ANNUAL REVIEW OF
FLUID MECHANICS

ANNUAL REVIEW OF FLUID MECHANICS

VOLUME 17, 1985

MILTON VAN DYKE, *Co-Editor*
Stanford University

J. V. WEHAUSEN, *Co-Editor*
University of California, Berkeley

JOHN L. LUMLEY, *Associate Editor*
Cornell University

ANNUAL REVIEWS INC.   4139 EL CAMINO WAY   PALO ALTO, CALIFORNIA 94306 USA

ANNUAL REVIEWS INC.
Palo Alto, California, USA

International Standard Serial Number: 0066-4189
International Standard Book Number: 0-8243-0717-8
Library of Congress Catalog Card Number: 74-80866

TYPESET BY A.U.P. TYPESETTERS (GLASGOW) LTD., SCOTLAND
PRINTED AND BOUND IN THE UNITED STATES OF AMERICA

Annual Review of Fluid Mechanics
Volume 17, 1985

CONTENTS

JAKOB ACKERET AND THE HISTORY OF THE MACH NUMBER, *N. Rott*   1

MULTICOMPONENT CONVECTION, *J. S. Turner*   11

RHEOMETRY OF POLYMER MELTS, *Joachim Meissner*   45

COATING FLOWS, *Kenneth J. Ruschak*   65

SEDIMENTATION OF NONCOLLOIDAL PARTICLES AT LOW REYNOLDS NUMBERS, *Robert H. Davis and Andreas Acrivos*   91

MATHEMATICAL MODELS OF DISPERSION IN RIVERS AND ESTUARIES, *P. C. Chatwin and C. M. Allen*   119

AERODYNAMICS OF SPORTS BALLS, *Rabindra D. Mehta*   151

BUOYANCY-DRIVEN FLOWS IN CRYSTAL-GROWTH MELTS, *W. E. Langlois*   191

SOUND TRANSMISSION IN THE OCEAN, *Robert C. Spindel*   217

FLUID MODELING OF POLLUTANT TRANSPORT AND DIFFUSION IN STABLY STRATIFIED FLOWS OVER COMPLEX TERRAIN, *William H. Snyder*   239

MATHEMATICAL MODELING FOR PLANAR, STEADY, SUBSONIC COMBUSTION WAVES, *D. R. Kassoy*   267

FLUID MECHANICS OF COMPOUND MULTIPHASE DROPS AND BUBBLES, *Robert E. Johnson and S. S. Sadhal*   289

THE RESPONSE OF TURBULENT BOUNDARY LAYERS TO SUDDEN PERTURBATIONS, *A. J. Smits and D. H. Wood*   321

MODELING EQUATORIAL OCEAN CIRCULATION, *Julian P. McCreary, Jr.*   359

THE KUTTA CONDITION IN UNSTEADY FLOW, *David G. Crighton*   411

TURBULENT DIFFUSION FROM SOURCES IN COMPLEX FLOWS, *J. C. R. Hunt*   447

GRID GENERATION FOR FLUID MECHANICS COMPUTATIONS, *Peter R. Eiseman*   487

COMPUTING THREE-DIMENSIONAL INCOMPRESSIBLE FLOWS WITH VORTEX ELEMENTS, *A. Leonard*   523

MANTLE CONVECTION AND VISCOELASTICITY, *W. R. Peltier*   561

INDEXES

   Subject Index   609

   Cumulative Index of Contributing Authors, Volumes 1–17   616

   Cumulative Index of Chapter Titles, Volumes 1–17   619

SOME RELATED ARTICLES IN OTHER *ANNUAL REVIEWS*

From the *Annual Review of Earth and Planetary Sciences*, Volume 12 (1984)

Double-Diffusive Convection Due to Crystallization in Magmas, Herbert E. Huppert and R. Stephen J. Sparks

Oceanography From Space, Robert H. Stewart

Theory of Hydrothermal Systems, Denis L. Norton

Sedimentation in Large Lakes, Thomas C. Johnson

From the *Annual Review of Physical Chemistry*, Volume 35 (1984)

Dielectric Properties of Polyelectrolyte Solutions, M. Mandel and T. Odijk

ANNUAL REVIEWS INC. is a nonprofit scientific publisher established to promote the advancement of the sciences. Beginning in 1932 with the *Annual Review of Biochemistry*, the Company has pursued as its principal function the publication of high quality, reasonably priced *Annual Review* volumes. The volumes are organized by Editors and Editorial Committees who invite qualified authors to contribute critical articles reviewing significant developments within each major discipline. The Editor-in-Chief invites those interested in serving as future Editorial Committee members to communicate directly with him. Annual Reviews Inc. is administered by a Board of Directors, whose members serve without compensation.

For the convenience of readers, a detachable order form/envelope is bound into the back of this volume.

Jakob Ackeret (1898–1981) at the age of 60. Taken from the anniversary volume of the *Zeitschrift für angewandte Mathematik und Physik*, Vol. 9b (1958).

Ann. Rev. Fluid Mech. 1985. 17 : 1–9

JAKOB ACKERET AND THE HISTORY OF THE MACH NUMBER[1]

N. Rott

Institut für Aerodynamik, ETH Zürich, CH-8092 Zürich, Switzerland[2]

I

When Jakob Ackeret was appointed privatdocent at the Federal Institute of Technology (ETH) in Zürich at the age of 30 in 1928, he already had a formidable list of credentials. After his graduation in mechanical engineering at the ETH in 1920, he was assistant to Aurel Stodola (1859–1942) for a year and then moved to Göttingen to work with Ludwig Prandtl (1875–1953). There he made essential contributions to the development of the aerodynamics institute. In 1925, he published the famous Ackeret formulas for the lift and drag of thin supersonic airfoils. A monograph on gasdynamics that he wrote for the *Handbuch der Physik* series appeared in 1927. In that year he also moved back to Zürich to become chief engineer for Escher Wyss, where he initiated the modern aerodynamic treatment of turbines and axial compressors.

His inaugural lecture at the ETH on 4 May 1929 was on drag at very high speeds. When defining the similarity properties of viscous compressible flow, he noted that it would be very convenient to have a special name for the important ratio of flow speed (or flight speed) v to sound speed a. He proposed the designation "Mach number." This was an immediate success, and the name is now known not only to people in the field (like "Reynolds number," a name proposed by Prandtl) but also to the general public.

Ackeret's inaugural lecture was later published (*Schweiz. Bauztg.*, 12

[1] Based, in part, on an eulogy held by the author at the ETH Zürich on 11 December 1982 and reprinted in the *Schweizer Ingenieur und Architekt* (Zürich) No. 21, 1983.
[2] Present address: Department of Aeronautics and Astronautics, Stanford University, Stanford, California 94305

1

0066–4189/85/0115–0001$02.00

October 1929); in it, he wrote (translated freely from German), "The well-known physicist Ernst Mach has recognized the sigificance of this ratio with particular clarity and has proved its importance with ingenious experiments; thus it appears to be very justifiable to call v/a the Mach number."

The work of Mach (1838–1916) to which Ackeret was referring is the well-known paper published in 1887 (with P. Salcher; *Sitzungsber. Akad. Wiss Wien* 95:41–50), where the head wave of a supersonic projectile was made visible and photographed for the first time. The experiment, based on the schlieren method invented in 1864 by August Toepler (1836–1912), was a breakthrough in its time; the pictures taken by Mach and Salcher were reproduced in the paper as woodcuts!

In addition to this experiment, Mach also gave a theoretical explanation of the head-wave phenomenon and showed the now standard figure of the Mach cone as the envelope of a series of spherical pulses emitted along the path of the supersonic projectile, each growing with the speed of sound. In Volume 15 (1983) of this series, an excellent and very detailed account of the contributions of Ernst Mach is given by H. Reichenbach, director of the Ernst-Mach-Institut in Freiburg, West Germany. However, it is not mentioned there—and it seems that this fact is headed for oblivion—that Mach's theoretical explanation of the "Mach cone" had already been given forty years earlier (1847) by Christian Doppler! Ackeret himself, however, was well aware of this fact and liked to comment on it when discussing the history of the Mach number. He used to say that the use of Doppler's name (instead of Mach's) could have caused confusion with the Doppler effect or with Doppler's principle. Also, Ackeret wanted to honor the experimentalist. Thus, Doppler was left with his effect, and Mach got his number.

It seems that an article devoted to the memory of Jakob Ackeret is also a suitable place to add a few remarks on the historical role of Doppler. In the next section, the contributions of Ackeret and of his Institute to high-speed aerodynamics are reviewed, including a few brief remarks on his many important contributions in other fields. In the final section, it is shown how the train of thought that initially led Doppler to the discovery of his principle also led him, systematically pursued, to the discovery of the Doppler-Mach cone. The material in this article was gathered by the author while working as a graduate student under the direction of Ackeret.

II

Ackeret was appointed professor at the ETH in 1931. He immediately started work on the construction of the Institut für Aerodynamik. The main facilities were two big wind tunnels, constructed by Ackeret (together with his trusted designer, J. Egli). Ackeret was probably the most successful practical engineer among the scientific pioneers of modern fluid dynamics. He maintained lifelong close connections with Escher Wyss and actively participated in actual designs (e.g. the construction of variable pitch propellers for ships and airplanes). His most important invention (together with C. Keller) is the gas turbine with a closed circuit, a machine that has not yet reached the practical significance that it potentially has.

Of the two wind tunnels constructed by Ackeret, one was a low-speed tunnel of conventional design but of unusual efficiency, a workhorse with many years of use still ahead. The second was the first supersonic wind tunnel built with a closed circuit. Ackeret had two main purposes in mind with his design. First, with the lower pressure level in the tunnel, high-speed runs could be realized with less power. Second, the changing of the pressure level allowed independent variation of the Reynolds number at a constant Mach number.

The construction of this tunnel was connected with important progress in the design of multistage axial compressors. The compressor used in the tunnel was built by Brown Boveri & Co. (BBC) in Baden. It absorbed 900 HP and provided $40 \, m^3 \, s^{-1}$ with a pressure ratio of 2.4; the efficiency of this 13-stage compressor was about 70%. The basic theory of the 1-stage axial compressor was enriched at that time by the thesis work of C. Keller, prepared under the direction of Ackeret. (It appeared in print in 1934.) As related by C. Seippel, who was head of the axial-compressor section of BBC at that time, only a 4-stage experimental engine existed when Ackeret decided to order the practically unproved multistage application of the advanced theory. Its immediate success profoundly influenced the spread of this engine type.

The first important application of the supersonic tunnel made full use of the independent variability of the Mach number and the Reynolds number. Ackeret had the idea to investigate the interaction of shock waves with boundary layers. The results were published in the series *Mitteilungen aus dem Institut für Aerodynamik* (No. 10, by J. Ackeret, F. Feldmann & N. Rott, 1946), as were the results of several other basic experiments using the wind tunnel. These included an examination of the problem of tunnel corrections for models investigated at high subsonic Mach numbers (No. 14, F. Feldmann, 1948); an experimental investigation of bodies of revolution, for which the theory at high subsonic Mach numbers was a matter of controversy before the appearance of the Göthert rule (No. 16, E. R. Van Driest, 1949); a study of the thermal effects in the wake of bluff bodies (No. 18, L. F. Ryan, 1951; No. 21, J. Ackeret, 1954); and experiments on grids in supersonic flow (No. 19, R. M. El Badrawy, 1952) and bodies of revolution at low supersonic Mach numbers (No. 24, H. R. Voellmy, 1958). A few

papers on high-speed flow did not appear in the Mitteilungen series—
in particular, an experimental verification of the transonic similarity
(*Z. Angew. Math. Phys.*, 1950) and measurements on inclined bodies
of revolution at high subsonic Mach numbers (*L'Aerotecnica*, 1951), both
by J. Ackeret, M. Degen & N. Rott.

By 1967, when Ackeret retired, 32 *Mitteilungen* volumes had appeared;
only those were mentioned above in which the Mach number played a role.
This is not the place to give a complete survey of this series, but No. 4/5 (H.
L. Studer & P. de Haller, 1934) should be mentioned because it included the
discovery of the stall flutter of single profiles and a treatment of ground
effects on wings. In addition, No. 13 (W. Pfenninger, 1946) should also be
noted for its report on important new experiments on boundary-layer
suction.

After his retirement, Ackeret remained active in many fields (e.g. wind
forces on buildings, ventilation of long tunnels). He also maintained his
lifelong interest in the history of science and technology. His most
important contribution in this field was the editing of the volume of Euler's
works on hydrodynamics. (In 1944, a turbine was built and tested at the
Institut according to ideas and sketches published by Euler in 1754.)

In 1973, Ackeret underwent a serious operation, after which he curtailed
many of his activities. His main interest remained the solution of the world
energy problem. He died on 26 March 1981, nine days after his eighty-third
birthday. His life work is an integral part of modern aerodynamics.

III

When Christian Doppler (1803–53) announced in 1842 the principle now
bearing his name, he was fully aware that he had to take into account the
relative motion of three things: the source, the observer, and the medium
(air, ether, etc.). This he did by examining two cases. In case 1, he considered
an observer moving toward a source at rest (relative to the medium at rest),
with the source emitting signals with its proper frequency ω_0. The
frequently measured by the observer is $\omega_I = \omega_0(1 + M)$, where M is
the Mach number of the approach velocity. In case 2, the observer is
at rest relative to the medium and is approached by a source, and thus
$\omega_{II} = \omega_0(1 - M)^{-1}$. Only to first order in M are the two results the same. It
is also clear that the two cases are vastly different in the level of difficulty
needed for their comprehension. Case 1 is almost trivial, while case 2
involves an understanding of the whole field generated by a moving source.
In 1842, Doppler restricted his attention to the part of the field lying in the
line of the source motion. In due course, however, he considered the whole
field, and his results were published (as in 1842) in the *Abhandlungen der*

Böhmischen Gesellschaft der Wissenschaften (Vol. 5, 1847). Doppler's own figures from this paper are reproduced here (Figures 1–6). First, Doppler constructed a subsonic (Figure 1) and a supersonic (Figure 2) field pattern; the latter figure is the first drawing showing a "Mach cone." He then proceeded to discuss the special case of sonic speed (Figure 3). Finally, he applied his construction to curved paths, again for subsonic, supersonic, and sonic speeds (Figures 4–6); these figures show how deeply Doppler explored the problem of a moving sound source. He even considered moving sources in dispersive media, albeit without conclusive results.

The involvement of Mach with Doppler's earlier work from 1842, when the "principle" was laid down, is presented in great detail in Reichenbach's article (mentioned above). Here only a brief outline of the main issues is given.

Mach's contribution to the understanding of the Doppler effect was both experimental and theoretical; his first paper on this subject was published in 1860, when he was a 22-year-old student, in the *Sitzungsberichte der Akademie der Wissenschaften in Wien* (1860; Reichenbach, loc. cit., p. 5). The work of Mach was a defense of Doppler's theories against (unjustified) criticism by Jozsef Petzval (1807–91), also of Vienna. Petzval was already well known for his contributions to geometrical optics; his lens design of 1840 revolutionized the early development of photography. In three papers presented to the Academy in Vienna in 1852, he proposed a theory that he tried to interpret as a refutation of Doppler's results. Basically, Petzval could not accept that a field of a moving source can be found without considering the interaction between source and medium. Doppler, however, came by sheer intuition to the (implicit) conclusion that this interaction only affects a near-field of negligible extension, and he found his results without resorting to any kind of calculations. Actually, Petzval was the first to propose that the field of a moving source could be determined by superposition of pulses distributed along its path, a method that can serve (as was pointed out by Mach) for a mathematical proof of Doppler's results. This method was used again much later by Prandtl (1938, *Schriften der deutschen Akademie für Luftfahrtforschung*).

Petzval remarked correctly that when source and observer are relatively at rest, then there is no frequency shift when the wind blows. From this he tried to construct a contradiction with Doppler's results; naturally, there is none. Mach made in 1862 (in *Annalen der Physik und Chemie*) the acrimonious remark, that in case that Professor Petzval would be serenaded (maybe for his contributions to this controversy), he obviously will hear the music in the correct tune, whether the wind blows or not.

The heated controversy had apparently cooled down considerably by 1887, when Mach made his famous experiments showing pictures of the

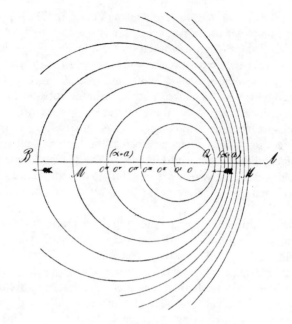

Figure 1

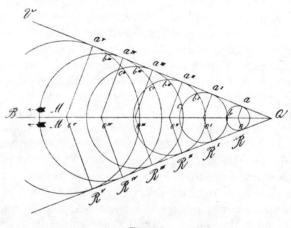

Figure 2

Figures 1–6 Original figures from Doppler's paper entitled "Ueber den Einfluss der Bewegung des Fortpflanzungsmittels auf die Erscheinungen der Aether-, Luft- und Wasserwellen." Figures 1–3: straight path. Figure 1: $M < 1$; Figure 2: $M > 1$; Figure 3: $M = 1$. Figures 4–6: curved paths. Figure 4: $M < 1$; Figure 5: $M > 1$; Figure 6: $M = 1$.

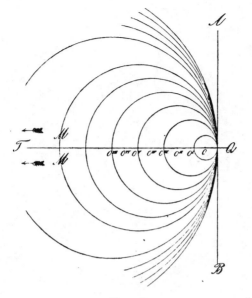

Figure 3

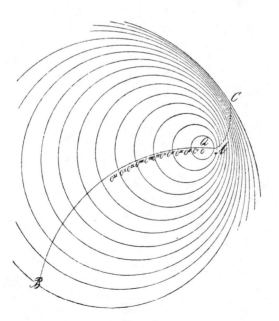

Figure 4

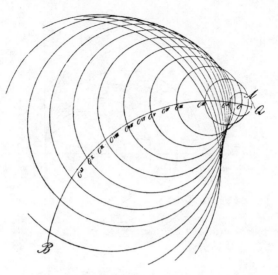

Figure 5

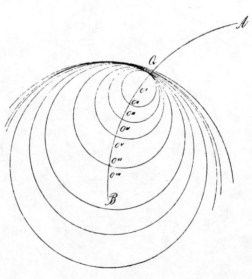

Figure 6

wave system of a supersonic bullet. A full account of these experiments is given by Reichenbach (loc. cit., p. 18); in Reichenbach's article, there is also (p. 25) a remarkable copy of a photograph taken by Mach, made from the original negative. Such pictures marked the beginning of a new era in the understanding of supersonic flow. In the theoretical explanations given by Mach, the legacy of Doppler is preeminently apparent. Mach, however, took the step of extending the ideas of Doppler from the sound field to the flow field.

The reader might ask why we should gloat over these old stories about the great men of the past. I think, however, that the Doppler-Petzval controversy is of interest as an elementary prelude to the difficult acceptance that the special theory of relativity received several decades later. Indeed, if two spaceships approached each other, carrying identical light sources, they could not distinguish the Doppler shifts that they observe. According to Einstein's theory, they would measure the geometric mean of Doppler's ω_I and ω_II.

The classical part of the history of the Doppler effect was brought to completion by a paper of Waldemar Voigt (1850–1919), best known for his fundamental contributions to the mechanics and optics of crystals. In 1887 he published a paper in the *Göttinger Nachrichten*, in which he showed that by a simple change of variables, the field of a moving source can be obtained from the field of a source at rest as a solution of the wave equation, and that this solution is in agreement with Doppler's predictions. Aerodynamicists know that such a change of variables is (almost) identical with a Lorentz transformation, except for a normalization factor that is trivial in the classical sense but essential for the explanation of the relativistic Doppler effect.

Today, Voigt's paper is largely forgotten and, in particular, not quoted by aerodynamicists. However, as was pointed out to me by Ackeret, the significance of this paper was known to Wolfgang Pauli (1900–58), whose monumental work on relativity appeared in the *Encyclopädie der mathematischen Wissenschaften* in 1921. Here, Voigt's work is the first reference. It represents the end of the classical era and gives a convenient point to establish a bridgehead in a new territory.

Ann. Rev. Fluid Mech. 1985. 17:11–44

MULTICOMPONENT CONVECTION

J. S. Turner

Research School of Earth Sciences, Australian National University, Canberra, A.C.T. 2601, Australia

1. INTRODUCTION

In two early reviews of this subject (Turner 1973, 1974), the phenomena described were adequately covered by the name *double-diffusive convection.* They were treated as generalizations of the process of thermal convection in a thin fluid layer, which arise when spatial variations of a second component with a different molecular diffusivity are added to the thermal gradients; the term *thermohaline convection* was introduced to describe the heat-salt system. The experimental and theoretical results then available could be described in terms of the differential diffusion of two components acting so as to release potential energy from a component that is "heavy at the top," even though the net density decreases upward. It was also shown that the form of the resulting motions depends on whether the driving energy comes from the component having the higher or lower diffusivity. Where one layer of fluid is placed above another (denser) layer having different diffusive properties, two basic types of convective instabilities arise, in the "diffusive" and "finger" configurations. These are illustrated in Figures 1 and 2, respectively. In both cases, the double-diffusive fluxes can be much larger than the vertical transport in a single-component fluid because of the coupling between diffusive and convective processes.

The main development of these ideas since 1974 has occurred in the field of oceanography, with interactions between theoreticians, laboratory experimenters, and sea-going oceanographers playing a vital role. The field has also broadened considerably, with new applications becoming apparent in addition to those outlined by Turner (1974). A rather personal historical review of the subject was presented by Huppert & Turner (1981a), in which they emphasized the importance of transfer of information from

11

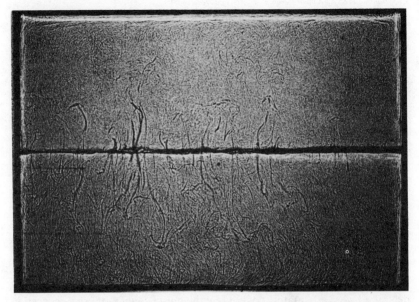

Figure 1 Shadowgraph picture of a "diffusive" interface formed by pouring a layer of NaCl solution ($\kappa = 1.5 \times 10^{-5}$ cm^2 s^{-1}) on top of a denser sucrose solution ($\kappa = 0.5 \times 10^{-5}$ cm^2 s^{-1}) in a tank 150 mm wide. Note the convective plumes on each side of the thin interface, evidence of strong interfacial transports.

one field to another. They concluded, however, that it is difficult to foster formal interactions of this kind, and that there must be a large element of chance in the spread of ideas across disciplines. In response to that review, an interdisciplinary conference was organized in March 1983 under the auspices of the Engineering Foundation, and summaries of the proceedings have been published (Huppert 1983, Chen & Johnson 1984). The present review has the related aim of making more widely known the increasing variety of topics to which these ideas can be applied.

It has become clear that not only can a wider range of phenomena be described in related terms, but that the previous physical descriptions, nomenclature, and even the basic definitions were unduly restrictive. There are situations (discussed below) where more than two components are significant—for example, in a star, the angular momentum, heat, magnetic field, and composition all diffuse at different rates, and they may vary spatially in dynamically important ways. Force fields other than gravity can be relevant, and there has been an increasing emphasis on two-dimensional effects due to a combination of horizontal and vertical property gradients. As disciplines that have developed relatively independently have begun to communicate with one another, differences of

Figure 2   Shadowgraph of a thickened "finger" interface formed by placing a layer of sucrose solution on top of a denser NaCl solution and leaving it for about three hours (cf. Figure 1).

language and ways of thinking have become more apparent. A major aim of the present review is to resolve some of these differences by setting out the basic principles in a way that will be accessible not only to those whose primary interest is fluid mechanics but also to specialists in other fields.

The historical development of the subject is not pursued here (see Turner 1974). The more recent results and publications are given more attention, though there is of course some repetition of material appearing in earlier reviews. Where possible, reference is made to recent specialized reviews of particular topics. Our starting point is a broader definition of the circumstances under which multicomponent convection can occur in a fluid subject to a gravitational field:

1. The fluid must contain at least two components with different molecular diffusivities, each of which affects the density of the fluid. Chemists prefer to use the terms "isothermal ternary system," referring to a solvent and two solutes or polymers, or "nonisothermal binary system," thus distinguishing between thermal and other diffusion processes.[1]

---

[1] Note that our use of the word "component" is also at variance with its technical significance in physical chemistry.

2. Differential or coupled diffusion can produce convective motions that are associated with a decrease in the gravitational potential energy of the system.

This second criterion represents an important generalization of the earlier ideas, as summarized in the first paragraph of this review. No longer is it assumed that one of the components must be distributed in a "hydrostatically unstable" manner; theoretical arguments suggest, for example, that when cross-diffusion terms are large, perturbations can grow in a ternary system even when both solutes are more concentrated at the bottom. Though the gravitational potential energy term is small compared to the free-energy change, it plays an essential role in the organization of the flow; this point is discussed in more detail in Section 3.2.

Another point, about which there is much confusion in the chemical diffusion literature, is worth making immediately. It is *not* necessary for the net vertical density gradient (across an interface, for example) to change sign and become statically unstable before convection can occur. Double-diffusive motions (and their equivalents in more complicated systems) can certainly develop under much less stringent conditions, which allow local density anomalies to develop while the horizontally averaged vertical density gradient ρ_z remains negative (i.e. the mean density decreases upward across an interface, but local anomalies cause local instabilities).

2.   VERTICAL GRADIENTS OF DIFFUSING COMPONENTS

2.1   *Prototype Laboratory Experiments*

The basic phenomena and underlying physical principles are still most easily introduced by describing several one-dimensional, two-component experiments in qualitative terms. Some recent one-dimensional results are mentioned immediately, but more subtle extensions of these experiments, and the associated theory, are treated separately.

SUPERIMPOSED FLUID LAYERS   A characteristic of multicomponent convection is that well-mixed layers tend to form, separated by interfaces across which there are large gradients of the several components. A single interface can be set up directly by pouring a layer of one solution carefully on top of a denser miscible layer, which either is at a different temperature or contains a different solute. For consistency with the notation used below, we denote by T (standing for temperature in the extensively studied heat-salt system) the component with the larger molecular diffusivity D_{11}, and by S the component ("salinity") with the smaller diffusivity D_{22}. The ratio D_{22}/D_{11} is defined as τ (< 1), and cross-diffusion terms are ignored for the present.

A "diffusive" interface results when the component T is heavy at the top and S is heavy at the bottom, with the lower layer being denser (see Figure 1). When T refers to temperature, this means that cold, dilute solution lies above hot, concentrated solution. As the name suggests, the vertical transports of T and S across the hydrostatically stable central core of the interface occur in this case entirely by molecular diffusion. But because $D_{11} > D_{22}$, the *edges* of the interface can become marginally unstable (cf. Linden & Shirtcliffe 1978). The resulting unstable buoyancy flux into the layers above and below drives large-scale convection that keeps the two layers well stirred and the interface sharpened.

In this case the downward flux of T [expressed in density terms as αF_T through the relation $d\rho = \rho_0(\alpha\, dT + \beta\, dS)$] is greater than the upward flux of $S(\beta F_S)$ (or in the heat-salt case, the upward flux of heat is greater than the flux of salt). It is appropriate to focus attention on the gravitational potential energy in the T field, which is continually decreasing, as is the potential energy of the system as a whole. The density difference between the two layers *increases* in time (in contrast to the slower evolution due to diffusion across an interface in a single solute system or that caused by mechanical mixing between two layers).

Over a wide range of conditions, $\beta F_S/\alpha F_T$ is observed to be nearly constant and close to $\tau^{1/2}$ for both two-solute and heat-solute systems. The reader is referred to earlier reviews for a fuller discussion of this point, and of the quantitative flux measurements in general. The experiments are described in Turner (1965) and Marmorino & Caldwell (1976).

In the opposite configuration, with a layer of S above a layer of T, the structure shown in Figure 2 is observed. Small disturbances can grow rapidly, and long, narrow, vertical convection cells called "salt fingers" are formed and extend through the interface. It is now the more rapid *horizontal* diffusion of T relative to S over the width of the fingers that makes possible the release of the potential energy stored in the S field. It can be shown [see Equation (2) below] that when τ is small (e.g. for heat-salt fingers), even a tiny fraction of salt in the warmer top layer will lead to the formation of persistent fingers. The finger mechanism of transport, in which gravity plays a vital role, is described in terms of this familiar system.

Each downward-moving finger is surrounded by upward-moving fingers, and vice versa (Shirtcliffe & Turner 1970). The downgoing fingers continually lose heat (by horizontal conduction) to the neighboring upgoing fingers; therefore the downgoing fingers become more dense and the upgoing fingers less dense. There is a slower transfer of salt sideways, which results in a small vertical salinity gradient (Figure 3). Thus the small-scale (finger) convective motions are driven by the *local* density anomalies between fingers and lead to $\beta F_S > \alpha F_T$. However, the horizontally

warm,salty

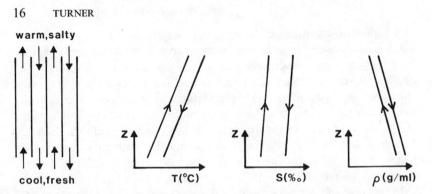

cool,fresh

Figure 3 Upgoing and downgoing salt fingers and their temperatures, salinities, and densities. Arrows denote the direction of motion. By this mechanism, both properties are transferred down their mean vertical gradients, but much more rapidly than by simple molecular diffusion.

averaged vertical density gradient through the interface remains stable and dominated by the T gradient. There is an unstable boundary layer at the *edge* of the salt-finger interface (cf. Figure 2) that drives convection in the layers. The potential energy of the whole system is again decreasing and the density difference between the layers is increasing, but now, since $\beta F_S > \alpha F_T$, this energy can be regarded as derived from the salt field.

The schematic distributions of S and T, and the net density at the beginning and end of a two-layer "finger" experiment, are shown in Figure 4. In the initial state (state 1), S and T are homogeneous in the upper and lower layers, respectively, and ρ is the same everywhere. Rapid convective transports due to fingers produce state 2, in which S is nearly uniformly

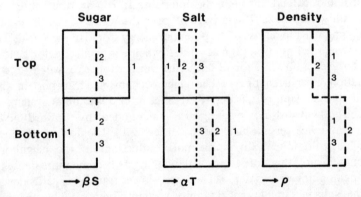

Figure 4 Distributions of S, T, and density at various stages of the "run-down" finger experiment described in the text. It has been assumed that $\alpha F_T/\beta F_S = 1/2$ for the purposes of illustration.

distributed with depth, while part of the T difference remains and there is a stable density step. The final state (state 3) of complete uniformity of both components would be achieved (by molecular diffusion) only after a much longer time.

The differences between the hydrostatic stability at the edge or the center of a "finger" interface and between static and dynamic instabilities have not always been clearly understood. Wendt (1962), for example, derived the necessary condition for *static* instability of an interface across which two solutes are diffusing as

$$R_\rho = \Delta\rho_1/\Delta\rho_2 < \tau^{-1/2}, \tag{1}$$

where $\Delta\rho_1$ and $\Delta\rho_2$ are the initial density excesses due to the faster-diffusing solute 1 in the lower layer and solute 2 in the upper layer. Huppert & Manins (1973) showed, however (see also the theoretical section below), that the center of such an interface can become *dynamically* unstable when

$$R_\rho < \tau^{-3/2}. \tag{2}$$

Since $\tau < 1$, (2) is less restrictive than (1), and dynamic instabilities in the form of the alternating finger motions sketched in Figure 3 can set in while the net density gradient remains statically stable at all times. For the heat-salt system, fingers can form when R_ρ is as large as 10^3, since $\tau \approx 10^{-2}$. Further laboratory experiments have unambiguously confirmed these results (Griffiths 1979b, Huppert & Hallworth 1984).

Fluxes have also been measured across finger interfaces. The most recent results for heat-salt fingers, as well as a critical survey of previous work, have been presented by McDougall & Taylor (1984), who extended the measurements to values of R_ρ as low as 1.2, into the range that is of most interest in the ocean.

FORMATION OF LAYERS FROM A GRADIENT   Layers can form when a smooth, statically stable gradient of one property has an opposing gradient of a second property superimposed on it, or when there is a vertical flux of the destabilizing component. The first case is the one for which most of the theory has been developed, but the formation of double-diffusive layers is most readily demonstrated in the second.

When a linear, stable salinity gradient is heated from below, for example, the fluid does not immediately convect from top to bottom. First, the heated layer immediately above the boundary breaks down to form a thin convecting layer. The depth d of this layer, and also the temperature and salinity steps ΔT and ΔS at its top, grow in time t like $t^{1/2}$, in such a way that $\alpha\Delta T = \beta\Delta S$ and there is no net density step (Turner 1968). The thickness δ of the diffusive thermal boundary layer growing ahead of the convecting layer is also proportional to $t^{1/2}$ and therefore to d, the multiplying constant being

$$Q = \frac{\delta}{d} = D_{11}S_*/H_*. \tag{3}$$

Here $H_* = -g\alpha F_T/\bar{\rho}C$ is the imposed buoyancy flux corresponding to the heat flux F_T into a fluid of specific C, and $S_* = -g\beta dS/dz$ is a measure of the initial salinity gradient. The behavior of the system depends only on Q and on the molecular properties through the Prandtl number σ and the diffusivity ratio τ. When Q and τ are small, the criterion for instability of the thermal boundary layer is a critical Rayleigh number based on δ and ΔT; after this criterion is achieved at a certain thickness, the depth d remains constant, and a new layer grows on top of the first. Huppert & Linden (1979) have reported a combined numerical and experimental study that treats the sequential formation of new layers at the top of the region while the lowest layers merge in pairs. Linden (1976) has used the sugar-salt system to study the case where there is a destabilizing salt (T) gradient, partly compensating the stabilizing sugar (S) gradient in the interior. He found that as the two gradients become nearly equal but opposite, the layer properties depend less and less on the boundary flux, which then just acts as a trigger for an internal instability.

Convecting layers can also be produced in the "finger" situation, with a flux of sugar (S) imposed at the top of a salt (T) gradient (Stern & Turner 1969). Fingers form, grow, and break up because of a collective instability having the form of an internal wave (Stern 1969, Holyer 1981); this process produces a convecting layer that deepens, bounded below by an interface containing shorter, stable fingers. These fingers in turn grow and become unstable, thus producing a second convecting layer. Viewed on the scale of the convecting layers, there is a strong similarity between finger and diffusive systems, since they are both driven by unstable buoyancy fluxes through the bounding interfaces.

2.2 Stability Analyses

Some of the statements made above in describing the experiments can only be properly justified by referring to the corresponding theoretical results. Only a few examples can be treated here, and these only in brief, but even the simplest stability arguments go some way toward describing and explaining what is observed, especially in the finger case.

LINEAR THEORIES WITH SIMPLE DIFFUSION OF TWO COMPONENTS   The early developments, initiated by Stern (1960), were reviewed by Turner (1974). Stern was the first to consider the stability of temperature and salinity profiles, both of which increase linearly with height between two free horizontal boundaries separated by a distance d and held at fixed temperatures and salinities. He thus concentrated on the "finger" case, though he mentioned the "diffusive" case in a footnote. His results are applicable to any pair of independently diffusing components, and they have since been extended (e.g. see Baines & Gill 1969) explicitly to the diffusive case and to other boundary conditions. Here we consider ΔT and ΔS to be the differences in solute concentrations between the boundaries; positive values imply that these are greater at the lower plate. Four nondimensional parameters are required to specify such systems. Using the notation already introduced, these can be defined as $R = g\alpha\Delta T d^3/D_{11}v$, the Prandtl number $\sigma = v/D_{11}$, $\tau = D_{22}/D_{11}$, and $R_\rho = |\alpha\Delta T/\beta\Delta S| = |\alpha T_z/\beta S_z|$. A frequently used combination is $Rs = R/R_\rho$.

Assuming small disturbances of the form $T' = e^{pt} \cos \pi ax \cdot \sin \pi bz$, for example, from the state of constant vertical property gradients T_z and S_z and no initial motion, we obtain a cubic characteristic equation in p, with coefficients that depend on R, Rs, σ, and τ. In a given fluid (fixed σ and τ) the stability boundaries may be calculated as lines in the R, Rs plane, and the character of the initial motion determined in each region. When $R > 0$ and $Rs < 0$ (the finger regime), the motion sets in as a direct, growing mode when

$$-R - Rs/\tau > \frac{27}{4}\pi^4. \tag{4}$$

When R and Rs are large (e.g. because the diffusivities are small), fingers can form provided

$$R_\rho < \tau^{-1}. \tag{5}$$

When τ is small, instability sets in at values of $Rs \ll -R$, i.e. while the system is still statically stable. In this limit, the fastest-growing motions have a horizontal length scale $\ell \sim (g\alpha\Delta T/dD_{11}v)^{-1/4}$, which is much smaller than d and represents a balance between the destabilizing diffusive effects and the dissipative viscous effects. This is in agreement with observations, and though the linear stability theory is not strictly applicable to fully developed fingers, a later finite-amplitude calculation by Huppert & Manins (1973) using the same force balance has verified that this is the relevant scale. We recall that these authors also derived the condition for dynamical instability at an interface, quoted in Equation (2) above. It can now be seen that this condition differs from the static condition (1) because of the factor τ^{-1} appearing in Equation (5).

Corresponding results have been obtained for the "diffusive" case ($R < 0$, $Rs > 0$), where it is found that instability to small perturbations occurs in an overstable or oscillatory mode and (at large $|Rs|$) when the net density distribution is statically stable. Many of the recent developments in this area relate to finite-amplitude calculations, which are discussed separately.

The calculations for the finger case have been extended by Schmitt (1979a, 1983) to cover general values of σ and τ. He has computed the maximum growth rates, the flux ratios of the two components (at particular values of R_ρ), the wave numbers of the fastest-growing fingers, and a "bandwidth," or measure of the spread of scales evolving from a white-noise spectrum. The fastest-growing fingers occur at low values of σ and τ. The flux ratios $\gamma = \alpha F_T/\beta F_S$ calculated by Schmitt (1979a) using similarity solutions and linear gradients are in good agreement with laboratory values for two-layer heat/salt and salt/sugar systems (Turner 1967, Schmitt 1979b, Lambert & Demenkow 1972, Griffiths & Ruddick 1980). With this encouragement, Schmitt (1983) produced the more general results shown in Figure 5. Note that for $\sigma > 1$ and $\tau < 10^{-1}$ the flux ratio lies between 0.5 and 0.7 at $R_\rho = 2$, and it is predicted (but not yet tested experimentally) that systems, such as liquid metals, with low σ and τ should have low γ (0.1–0.2).

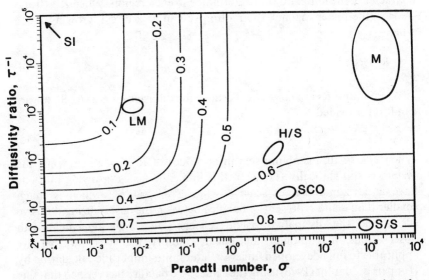

Figure 5   Contours of the flux ratio $\gamma = \alpha F_T/\beta F_S$ in fingers, calculated for general values of σ and τ with $R_\rho = 2$ by Schmitt (1983). The regions corresponding to a number of natural double-diffusive systems are denoted by the following abbreviations: LM = liquid metals and metallic semiconductors, SCO = semiconductor oxides, H/S = heat/salt, S/S = salt/sugar, M = magmas. The region for stellar interiors (SI) lies off the diagram at about $\sigma = 2 \times 10^{-6}$, $\tau = 10^{-7}$.

The growth-rate calculations in the relevant σ, τ ranges indicate that stellar interiors, liquid metals, and warm heat/salt systems should develop vigorous fingers, while salt/sugar and magma systems will be more sluggish.

COUPLED MOLECULAR DIFFUSION   When cross-diffusion terms are significant, further kinds of instabilities become possible. If the fluxes are parameterized in terms of the gradients by

$$- \text{Flux of } T = D_{11} \nabla T + D_{12} \nabla S, \tag{6}$$

$$- \text{Flux of } S = D_{22} \nabla S + D_{21} \nabla T, \tag{7}$$

with the density ρ of the fluid given by $\rho = \rho_0 (1 + \alpha T + \beta S)$, the cases already discussed correspond to $D_{12} = D_{21} = 0$. These relations apply strictly only to isothermal ternary systems, but when T and S represent heat and solute, D_{21} is proportional to the Soret coefficient. Hurle & Jakeman (1971) have shown that a large value of D_{21} produces a concentration gradient near a heated boundary, which can influence the form of the instability. For example, a homogeneous layer of water-methanol mixture (with a low concentration of methanol) can break down when heated from below in an oscillatory rather than direct mode because of the stabilizing concentration gradient set up in this way. Schechter et al. (1972) and Verlande & Schechter (1972) have investigated theoretically the effect of heating above and cooling below a thin layer of solution, and they have shown that double-diffusive instabilities can and do occur when the density gradient is statically stable (see also Platten & Legros 1983).

McDougall & Turner (1982) and McDougall (1983a) have taken a further step and investigated the conditions under which fingers can form in a ternary system where both D_{12} and D_{21} can be significant. On following through the linear stability analysis for small perturbations from linear gradients and no initial motion, the condition for finger instability with $\alpha T_z < 0$ can be written as

$$\left(\frac{\beta}{\alpha} \frac{D_{21}}{D_{22}} - 1 \right) + \frac{1}{\tau} \frac{\beta}{\alpha} \frac{S_z}{T_z} \left(\frac{\alpha}{\beta} \frac{D_{12}}{D_{11}} - 1 \right) > \frac{27}{4} \frac{\pi^4}{R} \left(1 - \frac{D_{12}D_{21}}{D_{11}D_{22}} \right). \tag{8}$$

This reduces to (4) when $D_{12} = D_{21} = 0$, but clearly considerable extra complexity is introduced by adding these two coefficients. McDougall (1983a) has sketched the stability boundaries as functions of $\beta D_{21}/\alpha D_{22}$ and $\alpha D_{12}/\beta D_{11}$ for specified ranges of αT_z and βS_z. He has also shown that essentially the same expression as (8), with the right-hand side zero, is obtained for finite-amplitude fingers if T_z and S_z are now taken to be the vertical gradients of the horizontal average of T and S through the fingers (cf. Huppert & Manins 1973).

Of special interest is the case where *both* components are hydrostatically stably stratified. From (8) it is seen that instability is assisted by large positive values of both D_{12} and D_{21}, and that at least one of $\beta D_{21}/\alpha D_{22}$ and $\alpha D_{12}/\beta D_{11}$ must exceed unity. Figure 6 illustrates the variations of properties in the upgoing and downgoing fingers when $D_{12} = 0$ and Equation (8) is satisfied; this should be compared with Figure 3, which shows the variations in "ordinary" fingers (when $D_{12} = D_{21} = 0$). In Figure 6, the T concentration of the downgoing fingers increases as a result of the diffusion of T from the upgoing into the downgoing fingers, in a manner that is unaffected by coupled diffusion. The flux of S between the fingers is, however, due mainly to the cross-diffusion flux driven by the spatial gradients of T; even though the concentration of S in a downgoing finger is already greater than that in the upgoing finger, it continues to increase because of the D_{21} term. Note that the horizontally averaged density gradient is statically stable, and that the downgoing fingers are more dense than the surrounding upgoing fingers. Thus, again denser fluid will be transported downward and less dense fluid upward, and the total gravitational potential energy of the fluid will decrease, in accordance with our extended definition of multicomponent convection.

McDougall (1983a) has generalized the above results to include nonzero values of both D_{12} and D_{21} and has also examined the instability of initially sharp interfaces with cross-diffusion included. When both components are stably stratified initially, or even when they are distributed in the "diffusive" sense, a large cross-diffusion term can lead to the formation of fingers before the density gradient at the center of the interface becomes statically unstable. Comparisons of this theory with experiments by polymer chemists are made in Section 3.2.

INFLUENCE OF A THIRD DIFFUSING COMPONENT   Griffiths (1979a,b) extended the linear stability analyses to include three solutes (without cross-diffusion effects). In this case, the characteristic equation is fourth order, rather than a

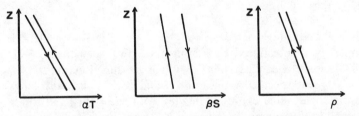

Figure 6   Sketch of the vertical profiles of αT, βS, and ρ in the centers of up- and downgoing fingers with both T and S stably stratified. Here $D_{12} = 0$ and D_{21} satisfies Equation (8), so that the cross-diffusive flux of S due to the T gradient sustains the density anomalies that drive the motion.

cubic, and the stability boundaries are planes in the space defined by three Rayleigh numbers. He gave special attention to systems with one diffusivity (say D_{11}) much larger than the other two (D_{22} and D_{33}) and with large Prandtl number, i.e. to the heat/two-solute system. Small concentrations of a third property with a smaller diffusivity can profoundly affect the nature of diffusive instabilities. For example, oscillatory and direct finger modes can be unstable simultaneously when the density gradients due to components with the largest D_{11} and smallest D_{33} diffusivities are of the same sign. What is actually seen experimentally depends on the relative growth rates of the two modes, and for three components of comparable diffusivities, fingers will be dominant.

Griffiths (1979b) has extended this analysis, using the technique that led to (2) in the two-solute case, to determine the type of interface formed between well-mixed layers containing three diffusing components with different relative concentrations. He tested his theory using three solutes (1: KCl, 2: NaCl, and 3: sucrose in water, listed in order of decreasing diffusivity), first with component 2 above 1 and 3, and then with 2 and 3 above 1. His experimental results are in good agreement with the theoretical criterion, which predicts the marginal concentration ratio $\beta_3 \Delta C_3 / \beta_1 \Delta C_1$ for fingers to form as a function of the overall density ratio R_ρ. Fingers forming away from the center, with a "diffusive" core, were also observed in a certain parameter range.

A final step in this series of related studies is mentioned now. Griffiths (1979c) measured the transport of multiple components through heat-solute diffusive interfaces, following some exploratory experiments by Turner et al. (1970). He also predicted theoretically that the ratio of transport coefficients K_i, defined as the mass flux divided by the corresponding concentration step, should be proportional to $\tau^{1/2} = (D_{33}/D_{22})^{1/2}$ at low total solute-heat density ratios R_ρ^T, and to τ at higher R_ρ^T, where a steady diffusive core dominates. The experiments with R_ρ^T between 2 and 4 were consistent with $K_3/K_2 = \tau$, but there was an even greater separation of components at higher R_ρ^T. This discrepancy has not been explained, but it is unlikely to result from the neglect of cross-diffusion terms. The separation of different salts during transport through a finger interface, on the other hand, is negligible when heat is the stabilizing property and the solutes are destabilizing.

NONLINEAR THEORIES Experiments such as the heating of a salinity gradient from below make it clear that a series of diffusive layers and interfaces can only be properly explained theoretically using nonlinear theories. The oscillatory instability predicted by linear theory is rapidly overtaken by monotonic motion and convective mixing within the layers, and there is not a close correspondence between the form of the growing motions and the final steady state, as there is in the finger case.

Huppert & Moore (1976) tackled the "diffusive" system numerically and found two different solution branches, one oscillatory and one steady. For $\tau = 0.1$, the smallest value for which it was convenient to run the numerical procedure, they found that nonlinear steady convection can occur at a lower value of the thermal Rayleigh number R than the linear prediction for the oscillatory mode. Recent numerical experiments by Toomre et al. (1982) have extended Huppert & Moore's (1976) calculations to $\tau = 0.01$ and $R \approx 10^6$. The solutions exhibit time-dependent behavior, including a sequence of period doublings, leading eventually to chaotic behavior. Such results have more general implications for the behavior of systems of differential equations describing other physical systems, and this aspect of the theoretical work has taken on a life of its own (e.g. see Marzec & Spiegel 1980).

Proctor (1981) has described the steady solutions near the minimum value of R, using an analysis valid as $\tau \to 0$. For small enough τ, salt can be swept to the boundaries of the convection cells, so that the S field is uniform in the interior and has large gradients only at the edges. A value of R sufficient to initiate purely thermal convection is thus able to drive steady motion, no matter how large the stabilizing salt field is initially.

There have been various attempts at steady and unsteady numerical models of a series of diffusive layers and interfaces, but none of these are really satisfactory yet. Piacsek & Toomre (1980) have produced a numerical model of salt fingers growing across an initially sharp interface. Though the ends of the fingers became unstable in this calculation, they did not evolve into convecting layers developing independently of the fingers, possibly because only a few fingers were included in the calculation, thereby restricting the horizontal scale of the convecting motions.

A totally different theoretical approach has been used by Stern (1982) to study fingers between two deep layers. He has applied a variational principle to place an upper bound on the salt flux, keeping the heat and momentum fluxes constant.

2.3   Direct Applications of One-Dimensional Results

Before we add extra physical effects, in other geometries, there are several examples of practical importance that can be discussed immediately.

LAKES AND SOLAR PONDS   Many lakes have now been described that are more saline at the bottom, where they are heated by the Sun or by geothermal energy. A good example is Lake Kivu (Newman 1976), which contains many well-mixed layers separated by diffusive interfaces. Griffiths (1979c) has shown that laboratory results on the transport of different components through these interfaces give a satisfactory explanation of the observed fluxes.

The main large-scale engineering application of double-diffusive concepts is to solar ponds, shallow artificial lakes that are density stratified. Radiant heat is trapped and stored in a bottom layer, with an insulating gradient layer above; the aim is to prevent convection that would increase the heat losses (Poplawsky et al. 1981). The salinity gradient is not constant with depth, and this has prompted theoretical studies (Walton 1982) of the breakdown, which is found to occur preferentially (in agreement with observations) in a thin layer where the salinity gradient is a minimum. The behavior of interfaces at the large values of R_ρ observed in solar ponds has been examined theoretically and experimentally (Newall 1984). The convective heat flux persists up to $R_\rho = 28$, well beyond the value given by theoretical models based on the stability of the interface, which predict a cutoff around $R_\rho = \tau^{-1/2} = 9$ (Linden & Shirtcliffe 1978). Allowance for extra entrainment due to the turbulence in the convecting layers extends the range over which the transports remain nonzero (McDougall 1984).

There are still many unsolved practical problems relating to long-term unsteady behavior of interfaces (for example, the erosion of thick interfaces) and to replenishment methods. Layering caused by sloping sidewalls may also be significant for the operation of solar ponds.

LIQUID NATURAL GAS TANKS   The phenomenon of *rollover* in LNG tanks has continued to receive attention. In storage tanks refilled from below with denser but warmer LNG than that already present, one or more layers, separated by diffusive interfaces, will tend to form. As the lighter fractions boil off at the top, the upper layer can increase in density more rapidly than the lower layer can by cooling through the interface, so that overturning occurs. This leads to an uncontrollable increase in the rate of release as warmer LNG is convected to the surface.

The earlier consideration of two-layer systems has been followed by calculations of the mixing between the new and resident LNG (Germeles 1975), but the influence of the multiple layering that can develop during this process has not yet been properly evaluated. Multi-diffusive convection is also likely to occur because of the different chemical compositions of LNG from various sources. The latter aspect is complicated by the fact that the diffusivities are not well known in this system.

3.   IMMEDIATE EXTENSIONS AND APPLICATIONS

A wide range of related phenomena are now described briefly in the context of their most common application (though some cross-references and overlap are inevitable).

3.1   *Oceanography*

The literature in this field, which has provided the main motivation for the development of the subject of double-diffusion over the past 25 years, is now very large. There is no doubt that double-diffusive transport mechanisms are significant in particular regions of the ocean, but their overall importance relative to other mixing mechanisms still awaits a systematic assessment (Turner 1981, Caldwell 1983). A selection of the more recent observational, laboratory, and theoretical developments of particular relevance to this field are presented in the remainder of this section.

SIGNATURES OF DOUBLE-DIFFUSIVE PROCESSES   New instruments with high resolution have produced many profiles of temperature and salinity showing a "finestructure" of well-mixed layers separated by sharp, nearly horizontal interfaces. When temperature and salinity both increase or both decrease with depth across these interfaces, in such a way that their density effects are opposing and nearly compensating, it is reasonable to presume that double-diffusive convection will be important. Two further generalizations have also emerged. First, conditions are especially favorable for double diffusion to occur when a layer with compensating T and S differences intrudes at its own density level into an environment with different properties. Layers are thus most prominent near regions of strong horizontal property contrasts, i.e. across fronts, and this has led to the studies reported in the next section. Second, the smallest scale fluctuations or microstructure should always be examined in relation to the finestructure. There are now many measurements (see, for example, Gargett 1976, Schmitt & Georgi 1982) that show that the regions of most intense activity are the upper and lower boundaries of intrusions. Once diffusive or finger interfaces have formed, by whatever mechanism, there is reason to believe that the fluxes through them are adequately treated using one-dimensional models.

The existence of salt fingers in the ocean has been demonstrated in several ways. Gargett & Schmitt (1982) have reported direct measurements of temperature fluctuations attributable to fingers, which have a quasi-periodic structure and limited amplitude. The observed spectral shape compares well with theoretical predictions, and the buoyancy flux estimated directly from the variance of temperature gradient is in fair agreement with predictions using laboratory flux laws. Fingers have also been observed directly, using an optical method (Williams 1975, Schmitt & Georgi 1982). The images can be quantified, and different structures can be related to fingering or mechanical mixing activity. The latter authors have shown that finger activity is strongest at $R_\rho \approx 1$, and shear instabilities associated with low Richardson numbers (due to shears created by internal wave motions) dominate near $R_\rho = 2$. Their correlations indicate that double diffusion is more common than shear instabilities in frontal interleaving zones, but they do not identify the mechanism responsible for creating the interleaving structures.

It has also been plausibly argued (Schmitt 1981) that the characteristic curvature of the T/S relation in central waters is consistent with $R_\rho = $ constant, a condition that can be maintained by (unequal) double-diffusive transports of T and S but that is inconsistent with equaly eddy diffusivities (which would produce a linear T/S relation). Gordon (1981) has also attributed observed large midthermocline mixing rates in the South Atlantic to salt-finger activity.

Two recent studies related to "diffusive" interfaces are worth reporting. McDougall (1981) has shown that the nonlinearity of the equation of state (e.g. in seawater near 0°C) modifies normal double-diffusive convection by causing asymmetric entrainment across an interface. This is more important than the previously discussed "cabbeling" process, which becomes possible where mixtures of two water masses are more dense than either parent mass. These results can be used to predict the signature imposed on a series of migrating interfaces by the asymmetrical entrainment. McDougall (1983b) has also developed a model for the formation of Greenland Sea bottom water, in which there is a balance between the advection of Atlantic water toward the center of the Sea and subsurface modification of its properties by double-diffusive convection.

HORIZONTAL GRADIENTS AND FRONTS   There have been many theoretical and laboratory studies following up early work on the effect of horizontal property gradients. Paliwal & Chen (1980) have generalized the problem of convection in a slot by considering stratified fluid contained in a small gap set at any angle to the vertical. Linear stability theory predicts the observed critical Rayleigh number and the scale of the steady convection cells. The case of a wide gap, with layers forming at a single heated or cooled sidewall, has been studied experimentally by Huppert & Turner (1980). They also examined vertical ice blocks melting into a salinity gradient (with an application to icebergs in mind) and showed that at Rayleigh numbers above about 10^5 the layer scale was

$$h = 0.65[\rho(T_w, S_\infty) - \rho(T_\infty, S_\infty)] \left/ \left|\frac{d\rho}{dz}\right.\right., \tag{9}$$

where the subscripts w and ∞ denote T and S values at the wall and in the far field, resepctively (i.e. the layer scale depends only on the temperature difference and not on the presence of a melt layer near the wall). Huppert et al. (1984a) have extended these experiments to include crystallization at the wall, or other solutes, and have shown that this conclusion appears also to be independent of σ and τ, i.e. of the molecular properties.

Tsinober et al. (1983) have described experiments in which layers were produced by heating at a small source in a linearly stratified salt solution. A plume penetrated to an increasing height above the source, with layers forming above and around it; in time, the evolving pattern shown in Figure 7 was set up, with features related to both bottom- and side-heating experiments. The convecting layers sloped downward as they moved radially away from a central plume into cooler surroundings, setting up systematic shearing motions with an outflow at the top and inflow at the bottom of each extending layer.

Intrusions of one fluid injected at its own density level into a gradient of another have been studied in the laboratory by Turner (1978), and the intrusive motions produced at a front (separating salt and sugar solutions with the same vertical density gradient) have been examined by Ruddick & Turner (1979). In the latter case a series of regular interleaving layers

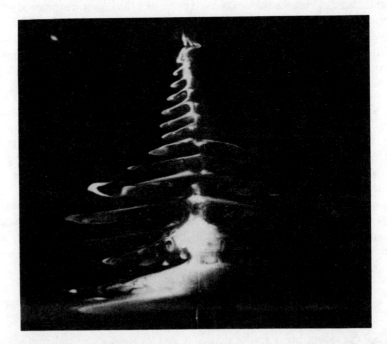

Figure 7   A series of double-diffusive layers produced after 35 min by heating a small region at the bottom of a salt-stratified tank 30 cm deep. Fluorescent dye placed near the bottom was lifted into convecting layers, which formed from the bottom up, and was also spread sideways because of the horizontal property gradients. (From Tsinober et al. 1983.)

develops, with both the vertical spacing and speed of advance approximately proportional to the horizontal property difference. There is a characteristic tilt of the layers across isopycnals as they advance that results from the net buoyancy flux through the finger interfaces. This tilt has now been documented in many ocean observations and supports fingering as the most probable driving mechanism; a particularly well-documented example has been described by Gregg (1980). Larson & Gregg (1983) have compared measured dissipation rates in profiles through intrusions with the production of turbulent energy by double-diffusive buoyancy fluxes. Their measurements are consistent with simple salt fingering in some regions, but they suggest some production of energy by shear in others; in addition, they point to the need for more accurate laboratory measurements of fluxes, especially at values of R_ρ near unity.

Note the remarkable general implication of the measurements that indicate that salt fingers are important in intrusions. Not only can molecular diffusion affect the motions on the layer scale, i.e. tens of meters in the vertical, but it can also, by driving the intrusions across fronts, influence large-scale mixing between water masses.

There is as yet no completely satisfactory theoretical description of the height scale or velocity of double-diffusive intrusions. Ruddick & Turner (1979) used a simple energy argument that is consistent with the vertical scale observed in their experiments with a sharp vertical front. For the oceanographically more realistic case of smooth, compensating horizontal gradients of T and S, Stern (1967) used a linear stability analysis to show that with vertical gradients in the finger sense, variations of T and S along isopycnals can be dissipated by interleaving motions that are driven by salt fingers. Toole & Georgi (1981) have extended Stern's results in a series of calculations (some of them including rotation) that give the scale and tilt of the fastest-growing mode. Their model is based on the assumption of eddy fluxes of heat and salt due to fingers, which depend only on the vertical gradient of S. Holyer (1983) has considered a similar problem (in the nonrotating case), but she parameterizes the flux divergences in a manner that uses molecular diffusivities and also includes horizontal derivatives of T and S. Her analysis is therefore indicative of the effect of horizontal gradients on the fingers themselves, rather than descriptive of the larger-scale horizontal interleaving that is caused by the fingers. It is still difficult to reconcile any of these theories with the laboratory results or with the limited field observations.

LANGMUIR CIRCULATIONS IN THE SURFACE LAYER   All the phenomena considered so far have depended on the presence of different thermal and solute diffusivities. Now we discuss a process that involves the diffusion of momentum, the two relevant transport parameters being a turbulent or eddy diffusivity K_T and an eddy viscosity v_T. Leibovich (1983) has recently reviewed the theories related to Langmuir cells and the formation of windrows. Their generation depends on an instability mechanism in which there is an interaction between a wind-driven current shear $U(z)$ and a Stokes drift produced by surface waves. Physically, this implies that the vertical vorticity associated with small lateral variations of the shear flow is twisted by the presence of the Stokes drift into that of the Langmuir circulations, with growing motions in a plane transverse to the wind direction.

When a stabilizing temperature gradient is added to this flow, the linearized equations governing the stability to rolls aligned with U are precisely those describing double-diffusive instabilities in a salinity gradient heated from below. In this case we have $v_T > K_T$, and the velocity plays the role previously taken by temperature, and temperature that of the stabilizing salt gradient in the heat/salt problem. The results have been extended to finite amplitude, including regions of monotonic growth in which the Langmuir cells distort the initial constant temperature gradient. These calculations produce a nearly uniform layer (in the horizontally averaged temperature) with a strong thermocline below it, but it is not clear if they properly describe vertical mixing, since the dynamical effects of the temperature field have not been considered.

Another instability involving angular momentum and horizontal density gradients has been applied in the oceanic context (Lambert 1982); this is treated in the astrophysics section (Section 4.1), since it was first discussed in that field.

3.2 Chemical Studies

Chemists measuring diffusion coefficients, and trying to understand the variations with concentration and temperature, have usually sought to avoid convective effects because they introduce unwanted complications. In some cases, however, they can be exploited [e.g. by Caldwell (1973) in his determination of the Soret coefficient for NaCl]. Furthermore, the deliberate exclusion of multicomponent convection from experiments might in fact make the measurements irrelevant to practical cases in which convection does occur. The influence of cross-diffusion effects in particular has already been discussed in general terms. Now we refer specifically to several observational studies carried out in a chemical or biochemical context that are consistent with this mechanism, but that have previously been described in a very different way.

OBSERVATIONS OF FINGERS IN POLYMER SOLUTIONS   In a series of papers, Preston and coworkers (Preston et al. 1980, Comper et al. 1983) have investigated transports between two layers containing various polymers in aqueous solution. An extensively studied case is a ternary system containing dextran and a small amount of a high-molecular-weight, long-chain polymer PVP in the bottom layer, and a less-concentrated dextran solution in the layer above it; in some experiments they added a trace of sorbitol to monitor the water in the bottom layer. The transport of either radioactively labeled dextran or PVP in this experiment is extremely rapid compared with ordinary diffusion in binary systems.

The PVP concentration in the upper layer increases linearly in time, rather than as $t^{1/2}$ (which is characteristic of a diffusion process), while the sorbitol (and water) do obey the $t^{1/2}$ relation. This rapid polymer transport is a consequence of the formation of fingers and the associated vertical convective motions. The phenomena continue to be observed when both components are more concentrated at the bottom, or even when the lower solution is supplemented with either low molecular weight solutes (such as NaCl) or dense solvents.

The theory of McDougall (1983a) has been applied to this case. He has argued (as described earlier) that cross-diffusion terms, if they are large enough for (8) to be satisfied, can allow fingers to form because of the property variations set up in the counterflowing fingers (see Figure 6). An extra (favorable) factor of $\tau^{-1/2}$ appears in the criterion for cross-diffusion fingers to form from an initially sharp interface [cf. the difference between Equations (2) and (5)]. There are as yet few quantitative experiments, and definitive comparisons between theory and experiment are difficult because of the lack of measurements of D_{12} and D_{21}. It would be especially valuable to know all four coefficients for a ternary system containing low-molecular-weight solutes with globular molecules so that the importance of coupled diffusion could be assessed and compared with possible effects due to networks of long-chain molecules in the polymer systems.

PREVIOUS DESCRIPTIONS AND INTERPRETATIONS   The motivation for much of the work just described has been biological, with the eventual aim of relating the observations to transport phenomena in connective tissues and membranes. In the laboratory model system, however, it seems clear that gravity has always been an important factor, since all the ordered structures (which we have identified with fingers) grow vertically. The weak dependence of fluxes on g observed in centrifuge experiments (Preston et al. 1980) is also consistent with this view (McDougall 1983a).

Explanations of the structure in terms of chemical potential gradients and other concepts familiar to physical chemists tend to obscure this dependence on gravity, and they give what I believe to be false hope that similar arguments might be extended to nonhorizontal interfaces. Consider again the simple two-layer system sketched in Figure 4. The major energy change (and increase in entropy) between states 1 and 2 are certainly associated with the free energy. Because the solute S is more uniformly distributed in state 2, there has been a decrease in the Gibbs free energy of $RT \ln 2$ per mole (where R is the gas constant and T the absolute temperature), which for 10-cm-deep layers of molar concentration amounts to approximately 17 J cm^{-2}. The corresponding change in gravitational potential energy is only about 10^{-4} times as large, and if the second component is included, the contrast between the gravitational potential energy and the free-energy changes is even greater. But the fact that the free-energy change is so much larger does *not* mean that it is the most important kinetically. It is the gravitational term that appears in the dynamical equations (i.e. the equations of motion) and provides the mechanism for organizing the flow into the efficient form shown in Figure 3. The coupling between vertical motions driven by gravity and diffusion acting on small scales produces much more rapid transports of properties than is possible with molecular diffusion alone.

Note that the final state 3 in Figure 4 represents a further large decrease in free energy (as the second component also becomes uniformly distributed), while there is an *increase* in gravitational potential energy (back to the original value in the specific example described). The process is very slow because only molecular diffusion is now acting, and there is no release of gravitational potential energy to organize a convective flow. Arguments based on fluxes driven by osmotic pressure, or more detailed thermo-dynamic considerations of the effect of one component on the chemical potential gradient of another, can go further than the free-energy calculations sketched above in describing the molecular processes that produce coupled transports. They do, however, all suffer from the same drawback: They cannot in themselves explain the origin of the regular vertically oriented structures and the very much larger fluxes. These organized flow structures depend essentially on gravity and cannot be produced by diffusion alone.

4.   EXTENSIONS TO OTHER DIFFUSIVE SYSTEMS

4.1   *Astrophysics*

As noted in Section 1, there are many diffusing properties in stars and, therefore, a wide variety of multicomponent convection effects that can be included in the broader definition. A direct analogue of heat/salt diffusive convection has been used to explain the properties of large stars with a helium-rich core, which is heated from below and thus convecting. Spiegel (1972) has shown that variations in the helium/hydrogen ratio can produce a density gradient that limits the helium transport by double-diffusive convection, though whether this may be in layers is still unclear. The process is called *semiconvection*, since it allows the motion outside the core to regulate itself and remain nearly convectively neutral. The potential temperature will be almost uniform (since this is the driving component), though there can still be a stabilizing gradient of He. The possibility of "helium fingers" in stellar atmospheres has been suggested too, following accretion of He-rich material onto a star.

ROTATION   An early example of an instability that is akin to salt fingers, but that is due to differential rotation in the radiation zone of a star (where heat transport is rapid), was discussed by Goldreich & Schubert (1967). If the angular momentum decreases radially outward, this provides a destabilizing distribution of the more slowly diffusing component (angular momentum) that can be superimposed on a stable temperature gradient. This could lead to the formation of "angular-momentum fingers," drawing on the energy of the differential rotation. McIntyre (1970) developed this theme further and showed both that a baroclinic circular vortex can be destabilized by diffusive effects in various ways and that oscillatory modes can also occur.

In the oceanic application where eddy viscosity and diffusivity may be appropriate and $\sigma > 1$, viscosity (now the faster transport process) can again be destabilizing and produce an overturning that leads to the formation of layers in a density stratification [see Lambert (1982) and the section on Langmuir circulations]. This type of instability occurs at Richardson numbers much greater than the critical value for nonrotating shear instability. Baker (1971) and Calman (1977) have described related laboratory experiments in which layers were produced by differential rotation of a disk in a rotating cylinder of water.

Worthem et al. (1983) have summarized relevant earlier work and carried out a linear stability analysis of a doubly diffusive (heat-salt) system on which rotation and shear have been superimposed. Their calculations reduce to the Baines & Gill (1969) results in one limit and the McIntyre (1970) rotating instability in another, and they indicate that shear and rotation in combination could cause instability and the formation of layers under oceanic conditions that would otherwise be stable. Recent theoretical work on rotating stars has added a compositional gradient (e.g. a H-He gradient) to temperature and angular momentum, where now the diffusivities have the relative magnitudes $\kappa_{He} \leq \nu \ll \kappa_T$.

MAGNETIC FIELDS   In the outer layers of stars like the Sun, thermal convection is affected by the presence of magnetic fields. The nature of magnetoconvection depends crucially on the ratio $\zeta = \eta/\kappa_T$ of the magnetic diffusivity η to the thermal diffusivity κ_T. In the astrophysical context ζ is small (because of radiative heat transport), and in this regime a rich variety of both steady and oscillatory solutions are possible. Proctor & Weiss (1982) have recently reviewed this subject, and they assert that "convection in a magnetic field is now the best studied example of double-diffusive convection and serves as a guide to the behavior of related systems." Magnetic fields can be stabilizing or destabilizing; they are also vector functions that can be concentrated by convective motions, so there are phenomena that are peculiar to this system. For example, theory can now shed light on the existence and properties of sunspots, which are maintained by strongly nonlinear processes.

The opposite case of ζ large is relevant to laboratory experiments and to studies of the Earth's core, and this work too is the subject of recent reviews (Busse 1983, Loper & Roberts 1983). A further reference to this case is made later, though neither of these vast subjects can be pursued much further here. Though particular combinations of effects have been isolated to make the studies manageable, all these systems are in principle quadruply diffusive and can involve angular momentum, magnetic field, temperature, and composition simultaneously. There are few finite-amplitude calculations, and even fewer that include horizontal gradients or spatial nonuniformities.

4.2   Metallurgy and Materials Science

The phenomenon of crystal growth has taken on a new practical importance with the increasing needs of the electronics industry for larger and more chemically homogeneous crystals. Crystal growth from a solution or melt involves both heat and mass transfer, and these usually lead to convection in the fluid, often double-diffusive in character. In common practice, convection was regarded as always deleterious and thus to be avoided (perhaps by mechanical stirring). Detailed fluid-dynamical studies have only recently begun to contribute to the understanding of the processes that lead to fluctuations in growth rate and consequent nonuniformities in crystals.

STABILITY OF A HORIZONTAL PLANAR INTERFACE   As a crystal grows upward from the cooled horizontal boundary of a region containing a solution or melt, an initially planar interface may become unstable in two ways: morphologically or hydrodynamically. Morphological instability pertains to the spontaneous change in shape of a two-phase interface during a phase change. The consequent influence on surface boundary conditions couples this to hydrodynamic instabilities in the fluid above, which depend on both temperature gradients and compositional gradients set up by the depletion or concentration in the melt of the crystallizing components [see Langer (1980), Coriell & Sekerka (1981), and Azouni (1981) for reviews of recent work]. The nature of the stability depends on the sign of the boundary concentration gradient, which decays exponentially away from the interface with length scale D/V, where D is the liquid diffusion coefficient and V the velocity of advance of the interface.

Linear stability analyses of the coupled problem (which have recently been extended numerically into the nonlinear regime) show that for large V and stabilizing boundary gradients of T and composition, the growing interface becomes morphologically unstable to short wavelength disturbances, which lead to the formation of dendrites. The mechanism has been understood qualitatively for many years; it is called "constitutional supercooling," and it depends on a local increase of solute concentration near the boundary. On the other hand, when a less dense component is rejected during crystallization and the T and compositional gradients are opposing, a truly double-diffusive instability is possible. The resulting fluid motions are described in the next section. The intermediate range, where there is close coupling between fluid and morphological effects, leads to more complicated behavior, including oscillatory modes. A noteworthy new numerical method of attack on this and related problems has been described by Ungar & Brown (1983).

Surface tension can also be important in this context, especially in a low-gravity environment. Convection phenomena produced, or partly driven, by gradients of interfacial surface tension are called "Marangoni effects." When these are combined with other driving forces, the problem becomes one involving multiple diffusivities, and the stability must be assessed accordingly. [Castillo & Velarde (1982) have used an energy argument to study this case.]

SOLIDIFICATION AND FLUID FLOW   The motions that develop above a cooled, rapidly crystallizing boundary at which lighter residual fluid is being released are closely related to salt fingers, though they are complicated by a "mushy zone" of crystals through which the convecting motions penetrate. These produce long, narrow segregated regions (called "freckles" because of their appearance in horizontal cross-sections), which can seriously weaken alloy castings. The phenomenon has recently attracted increased attention because of its relevance to solidification at the Earth's inner core (Roberts & Loper 1983).

It is still not possible to predict in detail the rate of growth of a dendritic interface: The presence of dendrites can retard the volume growth by a large factor compared with the theory for a planar interface. Associated with this growth is the rejection of solute laterally between the dendrite arms, and the behavior of even a single growing tip will depend strongly on the direction of the "compositional convection" set up by this process.

The large-scale effects of fluid flow in a solidifying ingot are better understood (Fisher 1981), at least in a qualitative way. Macrosegregation, or the gross separation of different components by the relative motion of crystals and remaining melt under gravity, is accomplished by mechanisms related to those described in the next section. They are complicated by the large fraction of a casting that contains dendrites (so that much of the convection is interdendritic) and by flows due to the shrinkage accompanying solidification. Existing theories give good predictions when the latter effect is dominant, but the convective case remains a major unsolved double-diffusive problem.

4.3 *Geology*

The rate of transfer of ideas from oceanographic to geological contexts has accelerated in recent years, following the review by Turner & Gustafson (1978). They considered the various flow phenomena that can result from the efflux of hot, salty water from vents in the seafloor, in relation to the genesis of massive sulfide ore deposits. This application subsequently has been pursued further (McDougall 1984, Campbell et al. 1984). It was soon realized, however, that double-diffusive effects can also be important during the solidification of large magma chambers (i.e. in magmas as well as aqueous solutions). It can now be confidently asserted that they probably play a role in all major igneous processes, and much of the recent work [reviewed by Huppert & Turner (1981a) and more particularly by Huppert & Sparks (1984)] has been carried out with these processes in mind. Aqueous solutions have also been used for the laboratory modeling of magma chambers, with crystallization added to the double-diffusive systems previously studied; these laboratory experiments are our main concern here.

CRYSTALLIZATION AND COMPOSITIONAL CONVECTION   Once thermal and compositional gradients have become established in a chamber, multicomponent convection will inevitably lead to layering. The starting point of the recent fluid-dynamical work was the idea that extensive quasi-horizontal layers are more likely to form in the fluid state, and that the crystallization and final mineral layering can subsequently be affected by these preexisting layers. This is very different from the long-held idea that crystal settling dominates layer formation in igneous rocks.

Most of the convective phenomena studied, including those responsible for setting up the compositional gradients from which layering can develop, depend on density changes produced by crystallization. In magmas (and many common solutions that can be used as laboratory analogues), crystallization causes much larger changes in melt (or solution) density than the associated temperature changes. In Na_2CO_3 solution, for example, the residual fluid is cold but less dense, and upward "compositional convection" can occur; its effects have now been studied in many different geometries.

Sidewall crystallization   Turner (1980) and Turner & Gustafson (1981) have shown how the cooling and crystallization of a homogeneous solution at a vertical wall generates a boundary-layer flow, which rises to the top to form a growing region that is thermally and compositionally stratified. Worster & Leitch (1984) have made a detailed theoretical and experimental study of the stratification produced by a laminar boundary layer rising along a plane vertical wall. Smaller-scale double-diffusive layering develops in the upper stratified region because of the cooling from the top and the side. When there are several components in the solution or melt, the same process leads to differentiation or fractionation, i.e. a vertical zonation of the several components.

McBirney (1980) has shown how this model can account for the changes with time in the composition of lava erupted from the same volcano, or the contrasting compositions from neighboring vents that tap different levels in a single stratified magma chamber. The striking trace-element gradients in ash-flow deposits derived from some high-silica rhyolite magma chambers (Hildreth 1981) can also be explained by convective fractionation. Highly fractionated fluids, with element concentrations depending systematically on their distribution coefficients, can be generated in boundary-layer flows that involve only a small proportion of the total chamber volume [see Sparks et al. (1984) for a fuller discussion]. In the future, the decisive tests of the dynamical ideas are likely to depend on a critical reevaluation of trace-element and isotopic data.

Top cooling   Chen & Turner (1980) also cooled and crystallized concentrated solutions from the top, with the roof of the container held below the eutectic temperature. These experiments have been followed up by Huppert et al. (1984b) with an application to komatiite (very high temperature) lavas in mind. Dendritic crystals were observed to grow down nearly vertically from the roof in a static fluid layer generated by the light residual fluid. The eutectic crystals advanced more slowly, with a horizontal front, and filled the interdendritic spaces (see Figure 8). The morphology of these crystals resembled the "spinifex texture" observed in komatiites, while the crystals growing on the bottom in a convecting layer were more solid and randomly oriented.

REPLENISHED MAGMA CHAMBERS   Experiments of a different kind have also been used to model chambers replenished from below with magma of different composition and temperature. Denser and less dense inputs, rapid and slower injection rates, and homogeneous or already stratified tanks have all been investigated with particular geological processes in mind. Three cases, all with hot, denser inputs, are given here as examples.

Huppert & Turner (1981b) placed a hot layer of KNO_3 solution under a cold, lighter $NaNO_3$ layer, taking care to minimize the mixing and to produce a sharp (triple-diffusive) interface. The rapid convective heat transfer through the interface led to the growth of crystals on the bottom, so that the density of the residual lower solution decreased. When it became equal to that in the upper layer, the interface broke down and the layers mixed thoroughly together. This is the behavior to be expected in a basaltic magma chamber, with nearly equal viscosities of the two layers. When the viscosity of the upper layer was much greater (as it would be with rhyolitic magma), then Huppert et al. (1983) showed, using glycerine as the upper fluid, that crystallization along the interface led to light residual KNO_3 fluid being released continuously into the viscous layer. There was little mixing, and a layer of light KNO_3 was deposited above the glycerine (with, incidentally, fingers forming on the interface with the glycerine below).

With a rapid input of either lighter or denser fluid, there can be extensive turbulent mixing with the originally homogeneous fluid already in the chamber. Campbell & Turner (1984) have examined the case of a dense, hot, salty fountain that is ejected upward from a source in the base of a tank of freshwater and falls back to produce a stratified region that breaks up into

Figure 8 Shadowgraph picture (taken 18 hr after the start) showing the effect of strong cooling at the top of a tank containing a stable gradient of Na_2CO_3 solution. Convecting double-diffusive layers formed first, and these influenced the subsequent growth of the crystals and the eutectic layer above. (From Chen & Turner 1980.)

thinner double-diffusive layers. Campbell & Murck (1984) have used this model to explain the properties and distribution of chromite seams in large igneous intrusions such as the Bushveld. This experiment is an extension of the "filling box" model of Baines & Turner (1969), which is also the basis for the work of McDougall (1984) on double-diffusive plumes in confined environments and their application to massive sulfide ore deposits.

4.4   Geophysics

There are many large-scale examples of processes in the Earth that can be influenced by multicomponent or compositional convection. The modification of the composition of midoceanic ridge basalts by convective processes in a magma chamber below, and the more general problem of differentiation of the crust from the mantle have been discussed in these terms. Two further examples are now described.

MANTLE CONVECTION   It is now believed that mantle convection must be the major driving mechanism for plate tectonics. There is still a debate about whether the convection occurs throughout the mantle or is confined to the upper 650 km (Spohn & Schubert 1982). A possible alternative, discussed by Richter (1979) and Stevenson & Turner (1979), for example, is that the mantle is chemically layered, with independent convective circulations in the upper and lower mantle and heat transfer (but little transfer of matter) across a double-diffusive interface between them. McKenzie & Richter (1981) concluded that if convection does occur in two layers, then the present surface heat flux is determined mainly by the initial thermal conditions of the Earth and not by the radioactive generation of heat.

Subducting plates are compositionally less dense than the mantle; they are made denser and plunge downward only because they are colder. As they sink, they equilibrate in temperature, and a small fraction of the lithosphere melts to form a water-rich component that is less dense than the surrounding mantle (see Ringwood 1982). This lighter material rises, and further melting occurs at shallower depths, to form andesitic volcanoes at an island arc or continental margin. This process too can be regarded as double-diffusive in character.

THE EARTH'S CORE   Several of the topics discussed in other contexts have implications for the dynamics of the Earth's core. Gubbins et al. (1979) and Loper & Roberts (1983) have convincingly argued that compositional convection, due to the solidification of dense metallic iron-nickel crystals on the inner core and the rejection of light components such as silicon, will be much more important than thermal effects in driving convection in the outer core. The latter authors further suggest that the boundary is likely to be dendritic, as observed for solidifying metal ingots; the evidence for this, however, is inconclusive.

An examination of the energetics of the core shows that compositional convection is in principle capable of sustaining a dynamo with a large toroidal magnetic field, whereas thermal driving is inefficient because of the heat conducted down the adiabatic gradient. The study of the magneto-hydrodynamic effects themselves still involves at least the magnetic field, rotation, and composition, all with different diffusion rates; this too has been recognized as a multicomponent convection problem.

5. CONCLUDING REMARKS

The first purpose of this article has been to show how the presence of several components in a fluid, each influencing the density and having different diffusive properties, can lead to dynamic instabilities, often well before a fluid system would become statically unstable. The convective phenomena described depend essentially on gravity or some other external force field. It is the coupling between this force field and diffusion that organizes the flow into a form that produces much more rapid transports than diffusion alone could achieve. Linear stability theories have now been extended to increasingly complicated multicomponent systems, and these give a first indication of the fully developed behavior; however, more nonlinear and numerical studies will certainly be needed in the future. Laboratory studies of idealized systems have been, and will continue to be, important in the development of this subject.

The second aim of this review has been to emphasize the very broad range of disciplines in which the effects of multicomponent convection should be taken into account. If these effects are ignored, a whole range of qualitatively different phenomena can be missed entirely. In particular, mixing rates will be seriously underestimated if only molecular diffusion is considered without taking into account the coupled convection that can occur in a gravitational field.

This review has been necessarily very selective, with space for only a few examples in each field. Other possible applications have not been discussed at all, and there are no doubt many more areas that readers will now be able to relate to the ideas presented here. The recent recognition that multicomponent convection is a fundamental phenomenon of importance in many disciplines should certainly lead to more rapid progress over the next few years.

ACKNOWLEDGMENTS

The author is grateful to I. H. Campbell, R. W. Griffiths, H. E. Huppert, G. N. Ivey, P. F. Linden, T. J. McDougall, and H. J. V. Tyrrell for their helpful comments on an earlier draft of this review.

Literature Cited

Azouni, M. A. 1981. Time-dependent natural convection in crystal growth systems. *PhysicoChem. Hydrodyn.* 2:295–309

Baines, P. G., Gill, A. E. 1969. On thermohaline convection with linear gradients. *J. Fluid Mech.* 37:289–306

Baines, W. D., Turner, J. S. 1969. Turbulent buoyant convection from a source in a confined region. *J. Fluid Mech.* 37:51–80

Baker, D. J. 1971. Density gradients in a rotating stratified fluid: experimental evidence for a new instability. *Science* 172:1029–31

Busse, F. H. 1983. Recent developments in the dynamo theory of planetary magnetism. *Ann. Rev. Earth Planet. Sci.* 11:241–68

Caldwell, D. R. 1973. Thermal and Fickian diffusion of sodium chloride in a solution of oceanic concentration. *Deep-Sea Res.* 20:1029–39

Caldwell, D. R. 1983. Small-scale physics of the ocean. *Rev. Geophys. Space Phys.* 21:1192–1205

Calman, J. 1977. Experiments on high Richardson number instability of a rotating stratified shear flow. *Dyn. Atmos. Oceans* 1:277–97

Campbell, I. H., Murck, B. W. 1984. A model for chromite seams in layered intrusions. *J. Petrol.* In press

Campbell, I. H., Turner, J. S. 1984. The fluid dynamics of fountains in magma chambers. *Proc. ISEM Field Conf. Open Magmat. Syst.*, ed. M. A. Dungan, T. L. Grove, W. Hildreth, pp. 23–25. Dallas: Inst. Study of Earth and Man, South. Methodist Univ.

Campbell, I. H., McDougall, T. J., Turner, J. S. 1984. A note on fluid dynamic processes which can influence the deposition of massive sulphides. *Econ. Geol.* In press

Castillo, J. L., Velarde, M. G. 1982. Buoyancy-thermocapillary instability: the role of interfacial deformation in one- and two-component fluid layers heated from below or above. *J. Fluid Mech.* 125:463–74

Chen, C. F., Johnson, D. H. 1984. Double-diffusive convection: a report on an Engineering Foundation Conference. *J. Fluid Mech.* 138:405–16

Chen, C. F., Turner, J. S. 1980. Crystallization in a double-diffusive system. *J. Geophys. Res.* 85:2573–93

Comper, W. D., Preston, B. N., Laurent, T. C., Checkley, G. J., Murphy, W. H. 1983. Kinetics of multicomponent transport by structured flow in polymer solutions. 4. Relationships between the formation of structured flows and kinetics of polymer transport. *J. Phys. Chem.* 87:667–73

Coriell, S. R., Sekerka, R. F. 1981. Effects of convective flow on morphological stability. *PhysicoChem. Hydrodyn.* 2:281–93

Fisher, K. M. 1981. The effects of fluid flow on the solidification of industrial castings and ingots. *PhysicoChem. Hydrodyn.* 2:311–26

Gargett, A. E. 1976. An investigation of the occurrence of oceanic turbulence with respect to finestructure. *J. Phys. Oceanogr.* 6:139–56

Gargett, A. E., Schmitt, R. W. 1982. Observations of salt fingers in the central waters of the eastern North Pacific. *J. Geophys. Res.* 87:8017–29

Germeles, A. E. 1975. Forced plumes and mixing of liquids in tanks. *J. Fluid Mech.* 71:601–23

Goldreich, P., Schubert, G. 1967. Differential rotation in stars. *Astrophys. J.* 150:571–87

Gordon, A. L. 1981. South Atlantic thermocline ventilation. *Deep-Sea Res.* 28:1239–64

Gregg, M. C. 1980. The three-dimensional mapping of a small thermohaline intrusion. *J. Phys. Oceanogr.* 10:1468–92

Griffiths, R. W. 1979a. The influence of a third diffusing component upon the onset of convection. *J. Fluid Mech.* 92:659–70

Griffiths, R. W. 1979b. A note on the formation of "salt-finger" and "diffusive" interfaces in three-component systems. *Int. J. Heat Mass Transfer* 22:1687–93

Griffiths, R. W. 1979c. The transport of multiple components through thermohaline diffusive interfaces. *Deep-Sea Res.* 26A:383–97

Griffiths, R. W., Ruddick, B. R. 1980. Accurate fluxes across a salt-sugar finger interface deduced from direct density measurements. *J. Fluid Mech.* 99:85–95

Gubbins, D., Masters, T. G., Jacobs, J. A. 1979. Thermal evolution of the Earth's core. *Geophys. J. R. Astron. Soc.* 59:57–99

Hildreth, W. 1981. Gradients in silicic magma chambers: implications for lithospheric magmatism. *J. Geophys. Res.* 86:10153–92

Holyer, J. Y. 1981. On the collective instability of salt fingers. *J. Fluid Mech.* 110:195–207

Holyer, J. Y. 1983. Double-diffusive interleaving due to horizontal gradients. *J. Fluid Mech.* 137:347–62

Huppert, H. E. 1983. Multicomponent convection: turbulence in Earth, Sun and sea. *Nature* 303:478–79

Huppert, H. E., Hallworth, M. A. 1984. Static and dynamic stability criteria during free diffusion in a ternary system. *J. Phys. Chem.* In press

Huppert, H. E., Linden, P. F. 1979. On heating a stable salinity gradient from below. *J. Fluid Mech.* 95:431–64

Huppert, H. E., Manins, P. C. 1973. Limiting conditions for salt-fingering at an interface. *Deep-Sea Res.* 20:315–23

Huppert, H. E., Moore, D. R. 1976. Nonlinear double-diffusive convection. *J. Fluid Mech.* 78:821–54

Huppert, H. E., Sparks, R. S. J. 1984. Double-diffusive convection due to crystallization in magmas. *Ann. Rev. Earth Planet. Sci.* 12:11–37

Huppert, H. E., Turner, J. S. 1980. Ice blocks melting into a salinity gradient. *J. Fluid Mech.* 100:367–84

Huppert, H. E., Turner, J. S. 1981a. Double-diffusive convection. *J. Fluid Mech.* 106:299–329

Huppert, H. E., Turner, J. S. 1981b. A laboratory model of a replenished magma chamber. *Earth Planet. Sci. Lett.* 54:144–72

Huppert, H. E., Sparks, R. S. J., Turner, J. S. 1983. Laboratory investigations of viscous effects in replenished magma chambers. *Earth Planet. Sci. Lett.* 65:377–81

Huppert, H. E., Kerr, R. C., Hallworth, M. A. 1984a. Heating or cooling a stable compositional gradient from the side. *Int. J. Heat Mass Transfer.* In press

Huppert, H. E., Sparks, R. S. J., Turner, J. S., Arndt, N. T. 1984b. The emplacement and cooling of komatiite lavas. *Nature* 309:19–22

Hurle, D. T. J., Jakeman, E. 1971. Soret-driven thermosolutal convection. *J. Fluid Mech.* 47:667–87

Lambert, R. B. 1982. Lateral mixing processes in the Gulf Stream. *J. Phys. Oceanogr.* 12:851–61

Lambert, R. B., Demenkow, J. W. 1972. On the vertical transport due to fingers in double-diffusive convection. *J. Fluid. Mech.* 54:627–40

Langer, J. S. 1980. Instabilities and pattern formation in crystal growth. *Rev. Mod. Phys.* 52:1–28

Larson, N. G., Gregg, M. C. 1983. Turbulent dissipation and shear in thermohaline intrusions. *Nature* 306:26–32

Leibovich, S. 1983. The form and dynamics of Langmuir circulations. *Ann. Rev. Fluid Mech.* 15:391–427

Linden, P. F. 1976. The formation and destruction of fine-structure by double-diffusive processes. *Deep-Sea Res.* 23:895–908

Linden, P. F., Shirtcliffe, T. G. L. 1978. The diffusive interface in double-diffusive convection. *J. Fluid Mech.* 87:417–32

Loper, D. E., Roberts, P. H. 1983. Compositional convection and the gravitationally powered dynamo. In *Stellar and Planetary Magnetism*, ed. A. M. Soward, pp. 297–327. New York: Gordon & Breach

Marmorino, G. O., Caldwell, D. R. 1976. Heat and salt transport through a diffusive thermohaline interface. *Deep-Sea Res.* 23:59–67

Marzec, C. J., Spiegel, E. A. 1980. Ordinary differential equations with strange attractors. *SIAM J. Appl. Math.* 38:403–21

McBirney, A. R. 1980. Mixing and unmixing of magmas. *J. Volcanol. Geotherm. Res.* 7:357–71

McDougall, T. J. 1981. Double-diffusive convection with a nonlinear equation of state. II. Laboratory experiments and their interpretation. *Prog. Oceanogr.* 10:71–89

McDougall, T. J. 1983a. Double-diffusive convection caused by coupled molecular diffusion. *J. Fluid Mech.* 126:379–97

McDougall, T. J. 1983b. Greenland Sea bottom water formation: a balance between advection and double-diffusion. *Deep-Sea Res.* 30:1109–17

McDougall, T. J. 1984. Fluid dynamic implications for massive sulphide deposits of hot saline fluid flowing into a submarine depression from below. *Deep-Sea Res.* 31:145–70

McDougall, T. J., Taylor, J. R. 1984. Flux measurements across a finger interface at low values of the stability ratio. *J. Mar. Res.* 42:1–14

McDougall, T. J., Turner, J. S. 1982. Influence of cross-diffusion on "finger" double-diffusive convection. *Nature* 299:812–14

McIntyre, M. E. 1970. Diffusive destabilization of a baroclinic circular vortex. *Geophys. Fluid Dyn.* 1:19–57

McKenzie, D. P., Richter, F. M. 1981. Parameterized thermal convection in a layered region and the thermal history of the Earth. *J. Geophys. Res.* 86:11667–80

Newall, T. A. 1984. Characteristics of a double diffusive interface at high density stability ratios. *J. Fluid Mech.* In press

Newman, F. C. 1976. Temperature steps in Lake Kivu, a bottom heated saline lake. *J. Phys. Oceanogr.* 6:157–63

Paliwal, R. C., Chen, C. F. 1980. Double-diffusive instability in an inclined fluid layer. 2. Theoretical investigation. *J. Fluid Mech.* 98:769–85

Piacsek, S. A., Toomre, J. 1980. Nonlinear evolution and structure of salt fingers. In *Marine Turbulence*, ed. J. C. J. Nihoul, pp. 193–219. Amsterdam: Elsevier

Platten, J. K., Legros, J. C. 1983. *Convection in Liquids*. Berlin: Springer-Verlag. 700 pp.

Poplawsky, C. J., Incropera, F. P., Viskanta, R. 1981. Mixed layer development in a double-diffusive, thermohaline system. *ASME J. Sol. Energy Eng.* 103:351–59

Preston, B. N., Laurent, T. C., Comper, W. D., Checkley, G. J. 1980. Rapid polymer transport in concentrated solutions through the formation of ordered structures. *Nature* 287:499–503

Proctor, M. R. E. 1981. Steady subcritical thermohaline convection. *J. Fluid. Mech.* 105:507–21

Proctor, M. R. E., Weiss, N. O. 1982. Magneto-convection. *Rep. Prog. Phys.* 45: 1317–79

Richter, F. M. 1979. Focal mechanisms and seismic energy release of deep and intermediate earthquakes in the Tonga-Kermadec region and their bearing on the depth extent of mantle flow. *J. Geophys. Res.* 84:6783–95

Ringwood, A. E. 1982. Phase transformations and differentiation in subducted lithosphere: implications for mantle dynamics, basalt petrogenesis and crustal evolution. *J. Geol.* 90:611–43

Roberts, P. H., Loper, D. E. 1983. Towards a theory of the structure and evolution of a dendrite layer. In *Stellar and Planetary Magnetism*, ed. A. M. Soward, pp. 329–49. New York: Gordon & Breach

Ruddick, B. R., Turner, J. S. 1979. The vertical length scale of double-diffusive intrusions. *Deep-Sea Res.* 26:903–13

Schechter, R. S., Prigogine, I., Hamm, J. R. 1972. Thermal diffusion and convective stability. *Phys. Fluids* 15:379–86

Schmitt, R. W. 1979a. The growth rate of super-critical salt fingers. *Deep-Sea Res.* 26A:23–40

Schmitt, R. W. 1979b. Flux measurements on salt fingers at an interface. *J. Mar. Res.* 37:419–35

Schmitt, R. W. 1981. Form of the temperature-salinity relationship in the central water: evidence for double-diffusive mixing. *J. Phys. Oceanogr.* 11:1015–26

Schmitt, R. W. 1983. The characteristics of salt fingers in a variety of fluid systems, including stellar interiors, liquid metals, oceans and magmas. *Phys. Fluids* 26: 2373–77

Schmitt, R. W., Georgi, D. T. 1982. Fine-structure and microstructure in the North Atlantic Current. *J. Mar. Res.* 40:659–705 (Suppl.)

Shirtcliffe, T. G. L., Turner, J. S. 1970. Observations of the cell structure of salt fingers. *J. Fluid Mech.* 41:707–19

Sparks, R. S. J., Huppert, H. E., Turner, J. S. 1984. The fluid dynamics of evolving magma chambers. *Philos. Trans. R. Soc. London Ser. A.* 310:511–34

Spiegel, E. A. 1972. Convection in stars. II. Special effects. *Ann. Rev. Astron. Astrophys.* 10:261–304

Spohn, T., Schubert, G. 1982. Modes of mantle convection and the removal of heat from the Earth's interior. *J. Geophys. Res.* 87:4682–96

Stern, M. E. 1960. The "salt fountain" and thermohaline convection. *Tellus* 12:172–75

Stern, M. E. 1967. Lateral mixing of water masses. *Deep-Sea Res.* 14:747–53

Stern, M. E. 1969. Collective instability of salt fingers. *J. Fluid Mech.* 35:209–18

Stern, M. E. 1982. Inequalities and variational principles in double-diffusive turbulence. *J. Fluid Mech.* 114:105–21

Stern, M. E., Turner, J. S. 1969. Salt fingers and convecting layers. *Deep-Sea Res.* 34: 95–110

Stevenson, D. J., Turner, J. S. 1979. Fluid models of mantle convection. In *The Earth: Its Origin, Evolution and Structure*, ed. M. W. McElhinney, pp. 227–63. New York: Academic. 597 pp.

Toole, J. M., Georgi, D. T. 1981. On the dynamics and effects of double-diffusively driven intrusions. *Prog. Oceanogr.* 10: 123–45

Toomre, J., Gough, D. O., Spiegel, E. A. 1982. Time-dependent solutions of multimode convection equations. *J. Fluid Mech.* 125:99–122

Tsinober, A. B., Yahalom, Y., Shlien, D. J. 1983. A point source of heat in a stable salinity gradient. *J. Fluid Mech.* 135:199–217

Turner, J. S. 1965. The coupled turbulent transports of salt and heat across a sharp density interface. *Int. J. Heat Mass. Transfer* 38:375–400

Turner, J. S. 1967. Salt fingers across a density interface. *Deep-Sea Res.* 14:599–611

Turner, J. S. 1968. The behavior of a stable salinity gradient heated from below. *J. Fluid Mech.* 33:183–200

Turner, J. S. 1973. *Buoyancy Effects in Fluids.* Cambridge: Cambridge Univ. Press. 367 pp.

Turner, J. S. 1974. Double-diffusive phenomena. *Ann. Rev. Fluid Mech.* 6:37–56

Turner, J. S. 1978. Double-diffusive intrusions into a density gradient. *J. Geophys. Res.* 83:2887–2901

Turner, J. S. 1980. A fluid-dynamical model of differentiation and layering in magma chambers. *Nature* 285:213–15

Turner, J. S. 1981. Small-scale mixing processes. In *Evolution of Physical Oceanography*, ed. B. A. Warren, C. Wunsch, pp. 236–62. Cambridge, Mass: MIT Press

Turner, J. S., Gustafson, L. B. 1978. The flow of hot saline solutions from vents in the sea floor—some implications for exhalative massive sulfide and other ore deposits. *Econ. Geol.* 73:1082–1100

Turner, J. S., Gustafson, L. B. 1981. Fluid motions and compositional gradients produced by crystallization or melting at vertical boundaries. *J. Volcanol. Geotherm. Res.* 11:93–125

Turner, J. S., Shirtcliffe, T. G. L., Brewer, P. G. 1970. Elemental variations of transport coefficients across density interfaces in multiple-diffusive systems. *Nature* 228:1083–84

Ungar, L. H., Brown, R. A. 1983. The role of multiple parameters on the stability of a rotating liquid drop. *Philos. Trans. R. Soc. London Ser. A* 201:347–71

Velarde, M. G., Schechter, R. S. 1972. Thermal diffusion and convective instability. II. An analysis of the convected fluxes.

Phys. Fluids 15:1707–14

Walton, I. C. 1982. Double-diffusive convection with large variable gradients. *J. Fluid Mech.* 125:123–35

Wendt, R. P. 1962. The density gradient and gravitational stability during free diffusion in three-component systems. *J. Phys. Chem.* 66:1740–42

Williams, A. J. 1975. Images of ocean microstructure. *Deep-Sea Res.* 22:811–29

Worster, M. G., Leitch, A. M. 1984. Laminar free convection in confined regions. *J. Fluid Mech.* In press

Worthem, S., Mollo-Christensen, E., Ostapoff, F. 1983. Effects of rotation and shear on doubly diffusive instability. *J. Fluid Mech.* 133:297–319

Ann. Rev. Fluid Mech. 1985. 17:45–64

RHEOMETRY OF POLYMER MELTS

Joachim Meissner

Institut für Polymere, Eidgenössische Technische Hochschule (ETH), CH-8092 Zürich, Switzerland

1. INTRODUCTION

The rheological properties of polymers in the molten state are important for polymer engineering and science for the following reasons: (a) they depend sensitively on the chemical structure of the individual macromolecules and, therefore, are of interest for polymer characterization; (b) they are required for the development of a realistic fluid dynamics of polymer melts as the basis for any theory of polymer processing; and (c) they are responsible for large molecular orientations that are formed in the melt during processing and frozen into the final products. There they create the inhomogeneous anisotropy of all physical quantities and alter the technological end-use properties. In spite of this relevance, the central problem of polymer-melt rheology is not yet solved, viz. the formulation of a workable and *correct* constitutive equation that describes the stress at any instant for any deformation history. For monographs concerning this subject, see Lodge (1964, 1974), Truesdell & Noll (1965), Astarita & Marrucci (1974), Bird et al. (1977a,b), and Janeschitz-Kriegl (1983).

The difficulties are due to the complicated, nonlinear rheological behavior of polymer melts. These melts are not only highly viscous liquids, but in addition they are pseudoplastic (viscosity decreases with increasing shear rate) and viscoelastic (mechanical response is time- and frequency-dependent), and their viscous flow is often connected with large, rubberlike elastic deformations. Furthermore, polymer melts can be elongated remarkably without rupture.

In the following, methods of polymer-melt rheometry are reviewed and discussed. Most of these methods originated in the plastics and rubber industry or were transferred from other areas, like the methods of linear

45

0066–4189/85/0115–0045$02.00

viscoelasticity developed for polymer solutions and solids (Ferry 1980). Dealy (1982) has given a complete review of commercially available rheometers for polymer melts, and physical details of rheological instruments have been treated by Walters (1975) and Whorlow (1980). Most of the conventional methods consider only one aspect, e.g. the flow curve in capillary rheometry or the material functions in viscometric flows. But the melts show a time-dependent response, and several types of nonviscometric flows exist. At present, most instruments operate in unidirectional deformations only. The response to a more general deformation history, during which the main axes of shear or elongation are altered, can only be investigated with new instruments, the development of which began recently. Results from these instruments will also act as guidelines for further theoretical developments.

The discussion here is restricted to thermoplastic polymers. For amorphous polymers, the molten state is defined as the temperature range above the glass transition temperature T_g, and for partially crystalline polymers, it is the temperature range above the highest melting temperature T_m at which all the crystalline material is molten. We assume that the macromolecules are sufficiently stable during the rheological measurements and that polymer melts are incompressible, such that their rheological properties are not influenced by a superimposed hydrostatic pressure. This assumption is not exactly correct (Maxwell & Jung 1957, Westover 1961), but it is the usual simplification.

2. LINEAR VISCOELASTIC METHODS

For infinitesimally small strain or strain rate, polymeric liquids are linear viscoelastic with the constitutive equation (incompressibility assumption)

$$\mathbf{p}(t) = 2 \int_{-\infty}^{t} G(t - t')\dot{\varepsilon}(t')\, dt', \tag{1}$$

where $\mathbf{p}$ is the extra stress tensor, $\dot{\varepsilon}$ the strain-rate tensor, and $G(t)$ the relaxation modulus. Equation (1) expresses Boltzmann's superposition principle, with the consequence that the material response can be predicted from the excitation if any one of the linear material functions is known for a sufficiently broad range of time (or frequency).

The material functions are defined according to the excitation applied, which is often a step-function strain (*relaxation test*), a step-function stress (*creep test*), a step-function strain rate (*stressing test*; Giesekus 1965), or an excitation consisting of harmonic oscillations (*dynamic test*). Figure 1 summarizes the excitation of and material response to these tests and defines the following material functions: *relaxation modulus G(t), transient* or *stressing viscosity* $\eta(t)$, *complex modulus* $G^*(\omega) = G'(\omega) + iG''(\omega)$ (G' = *storage modulus*, G'' = *loss modulus*), *complex viscosity* $\eta^*(\omega)$, and *creep compliance* $J(t)$. The *relaxation spectrum* $H(\tau)$ is a distribution function of the strength of the relaxation processes corresponding to *relaxation times* τ, and $L(\tau')$ is the *retardation spectrum* with *retardation time* τ' (Ferry 1980).

Additional information can be obtained from combined tests, e.g. *creep recovery* with a sudden removal of the stress at instant t_1. The measured shear recovery γ_R allows the separation of $\gamma(t_1)$ into the recoverable, elastic and the irrecoverable, viscous portions : $\gamma(t_1) = \gamma_R + \gamma_v$. For a creep test long enough to achieve the linear part of the creep curve, the equilibrium compliance $J_e = \int_{-\infty}^{+\infty} L(\tau')\, d\ln \tau'$ is obtained; this is then equal to the recoverable compliance $J_R = \gamma_R/p_{21}$ and is another linear-viscoelastic material parameter, as is the zero-shear viscosity

$$\eta_0 = \lim_{\dot{\gamma} \to 0, t \to \infty} \eta(t).$$

The parameter η_0 depends sensitively on the weight-average molecular weight M_w of polymer melts : $\eta_0 \propto M_w^{3,4}$ for arbitrary unimodal molecular-weight distributions, whereas J_e is a function of this distribution (Graessley 1974, Zosel 1971). Zosel's measurements were in oscillation; from G' and

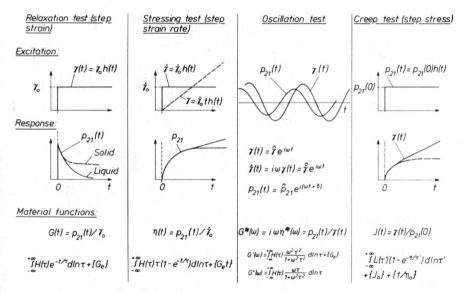

Figure 1 Basic tests in linear viscoelasticity (formulated for shear) and definition of material functions and their relations to the spectra $H(\tau)$ and $L(\tau')$. The unit step function is given by $h(t)$, and the quantities in braces denote optional constants (G_e: equilibrium shear modulus, J_0: instantaneous compliance).

G'', it follows that

$$J_e = \lim_{\omega \to 0} G'(\omega)/|G^*(\omega)|^2.$$

Instruments for linear-viscoelastic measurements have been reviewed by Ferry (1980), and Dealy (1982) has described commercial rheometers. For polymer melts, the excitations of Figure 1 and their combinations are utilized (Plazek 1968, denOtter 1969, Laun & Meissner 1980). For consistency tests, results from different modes have been compared, e.g. creep and constant strain rate in elongation (Münstedt 1979). After steady shear flow, the stress relaxes with $\int_{-\infty}^{+\infty} H(\tau)\tau \exp[-(t-t_1)/\tau]\, d\ln \tau$ more slowly than in simple relaxation, because $H(\tau)$ is enhanced by τ in the integral. For step-function strain, the problem of the finite rise time Δt exists (Laun 1978); a shear rheometer of short response time (< 10 ms) has been described by Laun (1982). These tests are characterized by one excitation parameter. The test conditions can be better optimized with the rheometer range and the melt properties in a step-function strain-rate test of duration t_1 followed by stress relaxation, with t_1 shorter than the time necessary to achieve steady flow (Meissner 1978, Smith 1979).

The test range of each rheometer can be extended by the time-temperature superposition principle (Ferry 1980). Polymer melts are thermo-rheologically simple (Schwarzl & Staverman 1952), i.e. on changing the temperature, all relaxation times change by the same factor a_T. This factor can be determined by shifting the viscoelastic quantity measured along the logarithmic time or frequency axis, such that, for the current temperature T and the reference temperature T_0, the curves overlap. The time and frequency reduced to T_0 are $t_r = t/a_T$ and $\omega_r = \omega a_T$, respectively, with t and ω corresponding to T. Figure 2 illustrates the application of this principle: On the left (a), the relaxation modulus $E(t)$ of polyisobutylene is shown as measured for $10 < t < 10^5$ s and $-80.8 \leq T \leq +50°C$. On the right (b), $T_0 = 25°C$ is selected as the reference temperature, and the curves are shifted such that they overlap, beginning with the curves for 50 and 0°C. The resulting master curve $E(t)$ extends over 16 decades. By means of $a_T(T)$ [shown in (c)], the master curve can be shifted from T_0 to other temperatures. However, the time-temperature superposition does not always work (Ferry 1980); for example, a few degrees above T_g, polystyrene can show different a_T for the viscous and the recoverable portions of the deformation (Plazek 1965).

3. EXTRUSION RHEOMETRY

The extrusion through capillaries and slits, and the correlation of output rate q and pressure gradient dp/dL along the die, is the oldest rheological

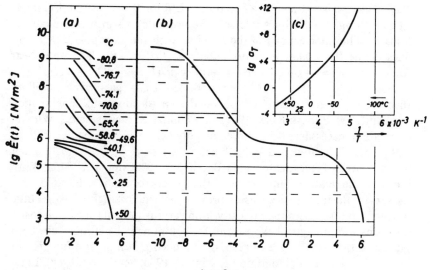

Figure 2 Time-temperature superposition of the elongational relaxation modulus $E(t)$ of polyisobutylene (Tobolsky & Catsiff 1956). Reference temperature is $T_0 = 25°C$.

test for molten polymeric systems (Marzetti 1923, Behre 1932, Nason 1945, Spencer & Dillon 1948, Merz & Colwell 1958). Extrusion rheometry is especially appropriate for high shear rates, and it represents a basic method for polymer fluid dynamics. However, it is difficult to measure at shear rates below 1 s^{-1}, and the resulting function $q(dp/dL)$ reflects neither time-dependent nor rubber-elastic rheological properties.

For laminar flows through a capillary of radius R, the shear stress at the wall is $p_{21}(R) = \frac{1}{2}R \, dp/dL$ and the *apparent shear rate* is $D = 4q/\pi R^3$, which is the magnitude of the true shear rate at the wall for a Newtonian liquid with zero velocity at the wall $[v(R) = 0]$. Polymer melts fulfill this boundary condition with few exceptions, e.g. polyvinylchloride (PVC) below 200°C (Münstedt 1975a). For any non-Newtonian fluid and for the condition $v(R) = 0$, the *true shear rate* $\dot{\gamma}(R)$ follows from the *Rabinowitsch-Weissenberg correction*: $\dot{\gamma}(R) = [3D + p_{21}(R) \, dD/dp_{21}(R)]/4$ (Rabinowitsch 1929). Chmiel & Schümmer (1971) and Laun (1983) have proposed simpler, approximate relations.

The pressure gradient dp/dL can be measured in slit dies by flush-mounted pressure transducers. Otherwise, *hole-pressure errors* follow that are characteristic for elastic liquids (Higashitani & Lodge 1975, Lodge & de Vargas 1983). In order to determine dp/dL for capillaries, functions $q(p)$ must be measured with several dies of different L/R ratios (Bagley 1957). In

Bagley plots, p is redrawn as a function of L/R for D = constant (Figure 3). The resulting straight lines intersect the ordinate axis in pressure corrections p_c. The intersections of the abscissa with the extrapolated straight lines define the *Bagley end correction* $e = |L/R(p \to 0)|$. The wall shear stress is $p_{21}(R) = -\Delta p/2(\Delta L/R) = -(p-p_0)/2(L/R) = p/2(e + L/R)$.

The results of Figure 3 were obtained by a gas-driven capillary rheometer; other systems are piston driven (deadweights, pneumatics, mechanically). Buck & Kerk (1969) have compared these systems. With all corrections necessary, flow curves of polymer melts can be accurate to within 5%. For branched polyethylenes, the flow curves from different instruments and investigators agreed to within 10% (Meissner 1975a). With long dies the relative contribution of the end correction is reduced, but for the same D the increased extrusion pressure acts increasingly as hydrostatic pressure and increases the viscosity. For various polymer melts, viscosity functions with analytical expressions for computer calculations have been published (Münstedt 1978). Often, the reduced plot η/η_0 as a function of log $\eta_0\dot{\gamma}$ is preferred (Vinogradov & Malkin 1966, Semjonow 1968). Laun (1983) describes the details of slit-die rheometry.

The large end or pressure corrections in Figure 3 indicate that the energy input into the rheometer is required for (*a*) viscous flow in the die, (*b*) energy dissipation in front of the die, and (*c*) elastic deformation, which to a large extent is transported with the melt through the die causing *extrudate swell* [i.e. the extruded strand diameter is larger than the die diameter (Bagley et al. 1963)]. The three energy portions could be separated experimentally (Meissner 1963, 1975a, Ramsteiner 1972). Malkin et al. (1970) considered a

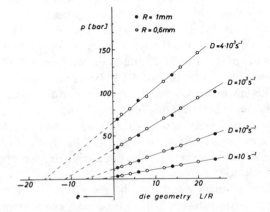

Figure 3   Bagley plots for a melt of high-impact polystyrene at 189°C (Meissner 1969a). The tests were performed with two sets of "flat" dies (180° entrance angle into the capillary) of 1.2 and 2 mm diameter (± 0.002 mm).

fourth energy portion because of the breakdown of the physical structure of the melt due to flow. The extrudate swell is taken as a measure for the melt elasticity and depends strongly on die length with short dies. Ramsteiner (1972) investigated the influence of different inlet geometries on extrudate swell. Reproducible results require a careful experimental procedure, including an additional heat treatment (Meissner 1967). Utracki et al. (1975) describe a semiautomatic device for measuring extrudate swell.

Bagley & Birks (1960) investigated the die-inlet flow cinematographically and observed a central melt stream surrounded by a secondary flow regime. Only the melt of the central stream entered the die, except when the flow became unstable. In the central stream, the melt is mainly deformed by elongation (Hürlimann & Knappe 1972). Kramer & Meissner (1980) proved this in a slit entrance by laser-Doppler velocimetry. The extrusion flow of polymer melts becomes unstable (onset of *melt fracture*) at flow rates far below the value corresponding to the critical Reynolds number (Tordella 1958, Petrie & Denn 1976). For some polymer melts, e.g. linear polyethylene, the onset of melt fracture is connected with a steplike increase in output rate q (Bagley et al. 1958, Tordella 1963); a comprehensive investigation of this phenomenon was performed by Uhland (1979).

4. VISCOMETRIC MATERIAL FUNCTIONS AND GENERAL SHEAR FLOWS

In *viscometric flows*, material surfaces, e.g. cylinders, slide along each other with a velocity difference proportional to their separation. From a rheological point of view, viscometric flows are equivalent to simple planar shear flow (Coleman et al. 1966). Let e_1 be the unit vector in the flow direction, e_2 the unit normal vector of the shear planes, and $e_3 = e_1 \times e_2$ the unit vector in the "third direction." Viscometric flows are defined by a step-function shear rate for the material particle under consideration, $\dot{\gamma}(t) = \dot{\gamma}_0 h(t)$ (Bird et al. 1977a). For incompressible liquids, the complete stress response consists of the shear stress $p_{21} = p_{12}$ and two normal-stress differences, $N_1 = p_{11} - p_{22}$ and $N_2 = p_{22} - p_{33}$. With p_{21} as odd and N_1 and N_2 as even functions of $\dot{\gamma}_0$, the normal-stress functions $\psi_1 = N_1/\dot{\gamma}_0^2$ and $\psi_2 = N_2/\dot{\gamma}_0^2$ are defined, in addition to the viscosity η. The parameters η, ψ_1, and ψ_2 are functions of $\dot{\gamma}_0$ and time t. The first normal-stress difference N_1 is connected with the recoverable portion γ_R of the shear strain (Weissenberg 1947). An analogous connection exists for solid rubber (Rivlin 1947).

For the simultaneous measurement of η, ψ_1, and ψ_2, the cone-and-plate rotational rheometer is especially appropriate. The shear history of the material in the gap is homogeneous, and torque T, normal force F (which tends to open the gap), and radial-pressure distribution $[-p_{22}(r)]$ are related to p_{21}, N_1, and N_2, as shown in Figure 4. The first simultaneous measurements of torque and normal force were performed in the Weissenberg rheogoniometer (Weissenberg 1948). The pressure distribution was reliably measured for polymer solutions by Christiansen & Leppard (1974). For the separation of N_1 and N_2, Kotaka et al. (1959) proposed a double measurement, in the cone-and-plate and the parallel-plates modes, and they derived the bottom equation of Figure 4.

A rheogoniometer for polymer melts was built by Pollett & Cross (1950). In addition to the total force F, the force acting on a separate central disk within the plate was measured, and N_1 and N_2 could be separated (Pollett 1955). Because of the high stresses, the rheometer must be very stiff to obtain the transients p_{21} and N_1 correctly, and only with large cone angles (6–8°) can the interaction between the sample in the gap and the rheometer compliance be eliminated (Meissner 1972a, Hansen & Nazem 1975, Nazem & Hansen 1976). Figure 5 shows as an example the resulting transients p_{21} and N_1, with remarkable maxima. The thermal expansion of polymer melts explains why the smallest temperature variations contribute to the noise of the normal-force signal and limit its resolution. With increasing shear, the melt often separates at the rim of the sample, causing an erroneous decrease of p_{21} and N_1. The onset of these irregularities can be reduced by screening rings (Gleissle 1978). More experimental difficulties involved in this method are listed and discussed by Lodge (1984).

With a servo-controlled drive, additional test modes can be performed, e.g. step-function shear stress, oscillations (also of large amplitudes), superimposed different unidirectional flows. Shear recovery (with the constraint of constant gap width) can be measured without loss due to

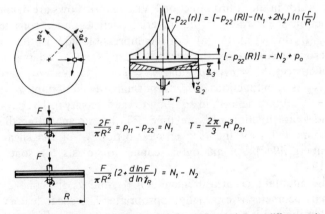

$$[-p_{22}(r)] = [-p_{22}(R)] - (N_1 + 2N_2)\ln\left(\frac{r}{R}\right)$$

$$[-p_{22}(R)] = -N_2 + p_o$$

$$\frac{2F}{\pi R^2} = p_{11} - p_{22} = N_1, \qquad T = \frac{2\pi}{3}R^3 p_{21}$$

$$\frac{F}{\pi R^2}\left(2 + \frac{d\ln F}{d\ln \dot{\gamma}_R}\right) = N_1 - N_2$$

Figure 4 Determination of the shear stress and the two normal-stress differences in the cone-and-plate and the parallel-plates rotational rheometers.

inertia of and friction in the instrument. In all these cases, $\gamma(t)$, $p_{21}(t)$, and $N_1(t)$ can be recorded (Meissner 1975b). As an example, the total recovery measured after simple shear flow up to total strains $\gamma_1 = \dot{\gamma}_0 t_1$ is added in Figure 5 and shows another maximum. Recoverable shear strains up to nine shear units were measured at higher $\dot{\gamma}_0$ by Laun (1982). The maxima of Figure 5 indicate that the network connectivity of the polymer melt is reduced by deformation (Wagner & Meissner 1980). Relaxation tests with shear steps up to $\gamma = 30$ confirm this statement, with the relaxation of p_{21} and N_1 yielding the same relaxation modulus with a factorization of the time and strain dependences (Laun 1978): $G(t, \gamma) = G(t)h(\gamma)$, where $G(t)$ is the linear viscoelastic modulus and $h(\gamma)$ is an attenuation function introduced by Wagner (1976). For polymer melts, $[-p_{22}(t)]$ and therefore N_2 could not yet be measured precisely. The method of Kotaka et al. (1959) gave negative N_2 for a polyethylene melt at $\dot{\gamma}_0 = 10 \text{ s}^{-1}$ and $\gamma = \dot{\gamma}_0 t = 50$: $N_2 = -0.22 \, N_1$ (calculated from F and correcting the sign error in Table 2 of Meissner 1967); however, the shear between parallel plates (Figure 4) is inhomogeneous, and therefore these results are only approximate.

Kataoka & Ueda (1969) superimposed oscillations on a steady shear rate flow of polyethylene melts. The superposition was "parallel"; a perpendicular superposition was performed with polyisobutylene solutions (Simmons 1968). Osaki et al. (1981) investigated theoretical aspects by double-step shear deformations of polystyrene solutions. In all these tests, the shear was unidirectional or reversed, except for the perpendicular superposition of oscillations with, however, small amplitudes.

Our present knowledge of the shear-flow behavior of polymer melts is limited because of the restriction to unidirectional test modes. For investigations with more general shear flows, a sandwich-type biaxial shear

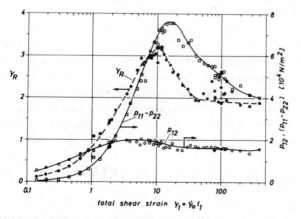

Figure 5   Transient response of a polyethylene melt to a step-function shear rate with $\dot{\gamma}_0 = 2 \text{ s}^{-1}$ at a temperature of 150°C (Meissner 1975b).

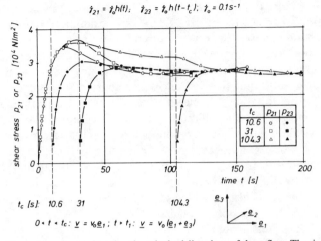

$$\dot{\gamma}_{21} = \dot{\gamma}_o h(t); \quad \dot{\gamma}_{23} = \dot{\gamma}_o h(t-t_c); \quad \dot{\gamma}_o = 0.1 s^{-1}$$

t_c	p_{21}	p_{23}
10.6	○	●
31	□	■
104.3	△	▲

t_c [s]:   10.6   31   104.3

$0 < t < t_c: \underline{v} = v_o\underline{e}_1 ; \quad t > t_1: \underline{v} = v_o(\underline{e}_1 + \underline{e}_3)$

Figure 6   Stress response on changing the principal directions of shear flow. The simple shear in the x_1 direction (with $\dot{\gamma}_0 = 0.1$ s^{-1}) was of duration t_c, at which time the additional shear flow in the x_3 direction was started. The tests were performed using polyisobutylene at room temperature (Meissner 1984).

rheometer was developed, in which two mutually perpendicular shear histories, independent of each other, can be imposed on the melt (H. P. Hürlimann & J. Meissner, in preparation). Figure 6 gives the shear stresses when, after a simple shear of duration t_c in the x_1 direction, a shear of equal magnitude is added in the x_3 direction. From the results it follows that there is a certain period of time (approximately 30 s) during which, at the onset of the second component, the two corresponding shear stresses are different, reflecting the melt anisotropy induced during the preliminary flow in the time interval $0 < t < t_c$ (Meissner 1984).

In the instruments reviewed so far, the stress components are measured mechanically, e.g. by forces and the torque. Using the stress-optical law, we can also measure these components by birefringence. Cone-and-plate rheometers utilizing this technique were developed by Janeschitz-Kriegl (1969, 1983). They can be built extremely stiff in order to eliminate the compliance of mechanical transducers. The coincidence of the transients $p_{21}(t)$ and $N_1(t)$ from optical and mechanical measurements was confirmed for a polystyrene melt (Gortemaker et al. 1976).

5.   RHEOMETERS FOR UNIAXIAL STRETCHING

In spite of its technical relevance, and excluding a few early investigations (Karam & Bellinger 1964, Ballman 1965), the development of elongational rheometry began around 1970 with emphasis on simple elongation

(Cogswell 1968, Meissner 1969b, 1971, 1972b, Vinogradov et al. 1970a,b 1972). The monograph by Petrie (1979) reviews the literature up to 1978. Only methods with nearly homogeneous elongations of the melt are considered here.

There are two different types of instruments, each complementing the other. The first is characterized by grips or clamps fixed to the ends of the rodlike sample, which is elongated by increasing the distance between the clamps. The advantage of this method is that the samples are small. But the rheometer length limits the maximum elongation ε_{max}. The second method utilizes rotary clamps and requires long samples, but ε_{max} is, in principle, infinite. In all cases the polymer-melt sample is floating in or on a supporting liquid to avoid its deformation by gravity.

The rheometers of the first type go back to Cogswell (1968) and were improved in several steps by Münstedt (1975b, 1979). Figure 7 gives the schematic diagrams of Münstedt's rheometers. The glass vessels contain silicone oil and have a heating jacket. In the left version, the lower sample end is fixed and the upper end is pulled by a force that is continuously reduced, such that the tensile stress σ is constant. For this purpose, a cam reduces the lever of the test load appropriately. The cam is shaped according to the initial sample length L_0 (30 or 50 mm). The position transducer measures the elongational creep curve $\varepsilon(t)$. Initially, polystyrene was investigated between 125 and 160°C; more recently, in an improved version of the instrument, polyethylene was tested up to 190°C (Laun & Münstedt 1976). The initial sample length L_0 can be reduced to 10 mm, which increases the maximum stretch $\lambda_{max} = L_{max}/L_0$ to 50. This is achieved by two small cams operating in series. At the end of the test, the sample is cut into small cutoffs, from which the homogeneity of the deformation is deduced.

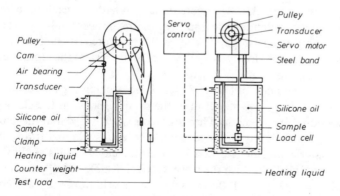

Figure 7 Mechanically (*left*) and electrically (*right*) driven rheometers for the elongation of polymer melts (Münstedt 1975b, 1979).

The same rheometer with a servo-controlled drive is shown on the right-hand side of Figure 7. The sample is fixed to a load cell, and the elongation is measured by an angular position transducer. With the servo drive, arbitrary stress or strain histories can be prescribed, provided that $\sigma > 0$. Münstedt (1979) performed multistep stress and strain-rate tests including recovery. The elongation is expressed in the Hencky measure $\varepsilon = \ln(L/L_0) = \ln \lambda$, where the current length is $L = L(t)$. For a constant strain rate test, L must increase exponentially with time. The samples are glued to the clamps after etching. M. Fleissner (personal communication) uses small metal cups as clamps and wraps the sample ends by a film of polyethylene that has a higher molecular weight and acts as an adhesive.

In contrast to the perpendicular instruments of Figure 7, horizontally positioned rheometers have the advantage that the densities of melt and supporting liquid need not match exactly, since the sample floats on this liquid (Vinogradov et al. 1970a, 1972). A servo drive provides constant strain rate or stress; Vinogradov et al. (1970b) have proposed different types of clamps. For short samples, the necking at the clamps causes an error of the true elongation of the melt deformed homogeneously in the center of the sample. When the test is terminated by cutting the sample, the cutoffs provide a correction of the drive control for the ensuing tests (Franck 1982, Franck & Meissner 1984).

The development of rheometers with rotary clamps was described previously (Meissner et al. 1981). Figure 8 shows a schematic diagram of the first version. The sample S (approximately 800 mm in length) is floating on silicone oil and is clamped between two pairs of gears (*rotary clamps*) Z_1, Z_2 that have a fixed distance L_0 and rotate in opposite directions. At constant speeds of rotation $n_1, n_2 = $ constant, the sample is elongated with $\dot\varepsilon = \dot\varepsilon_0$ = constant. At the end of the test, the cutters C_i cut the sample into cutoffs, the weight of which determines the true total elongation and its homo-geneity. The motors M_1, M_2 operate independently of each other; M_1 is mounted at the free end of a leaf spring LS, which is bent only by the tensile force F acting in the sample. The force F is measured by the transducer T. The tensile stress $\sigma(t)$ follows from the recorded $F(t)$ and the cross-sectional area $q(t) = q_0 \exp(-\dot\varepsilon_0 t) = q_0 \exp(-\varepsilon)$.

With the rheometer of Figure 8 there is no necking because each rotary clamp always grips new material of the sample. Between the clamps, the cross-sectional area is reduced homogeneously, provided that the temperature field is sufficiently homogeneous. The rotary clamps transport the sample material outside L_0; therefore, the total elongation is not limited by the rheometer length. With the newest model (Raible 1981, Meissner et al. 1981), $\varepsilon_{max} = 7$ ($\lambda_{max} = 1100$) was achieved. In addition, the transient and total recovery are measured as functions of the total elongation. By means of a servo control of motor M_2, several strain and stress histories can be imposed on the melt under test, e.g. constant strain rate with superimposed oscillations (Meissner 1972b). Raible et al. (1982) performed melt elongation with constant tensile force, a test mode of special interest for the melt-spinning problem.

For precise results at large strains, quality parameters for the test performance were established (Meissner et al. 1981). The strain-rate error $\Delta\dot{\varepsilon}/\dot{\varepsilon}_0$ produces a stress error $\Delta\sigma/\sigma = [\exp(\varepsilon\Delta\dot{\varepsilon}/\dot{\varepsilon}_0)] - 1$, which increases more than linearly with $\Delta\dot{\varepsilon}/\dot{\varepsilon}$ at large ε (where the error may become unacceptable). Repeated tests with $\varepsilon > 4$ may yield contradictory shapes of the stress-strain diagrams measured if the test quality is not controlled and excellent. Carefully made initial samples and a homogeneous temperature in the rheometer ($\Delta T < 0.1°C$ could be achieved at 150 and 190°C) are essential.

The first range of strain rates ($\dot{\varepsilon}_0 = 0.001-1$ s^{-1}) was extended up to 30 s^{-1} by reducing L_0 from 650 to 70 mm, and down to 2×10^{-5} s^{-1} by combining the two methods of Figures 7 and 8 (Laun & Münstedt 1978). One end of the sample was glued to a force-measuring device of high resolution ($< 10^{-4}$ N); the other end was extended by a rotary clamp. At $\dot{\varepsilon}_0 = 10^{-4}$ s^{-1}, the deformation is very small and its inhomogeneity negligible. But the interfacial tension between melt and supporting liquid must be considered, and for tests of long duration (as for low $\dot{\varepsilon}_0$) the chemical stability of the polymer melts may not be sufficient. Both problems have been analyzed carefully (Laun & Münstedt 1978). For recovery measurements, the interfacial tension must also be considered (Meissner et al. 1981). The combination of the two rheometers (one sample end fixed, the other extended by a rotary clamp or a roller) was also utilized by other authors for higher $\dot{\varepsilon}_0$ (Ide & White 1978, Ishizuka & Koyama 1980). Without a statement concerning the homogeneity of the elongation, erroneous results must be expected at large strains.

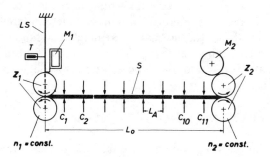

Figure 8   Rheometer with two rotary clamps for simple elongation and other uniaxial stretching of polymer melts (Meissner 1971).

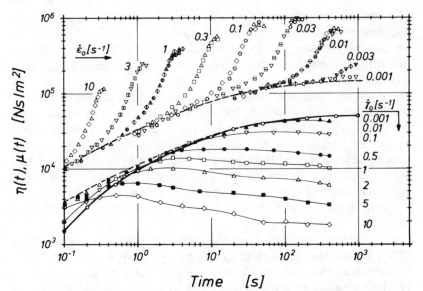

Figure 9 Transient elongational and shear viscosities of a melt of branched polyethylene (sample "A" of an extensive IUPAC investigation; Meissner 1975a). The temperature was 150°C, and the results are combined from the studies of Münstedt & Laun (1979) and Meissner (1972a).

For results from constant-strain-rate tests, the elongational viscosity $\mu(t) = \sigma(t)/\dot{\varepsilon}_0$ was introduced, analogously to the definition of $\eta(t)$. Both viscosities are shown in Figure 9 for a melt of branched polyethylene. For shear as well as for elongation, the behavior is linear viscoelastic at low strain rates with $\mu(t)/\eta(t) = 3$, reflecting the *Trouton* ratio. At higher strain rates, there is strain-hardening in elongation but strain-thinning in shear. At the highest strain $\varepsilon = 6$, $\mu(t)$ shows a maximum, which was confirmed more clearly when precise measurements could be made up to $\varepsilon = 7$ (Raible 1981, Meissner et al. 1981). This maximum in $\varepsilon(t)$ is accompanied by a maximum in the recovery. Figure 9 gives the complete material response for elongation; for shear, the normal-stress functions should be added. A similar strain-hardening in elongation was found for polyisobutylene (Meissner et al. 1982), but for linear polyethylene this phenomenon is much smaller (Münstedt & Laun 1981).

6.   MULTIAXIAL STRETCHING

Except for film-blowing studies, the method utilized up to 1980 for multiaxial stretching was bubble inflation. In this method a sheet of polymeric material is clamped at the rim, heated to measuring temperature, and inflated by air pressure. The pressure can be controlled such that at the center of the growing bubble, the strain rate or the stress is constant. The deformation is determined from a grid printed on the sample and photographed during the inflation from the top and side (Joye et al. 1972, 1973, Denson & Crady 1974, de Vries & Bonnebat 1977). At the pole of the bubble, the elongation is equibiaxial with a circular sample and planar with a rectangular sample (no deformation in one direction). Equivalent to equibiaxial elongation is simple compression, a technique introduced by Chartraei et al. (1981). In simple compression, a disk of polymer melt is compressed between two flat plates; compression force and sample thickness are then recorded. A lubricant between the plates and the melt prevents any shear flow.

The methods described so far restrict the multiaxial elongation to few modes. By using new rotary clamps, more general test modes can be performed. Each clamp consists of two grooved cylinders (77 mm long, 74 mm diameter) with a stepping motor inside. A bandlike sample is drawn into the nipline between the two cylinders and wound up on a third cylinder. The three cylinders are mounted on the same frame, which is suspended by leaf springs. The displacements of the springs correspond to the force components acting on the clamp in and normal to the direction of the sample motion. The clamps form modules that introduce the sample deformation and measure the force components acting on the sample at the rim. A versatile elongational rheometer is built by arranging the modules in different positions. Two modules in opposed positions perform simple elongation, as in Figure 8. Eight modules positioned on a circle elongate a disklike sample equibiaxially. Such tests were performed with poly-isobutylene (initial sample diameter of 350 mm). The sample could be extended homogeneously, with the original thickness of 5 mm drawn down to 12 μm. The sample has to be cut intermittently at the rim into bands, which are wound up on each module. This rheometer is shown in Figure 10. The white circle in the middle is talcum powder, which supports the sample before the deformation is started. The tubes around the apparatus are for the pneumatic operation of the cutting elements between the clamps (Stephenson 1980, Meissner et al. 1981).

For planar elongation, the positions of the eight modules form a rectangle, with three modules on each long side and one module on each short side. Only the clamps on the long sides rotate with constant speed, whereas the clamps on the short sides do not rotate but fix the sample in the direction of constant dimension. With this arrangement the elongation in the central area of a polyisobutylene sample is planar and homogeneous, and the stresses in the direction of elongation and in the perpendicular direction of constant dimension can be measured simultaneously

Figure 10   Rheometer with eight rotary clamps for equibiaxial elongation of polymer melts at room temperature (Stephenson 1980, Meissner et al. 1981).

(Demarmels 1983, Meissner et al. 1982). With appropriate positions and speed controls of the rotary clamps, elongational tests with arbitrary, constant strain rates can be performed. However, such tests do not represent the general case with a variation of strain-rate ratios during a test. For this purpose, the clamps are mounted on pivots to adjust their mutual orientations with the variation of the strain-rate ratios (Demarmels 1983, Meissner 1984).

For multiaxial elongations with constant strain rates, the following classification was developed (Stephenson 1980, Meissner et al. 1982). In a Cartesian coordinate system, the axes x_i are oriented such that $\dot{\varepsilon}_0 \equiv \dot{\varepsilon}_{11} \geq \dot{\varepsilon}_{22} \geq \dot{\varepsilon}_{33}$, where $\dot{\varepsilon}_0$ is the maximum strain rate. The ratio $m = \dot{\varepsilon}_{22}/\dot{\varepsilon}_{11}$ ($-0.5 \leq m \leq 1$) denotes three special cases by the values $m = -0.5$ for simple elongation, $m = 1$ for equibiaxial elongation, and $m = 0$ for planar elongation. As $\dot{\varepsilon}_{33}$ is determined from $\dot{\varepsilon}_{11}$ and $\dot{\varepsilon}_{22}$ due to the incompressibility of the melt, any multiaxial elongation with constant strain rates is uniquely characterized by $\dot{\varepsilon}_0$ and m, with $\dot{\varepsilon}_0$ representing the intensity and m the mode of the test.

From the force components acting on the rotary clamps, the following differences of the stress components can be determined:

$$\sigma_1 \equiv p_{11} - p_{33}, \qquad \sigma_2 \equiv p_{22} - p_{33}, \qquad \sigma_3 \equiv p_{11} - p_{22} = \sigma_1 - \sigma_2. \qquad (2)$$

Only two σ_i are independent of each other. From the linear-viscoelastic equation (1), it follows for such a flow with $\dot{\varepsilon}_{11}(t) = \dot{\varepsilon}_0 h(t)$ that

$$\int_0^t G(t-t')\,dt' = \sigma_i(t)A_i(m)/\dot{\varepsilon}_0 \qquad (i = 1, 2, 3). \qquad (3)$$

The front factors $A_i(m)$ are easily calculated (e.g. for simple elongation $A_1 = A_3 = 1/3$; for planar elongation $A_1 = 1/4$, $A_2 = A_3 = 1/2$). The integral of Equation (3) is the same as the transient linear-viscoelastic shear viscosity $\eta(t)$. For multiaxial elongations we denote the right-hand side of Equation (3) as elongational viscosity μ_i. Hence, for the linear-viscoelastic limit, all the $\mu_i(t)$ are equal to $\eta(t)$. By this new definition of the three elongational viscosities ($i = 1, 2, 3$), the nonlinear behavior can be described efficiently because the linear $\eta(t)$ serves as a reference that is easily measured by simple shear and $\dot{\gamma}_0 \to 0$. For simple, equibiaxial, and planar elongations of polyisobutylene, the nonlinear deviations from $\eta(t)$ were investigated, with surprising results. As an example, for planar elongations the material response starts linear-viscoelastically followed by strain-hardening for μ_1 [i.e. the curves $\mu_1(t)$ increase above $\eta(t)$], whereas $\mu_2(t)$ shows strain-thinning [i.e. the curves $\mu_2(t)$ increase more slowly than $\eta(t)$] (Demarmels 1983, Meissner et al. 1982).

Literature Cited

Astarita, G., Marrucci, G. 1974. *Principles of Non-Newtonian Fluid Mechanics*. London: McGraw-Hill. 289 pp.

Bagley, E. B. 1957. End corrections in the capillary flow of polyethylene. *J. Appl. Phys.* 28:624–27

Bagley, E. B., Birks, A. M. 1960. Flow of polyethylene into a capillary. *J. Appl. Phys.* 31:556–61

Bagley, E. B., Cabot, I. M., West, D. C. 1958. Discontinuity in the flow curve of polyethylene. *J. Appl. Phys.* 29:109–10

Bagley, E. B., Storey, S. H., West, D. C. 1963. Post extrusion swelling of polyethylene. *J. Appl. Polym. Sci.* 7:1661–72

Ballman, R. L. 1965. Extensional flow of polystyrene melt. *Rheol. Acta* 4:137–40

Behre, J. 1932. Ueber Plastizitätsmessungen in der Gummi-Industrie. *Kautschuk* 8:2–5, 167–71

Bird, R. B., Armstrong, R. C., Hassager, O. 1977a. *Dynamics of Polymeric Liquids. Vol. 1. Fluid Mechanics*. New York: Wiley. 470 + 89 pp.

Bird, R. B., Hassager, O., Armstrong, R. C., Curtiss, C. F. 1977b. *Dynamics of Polymeric Liquids. Vol. 2. Kinetic Theory*. New York: Wiley. 256 + 31 pp.

Buck, M., Kerk, K. 1969. Vergleichsmessungen der Schmelzviskosität an vier verschiedenen Viskosimetern von zum Teil sehr unterschiedlicher Bauart. *Rheol. Acta* 8:372–81

Chartraei, S., Macosco, C. W., Winter, H. H. 1981. Lubricated squeezing flow: a new biaxial extensional rheometer. *J. Rheol.* 25:433–43

Chmiel, H., Schümmer, P. 1971. Eine neue Methode zur Auswertung von Rohrrheometer-Daten. *Chem. Ing. Tech.* 43:1257–59

Christiansen, E. B., Leppard, W. R. 1974. Steady-state and oscillatory flow properties of polymer solutions. *Trans. Soc. Rheol.* 18:65–86

Cogswell, F. N. 1968. The rheology of polymer melts under tension. *Plast. Polym.* 36(4):109–11

Coleman, B. D., Markovitz, H., Noll, W. 1966. *Viscometric Flows of Non-Newtonian Fluids*. Berlin: Springer. 130 pp.

Dealy, J. M. 1982. *Rheometers for Molten Plastics.* New York: Van Nostrand Reinhold. 302 pp.

Demarmels, A. 1983. *Das rheologische Verhalten von Polyisobutylen bei mehrachsigen Dehnströmungen und seine Beschreibung in den Netzwerktheorien.* PhD thesis No. 7345. ETH Zürich. 125 pp.

denOtter, J. L. 1969. Dynamic measurements on polymer solutions and melts. *Rheol. Acta* 8 : 355–63

Denson, C. D., Crady, D. L. 1974. Measurements on the planar extensional viscosity of bulk polymers: the inflation of a thin rectangular polymer sheet. *J. Appl. Polym. Sci.* 18 : 1611–17

de Vries, A. J., Bonnebat, C. 1977. Uni- and biaxial orientation of polymer films and sheets. *J. Poly. Sci., Polym. Symp.* 58 : 109–56

Ferry, J. D. 1980. *Viscoelastic Properties of Polymers.* New York: Wiley. 641 pp. 3rd ed.

Franck, A. 1982. *Untersuchungen über das Kriechverhalten von Polystyrol-Schmelzen in Abhängigkeit von Molekulargewicht und Molekulargewichtsverteilung mit einem neuentwickelten Dehnungsrheometer.* PhD thesis No. 7158. ETH Zürich. 156 + 17 pp.

Franck, A., Meissner, J. 1984. The influence of blending polystyrenes of narrow molecular weight distributions on melt creep flow and creep recovery in elongation. *Rheol. Acta* 23 : 117–23

Giesekus, H. 1965. A symmetric formulation of the linear theory of viscoelastic materials. *Proc. Int. Congr. Rheol., 4th, 1963,* 3 : 15–28. New York: Interscience

Gleissle, W. 1978. *Ein Kegel-Platte Rheometer für sehr zähe viskoelastische Flüssigkeiten bei hohen Schergeschwindigkeiten, Untersuchung des Fliessverhaltens von hochmolekularem Siliconöl und Polyisobutylen.* PhD thesis. Univ. Karlsruhe. 168 pp.

Gortemaker, F. H., Hansen, M. G., deCindio, B., Laun, H. M., Janeschitz-Kriegl, H. 1976. Flow birefringence of polymer melts: application to the investigation of time dependent rheological properties. *Rheol. Acta* 15 : 256–67

Graessley, W. W. 1974. The entanglement concept in polymer rheology. *Adv. Polym. Sci.* 16 : 1–179

Hansen, M. G., Nazem, F. 1975. Transient normal force transducer response in a modified Weissenberg rheogoniometer. *Trans. Soc. Rheol.* 19 : 21–36

Higashitani, K., Lodge, A. S. 1975. Hole pressure error measurements in pressure-generated shear flow. *Trans. Soc. Rheol.* 19 : 307–35

Hürlimann, H. P., Knappe, W. 1972. Der Zusammenhang zwischen der Dehnspannung von Kunststoff-Schmelzen im Düseneinlauf und im Schmelzbruch. *Rheol. Acta* 11 : 292–301

Ide, Y., White, J. L. 1978. Experimental study of elongational flow and failure of polymer melts. *J. Appl. Polym. Sci.* 22 : 1061–79

Ishizuka, O., Koyama, K. 1980. Elongational viscosity at a constant elongational strain rate of polypropylene melt. *Polymer* 21 : 164–70

Janeschitz-Kriegl, H. 1969. Flow birefringence of elastico-viscous polymer systems. *Adv. Polym. Sci.* 6 : 170–318

Janeschitz-Kriegl, H. 1983. *Polymer Melt Rheology and Flow Birefringence.* Berlin: Springer. 524 pp.

Joye, D. D., Poehlein, G. W., Denson, C. D. 1972, 1973. A bubble inflation technique for the measurement of viscoelastic properties in equal biaxial extensional flow. *Trans. Soc. Rheol.* 16 : 421–45, 17 : 287–302

Karam, H. J., Bellinger, J. C. 1964. Tensile creep of polystyrene at elevated temperatures. Part I. *Trans. Soc. Rheol.* 8 : 61–72

Kataoka, T., Ueda, S. 1969. Influence of superimposed steady shear flow on the dynamic properties of polyethylene melts. *J. Polym. Sci. Part A-2* 7 : 475–81

Kotaka, T., Kurata, M., Tamura, M. 1959. Normal stress effects in polymer solutions. *J. Appl. Phys.* 30 : 1705–12

Kramer, H., Meissner, J. 1980. Application of the laser Doppler velocimetry to polymer melt flow studies. In *Rheology,* ed. G. Astarita, G. Marrucci, L. Nicolais, 2 : 463–68. New York: Plenum

Laun, H. M. 1978. Description of the nonlinear shear behaviour of a low density polyethylene melt by means of an experimentally determined strain dependent memory function. *Rheol. Acta* 17 : 1–15

Laun, H. M. 1982. Elastic properties of polyethylene melts at high shear rates with respect to extrusion. *Rheol. Acta* 21 : 464–69

Laun, H. M. 1983. Polymer melt rheology with a slit die. *Rheol. Acta* 22 : 171–85

Laun, H. M., Meissner, J. 1980. A sandwich-type creep rheometer for the measurement of rheological properties of polymer melts at low shear stresses. *Rheol. Acta* 19 : 60–67

Laun, H. M., Münstedt, H. 1976. Comparison of the elongational behaviour of a polyethylene melt at constant stress and constant strain rate. *Rheol. Acta* 15 : 517–24

Laun, H. M., Münstedt, H. 1978. Elongational behaviour of a low density polyethylene melt. I. Strain rate and stress dependence of viscosity and recoverable strain in the steady-state. Comparison with shear data. Influence of interfacial tension. *Rheol. Acta* 17:415–25

Lodge, A. S. 1964. *Elastic Liquids*. London: Academic. 389 pp.

Lodge, A. S. 1974. *Body Tensor Fields in Continuum Mechanics*. New York: Academic. 319 pp.

Lodge, A. S. 1984. A classification of constitutive equations based on stress relaxation predictions for the single-jump shear strain experiment. *J. Non-Newtonian Fluid Mech.* 14:67–83

Lodge, A. S., de Vargas, L. 1983. Positive hole pressures and negative exit pressures generated by molten polyethylene flowing through a slit die. *Rheol. Acta* 22:151–70

Malkin, A. Ya., Yarlykov, B. V., Vinogradov, G. V. 1970. Most important features of prestationary deformation regimes and stress relaxation of polymers in the viscous state. *Rheol. Acta* 9:329–38

Marzetti, B. 1923. La plasticita della gomma cruda. *G. Chim. Ind. Appl.* 5:342–43

Maxwell, B., Jung, A. 1957. Hydrostatic pressure effect on polymer melt viscosity. *Mod. Plast.* 35(November):174–82, 276

Meissner, J. 1963. Untersuchungen über das Fliessverhalten von geschmolzenem Polyäthylen mit dem Kapillarviskosimeter. *Materialprüfung* 5:107–13

Meissner, J. 1967. Die Kunststoff-Schmelze als elastische Flüssigkeit. *Kunststoffe* 57:397–400, 702–10

Meissner, J. 1969a. Rheologische Grundlagen (der Verarbeitung von Polystyrol). In *Kunststoff-Handbuch*, ed. R. Vieweg, G. Daumiller, 5:162–87. München: Hanser

Meissner, J. 1969b. Rheometer zur Untersuchung der deformationsmechanischen Eigenschaften von Kunststoff-Schmelzen unter definierter Zugbeanspruchung. *Rheol. Acta* 8:78–88

Meissner, J. 1971. Dehnungsverhalten von Polyäthylen-Schmelzen. *Rheol. Acta* 10:230–42

Meissner, J. 1972a. Modifications of the Weissenberg rheogoniometer for measurement of transient rheological properties of molten polyethylene under shear. Comparison with tensile data. *J. Appl. Polym. Sci.* 16:2877–99

Meissner, J. 1972b. Development of a universal extensional rheometer for the uniaxial extension of polymer melts. *Trans. Soc. Rheol.* 16:405–20

Meissner, J. 1975a. Basic parameters, melt rheology, processing and end-use properties of three similar low density polyethylene samples. *Pure Appl. Chem.* 42:553–612

Meissner, J. 1975b. Neue Messmöglichkeiten mit einem zur Untersuchung von

Kunststoff-Schmelzen geeigneten modifizierten Weissenberg-Rheogoniometer. *Rheol. Acta* 14:201–18

Meissner, J. 1978. Combined constant strain rate and stress relaxation test for linear viscoelastic studies. *J. Polym. Sci., Polym. Phys. Ed.* 16:915–19

Meissner, J. 1984. Alte und neue Wege in der Rheometrie der Polymer-Schmelzen. *Chimia* 38:35–45, 65–75

Meissner, J., Raible, T., Stephenson, S. E. 1981. Rotary clamp in uniaxial and biaxial extensional rheometry of polymer melts. *J. Rheol.* 25:1–28

Meissner, J., Stephenson, S. E., Demarmels, A., Portmann, P. 1982. Multiaxial elongational flows of polymer melts—classification and experimental realization. *J. Non-Newtonian Fluid Mech.* 11:221–37

Merz, E. H., Colwell, R. E. 1958. A high shear rate capillary rheometer for polymer melts. *ASTM Bull.* 232:63–67

Münstedt, H. 1975a. Rheologische Eigenschaften von Polyvinylchlorid im Vergleich mit Styrolpolymerisaten. *Angew. Makromol. Chem.* 47:229–42

Münstedt, H. 1975b. Viscoelasticity of polystyrene melts in tensile creep experiments. *Rheol. Acta* 14:1077–88

Münstedt, H. 1978. Viskositätsdaten von Kunststoff-Schmelzen. *Kunststoffe* 68:92–98

Münstedt, H. 1979. New universal extensional rheometer for polymer melts. Measurements on a polystyrene sample. *J. Rheol.* 23:421–36

Münstedt, H., Laun, H. M. 1979. Elongational behaviour of a low density polyethylene melt. II. Transient behaviour in constant stretching rate and tensile creep experiments. Comparison with shear data. Temperature dependence of the elongational properties. *Rheol. Acta* 18:492–504

Münstedt, H., Laun, H. M. 1981. Elongational properties and molecular structure of polyethylene melts. *Rheol. Acta* 20:211–21

Nason, H. K. 1945. A high temperature, high pressure rheometer for plastics. *J. Appl. Phys.* 16:338–43

Nazem, F., Hansen, M. G. 1976. Stress growth and relaxation of a molten polyethylene in a modified Weissenberg rheogoniometer. *J. Appl. Polym. Sci.* 20:1355–70

Osaki, K., Kimura, S., Kurata, M. 1981. Relaxation of shear and normal stresses in double-step shear deformations for a polystyrene solution. A test of the Doi-Edwards theory for polymer rheology. *J. Rheol.* 25:549–62

Petrie, C. J. S. 1979. *Elongational Flows*. London: Pitman. 254 pp.

Petrie, C. J. S., Denn, M. M. 1976. Instabilities in polymer processing. *AIChE J.* 22:209–36

Plazek, D. J. 1965. Temperature dependence of the viscoelastic behavior of polystyrene. *J. Phys. Chem.* 69:3480–87

Plazek, D. J. 1968. Magnetic bearing torsional creep apparatus. *J. Polym. Sci. Part A-2* 6:621–38

Pollett, W. F. O. 1955. Rheological behaviour of continuously sheared polythene. *Br. J. Appl. Phys.* 6:199–206

Pollett, W. F. O., Cross, A. H. 1950. A continuous-shear rheometer for measuring total stress in rubber-like materials. *J. Sci. Instrum.* 27:209–12

Rabinowitsch, B. 1929. Ueber die Viskosität und Elastizität von Solen. *Z. Physikal. Chem. Abt. A* 145:1–26

Raible, T. 1981. *Deformationsverhalten von geschmolzenem Polyäthylen im Zugversuch bei grossen Gesamtdehnungen*. PhD thesis No. 6751. ETH Zürich. 104 pp.

Raible, T., Stephenson, S. E., Meissner, J., Wagner, M. H. 1982. Constant force elongational flow of a low-density polyethylene melt—experiment and theory. *J. Non-Newtonian Fluid Mech.* 11:239–56

Ramsteiner, F. 1972. Einfluss der Düsengeometrie auf Strömungswiderstand, Strangaufweitung und Schmelzbruch von Kunststoff-Schmelzen. *Kunststoffe* 62:766–72

Rivlin, R. S. 1947. Torsion of a rubber cylinder. *J. Appl. Phys.* 18:444–49

Schwarzl, F., Staverman, A. J. 1952. Time-temperature dependence of linear viscoelastic behavior. *J. Appl. Phys.* 23:838–43

Semjonow, V. 1968. Schmelzviscositäten hochpolymerer Stoffe. *Adv. Polym. Sci.* 5:387–450

Simmons, J. M. 1968. Dynamic modulus of polyisobutylene solutions in superposed steady shear flow. *Rheol. Acta* 7:184–88

Smith, T. L. 1979. Evaluation of the relaxation modulus from the response to a constant rate of strain followed by a constant strain. *J. Polym. Sci., Polym. Phys. Ed.* 17:2181–88

Spencer, R. S., Dillon, R. E. 1948. The viscous flow of molten polystyrene. *J. Colloid Sci.* 3:163–80

Stephenson, S. E. 1980. *Biaxial elongational flow of polymer melts and its realization in a newly developed rheometer*. PhD thesis No. 6664. ETH Zürich. 105 pp.

Tobolsky, A. V., Catsiff, E. 1956. Elastoviscous properties of polyisobutylene (and other amorphous polymers) from stress relaxation studies. IX. A sum-

mary of results. *J. Polym. Sci.* 19:111–21

Tordella, J. P. 1958. An instability in the flow of molten polymers. *Rheol. Acta* 1:216–21

Tordella, J. P. 1963. Unstable flow of molten polymers: a second site of melt fracture. *J. Appl. Polym. Sci.* 7:215–29

Truesdell, C., Noll, W. 1965. The non-linear field theories of mechanics. In *Handbuch der Physik*, ed. S. Flügge, Vol. III/3. Berlin: Springer. 602 pp.

Uhland, E. 1979. Das anomale Fliessverhalten von Polyäthylen hoher Dichte. *Rheol. Acta* 18:1–24

Utracki, L. A., Bakerdjian, Z., Kamal, M. R. 1975. A method for the measurement of the true die swell of polymer melts. *J. Appl. Polym. Sci.* 19:481–501

Vinogradov, G. V., Malkin, A. Ya. 1966. Rheological properties of polymer melts. *J. Polym. Sci. Part A-2* 4:135–54

Vinogradov, G. V., Radushkevich, B. V., Fikhman, V. D. 1970a. Extension of elastic liquids: polyisobutylene. *J. Polym. Sci. Part A-2* 8:1–17

Vinogradov, G. V., Fikhman, V. D., Radushkevich, B. V., Malkin, A. Ya. 1970b. Viscoelastic and relaxation properties of a polystyrene melt in axial extension. *J. Polym. Sci. Part A-2* 8:657–78

Vinogradov, G. V., Fikhman, V. D., Radushkevich, B. V. 1972. Uniaxial extension of polystyrene at true constant stress. *Rheol. Acta* 11:286–91

Wagner, M. H. 1976. Analysis of time-dependent non-linear stress-growth data for shear and elongational flow of a low-density branched polyethylene melt. *Rheol. Acta* 15:136–42

Wagner, M. H., Meissner, J. 1980. Network disentanglement and time-dependent flow behaviour of polymer melts. *Makromol. Chem.* 181:1533–50

Walters, K. 1975. *Rheometry*. London: Chapman & Hall. 278 pp.

Weissenberg, K. 1947. A continuum theory of rheological phenomena. *Nature* 159:310–11

Weissenberg, K. 1948. Some new anisotropic time effects in rheology. *Nature* 161:324–25

Westover, R. F. 1961. Effect of hydrostatic pressure on polyethylene melt rheology. *SPE Trans.* 1:14–20

Whorlow, R. W. 1980. *Rheological Techniques*. Chichester: Horwood. 447 pp.

Zosel, A. 1971. Der Einfluss von Molekulargewicht und Molekulargewichtsverteilung auf die viskoelastischen Eigenschaften von Polystyrolschmelzen. *Rheol. Acta* 10:215–24

Ann. Rev. Fluid Mech. 1985. 17:65–89

COATING FLOWS

Kenneth J. Ruschak

Research Laboratories, Eastman Kodak Company, Rochester, New York 14650

Introduction

A coating flow is a fluid flow that is useful for covering a large surface area with one or more thin, uniform liquid layers. The liquid film is subsequently dried or cured and often serves to protect or decorate the substrate. The film may also serve a more active function, such as the recording of information.

A familiar coating flow is that associated with the application of paint by brush or roller. Although everyday painting experience may suggest that the problem of applying a thin liquid layer to a surface is a trivial one, this is not the case. In industrial applications, the layer thickness may have to be very small and at the same time highly accurate. Moreover, a wide range of rheologies are encountered, and often the properties of the liquid are not adjustable for the purpose of coating. For productivity reasons a high speed of application may be required, and several discrete layers may have to be applied simultaneously. As a result of such demands, attempts to use a specific coating process for a given application frequently fail. The liquid layer may not be continuous, and if it is, longitudinal or transverse waves or streaks may be observed. It is also common for air to be entrained by the substrate along with the liquid.

The theoretical fluid-mechanics problem is first to identify a prospective steady, two-dimensional, film-forming flow, given the liquid properties, the coating thickness, and the application speeds of interest. This step is complicated by the fact that coating flows are free-boundary flows and therefore inherently nonlinear; that is, the region in space occupied by the flowing liquid is not known at the start but is in fact part of the solution to the hydrodynamic equations. Once a flow field is found, its stability to disturbances must be evaluated. Finally, it is necessary to predict whether the flow will effectively displace air from the surface to be coated. Stated in this way, the theoretical coating problem is anything but trivial, and the fact

65

0066–4189/85/0115–0065$02.00

that coating technology has developed largely as an art can perhaps be more readily appreciated.

The dozens of coating devices in use attest to the practical difficulty of the coating problem. Higgins (1965) and Booth (1970) have described many of these. Even a cursory review of this material makes it evident that basic science has not had a great deal to do with the development of the technology. The outline by Kaulakis (1974) of the development of coating technology in the paper industry conveys much the same message.

This is not to say that useful research on coating flows has not been done. Rather, as in many other areas of technology, largely empirical developments have outpaced scientific capabilities and understanding. Theoretical and experimental studies have led to many insights into the structure and stability of film-forming flows. For the most part, however, flow geometries have been highly simplified and the mathematical analyses quite limited in their range of applicability. Methods for computer simulation of coating flows under very general conditions have been maturing rapidly, however, and there is reason to expect that scientific studies will have a greater effect on the development of the technology in the future.

Nearly Rectilinear Flow Fields

Consider the steady, two-dimensional flow of a Newtonian liquid of viscosity μ, density ρ, and surface tension σ. Frequently the flow fields that have been studied in connection with coating flows exhibit nearly parallel streamlines. In such cases the governing equations can be substantially simplified. As will be seen, however, there has been no general agreement on which terms are properly discarded, and frequently the limitations of the results have not been clearly set forth. It therefore seems worthwhile to present the equations scaled for nearly rectilinear or quasi-one-dimensional flow. Higgins et al. (1977) and Higgins & Scriven (1979) have developed the equations and solution strategies more generally.

Suppose that the streamlines are all nearly parallel to the X axis, with a small characteristic slope δ. The characteristic length in the Y direction, T, will be a measure of the film thickness, and it follows that the characteristic length in the X direction is T/δ. The characteristic speed in the X direction, S, is usually the speed of the substrate. For a nontrivial result, the two terms in the continuity equation must be of the same size; this implies that the characteristic speed in the Y direction is δS. In strictly rectilinear flow the streamwise pressure gradient is balanced by the transverse gradient of the shear stress. The scaling for the pressure in nearly rectilinear flow, $\mu S/\delta T$, follows when the same two terms are balanced. With x, y as the dimensionless coordinates, u, v as the corresponding dimensionless velocity components, and p as the dimensionless pressure, the momentum and continuity equations scaled for nearly rectilinear flow are

$$r\delta(uu_x + vu_y) = -p_x + u_{yy} + \delta^2 u_{xx} + g\cos\theta, \tag{1}$$

$$r\delta^3(uv_x + vv_y) = -p_y + \delta^2 v_{yy} + \delta^4 v_{xx} + \delta g\sin\theta,$$

$$u_x + v_y = 0.$$

In these equations, $r = \rho S T/\mu$ is the Reynolds number, $g = \rho G T^2/\mu S$ is the inverse Stokes number, and the components of the gravitational vector are $G\cos\theta$, $G\sin\theta$. When $\delta^2 \ll 1$, all but one of the viscous terms can be neglected, and the simplified equations for nearly rectilinear flow result. It is also commonly assumed that $r\delta \ll 1$ so that the inertial terms are negligible. These simplifications are referred to as the lubrication approximation because they are frequently made in lubrication problems. The resulting equations lead to a parabolic velocity profile as in rectilinear flow; however, the profile is slowly varying in the X direction, unlike rectilinear flow.

In addition to the customary no-slip boundary condition at solid boundaries, traction and kinematic boundary conditions must be applied at the liquid/air interface $y = h(x)$. Usually the dynamic effects of the ambient air are negligible, and the traction exerted by the flowing liquid on the curved meniscus is balanced by the action of surface tension. The traction boundary condition can be resolved into components normal and tangential to the interface, and the tangential component can be used to simplify the normal component to obtain

$$0 = p + 2\delta^2 u_x(1 + \delta^2 h_x^2)/(1 - \delta^2 h_x^2) + (\delta^3/c)h_{xx}/(1 + \delta^2 h_x^2), \tag{2}$$

$$0 = (u_y + \delta^2 v_x)(1 - \delta^2 h_x^2) - 4\delta^2 h_x u_x,$$

where $c = \mu S/\sigma$ is the capillary number. These boundary conditions also simplify greatly when the terms estimated to be of order δ^2 are neglected. The kinematic boundary condition $v = h_x u$ expresses the fact that the liquid does not cross the interface.

Film Formation at Small Capillary Number

In general, film-forming flows are strongly two dimensional, and simple relationships between the parameters of the flow field are not readily obtainable. However, Landau & Levich (1942) have derived a simple relationship for film formation at small capillary number (see also Levich 1962).

Suppose that the final film thickness D is much less than the overall length scale of the flow field when the capillary number is small (Figure 1). Viscous effects are then important over the length scale D of the region where the film is entrained, but they are negligible outside this region where the gross shape of the meniscus is controlled by surface tension and a hydrostatic pressure field. Because the film is relatively thin, this static meniscus appears to be tangent to the moving substrate. If the radius of curvature of the static meniscus at its apparent point of tangency is R, then the overall scale of the flow field can be taken as R, and the supposition is that $D \ll R$. Viscous effects combine with surface tension to determine meniscus shape in the film-entrainment region, where the flow is likely to be nearly rectilinear (since the static meniscus approaches the substrate at a shallow angle). By balancing viscous and surface-tension terms in Equation (2), it follows that $\delta \sim c^{1/3}$ in the film-entrainment region.

A major contribution of Landau & Levich was to identify a matching condition for the film-entrainment and static-meniscus regions. They found that if the meniscus curvature in the film-entrainment region is of order $1/R$ where the two regions overlap, then the two regions will blend together smoothly. In the film-entrainment region, where the slope of the meniscus relative to the substrate is small, the curvature is given by the second derivative of film thickness with respect to distance along the substrate and thus is of order δ^2/D. Equating this estimate to the curvature of the static meniscus near the film-entrainment region, $1/R$, gives a second estimate for δ, namely $(D/R)^{1/2}$. When the two estimates are equated, a relationship between the final film thickness and the radius of curvature of the static meniscus at its apparent point of tangency with the substrate results:

$$D/R = 1.34c^{2/3}. \tag{3}$$

The constant of proportionality is determined through a detailed treatment of the equations. Deryagin & Levi (1964, p. 37) derived this particular form of the result. Ruschak (1974, 1976) and Wilson (1982) have shown that Equation (3) is asymptotically valid as $c \to 0$.

For Equation (3) to be useful, the flow in the film-entrainment region must be nearly rectilinear, and inertial and gravitational effects must be negligible there. Referring to Equations (1) and (2) and using Equation (3) and the estimate for δ just derived, these restrictions are, respectively, $c^{2/3} \ll 1$, $\rho S^2 R/\sigma \ll 1$, and $Bc^{1/3} \cos \theta \ll 1$, where $B = \rho G R^2/\sigma$ is the Bond

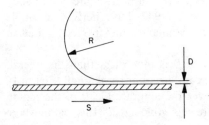

Figure 1 Film formation at small capillary numbers.

number. Although these restrictions are quite severe, the equation nevertheless helps to explain some of the basic characteristics of various coating flows, as will now be seen.

Premetered Coating Flows

In precision coating it is not desirable for the coating thickness to depend upon the viscosity of the liquid or other parameters that are difficult to control precisely. In premetered coating flows, the coating thickness is an independent parameter that can be varied within limits. Slide and curtain coating are examples of premetered coating flows.

Consider the simple slot coater shown in Figure 2 under the same limiting conditions as in the analysis of Landau & Levich (Ruschak 1976). More general geometries have been analyzed by Higgins & Scriven (1980). The two static menisci that must be considered are the coating meniscus, which forms the film, and the wetting meniscus, which displaces air from the substrate. It is presumed that these menisci are so small ($B \ll 1$) that gravity does not affect their shape, which is therefore circular. In practice, a pressure difference ΔP is maintained across the liquid in the gap between the slot and the substrate, where ΔP is the amount by which the ambient pressure downstream of the coating meniscus exceeds that upstream of the wetting meniscus. The coating meniscus must bridge the gap E between the slot and the substrate and must be tangent to the substrate at its apparent point of contact. On purely geometric grounds, then, its radius of curvature R, as computed from Equation (3) using the desired coating conditions, cannot be smaller than $E/2$. Further restrictions on R follow from the requirement that the wetting meniscus bridge the gap between the slot and the substrate and intersect the substrate at the apparent dynamic contact angle ζ (Dussan

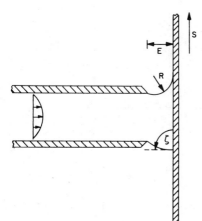

Figure 2   Slot coater at small capillary numbers.

V. 1979), as shown in Figure 2. The complex subject of dynamic wetting is covered in the final section of this review; here, the dynamic contact angle is presumed to be known.

The geometrical restrictions on the radius of curvature of the coating meniscus are summarized in Figure 3, which was drawn for a dynamic contact angle of 120°. The coating meniscus is free to adopt any radius of curvature within a range of values that depends upon the imposed pressure differential. When the flow rate is imposed at a given speed, thus determining the coating thickness D, the coating meniscus adopts the radius of curvature necessary to satisfy Equation (3), as long as this radius satisfies the inequalities depicted in Figure 3. If any of the inequalities is violated, then coating a uniform layer under the specified conditions is not

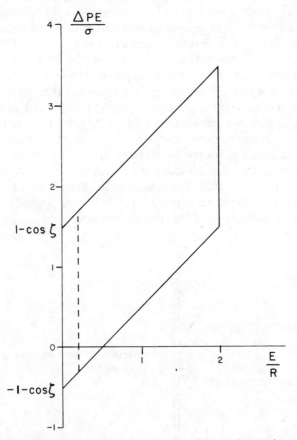

Figure 3   Possible values for the radius of curvature R of the coating meniscus for the slot coater at small capillary numbers.

possible. Figure 3 shows that a pressure difference across the gap increases the utility of the coating device by enabling the coating meniscus to adopt higher curvatures and thus produce thinner coatings.

According to Figure 3, the radius of curvature of the coating meniscus can be arbitrarily large. This is because the effect of gravity on the shape of the coating meniscus has been neglected. If gravity is retained in the analysis, there is an upper limit to the radius of curvature of the coating meniscus at its point of tangency with the substrate, as suggested by the dashed line in Figure 3.

Tallmadge et al. (1979) experimentally studied the slide coater, in which liquid flows down an inclined plane that ends a small distance from the substrate. They did not impose a pressure difference across the gap. As the speed was increased at constant flow rate, a point was reached where it became impossible to preserve a two-dimensional flow. The film would either narrow and resist being drawn to the proper width or, more commonly, split into two or more parts. This "split point" marks the crossing of the lower boundary in Figure 3. On the other hand, as coating speed was decreased, a point was reached where not all of the liquid supplied was taken up by the moving substrate, and dripping was observed below the point of application. This "drip point" marks the crossing of the left, dashed boundary in Figure 3.

Coating Flows Metered by the Coating Meniscus

DIP COATING AT LOW CAPILLARY NUMBER   In a self-metering coating flow, the coating thickness is a dependent variable. The best-known example of a self-metering coating flow is free withdrawal, or dip coating, in which the substrate is withdrawn from a large reservoir of liquid (Figure 4). This type of coating flow has been reviewed by Tallmadge & Gutfinger (1967). In Equation (3), R, the radius of curvature of the static meniscus at the coating point, is uniquely determined by the hydrostatic pressure field to be $[\sigma/2\rho G(1+\sin \alpha)]^{1/2}$. Unlike premetered coating, the radius of curvature of the coating meniscus is not free to vary, with the result that the coating thickness becomes the dependent variable in Equation (3). Thus the coating thickness depends upon the fluid properties and the substrate speed, which is usually a disadvantage. Ruschak (1974) compared the predictions of Equation (3) for vertical withdrawal ($\alpha = 0$) with the low-capillary-number data of Moray (1940) and found good agreement only at capillary numbers below about 0.01. The quantitative usefulness of Equation (3) is limited indeed.

DIP COATING AT HIGH CAPILLARY NUMBER   Many attempts have been made to improve upon the predictions of the equation of Landau & Levich for dip coating at capillary numbers higher than 0.01. One useful result is that there is an upper limit to the film thickness. Derjaguin (1945) and Van Rossum (1958) studied the fully rectilinear flow far above the liquid reservoir and found that there is a maximum possible flow rate that can be lifted by the moving surface against gravity. This maximum flow rate is $2SD'/3$, where the film thickness D' is given by $(\mu S/\rho G \cos \alpha)^{1/2}$. This result is consistent with $g \cos \theta$ being of order unity in Equation (1), i.e. consistent with viscous and gravitational effects being in balance. Derjaguin concludes that this will be the film thickness approached as the capillary number becomes large with inertial effects remaining negligible. More recent theoretical studies by Homsy & Geyling (1977) and by Tuck (1983) support this contention. Experiments, however, have generally yielded a value of $d = D/D'$ that is less than unity at high capillary number. For example, Van Rossum (1958) could not produce films thicker than $d = 0.68$, while Groenveld (1970) reports 0.66, and the highest value obtained by Spiers et al. (1974) is about 0.8. Moreover, after reaching its maximum value, d decreases slightly with further speed increases.

Marques et al. (1978) have performed numerical experiments on the boundary-layer equations that may account for this apparent lack of agreement between experiment and theory. The boundary-layer equations are in fact the equations for nearly rectilinear flow when the inertial and gravitational terms are retained but the surface-tension term is discarded. In the numerical experiments a film thickness greater than $d = 0.67$ could

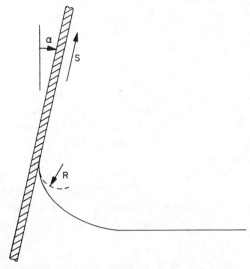

Figure 4   Free withdrawal from a pool of liquid.

not be produced, and so significant inertial effects may explain the experimental data at high capillary number.

DIP COATING AT INTERMEDIATE CAPILLARY NUMBER   To bridge the gap between the results at high and low capillary numbers, two-dimensional flow equations must be faced. By the estimates given earlier for the limits of applicability of Equation (3), the effect of gravity in the film-entrainment region becomes important at about the same capillary number that the assumption of nearly rectilinear flow breaks down. That this is so is also evident from the two-dimensional finite-element calculations of Ruschak (1982), which show Equation (3) failing at a capillary number of about 0.01, even though gravity is not included in the calculations.

Nonetheless, the assumption of nearly rectilinear flow has often been retained in analyses of dip coating. The effect of gravity in the film-entrainment region has been successfully incorporated into the analysis of the equations of nearly rectilinear flow by Lee & Tallmadge (1975), who solved the equations numerically, and also by Wilson (1982), who obtained an explicit result by the method of matched asymptotic expansions. Unfortunately, the inclusion of gravity alone does not greatly expand the range of validity of Equation (3). In analyses where nearly rectilinear flow has not been assumed, some, but not all, of the terms estimated to be of order δ^2 have been kept. The papers by Spiers et al. (1974) and by Esmail & Hummel (1975) are examples.

Another difficulty in extending the analysis of Landau & Levich to higher capillary numbers lies in matching the solution to the differential equation giving the meniscus shape of the film-entrainment region with the static meniscus associated with the reservoir. Whenever gravity or other higher-order terms are retained, the procedure of Landau & Levich for matching curvatures breaks down. That this problem can be rationally approached is evidenced by the analyses of Lee & Tallmadge and Wilson just mentioned. Unfortunately, ad hoc measures pervade the literature. In some cases the differential equation for the film shape has been illogically altered to produce a new differential equation that does permit the curvature matching of Landau & Levich; examples include the papers by White & Tallmadge (1965) and Spiers et al. (1974). In other cases the static meniscus is assumed to extend up to the stagnation point, which is always present on the meniscus. The differential equation for the static meniscus is applied as a boundary condition at the stagnation point. This procedure does not guarantee continuous slope and curvature along the entire meniscus, contrary to the claims of its first proponents, Esmail & Hummel (1975), and there would seem to be nothing special about the stagnation point on the meniscus that would permit one to deduce that dynamic effects are negligible below this point. What is more, the differential equation for the static meniscus that is applied at the stagnation point has been incorrectly taken from the book of Deryagin & Levi (1964, p. 36); a change in the positive direction of the X axis produces a change in the sign of the first derivative of the film thickness, which has apparently been overlooked. Although Nigam & Esmail (1980) later admitted the ad hoc nature of this boundary condition, it continues to appear in papers (Tekic & Jovanovic 1982), and in its incorrect form at that.

Although improved agreement with experimental data at moderate and/or high capillary numbers is claimed for many analyses subsequent to that of Landau & Levich, this does not really justify ad hoc mathematical procedures. Perhaps the best way to solve the dip-coating problem at capillary numbers higher than 0.01 is by a two-dimensional numerical method. Lee & Tallmadge (1974) used a finite-difference scheme and had some success for capillary numbers larger than about 1.0. Current finite-element methods can handle this problem more easily and without the restriction on the capillary number.

Coating Flows Metered Upstream of the Coating Meniscus

BLADE COATING   In dip coating the flow rate is metered by the coating meniscus, and as a result the coating thickness is sensitive to the fluid properties and the substrate speed. There is, however, an important class of self-metering coating flows in which the metering takes place upstream of the coating meniscus. The coating meniscus essentially responds to an imposed flow rate, much as in premetered coating. While these coating flows are not as accurate as a premetered coating flow, a substantial degree of independence of the coating thickness from the fluid properties and the substrate speed can be achieved.

One example of such a self-metering coating device is the blade coater, commonly used to apply coatings during the manufacture of paper (Gartaganis et al. 1978). A simple model of a blade coater by Greener & Middleman (1974) illustrates the principle. The face of the blade has a small slope δ with respect to the rigid substrate, so that the liquid is drawn into a slowly converging region (Figure 5). The blade is presumed to be flooded upstream (an excess of liquid is supplied), and surface-tension effects are presumed to be negligible so that the pressure can be taken to be sensibly atmospheric at each end of the blade. In terms of the dimensionless group $b = \delta L/E$, the ratio of the coating thickness to the gap is given by $D/E = (1+b)/(2+b)$. The most important feature of this result is that the coating thickness is independent of fluid properties and substrate speed.

Often the blade is not rigidly fixed with respect to the substrate. Rather, it is loaded by one means or another, and the loading controls the

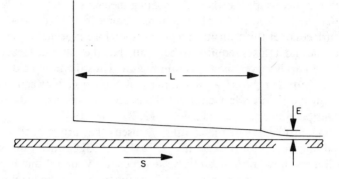

Figure 5   Blade coater flooded upstream.

coating gap. If A is the force per unit width applied to the blade, then $A\delta^2/6\mu S = \ln(1+b) - 2b/(2+b)$. Thus b and the gap E follow when A is specified. For the special case $b \ll 1$, $E = L(\delta\mu S/2A)^{1/3}$, and the larger the loading the smaller the gap.

ROLL COATING   An important variation of this type of flow field is that for the roll coater. As shown in Figure 6, two rotating rollers are separated by a gap E, which is much smaller than their radii R_1 and R_2. For the moment the ratio of the speeds of the rollers, a, may be any number. It is usually presumed that the important part of the flow field is near the nip, where the roller surfaces are close together and nearly parallel, so that the equations for nearly rectilinear flow can be used. If the characteristic film thickness T is taken to be E and the characteristic length in the X direction to be $(ER)^{1/2}$, where $1/R \equiv 1/R_1 + 1/R_2$, then the relative slope between the cylinder surfaces is on the order of $\delta \equiv (E/R)^{1/2}$, and the flow will be nearly rectilinear when $\delta^2 \ll 1$. Furthermore, when $\delta^2 \ll 1$ the distance $F(X)$ between the rollers may be approximated by the first terms of a power series, namely $f \equiv F/E = 1 + x^2/2$. It is also usually presumed that inertial and gravitational effects are negligible.

If the rollers are flooded on both sides (that is, if the liquid extends to $|x| \gg 1$), then the dimensionless flow rate is given by $\lambda \equiv Q/SE = 2(1+a)/3$, where Q is the volumetric flow rate per unit width (Gatcombe 1945). When the rollers are not flooded on one side and there is a coating meniscus, the problem is considerably more difficult to solve, because the flow field in the vicinity of the coating meniscus is two dimensional (Taylor 1963). If the meniscus forms at $x = m$, then the slope between the roller surfaces there will be δm. Furthermore, as long as inertial effects are negligible, the transition from the nearly rectilinear flow between the roller surfaces to fully formed liquid films will take place over a distance comparable to that between the roller surfaces where the coating meniscus forms, namely $W = Ef(m)$ (Coyne & Elrod 1970). The fractional change in distance between the roller surfaces over this transition region is therefore expected to be on the order of δm, and if this is required to be small, then the roller surfaces can be considered parallel over the transition region. Thus the flow field divides naturally into two parts: a two-dimensional portion between parallel surfaces in the immediate vicinity of the coating meniscus, and a nearly rectilinear portion elsewhere (Taylor 1963, Ruschak 1982).

The flow problem near the coating meniscus that arises in this way is sketched in Figure 7 for the case $a = 1$. Taylor (1963) pointed out that the dimensionless flow rate $q \equiv Q/SW$ depends only upon the capillary number. The asymptotic pressure gradient upstream can be expressed in terms of q as $dp/dx = 12(1 - q)/f^2$. Indeed, this pressure gradient provides one of the boundary conditions at $x = m$ that the solution to the equations for nearly rectilinear flow between the rollers must satisfy. Unfortunately, this boundary condition is not useful until q is determined as a function of the capillary number by solving the difficult flow problem of Figure 7. In

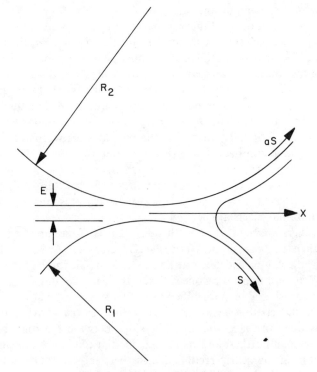

Figure 6   Roll coater flooded upstream.

fact, Taylor resorted to special experiments to measure q. Later, Coyne & Elrod (1970) determined q approximately (for a closely related flow problem) by a method that involves assuming a functional form for the velocity field, and Ruschak (1982) solved for q without making simplifications by using the finite-element method. When the capillary number is less than about 0.01, Equation (3) can be used to compute q. With $W/2$ as the appropriate radius of curvature for the coating meniscus, the result is $q = 1.34c^{2/3}$. According to Ruschak's results, q is given approximately by $0.54c^{1/2}$ for $0.01 < c < 0.1$, and q approaches a constant value of about 0.41 as $c \to \infty$. Thus the pressure gradient of the nearly rectilinear flow should be $12k/f^2$ where the coating meniscus forms, and k decreases from 1 to about 0.59 as the capillary number increases from 0 to ∞.

Frequently, approximate expressions have been used for the pressure gradient where the coating meniscus forms. Dowson & Taylor (1979), Coyne & Elrod (1970), Savage (1977a), Sullivan & Middleman (1979), and Bixler (1982) have critically discussed many of these. The Swift-Steiber approximation, for instance, is to set the pressure gradient to zero where the coating meniscus forms, but as has just been seen, this is never strictly true. According to the popular Prandtl-Hopkins condition, the interface forms at a stagnation point in the nearly rectilinear flow field (i.e. at a point where u and u_y are both zero). This leads to a pressure gradient of $12(2/3)/f^2$, which is close to the correct result at capillary numbers near unity.

Besides the pressure gradient, the pressure where the coating meniscus forms is also required as a boundary condition and may be estimated (Taylor 1963) or obtained by the solution of the flow field of Figure 7 (Ruschak 1982). When the capillary number is large, this pressure will be on the order of $\mu S/W$, which is small compared with the characteristic pressure of $\mu S/\delta E$ and may be taken as zero. When the capillary number is small, on the other hand, the pressure will be of the order of $2\sigma/W$, and the boundary condition becomes $p = -2\delta/fc$, which can be of some consequence if c is smaller than δ. For simplicity, this latter expression is used throughout what follows.

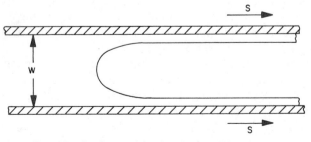

Figure 7   Coating meniscus between parallel surfaces.

The value of λ varies little with δ and c and is close to 4/3, as when the rollers are flooded both upstream and downstream. Benkreira et al. (1981) made 1500 experimental measurements for forward roll coating ($a > 0$) at capillary numbers between 0.03 and 15, and for δ between about 0.04 and 0.3. They found $\lambda = 1.31$ with a standard deviation of 0.4%. Pitts & Greiller (1961) measured flow rates for δ between about 0.06 and 0.3 and for $a = 1$. They found λ to lie between 1.26 and 1.38, and they noted that λ decreases slightly with increasing speed (capillary number). According to the theoretical results of Ruschak (1982), for capillary numbers at even multiples of 10 between 0.01 and 10, the predicted values of λ are, respectively, 1.36, 1.33, 1.30, and 1.29 for $\delta = 0.06$. Thus the flow rates predicted by Gatcombe (1945) for flooded rollers are expected to be correct within a few percent for typical values of c and δ, and as a result, the rollers meter the liquid in much the same way as the blade coater.

The position of the coating meniscus is simply related to λ and q. From the definition of q, the distance between the roller surfaces where the coating meniscus forms is $W = Q/Sq = E\lambda/q$, so that $f = \lambda/q$ and $m = [2(\lambda/q - 1)]^{1/2}$. When the capillary number is so small that q is much smaller than 1, the coating meniscus forms too far from the nip ($m \gg 1$) for the assumption of nearly rectilinear flow to be valid up to the vicinity of the coating meniscus, and a fully two-dimensional analysis is in order. As the capillary number increases, so too does q increase, and the meniscus is drawn toward the nip.

It is necessary to know in forward roll coating ($a > 0$) how the flow divides between the two rollers, or in other words what is the coating thickness on each roller? Pitts & Greiller (1961) observed that two regions of recirculation are usually (but not always) present upstream of the coating meniscus. Savage (1982) noted that the equations for nearly rectilinear flow predict the stagnation point marking the onset of these recirculations. On the assumption that the streamline passing through this point ultimately becomes the free-surface streamline (an assumption consistent with the observations of Pitts & Greiller), he was able to predict quantitatively how the flow divides. Actually, C.-S. Yih pursued this idea some years earlier (see Hintermaier & White 1965) and obtained a result that, although correct, is limited to the special case where a is nearly 1. With λ_1 and λ_2 the flow rates ultimately associated with rollers 1 and 2 so that $\lambda_1 + \lambda_2 = \lambda$, Savage's result may be expressed as $\lambda_2/\lambda_1 = a^{1.5}$. This form of the result was pointed out by D. J. Coyle (personal communication, 1983) and follows immediately from Savage's work. Savage presented experimental data that support his result, and Benkreira et al. (1981) used a least-squares analysis of their 1500 data points to obtain $\lambda_2/\lambda_1 = 0.87a^{1.65}$ (the asymmetry for $a = 1$ was attributed to gravitational effects). Equation (3) predicts an exponent of 1.67 for very small capillary numbers.

Instability of Coating Flows: Ribbing Lines

The calculation of the steady, two-dimensional flow field, and sometimes the determination of the range of conditions over which such a solution to the governing equations exists, is often just a prelude to a determination of the stability of the flow. The flow instability called ribbing lines has received a great deal of attention in the literature. This instability may be observed in roll coating and slide coating, for example. The flow field for ribbing lines is steady but three dimensional. More specifically, the flow field becomes periodic in the third dimension, and the coating itself becomes ribbed and uneven. Such an event is usually intolerable in a coating process, and so the study of this and other instabilities is of considerable importance.

The theoretical calculations and experimental observations of Saffman & Taylor (1958) provided the first insights into the ribbing instability. Saffman & Taylor used the equations for nearly rectilinear flow, expanded to include the third dimension, to evaluate the stability of a coating meniscus between parallel surfaces (Figure 7) to small disturbances. They found that the flow is unstable if the pressure gradient corresponding to the rectilinear flow upstream of the meniscus, namely $dP/dX = 12\mu S(1-q)/W^2$, is positive. Because q cannot exceed 0.41 (Ruschak 1982), this is always the case.

Arguments by Pitts & Greiller (1961) are useful in trying to understand this result. They consider the evolution of the disturbance to be quasi-steady-state, and they suppose that at some location in the third dimension the coating meniscus moves slightly upstream because of a disturbance. The linear pressure field of positive slope upstream of the displaced meniscus remains the same as that behind the undisturbed meniscus, but it is shifted along the X axis by the distance the displaced meniscus has moved. It follows that at any position X, the pressure behind the disturbed meniscus is more positive than that behind the undisturbed meniscus. Therefore, flow will occur from the region behind the disturbed meniscus toward the region behind the undisturbed meniscus, and the disturbance will be reinforced.

In forward roll coating with equal roller speeds, the position of the coating meniscus is near the nip where the roller surfaces are nearly parallel if the capillary number is sufficiently large, as was seen in the preceding section. In light of the findings of Saffman & Taylor, it is not surprising that forward roll coating is subject to the ribbing-line instability. According to the experimental observations of Pitts & Greiller (1961), Mill & South (1967), and Greener et al. (1980), forward roll coating is stable at sufficiently low capillary numbers but eventually becomes unstable as the capillary number is increased and the coating meniscus is drawn toward the nip. Thus the fact that the roller surfaces diverge to some extent evidently has a stabilizing effect on the flow field.

The mechanisms by which diverging surfaces stabilize the flow field were elucidated by Pearson (1960) in his analysis of a spreader and by Pitts & Greiller (1961) in their analysis of the forward roll coater for the case $a = 1$. Following Pitts & Greiller, suppose that at some location in the third dimension the meniscus moves slightly upstream because of a disturbance. Because the roller surfaces diverge slightly, the radius of curvature of the coating meniscus must decrease as it moves inward, and thus the pressure at the meniscus will become more negative by the amount $\varepsilon(2\sigma/W^2)dF/dX$, where ε is the distance the meniscus has moved and W is the distance between the roller surfaces at the position of the coating meniscus in the two-dimensional flow field. The pressure behind the undisturbed meniscus at the value of X of the displaced meniscus is more negative than the pressure at the undisturbed meniscus by the amount $\varepsilon \, dP/dX$, where the pressure gradient is that for the nearly rectilinear flow just upstream of the coating meniscus in the two-dimensional flow field. If, at this X, the pressure is higher behind the displaced meniscus than behind the un-displaced meniscus, then liquid will flow from behind the displaced meniscus toward the undisplaced meniscus and increase the disturbance. Thus, comparing the above two terms, the disturbance can grow whenever $dP/dX > (2\sigma/W^2)dF/dX$. For instability the pressure gradient must not only be positive but also sufficiently large. Savage (1977a,b) proved that this inequality is a necessary but not sufficient condition for instability, and Gokhale (1983) derived a stronger necessary condition.

The second stabilizing mechanism follows from the result of Taylor (1963) on the geometric similarity of the flow field near the coating meniscus. Taylor's result is again that the flow rate Q at the coating meniscus is given by SWq, where W is the distance between the surfaces and where q depends only upon the capillary number. Thus, if a disturbance causes a section of the coating meniscus to move slightly upstream, the flow rate past it is reduced because the surfaces are closer together. A simple mass-conservation argument at the meniscus shows that this response has a dampening effect on the disturbance.

Figure 8 shows a flow geometry similar to the spreader considered by Pearson (1960), which I have analyzed for this review. The dimensionless flow rate λ is taken to be 4/3 to approximate that for the forward roll coater. In Figure 9 the capillary number above which the flow field is unstable to small disturbances is plotted against α, the relative slope between the linearly diverging surfaces. The curve has purposely been extended to values of α that are not small. In light of the stabilizing mechanisms

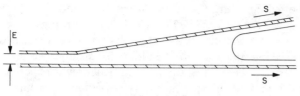

Figure 8   Coating meniscus between linearly diverging surfaces.

discussed, it is not surprising that the capillary number at which the flow becomes unstable to small disturbances increases with the slope between the surfaces. The rather substantial divergence of $10°$ ($\alpha = 0.176$) corresponds to the small capillary number of 0.054. The stabilizing mechanisms are weak indeed.

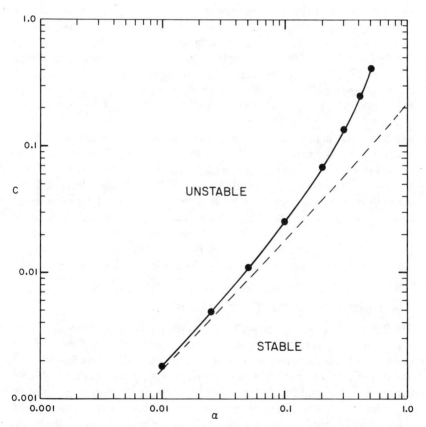

Figure 9   Stability of a coating meniscus between linearly diverging surfaces: ● linear-stability analysis following Pearson (1960); ———— necessary condition for instability following Pitts & Greiller (1961) and Savage (1977a,b).

Also plotted in Figure 9 is the necessary condition for instability determined by Pitts & Greiller (1961). This condition appears to approach the exact result asymptotically as the capillary number decreases. There is other evidence that the necessary condition for instability is useful only for small capillary numbers. By comparing predictions with experimental data, Savage (1977a) found that the necessary condition for instability gives a reasonable estimate of the region of stability for a roller and flat-plate geometry over the capillary number range of 0.06–0.11. On the other hand, when Greener et al. (1980) compared their stability data for forward roll coating for capillary numbers in the range 0.1–10, they found that the necessary condition for instability substantially underestimates the region of stability. However, under the experimental conditions of Greener et al., the coating meniscus forms so far from the nip that a fully two-dimensional treatment is probably in order anyway.

Although not strictly correct, it is instructive to use the stability results for flow in a linearly diverging region (Figure 9) to predict the stability of the forward roll coater. To do this it is necessary only to estimate the slope between the roller surfaces where the coating meniscus forms. Since the meniscus forms where $f = \lambda/q$, and since $f = 1 + x^2/2$, it follows that this slope is given by $\delta[2(\lambda/q - 1)]^{1/2}$. Moreover, λ is close to 4/3, and the value of q as a function of c is known (Ruschak 1982). Thus for a given capillary number, q can be looked up and Figure 9 can be used to estimate the slope between the roller surfaces at which instability occurs. The value of δ that gives this slope then follows immediately. The results are shown in Figure 10, along with the experimental data of Mill & South (1967) and the experimental correlations of Pitts & Greiller (1961) and Greener et al. (1980). The agreement between the predictions and the data is good. It therefore seems that in forward roll coating the stability of the flow depends primarily upon the divergence between the roller surfaces where the coating meniscus locates: the greater the divergence, the more likely the flow is to be stable.

Computer Methods for Flow Simulation

The flow near a coating meniscus is nearly always two dimensional. This is true even when, as in roll coating, the major portion of the flow field has nearly parallel streamlines. Moreover, viscous, inertial, surface tension, and gravitational effects may all be important; the flow geometry may be highly irregular and complicated (for example, some fluid boundaries may be compliant); and sometimes, as in the photographic industry, several liquid layers may be applied simultaneously. Often non-Newtonian effects are important. Moreover, an analysis of the two-dimensional, steady-state flow

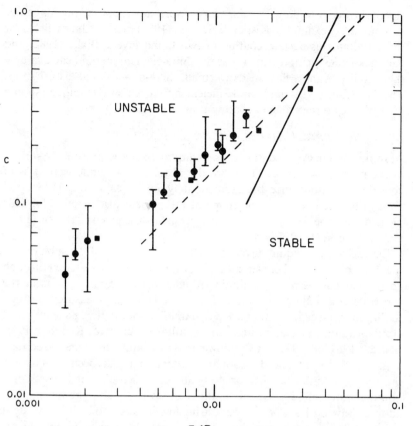

Figure 10 Stability of forward roll coating: ● data by Mill & South (1967); ———— experimental correlation of Pitts & Greiller (1961); ——— experimental correlation of Greener et al. (1980); ■ theoretical prediction based on the stability of a coating meniscus between linearly diverging surfaces (Figures 8 and 9).

is often of little value if the stability of the flow cannot be assessed. These requirements are largely beyond the means of classical analysis.

In recent years computer methods for flow simulation have advanced to the point where most steady-state, two-dimensional flow fields can be efficiently calculated. Kistler & Scriven (1983) recently reviewed these methods. They included as examples calculated flow fields for roll, extrusion, slide, and curtain coating. Kobayashi et al. (1982) did a parameter study on the slide-coater flow field and found good agreement between a computed and an experimentally obtained flow profile. Kistler

(1983) showed how flow fields for two or more superimposed liquids can be computed. As examples, Kistler & Scriven (1982) presented flow fields for the simultaneous curtain coating of two liquid layers. Bixler (1982) and Ruschak (1983) demonstrated that the finite-element method can evaluate the stability of these flow fields to small, three-dimensional disturbances. The availability of reliable and efficient finite-element techniques should greatly expand the scope of coating-flow studies.

Dynamic Contact Angles and Air Entrainment

At some point in every coating process, liquid displaces air from the surface of the moving substrate. Therefore the physics of dynamic wetting is of considerable importance, but unfortunately not a great deal is understood about this phenomenon. The scope of the present discussion is limited; more detailed reviews of contact-line problems are given by Davis (1983) and Dussan V. (1979).

During coating, liquid displaces air from the surface of the substrate at the three-phase line where air, liquid, and solid meet. When viewed through an optical microscope, the liquid/air interface appears to intersect the substrate at a well-defined angle. This angle is usually measured through the liquid and is called the apparent dynamic contact angle. The sequence of events leading to air entrainment is as follows (Burley & Kennedy 1976, Blake & Ruschak 1979). At very low speeds, the apparent dynamic contact angle equals the advanced stationary contact angle. As speed is increased, the apparent contact angle monotonically increases, ultimately reaching a nominal value of 180°. At this point the previously straight wetting line adopts a sawtooth shape (i.e. the wetting line breaks into inclined, straight sections). The flow becomes three dimensional and possibly also unsteady. At sufficiently high speeds, air bubbles are entrained from the vertices of the sawtooth wetting line that are the farthest downstream. In coating flows, therefore, it is generally necessary to avoid apparent dynamic contact angles of nominally 180°, and consequently the factors that determine the value of the apparent dynamic contact angle are of great interest.

A contact angle is required as a boundary condition at the wetting line when, as is generally the case, the effect of surface tension is important on the overall scale of the flow field. At high capillary numbers, however, surface tension can be discarded from the hydrodynamic equations outside a thin boundary layer near the wetting line, and the apparent dynamic contact angle becomes a dependent variable determined by the details of the flow field away from the three-phase line (Kistler & Scriven 1982).

An important attribute of the apparent dynamic contact angle is that its value depends upon more than just the properties of the liquid and substrate and the speed of wetting. In other words, its value depends upon the details of the flow field of which it is a part, and so it is not generally possible to measure the apparent dynamic contact angle in one flow geometry and then apply this angle as a boundary condition in another flow geometry. The apparent dynamic contact angles reported by Ngan & Dussan V. (1982) for liquid displacing air between glass slides at low capillary number are the most direct supporting evidence for this conclusion. Ngan & Dussan V. found that the spacing between the slides affects the value of the apparent dynamic contact angle. Kistler & Scriven (1982) provided further support by calculating the variation of the apparent dynamic contact angle with changes in flow parameters for curtain coating at high capillary numbers. Less direct support comes from the experimental finding that the speed of onset of air entrainment can be changed by altering the flow field (Perry 1967, Levi & Akulov 1964, Levi 1966). It is, in fact, a common observation in the coating art that the speed of air entrainment depends upon the coating method.

Modeling of the flow near the three-phase line is complicated by the breakdown of the classical hydrodynamic equations and/or boundary conditions in this region. Huh & Scriven (1971) showed that the equations and boundary conditions of classical hydrodynamics make the unacceptable prediction that the flow of liquid in the neighborhood of a contact line exerts an unbounded force on the substrate. They suggest that the classical no-slip boundary condition may not be valid near the wetting line. Although subsequent investigators have shown that allowing slip in the immediate vicinity of the three-phase line removes this force singularity (Dussan V. 1979), the problems remain of finding a realistic model for slip and specifying a contact angle at the three-phase line.

Hansen & Toong (1971) realized that hydrodynamic forces can cause the slope of the meniscus to change rapidly very near the three-phase line. Indeed, they predicted that significant changes in slope occur over distances too small to be resolved with an optical microscope. They conclude that, in general, the apparent dynamic contact angle is not equal to the actual contact angle at the three-phase line, and that much if not all of the variation of the apparent dynamic contact angle with speed can be attributed to hydrodynamic bending of the meniscus very near the substrate, rather than to changes in the actual contact angle. Some later investigators (e.g. Hocking & Rivers 1982) have assumed that the actual dynamic contact angle does not vary at all with speed, so that changes in apparent dynamic contact angle are entirely attributable to hydrodynamic bending of the meniscus over a region very close to the substrate.

By incorporating the disjoining pressure of Deryagin, Teletzke et al. (1984) modified the classical hydrodynamic equations to include the effects of fluid microstructure over distances from the substrate so small (less than

1 μm) that the liquid is no longer homogeneous. They predict that below a certain speed no air is entrained, but that above this speed an air film is entrained that is at first so thin that the residual air is not likely to be detected (Miyamoto & Scriven 1982). The thickness of the air film increases with further speed increases. Thus the question may not be whether air is entrained, but whether enough is entrained to produce detectable residual air. Even with an air film left on the substrate, Teletzke et al. found that the apparent dynamic contact angle was not 180°. The apparent angle became 180° only when the air film got so thick that disjoining pressures became negligible. The presence of a thin air film separating the substrate from the liquid circumvents the singularity at a three-phase line. Teletzke et al. suggest that this air film may justify slip near a three-phase line in macroscopic hydrodynamic models.

In a third approach, the displacement of air by liquid at a three-phase line is modeled as a molecular rate process by using the statistical-mechanical theory of rate processes developed by H. Eyring and others. The theory leads to an expression for the apparent dynamic contact angle as a function of speed that fits experimental data very well (Blake & Haynes 1969). Presently, however, the theory does not take account of hydrodynamic effects on the apparent dynamic contact angle.

Unfortunately, these theories are not developed to the point where definite judgments can be made about their utility. Even further away is the routine application of a proven theory for dynamic wetting to the modeling of coating flows.

Concluding Remarks

Considerable progress has been made at elucidating the structure and stability of coating flows. However, the limitations of classical analytical approaches have certainly limited progress on the modeling of coating flows. Until recently, nearly all coating flows studied were characterized by nearly parallel streamlines. This is doubtless the case because the equations of capillary hydrodynamics simplify greatly for nearly rectilinear flow. In general, however, the flow near a coating meniscus is highly two-dimensional. The failure to treat the two-dimensional areas of flow fields realistically has led in the past to the use of ad hoc boundary conditions on the nearly rectilinear portion of the flow field. The range of validity of the analysis is then in doubt, and the ad hoc boundary conditions become as much a subject of attention as the original problem.

There is a great need to look beyond coating flows that exhibit nearly parallel streamlines. It is similarly desirable to relax the usual restriction of negligible inertial effects. Non-Newtonian effects merit more attention, because many if not most of the liquids coated depart significantly from

Newtonian behavior. It is also important to recognize that an analysis of the steady, two-dimensional flow field is likely to have limited value unless the stability of the flow field to three-dimensional disturbances can be evaluated. Finite-element methods implemented on modern computers have demonstrated considerable promise for handling such complications.

Although many of the remaining problems are computational, the phenomenon of air entrainment stands out as one for which a proven model is not available. The phenomenon is a complex one, involving both macroscopic flow effects and microscopic events at a dynamic wetting line that are far from clearly understood.

Flow fields that can transform a liquid of given properties into a suitably thin, uniform film at high speeds are not easily identified and analyzed. Nonetheless, the interplay among viscous, inertial, gravitational, and capillary effects, as well as the nuances of free boundaries, makes this challenging class of flow problems an interesting one for fluid mechanicians.

ACKNOWLEDGMENT

I thank Eastman Kodak Company for supporting this work.

Literature Cited

Benkreira, H., Edwards, M. F., Wilkinson, W. L. 1981. Roll coating of purely viscous liquids. *Chem. Eng. Sci.* 36:429–34

Bixler, N. E. 1982. *Stability of a coating flow.* PhD thesis. Univ. Minn., Minneapolis

Blake, T. D., Haynes, J. M. 1969. Kinetics of liquid/liquid displacement. *J. Colloid Interface Sci.* 30:421–23

Blake, T. D., Ruschak, K. J. 1979. A maximum speed of wetting. *Nature* 282:489–91

Booth, G. L. 1970. *Coating Equipment and Processes.* New York: Lockwood. 453 pp.

Burley, R., Kennedy, B. S. 1976. An experimental study of air entrainment at a solid/liquid/gas interface. *Chem. Eng. Sci.* 31:901–11

Coyne, J. C., Elrod, H. G. 1970. Conditions for the rupture of a lubricating film. Part I: Theoretical model. *J. Lubr. Technol.* 92: 451–56

Davis, S. H. 1983. Contact-line problems in fluid mechanics. *J. Appl. Mech.* 50:977–82

Derjaguin, B. 1945. On the thickness of the liquid film adhering to the walls of a vessel after emptying. *Acta Physicochim. URSS* 20:349–52

Deryagin, B. M., Levi, S. M. 1964. *Film Coating Theory.* New York: Focal. 190 pp.

Dowson, D., Taylor, C. M. 1979. Cavitation in bearings. *Ann. Rev. Fluid Mech.* 11:35–66

Dussan V., E. B. 1979. On the spreading of liquids on solid surfaces: static and dynamic contact angles. *Ann. Rev. Fluid Mech.* 11:371–400

Esmail, M. N., Hummel, R. L. 1975. Nonlinear theory of free coating onto a vertical surface. *AIChE J.* 21:958–65

Gartaganis, P. A., Cleland, A. J., Wairegi, T. 1978. Blade mechanics of extended blade coaters. *Tappi* 61:77–81

Gatcombe, E. K. 1945. Lubrication characteristics of involute spur gears. *Trans. ASME* 67:177–88

Gokhale, V. V. 1983. Improved stability criterion for lubrication flow between counterrotating rollers. *AIChE J.* 29:865–66

Greener, J., Sullivan, T., Turner, B., Middleman, S. 1980. Ribbing instability of a two-roll coater: Newtonian fluids. *Chem. Eng. Commun.* 5:73–83

Greener, Y., Middleman, S. 1974. Blade-coating of a viscoelastic fluid. *Polym. Eng. Sci.* 14:791–96

Groenveld, P. 1970. High capillary number withdrawal from viscous Newtonian liquids by flat plates. *Chem. Eng. Sci.* 25:33–40

Hansen, R. J., Toong, T. Y. 1971. Dynamic contact angle and its relationship to forces of hydrodynamic origin. *J. Colloid Interface Sci.* 37:196–207

Higgins, B. G., Scriven, L. E. 1979. Interfacial

shape and evolution equations for liquid films and other viscocapillary flows. *Ind. Eng. Chem. Fundam.* 18:208–15

Higgins, B. G., Scriven, L. E. 1980. Capillary pressure and viscous pressure drop set bounds on coating bead operability. *Chem. Eng. Sci.* 35:673–82

Higgins, B. G., Silliman, W. J., Brown, R. A., Scriven, L. E. 1977. Theory of meniscus shape in film flows. A synthesis. *Ind. Eng. Chem. Fundam.* 16:393–401

Higgins, D. G. 1965. Coating methods. In *Encyclopedia of Polymer Science and Technology*, 3:765–807. New York: Wiley

Hintermaier, J. C., White, R. E. 1965. The splitting of a water film between rotating rolls. *Tappi* 48:617–25

Hocking, L. M., Rivers, A. D. 1982. The spreading of a drop by capillary action. *J. Fluid Mech.* 121:425–42

Homsy, G. M., Geyling, F. T. 1977. A note on instabilities in rapid coating of cylinders. *AIChE J.* 23:587–90

Huh, C., Scriven, L. E. 1971. Hydrodynamic model of steady movement of a solid/liquid/fluid contact line. *J. Colloid Interface Sci.* 35:85–101

Kaulakis, F. 1974. Evolution of coating machinery. *Tappi* 57:80–84

Kistler, S. F. 1983. *The fluid mechanics of curtain coating and related viscous free surface flows with contact lines.* PhD thesis. Univ. Minn., Minneapolis

Kistler, S. F., Scriven, L. E. 1982. Finite-element analysis of dynamic wetting for curtain coating at high capillary numbers. *AIChE Winter Meet., Orlando, Fla.*, Pap. 45d

Kistler, S. F., Scriven, L. E. 1983. Coating flows. In *Computational Analysis of Polymers*, ed. J. Pearson, S. Richardson, Chap. 8. Barking, Essex, Engl: Appl. Sci. 343 pp.

Kobayashi, C., Saito, H., Scriven, L. E. 1982. Study of slide coating by finite-element method. *AIChE Winter Meet., Orlando, Fla.*, Pap. 45e

Landau, L., Levich, B. 1942. Dragging of a liquid by a moving plate. *Acta Physicochim. URSS* 17:42–54

Lee, C. Y., Tallmadge, J. A. 1974. Dynamic meniscus profiles in free coating. III. Predictions based on two-dimensional flow fields. *AIChE J.* 20:1079–86

Lee, C. Y., Tallmadge, J. A. 1975. Meniscus shapes in withdrawal of flat sheets from liquid baths. II. A quasi-one-dimensional flow model for low capillary numbers. *Ind. Eng. Chem. Fundam.* 14:120–26

Levi, S. M. 1966. Effect of coating conditions on the kinetic wetting of moving emulsion support. *Zh. Nauchn. Prikl. Fotogr. Kinematogr.* 11:401–2

Levi, S. M., Akulov, V. I. 1964. Investigations of kinetics of wetting in photographic emulsions. *Zh. Nauchn. Prikl. Fotogr. Kinematogr.* 9:124–26

Levich, V. G. 1962. *Physicochemical Hydrodynamics.* Englewood Cliffs, N.J: Prentice-Hall. 700 pp.

Marques, D., Costanza, V., Cerro, R. L. 1978. Dip coating at large capillary numbers: an initial value problem. *Chem. Eng. Sci.* 33:87–93

Mill, C. C., South, G. R. 1967. Formation of ribs on rotating rollers. *J. Fluid Mech.* 28:523–29

Miyamoto, K., Scriven, L. E. 1982. Breakdown of air film entrained by liquid coated on web. *AIChE Ann. Meet., Los Angeles, Calif.*, Pap. 101g

Morey, F. C. 1940. Thickness of a liquid film adhering to a surface slowly withdrawn from the liquid. *J. Res. Natl. Bur. Stand.* 25:385–93

Ngan, C. G., Dussan, V., E. B. 1982. On the nature of the dynamic contact angle: an experimental study. *J. Fluid Mech.* 118:27–40

Nigam, K. D. P., Esmail, M. N. 1980. Liquid flow over a rotating dip coater. *Can. J. Chem. Eng.* 58:564–68

Pearson, J. R. A. 1960. The instability of uniform viscous flow under rollers and spreaders. *J. Fluid Mech.* 7:481–500

Perry, R. T. 1967. *Fluid mechanics of entrainment through liquid-liquid and liquid-solid junctures.* PhD thesis. Univ. Minn., Minneapolis

Pitts, E., Greiller, J. 1961. The flow of thin liquid films between rollers. *J. Fluid Mech.* 11:33–50

Ruschak, K. J. 1974. *The fluid mechanics of coating flows.* PhD thesis. Univ. Minn., Minneapolis

Ruschak, K. J. 1976. Limiting flow in a premetered coating device. *Chem. Eng. Sci.* 31:1057–60

Ruschak, K. J. 1982. Boundary conditions at a liquid/air interface in lubrication flows. *J. Fluid Mech.* 119:107–20

Ruschak, K. J. 1983. A three-dimensional linear stability analysis for two-dimensional free boundary flows by the finite-element method. *Comput. Fluids* 11:391–401

Saffman, P. G., Taylor, G. I. 1958. The penetration of a fluid into a porous medium or Hele-Shaw cell containing a more viscous liquid. *Proc. R. Soc. London Ser. A* 245:312–29

Savage, M. D. 1977a. Cavitation in lubrication. Part 1. On boundary conditions and cavity-fluid interfaces. *J. Fluid Mech.* 80:743–55

Savage, M. D. 1977b. Cavitation in lubrication. Part 2. Analysis of wavy interfaces. *J. Fluid Mech.* 80:757–67

Savage, M. D. 1982. Mathematical models for coating processes. *J. Fluid Mech.* 117:443–55

Spiers, R. P., Subbaraman, C. V., Wilkinson, W. L. 1974. Free coating of a Newtonian liquid onto a vertical surface. *Chem. Eng. Sci.* 29:389–96

Sullivan, T. M., Middleman, S. 1979. Roll coating in the presence of a fixed constraining boundary. *Chem. Eng. Commun.* 3:469–82

Tallmadge, J. A., Gutfinger, C. 1967. Entrainment of liquid films. *Ind. Eng. Chem.* 59:18–34

Tallmadge, J. A., Weinberger, C. B., Faust, H. L. 1979. Bead coating instability: a comparison of speed limit data with theory. *AIChE J.* 25:1065–72

Taylor, G. I. 1963. Cavitation of a viscous fluid in narrow passages. *J. Fluid Mech.* 16:595–619

Tekic, M. N., Jovanovic, S. 1982. Liquid coating onto a rotating roll. *Chem. Eng. Sci.* 37:1815–17

Teletzke, G. F., Davis, H. T., Scriven, L. E. 1984. Wetting hydrodynamics. *J. Fluid Mech.* In press

Tuck, E. O. 1983. Continuous coating with gravity and jet stripping. *Phys. Fluids* 26:2352–58

Van Rossum, J. J. 1958. Viscous lifting and drainage of liquids. *Appl. Sci. Res. Sect. A* 7:121–44

White, D. A., Tallmadge, J. A. 1965. Theory of drag out of liquids on flat plates. *Chem. Eng. Sci.* 20:33–37

Wilson, S. D. R. 1982. The drag-out problem in film coating theory. *J. Eng. Math.* 16:209–21

Ann. Rev. Fluid Mech. 1985. 17:91–118

SEDIMENTATION OF NONCOLLOIDAL PARTICLES AT LOW REYNOLDS NUMBERS

Robert H. Davis

Department of Chemical Engineering, University of Colorado, Boulder, Colorado 80309

Andreas Acrivos

Department of Chemical Engineering, Stanford University, Stanford, California 94305

1. INTRODUCTION

Sedimentation, wherein particles fall under the action of gravity through a fluid in which they are suspended, is commonly used in the chemical and petroleum industries as a way of separating particles from fluid, as well as a way of separating particles with different settling speeds from each other. Examples of such separations include dewatering of coal slurries, clarification of waste water, and processing of drilling and mining fluids containing rock and mineral particles of various sizes. The separation of different particles by sedimentation is also the basis of some laboratory techniques for determining the distribution of particle sizes in a particulate dispersion.

Owing to the significance of the subject, there have been numerous experimental and theoretical investigations of the sedimentation of particles in a fluid. One of the earliest of these is Stokes' analysis of the translation of a single rigid sphere through an unbounded quiescent Newtonian fluid at zero Reynolds number, which led to his well-known law

$$u^{(0)} = \frac{2a^2(\rho_s - \rho)g}{9\mu}, \tag{1.1}$$

91

0066–4189/85/0115–0091\$02.00

where $u^{(0)}$ is the settling velocity of the sphere, a is its radius, ρ_s is its density, ρ is the fluid density, μ is the fluid viscosity, and g is the gravitational constant. Since then, research has focused on extending Stokes' law by considering nonspherical rigid particles, drops and bubbles, non-Newtonian fluids, nonzero Reynolds numbers, the presence of a wall near the particle, and interactions between particles.

In this paper, we do not attempt to present a complete review of the general subject of sedimentation, but rather we discuss recent developments in three areas:

1. *The sedimentation of monodisperse suspensions.* Here we focus on the theoretical and experimental determination of the average settling velocity of identical spherical particles in a suspension that is not infinitely dilute.
2. *The sedimentation of polydisperse suspensions.* For suspensions containing more than one particle species (i.e. different sizes and densities), we discuss recent studies that deal with the prediction of the local average settling velocity of a spherical particle in the midst of unlike spheres, and also with the macroscopic description of the sedimentation process as particles with different fall speeds separate from one another.
3. *Enhanced sedimentation in inclined channels.* The curious observation that the clarification rate of a suspension is increased merely by inclining a settling vessel from the vertical has recently been analyzed using the appropriate equations of continuum mechanics. We summarize the findings for the flow profiles and sedimentation rates as well as the results of a linear stability analysis used to predict the formation and growth of waves in inclined settling channels; these waves lead to a decrease in the efficiency of such processes.

Before proceeding with our review, we wish to call attention to several other articles that have appeared in this series in recent years that focus on properties of particulate suspensions other than sedimentation. Batchelor (1974) reviewed the effective transport properties of two-phase materials with random structure and included a description of the sedimentation of a dilute dispersion of equal spheres. Herczyński & Pieńkowska (1980) discussed the statistical theory of a suspension and, in particular, its rheological properties. Leal (1980) considered general particle motions in a viscous fluid in the presence of small inertia effects or small non-Newtonian effects, or when the particles (or bubbles) are slightly nonspherical. The Brownian motion of small colloidal particles suspended in liquids was reviewed by Russel (1981). In general, Brownian motion is only significant for particles that are so small that sedimentation due to gravity is negligibly slow, but circumstances exist where both effects are important. Finally,

Schowalter (1984) discussed the stability and coagulation of particulate dispersions subjected to shear flows. Included in his review is a discussion of the interparticle forces, such as attractive van der Waals forces and repulsive electric double-layer forces, that characterize colloidal dispersions.

2. SEDIMENTATION OF MONODISPERSE SUSPENSIONS

We consider a suspension composed of rigid spherical particles of equal size and density sedimenting in a Newtonian fluid at very small particle Reynolds numbers. For typical hydrosol dispersions, the Reynolds number $\rho u^{(0)} a / \mu$ is small compared with unity when the spheres are less than 0.1 mm in diameter. Furthermore, let us suppose that the suspension is stable, i.e. that any Brownian motion and attractive van der Waals forces are too weak to cause the suspension to coagulate [see Spielman (1970) and Valoulis & List (1984) for detailed criteria for the onset of coagulation of suspended particles]. Then, an initially well-mixed suspension will separate into three regions when subjected to batch sedimentation in a vessel with vertical sidewalls. A layer of clarified fluid will form at the top of the settling vessel, whose depth increases with time as the particles settle. Below this layer lies the suspension region, sometimes referred to as the *clarification zone*. Since the spheres are identical, the interface between the clear fluid and the suspension will remain fairly sharp, though it may spread somewhat owing to particle diffusion. If the suspension is infinitely dilute, the particles settle with their Stokes velocity given by (1.1). On the other hand, for particle volume fractions as small as 1%, the average settling velocity of the spheres is noticeably lower than that given by Stokes' law. This phenomenon can be represented by a hindered settling function $f(c)$ such that the average fall velocity of a sphere in the suspension is given by $v = u^{(0)} f(c)$. It is generally assumed that $f(c)$ depends only on the solids volume fraction c and that it is a monotonically decreasing function with $f(0) = 1$. However, for colloidal dispersions, this correction to Stokes' law will also depend on the nature of any interparticle forces present. Strong double-layer repulsive forces will tend to keep the particles well separated, and hence will decrease the average fall speed—whereas strong attractive van der Waals forces will lead to an excess of spheres that are close partners to another sphere, and hence will increase the average settling velocity of the suspension. It is conceivable that if the attractive forces are sufficiently strong, $f(c)$ may exceed unity; however, it is expected that in this case the suspension would be unstable, and that the increased settling rate would then be accompanied by the formation of doublets, triplets, etc.

The third region within the settling vessel is a sediment layer that forms on the bottom of the vessel. Often it is assumed that c_s, the solids concentration in this layer, is a constant and is equal to the maximum random packing density of the spheres, which is about 0.6. In reality, the solids gradually compact in this zone as the liquid in the interstices between the spheres is slowly squeezed out. Thus, the sediment layer at the bottom of the vessel is sometimes called the *compression zone* or *thickening zone*. Mathematical models of the gravity thickening process have been reviewed by Dixon (1979). In general, the settling or compaction rate of the particles depends not only on the local solids concentration in the sediment but also on the concentration gradient, the depth of the sediment layer, and the interparticle forces between the closely packed spheres.

Discontinuities in the solids volume fraction c occur at the interfaces that separate the suspension from the clear fluid at the top and from the sediment layer at the bottom of the vessel. If the suspension is sufficiently concentrated, gradients or discontinuities in the solids volume fraction may also appear within the interior of the suspension. The well-known kinematic-wave theory describing this phenomenon was first given by Kynch (1952), and it is not repeated here. A cohesive account of the theory and detailed criteria for the presence of internal concentration gradients can be found in the book by Wallis (1969). Recent developments include the application of one-dimensional kinematic-wave theory to settling vessels with nonconstant cross-section (Baron & Wajc 1979, Schneider 1982) and also the modification of Kynch's theory of sedimentation in order to account for the layer of compacting sediment that forms at the vessel bottom (Fitch 1983).

Application of kinematic-wave theory to sedimentation, as well as the design of both batch and continuous settling vessels, requires knowledge of the hindered settling function $f(c)$. Thus, considerable effort has been expended in the past to determine this function either experimentally or theoretically.

Empirical Formulae for the Hindered Settling Function

The results of many sedimentation and fluidization experiments reported in the literature have been summarized by Barnea & Mizrahi (1973) and Garside & Al-Dibouni (1977). The former authors proposed that the semiempirical formula

$$f(c) = \frac{(1-c)^2}{(1+c^{1/3})\exp(5c/3(1-c))} \tag{2.1}$$

provides the best fit to existing experimental data for very small particle Reynolds numbers. On the other hand, the most commonly used empirical

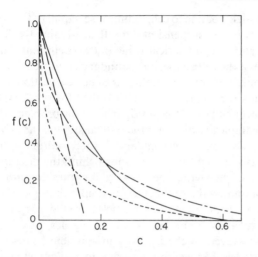

Figure 1   The hindered settling function $f(c)$; ———— the empirical correlation given by Equation (2.2); ———— the empirical correlation given by Equation (2.1); —— Batchelor's (1972) theory for dilute suspensions; ---- the exact theoretical results of Sangani & Acrivos (1982) for simple cubic arrays.

correlation is that attributed to Richardson & Zaki (1954),

$$f(c) = (1-c)^n,$$ (2.2)

where, according to Garside & Al-Dibouni (1977), a value of $n = 5.1$ most accurately represents their data for small Reynolds number. These two functions are plotted in Figure 1 for comparison, and it is evident that their behavior is quite different, especially for small c. In the dilute limit, Equation (2.1) has the form

$$f(c) \sim 1 - c^{1/3},$$ (2.3)

whereas Equation (2.2), with $n = 5.1$, becomes,

$$f(c) \sim 1 - 5.1c.$$ (2.4)

The experimental data reported by Garside & Al-Dibouni (1977) generally fall between the predictions of (2.1) and (2.2).

Theoretical Formulae for the Hindered Settling Velocity

We now direct our attention to the theoretical models that have been proposed for computing the settling function $f(c)$. Among the earlier and more widely used are the so-called "cell models" or "self-consistent" schemes, which involve solving the creeping-flow equations within a fluid cell (typically assumed spherical) encasing a representative particle. The ratio of the particle volume to cell volume is set equal to the volume fraction of the particles in the suspension. One then obtains the formula $f(c) = 1 - \beta c^{1/3}$, but since the numerical value of the coefficient β depends on an ad hoc choice of the cell shape and boundary conditions on the surface of this cell, the reliability of such models is open to question.

From a fundamental point of view, it is therefore preferable to compute $f(c)$ via an exact analysis of the multiparticle flow problem. Unfortunately, owing to the complexity of the mathematical system, the analysis has been restricted so far to dilute systems and is therefore of limited usefulness, although it has shed light on the reasons behind the discrepancy between (2.3) and (2.4). To begin with, when $c \to 0$, it is natural to suppose that for the purpose of computing their influence on a representative or test particle, it suffices to replace all the other particles in the suspension by point singularities such as point forces, point dipoles, etc. The analysis is complicated, however, by the fact that in a system containing an infinite number of particles, the resulting sums or integrals are often divergent. This difficulty can be avoided in a number of ways; at present, the most popular is the renormalization technique first introduced by Batchelor (1972) in his analysis of the sedimentation of a statistically homogeneous, dilute suspension of monodisperse spheres. This technique has been elaborated on by Jeffrey (1974) and O'Brien (1979), and its salient features have also been summarized by Batchelor (1974). This method has been modified recently by Feuillebois (1984) in order to investigate the sedimentation of monodisperse spheres in a dilute suspension that is homogeneous in any horizontal plane but in which a vertical concentration profile is prescribed. Another approach is that of Hinch (1977), who constructs a hierarchy of equations describing the behavior of a two-phase macroscopically homogeneous material and shows how bulk parameters such as the sedimentation velocity can be computed in a systematic way. Another method is to use Fourier integrals and generalized functions (Saffman 1973). At any rate, it is generally accepted that the earlier difficulties involving nonabsolutely convergent sums or integrals can be overcome, and that, in principle at least, it should be possible to compute the settling function $f(c)$ from first principles.

Even in dilute systems, however, the departure of $f(c)$ from unity depends on the configuration of the particles. Saffman (1973), who appears to have been the first to discuss this point in detail, considers three distinct possibilities: (I) regular periodic arrays of particles, (II) random fixed arrays, and (III) random free arrays. In a regular array, the force and the velocity are the same for each particle and, as first shown by Hasimoto (1959) and in more detail by Sangani & Acrivos (1982), $f(c) \sim 1 - \beta c^{1/3} + O(c)$, where β is a constant whose value depends on the assumed lattice configuration, e.g. $\beta = 1.76$ for a simple cubic lattice. On the other hand, for case II, first studied by Brinkman (1947), the corresponding expression for f is, according to Hinch (1977),

$$f(c) \sim 1 - \frac{3}{\sqrt{2}} c^{1/2} - \frac{135}{64} c \ln c - 12.0c + \ldots .$$

The most relevant case for sedimentation is, however, that of random free arrays, but here the probability density function $P(\mathbf{r})$, which denotes the probability of finding a sphere with its center at $\mathbf{r}$ given that there is a sphere with its center at the origin, is not known a priori and can vary from system to system, depending on the conditions of the experiment. As a consequence, the functional dependence of the hindered settling function on c can assume different forms, since $f(c)$ is determined in part by $P(\mathbf{r})$. A simple explanation of this dependence of the settling velocity on the structure of the suspension is as follows. The first correction to the Stokes settling velocity from the presence of a second sphere is of order $u^{(0)}a/d$, where d is the distance between the two sphere centers. In a random dispersion, for which $P(\mathbf{r})$ is a constant for $|\mathbf{r}| > 2a$ and vanishes for $0 \le |\mathbf{r}| \le 2a$, there is an $O(c)$ number of spheres, on the average, within a distance $d = O(a)$. Since these near neighbors exert the greatest influence, we expect the correction to Stokes' law to be of the form $f(c) \sim 1 - O(c)$, and in fact Batchelor's (1972) exact analysis yields $f(c) \sim 1 - 6.55c + O(c^2)$. If, on the other hand, the nearest neighbors are at a distance approximately equal to the average separation, i.e. $d = O(ac^{-1/3})$, and there are an $O(1)$ number of such neighbors, then we should expect that $f(c) \sim 1 - O(c^{1/3})$. Thus, which of the two empirical formulas (2.3) and (2.4) is predicted by theory depends on the assumptions made regarding the statistical structure of the suspension.

The question that needs to be resolved now is whether a sedimenting suspension would be expected to have a random or an ordered microscale structure. Certainly, if Brownian motion is strong, the suspension will remain random. Indeed, in the experiments reported by Cheng & Schachman (1955), Kops-Werkhoven & Fijnaut (1981), Buscall et al. (1982), and Tackie et al. (1983) using micron- and submicron-sized uniform colloidal spheres, the sedimentation results at low concentrations were of the form $f(c) \sim 1 - \gamma c + O(c^2)$. The observed values of γ were generally in the range 4 to 6, and the linear form of the hindered settling function was accurate provided that c was less than about 0.08. Batchelor & Wen (1982) discuss these experiments more fully and conclude that the difference between the observed values of γ and Batchelor's (1972) predicted value of $\gamma = 6.55$ could be a consequence of van der Waals attractive forces causing an excess number of close pairs whose common speed of fall exceeds the fall speed of an isolated sphere. On the other hand, for the suspensions containing larger particles listed by Barnea & Mizrahi (1973), the correction of Stokes' settling law was more closely represented by (2.3), which, from our discussion of the theory, indicates that the particles were not randomly distributed. Although Batchelor (1972) has shown that pairwise interactions alone will not give rise to the prediction of order in the suspension, this may not be the case when interactions between many particles are involved.

Lynch & Herbolzheimer (1983) recently reported the first results of a numerical simulation designed to test this possibility. Although their results suggested that near-pairs of particles are less likely to persist in a sedimenting suspension, such a conclusion is only tentative at this stage. These authors also noted on theoretical grounds, however, that if multiparticle hydrodynamic interactions do cause the particles to become well spaced, then when the suspension is sheared the microscale arrangement should be altered so as to cause the sedimentation velocity to increase. This prediction was verified in their Couette device experiments in that, at asymptotically high shear rates, the settling function was found to decrease as $1 - 4c$, whereas when the shear was turned off, it returned to the form $1 - c^{1/3}$. These results tend to further support the notion that there exists in sedimenting suspensions a microscale "structure" whose origin is at present unknown.

3.   SEDIMENTATION OF POLYDISPERSE SUSPENSIONS

The majority of theoretical and experimental investigations of the sedimentation process have focused on monodisperse suspensions, whereas in most cases of practical importance, the suspensions contain particles having a range of settling velocities due to variations in their size, shape, and density. Again, we restrict our discussion to rigid spheres that have very small particle Reynolds numbers. We also assume that the suspension does not coagulate. First, let us note, however, that in contrast to monodisperse suspensions, the particles in a polydisperse suspension move relative to one another as a result of gravity. If this relative motion brings two particles sufficiently close to one another, then they will form a permanent doublet as a result of the attractive van der Waals force acting between them. The conditions governing the stability of a sedimenting polydisperse suspension and the rate of coagulation of an unstable suspension have been presented in the recent works of Melik & Fogler (1984), Davis (1984), and Wen & Batchelor (1984).

Preliminary investigations of stable, i.e. noncoagulating, suspensions containing a small number of distinct species indicate that the particle concentration does not remain uniform. Instead, as the settling develops, the faster-falling particles move away from the others, thereby creating different regions in the interior of the suspension. The lower region, i.e. that just above the sediment layer, contains all the particle species at their initial concentrations, whereas the region immediately above it is devoid of the fastest-settling species. Each successive region contains one fewer species than the region below, with the upper region of the suspension containing only particles of the slowest-settling species (see Figure 2). In most cases, these distinct regions are separated by a near-discontinuity, or "shock," in the species concentration distribution; these shocks can be quite steep, but diffusion of the particles ultimately limits the steepening so that there is a sharp but continuous transition across each shock separating two regions of uniform concentration. For continuous distributions of particle sizes and densities, no distinct shocks develop, but rather the species concentrations vary continuously as a function of height in the settling vessel.

In describing the above behavior, we have assumed that the spatial distribution of particles in any horizontal plane is uniform, and this is indeed true for dilute suspensions. However, in the settling of bidisperse suspensions at high enough solids concentrations, the remarkable observation has been made that vertical streams develop in the suspension. Whitmore (1955) was the first to describe this phenomenon after perform-

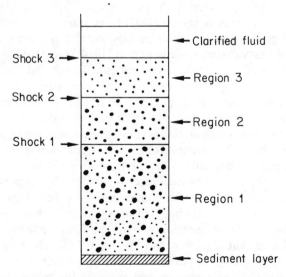

Figure 2 The regions that develop during the sedimentation of a mixture of three distinct species of particles. Region 1 contains all three species of particles, region 2 is devoid of the fastest-settling species, and region 3 contains only the slowest-settling species.

ing experiments with a suspension containing a mixture of heavy particles and neutrally buoyant particles. He reported that, within a few moments after the well-mixed suspension was allowed to stand, a lateral segregation of the particles set in that evolved into vertical fingers of a few millimeters in width. These fingers were composed of fluid and only the particle species that was present in lower concentration embedded within a continuum composed of fluid and only the other particle species. The fingers containing the neutrally buoyant particles were convected upward, whereas the streams containing the heavy particles descended toward the bottom of the vessel at a rate appreciably greater than that prevailing in the absence of the neutrally buoyant particles.

More recently, Whitmore's discovery of lateral segregation and subsequent convection in bidisperse suspensions has been explored in more detail by R. H. Weiland and coworkers. Upon adding positively buoyant particles to an otherwise uniformly settling suspension, Weiland & McPherson (1979) also observed a rapid lateral segregation of the two species of particles into vertically directed streams. This phenomenon is depicted in Figure 3. The suspending fluid is aqueous glycerol with density $1.12 \, \text{g cm}^{-3}$ and viscosity 4.95 cP, the heavy particles are polyvinylchloride spheres having a density of $1.41 \, \text{g cm}^{-3}$ and a mean diameter of 137 μm, and the buoyant particles are hollow ceramic spheres with a density of 0.595 g cm^{-3} and a mean diameter also of 137 μm. The heavy particles were dyed and appear as the light areas in the photograph, whereas fluid containing only buoyant particles is represented by the dark areas. Although the observed fingering structure develops most rapidly in systems containing buoyant and heavy particles, Weiland et al. (1984) report that fingering can also result when all the particles are more dense than the fluid but are of two distinct species, distinguishable by either density or size.

The phenomenon described above is of course fascinating from the point of view of fluid mechanics, but in fact (contrary to some of the statements that have been made in the literature) it is of relatively minor importance as a means of increasing the overall settling rate or, equivalently, of decreasing the total time required to settle the suspension. This is because, as a result of the vigorous agitation and convection brought about by the presence of the vertical streamers, a suspension containing two types of particles (say, heavy particles and neutrally buoyant ones) will quickly stratify into two layers: a bottom layer containing only the heavy particles but at a concentration c_H^* that is greater than the original concentration c_H, and a top layer with all the lighter particles. The heavy particles then continue to settle as in vertical sedimentation, and although the volume occupied by this more concentrated suspension is smaller than that of the initial suspension, the fact that $f(c_H^*) < f(c_H)$ implies that the heavy particles will settle slower than would have been the case before. Thus, the total settling time may not be greatly reduced, and under some conditions it may even increase. In fact, in a recent paper that presents both a model of the vertical streams and also quantitative results of two-species sedimentation experiments, Fessas & Weiland (1984) have reported that the settling rate of glass spheres in carbon tetrachloride can be accelerated up to sixfold upon the addition of PVC particles; however, the total settling time, which includes both the accelerated convection of the vertical streamers and the subsequent hindered settling of the concentrated heavy particles, has been observed to be reduced by no more than a factor of two (Weiland & McPherson 1979). Often the observed enhancement is considerably less than this. Finally, we note that if the total solids concentration is too low

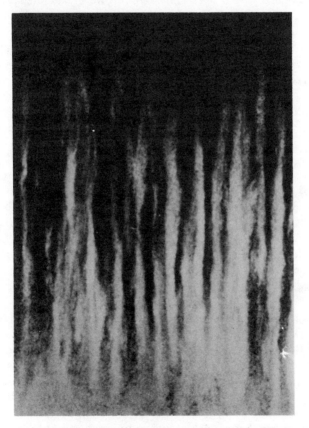

Figure 3 Vertical finger formation during sedimentation of a mixture of buoyant and heavy particles. The photograph was taken 60 seconds after mixing was completed, and the heavy particles are dyed. (From Weiland et al. 1984.)

(less than about 10% by volume), lateral segregation and fingering are not observed, and both solids components settle at reduced rates. Moreover, the cause of the lateral segregation of two species has heretofore not been addressed in the literature. This appears to be a question of hydrodynamic stability involving multiparticle interactions, certainly a fascinating area for future research.

We now return to our original description of the polydisperse sedimentation process and restrict our attention to suspensions where the spatial distribution of particles in any horizontal plane remains uniform. When a dispersion is extremely dilute, each individual particle falls with its Stokes velocity given by (1.1). Stokes' law provides a quantitative description of the sedimentation process for such dilute suspensions and forms the basis for the design of many particle-size analyzers. However, as stated earlier, even when the total volume fraction of particles exceeds only about 0.01, the behavior differs significantly from the prediction of Stokes' law as a consequence of interactions between particles of the various species. In particular, when a polydisperse suspension is not extremely dilute, the volume fraction of any one particle species differs in the various regions in the suspension. For a discrete distribution of N distinct particle species, Smith (1966) has derived from the requirement of particle flux continuity that

$$c_i^{(k+1)}(v_k^{(k)} - v_i^{(k+1)}) = c_i^{(k)}(v_k^{(k)} - v_i^{(k)}), \qquad i = k+1, \ldots, N, \tag{3.1}$$

where the superscript (k) refers to the conditions in region k (the regions are numbered sequentially starting at the bottom of the vessel—see Figure 2), and the subscripts refer to the particle species, which are numbered such that the fastest-settling species in region k is denoted as species k, i.e. $c_i^{(k)} = 0$ for $i < k$. A particle of species i has radius a_i, density ρ_i, volume fraction c_i, and a Stokes settling velocity $u_i^{(0)}$. Its average fall speed v_i depends on the local concentrations of all of the particles present, and we express this as

$$v_i = u_i^{(0)} f_i(\mathbf{c}), \tag{3.2}$$

where $\mathbf{c}$ is the vector of the volume fractions of all of the particle species. In general, f_i will differ for each particle species, and it will depend also on the effects of Brownian motion and interparticle forces, as discussed below.

In view of the fact that the average settling velocities are related to the volume fraction of all species present, Equation (3.1) represents a system of coupled, nonlinear algebraic equations for the unknowns $c_i^{(k+1)}$ that can be solved numerically using standard iterative techniques. Knowledge of the $c_i^{(k+1)}$ values, together with (3.2), suffices to give a complete description of the separation process. For continuous distributions of particles, Davis et al. (1982) and Greenspan & Ungarish (1982) have shown that the difference equations given by (3.1) are replaced by differential equations. However, these differential equations can be solved by finite-difference techniques that yield a set of coupled equations similar to (3.1) with N large.

In order to carry out the calculation described above, it is of course necessary to know the hindered settling function $f_i(\mathbf{c})$ for polydisperse suspensions. Smith (1965, 1966, 1967) has presented a theoretical cell model for the sedimentation of multisized particles that is an extension of the fluid cell model originally proposed by Happel (1958) for the settling of equisized particles. Smith's theory successfully predicts the trends of measured settling velocities with changing solids concentration, but it underestimates the velocities, just as does the cell theory for the settling of a single species. Lockett & Al-Habbooby (1973, 1974) carried out sedimentation and counter-current solid-liquid vertical-flow experiments with binary particle mixtures. They proposed that the monodisperse Richardson & Zaki (1954) correlation [Equation (2.2)] could be applied to particles of each of the two sizes in a binary suspension simply by using the total local solids volume fraction $c = \sum c_i$ without any reference to the individual particle species. Mirza & Richardson (1979) extended the Lockett & Al-Habbooby model to the sedimentation of multisized particle systems. The models of both of these sets of authors were found to predict sedimentation velocities that were larger than those found experimentally. In order to obtain a better representation of their experimental data, Mirza & Richardson (1979) applied an additional empirical correction factor of $(1-c)^{0.4}$ to the predicted sedimentation velocities. In a similar study, Selim et al. (1983) treated equidensity particles and used the total local volume fraction of solids in the Richardson & Zaki correlation for the slip velocity of the particles relative to the fluid; in addition, they proposed that the Stokes settling velocity of a particle of species i be modified by replacing the fluid density ρ in (1.1) with the average density of a suspension consisting of fluid and of particles of sizes smaller than that of species i. Clearly, this approach represents only an ad hoc model for the effects of particle interactions between different species; nonetheless, using concentrated suspensions with $0.12 \leq c \leq 0.45$, Selim et al. (1983) found very good agreement between their model and experiments.

In an application of steady-state continuous sedimentation of poly-disperse suspensions in an inclined settling channel (see the following section), Davis et al. (1982) measured the total solids concentration as a function of height in the channel for a feed suspension containing a known "bell-shaped" particle-size distribution. These authors compared their data with theoretical predictions based on the assumption of Lockett & Al-

Habbooby (1973, 1974) that the hindered settling effect is adequately represented by empirical correlations for monodisperse suspensions. For total solids volume fractions in the feed of 2 and 5%, the experimental data lay in-between the theoretical predictions using (2.1) and (2.2) for the hindered settling function.

In each of the theoretical models discussed above, the results for monodisperse suspensions were used to predict the effects of particle-particle interactions in polydisperse systems, or else an ad hoc modification of the monodisperse results was introduced. However, for general systems containing large variations in particle sizes or densities, we would not expect that polydisperse settling phenomena could be accurately described by this type of approach. On the other hand, for suspensions that are sufficiently dilute that pairwise particle interactions are dominant, the hindered settling effect can be calculated exactly. Such a calculation has been presented recently by Batchelor (1982). Following his earlier investigation of monodisperse suspensions, Batchelor showed that the mean velocity of a particle of species i in a stable dispersion of N distinct species is

$$v_i = u_i^{(0)} \left\{ 1 + \sum_{j=1}^{N} S_{ij} c_j \right\}, \qquad i = 1, 2, \ldots, N, \tag{3.3}$$

correct to order c. The dimensionless sedimentation coefficient S_{ij} is a function of (a) the size ratio $\lambda = a_j/a_i$, (b) the reduced density ratio $\gamma = (\rho_j - \rho)/(\rho_i - \rho)$, (c) the Peclet number of the relative motion of an i particle and a j particle, and (d) a dimensionless measure of the interparticle force potential. The Peclet number, whose inverse measures the effect of Brownian diffusion compared with the effect of relative motion due to gravity, is typically of order unity for micron-sized spheres in water, and it increases in proportion to $(a_i + a_j)^4$ for a given λ.

The computation of the sedimentation coefficient S_{ij} for a pair of unlike spheres is much more complicated than for equal spheres, because the pair-distribution function—defined as the probability of finding the center of a particle of species j in unit volume at a position $\mathbf{r}$ relative to the center of a particle of species i—is nonuniform and must be determined as part of the solution. This distribution function is governed (a) by a particle-pair conservation equation that in a dilute dispersion depends on the relative motion of two particles through the suspending fluid due both to gravity and to any interactive force that may act between the two particles; and (b) by Brownian diffusion of the two particles. For details of the calculation of the pair-distribution function and the sedimentation coefficients, the reader is referred to Batchelor (1982) and to a sequel paper by Batchelor & Wen (1982), in which numerical results for these quantities have been compiled.

4. ENHANCED SEDIMENTATION IN INCLINED CHANNELS

The sedimentation process is often very slow, especially when the particles are small, and hence there exists a need for constructing simple devices that could accomplish this solid-liquid separation more rapidly. One such class of devices, known commercially as "supersettlers" or "lamella settlers," consists of settling vessels having inclined walls and in which the retention times can be reduced by an order of magnitude or more below those in corresponding vertical settlers. These settlers may be composed either of a narrow tube or channel inclined from the vertical or of a large tank containing several closely spaced tilted plates.

The phenomenon of enhanced sedimentation in inclined channels has a long history, and one of the first references to appear in the literature is attributed to the physician Boycott (1920), who observed that "when . . . blood is put to stand in narrow tubes, the corpuscles sediment a good deal faster if the tube is inclined than when it is vertical." Subsequently, many investigators have studied the phenomenon for a variety of suspensions and have reported that a severalfold increase in the sedimentation rate could indeed be achieved; an excellent summary of the early papers on the subject is given by Hill (1974).

The enhancement in the sedimentation rate can be viewed as resulting from the fact that whereas particles can only settle onto the bottom in a channel with vertical walls, such particles can also settle onto the upward-facing wall in a tilted channel, as shown in Figure 4. These particles then

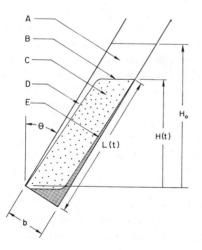

Figure 4 The different regions of the flow field during sedimentation in an inclined channel: (*A*) region of particle-free fluid above the suspension, (*B*) interface between the particle-free fluid and the suspension, (*C*) suspension, (*D*) thin particle-free fluid layer beneath the downward-facing surface, (*E*) concentrated sediment.

form a thin sediment layer that rapidly slides down toward the bottom of the vessel under the action of gravity. Thus, one can interpret the increase in the settling rate as due to an increase in the surface area available for settling. Of course, since the suspension is incompressible, coincident with the removal of particles from the suspension, there must also occur an accompanying production of clarified fluid. This takes place at the top of the suspension and along the downward-facing wall, where a thin clear-fluid layer is formed (cf. Figure 4). This particle-free layer is buoyant compared with the bulk suspension, and the fluid within it flows rapidly to the top of the vessel. Also, this layer reaches a steady-state thickness in a relatively short time, because the fluid that is being entrained into the layer exerts a drag force on the particles at the interface that counterbalances the normal component of the gravitational force on these particles.

Although this so-called Boycott effect of enhanced sedimentation has been known for many years, it has only recently been analyzed using first principles. Hill et al. (1977) were the first to study the phenomenon theoretically using the appropriate equations of continuum mechanics. These authors used a numerical solution to obtain the flow profiles and sedimentation rates in various upward-pointing cones. Subsequently, Probstein et al. (1981), Rubinstein (1980), and Leung & Probstein (1983) performed an analytical analysis of continuous steady-state settling in relatively long and narrow inclined channels and treated the problem as a two-dimensional viscous channel flow. Their analysis includes three stratified layers: a clarified layer, a suspension layer, and a sediment layer. Equations were developed for the velocity profiles and fluxes in each of these layers, as well as for the location of the interfaces separating them. Concurrently, Acrivos & Herbolzheimer (1979) and Herbolzheimer & Acrivos (1981) presented analytical derivations of the laminar-flow profiles in two-dimensional inclined channels for both batch and continuous settling. The former paper applies to low-aspect-ratio vessels in which the clarified-fluid layer that forms beneath the downward-facing wall of the channel is much smaller than the spacing between the walls (see Figure 4), whereas the latter paper considers very tall and narrow vessels in which the clarified-fluid layer occupies an appreciable portion of the channel.

A result common to each of the analyses discussed above is the verification of the expression for the clarification rate given via an elementary kinematic model by Ponder (1925) and independently by Nakamura & Kuroda (1937)—the so-called PNK theory. This theory states that the rate of production of clarified fluid is equal to the vertical settling velocity of the particles multiplied by the horizontal projection of the channel area available for settling. For the parallel-plate geometry of

Figure 4, the expression is

$$S^*(t) = v_0 b \left(\cos \vartheta + \frac{L}{b} \sin \vartheta \right), \tag{4.1}$$

where $S^*(t)$ is the volumetric rate of production of clarified fluid per unit depth in the third dimension of the vessel (hereafter referred to as the settling rate), v_0 is the vertical settling velocity [i.e. (1.1) multiplied by the hindered settling function $f(c)$ discussed earlier], ϑ is the angle inclination of the plates from the vertical, b is the spacing between the plates, and L is the length of the vessel. It is evident from (4.1) that the augmentation in the rate of settling should be $O(L/b)$, which, in principle, could be made arbitrarily large by decreasing the spacing between the inclined walls.

The enhancement in the settling rate predicted by (4.1) holds true only so long as the flow in the channel remains laminar. Under some conditions, however, waves have been observed to form at the interface between the suspension and the layer of particle-free fluid underneath the downward-facing wall (Pearce 1962, Probstein & Hicks 1978, Leung 1983, Herbolzheimer 1983, Davis et al. 1983). These waves grow as they travel up the vessel and often break before reaching the top of the suspension. Evidently, the occurrence of such an instability limits the efficiency of the inclined settling process, because when the waves break, fluid that has already been clarified is remixed with the suspension. A photograph depicting this phenomenon is shown in Figure 5. The particles are glass spheres having a density of 2.44 g cm^{-3} and an average diameter of $130 \ \mu\text{m}$. The suspending fluid is a synthetic Newtonian lubricant having a density of 1.05 g cm^{-3} and a viscosity of 10 cP.

Recently, classical linear stability theory has been applied to the inclined settling process by Herbolzheimer (1983), Davis et al. (1983), and Leung (1983) in order to elucidate the conditions under which the flow becomes unstable and to determine the growth rate of the waves. In what follows, we present the findings of these stability analyses, preceded by a brief discussion of the laminar-flow profiles.

Theory and Experiments for Laminar Flow

We consider here the suspension to be composed of a Newtonian fluid and identical solid spheres of vanishingly small particle Reynolds number. The sedimentation of polydisperse suspensions in inclined channels has been treated by Davis et al. (1982), to which the reader is referred for details. A key feature of our analysis is to treat the suspension as an effective fluid whose properties depend on the solids volume fraction. The particles are assumed to move with the bulk average velocity plus a slip velocity in the direction of gravity equal to the vertical settling velocity of the suspension. The latter is determined either from a vertical sedimentation experiment or from an empirical correlation. Aside from the vessel geometry and solids volume fraction, a complete description of the settling process is governed by two dimensionless parameters: R, a sedimentation Reynolds number; and Λ, the ratio of a sedimentation Grashof number to R, where $R \equiv \rho H v_0/\mu$ and $\Lambda \equiv H^2 g(\rho_s - \rho)c_0/v_0\mu$, with H being the vertical height of the suspension and c_0 the volume fraction of solids (which remains uniform everywhere within the interior of the suspension). Since in most cases of interest, Λ is $O(10^6)$–$O(10^8)$ while R is only $O(1)$–$O(10)$, an asymptotic analysis for determining the details of the motion was developed by Acrivos & Herbolzheimer (1979) and Herbolzheimer & Acrivos (1981) by taking $\Lambda \gg 1$ with $R\Lambda^{-1/3} \ll 1$. In this limit, the flow along the channel in the thin clarified-fluid layer to leading order is parallel and fully developed. In particular, if a coordinate system is defined with x directed along the downward-facing wall of the channel and y perpendicular to it, then

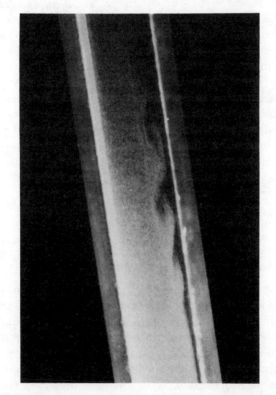

Figure 5   Wave formation and breakage during sedimentation in an inclined channel.

provided that the aspect ratio of the channel is not too large, the velocity in the x direction is given by

$$u = \Lambda^{1/3}(\tilde{y}\tilde{\delta} - \tfrac{1}{2}\tilde{y}^2) \cos \vartheta, \tag{4.2}$$

where the velocity u has been rendered dimensionless by v_0, $\tilde{\delta} \equiv \Lambda^{1/3}\delta$, $\tilde{y} \equiv \Lambda^{1/3}y$, and y is the distance from the wall made dimensionless by H. Also, after a short initial time period, the thickness of the clear-fluid layer reaches a steady-state thickness, which for the parallel-plate geometry is

$$\delta(x) = \Lambda^{-1/3}(3x \tan \vartheta)^{1/3}, \tag{4.3}$$

where both δ and x have been made dimensionless by H. An experimental determination of the clarified-fluid layer thickness provides a sensitive test of the theory, and in a series of experiments reported by Acrivos & Herbolzheimer (1979), measurements of $\delta(x)$ and of the sedimentation rate were in excellent agreement with the predictions. Since the thickness of the clear-fluid layer scales as $\Lambda^{-1/3}$, it will remain thin relative to the channel width so long as $b/H \gg \Lambda^{-1/3}$. Also, in view of the fact that the clarified-fluid layer has a very large velocity at its interface with the adjacent suspension, it drives a boundary-layer flow in the suspension region. The details of the boundary-layer analysis that describes this flow are given in Acrivos & Herbolzheimer (1979).

When the aspect ratio of the channel, H/b, is of order $\Lambda^{1/3}$ or greater, the clear-fluid layer occupies an appreciable fraction of the channel. Under these conditions, the flow along the channel to leading order is parallel and fully developed in both the clear-fluid layer and the adjacent suspension layer. In particular, the velocity in the up-channel direction is given by (Rubinstein 1980, Herbolzheimer & Acrivos 1981)

$$u = \Lambda^{1/3}\left\{\frac{6Q}{\tilde{b}}\,\bar{y}(1-\bar{y}) - \bar{y}[(\bar{\delta}+1/2)\bar{y}-\bar{\delta}](1-\bar{\delta})^2\tilde{b}^2 \cos \vartheta\right\}, \tag{4.4a}$$

in the pure-fluid layer (i.e. for $0 \leq \bar{y} \leq \bar{\delta}$), and by

$$u_s = \Lambda^{1/3}\left\{\frac{6Q}{\tilde{b}}\,\bar{y}(1-\bar{y}) - (1-\bar{y})[(3/2-\bar{\delta})\bar{y}-1/2]\bar{\delta}^2\tilde{b}^2 \cos \vartheta\right\}, \tag{4.4b}$$

in the suspension layer (i.e. for $\bar{\delta} \leq \bar{y} \leq 1$). The variables have been made dimensionless using H_0 and v_0 as the characteristic length and velocity, respectively. The variable Q is the net flow rate of material across any plane of constant x, $\tilde{b} \equiv \Lambda^{1/3}b/H_0$ is an $O(1)$ quantity by definition, $\bar{y}$ is the fractional distance across the channel (i.e. $\bar{y} \equiv \Lambda^{1/3}y/\tilde{b}$), and $\bar{y} = \bar{\delta}(x,t)$ is the equation for the interface separating the pure fluid and the suspension. For simplicity, the presence of the sediment layer on the upward-facing wall has been neglected, and the viscosity of the suspension has been set equal to that of the pure fluid. These approximations are accurate in the dilute limit, and removing them is relatively straightforward (Herbolzheimer 1980, Leung & Probstein 1983).

In order to complete the description of the laminar-flow field in these high-aspect-ratio vessels, it is necessary to determine the thickness of the clear-fluid layer $\bar{\delta}(x, t)$ from the kinematic condition for the interface. For batch settling, one surprising result is that the clear-fluid layer attains a steady shape only along the lower portion of the channel given by

$$\bar{\delta} = \tfrac{1}{2}\{1 - [1 - 4(3x \tan \vartheta)^{1/3}/\bar{b}]^{1/2}\}, \tag{4.5a}$$

while, in contrast, its thickness is independent of position and sweeps across the upper portion of the vessel linearly with time above some critical point

$$\bar{\delta} = \Lambda^{1/3} t \sin \vartheta/\bar{b}. \tag{4.5b}$$

In other words, when viewed as a function of distance along the channel, the thickness of the clear-fluid layer becomes discontinuous. The equation for the point of discontinuity separating the two parts of the interface is given by Equation (3.7) of Herbolzheimer & Acrivos (1981), who also report experimental verification of this interesting behavior.

Of greater practical interest than batch settling is the continuous operation of inclined settlers. Even though (as mentioned earlier) the thickness of the clear-fluid layer remains transient in the upper portion of the vessel for batch systems, the analyses of Herbolzheimer & Acrivos (1981) and Leung & Probstein (1983) predict that high-aspect-ratio settlers can be operated continuously provided the feed and widthdrawal locations are chosen properly. A thorough discussion of the feed and withdrawal conditions that should lead to a steady operation of such settlers is found in Section 4 of Herbolzheimer & Acrivos (1981). We focus here on two such regimes that are of the most widespread interest. In the first regime (regime 1), the feed is introduced at the bottom of the vessel while clarified fluid is removed from the top, so that a net flow of fluid up the channel is maintained. In this case, $\bar{\delta}(x)$ is governed by

$$(3 - 2\bar{\delta})\bar{\delta}^2 Q + \frac{\bar{b}^3}{3} \bar{\delta}^3(1 - \bar{\delta})^3 \cos \vartheta = x \sin \vartheta, \tag{4.6}$$

where [from (4.1)] $Q = \tan \vartheta + \Lambda^{-1/3}\bar{b} \sec \vartheta \cong \tan \vartheta$. Numerical solutions of (4.6) are possible for all values of $\bar{b}$, and experimental verification of these solutions has been reported by Davis et al. (1983).

In regime 2 both the feed and withdrawal are located at the top of the vessel. Thus $Q = 0$ everywhere along the channel, and (4.6) has two steady solutions,

$$\bar{\delta}(x) = \tfrac{1}{2}[1 \pm (1 - 4(3x \tan \vartheta)^{1/3}/\bar{b})^{1/2}]. \tag{4.7}$$

The existence of these two solutions was first predicted by Probstein et al. (1981). In the solution with the negative root, termed the "subcritical mode," the thickness of the clear-fluid layer vanishes at the bottom of the vessel and then grows with x—as is always observed in batch settling—whereas in the second solution, termed the "supercritical mode," the suspension layer vanishes at the bottom of the vessel and then grows with x. Which of these two modes is realized in practice cannot be predicted from the fully developed flow theory discussed thus far. However, Rubinstein (1980) has presented a two-dimensional model for the entry-flow problem that enables one to relate the realization of a particular mode to the feed and withdrawal conditions at the top of the channel. The supercritical mode of operation was first observed experimentally by Probstein & Hicks (1978), and later by Leung & Probstein (1983), who argued that this mode may be advantageous in suppressing the effects of any remixing of the particle-free fluid and the suspension; such remixing might occur if the interface between these regions became unstable. However, as can be seen from (4.7), real solutions for regime 2 exist only for $\bar{b} > 4(3 \sec \vartheta \tan \vartheta)^{1/3}$; this restriction limits the maximum aspect ratio that can be used for the top-feeding modes. On the other hand, a vessel with the feed at the bottom can be operated, in principle, for arbitrarily large values of the aspect ratio. Hence, if the flow remains stable, the highest possible enhancement in the settling rate can be achieved using regime 1. The aim of the stability analysis that follows is to predict the conditions under which the flow will remain stable.

Theory and Experiments for Unstable Flow

It has been noted earlier that in certain cases the interface between the clear-fluid slit and the suspension becomes wavy. Evidently, the occurrence of such an instability limits the efficiency of the inclined settling process, especially when the waves break. This phenomenon of wave formation and growth is similar to the well-known problem of a falling liquid film on an inclined surface, which has also been observed to become unstable. In order to investigate the origins of this problem, a linear-stability analysis has been performed for the laminar-flow profiles presented in the previous section. The details of this analysis can be found in the recent papers by Herbolzheimer (1983) and Davis et al. (1983).

The theoretical development proceeds along the lines of the classical works by Yih (1963) and by Benjamin (1957), who studied the instability of the falling liquid film. The disturbances to the base-state flow are assumed to be two dimensional and infinitesimally small and are expressed as superpositions of normal modes of the general form

$$\Psi(x, y, t) = \psi(y) \exp i(\alpha x - \omega t), \tag{4.8a}$$

$$\Phi(x, y, t) = \varphi(y) \exp i(\alpha x - \omega t), \tag{4.8b}$$

where, since the waves are spatially (rather than temporally) growing, the wave number of the disturbance α is treated as complex, while the frequency ω is a real constant. Also, Ψ and Φ are the normal-mode disturbance stream functions in the clear-fluid layer and in the adjacent suspension layer, respectively. A similar expression applies for the displaced position of the interface separating these two regions. The analysis is a local (quasi-parallel and quasi-steady) one, because the base-state profiles vary parametrically with x and t but only slowly compared with the variations in the disturbances.

When (4.8a,b) are substituted into the linearized momentum equations for the disturbances, the well-known Orr-Sommerfeld equation results, together with the associated boundary conditions of no-slip at the walls and matching of velocity and stress at the interface. This system represents an eigenvalue problem with α as the complex eigenvalue. If the solution of this problem yields a positive value for α_I, the imaginary part of the eigenvalue, then the disturbances are damped as they travel up the vessel; whereas, if α_I is negative, then these same disturbances grow and the system is unstable.

The Orr-Sommerfeld equation was first solved asymptotically for small frequencies. [This corresponds to long wavelengths, since the wave speed, scaled with the average velocity in the clear-fluid layer, remained $O(1)$.] These results were then supplemented with a numerical analysis because the most highly amplified waves were found to occur for moderate values of the frequency and the Reynolds number. For sedimentation in low-aspect-ratio vessels, the asymptotic results of the linear theory show that small disturbances always grow, i.e. that the presumably laminar flow in the narrow slit is always unstable, provided that the local Reynolds number Rx is greater than $140\bar{\delta}/(57 \cos \vartheta)$, which is generally very small. An equivalent result was found by Leung (1983), who performed a long-wavelength analysis of the stability of the "supercritical" mode having a very thin suspension layer. However, the rate of growth of the disturbances in each case depends on R, Λ, and the geometrical and physical properties of the system, so that under many circumstances these disturbances do not grow to a visible size over the length of the vessel. Thus, in many of the experiments described earlier, the flow appeared to be stable. The local exponential amplification factor for the wave growth is $-\alpha_I$, which is seen plotted in Figure 6 for the most unstable mode as a function of ϑ and for various values of the local Reynolds number Rx. Clearly, according to the predictions of the theory, the waves have a maximum amplification that generally occurs in the range $\vartheta = 5-15°$.

Using the linear-stability analysis as a guide, experiments for which a visible observation of the instability was expected were performed by Herbolzheimer (1983), both on a batch and a continuous basis, under the

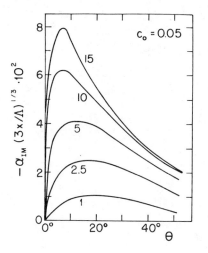

Figure 6 The parameter $-\alpha_{IM}\,(3x/\Lambda)^{1/3}$ for the most unstable mode versus the angle of inclination ϑ for a low-aspect-ratio vessel with $c_0 = 0.05$ and $Rx = 1, 2.5, 5, 10, 15$. (From Herbolzheimer 1983.)

following set of conditions: $8 \le H_0/b \le 15$, $5 \le \vartheta \le 50°$, $4 \le R_0 \le 25$, $7 \times 10^6 \le \Lambda_0 \le 2 \times 10^8$, $0.005 \le c_0 \le 0.15$ (where H_0 refers to the initial height of the suspension, and R_0 and Λ_0 are equivalent to R and Λ, respectively, only with L replacing H in their definitions). The experimental data were found to follow the general trends predicted by the linear theory. In particular, the distance along the channel at which waves first become visible, x_i, was measured because the inverse of this distance provides an experimental indication of the average growth rate of the waves (Herbolzheimer 1983, Davis et al. 1983). As an example of these measurements, $1/x_i$ is plotted in Figure 7 as a function of ϑ for $c_0 = 0.05$ and various values of R_0. From the figure, it is evident that the shapes of the curves are very similar to the corresponding theoretical predictions for the maximum amplification rates shown in Figure 6.

The linear stability analysis and experiments described above have been extended to determine the amplification rates of waves in tall, narrow channels with $H/b \ge O(\Lambda^{1/3})$. This case is of particular interest, since in light of (4.1), such vessels give very high enhancement in the clarification rate. It is important to know, therefore, under what conditions, if any, a stable operation of these vessels can be achieved. The analysis for these high-aspect-ratio vessels led to a critical Reynolds number R_c, below which the system can be expected to be stable to infinitesimal disturbances, unlike the case of low-aspect-ratio vessels, which were shown to be unstable for all but very small values of the Reynolds number. In general, it was shown by Davis et al. (1983) that R_c increased with increasing H/b and increasing ϑ. Thus, the same geometric factors that lead to the highest enhancement in the settling rate also give the most stable operating conditions—a

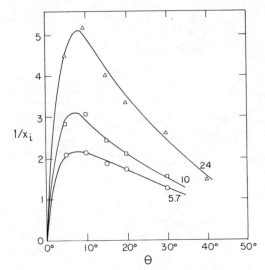

Figure 7   The inverse of the visually determined wave inception distance $1/x_i$ versus the angle of inclination ϑ for a low-aspect-ratio vessel, with $c_0 = 0.05$, $\Lambda_0 = 8.7 \times 10^7$, and $R_0 = 5.7, 10$, and 24. The solid lines are visual best-fit estimates through the data. (From Herbolzheimer 1983.)

surprising result that can be very useful in the future design of "supersettlers."

This exciting prediction was tested in the laboratory using a bottom-feeding continuous vessel with an aspect ratio of approximately 100, and the results were in very good agreement with the linear theory. This can be seen in Figures 8 and 9, where the amplification rates of the waves are plotted for theory and experiment, respectively, as functions of ϑ under the conditions $\Lambda_0 = 1.3 \times 10^7$, $\bar{b}_0 = 2.4$, $c_0 = 0.01$, and for various values of the Reynolds number. The theoretical amplification rates are computed at the location midway up the channel. Compared with the low-aspect-ratio case (Figures 6 and 7), there is dramatic stabilization for $\vartheta > 20°$.

5.   CONCLUSIONS

In this review, we have considered recent advances in the understanding of the sedimentation of monodisperse suspensions, the sedimentation of polydisperse suspensions, and enhanced sedimentation in inclined channels. Although sedimentation is a classical subject with a long history, it is our hope that the reader will be convinced that many opportunities for future research still very much exist. From our discussion, which has been limited to dispersions of rigid spherical particles having zero particle

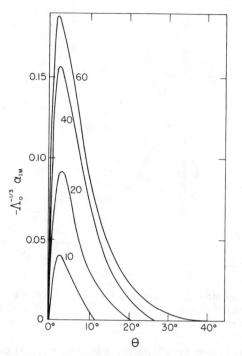

Figure 8 The scaled maximum amplification factor $-\Lambda_0^{-1/3}\alpha_{\mathrm{IM}}$ versus the angle of inclination ϑ for regime 1, with $x_0 = 1/2$, $\bar{b}_0 = 2.365$, and $R_0 = 10, 20, 40$, and 60. (From Davis et al. 1983.)

Reynolds numbers, it is evident that continued research is needed in many areas. Certainly, the problem of determining the effects of multiparticle interactions during the sedimentation of random, concentrated suspensions is both an important and a challenging one. Moreover, for dilute dispersions, the question still remains as to whether or not a settling suspension develops a structure having a length scale comparable to the mean particle spacing. Also, further experimental and theoretical research is needed in order to relate the macroscopic properties of a sedimenting polydisperse suspension to the local microphysical interactions between the different particles. The augmentation of the sedimentation rate in inclined channels holds great promise for the design of high-capacity settling vessels; continued work needed in this area includes studying the effects of inertia on the laminar-flow profiles in the channel and performing a nonlinear analysis in order to investigate the breaking of the waves at the liquid-suspension interface and to determine the extent of the subsequent remixing of the clarified fluid with the suspension.

Beyond this, there is considerable scope for further research on the

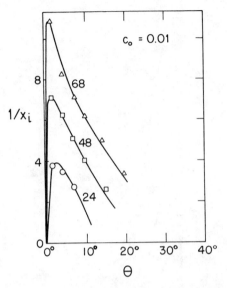

Figure 9   The inverse of the visually determined wave inception distance $1/x_i$ versus the angle of inclination ϑ for regime 1, with $c_0 = 0.01$, $\Lambda_0 = 1.3 \times 10^7$, $\bar{b}_0 = 2.4$, and $R_0 = 24, 48$, and 68. The solid lines are visual best-fit estimates through the data. (From Davis et al. 1983.)

sedimentation of nonspherical particles, of nonrigid particles such as drops and bubbles, and of particles having nonzero inertia. Also, as shown by the pioneering studies of Greenspan (1983) and Ungarish & Greenspan (1983, 1984), the effects of rotation and of the associated centrifugal forces on the separation of particles from a suspension give rise to a host of fascinating flow phenomena that are of interest from both the fundamental and practical points of view; these deserve a great deal more attention.

ACKNOWLEDGMENTS

This review was supported in part by the National Science Foundation under Grant No. CPE 81-08200. We are grateful to Professor Eric Herbolzheimer for his constructive comments and to Professor Ralph Weiland for providing us with copies of his work prior to its publication.

Literature Cited

Acrivos, A., Herbolzheimer, E. 1979. Enhanced sedimentation in settling tanks with inclined walls. *J. Fluid Mech.* 92:435–57

Barnea, E., Mizrahi, J. 1973. A generalized approach to the fluid dynamics of particulate systems: Part 1. General correlation for fluidization and sedimentation in solid multiparticle systems. *Chem. Eng. J.* 5:171–89

Baron, G., Wajc, S. 1979. Behinderte Sedimentation in Zentrifugen. *Chem. Ing. Tech.* 51:333

Batchelor, G. K. 1972. Sedimentation in a dilute dispersion of spheres. *J. Fluid Mech.* 52:245–68

Batchelor, G. K. 1974. Transport properties of two-phase materials with random structure. *Ann. Rev. Fluid Mech.* 6:227–55

Batchelor, G. K. 1982. Sedimentation in a dilute polydisperse system of interacting spheres. Part 1. General theory. *J. Fluid Mech.* 119:379–408

Batchelor, G. K., Wen, C.-S. 1982. Sedimentation in a dilute polydisperse system of interacting spheres. Part 2. Numerical results. *J. Fluid Mech.* 124:495–528

Benjamin, T. B. 1957. Wave formation in laminar flow down an inclined plane. *J. Fluid Mech.* 2:554–74

Boycott, A. E. 1920. Sedimentation of blood corpuscles. *Nature* 104:532

Brinkman, H. C. 1947. A calculation of the viscous force exerted by a flowing fluid on a dense swarm of particles. *Appl. Sci. Res.* A 1:27–34

Buscall, R., Goodwin, J. W., Ottewill, R. H., Tadros, T. F. 1982. The settling of particles through Newtonian and non-Newtonian media. *J. Colloid Interface Sci.* 85:76–86

Cheng, P. Y., Schachman, H. K. 1955. Studies on the validity of the Einstein viscosity law and Stokes' law of sedimentation. *J. Polym. Sci.* 16:19–30

Davis, R. H. 1984. The rate of coagulation of a dilute polydisperse system of sedimenting spheres. *J. Fluid Mech.* 145:179–99

Davis, R. H., Herbolzheimer, E., Acrivos, A. 1982. The sedimentation of polydisperse suspensions in vessels having inclined walls. *Int. J. Multiphase Flow* 8:571–85

Davis, R. H., Herbolzheimer, E., Acrivos, A. 1983. Wave formation and growth during sedimentation in narrow tilted channels. *Phys. Fluids* 26:2055–64

Dixon, D. C. 1979. Theory of gravity thickening. In *Progress in Filtration and Separation*, ed. R. J. Wakeman, pp. 131–78. London: Elsevier

Fessas, Y. P., Weiland, R. H. 1984. The settling of suspensions promoted by rigid buoyant particles. *Int. J. Multiphase Flow.* In press

Feuillebois, F. 1984. Sedimentation in a dispersion with vertical inhomogeneities. *J. Fluid Mech.* 139:145–72

Fitch, B. 1983. The Kynch theory and compression zones. *AIChE J.* 29:940–46

Garside, J., Al-Dibouni, M. R. 1977. Velocity-voidage relationship for fluidization and sedimentation in solid-liquid systems. *Ind. Eng. Chem. Process Des. Dev.* 16:206–14

Greenspan, H. P. 1983. On centrifugal separation of a mixture. *J. Fluid Mech.* 127:91–101

Greenspan, H. P., Ungarish, M. 1982. On hindered settling of particles of different sizes. *Int. J. Multiphase Flow* 8:587–604

Happel, J. 1958. Viscous flow in multiparticle systems: slow motion of fluids relative to bed of spherical particles. *AIChE J.* 4:197–201

Hasimoto, H. 1959. On the periodic fundamental solutions of the Stokes equations and their application to viscous flow past a cubic array of spheres. *J. Fluid Mech.* 5:317–28

Herbolzheimer, E. 1980. *Enhanced sedimentation in settling vessels having inclined walls*. PhD thesis. Stanford Univ., Stanford, Calif.

Herbolzheimer, E. 1983. The stability of the flow during sedimentation beneath inclined surfaces. *Phys. Fluids* 26:2043–54

Herbolzheimer, E., Acrivos, A. 1981. Enhanced sedimentation in narrow tilted channels. *J. Fluid Mech.* 108:485–99

Herczyński, R., Pieńkowska, I. 1980. Toward a statistical theory of suspension. *Ann. Rev. Fluid Mech.* 12:237–69

Hill, W. D. 1974. *Boundary-enhanced sedimentation due to settling convection.* PhD thesis. Carnegie-Mellon Univ., Pittsburgh, Pa.

Hill, W. D., Rothfus, R. R., Li, K. 1977. Boundary-enhanced sedimentation due to settling convection. *Int. J. Multiphase Flow* 3:561–83

Hinch, E. J. 1977. An averaged-equation approach to particle interactions in a fluid suspension. *J. Fluid Mech.* 83:695–720

Jeffrey, D. J. 1974. Group expansions for the bulk properties of a statistically homogeneous random suspension. *Proc. R. Soc. London Ser. A* 338:503

Kops-Werkhoven, M. M., Fijnaut, H. M. 1981. Dynamic light scattering and sedimentation experiments on silica dispersions at finite concentrations. *J. Chem. Phys.* 74:1618–25

Kynch, G. J. 1952. A theory of sedimentation. *Trans. Faraday Soc.* 48:166–76

Leal, L. G. 1980. Particle motions in a viscous fluid. *Ann. Rev. Fluid Mech.* 12:435–76

Leung, W. F. 1983. Lamella and tube settlers. II. Flow stability. *Ind. Eng. Chem. Process Des. Dev.* 22:68–73

Leung, W. F., Probstein, R. F. 1983. Lamella and tube settlers. I. Model and operation. *Ind. Eng. Chem. Process Des. Dev.* 22:58–67

Lockett, M. J., Al-Habbooby, H. M. 1973. Differential settling by size of two particle species in a liquid. *Trans. Inst. Chem. Eng.* 51:281–92

Lockett, M. J., Al-Habbooby, H. M. 1974. Relative particle velocities in two-species settling. *Powder Technol.* 10:67–71

Lynch, E. D., Herbolzheimer, E. 1983. Formation of microscale structure in sedimenting suspensions. *Bull. Am. Phys. Soc.* 28:1365

Melik, D. H., Fogler, H. S. 1984. Kinetics of gravity-induced flocculation. *J. Colloid Interface Sci.* In press

Mirza, S., Richardson, J. R. 1979. Sedimentation of suspensions of particles of two or more sizes. *Chem. Eng. Sci.* 34: 447–54

Nakamura, N., Kuroda, K. 1937. La cause de l'acceleration de la vitesse de sedimentation des suspensions dans les recipients inclines. *Keijo J. Med.* 8:256–96

O'Brien, R. W. 1979. A method for the calculation of the effective transport properties of suspensions of interacting particles. *J. Fluid Mech.* 91:17–39

Pearce, K. W. 1962. Settling in the presence of downward facing surfaces. *Proc. Congr. Eur. Fed. Chem. Eng., 3rd, London*, pp. 30–39

Ponder, E. 1925. On sedimentation and rouleaux formation. *Q. J. Exp. Physiol.* 15:235–52

Probstein, R. F., Hicks, R. 1978. Lamella settlers: a new operating mode for high performance. *Ind. Water Eng.* 15:6–8

Probstein, R. F., Yung, R., Hicks, R. 1981. A model for lamella settlers. In *Physical Separations*, ed. M. P. Freeman, J. A. Fitzpatrick, pp. 53–92. New York: Eng. Found.

Richardson, J. F., Zaki, W. N. 1954. Sedimentation and fluidization: Part I. *Trans. Inst. Chem. Eng.* 32:35–53

Rubinstein, I. 1980. A steady laminar flow of suspension in a channel. *Int. J. Multiphase Flow* 6:473–90

Russel, W. B. 1981. Brownian motion of small particles suspended in liquids. *Ann. Rev. Fluid Mech.* 13:425–55

Saffman, P. G. 1973. On the settling speeds of free and fixed suspensions. *Stud. Appl. Math.* 52:115–27

Sangani, A. S., Acrivos, A. 1982. Slow flow through a periodic array of spheres. *Int. J. Multiphase Flow* 8:343–60

Schneider, W. 1982. Kinematic-wave theory of sedimentation beneath inclined walls. *J. Fluid Mech.* 120:323–46

Schowalter, W. R. 1984. Stability and coagulation of colloids in shear fields. *Ann. Rev. Fluid Mech.* 16:245–61

Selim, M. S., Kothari, A. C., Turian, R. M. 1983. Sedimentation of multisized particles in concentrated suspensions. *AIChE J.* 29:1029–38

Smith, T. N. 1965. The differential sedimentation of particles of two different species. *Trans. Inst. Chem. Eng.* 43:T69–73

Smith, T. N. 1966. The sedimentation of particles having a dispersion of sizes. *Trans. Inst. Chem. Eng.* 44:T153–57

Smith, T. N. 1967. The differential sedimentation of particles of various species. *Trans. Inst. Chem. Eng.* 45:T311–13

Spielman, L. A. 1970. Viscous interactions in Brownian coagulation. *J. Colloid Interface Sci.* 33:562–71

Tackie, E., Bowen, B. D., Epstein, N. 1983. Hindered settling of uncharged and charged submicrometer spheres. *Ann. NY Acad. Sci.* 404:366–67

Ungarish, M., Greenspan, H. P. 1983. On two-phase flow in a rotating boundary layer. *Stud. Appl. Math.* 69:145–75

Ungarish, M., Greenspan, H. P. 1984. On centrifugal separation of particles of two sizes. *Int. J. Multiphase Flow* 10:133–48

Valoulis, I. A., List, E. J. 1984. Collision efficiencies of diffusing spherical particles: hydrodynamics, van der Waals and electrostatic forces. *Adv. Colloid Interface Sci.* In press

Wallis, G. B. 1969. *One-Dimensional Two-Phase Flow.* New York: McGraw-Hill. 408 pp.

Weiland, R. H., McPherson, R. R. 1979. Accelerated settling by addition of buoyant particles. *Ind. Eng. Chem. Fundam.* 18:45–49

Weiland, R. H., Fessas, Y. P., Ramarao, B. V. 1984. On instabilities during sedimentation of two-component mixtures of solids. *J. Fluid Mech.* 142:383–89

Wen, C.-S., Batchelor, G. K. 1984. The rate of coagulation in a dilute suspension of small particles. *Sci. Sinica.* In press

Whitmore, R. L. 1955. The sedimentation of suspensions of spheres. *Br. J. Appl. Phys.* 6:239–45

Yih, C.-S. 1963. Stability of liquid flow down an inclined plane. *Phys. Fluids* 6:321–34

Ann. Rev. Fluid Mech. 1985. 17 : 119–49

MATHEMATICAL MODELS OF DISPERSION IN RIVERS AND ESTUARIES

P. C. Chatwin and C. M. Allen

Department of Applied Mathematics and Theoretical Physics, University of Liverpool, P.O. Box 147, Liverpool L69 3BX, England

INTRODUCTION

In rivers and (especially) estuaries, the processes controlling the dispersion of dissolved and suspended pollutants are numerous and complicated. Among the factors that make a quantitative description hard are the turbulence; the effects of topography, buoyancy, and tides; and the abundant nonlinear interactions. However, despite such difficulties, there has been much research on dispersion in rivers and estuaries. This research has been, of course, primarily motivated by the practical importance of these flows, but it must be noted also that there are many basic scientific problems involved—problems whose better understanding would be valuable in predicting dispersion not only in rivers and estuaries but also in other important flows.

In the last few years there have been many books, reviews, and conferences dealing with themes that are highly relevant to that of the present article. Thus, a comprehensive and relatively recent account of flow and dispersion in rivers and estuaries is given in the book by Fischer et al. (1979), while recent research on estuarine mixing and dispersion is also summarized, for example, in reviews by Bowden (1981, 1982), and by Fischer (1976) in an earlier volume in this series. The present review is not an attempt to bring up to date the accounts of the whole field of dispersion given in such references. Space would not permit us to provide a coverage that was both comprehensive and at the degree of depth that we wish to achieve. Furthermore, in some aspects of the subject, including especially many basic physical processes, there is little to be added to existing accounts, at least at the time of this writing (January 1984).

119

0066–4189/85/0115–0119$02.00

Consequently, in the hope that more value may thereby derive from this review, we have chosen to cover in some detail only one aspect of the field, namely the different types of mathematical models used to describe dispersion in rivers and estuaries. This choice, of course, reflects our own research experience, but we think that it also has merit for other, perhaps more objective, reasons. While many different models are currently being used, developed, and investigated, there are few signs yet that a consensus has been reached on either (a) the precise flow characteristics that need to be known before an appropriate model can be chosen, or (b) how the choice of model will be affected by the detailed practical questions being investigated. It is also important to consider the physical principles on which the models are based; for instance, such knowledge is essential in identifying the inherent limitations of certain models. In brief, therefore, the aim of this article is to provide a balanced assessment of current knowledge and beliefs regarding the matching together of practical dispersion problems with suitable mathematical models.

We begin with a relatively full account of the scientific basis of the most commonly used mathematical models of dispersion. Particular attention is paid to the different types of averaging that are employed. This section serves as an introduction to the next two sections, which deal with dispersion in rivers and estuaries, respectively, with the emphasis in each section on recent work that is both promising and original. We conclude with a summary, which includes suggestions for future research work of potential value.

In writing this review, we have inevitably been influenced by our own prejudices, of which three ought to be admitted here.

1. There is knowledge of direct relevance to dispersion in rivers and estuaries to be gained from research in other fields, such as atmospheric dispersion. Unfortunately, researchers do not always take full advantage of this good fortune; thus, we cite such "outside" work where it seems helpful in achieving this article's purpose.
2. Some important papers dealing with mathematical models of dispersion have not received the attention they perhaps merit, mainly because their style is too mathematical and/or insufficiently practical. Here, while complicated mathematics is avoided as far as possible, we nevertheless attempt to provide convincing justification of the potential practical relevance of such papers.
3. In our view, consideration of recent research suggests that somewhat different emphases ought to be placed on the roles of certain classical papers than has hitherto been done. These papers are considered in many accounts of dispersion in rivers and estuaries; therefore, some

topics considered in such accounts may appear to be dealt with here from a different point of view.

BASIC PRINCIPLES OF SOME DISPERSION MODELS

The Fundamental Equation

We emphasize at the outset that the terms concentration and velocity, denoted by the symbols Γ and Υ, respectively, refer exclusively throughout this review to quantities defined in accord with the continuum hypothesis (Batchelor 1967, pp. 4–6, Fischer et al. 1979, p. 16). Both Γ and Υ are functions of position x and time t. Consider first the case when Γ is the concentration of a conserved substance like salt in an estuary. Except at places where there are sources and sinks, the equation governing Γ in any single realization is then (Fischer et al. 1979, pp. 50–51)

$$\frac{\partial \Gamma}{\partial t} + \Upsilon \cdot \nabla \Gamma = D \nabla^2 \Gamma, \tag{1}$$

where D (assumed constant) is the molecular diffusivity. Typical values of D for salt and heat in water are 1.1×10^{-9} and 1.4×10^{-7} m^2 s^{-1}, respectively.

While there is no evidence that Equation (1) is not (in practice) an exact description of the physics, it is not currently used, as it stands, in investigating practical dispersion problems in turbulent flows. Essentially, this is because it is impossible to specify completely the velocity field $\Upsilon(x, t)$, containing as it does not only the effects of topography, buoyancy forces, tidal forcing, etc., but also the random turbulent fluctuations. Even were such a specification possible, the numerical solution of Equation (1) for Γ would itself be extremely difficult to obtain and would, of course, contain much unwanted statistical noise.

One further important point about the physics represented by Equation (1) should be noted. On the basis of simple arguments (Tennekes & Lumley 1972, pp. 14–24, 240), it is known that at the high Reynolds numbers characteristic of flows in rivers and estuaries, the random concentration field generated by the random velocity field has spatial structure over a vast range of length scales, extending from a dimension characteristic of the overall flow geometry (e.g. the width of an estuary) downward to the contaminant microscale (often known as the conduction cutoff length). This microscale is estimated to be of order 10^{-4} m in the ocean (Chatwin & Sullivan 1979b), and it is of the same order in rivers and estuaries. Thus the distribution of dispersing contaminant in any realization will normally contain substantial, but randomly distributed, regions of patchiness in which large variations of concentration occur over distances as small as about 10^{-4} m. Associated with this range of length scales is a corresponding range of time scales extending down to about 10^{-2} s.

An analogous spatial and temporal structure exists for the velocity field $\Upsilon(\mathbf{x}, t)$. A relatively recent development in calculating turbulent velocity fields has been the direct use of the full unsteady Navier-Stokes equations to determine the large-eddy structure of Υ, accompanied by semiempirical modeling of the fine-scale structure (subgrid scales). This approach explicitly recognizes the vast range of length scales present in the structure of Υ, particularly the impossibility of computing the smallest scales (Reynolds & Cebeci 1978). Should it ever be applied successfully to velocity fields in real rivers and estuaries, it would then be natural to apply a similar approach to Equation (1) for the concentration field. Work by Antonopoulos-Domis (1981) is relevant in this regard.

Ensemble Means and Time Averages

Essentially all mathematical models of dispersion in rivers and estuaries deal with some sort of average concentration, although it is not always made clear precisely what average is meant. Exclusive attention on an average concentration implies no interest in the fluctuations of the actual concentration $\Gamma(\mathbf{x}, t)$ about that average or, for example, in the peak concentration. This restriction has potentially serious consequences for the investigation of some practical problems, such as the assessment of possible toxic effects. For this reason, increasing attention is being devoted in some related fields to assessing magnitudes of concentration fluctuations (see, for example, Chatwin & Sullivan 1978, 1979a,b, Chatwin 1982), and, even more ambitiously perhaps, to calculating the probability density function of concentration (see, for example, Pope 1979). It seems inevitable that such approaches will eventually be applied also to appropriate problems in rivers and estuaries. It also ought to be noted that no matter how the average concentration is defined, it is not (a) a concentration that is ever observed (just as the average score with one throw of a die—namely $3\frac{1}{2}$—is never observed), nor (b) is it usually a typical value of the actual concentration Γ. The explanation of (b) is, of course, that the spatial patchiness referred to above, together with (possibly large) meandering, generally causes the average concentrations to be much smaller than typical concentrations (Fischer et al. 1979, p. 269). These shortcomings of dealing with dispersion exclusively in terms of an average concentration arise inevitably because of the impossibility of giving an adequate description of a phenomenon as complicated as turbulent diffusion in terms of a single scalar field, even when this is an average concentration.

Nevertheless, a properly defined average concentration is, arguably,[1] the single most important scalar field associated with any dispersion process in turbulent flow; it is also undoubtedly the simplest property of such a process that can both be measured and modeled with some hope of reasonable accuracy. For this reason, nearly all research, and therefore the remainder of this review, considers only an average concentration; thus it is necessary to discuss some different meanings given to this term.

The basic average in all work on turbulent diffusion (and turbulence) is the ensemble mean. An ensemble, stated to be "probably the most fundamental idea" by Lumley & Panofsky (1964, p. 6), is a well-defined collection of different possible realizations of $\Gamma(x, t)$. The ensemble must be sufficiently precisely described so that it can be decided unambiguously whether any single realization does or does not belong to the ensemble. Many different ensembles can be defined for any turbulent-diffusion process. As a simple illustrative example, let $\Gamma(x, t)$ denote the concentration of sewage from a particular outfall in a particular estuary, and consider two possible ensembles.

1. Ensemble A: Interest is in the maximum upstream concentration of sewage on the flood tide. In this case it is appropriate to admit only those realizations in which $\Gamma(x, t)$ is measured at positions x upstream of the outfall and at times t of maximum flood.
2. Ensemble B: It is required to obtain an overall picture of the distribution of sewage throughout the estuary without any reference to special times. In this case all observations of $\Gamma(x, t)$ would be relevant, although it must obviously be ensured that the whole set of data collected is a fair representation of conditions throughout the estuary, i.e. that the data collected are a random sample of all possible observations.

This example illustrates that the phrase "best ensemble" has no meaning without further qualification, and that in each case, the selection of the appropriate ensemble must be determined by the problem under investigation. It is also evident from this example that the statistical properties of $\Gamma(x, t)$ will differ from ensemble to ensemble. For example, the spatial average over a particular estuarine cross section of the ensemble mean concentration for Ensemble A will differ from the same average for Ensemble B, since in the former case, only measurements made at times of maximum flood are admitted. Further discussion of the concept of an ensemble in a different context—namely the atmospheric dispersion of heavy gases—is given by Chatwin (1982).

It is unfortunately, but obviously, true that there are severe, often

---

[1] The other obvious candidate is the probability density function of concentration.

practically insurmountable difficulties associated with using the theoretical concept of an ensemble. All such difficulties derive essentially from the fact that, in general, the statistical properties of $\Gamma(\mathbf{x}, t)$ (and of all other random fields including the velocity) can, in principle, be estimated satisfactorily only by taking the arithmetical mean of the results of a number of separate realizations (replications), all within the specified ensemble. The statistical property of concern here is the ensemble mean concentration $C(\mathbf{x}, t)$. Suppose the results of n separate realizations of $\Gamma(\mathbf{x}, t)$ are available, and that $\Gamma^{(r)}(\mathbf{x}, t)$ is the value of $\Gamma(\mathbf{x}, t)$ in the rth realization. Then, the approximation

$$\frac{1}{n} \sum_{r=1}^{n} \Gamma^{(r)}(\mathbf{x}, t) \approx C(\mathbf{x}, t) \tag{2}$$

has to be used to estimate $C(\mathbf{x}, t)$. Equation (2) is an approximation because $C(\mathbf{x}, t)$ is not random, unlike the values of $\Gamma^{(r)}(\mathbf{x}, t)$. However, the net effect of the random fluctuations on the estimate of $C(\mathbf{x}, t)$ obtained by Equation (2) becomes less as n increases. The ensemble mean velocity field $\mathbf{U}(\mathbf{x}, t)$ also has to be estimated from separate realizations by a relation analogous to Equation (2). The governing equation for $C(\mathbf{x}, t)$ is obtained by writing $\Gamma = C + c$ and $\Upsilon = \mathbf{U} + \mathbf{u}$ (the Reynolds decomposition) in Equation (1), where c and $\mathbf{u}$ are the fluctuations in concentration and velocity, respectively. On taking the ensemble mean, the resulting equation can be rearranged to give

$$\frac{\partial C}{\partial t} + \mathbf{U} \cdot \nabla C = \nabla \cdot (-\overline{\mathbf{u}c}) + D\nabla^2 C, \tag{3}$$

where the overbar denotes the ensemble mean, again estimated (in theory) by a relation like Equation (2). In the rearrangement it has been assumed that the velocity field is incompressible, so that $\nabla \cdot \mathbf{U} = \nabla \cdot \mathbf{u} = 0$; accordingly, the second term on the left-hand side of Equation (3) can also be rearranged in the form $\nabla \cdot (\mathbf{U}C)$.

The notation in Equation (2) emphasizes that, in general, the ensemble mean concentration depends on position $\mathbf{x}$ and time t. But in certain situations it may happen that the statistical properties are independent of one or more components of $\mathbf{x}$ or of t. In this case, ensemble means can be estimated by integration of the record of a single realization of the ensemble (the ergodic hypothesis). Suppose, for example, that $C(\mathbf{x}, t)$ is independent of t, so that $C(\mathbf{x}, t) \equiv C(\mathbf{x})$. Then, instead of Equation (2), $C(\mathbf{x})$ can be estimated by

$$\frac{1}{T} \int_{\tau}^{\tau+T} \Gamma(\mathbf{x}, t') \, dt' \approx C(\mathbf{x}), \tag{4}$$

where $\Gamma(\mathbf{x}, t')$ is the concentration in one realization, and τ is any time. The accuracy of this estimate increases as T increases, just as the accuracy of Equation (2) increases as n increases. Use of Equation (4) is justified, for example, when Γ is the concentration of effluent from a steady discharge into a river whose flow is unaffected by the tide or by any other causes of systematic temporal variation. In this case, i.e. when C is independent of t, the governing equation for C is

$$\mathbf{U} \cdot \nabla C = \nabla \cdot (-\overline{uc}) + D\nabla^2 C, \tag{5}$$

obtained directly from Equation (3) by putting $\partial C/\partial t \equiv 0$.

It is obvious that C cannot strictly be independent of t unless (a) the discharge is steady and (b) the mean velocity $\mathbf{U}$ is also independent of t. Problems occur when these conditions are violated, as they frequently are in practical problems; in estuaries, for example, the tides cause $\mathbf{U}$ to have time-dependent periodic components.

The problems occurring when C and $\mathbf{U}$ depend on t derive from one simple fact: Ensemble means cannot then be estimated by integration over time, as in Equation (4). Tennekes & Lumley (1972, p. 34) note that "time averages would not make sense in an unsteady situation." Two very different attitudes are taken to this problem.

Theoreticians insist that only ensemble means, estimated by Equation (2), can be used, in which case Equation (3) holds without approximation. This approach is practically naive, since there are two severe drawbacks. Firstly, the cost of performing sufficient replications to obtain stable ensemble means is prohibitive in almost all field trials.[2] Secondly, it is difficult to attach practical meaning to the concept of an ensemble mean in estuaries (and even more so in the oceans and atmosphere) in circumstances where it is impossible (irrespective of cost) to perform exact replications because of systematic trends on long and enormously long time scales.[3] Some causes of such trends are bed erosion, man-made constructions, seasonal changes in weather, and even variations in climate over periods of many years.

[2] Perhaps this restriction will eventually be overcome in certain circumstances by laboratory experiments on reliable small-scale models, a possibility suggested by work in a related field of comparable difficulty (Hall et al. 1982, Meroney & Lohmeyer 1982). Note also that ensemble averages can be estimated by repetitions of numerical simulations, as evidenced by work discussed later (Sullivan 1971, Allen 1982).

[3] For such reasons, the basic concept of an ensemble is sometimes regarded as valueless. Such judgments are unjustified. In many situations, long-term trends are not present; in other cases, such as estuaries, it is perfectly proper (from a theoretical point of view at least) to consider a hypothetical, albeit unrealizable, ensemble. This is standard practice in statistical analysis (Kendall & Stuart 1977, p. 221).

The other approach to the problem is to measure and consider only average values obtained by time integration. The quantity considered is $C_T(\mathbf{x}, t)$, defined by (Fischer et al. 1979, p. 16)

$$C_T(\mathbf{x}, t) = \frac{1}{T} \int_{t-\frac{1}{2}T}^{t+\frac{1}{2}T} \Gamma(\mathbf{x}, t')\, dt', \tag{6}$$

i.e. by the left-hand side of Equation (4) with $\tau = t - \frac{1}{2}T$, for some assigned value of T. This approach has the great merit of being practical; there are no conceptual difficulties in determining C_T. However, problems still remain, which often are not explicitly mentioned. As noted earlier, C_T in Equation (6) remains a random variable whose statistical properties depend on T (Hino 1968, Csanady 1980, pp. 75–76). It is usually assumed in estuarine research that there is a range of values of T of the order of 60–90 s (Dyer 1973, p. 29) for which C_T is essentially stable, i.e. independent of T. This assumption supposes the existence of a spectral gap in the velocity spectrum; however, the gap is not completely empty, so that measured values of C_T will sometimes be influenced by eddies whose characteristic time scale is of order T. Difficulties of interpretation may also be caused by fluctuations of this scale in the discharge rate of a source of contaminant, especially near the source.

By analogy with Equation (3), the equation governing $C_T(\mathbf{x}, t)$ is normally taken to be

$$\frac{\partial C_T}{\partial t} + \mathbf{U}_T \cdot \nabla C_T = \nabla \cdot \{ -(\mathbf{u}_T c_T)_T \} + D\nabla^2 C_T, \tag{7}$$

where $\mathbf{U}_T$ and $(\mathbf{u}_T c_T)_T$ are defined in terms of Υ and $\mathbf{u}_T c_T$ by the obvious parallels of Equation (6), and $\mathbf{u}_T$ and c_T are the new fluctuations, different of course from $\mathbf{u}$ and c defined earlier. It is usually claimed, or implied, that Equation (7) can be derived rigorously from Equation (1), but this is not so. The difficulty is that terms like $(\mathbf{U}_T c_T)_T$ are not identically zero, and therefore they are not obviously negligible compared with the term involving $(\mathbf{u}_T c_T)_T$, which is retained on the right-hand side of Equation (7). It is not even true that $(c_T)_T$ and $(\mathbf{u}_T)_T$ are identically zero, for use of $c_T = \Gamma - C_T$ in Equation (6) gives

$$(c_T)_T = C_T(\mathbf{x}, t) - \frac{1}{T} \int_{t-\frac{1}{2}T}^{t+\frac{1}{2}T} C_T(\mathbf{x}, t')\, dt', \tag{8}$$

where the dependence of $(c_T)_T$ on $\mathbf{x}$ and t has been suppressed for clarity. Since (except when $t' = t$) the values of $C_T(\mathbf{x}, t')$ depend on values of Γ at times outside the interval $(t-\frac{1}{2}T, t+\frac{1}{2}T)$, whereas the value of $C_T(\mathbf{x}, t)$ depends only on values of Γ for times within this interval, nothing further can be said rigorously about the value of $(c_T)_T$. Note that this problem does not arise when C is independent of t, even when it is estimated by Equation (4). The point is that T in Equation (4) can be made arbitrarily large, unlike T in Equations (6) and (8).

The difficulties with using C_T have been discussed above at some length because even though they relate to the basic principles of dispersion analysis, they are often ignored without justification. While these difficulties will inevitably be significant for some practical problems, they may well not affect the treatment of many others. In the remainder of this review, Equation (7) is usually regarded as an acceptable model equation for C_T, with the provisos that the results obtained with it will, on some random occasions, disagree with observations, and that the value of T must be appropriately chosen. Also, since Equations (3) and (7) are structurally identical, there is no need to retain the subscripts T in C_T and $(\mathbf{u}_T c_T)_T$; hence, Equation (3) is regarded both as the exact equation for the ensemble mean concentration and as a model equation for the time-averaged concentration.

Effects of Instrumentation

Observations require instruments, and all instruments inevitably introduce averaging, which in the general case amounts to a smoothing over both space and time. The precise details of this smoothing vary from instrument to instrument and will generally be influenced by an instrument-dependent weighting function (Chatwin 1982). Many textbooks, including Lumley & Panofsky (1964, pp. 35–58) and Pasquill (1974, pp. 11–22), contain detailed discussions of some aspects of this problem.

Here there is space only for a few brief comments on this important question, which is attracting increasing attention, especially when concentration fluctuations are involved. Note first that some instrument smoothing is unavoidable in field observations, because the smallest length scales present in the structure of Γ in rivers and estuaries are of order 10^{-4} m (as noted earlier). Inevitably, therefore, the measured concentration Γ_m differs from Γ and does not obey Equation (1). This provides a further reason, and a strong one, why Equation (3) must be regarded in practice as a model (and not an exact) equation, with C in Equation (3) as the time-averaged measured concentration. The greatest problems associated with instrument smoothing seem likely to occur when the small-scale structure of Γ plays an important role, and this is not so for the mean concentration C. However, in view of the increasing interest in problems involving the fluctuations of Γ about its mean, it is important for reports of experimental data to give full details of the resolution characteristics of the concentration sensors.

A different problem concerned with measurements has been studied by

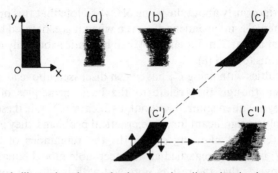

Figure 1   Sketch illustrating the mechanisms causing dispersion in the x direction. (a) Molecular diffusion. (b) Turbulent diffusion. (c) Initial advection, which enhances transverse diffusion (c') and hence is smeared (c'').

Kjerfve et al. (1981, 1982). They considered the number of separate readings needed to obtain reliable estimates of cross-sectional and tidal averages.

Mechanisms of Dispersion

In this review the term dispersion refers to any tendency for the distribution of $C(\mathbf{x}, t)$ to spread. There are three obvious mechanisms, represented by terms in Equation (3), that cause dispersion in a given direction, say that parallel to the x axis. These are (a) direct molecular diffusion, represented by $D(\partial^2 C/\partial x^2)$; ($b$) direct turbulent diffusion,[4] represented by $(\partial/\partial x)(-\overline{uc})$; and ($c$) advection, represented by $U(\partial C/\partial x)$. Here U and u are the x components of $\mathbf{U}$ and $\mathbf{u}$, respectively. These mechanisms are illustrated schematically in the top part of Figure 1.

Molecular diffusion and turbulent diffusion, while qualitatively similar, differ in two respects. In general the intensity of turbulent diffusion varies in directions such as Oy transverse to the direction of dispersion, unlike that of molecular diffusion. Also, turbulent diffusion is much more vigorous, so much so in fact that the effect of molecular diffusion on C (but not on some other statistical properties of Γ) can be ignored.

Advection by the mean current [mechanism (c) in Figure 1] is at first sight far more important than the other mechanisms, since it would appear to lead to spreading at a constant rate (proportional to the change in U in a direction transverse to the direction of dispersion); turbulent and molecular diffusion, on the other hand, generate rates of spread that decrease

[4] Note that the mechanism represented by $\nabla \cdot (-\overline{uc})$, and here called turbulent diffusion, can legitimately be regarded as hypothetical, since, unlike molecular diffusion and advection, it does not appear in Equation (1) for the actual concentration Γ and is a direct consequence of the averaging process.

with time (proportional to $t^{-1/2}$ in some common circumstances) because of the continual decrease of gradients of C in the x direction. That this explanation is too superficial was first shown by Taylor (1953, 1954) in two profound papers.

The mechanism discovered by Taylor is illustrated, again schematically, in the bottom part of Figure 1. In the presence of any variation of U in a transverse direction (and without such variation there is no effect whatsoever of advection on dispersion, since the distribution of C is simply transported as a whole), advection increases (or, as in Figure 1, even generates) transverse gradients of C (e.g. $\partial C/\partial y$). Consequently transverse diffusion, both molecular [e.g. $D(\partial^2 C/\partial y^2)$] and turbulent [e.g. $(\partial/\partial y)$ $(-\overline{vc})$], is enhanced as indicated in (c') in Figure 1. Thus the distribution of C tends to become smeared out laterally, with the result that advection in the x direction [$U(\partial C/\partial x)$] is less effective than suggested in the previous paragraph at dispersing the distribution of C in this direction [see (c'') in Figure 1]. The term *shear dispersion* (or *longitudinal dispersion*) is used to denote this contribution to dispersion, arising from an interaction between advection by a transversely sheared mean velocity and transverse diffusion.

Before Equation (3) can be used to estimate dispersion, including shear dispersion, a prescription is needed for dealing with the turbulent diffusion term $\nabla \cdot (-\overline{uc})$. This term can be rewritten

$$\frac{\partial}{\partial x}(-\overline{uc}) + \frac{\partial}{\partial y}(-\overline{vc}) + \frac{\partial}{\partial z}(-\overline{wc}),$$ (9)

and it is almost invariably assumed that there exist (variable) eddy diffusivities ε_x, ε_y, ε_z such that, by analogy with molecular diffusion,

$$-\overline{uc} = \varepsilon_x \frac{\partial C}{\partial x}, \qquad -\overline{vc} = \varepsilon_y \frac{\partial C}{\partial y}, \qquad -\overline{wc} = \varepsilon_z \frac{\partial C}{\partial z}.$$ (10)

Expression (9) then becomes

$$\frac{\partial}{\partial x}\left(\varepsilon_x \frac{\partial C}{\partial x}\right) + \frac{\partial}{\partial y}\left(\varepsilon_y \frac{\partial C}{\partial y}\right) + \frac{\partial}{\partial z}\left(\varepsilon_z \frac{\partial C}{\partial z}\right).$$ (11)

The literature on turbulent-diffusion theory contains many discussions of the validity, or otherwise, of the step from (9) to (11) [see, for example, Tennekes & Lumley (1972, pp. 11, 40–52)]. Only some brief comments are made here. First, the step from (9) to (11) does not have a sound theoretical justification, especially in flows like estuaries where stratification is important. The case for its use is therefore primarily empirical—namely, that satisfactory agreement with experimental results can be obtained for suitably chosen ε_x, ε_y, ε_z. It has to be stated, however, that unfortunately the practical adequacy, or otherwise, of modeling $\nabla \cdot (-\overline{\mathbf{u}c})$ by (11) is a question that seems to be of little or no interest to experimenters. Thus, data are analyzed to obtain formulae and to investigate scaling laws for ε_x, ε_y, ε_z, but they are not used to test whether a different representation from Equation (10) could be practically preferable. Hopefully, this limited outlook will soon broaden, stimulated perhaps by recent rapid progress in numerical modeling of turbulent flows using sophisticated closure schemes. Work by Smith & Takhar (1981) and T. J. Smith (1982), while using eddy diffusivities, indicates that some of these schemes at least can be used successfully in flows, like estuaries, with complicated geometries. In view of earlier comments about averaging, it ought to be noted also that measured eddy diffusivities are functions of the time interval T in Equation (6); therefore, $\mathbf{u}_T$ and c_T are not turbulent fluctuations in the normal sense. A final comment is that the step from (9) to (10) ignores, as noted by Pritchard (1958), the possibility of so-called mixed terms in (10) and hence (11). Such terms occur if the first equation in (10) is generalized to $-\overline{uc} = \varepsilon_x(\partial C/\partial x)$ $+ \varepsilon_{xy}(\partial C/\partial y) + \varepsilon_{xz}(\partial C/\partial z)$, with similar expressions for $-\overline{vc}$ and $-\overline{wc}$. While the neglect of these extra terms is probably justified at present from a practical point of view in rivers and estuaries, given the data that are available, it is worth noting that the analogous terms in atmospheric dispersion are now thought to be important in some circumstances (see, for example, Yaglom 1976).

When molecular diffusion is ignored and turbulent diffusion is represented by (11), Equation (3) becomes

$$\frac{\partial C}{\partial t} + U \frac{\partial C}{\partial x} + V \frac{\partial C}{\partial y} + W \frac{\partial C}{\partial z} = \frac{\partial}{\partial x}\left(\varepsilon_x \frac{\partial C}{\partial x} \right)$$
$$+ \frac{\partial}{\partial y}\left(\varepsilon_y \frac{\partial C}{\partial y} \right) + \frac{\partial}{\partial z}\left(\varepsilon_z \frac{\partial C}{\partial z} \right), \quad (12)$$

where (U, V, W) are the components of $\mathbf{U}$, to be determined from the Reynolds equations for the velocity field. Equation (12) has formed the basis of many investigations, some of which are described in this paper. The axes used in the remainder of this paper are shown in Figure 2 and are the same as those used by Fischer et al. (1979). The x axis is always in the direction of the mean discharge in rivers and is toward the sea in estuaries. Unfortunately, no choice of axes can satisfy the normal conventions in both hydraulics and oceanography!

Taylor's Model of Longitudinal Dispersion

Interest often focuses on the dispersion of contaminant in the x direction, i.e. in the streamwise or longitudinal direction. It is then common (and

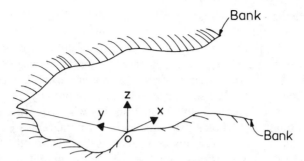

Figure 2   The axes used in this paper. Axes Ox and Oy are horizontal, and axis Oz is vertically upward.

natural) to consider only the variation of C_m with x and t, where

$$C_m(x, t) = \frac{1}{A} \int \int C(\mathbf{x}, t) \, dy \, dz \tag{13}$$

is the cross-sectional average of C, A is the cross-sectional area of the stream (depending in general on x and t), and the integration is over the complete stream cross section. Analogous definitions are used for U_m, V_m, and W_m.

Taylor (1954) showed that in certain circumstances the equation governing $C_m(x, t)$ is

$$\frac{\partial C_m}{\partial t} + U_m \frac{\partial C_m}{\partial x} = K_m \frac{\partial^2 C_m}{\partial x^2}, \tag{14}$$

where K_m is a constant, conveniently called the *effective longitudinal dispersion coefficient*. The conditions that ensure that Equation (14) holds are important enough to be listed here explicitly (Chatwin 1980):

1. The velocity field Υ is statistically steady, i.e. U, V, W, and the eddy diffusivities are independent of t. It is obvious that this condition will not normally be satisfied when the dispersing contaminant is not passive, i.e. when it generates buoyancy forces.
2. The cross-sectional area A is a constant, independent of x and t.
3. The time that has elapsed since the dispersion started is sufficiently large compared with the time taken for thorough mixing of the contaminant over the cross section.

In accordance with the earlier discussion, the value of K_m in Equation (14) depends on the details of the variation of U, ε_y, and ε_z with the transverse coordinates y and z; this value can be estimated most efficiently by the method given by Taylor (1953, 1954). This method was later applied to open channels by Elder (1959) and to rivers by Fischer (1967). Aris (1956) used the method of moments to confirm the validity of Taylor's method and also to show that streamwise turbulent diffusion (neglected by Taylor) made a small additive contribution to K_m. It should also be noted that conditions (1) and (2) above are precisely those that ensure that the Lagrangian streamwise velocity of a contaminant particle is a stationary random function of time, so that the famous analysis by Taylor (1921) applies. It can be deduced from this analysis that when condition (3) is also satisfied, Equation (14) holds. The remarks in this paragraph are developed in more detail in such accounts as Fischer (1973), R. Smith (1979), and Chatwin (1980). However, as is appropriate in nearly all practical problems in rivers and estuaries, the remainder of this paper emphasizes situations where conditions (1), (2), and (3) above are not all satisfied.

Other Types of Dispersion Models for Conserved Substances

The differential equations presented above form the basis for most investigations of dispersion, but there are other types of models that are used.

Pritchard (1969) developed a simple model for the dispersion of pollutants in estuaries, a model that extends the earlier tidal prism concept (Ketchum 1955). This model, termed a *box model*, is purely algebraic and (in its basic version) assumes that the pollutant is uniformly mixed over the cross section and that its concentration is time independent, so that the model can apply only to tidally averaged concentrations. The estuary is divided longitudinally into boxes, with mixing across the boundaries between pairs of adjacent boxes described by exchange coefficients. It is supposed that these coefficients are the same for all dispersing substances; accordingly, they can be determined by measuring the longitudinal distribution of salinity. In most cases, the model is refined by also dividing the estuary vertically into upper and lower sections, with different concentrations in each section. In this way some account is taken of the effects of stratification. A detailed summary of box models is given by Officer (1980), who also cites some successful applications. The box-model approach is compared with one based on differential equations in Officer & Lynch (1981). It is obvious that box models do not take account of all the physical processes affecting dispersion; in that sense they are less scientifically justified than models based on Equation (3). On the other hand, they are quick to use and they are robust. These are important practical considerations, especially when it is realized that models based on Equation (3) may give far more information than is required in an engineering application, and that this information may be sensitively dependent on experimental errors in the measurements that are needed as input.

In the earlier discussion it was pointed out that use of equations like (3) or (14) could lead to difficulties in interpretation because of the different averaging processes involved, and that such use denied any interest in concentration fluctuations. In a numerical simulation of dispersion in open-channel flow, Sullivan (1971) avoided these shortcomings by regarding a source of pollutant as composed of very many (usually 5000) particles, each of which undergoes a random walk. In each small discrete time step of the walk, the particle is advected downstream at the local value of U and is given, simultaneously, a random lateral displacement of magnitude determined by the measured turbulence characteristics. At any time after the start of dispersion, the statistical properties of the distribution of concentration are obtained by straightforward ensemble averaging. The results of this simulation were in good agreement with laboratory measurements. Sullivan's method was applied by Allen (1982) to simulate dispersion in time-periodic flow, again in an open channel. In this case up to 10,000 particles were used, and the results were consistent with the calculations of R. Smith (1983a) based on Equation (3). Further investigation is needed to see how readily this simulation method can be developed to deal, for example, with complicated geometries. Bowden (1982) notes that "the random walk method is an interesting new technique but it is too early to say yet what advantages it may have in practical problems of dispersion in estuaries."

Finally, we mention the Lagrangian transport method (Fischer 1972b), described in Fischer et al. (1979, pp. 289–91). This method has elements in common with both the box model and Sullivan's simulation method in that the estuary is divided longitudinally into advecting elements, between which mixing takes place. As operated, the method is computationally more efficient than any finite-difference scheme and minimizes numerical diffusion (Fischer et al. 1979, p. 289). Its great potential advantage is its adaptability to different flow situations, including those as complicated as deltas; however, further comparison with data is needed before a final verdict can be given on its accuracy.

Dispersion of Nonconserved Substances

While many practical problems involve nonconserved substances like dissolved biochemical oxygen, heavy metals, and heat, these problems seem to have received less attention from modelers than their importance merits. Most methods discussed in textbooks (e.g. Dyer 1973, pp. 119–21, Officer 1976, pp. 230–35) deal with estuaries and use a first-order decay term in a model equation for the average of $C_m(x, t)$ over the tidal cycle, where C_m is defined in Equation (13). Such averages are denoted here by angle brackets, e.g. $\langle C_m \rangle$. A typical model equation is therefore

$$\langle U_m \rangle \frac{\partial \langle C_m \rangle}{\partial x} = \langle K_m \rangle \frac{\partial^2 \langle C_m \rangle}{\partial x^2} - k \langle C_m \rangle, \tag{15}$$

where k is a decay constant. Solutions of this equation for values of $\langle U_m \rangle$ and $\langle K_m \rangle$ that are independent of x led Fischer et al. (1979, pp. 145–47) to an interesting observation. According to the discussion following Equation (14), longitudinal dispersion can validly be modeled by a term like $\langle K_m \rangle \partial^2 \langle C_m \rangle / \partial x^2$ in Equation (15) only after the dispersing substance is thoroughly mixed over the cross section. The distance downstream from the source needed to achieve this mixing is of order $x_1 = \langle U_m \rangle b^2 / \langle \varepsilon_{ym} \rangle$, where b is the breadth of the flow and ε_{ym} is the cross-sectional mean of ε_y, the transverse eddy diffusivity. The solutions show that when $x_1 \lesssim x_2$ (where $x_2 = \langle U_m \rangle / k$ is the order of magnitude of the decay distance), the term involving $\langle K_m \rangle$ in Equation (15) has almost negligible effect on the variation of $\langle C_m \rangle$ with x, whereas when $x_1 \gtrsim x_2$, the solution of Equation (15) has no practical significance because almost all the dispersing substance has disappeared through decay before cross-sectional mixing has occurred.[5] In either case, Equation (15) is of little or no value. But note that these conclusions are based on the frequently incorrect restrictions that $\langle U_m \rangle$ and $\langle K_m \rangle$ are independent of x and that the (tidally averaged) source conditions are steady.

Two other approaches to modeling the dispersion of nonconserved substances in estuaries have been developed recently, largely by C. B. Officer and his coworkers. It must be noted, however, that so far these approaches have been applied mainly under the restrictive assumption of steadiness. The simplest version of the first method (Officer 1979) supposes, assuming the equality of $\langle K_m \rangle$ for salinity and concentration, that the value of $\langle C_m \rangle$ for a conserved substance varies linearly with the corresponding value for the salinity; it then shows how quantitative estimates for the loss of a nonconserved substance can be obtained in terms of the deviations from linearity observed in the graph of its $\langle C_m \rangle$ against salinity. Extensions of the method to cover, for example, situations where $\langle U_m \rangle$ is dependent on x are described in Rattray & Officer (1981), Officer & Lynch (1981), and Rattray & Uncles (1983). The second method extends box models to include a decay term (Officer 1980, Officer & Lynch 1981); this approach is less developed than the first. Both of these methods are relatively easy to apply, but there is a need for more comparisons, both with experiments and with

---

[5] Essentially the same argument applies when loss of the dispersing substance occurs through absorption at the boundaries. Such problems are discussed by R. Smith (1983d), who notes, however, that other model equations, proposed by him and considered later in this article, do not have the disadvantages of Equation (15) in this regard.

the results of models, like that of Equation (15) or that obtained by modifying Fischer's numerical Lagrangian transport method to include decay.

DISPERSION IN RIVERS

Introduction

In this section we consider only situations for which, in the absence of the dispersing substance, the flow is unstratified and the mean velocity field U is steady. Most of our attention is paid to cases where the contaminant is effectively passive. Under these circumstances there are many similarities between dispersion in rivers on the one hand, and dispersion in pipes and open channels on the other; thus, work in the latter areas is considered when appropriate, but additional complications due (mainly) to geometry are, of course, also discussed.

Much work on dispersion in rivers (and in pipes and open channels) has been based on Taylor's equation (14) and, in particular, has involved estimating a value for K_m, the longitudinal dispersion coefficient. Useful summaries of such work are given by Fischer (1973) in an earlier volume of this series, and in Chapter 5 of Fischer et al. (1979). While these accounts make it clear that Equation (14) is not universally applicable, it is not until fairly recently that the limitations of this equation have been widely recognized; accordingly, and consistently with recent trends, the present account emphasizes practical problems for which Equation (14)—in the form given above—does not seem likely to be the most satisfactory possible mathematical model.

Describing Deviations From Gaussianity

A solution of Equation (14) for the dispersion of a finite quantity Q is the Gaussian expression

$$C_m = \frac{Q}{2A\{K_m\pi(t-t_0)\}^{1/2}} \exp\left\{ -\frac{[(x-x_0)-U_m(t-t_0)]^2}{4K_m(t-t_0)} \right\}, \tag{16}$$

where x_0 and t_0 are constants, and A is the cross-sectional area of the river. According to Equation (16), the center of mass X_m and the variance σ_m^2 are given by

$$X_m = \int_{-\infty}^{\infty} x(AC_m/Q)\,dx = U_m t + (x_0 - U_m t_0), \tag{17}$$

$$\sigma_m^2 = \int_{-\infty}^{\infty} (x-X_m)^2(AC_m/Q)\,dx = 2K_m(t-t_0). \tag{18}$$

Observations in rivers and open channels normally record values of C_m at a fixed value of x. The resulting curves of C_m vs. t are then bound to be asymmetrical, simply because the cloud is evolving as it moves past the measuring station. Nevertheless, this asymmetry can be catered for, either by replotting in the way proposed by Chatwin (1971) or by considering the moments of C_m with respect to time (Tsai & Holley 1978, R. Smith 1984); the observed profiles are found to be inconsistent with Equation (16). Day & Wood (1976) state that they "are unaware of any Gaussian concentration distribution ever being recorded from flow in an open channel." It is important to note also (Chatwin 1972) that the restrictions needed to derive Taylor's equation (14) are such that deviations of observed profiles from that in Equation (16) are not describable by Equation (14). Thus the apparently greater generality of the differential equation (14) compared with the simple expression in Equation (16) is spurious.

This failure of observed profiles to agree with Equation (16) has been known for many years (Elder 1959, Fischer 1967). All authorities agree that the principal cause of the disagreement is that the elapsed time t (or the equivalent distance x) since release is less than t_1, the time required for Taylor's equation (14) [or the equivalent Equation (16)] to be a valid approximation. Several estimates of the order of magnitude of t_1 have been made, some based on the time needed before Equation (16) holds and others on the requirement, less stringent in practice, that σ_m^2 [defined in Equation (18)] be linear in t. As noted earlier, t_1 is the time taken for thorough mixing of the dispersing substance over the flow cross section, and Fischer (1967) suggested that a reasonable estimate in practice for a source at the side of the river is $t_1 \sim 0.4 b^2 / \varepsilon_{ym}$, where b is the breadth of the river (assumed much greater than its depth d). This estimate is consistent with theoretical work (Chatwin 1970, Chatwin & Sullivan 1982). For the Green-Duwamish River, Washington (where $b \approx 22.25$ m and $b/d \approx 16$), and for the Missouri River near Omaha (where $b \approx 180$ m and $b/d \approx 60$), Fischer's formula gives $t_1 \sim 3\frac{1}{4}$ hr and $t_1 \sim 27$ hr, respectively.

Fischer's estimate of t_1 is a bulk estimate in the sense that it ignores mixing in any regions of very weak turbulence (such as viscous sublayers in smooth-walled pipes and open channels, and quasi-stagnant "dead zones" randomly occurring in the bottoms and sides of rivers and natural streams). Such regions are also associated with low values of U and, hence, with "tails" in the observed profiles of C_m. In analyzing dispersion experiments, it is possible to ignore such tails, as done by Elder (1959), for example, and to use Taylor's equation (14) only to describe the remaining faster-moving parts of the cloud[6] (provided, of course, that the elapsed time still exceeds

---

[6] However, Day (1975) reports that Taylor's analysis did not apply to his profiles of C_m, even when the tails were ignored and even for values of t greater than t_1. This disagreement led Day & Wood (1976) to develop a new mathematical model based on self-similarity in time, and this model was extended by Beltaos (1980). Fischer et al. (1979, p. 134) attribute the disagreement to dead zones in the bottoms of the mountain streams where the data were obtained.

Fischer's estimate of t_1 given above). Determination of suitable values of U_m and K_m in such an analysis is discussed by Sullivan (1971), with particular reference to an open channel. The problem with this viewpoint is that these values of U_m and K_m, and the total quantity of material in the part of the cloud being analyzed, will change with time as the tail merges into the rest of the cloud. This merging process will be essentially complete after a time t_2, much greater than t_1 estimated above, where the magnitude of t_2 depends on the geometry and dynamics of the regions in the flow that give rise to the tails. Chatwin (1971) and Dewey & Sullivan (1977) estimate t_2 and the corresponding value of K_m for viscous sublayers in smooth pipes and open channels, and Valentine & Wood (1977) consider the case of dead zones in natural streams.

The inevitability that observed distributions of C_m will not be Gaussian has led to many attempts to investigate dispersion before Equation (16) can be applied. Two different approaches have been adopted: Either (a) the aim is to describe and predict how the Gaussian curve in Equation (16) is reached, or (b) the work considers how the dispersion process evolves from release. In case (a) the concern is with times t less than, but comparable with t_1, while in case (b) it is with small times only, starting with release at $t = 0$. Before discussing work in each of these areas, we emphasize that in both cases Taylor's description of the basic physics of shear dispersion (as illustrated in Figure 1c–c″) still applies, even though the net effect is no longer described by the simple equation (14).

For case (a), it has been known since the work of Aris (1956) that the skewness λ_3 and the kurtosis λ_4 of the C_m–x curve, the simplest measures of its deviation from Gaussianity, are proportional to $t^{-1/2}$ and t^{-1}, respectively, and so decay to zero rather slowly. Chatwin (1970) showed how C_m could be described in terms of parameters like λ_3 and λ_4 by means of an asymptotic series whose first few terms are

$$C_m = \frac{Q}{A\sigma_m(2\pi)^{1/2}} e^{-\frac{1}{2}\xi^2} \left\{ 1 + \frac{\lambda_3}{6} H_3(\xi) \right.$$

$$\left. + \left[\frac{\lambda_4}{24} H_4(\xi) + \frac{\lambda_3^2}{72} H_6(\xi) \right] + \cdots \right\}, \quad (19)$$

where σ_m is given by Equation (18), $\xi = [(x - x_0) - U_m(t - t_0)]/\sigma_m$, and the H_n are Hermite polynomials with $H_3 = \xi^3 - 3\xi$, $H_4 = \xi^4 - 6\xi^2 + 3$, and $H_6 = \xi^6 - 15\xi^4 + 45\xi^2 - 15$. It will be seen that Equation (19) is exactly Equation (16) when $\lambda_3 = \lambda_4 = 0$. While Equation (19) was originally derived directly from Equation (12), it has an easier derivation as Edgeworth's series (Kendall & Stuart 1977, pp. 169–72) representing the deviation of any profile from Gaussianity. There is therefore little doubt that Equation (19) is formally correct in all cases in which C_m eventually becomes Gaussian. Practical use of this equation, as described by Chatwin (1970, 1980), requires the estimation of parameters like λ_3 and λ_4, which are sensitively dependent on the tails of the observed profiles.

Formally, the series for C_m in Equation (19) is the solution of the infinite-order differential equation (Chatwin 1970)

$$\frac{\partial C_m}{\partial t} + U_m \frac{\partial C_m}{\partial x} = K_m^{(2)} \frac{\partial^2 C_m}{\partial x^2} + K_m^{(3)} \frac{\partial^3 C_m}{\partial x^3} + \cdots, \tag{20}$$

where the constants $K_m^{(r)}$ (with $K_m^{(2)} \equiv K_m$) depend on the flow properties but not on the way in which the contaminant is released at the source. Equations (19) and (20) are asymptotic representations whose accuracy increases as t increases; they are not convergent as t approaches the time of release and are therefore less accurate as t decreases. In an attempt to avoid this shortcoming, Gill & Sankasubramanian (1970) proposed that the $K_m^{(r)}$ in Equation (20) should be functions of t. However, this use of time-dependent coefficients is unphysical because, for example, it violates the fundamental requirement that solutions from different sources should be superposable.

In using equations like (20), the detailed source properties (which have a lasting influence on parameters like λ_3 and λ_4) affect only the temporal boundary conditions (Chatwin 1970, 1972). R. Smith (1981a, 1982a) devised an ingenious model equation in which these source properties appear explicitly. This equation was called a delay-diffusion equation by its inventor because it recognizes that shear dispersion is a gradual process, and that what happens at time t depends on the concentration distribution at earlier times. The full delay-diffusion equation is an infinite-order equation like Equation (20), and it can also be truncated after any finite number of terms. For reasons of space and simplicity, only the lowest-order truncation is given here, and only for the case of a cross-sectionally uniform source of strength $Q_m(x, t)$. If we neglect longitudinal turbulent diffusion, this truncated equation is (R. Smith 1981a)

$$\frac{\partial C_m}{\partial t} + U_m \frac{\partial C_m}{\partial x} = \int_0^\infty \frac{\partial K(\tau)}{\partial \tau} \frac{\partial^2 C_m(x - x_0(\tau), t - \tau)}{\partial x^2} d\tau + Q_m, \tag{21}$$

where $K(\tau)$ tends to K_m as $\tau \to \infty$. The flow properties determine $K(\tau)$ and also the form of the spatial decay $x_0(\tau)$, which can be selected so that the solution of Equation (21) has the exact values of X_m, σ_m^2, and λ_3. Numerical solutions of Equation (21) are in good agreement with numerical solutions of the exact equation for C in Poiseuille flow for values of t comparable with t_1, and also with Elder's (1959) observations in open-channel flow. Smith (1982c) shows how Equation (21) can be extended to deal with all sources, and he emphasizes that the delay-diffusion equation (and all truncations) possess the superposition property. It is still not clear how the results of Smith's equation compare with the expansion in Equation (19) or with solutions of the exact equation for small values of t. Answers to such questions would be desirable.

Dispersion Soon After Release

It has already been noted that equations like (19) and (21) are not valid for $t \ll t_1$. As suggested by the typical values of t_1 quoted earlier, there are many important problems in which such times are of primary interest, and new methods are needed to deal with them. McQuivey & Keefer (1976a) developed a convective model and applied it to dispersion experiments in the Mississippi River (McQuivey & Keefer 1976b). Most of the other models have been based on Equation (12) for C, since there is incomplete cross-sectional mixing and it is no longer appropriate to work directly with C_m. Expansions of C in powers of t can be obtained (Chatwin 1976b, 1977, Barton 1978), but their range of usefulness in rivers is very restricted.

A much more promising method has been developed by R. Smith (1981c), in which the rapidly varying concentration $C(\mathbf{x}, t)$ is expressed in terms of two slowly varying functions $a(\mathbf{x}, t)$ and $\phi(\mathbf{x}, t)$ by the equation $C = a \exp(\pm \phi)$, where the $\pm$ sign enables obvious symmetries to be treated accurately. Smith interprets this equation as a ray expansion, with dispersion information being propagated by diffusion along rays. He develops equations for a and ϕ from Equation (12) and shows how to deal with reflection at boundaries. The results obtained from some simple flows are good approximations for times much greater than those achieved by other methods, and applications to unsteady discharges in rivers are awaited eagerly.

Dispersion From Steady Sources

The ray-expansion method has been applied (R. Smith 1981b, 1983b) to steady discharges in rivers, with emphasis on problems such as the dependence of the shoreline concentration on the discharge location. The potential power of the method is apparent from these papers, which include variations of river depth and channel meandering. These factors are also catered for in another paper (R. Smith 1982b), which considers the levels of concentration far downstream from the discharge and estimates the optimum discharge location, defined as that for which the peak shoreline concentration is least.

The technique used in the last paper appears to be similar in spirit to one employed by McNulty (1983) and Nokes et al. (1984); the details are different because Smith incorporated effects like depth variation. Nokes et al. considered dispersion from a steady transverse line source in a channel of uniform depth and used the equation

$$U\frac{\partial C}{\partial x} = \frac{\partial}{\partial z}\left(\varepsilon_z \frac{\partial C}{\partial z}\right), \tag{22}$$

which is obtained from Equation (12) under steady conditions when longitudinal turbulent diffusion is negligible and when the flow is unidirectional. A formal solution of Equation (22) as an eigenfunction expansion is given by Nokes et al. for the case when $U(z)$ is logarithmic and ε_z varies with z in the way prescribed by Reynolds analogy. The results appear to agree quite well with experiments described in McNulty (1983). Nokes et al. emphasize the practical importance of steady discharges, which have received far less attention from theoreticians than have instantaneous sources. Hopefully, this omission is now being remedied.

Some Geometrical Complications

When the cross section of the river channel is not uniform, the velocity of a contaminant particle is never a stationary random function of t, and so Taylor's equation (14) cannot strictly be correct even when $t \gg t_1$. Nevertheless, in cases where the length scale characterizing variations in channel geometry is not too small, it seems plausible that an analogue of Equation (14) might still eventually apply. Since rivers are much wider than they are deep in the majority of cases, such analogues have been developed from equations like

$$\frac{\partial C_d}{\partial t} + U_d\frac{\partial C_d}{\partial x} + V_d\frac{\partial C_d}{\partial y} = \frac{1}{d}\left\{\frac{\partial}{\partial x}\left(d\varepsilon_{xd}\frac{\partial C_d}{\partial x}\right) + \frac{\partial}{\partial y}\left(d\varepsilon_{yd}\frac{\partial C_d}{\partial y}\right)\right\}, \tag{23}$$

where the axes are shown in Figure 2, $d = d(x, y)$ is the depth, and the subscript d denotes a depth average, e.g.

$$C_d(x, y) = \frac{1}{d(x, y)}\int_{-d}^{0} C(x, y, z)\,dz. \tag{24}$$

Following earlier work by Yotsukura & Cobb (1972) and Yotsukura & Sayre (1976), R. Smith (1983c) transformed the coordinates in Equation (23) to flow-following coordinates, so that (in these coordinates) changes in C_{dm} (the transverse average of C) are due entirely to shear dispersion, not per se to the nonuniform advection. He was then able to show that Equation (14) applied to C_{dm}, but only in these flow-following coordinates. The situation in the normal coordinates is more complicated, with (for example) the possibility that the analogue of K_m can be negative for sufficiently rapid variations! Smith's analysis includes the effect of meandering, which is also considered at length in Chapter 5 of Fischer et al. (1979), where particular emphasis is put on the profound effects that meandering (and small irregularities in channel geometry) can have on the effective value of ε_{yd}, and hence on the intensity of the longitudinal dispersion.

Earlier it was emphasized that use of Taylor's equation (14) is limited to times t such that $t > t_1$, a condition that is often not satisfied in practical applications; the same restrictions apply of course to the analysis just described.

DISPERSION IN ESTUARIES

Introduction

Dispersion in estuaries is obviously a much more difficult process to model than dispersion in rivers. It is influenced by all the complications discussed above, by tidal forcing (causing both the velocity distribution and the flow geometry to vary with time), and, most seriously [when the dispersing substance is buoyant (like heat) or heavy (like salinity)], by the dependence of the velocity field and the eddy diffusivities on the dispersion itself. The process is then highly nonlinear.

There is not enough space here even to attempt an adequate summary of all recent important work; that would require at least one book. Instead, only some of the work directly related to mathematical models is considered. For other, complementary accounts, the reader is urged to consult, for example, Bowden (1981, 1982), Fischer et al. (1979, especially Chapter 7), and Officer (1983). There are also many conference proceedings devoted largely or exclusively to dispersion and mixing in estuaries, such as Hamilton & Macdonald (1980), Harris (1979), Kjerfve (1978), and Nihoul (1978).

Our emphasis throughout is on longitudinal dispersion, i.e. the variation with x (and t) of quantities like C, C_m, or C_{dm}, where (it is recalled) C is the ensemble mean concentration of the dispersing substance (usually salt in this section). At a fixed downstream location x and a particular time t, the total longitudinal advective flux $\Phi(x, t)$ is given by

$$\Phi(x,t) = F(x,t) + f(x,t) = \int \int (U+u)(C+c) \, dy \, dz, \qquad (25)$$

where F is the ensemble mean of Φ (so that $\bar{f} = 0$) and the integration is over the whole cross section. Since the cross-sectional area varies with time, F is not (in general) expressible simply in terms of integrals of UC and $\overline{uc}$. For similar reasons, it is not easy to derive a simple equation for the longitudinal mass balance from either Equation (1) or Equation (12), a point that is carefully discussed by Dronkers (1982).

One-Dimensional Models of Longitudinal Dispersion in Estuaries

Nevertheless, the most frequently used models suppose that an effective longitudinal dispersion coefficient $K_m = K_m(x, t)$ can be used to describe the ensemble-mean mass flux, so that

$$F = A\left(U_m C_m - K_m \frac{\partial C_m}{\partial x} \right),$$ (26)

where $A = A(x, t)$ is the (ensemble mean) cross-sectional area. Correspondingly, an equation like

$$\frac{\partial}{\partial t}(A C_m) + \frac{\partial}{\partial x}(A U_m C_m) = \frac{\partial}{\partial x}\left(A K_m \frac{\partial C_m}{\partial x} \right)$$ (27)

(Fischer 1976, R. Smith 1980, Bowden 1982) is usually assumed to be a valid representation of the longitudinal mass balance. Many workers consider tidally averaged values only and use equations for $\langle C_m \rangle$ analogous to Equations (26) and (27), with A, U_m, and K_m replaced by corresponding tidal averages. In this case it is often appropriate to neglect the time derivative of $\langle A \rangle \langle C_m \rangle$, especially when C_m refers to salinity.

It is obvious from earlier remarks that Equation (27) can be a good approximation for the salinity only when certain conditions are met. To begin with, note that a description in terms of C_m makes practical sense only when the estuary is long compared with the cross-sectional dimensions and the tidal excursion; for the same reason, changes of geometry in the x direction must be sufficiently slow. For salinity in estuaries, essentially all time variation is due to tidal forcing, and the concept (crucial earlier) of "time since release" becomes irrelevant. However, the values of time scales such as the tidal period and that characteristic of cross-sectional mixing are naturally important in determining both the magnitude of F in Equation (26) and how this depends on the many different processes contributing to longitudinal dispersion in estuaries. In the case of salinity, this question was the principal theme of an earlier review in this series (Fischer 1976) and continues to be of central importance. Therefore, some of the main ways in which this topic has been investigated recently are now briefly summarized.

Modeling the Longitudinal Flux of Salinity

One method of research has been to concentrate only on single effects, such as the temporal oscillations (Bowden 1965, Holley et al. 1970, Chatwin 1975), the vertical shear in the buoyancy-driven residual current (Chatwin 1976a), and the transverse shear resulting from a nonrectangular cross section (Imberger 1976). R. Smith (1982a, 1983a) has considered the first of these effects, addressing (in particular) problems arising because the salinity distribution (or contaminant cloud) may actually be contracting during substantial periods in the tidal cycle, a situation not naturally modeled by an equation like (27), to say the least! In R. Smith (1982a) it is shown that the delay-diffusion equation (21) can cope with such contraction, giving results for the long-term-averaged dispersion that agree well with those of Holley et al. (1970), but that have more generality.

Such work has served principally as an aid to understanding and (except in very special cases) does not model what occurs in real estuaries, where several effects are likely to be simultaneously important. In an attempt to assess the relative importance of these different effects, authors such as Fischer (1972a) and Dyer (1973) expressed $\langle F \rangle$, the tidally averaged value of the longitudinal flux, as the sum of different terms. Such expressions occur inevitably once the velocity, concentration, and cross-sectional area are themselves decomposed into sums of various contributions (the tidal average of the cross-sectional mean, a term representing the transverse variation of the deviation from this average, etc.). However, Rattray & Dworski (1980) demonstrated in an important paper that the conclusions to be derived from this method of analysis (such as the relative importance of vertical and transverse variations to the flux) depend on the details of the decomposition, details which are of course chosen by the analyst. Possible causes of Rattray & Dworski's results, which are perhaps surprising at first sight, are (a) unjustified assumptions related to the different sorts of averaging processes involved in the decomposition [cf. the earlier comments following Equation (7) on an analogous point] and (b) the artificial oversimplification of the physics that such decompositions imply, granted the presence of abundant and strong nonlinear interactions.

In the light of the above remarks, the continuing controversy about the relative importance of transverse and vertical variations on the value of $\langle F \rangle$ may be less fundamental than was once believed, in that the issue is to some extent prejudged by the method of analysis chosen.[7] With this proviso, it is important nevertheless to note work on the effects on longitudinal dispersion of the residual gravitational circulation by, for example, Dronkers & Zimmerman (1982), Holloway (1981), Lewis & Lewis (1983), Uncles (1982), and Uncles et al. (1983). Each of these papers contains interesting analyses of data from one or more estuaries. R. Smith (1976, 1977, 1980) has used the technique of maximum-generality scaling to derive Equation (27) for several different sets of estuarine conditions and to

---

[7] For example, Wilson & Okubo (1978) analyzed data on dye dispersion from a point release in the lower York Estuary using a model equation with vertical advection and no transverse variation. Reasonable correlations were achieved.

establish corresponding quantitative estimates of K_m and $\langle K_m \rangle$. In the last of these papers, there is also a lucid account of how the values of $\langle K_m \rangle$ in estuaries, compared with rivers of comparable cross-sectional dimensions, are limited when there is insufficient time for cross-sectional mixing during one tidal cycle. Thus, Fischer et al. (1979, pp. 262–63) observe that typical values of $\langle K_m \rangle$ in estuaries are 100–300 m^2 s^{-1}, compared with a value of 1500 m^2 s^{-1} in a 200-m-wide reach of the Missouri River.

The Dependence of K_m on Salinity in Estuaries

Since the gravitational circulation in estuaries depends on the salinity distribution, it is clear that the value of K_m will depend on the salinity, if we assume Equation (27) to be an adequate model. Equation (27) is thus, in general, a nonlinear equation, and obviously this should be important when analyzing data to determine the value of K_m and its dependence on external parameters. Remarks by Prandle (1982) on a paper by West & Broyd (1981) are pertinent in this context.

Several authors (Chatwin 1976a, Imberger 1976, R. Smith 1976, 1980) have derived a dependence of $\langle K_m \rangle$ on $\langle C_m \rangle$, such as

$$\langle K_m \rangle = \alpha_0 + \alpha_2 \left\{ \frac{\partial \langle C_m \rangle}{\partial x} \right\}^2, \tag{28}$$

where α_0 and α_2 are constants independent of $\langle C_m \rangle$ but dependent on parameters like the flow geometry and the freshwater discharge. This was first obtained by Erdogan & Chatwin (1967) for the longitudinal dispersion of a slightly buoyant contaminant in laminar flow in a tube. A different model for the dependence of $\langle K_m \rangle$ on $\langle C_m \rangle$ was proposed by Godfrey (1980) in a numerical model of the James River estuary in Virginia. In a discussion of salinity intrusion in estuaries, Prandle (1981) showed that data from eight estuaries could be fitted reasonably well with each of three assumptions about $\langle K_m \rangle$, namely $\langle K_m \rangle = \alpha_0$, $\langle K_m \rangle = \alpha_1 \partial \langle C_m \rangle / \partial x$, and $\langle K_m \rangle = \alpha_2 \{ \partial \langle C_m \rangle / \partial x \}^2$; this interesting conclusion has important consequences from the point of view of practical prediction.

While much of the earlier discussion in this article, including that on dispersion in rivers, is directly relevant to the longitudinal dispersion of passive contaminants in estuaries, a different situation occurs when heat is the dispersing substance. This is particularly important in the context of using river or estuary water for cooling purposes in industrial processes. Since heat also generates buoyancy forces, the value of $\langle K_m \rangle$ for its longitudinal dispersion in estuaries will be a nonlinear function of the total density gradient, with contributions to the density being generated by both heat and salinity. This problem is discussed by R. Smith (1978a), who later

(R. Smith 1978b) also considers the lateral dispersion of heat, a question first investigated by Prych (1970) and particularly relevant for conditions near the source of heat.

Closing Comments on Dispersion in Estuaries

Near the end of his review in this series, Fischer (1976) states that "it is not yet possible to look at a given estuary, compute the values of some appropriate dimensionless parameters, and say with certainty which mass-transport mechanisms are the most important or what factors control the intrusion of salinity." Although R. Smith (1980) has solved this problem for wide estuaries in which the vertical mixing is rapid but the transverse mixing is slow (compared with the total period in each case), Fischer's statement, unfortunately, is still true when applied to the class of all estuaries; more seriously, we are not optimistic that current research is leading toward an eventual solution. The complexity of Smith's calculations shows that a theoretical solution to Fischer's problem will be extremely difficult, if not impossible. Success is more likely to come using well-designed experiments not only in the field but also in laboratories, where accurate measurements can be made more readily (Prandle 1984). However, the data must be analyzed in a sensible way, and there seems to be no consensus yet on an optimum method (assuming one exists). Papers discussing apparently similar data sets are often difficult to compare because of marked differences in the way the data are used. It is our opinion that this matter should receive urgent and intense attention.

CONCLUSIONS

It seems that there has been good progress recently on dispersion in rivers and open channels, evidenced by the frequent availability of potentially useful models for practically important questions. Progress on the more difficult question of dispersion in estuaries has been less satisfactory, as noted immediately above.

We wish finally to emphasize a few general conclusions that we have reached while writing this review. First, there should be more recognition of the potential shortcomings of standard techniques like the use of time-averaged concentrations and eddy diffusivities, and of the limitations of standard models like that in Equation (14). Also, theoreticians should make more effort to present mathematical results in forms that make their importance obvious and to give detailed numerical examples of their application to real flows. Above all, knowledge about dispersion could be significantly advanced by more well-designed laboratory experiments.

ACKNOWLEDGMENTS

During the period when this article was written, C. M. Allen was supported under M.O.D. Agreement No. AT 2067/046. We are grateful to many people for their help, especially David Prandle, Ron Smith, and Ian Wood. Finally, we wish to acknowledge the debt that each of us owes to Hugo Fischer, in whose memory we dedicate this review.

Literature Cited

Allen, C. M. 1982. Numerical simulation of contaminant dispersion in oscillatory flows. *Proc. R. Soc. London Ser. A* 381: 179–94

Antonopoulos-Domis, M. 1981. Large-eddy simulation of a passive scalar in isotropic turbulence. *J. Fluid Mech.* 104: 55–79

Aris, R. 1956. On the dispersion of a solute in a fluid flowing through a tube. *Proc. R. Soc. London Ser. A* 235: 67–77

Barton, N. G. 1978. The initial dispersion of soluble matter in 3D flow. *J. Austral. Math. Soc. Ser. B* 20: 265–79

Batchelor, G. K. 1967. *An Introduction to Fluid Dynamics.* Cambridge: Cambridge Univ. Press. 615 pp.

Beltaos, S. 1980. Longitudinal dispersion in rivers. *J. Hydraul. Div. Proc. ASCE* 106: 151–72

Bowden, K. F. 1965. Horizontal mixing in the sea due to a shearing current. *J. Fluid Mech.* 21: 83–95

Bowden, K. F. 1981. Turbulent mixing in estuaries. *Ocean Manage.* 6: 117–35

Bowden, K. F. 1982. *Theoretical and practical approaches to studies of turbulent mixing processes in estuaries.* Presented at NERC Workshop on Estuarine Processes, Univ. East Anglia, September

Chatwin, P. C. 1970. The approach to normality of the concentration distribution of a solute in a solvent flowing along a straight pipe. *J. Fluid Mech.* 43: 321–52

Chatwin, P. C. 1971. On the interpretation of some longitudinal dispersion experiments. *J. Fluid Mech.* 48: 689–702

Chatwin, P. C. 1972. The cumulants of the distribution of concentration of a solute dispersing in solvent flowing through a tube. *J. Fluid Mech.* 51: 63–67

Chatwin, P. C. 1975. On the longitudinal dispersion of passive contaminant in oscillatory flows in tubes. *J. Fluid Mech.* 71: 513–27

Chatwin, P. C. 1976a. Some remarks on the maintenance of the salinity distribution in estuaries. *Estuarine Coastal Mar. Sci.* 4: 555–66

Chatwin, P. C. 1976b. The initial dispersion of contaminant in Poiseuille flow and the smoothing of the snout. *J. Fluid Mech.* 77: 593–602

Chatwin, P. C. 1977. The initial development of longitudinal dispersion in straight tubes. *J. Fluid Mech.* 80: 33–48

Chatwin, P. C. 1980. Presentation of longitudinal dispersion data. *J. Hydraul. Div. Proc. ASCE* 106: 71–83

Chatwin, P. C. 1982. The use of statistics in describing and predicting the effects of dispersing gas clouds. *J. Hazard. Mater.* 6: 213–30

Chatwin, P. C., Sullivan, P. J. 1978. How some new fundamental results on relative turbulent diffusion can be relevant in estuaries and other natural flows. See Nihoul 1978, pp. 233–42

Chatwin, P. C., Sullivan, P. J. 1979a. The relative diffusion of a cloud of passive contaminant in incompressible turbulent flow. *J. Fluid Mech.* 91: 337–55

Chatwin, P. C., Sullivan, P. J. 1979b. Measurements of concentration fluctuations in relative turbulent diffusion. *J. Fluid Mech.* 94: 83–101

Chatwin, P. C., Sullivan, P. J. 1982. The effect of aspect ratio on longitudinal diffusivity in rectangular channels. *J. Fluid Mech.* 120: 347–58

Csanady, G. T. 1980. *Turbulent Diffusion in the Environment.* Dordrecht: Reidel. 248 pp.

Day, T. J. 1975. Longitudinal dispersion in natural channels. *Water Resour. Res.* 11: 909–18

Day, T. J., Wood, I. R. 1976. Similarity of the mean motion of fluid particles dispersing in a natural channel. *Water Resour. Res.* 12: 655–66

Dewey, R., Sullivan, P. J. 1977. The asymptotic stage of longitudinal turbulent dispersion within a tube. *J. Fluid Mech.* 80: 293–303

Dronkers, J. 1982. Conditions for gradient-type dispersive transport in one-dimensional, tidally-averaged transport models. *Estuarine Coastal Shelf Sci.* 14:599–621

Dronkers, J., Zimmerman, J. T. F. 1982. Some principles of mixing in tidal lagoons. *Oceanol. Acta Spec. Vol., Dec., 1982*, pp. 107–17

Dyer, K. R. 1973. *Estuaries: A Physical Introduction.* London: Wiley. 140 pp.

Elder, J. W. 1959. The dispersion of marked fluid in turbulent shear flow. *J. Fluid Mech.* 5:544–60

Erdogan, M. E., Chatwin, P. C. 1967. The effects of curvature and buoyancy on the laminar dispersion of solute in a horizontal tube. *J. Fluid Mech.* 29:465–84

Fischer, H. B. 1967. The mechanics of dispersion in natural streams. *J. Hydraul. Div. Proc. ASCE* 93:187–216

Fischer, H. B. 1972a. Mass transport mechanisms in partially stratified estuaries. *J. Fluid Mech.* 53:671–87

Fischer, H. B. 1972b. A Lagrangian method for predicting pollutant dispersion in Bolinas Lagoon, Marin County, California. *US Geol. Surv. Prof. Pap. 582-B.* 32 pp.

Fischer, H. B. 1973. Longitudinal dispersion and turbulent mixing in open-channel flow. *Ann. Rev. Fluid Mech.* 5:59–78

Fischer, H. B. 1976. Mixing and dispersion in estuaries. *Ann. Rev. Fluid Mech.* 8:107–33

Fischer, H. B., List, E. J., Koh, R. C. Y., Imberger, J., Brooks, N. H. 1979. *Mixing in Inland and Coastal Waters.* New York: Academic. 483 pp.

Gill, W. N., Sankasubramanian, R. 1970. Exact analysis of unsteady convective diffusion. *Proc. R. Soc. London Ser. A* 316: 341–50

Godfrey, J. S. 1980. A numerical model of the James River estuary, Virginia, U.S.A. *Estuarine Coastal Mar. Sci.* 11:295–310

Hall, D. J., Hollis, E. J., Ishaq, H. 1982. A wind tunnel model of the Porton dense gas spill field trials. *Rep. No. LR 394(AP),* Warren Spring Lab., Stevenage, Herts., Engl.

Hamilton, P., Macdonald, K. B., eds. 1980. *Estuarine and Wetland Processes.* New York: Plenum. 653 pp.

Harris, C. J., ed. 1979. *Mathematical Modelling of Turbulent Diffusion in the Environment.* London: Academic. 500 pp.

Hino, M. 1968. Maximum ground-level concentration and sampling time. *Atmos. Environ.* 2:149–65

Holley, E. R., Harleman, D. R. F., Fischer, H. B. 1970. Dispersion in homogeneous estuary flow. *J. Hydraul. Div. Proc. ASCE* 96:1691–1709

Holloway, P. E. 1981. Longitudinal mixing in the upper reaches of the Bay of Fundy. *Estuarine Coastal Shelf Sci.* 13:495–515

Imberger, J. 1976. Dynamics of a longitudinally stratified estuary. *Proc. Coastal Eng. Conf. ASCE, 15th, Honolulu,* 4:3108–17

Kendall, M. G., Stuart, A. 1977. *The Advanced Theory of Statistics.* Vol. 1, *Distribution Theory.* London: Charles Griffin. 4th ed. 472 pp.

Ketchum, B. H. 1955. Distribution of coliform bacteria and other pollutants in tidal estuaries. *Sewage Ind. Wastes* 27:1288–96

Kjerfve, B. J., ed. 1978. *Estuarine Transport Processes, Belle W. Baruch Libr. Mar. Sci. No. 7.* Columbia: Univ. S.C. Press. 331 pp.

Kjerfve, B. J., Stevenson, L. H., Proehl, J. A., Chrzanowski, T. H., Kitchens, W. M. 1981. Estimation of material fluxes in an estuarine cross-section: a critical analysis of spatial measurement density and errors. *Limnol. Oceanogr.* 26:325–35

Kjerfve, B. J., Proehl, J. A., Schwing, F. B., Seim, H. E., Marozas, M. 1982. Temporal and spatial considerations in measuring estuarine water fluxes. In *Estuarine Comparisons,* ed. V. S. Kennedy, pp. 37–51. New York: Academic

Lewis, R. E., Lewis, J. O. 1983. The principal factors contributing to the flux of salt in a narrow, partially stratified estuary. *Estuarine Coastal Shelf Sci.* 16:599–626

Lumley, J. L., Panofsky, H. A. 1964. *The Structure of Atmospheric Turbulence.* New York: Interscience. 239 pp.

McNulty, A. J. 1983. *Dispersion of a continuous pollutant source in open channel flow.* PhD thesis. Dept. Civ. Eng., Univ. Canterbury, Christchurch, N.Z. 210 pp.

McQuivey, R. S., Keefer, T. N. 1976a. Convective model of longitudinal dispersion. *J. Hydraul. Div. Proc. ASCE* 102: 1409–24

McQuivey, R. S., Keefer, T. N. 1976b. Dispersion—Mississippi River below Baton Rouge, La. *J. Hydraul. Div. Proc. ASCE* 102:1425–37

Meroney, R. N., Lohmeyer, A. 1982. Gravity spreading and dispersion of dense gas clouds released suddenly into a turbulent boundary layer. *Gas Res. Inst. Res. Rep. GRI-82/0025,* Chicago, Ill. 220 pp.

Nihoul, J. C. J., ed. 1978. *Hydrodynamics of Estuaries and Fjords, Proc. Int. Liège Colloq. Ocean Hydrodyn., 9th.* Amsterdam: Elsevier. 546 pp.

Nokes, R. I., McNulty, A. J., Wood, I. R. 1984. Dispersion from a steady two-dimensional horizontal source. Submitted for publication

Officer, C. B. 1976. *Physical Oceanography of Estuaries (and Associated Coastal Waters).* New York: Wiley. 465 pp.

Officer, C. B. 1979. Discussion of the behaviour of non-conservative dissolved

148   CHATWIN & ALLEN

constituents in estuaries. *Estuarine Coastal Mar. Sci.* 9:91–94

Officer, C. B. 1980. Box models revisited. See Hamilton & Macdonald 1980, pp. 65–114

Officer, C. B. 1983. Physics of estuarine circulation. In *Estuaries and Enclosed Seas*, ed. B. H. Ketchum, pp. 15–41. Amsterdam: Elsevier. 500 pp.

Officer, C. B., Lynch, D. R. 1981. Dynamics of mixing in estuaries. *Estuarine Coastal Shelf Sci.* 12:525–33

Pasquill, F. 1974. *Atmospheric Diffusion*. Chichester: Ellis Horwood. 429 pp.

Pope, S. B. 1979. The statistical theory of turbulent flows. *Philos. Trans. R. Soc. London Ser. A* 291:529–68

Prandle, D. 1981. Salinity intrusion in estuaries. *J. Phys. Oceanogr.* 11:1311–24

Prandle, D. 1982. Comment on "Dispersion coefficients in estuaries. West, J. R., Broyd, T. W." *Proc. Inst. Civ. Eng.* 73:253–54

Prandle, D. 1984. On salinity regimes and the vertical structure of residual flows in narrow tidal estuaries. Submitted for publication

Pritchard, D. W. 1958. The equations of mass continuity and salt continuity in estuaries. *J. Mar. Res.* 17:412–23

Pritchard, D. W. 1969. Dispersion and flushing of pollutants in estuaries. *J. Hydraul. Div. Proc. ASCE* 95:115–24

Prych, E. A. 1970. Effects of density differences on lateral mixing in open channel flows. *Rep. No. KH-R-21*, W. M. Keck Lab., Calif. Inst. Technol., Pasadena. 225 pp.

Rattray, M. Jr., Dworski, J. G. 1980. Comparison of methods for analysis of the transverse and vertical circulation contributions to the longitudinal advective salt flux in estuaries. *Estuarine Coastal Mar. Sci.* 11:515–36

Rattray, M. Jr., Officer, C. B. 1981. Discussion of trace metals in the waters of a partially-mixed estuary. *Estuarine Coastal Shelf Sci.* 12:251–66

Rattray, M. Jr., Uncles, R. J. 1983. On the predictability of the ^{137}Cs distribution in the Severn estuary. *Estuarine Coastal Shelf Sci.* 16:475–87

Reynolds, W. C., Cebeci, T. 1978. Calculation of turbulent flows. In *Turbulence, Top. Appl. Phys.*, ed. P. Bradshaw, 12:193–229. New York: Springer-Verlag

Smith, R. 1976. Longitudinal dispersion of a buoyant contaminant in a shallow channel. *J. Fluid Mech.* 78:677–88

Smith, R. 1977. Long-term dispersion of contaminants in small estuaries. *J. Fluid Mech.* 82:129–46

Smith, R. 1978a. Effect of salt upon hot-water dispersion in well-mixed estuaries. *Estuarine Coastal Mar. Sci.* 7:445–54

Smith, R. 1978b. Effect of salt upon hot-water dispersion in well-mixed estuaries. Part 2. Lateral dispersion. *Proc. Conf. Waste Heat Manage. Util., 2nd, Univ. Miami, Fla.*, ed. S. S. Lee, S. Sengupta, pp. 91–108

Smith, R. 1979. Calculation of shear-dispersion coefficients. See Harris 1979, pp. 343–62

Smith, R. 1980. Buoyancy effects upon longitudinal dispersion in wide well-mixed estuaries. *Philos. Trans. R. Soc. London Ser. A* 296:467–96

Smith, R. 1981a. A delay-diffusion description for contaminant dispersion. *J. Fluid Mech.* 105:469–86

Smith, R. 1981b. Effect of non-uniform currents and depth variations upon steady discharges in shallow water. *J. Fluid Mech.* 110:373–80

Smith, R. 1981c. The early stages of contaminant dispersion in shear flows. *J. Fluid Mech.* 111:107–22

Smith, R. 1982a. Contaminant dispersion in oscillatory flows. *J. Fluid Mech.* 114:379–98

Smith, R. 1982b. Where to put a steady discharge in a river. *J. Fluid Mech.* 115:1–11

Smith, R. 1982c. Non-uniform discharges of contaminants in shear flows. *J. Fluid Mech.* 120:71–89

Smith, R. 1983a. The contraction of contaminant distributions in reversing flows. *J. Fluid Mech.* 129:137–51

Smith, R. 1983b. The dependence of shoreline contaminant levels upon the siting of an effluent outfall. *J. Fluid Mech.* 130:153–64

Smith, R. 1983c. Longitudinal dispersion coefficients for varying channels. *J. Fluid Mech.* 130:299–314

Smith, R. 1983d. Effect of boundary absorption upon longitudinal dispersion in shear flows. *J. Fluid Mech.* 134:161–77

Smith, R. 1984. Temporal moments at large distances downstream of contaminant releases in rivers. *J. Fluid Mech.* 140:153–74

Smith, T. J. 1982. On the representation of Reynolds stress in estuaries and shallow coastal seas. *J. Phys. Oceanogr.* 12:914–21

Smith, T. J., Takhar, H. S. 1981. A mathematical model for partially mixed estuaries using the turbulence energy equation. *Estuarine Coastal Shelf Sci.* 13:27–45

Sullivan, P. J. 1971. Longitudinal dispersion within a two-dimensional turbulent shear flow. *J. Fluid Mech.* 49:551–76

Taylor, G. I. 1921. Diffusion by continuous movements. *Proc. London Math. Soc. Ser. A* 20:196–211

Taylor, G. I. 1953. Dispersion of soluble matter in solvent flowing slowly through a tube. *Proc. R. Soc. London Ser. A* 219:186–203

Taylor, G. I. 1954. The dispersion of matter in turbulent flow through a pipe. *Proc. R. Soc. London Ser. A* 223:446–68

Tennekes, H., Lumley, J. L. 1972. *A First Course in Turbulence.* Cambridge, Mass: MIT Press. 300 pp.

Tsai, Y. H., Holley, E. R. 1978. Temporal moments for longitudinal dispersion. *J. Hydraul. Div. Proc. ASCE* 104:1617–34

Uncles, R. J. 1982. Residual currents in the Severn Estuary and their effects on dispersion. *Oceanol. Acta* 5:403–10

Uncles, R. J., Bale, A. J., Howland, R. J. M., Morris, A. W., Elliott, R. C. A. 1983. Salinity of surface water in a partially-mixed estuary, and its dispersion at low run-off. *Oceanol. Acta* 6:289–96

Valentine, E. M., Wood, I. R. 1977. Longitudinal dispersion with dead zones. *J. Hydraul. Div. Proc. ASCE* 103:975–90

West, J. R., Broyd, T. W. 1981. Dispersion coefficients in estuaries. *Proc. Inst. Civ. Eng.* 71:721–37

Wilson, R. E., Okubo, A. 1978. Longitudinal dispersion in a partially mixed estuary. *J. Mar. Res.* 36:427–47

Yaglom, A. M. 1976. Diffusion of an impurity emanating from an instantaneous point source in a turbulent layer. *Fluid Mech.–Sov. Res.* 5:73–87

Yotsukura, N., Cobb, E. D. 1972. Transverse diffusion of solutes in natural streams. *US Geol. Surv. Prof. Pap. 582-C.* 19 pp.

Yotsukura, N., Sayre, W. W. 1976. Transverse mixing in natural channels. *Water Resour. Res.* 12:695–704

Ann. Rev. Fluid Mech. 1985. 17:151–89

AERODYNAMICS OF SPORTS BALLS

Rabindra D. Mehta[1]

Aerodynamics Research Branch, NASA Ames Research Center, Moffett Field, California 94035

1. INTRODUCTION

Aerodynamics plays a prominent role in almost every sport in which a ball is either struck or thrown through the air. The main interest is in the fact that the ball can be made to deviate from its initial straight path, resulting in a curved flight path. The actual flight path attained by the ball is, to some extent, under the control of the person striking or releasing it. It is particularly fascinating that not all the parameters that affect the flight of a ball are under human influence. Lateral deflection in flight (variously known as *swing, swerve,* or *curve*) is well recognized in cricket, baseball, golf, and tennis. In most of these sports, the swing is obtained by spinning the ball about an axis perpendicular to the line of flight, which gives rise to what is commonly known as the *Magnus effect.*

It was this very effect that first inspired scientists to comment on the flight of sports balls. Newton (1672), at the advanced age of 23, had noted how the flight of a tennis ball was affected by spin, and he gave this profound explanation: "For, a circular as well as a progressive motion ..., its parts on that side, where the motions conspire, must press and beat the contiguous air more violently than on the other, and there excite a reluctancy and reaction of the air proportionably greater." Some 70 years later, in 1742, Robins showed that a transverse aerodynamic force could be detected on a rotating sphere. However, Euler completely rejected this possibility in 1777 (see Barkla & Auchterlonie 1971). The association of this effect with the name of Magnus was due to Rayleigh (1877), who, in his paper on the irregular flight of a tennis ball, credited him with the first "true explanation" of the effect. Magnus had found that a rotating cylinder

---

[1] Present address: Department of Aeronautics and Astronautics, Joint Institute for Aeronautics and Acoustics, Stanford University, Stanford, California 94305.

151

0066–4189/85/0115–0151$02.00

moved sideways when mounted perpendicular to the airflow. Rayleigh also gave a simple analysis for a "frictionless fluid," which showed that the side force was proportional to the free-stream velocity and the rotational speed of the cylinder. Tait (1890, 1891, 1893) used these results to try to explain the forces on a golf ball in flight by observing the trajectory and time of flight. This was all before the introduction of the boundary-layer concept by Prandtl in 1904. Since then, the Magnus effect has been attributed to asymmetric boundary-layer separation. The effect of spin is to delay separation on the retreating side and to enhance it on the advancing side. Clearly, this would only occur at postcritical Reynolds numbers (Re = Ud/v, where U is the speed of the ball or the flowspeed in a wind tunnel, d is the ball diameter, and v is the air kinematic viscosity), when transition has occurred on both sides. A smooth sphere rotating slowly can experience a negative Magnus force at precritical Reynolds numbers, when transition occurs first on the advancing side.

Most of the scientific work on sports ball aerodynamics has been experimental in nature and has concentrated on three sports balls: the cricket ball, baseball, and golf ball. Details of these three balls, together with typical operating conditions, are given in Figure 1.

The main aim in cricket and baseball is to deliberately curve the ball through the air in order to deceive the batsman or batter. However, the tools and techniques employed in the two sports are somewhat different, which results in the application of slightly different aerodynamic principles. An interesting comparison of the two sports is given by Brancazio (1983). In golf, on the other hand, the main aim generally is to obtain the maximum distance in flight, which implies maximizing the lift-to-drag ratio. In this article, the more significant research performed on each of the three balls is reviewed in turn, with emphasis on experimental results as well as the techniques used to obtain them. While many research papers and articles were consulted in preparing this review, only those that have made relevant and significant contributions to the subject have been cited. For an overview of the physics of many ball games, see Daish (1972).

2.   CRICKET BALL AERODYNAMICS

2.1   Basic Principles

The actual construction of a cricket ball and the principle by which the faster bowlers swing the ball is somewhat unique to cricket. A cricket ball has six rows of prominent stitching, with typically 60–80 stitches in each row (primary seam). The stitches lie along the equator holding the two leather hemispheres together. The better quality cricket balls are in fact made out of four pieces of leather, so that each hemisphere has a line of internal stitching forming the "secondary seam." The two secondary seams,

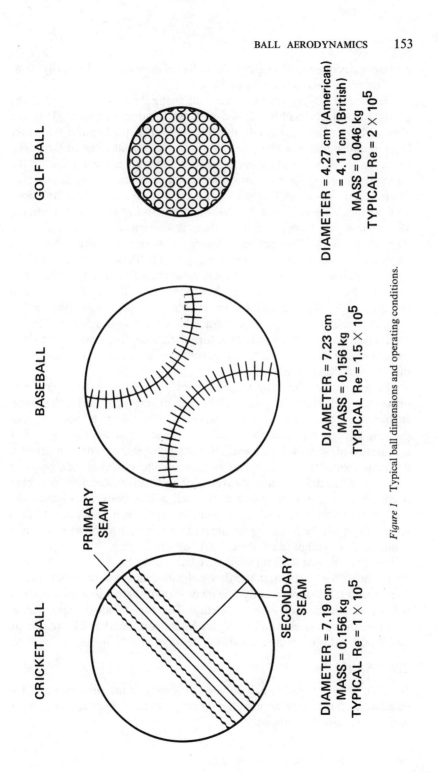

Figure 1 Typical ball dimensions and operating conditions.

primarily designed to strengthen the ball structure, are traditionally set at right angles to each other (Figure 1).

Fast bowlers in cricket make the ball swing by a judicious use of the primary seam. The ball is released with the seam at a small angle to the airflow. Under the right conditions, the seam trips the laminar boundary layer into turbulence on one side of the ball. This turbulent boundary layer, by virtue of its increased energy, separates relatively late compared to the boundary layer on the nonseam side, which separates in a laminar state. Figure 2 shows a cricket ball held stationary in a wind tunnel with the seam set at an incidence angle $\alpha = 40°$ to the airflow. Smoke was injected into the separated region behind the ball, where it was entrained right up to the separation points. The boundary layer on the lower surface has been tripped by the seam into turbulence, evidenced by the chaotic nature of the smoke edge just downstream of the separation point. On the upper surface a smooth, clean edge confirms that the separating boundary layer was in a laminar state. The laminar boundary layer on the upper surface has separated relatively early, at a point which makes an angle θ with the horizontal of about 90°, whereas the turbulent boundary layer on the lower surface separates at $\theta = 120°$. This asymmetry is further confirmed by the upward deflection of the wake flow.

The asymmetric boundary-layer separation results in an asymmetric pressure distribution that produces the side force responsible for the swing. In practice, some spin is also imparted to the ball, but about an axis perpendicular to the seam plane (i.e. along the seam) so that the asymmetry is maintained. A prominent seam obviously helps the transition process, whereas a smooth and polished surface on the nonseam side helps to maintain a laminar boundary layer. For this reason bowlers in cricket normally keep one hemisphere of the ball highly polished, whereas the other hemisphere is allowed to roughen, during the course of play. While it is legal to polish the ball using natural substances such as sweat or saliva, it is not legal to scuff or mark the cricket ball deliberately.

The basic principles behind cricket ball swing have been understood by scientists for years, and the first published discussion is that due to Lyttleton (1957). More recently, Mehta & Wood (1980) discussed the whole subject of cricket ball swing in detail. The two detailed experimental investigations on cricket ball swing are by Barton (1982) and Bentley et al. (1982); the latter is covered briefly in Mehta et al. (1983).

2.2 Static Tests

In static tests the cricket ball is held stationary in an airstream and the continuous forces measured, either directly or through integration of the surface pressure distributions.

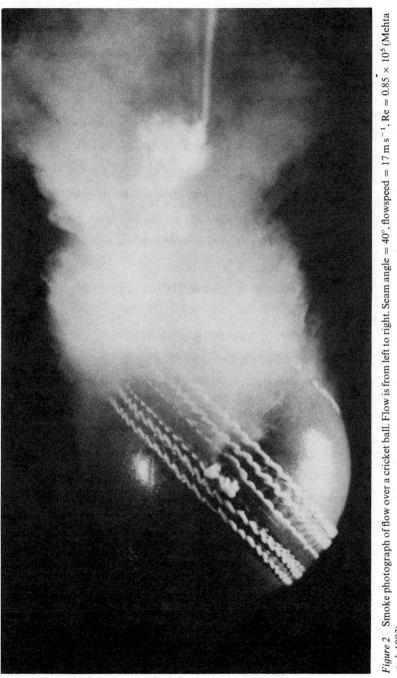

Figure 2 Smoke photograph of flow over a cricket ball. Flow is from left to right. Seam angle $= 40°$, flowspeed $= 17 \text{ m s}^{-1}$, $\text{Re} = 0.85 \times 10^5$ (Mehta et al. 1983).

Detailed pressure distributions along the equator of a ball supplied with 24 pressure tappings were measured by Bentley et al. (1982). The pressure tappings were installed in the horizontal plane, perpendicular to the seam plane. Figure 3 shows the measured pressures on the ball mounted in a wind tunnel at $\alpha = 20°$. The measurements on the seam side of the ball are represented by those shown on the right-hand side in Figure 3. At low values of Re or U, the pressure distributions on the two hemispheres are equal and symmetrical, so there would be no side force. At $U = 25$ m s^{-1}, the pressure dip on the right-hand face of the ball is clearly lower than that on the left-hand face. This would result in the ball swinging toward the right. The maximum pressure difference between the two sides occurs at $U = 29$ m s^{-1}, when presumably the boundary layer on the seam side is fully turbulent while that on the nonseam side is still laminar. Even at the highest flowspeed achieved in this test ($U = 36.5$ m s^{-1}), the asymmetry in pressure distributions is still clearly indicated, although the pressure difference is reduced. The actual (critical) flowspeeds at which the asymmetry appears or disappears were found to be a function of the seam angle, surface roughness, and free-stream turbulence—in practice it also depends on the spin rate of the ball. In general, though, a roughness height of the same order as the boundary-layer thickness is required for the successful transition of a

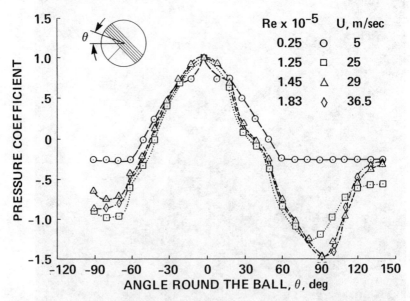

Figure 3   Pressure distributions on a cricket ball held at an incidence angle of 20° (Mehta et al. 1983).

laminar boundary layer into turbulence. Typically, the primary seam height is about 1 mm, and the laminar boundary-layer thickness at, say, $\theta = 20°$ is less than 0.5 mm. In terms of Reynolds number, Graham (1969) suggested that the value of Re based on roughness height should be at least 900. For a cricket ball seam, this corresponds to $U \sim 14$ m s^{-1}. Note that the magnitude of the suction peaks in Figure 3 (~ -1.5) often exceeds that given by potential-flow theory for spheres (-1.25). Bentley et al. (1982) attributed this to the higher induced velocities due to "tip vortices" produced on lifting bodies of finite span.

Barton (1982) measured side forces on cricket balls by using a compound pendulum system where the ball was allowed to swing transversely. The side force can be evaluated once the transverse deflection is measured. Figure 4 shows some of Barton's results obtained using this technique. Except for the ball that had been used in a cricket match for 40 overs (240

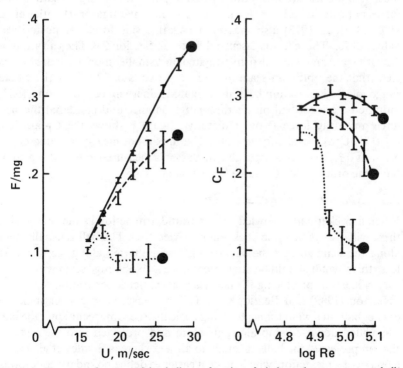

Figure 4 Transverse force on cricket balls as a function of wind speed: ————, new ball; ————, 10-over ball; ...... 40-over ball. Bars indicate fluctuations. The seams were fixed at about 30° to the airflow. The transverse forces became intermittent at the points marked ●, and they dropped quickly to zero at higher speeds (Barton 1982).

deliveries), the normalized side force (F/mg, where F is the side force and m is the mass of the ball) is seen to increase with flowspeed up to about 30 m s^{-1}. A nonspinning cricket ball can experience a side force equivalent to almost 40% of its own weight. The side-force coefficient [$C_F = F/(\frac{1}{2}\rho U^2 S)$, where ρ is the air density and S is the ball projected area] also varies with Re and has a maximum value of about 0.3. In a similar setup to Barton's, Hunt (1982) and Ward (1983) made measurements on a large-scale model of a cricket ball. Some typical averaged results for $\alpha = 0°$, but with one side roughened and the other side smooth, are shown in Figure 5. The main advantage here was in being able to investigate the behavior at high Reynolds numbers. The critical value at which C_F starts to decrease is about Re $= 1.5 \times 10^5$, which corresponds to a flowspeed of about 30 m s^{-1}. The side force starts to decrease when "natural" transition occurs on the nonseam side, which leads to a reduction in the flow asymmetry. It should be noted that fast bowlers in cricket achieve bowling speeds of up to 40 m s^{-1} (Re $= 1.9 \times 10^5$). The most striking feature in Figure 5 is the appearance of large negative side forces at postcritical Reynolds numbers. Other investigators (Bentley et al. 1982, Horlock 1973) also measured negative side forces at postcritical values of Re. This effect is discussed below in Section 2.4. The variation of C_F with seam angle was also investigated on a smaller model (Figure 6). It is clear that the optimum seam angle is approximately 20° over the whole range of Reynolds numbers investigated. Sherwin & Sproston (1982) mounted a cricket ball on a strain-gauge balance and measured the side force and drag force D on it directly. Figure 7 shows the C_F and C_D [$= D/(\frac{1}{2}\rho U^2 S)$] results for $\alpha = 30°$. The maximum measured value of C_F is about 0.3, which is comparable to Barton's measurements and is also of the same order as the C_D value for the ball (~ 0.45).

2.3   *Spinning Cricket Ball Tests*

When a cricket ball is bowled, with a round arm action as the rules insist, there will always be some backspin imparted to it. The ball is usually held along the seam so that the backspin is also imparted along the seam. At least this is what should be attempted, since a "wobbling" seam will not be very efficient at producing the necessary asymmetric separation.

Barton (1982) and Bentley et al. (1982) measured forces on spinning cricket balls in a wind tunnel by using basically the same technique; the ball was rolled down a ramp along its seam and projected into the airflow. Barton projected the balls through an air jet, whereas Bentley et al. used a closed working section, which made it easier to define boundary conditions for the computation of the forces. The spin rate ω was varied by changing the starting point along the track, and the seam angle was varied by adjusting the alignment of the ramp with the airflow. Once the conditions at the entry to the wind tunnel and the deflection from the datum are known, the forces due to the airflow can be easily evaluated. Barton calculated the spin rate at the end of the ramp, whereas Bentley et al. measured it by photographing the ball path using a stroboscope. They found that the energy equation used by Barton may have slightly overestimated the spin

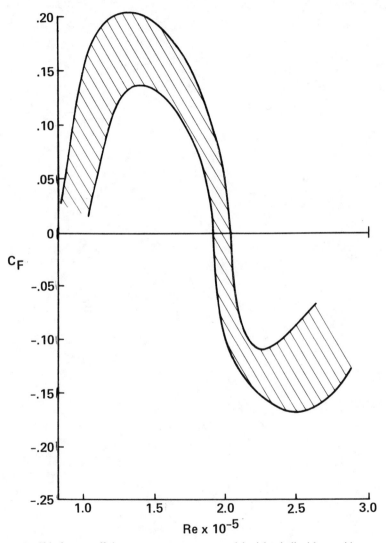

Figure 5   Side-force coefficient measurements on a model cricket ball with one side smooth and the other roughened. Seam angle = 0°; results averaged over eight runs (Hunt 1982).

rate because of the exclusion of the "friction-loss" terms. No account was taken in either investigation of the Magnus force experienced by the spinning balls, although Bentley et al. included a brief discussion on how the measurements may have been affected. The overall accuracy of Bentley et al.'s measurement technique was verified by comparing the C_D results with existing data on spheres.

Barton (1982) found considerable scatter in his measurements, which he

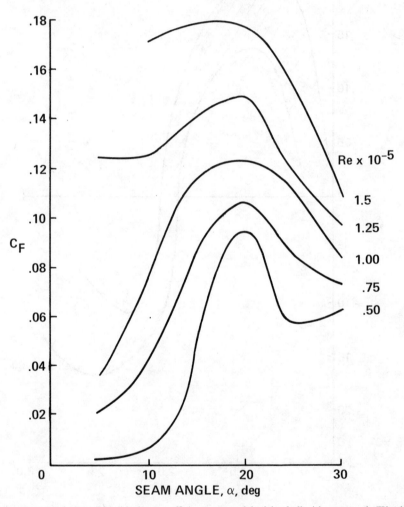

Figure 6   Variation of the side-force coefficient on a model cricket ball with seam angle (Ward 1983).

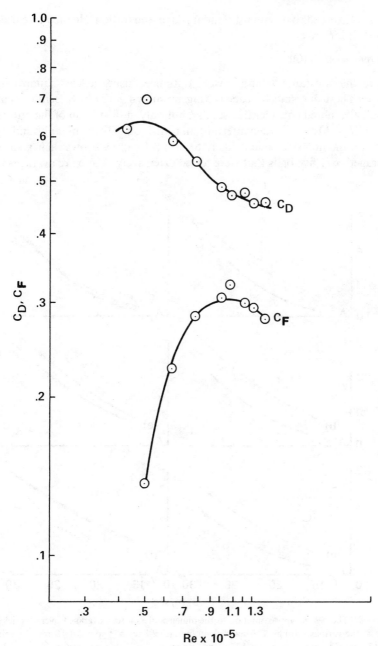

Figure 7 Variation of the side force and drag coefficients on a cricket ball with Reynolds number. Seam angle = 30° (Sherwin & Sproston 1982).

partly attributed to irregular boundary-layer separation. He therefore used a best-fit formula:

$$F/mg = A(U/10)^\gamma,$$

where the constants A and γ were optimized using a least-squares fit. Barton's best-fit results for each setting are shown in Figure 8. Bentley et al. (1982) also noted considerable scatter, but only in the region of the critical speed ($U \sim 30\,\mathrm{m\,s^{-1}}$) when intermittent transition of the laminar boundary layer on the nonseam side is likely to occur. Figure 9 shows their results, averaged over five balls that were tested extensively. These are averages of

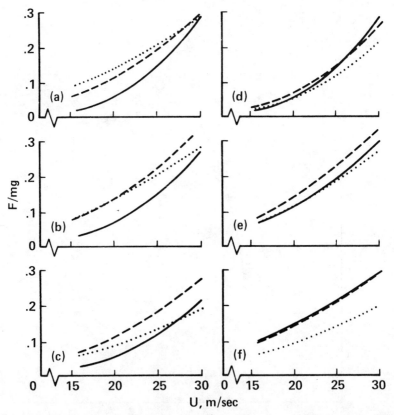

Figure 8 The best-fit approximation to the dimensionless forces, averaged over all cricket balls at the various settings. Left-hand diagrams: (a) 2.1 rev s^{-1} spin, (b) 4.9 rev s^{-1} spin, (c) 9.3 rev s^{-1} spin; ————, 10° seam angle; ————, 20° seam angle; ......, 30° seam angle. Right-hand diagrams: (d) 10° seam angle, (e) 20° seam angle, (f) 30° seam angle; ———— 2.1 rev s^{-1} spin; ————, 4.9 rev s^{-1} spin; ......, 9.3 rev s^{-1} spin (Barton 1982).

the actual measurements and are not an averaged best-fit. Side forces equivalent to about 30% of the balls's weight are experienced by spinning cricket balls in flight, slightly lower than in the static tests. There is also some dependence of the side force on spin rate. As with the static tests, the critical flowspeed at which the side force starts to decrease in these tests is about 30 m s^{-1}. Therefore, the main results from the two types of testing techniques are comparable.

The actual trajectory of a cricket ball can be computed using the measured forces. Figure 10 shows the computed trajectories at five bowling speeds for the ball exhibiting the best swing properties ($F/mg \sim 0.4$ at $U = 32$ m s^{-1}, $\alpha = 20°$, $\omega = 14$ rev s^{-1}). The results illustrate that the flight path is almost independent of speed in the range $24 < U < 32$ m s^{-1}. The trajectories were computed using a simple relation, which assumes that the side force is constant and acts perpendicular to the initial trajectory. This gives a lateral deflection that is proportional to time squared and hence a parabolic flight path. By using a more accurate step-by-step model, in

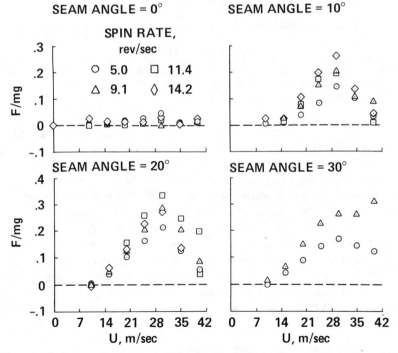

Figure 9   Variation with flowspeed of the normalized side force, averaged over five cricket balls (Mehta et al. 1983).

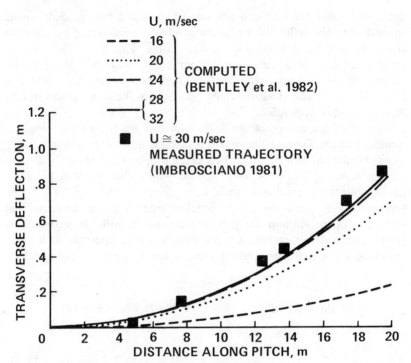

Figure 10 Computed flight paths using measured forces for the cricket ball with the best swing properties. Seam angle = 20°, spin rate = 14 rev s^{-1} (Bentley et al. 1982).

which the side force acted perpendicular to the instantaneous flight direction, R. D. Mehta & T. T. Lim (unpublished results) found that the final deflection was only reduced by about 6%. This calculation also included a semiempirical relation that modeled the change in α due to the movement of the stagnation point as the ball follows a curved flight path. In some photographic studies of a swing bowler, it was confirmed that the trajectories were indeed parabolic (Imbrosciano 1981). These studies also showed that the final predicted deflections of over 0.8 m were not unreasonable. One of the photographed sequences was later analyzed (R. D. Mehta & T. T. Lim, unpublished results), and the actual flight path for this sequence is also plotted in Figure 10. The agreement is excellent considering the crudity of the experimental and analytical techniques.

2.4 Optimum Conditions for Swing

It is often suggested that the actual construction of a cricket ball is important in determining the amount of swing obtained. Since some balls are stitched by hand, these differences can be significant. A two-piece ball is in general found to have better swing properties than a four-piece ball, where the secondary seam produces an effective roughness that helps to cause transition of the laminar boundary layer on the nonseam side. Barton (1982) concluded that a ball with a more pronounced primary seam than average (>1 mm) swung more. However, Bentley et al. (1982) investigated the seam structure on a variety of balls, and while small differences in size and shape were noted, these could not be correlated with the amount of swing. They concluded that, on the whole, the seam on all new balls is efficient at tripping the boundary layer in the speed range $15 < U < 30$ m s^{-1}. It was also apparent that for good swing properties it is perhaps more important to have a perfectly smooth surface on the nonseam side. Both these investigations confirmed that the swing properties of a ball deteriorate with age as the seam is worn and the surface scarred. (In cricket a ball is only replaced by a new one after about 450 deliveries.) There are two other parameters that the bowler can control to some extent: the ball seam angle and the spin rate.

The optimum seam angle for $U \sim 30$ m s^{-1} is found to be about $20°$ (Figures 6, 8, and 9). At lower speeds (especially for $U < 15$ m s^{-1}) a bowler should select a larger seam angle ($\alpha \sim 30°$), so that by the time the flow accelerates around to the seam, the critical speed has been reached. It is better not to trip the boundary layer too early (low α), since a turbulent boundary layer grows at a faster rate and will therefore separate relatively early (compared with a later tripping). At the same time, the seam angle should not be so large that the boundary layer separates before reaching the seam, since this would result in symmetrical separation on the ball and hence zero side force. In a case like this, if transition occurs in the boundary layer upstream of the seam, then the effect of the seam will be to act as a boundary-layer "fence" that thickens the boundary layer even further. This asymmetry would lead to a negative side force such as that shown in Figure 5 for postcritical Reynolds numbers. This effect can be produced even at low seam angles by inducing early transition of the laminar boundary layer through an increase in the free-stream turbulence. This was confirmed in some experiments performed by Bentley et al. (1982), where significant negative side forces ($F/mg \sim -0.15$) were measured when free-stream turbulence of the same scale as the ball radius was introduced in the wind tunnel.

Spin on the ball helps to stabilize the seam orientation. Basically, for stability, the angular momentum associated with the spin should be greater than that caused by the torque about the vertical axis due to the flow asymmetry. The combination of these two moments can lead to what is commonly known as gyroscopic precession—a moment about the third (horizontal) axis in the plane of the seam that attempts to tumble the ball over. Although this effect was sometimes observed (Barton 1982, Bentley et al. 1982), it proved rather difficult to correlate with seam angle or spin rate. Too much spin is of course also detrimental, since the effective roughness on the ball's surface is increased (i.e. the critical Reynolds number is reached sooner). This would obviously be more relevant at the higher speeds ($U \sim 25$ m s^{-1}). Barton's (1982) results (Figure 8) seem to indicate that the optimum spin rate is about 5 rev s^{-1}, whereas Bentley et al.'s (1982) results (Figure 9) suggest a much higher optimum spin rate of about 11 rev s^{-1}. The maximum spin rate investigated by Barton was 9.3 rev s^{-1}. While some of this discrepancy may be due to the relatively high turbulence level ($\sim 1\%$) in Barton's wind tunnel (compared with 0.2% in Bentley et al.'s tunnel), the level of the discrepancy is still somewhat surprising. In practice, a bowler can impart spin of up to 14 rev s^{-1}, but it should be noted that this is not an easy parameter to control.

2.5   Effect of Meteorological Conditions

The effect of weather conditions is by far the most discussed and controversial topic in cricket. It is widely believed that humid or damp days are conductive to swing bowling, but there has been no scientific proof of this.

The flow pattern around a cricket ball depends only on the properties of the air and the ball itself. The only properties of air that may conceivably be influenced by a change of meteorological conditions are the viscosity or density. Such changes would then affect the ball Reynolds number. However, Bentley et al. (1982) showed that the average changes in temperature and pressure encountered in a whole day would not change the air density and viscosity, and hence the ball Reynolds number, by more than about 2%.

Several investigators (Barton 1982, Horlock 1973, Sherwin & Sproston 1982) have confirmed that no change was observed in the measured pressures or forces when the relative humidity changed by up to 40%.

It has been suggested (Sherwin & Sproston 1982) that humid days are perhaps associated with general calmness in the air and thus less atmospheric turbulence. However, there is no real evidence for this, and even if it were the case, the turbulence scales would be too large to have any significant effect on the flow regime over the ball. Binnie (1976) suggested that the observed increase in swing under conditions of high humidity is caused by "condensation shock" near the point of minimum pressure. This shock would then assist the seam in tripping the laminar boundary layer. However, his calculations showed that this effect could only occur when the relative humidity was nearly 100%. Also, as discussed previously, the seam (on new balls at least) is adequate in tripping the boundary layer in the Reynolds-number range of interest.

The only investigation where the effect of humidity on swing was studied directly is that due to Bentley et al. (1982). They measured surface contours and side forces both on balls that were exposed to high levels of humidity and on balls that were wet completely. As shown in Figures 11a and b, no significant effect was noted. Photographic tests also showed no discernible effect on the seam or ball surface. Bentley et al. hypothesized that the spin rate imparted to a cricket ball may be affected by damp conditions, since the surface on a new ball becomes tacky and thus allows the bowler a better grip. However, there is no real evidence for this, and so this aspect of cricket ball aerodynamics still remains a mystery.

3. BASEBALL AERODYNAMICS

3.1 *Basic Principles*

Although a baseball has virtually the same size and weight as a cricket ball, there are major differences in the cover design that affect the aerodynamics. The cover of a baseball consists of two hourglass-shaped segments of white leather seamed together by a single row of about 216 stitches (Figure 1). Two basic aerodynamic principles are used to make a baseball curve in flight: spin, and asymmetric boundary-layer tripping due to seam position.

First, consider a curveball. The ball is released so that it acquires top spin about the horizontal axis. The spin produces the Magnus force that makes it curve downward, faster than it would under the action of gravity alone. In this case, the seam produces an overall roughness that helps to reduce the

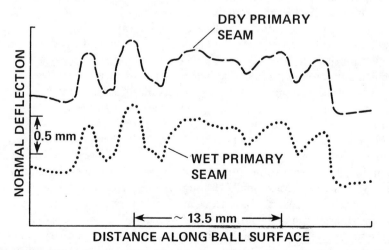

Figure 11a Talysurf contour plots of the primary seam on a cricket ball to investigate effects of humidity (Mehta et al. 1983).

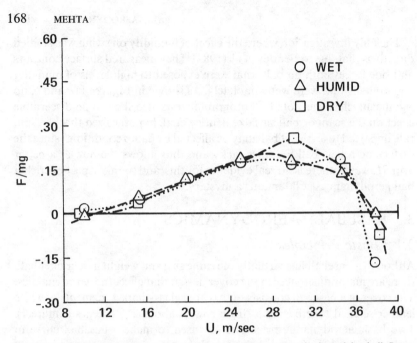

Figure 11b   Effect of humidity on the measured side forces on a spinning cricket ball. Seam angle = 20°, spin rate = 5 rev s^{-1} (Bentley et al. 1982).

critical Reynolds number. The pitcher, by varying the angles of his arm and wrist, generates spin about different axes, which produces different rates and directions of curvature. Spin rates of up to 30 rev s^{-1} and speeds of up to 45 m s^{-1} are achieved by pitchers in baseball.

Figure 12 shows a flow-visualization photograph (using smoke) of a baseball held stationary, but spinning counterclockwise at 15 rev s^{-1}, in a wind tunnel with $U = 21$ m s^{-1}. The "crowding together" of smoke filaments over the bottom of the ball shows an increased velocity in this region and a corresponding decrease in pressure. This would tend to deflect the ball downward. Note also the upward deflection of the wake. Thus, a postcritical flow is obtained at these operating conditions.

Second, consider an ideal knuckleball. The ball is released so that it has no spin at all. Then, depending on the position of the seam, asymmetric boundary-layer separation and hence swing can be obtained, much in the same way as with a cricket ball. Although banned over 60 years ago, spit or its modern counterpart, Vaseline, is still sometimes used so that the ball may be squirted out of the fingers at high speed. In Figure 13, the ball is not spinning, but it is so oriented that the two seams trip the boundary layer on the upper side of the ball. The boundary layer on the lower surface is seen to

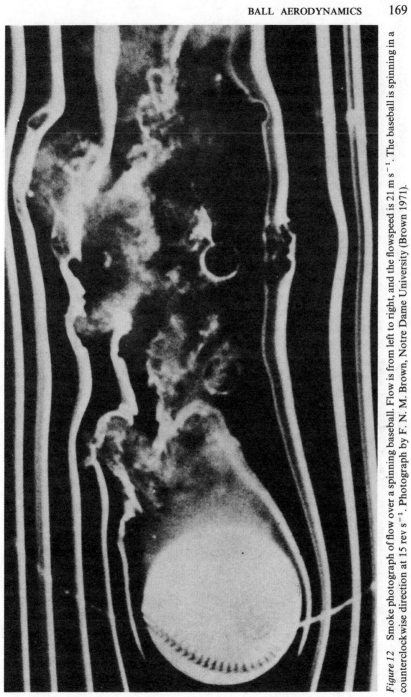

Figure 12   Smoke photograph of flow over a spinning baseball. Flow is from left to right, and the flowspeed is 21 m s^{-1}. The baseball is spinning in a counterclockwise direction at 15 rev s^{-1}. Photograph by F. N. M. Brown, Notre Dame University (Brown 1971).

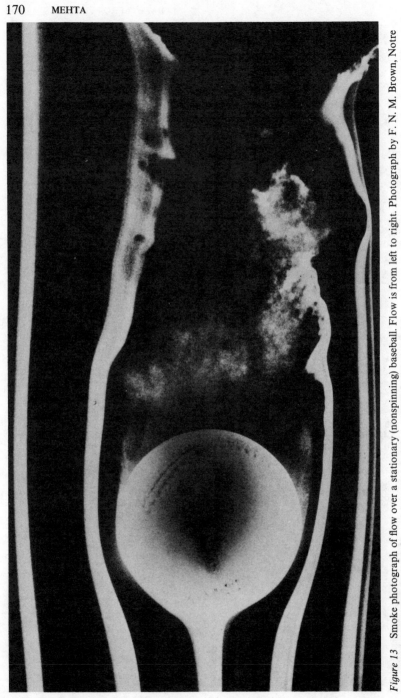

Figure 13   Smoke photograph of flow over a stationary (nonspinning) baseball. Flow is from left to right. Photograph by F. N. M. Brown, Notre Dame University (Brown 1971).

separate relatively early in a laminar state. Once again, the downward deflection of the wake confirms the presence of the asymmetric boundary-layer separation, which would deflect the ball upward.

3.2   Spinning Baseball Tests (Curveball)

The effects of spin and speed on the lateral deflection of a baseball were investigated in detail by Briggs (1959). In his first set of experiments, he fired spinning baseballs from an airgun and measured lateral deflections, but this technique did not prove very successful.

In Briggs' (1959) second set of measurements, spinning baseballs were dropped through a 1.8-m wind tunnel. The spinning mechanism was mounted on top of the tunnel and consisted of a suction cup mounted on the shaft that supported the ball. The spinning ball was released by a quick-acting valve that cut off the suction. The lateral deflection was taken as one half of the measured spread of the two points of impact, with the ball spinning first clockwise and then counterclockwise. The spin rate of the spinning mechanism was measured using a stroboscope. The mechanism was aligned so that a baseball spinning about a vertical axis was dropped through a horizontal airstream. Hence, a lateral deflection, perpendicular to the effects due to the drag force and gravity, was produced as a result of the effects of the spin. And since the initial velocity for each run was the same, the deflection gave a direct representation of the lateral force.

In Figure 14, Briggs' measured lateral deflections are illustrated. The straight lines through the origin show that within the limits of experimental accuracy, the lateral deflection is proportional to spin rate. However, Briggs' extrapolation to zero deflection at zero spin is not accurate, since a nonspinning baseball can also develop a lateral force. Hence, the behavior of the lateral force at low spin rates must be nonlinear. In Figure 15, the ratio of deflections is plotted against the ratio of flowspeed, for a given spin rate. Briggs concluded, erroneously, that the implication of the linear relation was that the lateral deflection was proportional to the square of the flowspeed. The graph actually confirms that C_F is independent of Re, which implies that the final deflection is independent of speed. But, once again, this linear relationship is not likely to hold as $U \to 0$ and Reynolds-number effects start to become important. So to summarize Briggs' (1959) results, the lateral deflection of a baseball is directly proportional to spin rate and is independent of the speed for $20 < U < 40$ m s^{-1} and $20 < \omega < 30$ rev s^{-1}.

Briggs (1959) also evaluated the final deflections that would be obtained in practice over the distance of 18.3 m (60 ft). He used a simple model, which assumed that the side force was constant and the distance traveled proportional to the square of the elapsed time, thus giving a parabolic flight path, as for the cricket ball. For $U = 23$ m s^{-1} at $\omega = 20$ rev s^{-1}, a lateral

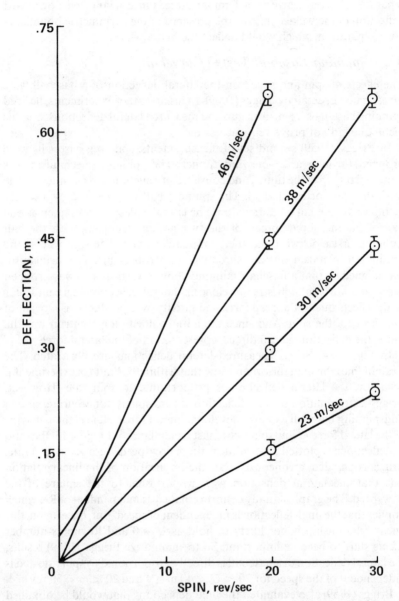

Figure 14 Lateral deflection of a baseball, spinning about a vertical axis, when dropped across a horizontal windstream. These values are all for the same time interval, 0.6 s, the time required for the ball to cross the stream (Briggs 1959).

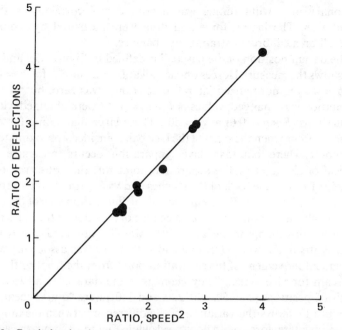

Figure 15 Graph showing that the lateral deflection of a spinning baseball is independent of the flowspeed (Briggs 1959).

deflection of about 0.28 m was obtained, whereas at $\omega = 30$ rev s^{-1}, it was about 0.43 m (corresponding to $F/mg \sim 0.2$). In the case of a curveball, the gravitational force would add to the force due to spin, and so the final deflection would be much greater.

In experiments described by Allman (1982), the actual flight paths of curveballs pitched by a professional were photographed using the strobo-scopic technique. Analysis of the flight paths confirmed that the ball travels in a smooth parabolic arc from the pitcher's hand to the catcher, and deflections of over a meter were observed. Therefore, the assumption of a constant side force for a spinning baseball seems to be valid. Since most of the deflection takes place in the second half of the flight, a batter often gets the impression that the ball "breaks" suddenly as it approaches home plate (a slider). However, changes in magnitude and direction of the side force on a ball in flight are not unknown in baseball. This effect is now discussed with reference to knuckleballs.

3.3 Nonspinning Baseball Tests (Knuckleball)

Watts & Sawyer (1975) investigated the nature of the forces causing the "erratic motions" of a knuckleball. The lateral force on a baseball held stationary in a wind tunnel was measured and correlated with seam orientation. The lateral force and drag were measured by mounting a baseball on a calibrated strain-gauge balance.

The datum position of the baseball is defined in Figure 16a, and Figure 16b shows the measured forces on the ball at $U = 21$ m s^{-1}, for $\phi = 0$–360°. At $\phi = 0°$, the normalized lateral force (F/mg) was zero, but as the ball orientation was changed, values of $F/mg = \pm 0.3$ were obtained with large fluctuations ($F/mg \sim 0.6$) at $\phi = 50°$. These large fluctuating forces were found to occur when the seam of the baseball coincided approximately with the point where boundary-layer separation occurs, an angle to the vertical of about 110°. The separation point was then observed to jump from the front to the back of the stitches and vice versa, thereby producing an unsteady flow field. The frequency of the fluctuation was of the order of 1 Hz, a value low enough to cause a change of direction in the ball's flight. A discontinuous change in the lateral force was also observed at $\phi = 140$ and 220°. Watts & Sawyer (1975) concluded that this was associated with the permanent movement of the separation point from the front to the rear of the seam (or vice versa). They claim that the data near all four of the "critical" positions ($\phi = 52, 140, 220$, and 310°) were "quite repeatable."

Figure 17 shows the variation of the difference between maximum and minimum lateral forces with flowspeed, while Figure 18 shows the variation

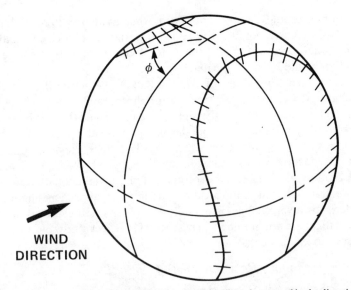

Figure 16a   Datum position of baseball at $\phi = 0°$. The ball can be rotated in the direction ϕ to a new position (Watts & Sawyer 1975).

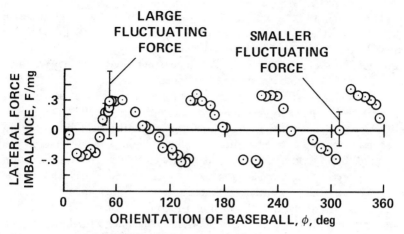

Figure 16b   The variation of the lateral force imbalance with orientation of the baseball—see Figure 16a for definition of ϕ (Watts & Sawyer 1975).

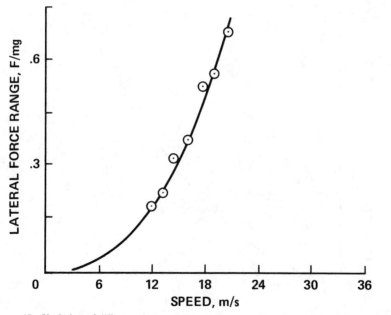

Figure 17   Variation of difference between the maximum and minimum lateral forces on a baseball (Watts & Sawyer 1975).

of the fluctuating force near $\phi = 52$ and $310°$. Since the magnitudes of the forces increase approximately as U^2, the lateral deflection for the knuckleball would also be independent of flowspeed within the range considered ($12 < U < 20$ m s^{-1}). The fluctuating force frequency was also found to be almost independent of flowspeed.

Watts & Sawyer (1975) computed some trajectories using their measured forces and the same simple assumption as Bentley et al. (1982) and Briggs (1959)—namely, that the lateral force on the baseball acts in a direction perpendicular to the original direction of level flight. The lateral force was assumed to be periodic in time, which implied that the lateral deflection would decrease with increasing spin. Some computed trajectories for the cases when the ball was initially oriented at $\phi = 90°$ and spin imparted such that the ball rotated a quarter- or a half-revolution during its flight to the home plate are illustrated in Figure 19. Clearly, the pitch with the lower spin rate has the maximum deflection and change of direction and would therefore be the more difficult one to hit. For the erratic flight of a knuckleball, Watts & Sawyer (1975) suggest that this could happen when

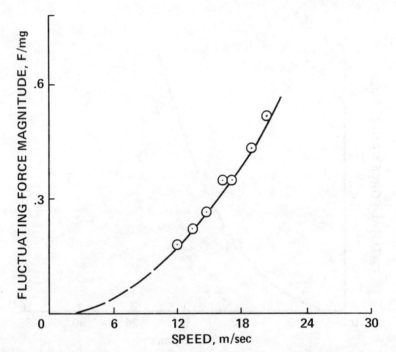

Figure 18   Variation of the magnitude of the fluctuating force on a baseball (Watts & Sawyer 1975).

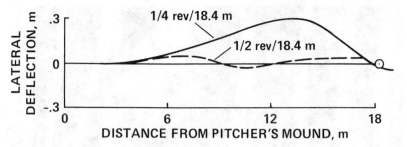

Figure 19   Typical computed trajectories for a slowly spinning baseball, with $U = 21$ m s^{-1} (Watts & Sawyer 1975).

the ball is so released that the seam lies close to the separation point. However, as Weaver (1976) rightly points out, a baseball thrown with zero or near-zero spin will experience a torque due to the flow asymmetry that will cause the ball to rotate. This is similar to the stability problem discussed in Section 2.4 for the cricket ball. It therefore seems that in practice it would be difficult to pitch a baseball that maintains, for the whole flight, an attitude where the erratic lateral force occurs. Thus, sudden changes in flight path are probably not as common as baseball players claim. And indeed, on studying the actual flight paths of professionally pitched knuckleballs (Allman 1983), it was found that while the direction of lateral deflection was unpredictable, there were no sudden, erratic changes in flight path.

4.   GOLF BALL AERODYNAMICS

4.1   *Basic Principles*

Golfing legend has it that in about the mid-nineteenth century a professor at Saint Andrews University discovered that a gutta-percha ball flew farther when its surface was scored (Chase 1981). This discovery soon became common knowledge, and it sparked off numerous cover designs, chosen more or less by intuition. Balls with raised patterns were not very successful, since they tended to collect mud and presumably also experienced a larger drag force. Covers with square and rectangular depressions were also tried, but by about 1930, the ball with round dimples had become accepted as the standard design. A conventional golf ball has either 330 (British ball) or 336 (American ball) round dimples arranged in regular rows (Figure 1).

In golf ball aerodynamics, apart from the lift force (which is generated by the backspin imparted to the ball), the drag and gravitational forces are also important, since the main objective is to obtain maximum distance in flight.

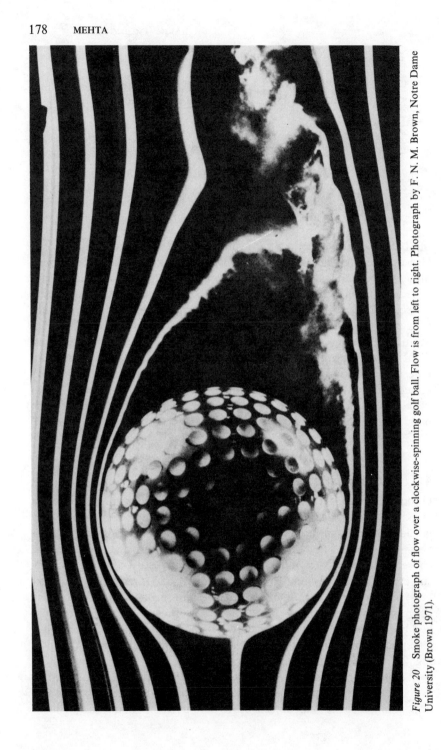

Figure 20 Smoke photograph of flow over a clockwise-spinning golf ball. Flow is from left to right. Photograph by F. N. M. Brown, Notre Dame University (Brown 1971).

At postcritical Reynolds numbers, the effect of spin is to delay separation on the upper (retreating) side and to advance it on the lower (advancing) side. In the smoke photograph of a clockwise spinning golf ball (Figure 20), the asymmetric separation and the downward-deflected wake are clearly illustrated, thus indicating a normal (upward) force on the ball. The effect of the dimples is to lower the critical Reynolds number. Figure 21 illustrates how dimples are more effective than, for example, sand-grain roughness at reducing the critical Reynolds number. The rapid rise in C_D for the sphere with sand-grain roughness is due to the forward movement of the transition point and the artificial thickening of the boundary layer by the roughness elements. The golf ball C_D value does not rise in this way, which indicates that dimples are effective at tripping the boundary layer without causing the thickening associated with positive roughness. The lift and drag coefficients are functions of Reynolds number and a spin parameter—the one normally used is V/U, where V is the equatorial speed of the ball. Unlike cricket and baseball, in golf both Re and V/U will change significantly during a typical flight, which means that C_L and C_D must also be expected to vary.

Typically, a golf ball is driven at an initial velocity of 75 m s^{-1} with the ball spinning backward at 3500 rpm, or about 60 rev s^{-1}. (Since the spin rates in golf are relatively high, they are conventionally quoted in rpm.) This gives a spin parameter of about 0.1 and a Reynolds number based on ball diameter (41.1 mm) of about 2.1×10^5.

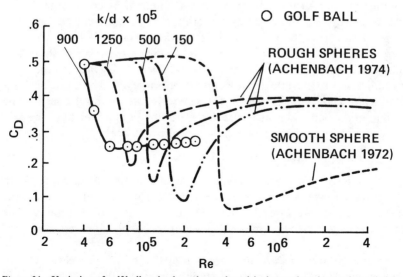

Figure 21 Variation of golf ball and sphere drag, where k is the sand-grain roughness height and d is the ball diameter (Bearman & Harvey 1976).

4.2   *Measurements of Forces on Golf Balls*

Davies (1949) measured the aerodynamic forces on golf balls by dropping spinning balls through a wind-tunnel working section. The ball was held between two shallow cups that were rotated with a variable speed motor (equipped with a tachometer) about a horizontal axis perpendicular to the airstream. A trigger released the springs on the cups, and the ball was allowed to drop through the airstream. The landing spot was marked on waxed paper, and by spinning the ball in one direction and then the other, the drag and lift forces could be evaluated. Davies claimed that the overall error in the measured forces was less than 10%. Spin rates of up to 8000 rpm were investigated at a flowspeed of about 32 m s^{-1}. This gave a Reynolds number based on ball diameter (42.7 mm) of about 9.4×10^4, somewhat lower than that attained by a ball leaving the tee ($\sim 2.1 \times 10^5$).

Bearman & Harvey (1976) measured the aerodynamic forces on model balls over a wide range of Reynolds numbers (0.4×10^5–2.4×10^5) and spin rates (0–6000 rpm). The two and one-half times full-scale models were constructed as hollow shells. A motor and bearing assembly on which the ball revolved was installed within the model. The model was supported from a wire, 0.5 mm in diameter ($\equiv 0.5\%$ of the model diameter). The upper support wire was attached to a strain-gauged arm, which in turn was mounted on a three-component force balance. The spin rate of the ball was measured using a stroboscope. Bearman & Harvey confirmed that the interference due to the support wires was minimal by comparing the C_D measurements on a nonrotating smooth sphere with previous data.

The results from both of these investigations (Davies 1949, Bearman & Harvey 1976) are shown in Figure 22 for a Reynolds number of about 10^5. The variation of C_L and C_D with V/U obtained by Davies has the same overall trends as the data due to Bearman & Harvey. At a spin rate of about 8000 rpm, Davies measured lift forces equivalent to about one half the weight of the ball and drag forces that were equal to the weight of the ball. He proposed a semiempirical relation for the lift force:

$$L = 0.029 \, [1 - \exp(-0.00026N)],$$

where N is the spin rate in revolutions per minute and L is the lift in kilograms. However, as Bearman & Harvey (1976) point out, this nonlinear behavior most likely results from Davies' low Reynolds number. The measurements due to Bearman & Harvey are probably more representative and accurate.

Figures 23 and 24 show the variation of C_L and C_D, respectively, with varying spin. On the whole, C_L is found to increase with spin, as one would expect, and C_D also increases as a result of induced drag effects associated

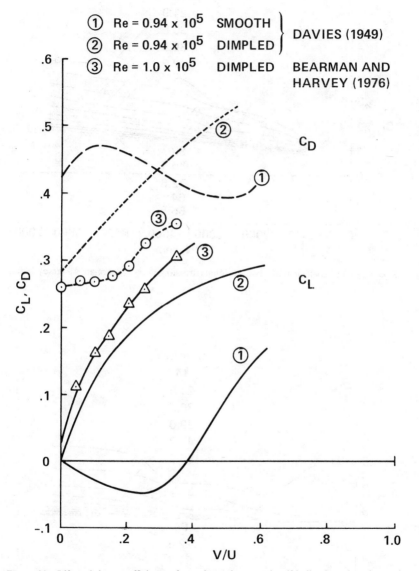

Figure 22 Lift and drag coefficients of rotating spheres and golf balls plotted against spin parameter (Bearman & Harvey 1976).

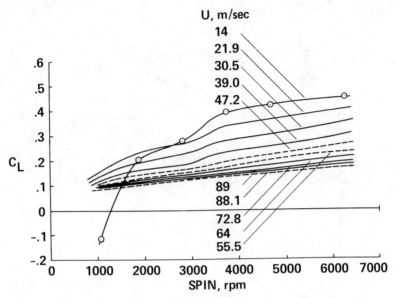

Figure 23   Lift coefficient of a conventional (British) golf ball (Bearman & Harvey 1976).

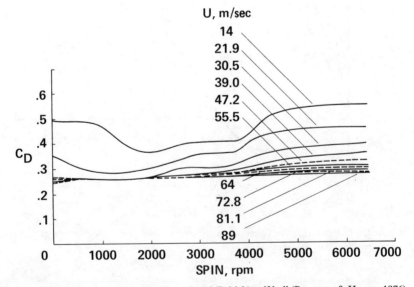

Figure 24   Drag coefficient of a conventional (British) golf ball (Bearman & Harvey 1976).

with lifting bodies. At a given spin rate, increasing U decreases the spin parameter, and hence C_L and C_D are also reduced. At postcritical Reynolds numbers, the relation between lift and spin rate is almost linear, as in the case of a baseball. However, this linear relationship does not hold as $N \to 0$; and for $U \sim 14$ m s^{-1} and $N < 1200$ rpm, a negative lift is obtained at this precritical Reynolds number. Bearman & Harvey (1976) conclude that the lift in this regime "cannot be explained by any simple attached flow circulation theory." This situation mainly results from the fact that transition is a very unstable phenomenon, which is easily influenced by parameters such as the details of the local surface roughness and free-stream turbulence.

4.3   Computation of Golf Ball Trajectories

Bearman & Harvey (1976) computed complete golf ball trajectories using the measured aerodynamic forces. The computation involved a step-by-step calculation procedure of the two components of the equation of motion:

$$\ddot{x} = -\frac{\rho S}{2m}(\dot{x}^2 + \dot{y}^2)(C_D \cos \beta + C_L \sin \beta),$$

$$\ddot{y} = \frac{\rho S}{2m}(\dot{x}^2 + \dot{y}^2)(C_L \cos \beta - C_D \sin \beta) - g,$$

where x and y are measured in the horizontal and vertical directions, respectively, and β is the inclination of the flight path to the horizontal [i.e. $\beta = \tan^{-1}(\dot{y}/\dot{x})$]. At each time step, the measured values (or interpolations) of C_L and C_D were used. The amount of spin decay during flight is difficult to predict. However, Bearman & Harvey tried some realistic assumptions, including the one wherein the decay was assumed to be proportional to N^2, and found that the computed trajectories were not affected significantly.

Figure 25 shows a computed trajectory for initial conditions typical of a professional golfer's drive. Note that toward the end of the flight, as the velocity decreases, the aerodynamic forces lose importance to the gravitational force.

The effect on range of the three main initial parameters (spin rate,

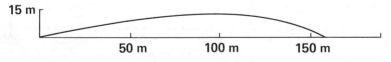

Figure 25   Computed golf ball trajectory. Initial conditions: velocity $= 57.9$ m s^{-1}, elevation $= 10°$, spin $= 3500$ rpm (Bearman & Harvey 1976).

velocity, and elevation) are shown in Figures 26*a*, *b*, and *c*, respectively. In Figure 26*a*, the maximum range for the initial conditions given is obtained for a spin rate just over 4000 rpm. In Figure 26*b*, the range increases rapidly with velocity for initial velocities greater than about 30 m s^{-1}. The range seems to be a relatively weak function of the initial elevation, although it is still increasing at an initial elevation angle of 15°. In practice, as Bearman & Harvey (1976) point out, hitting the ball harder increases both the initial velocity and spin. Bearman & Harvey (1976) also compared the ranges computed from the wind-tunnel results with actual measured values using golf balls launched by a driving machine. In general, the ball traveled slightly farther than predicted. This was attributed to incorrect scaling of the dimple edge radius.

4.4   Effect of Dimple Geometry

The results discussed in Section 4.3 clearly show that the aerodynamics of a golf ball depend critically on the flow induced by the dimples. The actual geometry of the dimples must therefore be expected to affect the flow regime

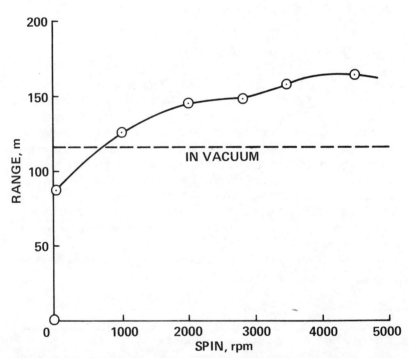

Figure 26a   Effect of spin on range for a conventional (British) golf ball. Initial conditions: velocity = 57.9 m s^{-1}, elevation = 10° (Bearman & Harvey 1976).

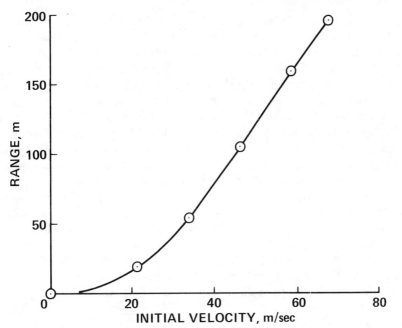

Figure 26b Effect of initial velocity on range for a conventional (British) golf ball. Initial conditions: elevation = 10°, spin = 3500 rpm (Bearman & Harvey 1976).

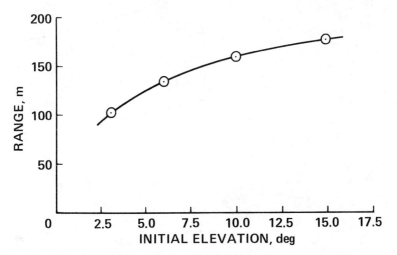

Figure 26c Effect of initial elevation on range for a conventional (British) golf ball. Initial conditions: velocity = 57.9 m s^{-1}, spin = 3500 rpm (Bearman & Harvey 1976).

and hence the aerodynamic forces on the golf ball. While this particular effect is difficult to understand and to quantify accurately, some experiments (described below) have been performed to establish its importance.

In general, for a given geometry and Reynolds number, the dimples would have to be deep enough to cause a disturbance in the laminar boundary layer. However, if the dimples are too deep, this may contribute to the drag force, although the actual mechanism causing it is not obvious. So there must be an optimum depth for the dimples at a given Reynolds number. This is shown to be the case in Figure 27, where the effect of square dimple depth on range is illustrated. The golf ball was launched by a driving machine. Clearly there is an optimum depth of about 0.25 mm at which the range is maximum.

Bearman & Harvey (1976) investigated the effects of changing the dimple shape. Apart from the model ball of conventional design, they also measured forces on a "hex-dimpled" ball, which had 240 hexagonal dimples and 12 pentagonal dimples arranged in a triangular pattern. The results for the postcritical regime are compared in Figure 28. In general, the hex-dimpled ball is superior to the conventional ball: it exhibits higher lift and

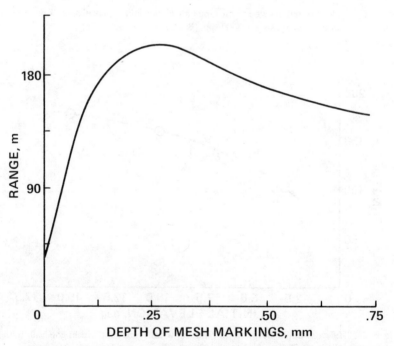

Figure 27   Effect of square dimple depth on range (Cochran & Stobbs 1968).

lower drag properties. Tests using a driving machine showed that under normal driving conditions the hex-dimpled ball traveled approximately 6 m farther than a conventional ball. Bearman & Harvey conclude that "hexagonal shaped dimples act as even more efficient trips than round dimples, perhaps by shedding into the boundary layer more discrete (horseshoe) vortices from their straight edges."

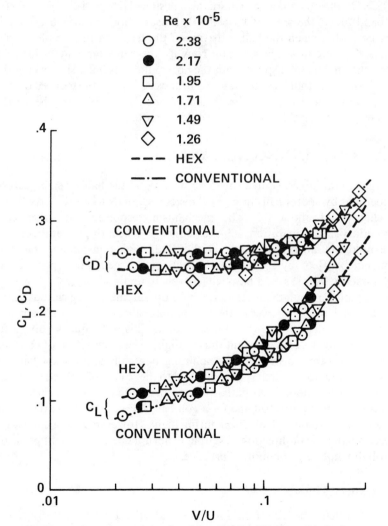

Figure 28 Comparison of conventional and hex-dimpled (British) golf balls (Bearman & Harvey 1976).

Apart from the dimple size and shape, the arrangement of dimples also seems to be a relevant parameter. A conventional golf ball has a "seam" with no dimples along the equator, where the two molded halves join together. At the relatively high Reynolds numbers and spin rates encountered in golf, this seam could produce an asymmetric flow, as on a cricket ball. This is of concern to (professional) golfers, and one of the new designs has 10 seams, which makes the ball more isotropic (Cavendish 1982). On the other hand, a ball was also designed that had dimples only on a band along the seam (Chase 1981). Presumably, the idea behind this design is that when the ball is driven off the tee, with the dimples in the vertical plane, it would generate roughly the same amount of lift as a conventional ball if it spins about the horizontal axis only. However, if the ball is sliced, so that it rotates about a near-vertical axis, the reduced overall roughness would increase the critical Reynolds number, and hence the sideways (undesirable) deflection would be reduced.

5. CONCLUDING REMARKS

Large lateral forces, equivalent to about 50% of the ball's weight, can be generated by sports balls in flight. This corresponds to a side force or lift coefficient of about 0.3. The mechanism responsible is asymmetric boundary-layer separation, achieved either by fence tripping at precritical Reynolds numbers [as on a cricket ball and nonspinning baseball (knuckleball)] or by spin-induced effects, which generate a postcritical, asymmetric flow [as on a baseball (curveball) and golf ball].

The most popular and effective method for measuring aerodynamic forces on balls is to release them (spinning) through an airstream and measure the deflection due to the side force. This technique is only useful when the side force is constant during flight. The assumption of constant side force seems to be valid for spinning baseballs and cricket balls; it results in a deflection that is proportional to the square of elapsed time, and hence in a parabolic flight path. More sophisticated models are necessary for a nonspinning baseball and for a golf ball, since the forces on these balls change magnitude and direction during flight. However, abrupt changes in force magnitude or direction, resulting in abrupt changes in flight path, are probably not very common in practice.

ACKNOWLEDGMENTS

I am grateful to Tee Lim for help with the cricket ball trajectory computations and to Jim Walton for helpful discussions on baseball aerodynamics. I would also like to thank all my colleagues at the Ames Research Center who reviewed an earlier draft of this article.

This article is dedicated to my late father.

Literature Cited

Achenbach, E. 1972. Experiments on the flow past spheres at very high Reynolds numbers. *J. Fluid Mech.* 54:565–75

Achenbach, E. 1974. The effects of surface roughness and tunnel blockage on the flow past spheres. *J. Fluid Mech.* 65:113–25

Allman, W. F. 1982. Pitching rainbows. *Sci.* 82 3(8):32–39

Allman, W. F. 1983. Flight of the knuckler. *Sci.* 83 4(5):92–93

Barkla, H. M., Auchterlonie, L. J. 1971. The Magnus or Robins effect on rotating spheres. *J. Fluid Mech.* 47:437–47

Barton, N. G. 1982. On the swing of a cricket ball in flight. *Proc. R. Soc. London Ser. A* 379:109–31

Bearman, P. W., Harvey, J. K. 1976. Golf ball aerodynamics. *Aeronaut. Q.* 27:112–22

Bentley, K., Varty, P., Proudlove, M., Mehta, R. D. 1982. An experimental study of cricket ball swing. *Aero Tech. Note 82–106*, Imperial Coll., London, Engl.

Binnie, A. M. 1976. The effect of humidity on the swing of cricket balls. *Int. J. Mech. Sci.* 18:497–99

Brancazio, P. J. 1983. The hardest blow of all. *New Sci.* 100:880–83

Briggs, L. J. 1959. Effect of spin and speed on the lateral deflection of a baseball; and the Magnus effect for smooth spheres. *Am. J. Phys.* 27:589–96

Brown, F. N. M. 1971. See the wind blow. *Dept. Aerosp. Mech. Eng. Rep.*, Univ. Notre Dame, South Bend, Ind.

Cavendish, M. 1982. Balls in flight. *Sci. Now* 1:10–13

Chase, A. 1981. A slice of golf. *Sci. 81* 2(6):90–91

Cochran, A., Stobbs, J. 1968. *Search for the Perfect Swing*, pp. 161–62. Philadelphia/New York: Lippincott.

Daish, C. B. 1972. *The Physics of Ball Games.* London: Engl. Univs. Press. 180 pp.

Davies, J. M. 1949. The aerodynamics of golf balls. *J. Appl. Phys.* 20:821–28

Graham, J. M. R. 1969. The development of the turbulent boundary layer behind a transition strip. *Aeronaut. Res. Counc. Rep. 31492*

Horlock, J. H. 1973. The swing of a cricket ball. In *Mechanics and Sport*, ed. J. L. Bleustein, pp. 293–303. New York: ASME

Hunt, C. 1982. *The aerodynamics of a cricket ball.* BSc dissertation. Dept. Mech. Eng., Univ. Newcastle, Engl.

Imbrosciano, A. 1981. The swing of a cricket ball. *Proj. Rep. 810714*, Newcastle Coll. of Adv. Educ., Newcastle, Austral.

Lyttleton, R. A. 1957. The swing of a cricket ball. *Discovery* 18:186–91

Mehta, R. D., Wood, D. H. 1980. Aerodynamics of the cricket ball. *New Sci.* 87:442–47

Mehta, R. D., Bentley, K., Proudlove, M., Varty, P. 1983. Factors affecting cricket ball swing. *Nature* 303:787–88

Newton, I. 1672. New theory of light and colours. *Philos. Trans. R. Soc. London* 1:678–88

Rayleigh, Lord. 1877. On the irregular flight of a tennis ball. *Messenger of Mathematics* 7:14–16. Reprinted in *Scientific Papers* (Cambridge, 1899) 1:344–46

Sherwin, K., Sproston, J. L. 1982. Aerodynamics of a cricket ball. *Int. J. Mech. Educ.* 10:71–79

Tait, P. G. 1890. Some points in the physics of golf. Part I. *Nature* 42:420–23

Tait, P. G. 1891. Some points in the physics of golf. Part II. *Nature* 44:497–98

Tait, P. G. 1893. Some points in the physics of golf. Part III. *Nature* 48:202–5

Ward, C. W. 1983. *The aerodynamics of a cricket ball.* BSc dissertation. Dept. Mech. Eng., Univ. Newcastle, Engl.

Watts, R. G., Sawyer, E. 1975. Aerodynamics of a knuckleball. *Am. J. Phys.* 43:960–63

Weaver, R. 1976. Comment on "Aerodynamics of a knuckleball." *Am. J. Phys.* 44:1215

Ann. Rev. Fluid Mech. 1985. 17 : 191–215
Copyright © 1985 by Annual Reviews Inc.

BUOYANCY-DRIVEN FLOWS IN CRYSTAL-GROWTH MELTS

W. E. Langlois

IBM Research Laboratory, San Jose, California 95193

INTRODUCTION

In most of the methods for synthetic production of single crystals, the crystal grows slowly from a fluid nutrient. Several different mechanisms can drive motions in this fluid, and these motions are of concern to the crystal grower because they influence the transport of dopant, impurities, and heat to the growth interface. The seriousness of this concern is reflected by the commissioning of at least seven review articles on the subject during the past decade (Carruthers 1975, 1979, Hurle 1977, Langlois 1981a, Pimputkar & Ostrach 1981, Kobayashi 1981, Jones 1984).

The nutrient fluid may be a vapor or a supersaturated solution, but we focus here upon systems in which a crystal is grown from its melt. Actually, we deal mostly with one specific method of growing crystals, viz. the Czochralski process. There are several reasons for this. The first is its importance. Four of the references cited in this review are from a single issue of *PhysicoChemical Hydrodynamics*, which was devoted to the role of convection and fluid flow in solidification and crystal growth. In the introductory article of that issue (Hurle & Jakeman 1981), the Czochralski process was described as "the most important and widely used technique." Indeed, Czochralski growth is ideally suited for producing the large quantities of pure crystal required by present technology. For example, single crystals of elemental silicon tens of kilograms in mass, and with diameters exceeding 100 mm, are now routinely produced. The second reason for dwelling on Czochralski growth follows from the first: Because of its importance, it has attracted far more attention from hydrodynamicists than have other melt growth systems. Finally, the flow in Czochralski melts is possessed of a rich structure with regard to both driving mechanisms and streaming patterns. The results obtained for this system provide at least a qualitative understanding of the other systems as well.

191

0066–4189/85/0115–0191$02.00

Czochralski Growth

In the Czochralski process, pictured in Figure 1, the crucible is initially loaded with the material to be crystallized. When this is melted, a seed crystal held at the end of a pull rod is dipped in. After appropriate start-up procedures, the growing crystal is slowly extracted from the melt; pull rates in the range of millimeters or centimeters per hour are typical.

To keep the nutrient molten, the crucible is maintained above the freezing point, which of course is the temperature at the crystal-growth interface. The resulting radial temperature gradient within the melt is itself sufficient to generate convective flow.

Other driving mechanisms often complicate matters. For example, the growing crystal is sometimes rotated as it is pulled. The objective is to improve uniformity by providing a viscous shear layer that tends to isolate the growth interface from the turmoil deeper in the melt. In the ideal limit, the rotating crystal would generate the von Kármán-Cochran centrifugal

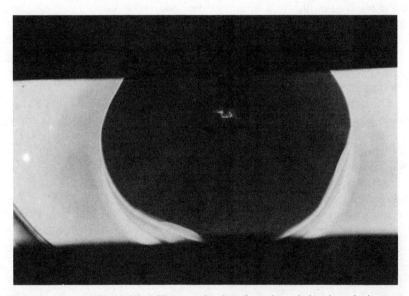

Figure 1 (Courtesy K. M. Kim) Photograph taken through a window into the inert gas atmosphere of a Czochralski furnace, showing a large-diameter (133 mm) crystal of silicon growing from melt contained in a fused silica crucible (25.4 cm in diameter). Note that the seed crystal, seen in the upper middle view field, is necked down considerably (to 3.4 mm). This is done deliberately, to suppress formation of dislocations, by temporarily pulling at a high rate (6.4 mm min^{-1}). During growth, the entire weight of the rotating crystal (about 10 kg at the end of the run) is supported by the neck. A typical pull rate for the subsequent full-diameter crystal is 1 mm min^{-1} (Kim et al. 1983).

flow, with its constant boundary-layer thickness. In practice, the finite crystal size and heterogeneous free-stream conditions lead to a less friendly flow. Nevertheless, crystal rotation does help in many cases.

In some applications, the crucible is also rotated to smooth out thermal asymmetries that might arise from irregularities in the heating. This too generates a centrifugal flow. Moreover, it imparts rotational velocity to most of the melt mass, causing it to exhibit some of the behavior associated with rotating fluid systems. In particular, eddies that arise within the melt show a pronounced tendency to form themselves into Taylor columns. Crystal and crucible rotation rates are typically in the range of radians per second.

Finally, there is thermocapillary flow. The surface-tension coefficients of many Czochralski melts vary strongly with temperature. Since there is a pronounced radial temperature gradient along the free surface, from the freezing point at the growth interface to a significantly higher temperature at the crucible wall, the surface tension diminishes with increasing radius. The nonuniform surface tension generates a finite shear stress at the free surface (Landau & Lifshitz 1959, p. 234), which pulls the fluid toward the crystal. Motions induced in this manner are called *Marangoni flows*, and their role in crystal growth has been the object of much study (Schwabe 1981). When the surface-tension variation results from temperature variation, the term *thermocapillary flow* is used to describe the concomitant motion.

When all of these driving mechanisms act in concert, the resulting flow can be quite complicated. In certain parameter ranges, which do occur in practice, the motion can lose its axial symmetry (Jones 1983b) or even become turbulent. Fortunately for the hydrodynamicist, however, the crystal grower usually wants to keep everything as smooth and symmetric as possible. Hence the regime of laminar, rotationally symmetric flow is of practical importance.

Czochralski Bulk Flow

Theoretical investigations of the convective flow in a Czochralski crucible employ a somewhat idealized configuration in which the top surface is flat and stationary, as shown in Figure 2. The term "Czochralski bulk flow" was coined to describe this model (Langlois & Shir 1977).

The bulk-flow idealization ignores the ripples that are usually present on a real melt. This simplification is not too bad : In most cases, the amplitude of the ripples is quite small, and domain perturbation can be used to move the boundary conditions down to the nominal level of the melt free surface (Langlois 1977a).

Of more concern is the departure of the bulk-flow idealization from

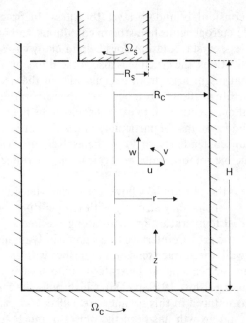

Figure 2   The geometrical configuration for the bulk-flow idealization.

reality in the neighborhood of the growth interface because the local flow in this region directly influences the growing crystal. In practice, the interface may be concave or convex, and it is somewhat elevated atop a meniscus rising from the surface of the melt. Fluid motion in the region bounded by the nominal surface level, the meniscus, and the growth interface has been termed the *near-field flow* (Pimputkar & Ostrach 1981). Studies that deal with solute transport across the near-field region require subsidiary conditions at the far boundary that are determined by the bulk flow. At the time of this writing, these have always been chosen more-or-less arbitrarily, but the analysis could presumably be improved by taking them from bulk-flow computations. The idea is that the bulk flow provides part of the environment that determines the character of the near-field flow, which in turn causes (one hopes) only a minor perturbation of the bulk flow.

FLOW VISUALIZATION

Direct observation of the flow in a Czochralski melt is by and large limited to the surface patterns. The melt is often opaque, and even if it is transparent, there is the problem of finding suitable tracer particles for a high-temperature liquid, not to mention the difficulties in photographing them within the furnace. Probes disturb the flow and are often physically or chemically attacked by the melt as well. Consequently, investigators have at various times sought to obtain a better understanding of melt convection by looking to models that resemble the Czochralski geometry, but that employ a liquid which is transparent and cool enough for accurate observation.

The Early Studies

At a time when hydrodynamicists had not yet discovered the subject, some experimental modeling of Czochralski flow was carried out by crystal growers themselves (Goss & Adlington 1959, Turovskii & Mil'vidskii 1962, Robertson 1966). Their results, which were obtained in an era when the significance of convective transport in Czochralski growth was not clearly perceived, illustrated qualitatively that such transport cannot be ignored.

Later Work

Semiquantitative experiments began soon thereafter (Carruthers 1967). Carruthers recognized that different flow regimes could occur and succeeded in modeling some some of them. He also realized that some of the patterns could be cellular, which would have severe implications about solute segregation in the growth process. One of his photographs is reproduced in Figure 3. Carruthers experiments, as well as his writings and

Figure 3   (Courtesy J. R. Carruthers) Flow pattern in a water-glycerol model of Czochralski bulk flow.

lectures, were a major factor in bringing the techniques of experimental and theoretical hydrodynamics to the problem. Later experiments by Carruthers and others (Carruthers & Nassau 1968, Brandle 1977, Shiroki 1977, Miller 1982) reflect this increased sophistication.

The experiments carried out during this period have served a purpose that may have been unexpected: They have provided partial checks for the computational models that were developed later. Even though these models are mainly intended for use in a different parameter range, they can easily be run with the parameters temporarily set to those of the experiments. Successful agreement gives some assurance that a model is correct. The check is not complete, however, because the model may have shortcomings that manifest themselves in the parameter range of the prototype but not in that of the experiment.

Similarity Parameters

I am sometimes asked why my own work on Czochralski flow makes such scant use of dimensionless variables, when their value in hydrodynamics has been proven many times over. I can only answer that few things in life are all good or all bad.[1] A dimensionless formulation can indeed be of some help in understanding the flow: For relating experimental models to operating prototypes, it is indispensable. However, it does have some drawbacks, which are not always fully appreciated.

First, there is the task of selecting meaningful scales. If one is content merely to express everything in terms of dimensionless quantities, elementary algebra suffices. However, to obtain a dimensionless formulation that leads to physical insight, more is required. As others have pointed out (Pimputkar & Ostrach 1981, Balasubramaniam & Ostrach 1984), the normalization should be carried out in such a way that the dimensionless variables are of unit order. This is less trivial. Consider, for example, the problem of choosing a meaningful velocity scale for Czochralski flow. Since the azimuthal velocity at the crystal edge is $R_s\Omega_s$, this is a plausible choice. However, if the crucible also rotates, $R_c\Omega_c$ may be a representative velocity for more of the melt mass. Worse still, either buoyant convection or thermocapillary flow might dominate the flow. Each of these has its own velocity scale, which is not so easy to express in terms of the external parameters of the problem.

Often the best one can do in practice is to choose scales using intuition and experience as guides, check a posteriori to evaluate the extent and

---

[1] For example, the Chicago sewer system is an engineering marvel but a terrible place to lose a contact lens.

limitations of their suitability, and issue appropriate warnings. It is largely a matter of taste, but I believe it is often preferable to focus on a particular system, e.g. silicon growth in a specified crucible, and formulate the problem in terms of the parameters that the crystal grower can adjust, even though they may bear physical dimensions.

In the case of Czochralski growth, one must also contend with a profusion of dimensionless constants. The simplest formulation of Czochralski bulk flow takes the crystal and crucible to be at uniform temperatures T_s and T_c, respectively, and assumes Stefan-Boltzmann radiation from the free surface. Even this simplified formulation involves nine independent scaling parameters. These can be chosen in various ways, but a convenient list is the following:

1. Ratio of crucible radius to crystal radius $\rho_R = R_c/R_s$
2. Melt aspect ratio $\mathcal{R} = H/R_c$
3. Ratio of crucible rotation rate to crystal rotation rate $\rho_\Omega = \Omega_c/\Omega_s$
4. Prandtl number $\mathrm{Pr} = v/\kappa$
5. Reynolds number $\mathrm{Re} = R_s^2\Omega_s/v$
6. Grashof number based on crucible radius $\mathrm{Gr} = \alpha g R_c^3\Delta T/v^2$
7. Marangoni number $\mathrm{Ma} = c_{tc}R_s\Delta T/\kappa$
8. Scaled emissivity $\varepsilon^* = (\varepsilon\sigma/\rho c_v)T_a^4/R_s$
9. Relative temperature difference $\Delta_r T = \Delta T/T_a$.

In these definitions, v is the kinematic viscosity, κ the thermal diffusivity, α the volumetric expansion coefficient, c_{tc} the thermocapillary coefficient, ρ the density, c_v the specific heat, σ the Stefan-Boltzmann constant, ε the emissivity, $\Delta T = T_c - T_s$, and $T_a = (T_c + T_s)/2$.

With so many parameters involved, one must be careful when reading— or making—a statement to the effect, say, that a flow transition takes place at such-and-such a Grashof number. The statement may well not extend to cases where some of the other parameters are out of scale. An example is discussed in the section on numerical simulation of Czochralski bulk flow.

More germane to the present section, it is evidently quite difficult to build a moderate-temperature model of Czochralski flow with all nine parameters in scale. For a crystal with a very high melting point, such as sapphire, correct modeling of the overall radiative heat loss (via the scaled emissivity), would require R_s to be so small that scaling the other parameters would probably be impossible. For a low-temperature melt, however, radiation may be negligible even in the prototype. The scaled emissivity and the relative temperature difference can then be ignored as scaling parameters. A similar hope can be entertained with respect to the Marangoni number.

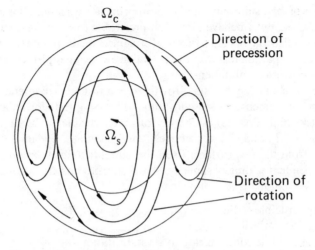

Figure 4   (Jones 1983a) Precessing pattern in a water-glycerol model.

Experiments Based on Similarity Considerations

A recent set of experiments (Jones 1983a) used glycerol and water to model Czochralski flow of silicon and of bismuth-silicon-oxide. As the latter material melts at 1168 K, radiative heat loss is not large. Hence, as indicated above, ε^* and $\Delta_r T$ become unimportant. The Prandtl number scaling was not ideal (4.5 to 7 for glycerol/water versus 23 for $Bi_{12}SiO_{20}$), but it was close enough for the results to be useful. A similar statement probably holds for the Marangoni number, although the point was not discussed. This left Jones with five parameters to scale properly. (He assigned names to only four of them, but the text makes it clear that he scaled the ratio of radii as well.) The set he chose was different from, but equivalent to, that introduced above.

Jones examined six different flow situations that involved various combinations of crystal rotation, crucible rotation, and differential heating. Various flow patterns resulted, some axisymmetric, some not. In some of the nonaxisymmetric cases, the departure from symmetry takes the form of the rather curious flow pattern shown in Figure 4. The pattern is not stationary but instead precesses in the direction of crucible rotation.

NUMERICAL SIMULATION OF CZOCHRALSKI BULK FLOW

There is now a large body of literature on numerical modeling of the rotationally symmetric regime of Czochralski bulk flow; these models use the Navier-Stokes equations with Boussinesq approximation. The transport coefficients are taken to be constant, which is not a fully satisfactory assumption, since they are really temperature dependent. However, to introduce variable coefficients would overresolve the accuracy of the data currently available for most Czochralski melts. Radiative heat transfer within the melt is neglected. This is reasonable for melts that are either cool or opaque. It is not too bad even for hot melts that are transparent to visible light, because the peak of the emission is in the infrared, where most melts have a relatively short mean free path for travel of radiation. The Rosseland approximation (Carruthers 1979) then permits treatment of internal radiative transfer by use of a modified thermal diffusivity. In principle, this depends on temperature, but a reasonable approximation is given by using an average temperature of the melt. Heating by viscous dissipation is insignificant compared with that supplied externally.

Primitive Equations Formulation

The fundamental dependent variables of Czochralski bulk flow are the three velocity components u, v, w, the pressure p, and the temperature T. These are governed by the continuity equation, the rotationally symmetric Navier-Stokes equation, and the energy equation. In the Boussinesq approximation, the last reduces to an advection-diffusion equation for the temperature.

It is better, however, to replace the azimuthal component of the Navier-Stokes equation by an equation governing the swirl $\Omega = rv$, which is a physically conserved quantity, viz the angular momentum per unit mass. Failure to do this is risky, although the early work was fortunate enough to avoid trouble. A problem does arise (Langlois 1981b) if the azimuthal motion is accompanied by a strong inward radial velocity. The coupling term uv/r in the discretized Navier-Stokes equation then leads to a false production of angular momentum that is difficult to eliminate. This is automatically avoided if one instead uses the swirl equation with a conservative differencing scheme. Thus, the governing differential equations are

$$\frac{1}{r}\frac{\partial}{\partial r}(ru) + \frac{\partial w}{\partial z} = 0, \tag{1}$$

$$\frac{\partial \Omega}{\partial t} + \frac{1}{r}\frac{\partial}{\partial r}(ru\Omega) + \frac{\partial}{\partial z}(w\Omega) = \frac{v}{r}\frac{\partial}{\partial r}\left[r^3\frac{\partial}{\partial r}\left(\frac{\Omega}{r^2}\right)\right] + v\frac{\partial^2\Omega}{\partial z^2}, \tag{2}$$

$$\frac{\partial u}{\partial t} + \frac{1}{r}\frac{\partial}{\partial r}(ru^2) + \frac{\partial}{\partial z}(wu) - \frac{\Omega^2}{r^3} = -\frac{1}{\rho}\frac{\partial p}{\partial r} + v\left(\nabla^2 u - \frac{u}{r^2}\right), \tag{3}$$

$$\frac{\partial w}{\partial t} + \frac{1}{r}\frac{\partial}{\partial r}(ruw) + \frac{\partial}{\partial z}(w^2) = -\frac{1}{\rho}\frac{\partial}{\partial z}(p+gz) + \alpha g(T-T_s) + \nu\nabla^2 w, \qquad (4)$$

$$\frac{\partial T}{\partial t} + \frac{1}{r}\frac{\partial}{\partial r}(ruT) + \frac{\partial}{\partial z}(wT) = \kappa\nabla^2 T. \qquad (5)$$

If one wishes to compute the transport of a trace component (either dopant or impurity), the concentration of the component will obey an equation much like (5), with its diffusion coefficient replacing the thermal diffusivity (Lee et al. 1984).

BOUNDARY CONDITIONS On the crucible wall and bottom, the no-slip velocity condition is applied in an obvious fashion. It is reasonably consistent with the bulk-flow approximation (which assumes that the crystal pull rate is slow enough that the change in melt depth can be neglected) to equate the melt velocity at the growth interface to the rotational velocity of the crystal. In some cases, however, the growth rate fluctuates enough that its instantaneous value is significantly higher than its time average. A correction for this can be incorporated by taking the view that melt "disappears" into the crystal at the growth interface, which is then modeled as a porous boundary with suction (Wilson 1978).

Various thermal boundary conditions can be supplied on the crucible wall and bottom. The most commonly used approach is to specify that both surfaces of the crucible are at a fixed temperature T_c. A minor variation, which slightly improves the realism when the crucible is heated only from the side, is to specify a nonuniform temperature distribution on the bottom (Langlois 1982, 1983, Langlois & Lee 1983a,b, Lee et al. 1984). Alternatively, it is possible to specify a heat flux from the crucible into the melt. The temperature at the growth interface is specified as the crystalline melting point.

Since no externally applied tractions act on the free surface, the mechanical boundary conditions on that surface are determined entirely by thermocapillarity. The normal derivative of the radial velocity is proportional to the radial derivative of the temperature, and the axial velocity vanishes (bulk-flow approximation). Because of axial symmetry, thermocapillarity does not directly drive the azimuthal flow; the normal derivative of the swirl vanishes on the free surface.

Complete modeling of the thermal boundary conditions on the free surface would be exceedingly complicated. The melt loses heat both by radiation and by sensible heat exchange with the overlying gas. Moreover, the radiative loss is not simply Stefan-Boltzmann radiation to infinity, because the melt faces upon solid surfaces (the furnace walls, the completed part of the crystal, etc.) that are at various temperatures. Many formulas that could be used have been set out (Kobayashi 1981). In practice, however, the difficulty of determining all the appropriate details leads to the use of some type of compromise. Stefan-Boltzmann radiation to surroundings that are at the crystalline melting point is a popular choice. Heat loss from the free surface should not be completely ignored, even for relatively cool melts, because it can lead to an undesirable phenomenon called *flaring*: If the portion of the free surface nearest the crystal becomes supercooled, the crystal may begin to grow laterally as a thin film of "ice," with disastrous results. Such supercooled regions do sometimes appear in simulations (Langlois 1977b).

With one exception, subsidiary conditions at the axis of symmetry are quite straightforward: All fluxes of momentum, heat, and trace solutes are in the axial direction. The exception pertains to the swirl. If an exact treatment were feasible, one would require Ω and its first radial derivative to vanish at the axis. In finite-difference approximation, however, the axial grid cells have finite volume and hence contain a small but finite residue of angular momentum. In order to ensure that truncation error does not lead to a false production or destruction of angular momentum, one should account for that residue; this is easily done (Langlois 1981b).

Some numerical modeling of Czochralski flow has been carried out with the primitive equations. Finite-difference methods (Mihelčič et al. 1981, 1982) and finite-element methods (Crochet et al. 1983a) have both been used. The main advantage of primitive equations modeling is that it can be extended to the nonaxisymmetric case in a more straightforward way. Most simulations of Czochralski flow, however, have employed the more frugal vorticity–stream function method, which we review next.

Vorticity–Stream Function Formulation

For rotationally symmetric flow, computation time can be reduced if the problem is reformulated so that the three variables u, v, p are eliminated in favor of the vorticity ω and Stokes stream function ψ, which are defined so that

$$\omega = \frac{\partial w}{\partial r} - \frac{\partial u}{\partial z}, \qquad u = \frac{1}{r}\frac{\partial \psi}{\partial z}, \qquad w = -\frac{1}{r}\frac{\partial \psi}{\partial r}. \tag{6}$$

These variables were employed in many of the studies described below (all of the Kobayashi-Arizumi work and my own work prior to 1981). However, it has been observed (Kyrazis 1977) that for numerically stable modeling of physically unstable flows, it is important to use ω/r rather than ω as a dependent variable. There is a physical reason for this (Langlois 1981b): A very old theorem due to A. V. Svanberg states that rotationally symmetric motion of an incompressible fluid is circulation preserving if and only if ω/r is constant for each fluid particle. This theorem is the rotationally symmetric analog of D'Alembert's vorticity theorem, which establishes ω as the corresponding quantity in plane flow. Thus ω/r is pseudo-conserved in rotationally symmetric flow in the same sense that ω is pseudo-conserved in plane flow. I have suggested (Langlois 1981b) that ω/r be termed the *Svanberg vorticity*, denoted by the symbol S. There is no evidence that the use of ω instead of S has led to incorrect results in Czochralski flow simulation, but in the interest of aesthetics and safety, the Svanberg vorticity should be used in future work.

In the previous subsection, the primitive equations were set out in physical form. In order to exhibit the way that the various scaling parameters enter the problem, we now exhibit the vorticity–stream function formulation in dimensionless form. The scaling parameters introduced earlier are concomitant with a choice of R_s as length scale, $1/\Omega_s$ as time scale, and $(T_c + T_s)/2$ and ΔT as scales of absolute and relative temperature, respectively. With the various symbols now denoting dimensionless variables, the governing equations (and a set of boundary conditions favorable for finite-difference approximation) are given by

$$\frac{\partial \Omega}{\partial t} + \frac{1}{r}\frac{\partial}{\partial r}(ru\Omega) + \frac{\partial}{\partial z}(w\Omega) = \frac{1}{Re}\left\{\frac{1}{r}\frac{\partial}{\partial r}\left[r^3\frac{\partial}{\partial r}\left(\frac{\Omega}{r^2}\right)\right] + \frac{\partial^2 \Omega}{\partial z^2}\right\}, \tag{7}$$

$$\frac{\partial S}{\partial t} + \frac{1}{r}\frac{\partial}{\partial r}(ruS) + \frac{\partial}{\partial z}(wS) + \frac{\partial}{\partial z}\left(\frac{\Omega^2}{r^4}\right)$$

$$= \frac{Gr}{\rho_R^3 Re^2}\frac{1}{r}\frac{\partial T}{\partial r} + \frac{1}{Re}\left\{\frac{1}{r}\frac{\partial}{\partial r}\left[\frac{1}{r}\frac{\partial}{\partial r}(r^2 S)\right] + \frac{\partial^2 S}{\partial z^2}\right\}, \tag{8}^2$$

$$\frac{\partial T}{\partial t} + \frac{1}{r}\frac{\partial}{\partial r}(ruT) + \frac{\partial}{\partial z}(wT) = \frac{1}{RePr}\left[\frac{1}{r}\frac{\partial}{\partial r}\left(r\frac{\partial T}{\partial r}\right) + \frac{\partial^2 T}{\partial z^2}\right], \tag{9}$$

$$\frac{\partial}{\partial r}\left(\frac{1}{r}\frac{\partial \psi}{\partial r}\right) + \frac{1}{r}\frac{\partial^2 \psi}{\partial z^2} = -rS, \tag{10}$$

$$S = -\frac{2}{r^2\Delta z^2}\,\psi(r, \rho_R R - \Delta z, t), \qquad \Omega = r^2, \qquad T = -\frac{1}{2},$$

$$\psi = 0 \text{ on the growth interface}, \tag{11}$$

$$S = -\frac{2}{\rho_R^2 \Delta r^2}\,\psi(\rho_R - \Delta r, z, t), \qquad \Omega = \rho_\Omega \rho_R^2, \qquad T = \frac{1}{2},$$

$$\psi = 0 \text{ on the crucible wall}, \tag{12}$$

---

[2] In an earlier review article (Langlois 1981a), the factor $1/\rho_R^3$ was incorrectly omitted from the buoyancy term.

$$S = -\frac{2}{r^2 \Delta r} \psi(r, \Delta z, t), \qquad \Omega = \rho_\Omega r^2, \qquad t = \frac{1}{2},$$

$\psi = 0$ on the crucible bottom, (13)

$$S = \frac{Ma}{\rho_\Omega Re} \frac{1}{r} \frac{\partial T}{\partial z}, \qquad \frac{\partial \Omega}{\partial z} = 0,$$

$$\left(\frac{\partial T}{\partial t}\right)_{rad} = -\frac{2\varepsilon^*}{\Delta z} \{[1 + (\Delta_r T)T]^4 - [1 - \tfrac{1}{2}(\Delta_r T)]^4\},$$

$\psi = 0$ on the free surface, (14)

where $(\partial T/\partial t)_{rad}$ signifies the contribution of radiation cooling to the temperature tendency at free-surface grid points. If the computation is carried out on a nonuniform grid, as is normally the case, Δr and Δz denote the grid spacings at the boundary point in question.

Subsidiary conditions on S at the axis of symmetry are not obvious, but finite-difference procedures can be derived by requiring that the flow have no singularities at the axis (Langlois 1981b).

STEADY-STATE SIMULATION   The pioneering work on digital simulation of Czochralski flow was carried out by Noboyuki Kobayashi and Tetsuya Arizumi using a model (Kobayashi 1980) that employs the steady-state form of the vorticity–stream function formulation. Some confidence in digital techniques was established (Kobayashi & Arizumi 1975) when their computed flow patterns closely resembled those found in one of the more extensive flow-visualization studies (Carruthers & Nassau 1968).

This model also yielded the first computed transition between qualitatively different regimes in Czochralski flow (Kobayashi 1978). With other parameters held fixed ($\rho_R = 2.5$, $\mathcal{R} = 2.0$, $\rho_\Omega = 0$, $Pr = 0.01$, $Re = 40$, $Ma = 0$, $\varepsilon^* = 0$, $\Delta_r T$ immaterial), the flow at low Grashof number in a nonrotating crucible is, in essence, a tip vortex generated by the rotating crystal. Above a transition range near $Gr = 3000$, buoyant convection dominates, so that the direction of circulation is actually reversed throughout much of the flow field.

Steady-state modeling of Czochralski flow is limited to configurations with low to moderate values of the Reynolds and Grashof numbers. At higher values the flow tends to be unsteady, so that convergence of a steady-state computation is somewhat chancy.

TIME-DEPENDENT SIMULATION   The first time-dependent code for Czochralski flow (Langlois & Shir 1977) was constructed to permit modeling of a variant of Czochralski growth (discussed later in this review) in which the crucible is rotated at a time-varying rate. However, it was soon learned (Langlois 1977b) that even when all boundary conditions are independent of time, steady flow is the exception rather than the rule for commercial growth of semiconductor crystals. More typically, the long-term flow is a slow oscillation.

The model was also used (Langlois 1979a) to explore the transition range of a configuration that exhibits either a tip vortex or strong buoyant convection, depending on the Grashof number. The investigation was motivated by a conversation with Phillip Yin, a crystal grower who had observed that the flow in gadolinium-gallium-garnet melt was quite different from that in a closely related compound, viz. neodymium-gallium-garnet, presumably because the latter has a much lower volumetric expansion coefficient. The Grashof number was systematically varied with the other scaling of parameters held fixed ($\rho_R = 1.5$, $\mathcal{R} = 0.6$, $\rho_\Omega = 0$, $Pr = 0.035$, $Re = 8000$, $Ma = 0$, $\varepsilon^* = 3.6$, $\Delta_r T = 0.08$), which led to the streamline plots shown in Figure 5.

Parts a and d of Figure 5 illustrate the two extreme cases of flow in a nonrotating crucible. The former shows the tip vortex that fills the flow region at zero Grashof number. The latter corresponds to $Gr = 1.2 \times 10^8$. At this large value, the remnant of the tip vortex is quite weak and is completely encapsulated by a strong buoyancy-driven eddy. There are two other minor vortices. One nests in the corner where the crucible wall and bottom meet and occurs frequently in Czochralski growth simulations. Real crucibles are made with slightly rounded corners, so it probably does not occur in practice. The other minor vortex lies just off the downward fetch of the convective eddy. Its presence in Czochralski flow had not been suspected before, although a similar vortex was found in an analysis of free convection in vertical, closed-bottom tubes (Lighthill 1953). A vortex such as this is unfavorable for crystal growth because solutes become trapped in it for long periods of time. For growth in a nonrotating crucible, it may be desirable to fashion the crucible bottom in a shape that discourages the formation of the vortex (Langlois 1979b).

Part b of Figure 5 illustrates the flow in the low end of the transition range ($Gr = 8 \times 10^6$). The tip vortex dominates (but does not overwhelm) the buoyancy-driven eddy. The contest between the two prevents the flow from settling down into a firm steady state. Part c shows the high end of the transition range ($Gr = 3.2 \times 10^7$). The buoyancy-driven eddy is now strong enough to generate a Lighthill vortex but not strong enough to keep it well separated from the tip vortex. Since both vortices rotate clockwise, they tend to devour each other, and the resulting flow is decidedly unsteady.

Note that the transition range of Grashof number in this study is about 10,000 times higher than that of Kobayashi's work. As indicated in the

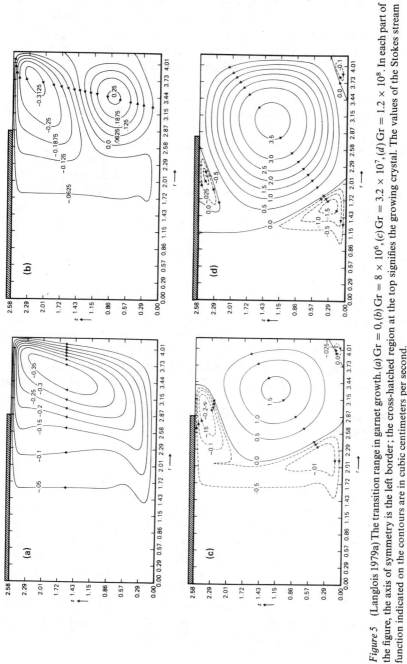

Figure 5 (Langlois 1979a) The transition range in garnet growth. (*a*) Gr = 0, (*b*) Gr = 8×10^6, (*c*) Gr = 3.2×10^7, (*d*) Gr = 1.2×10^8. In each part of the figure, the axis of symmetry is the left border; the cross-hatched region at the top signifies the growing crystal. The values of the Stokes stream function indicated on the contours are in cubic centimeters per second.

section on flow visualization, this is not surprising, because the other similarity parameters are not in scale. The equation governing the Svanberg vorticity suggests that the dimensionless grouping most strongly influencing the transition should be $Gr/Re^2 \rho_R^3$. The transition range of this quantity is of the same order of magnitude for the two studies; the remaining difference, a factor of four or so, is to be expected because the Reynolds number also enters the equations independently of the Grashof number, and because other parameters are somewhat out of scale as well.

VARIANTS OF CZOCHRALSKI GROWTH

Some significant modifications of the basic Czochralski technique have been introduced to deal with special problems or to attempt improvement of crystal quality. In the latter case, one of the main objectives is to minimize "striations," i.e. bands of high impurity or dopant concentration that result from flow irregularities such as oscillation, pluming, or vortex breakup. As the fluid dynamics of these variants requires special treatment, we review them here briefly.

Accelerated Crucible Rotation

Some time ago it was realized that crystal growth from a high-temperature solution could be enhanced if the solution were kept more thoroughly stirred by periodically reversing the direction of crystal rotation (Scheel & Schulz-Dubois 1971a, Scheel et al. 1971, Schulz-Dubois 1972, Scheel 1972, Mueller-Krumbhaar & Scheel 1974). This has been termed the accelerated crucible rotation technique (ACRT). Accelerated crystal rotation is sometimes also employed. More recently (Patrick & Westdorp 1977, Scheel & Mueller-Krumbhaar 1980) the technique has been applied to melt growth. The acronym is then sometimes modified to CACRT, signifying "Czochralski ACRT." The possibility that melt growth could be improved in this manner had, in fact, been anticipated some time earlier (Scheel & Schulz-Dubois 1971b).

The rationale behind CACRT is distinctive. Most techniques for alleviating striations seek to make the flow as quiescent as possible, so that transport of impurities and dopant is effected largely by diffusion rather than convection. Instead, CACRT seeks to "homogenize" the flow, breaking up vortices and other irregularities before significant pockets of trace constituents can build up.

Digital simulation of ACRT has been carried out (Mihelčič et al. 1981, 1982). As part of the study, the computer output was used to generate motion pictures that vividly depict ACRT at work.

Liquid-Encapsulated Czochralski (LEC)

The important III–V semiconductors present a problem for crystal growers because they dissociate when heated. In straightforward Czochralski growth, the more volatile component would therefore distill off, leaving a stoichiometrically incorrect melt. This can be avoided by covering the melt with a layer of immiscible liquid encapsulant, e.g. molten boric oxide, and operating with an external pressure above the dissociating pressure.

Liquid-encapsulated Czochralski growth was developed some time ago (Metz et al. 1962) and passed through a period when it was only of minor importance. However, the recent surge of interest in electronics applications of gallium arsenide has revitalized the subject (Burggraaf 1982).

It is risky to use the bulk-flow approximation for LEC, because the interface between the melt and encapsulant (liquid-liquid rather than liquid-gas) can support large-amplitude gravity waves. This is probably not a problem if the encapsulant layer is sufficiently thin, but some caution is certainly recommended.

Czochralski Growth in an External Magnetic Field

Some important crystal materials are good electrical conductors in their liquid state. Molten silicon, for example, conducts nearly as well as mercury. This opens the possibility of using magnetohydrodynamic effects to modify the fluid motion. Preliminary experiments suggesting that this could be beneficial were carried out 15 years ago for Czochralski growth (Witt et al. 1970), and even earlier for other growth systems (Chedzey & Hurle 1966, Utech & Flemings 1966). More recently, significant improvements in the quality of silicon crystals have been obtained by Czochralski growth in a strong magnetic field (Hoshikawa et al. 1980, 1981, Suzuki et al. 1981, Kim et al. 1981, Kim 1982).

The specific case of Czochralski flow in an axial magnetic field (Kim et al. 1981) can be simulated with relatively minor adaptations of the codes (described in the preceding section) that deal with the rotationally symmetric case. It is not necessary to bring the full weight of MHD theory to bear on the problem, because the magnetic Reynolds number (analogous to the hydrodynamic Reynolds number, with electrical resistivity playing the role of kinematic viscosity) is small compared with unity (about 0.04 for commercial growth of silicon). This leads (Langlois & Walker 1982) to a twofold simplification: (a) The induced magnetic field is negligible (no need to advance the induction equations prognostically!), so the induced current can be calculated from Ohm's law for a moving medium by use of the applied field in the $\mathbf{v} \times \mathbf{B}$ term; (b) to first order, the electric field is irrotational and can therefore be derived from a potential. Moreover, the potential can be eliminated from the problem in the same way that pressure is eliminated in the vorticity–stream function method. The rotational symmetry implies that there is no potential gradient in the azimuthal direction, so the azimuthal current is simply proportional to the product of the radial velocity and the applied field. The radial and axial currents can be derived from a current function, which is completely analogous to the Stokes stream function. It obeys a diagnostic equation that can be derived from the meridional components of Ohm's law by cross-differentiation and subtraction. The differential operator in that equation is precisely that of (10). Hence, it can be solved by the same computational method (Golub & Langlois 1979). For silicon at least, and probably for many other useful materials as well, the equation governing the temperature remains unchanged, because Joule heating is negligible (Langlois & Lee 1983b).

At very large values of the magnetic Reynolds number, e.g. for superconductors or for MHD on the astrophysical scale, the magnetic field tends to become "frozen in," i.e. each line of force behaves as if it were permanently attached to the fluid (Chandrasekhar 1961). If the external field is a strong one, this places severe constraints upon the possible motions. "Freezing in" has no bearing on most instances of liquid-metal MHD on the laboratory scale, including magnetic Czochralski flow, since the magnetic Reynolds number is quite small, as indicated above. In this case, the MHD influence on the motion is induction drag, which tends to damp out motion across the (fixed) lines of force. Figure 6 illustrates the effect for molten silicon (Langlois & Lee 1983a). When no field is applied, the flow is unsteady, but at large time it never differs much from the circulation shown on the left. The parameters of the motion are $\rho_R = 1.851$, $R = 0.656$, $\rho_\Omega = -0.680$, $\mathrm{Pr} = 0.0056$, $\mathrm{Re} = 38{,}000$, $\mathrm{Gr} = 10^{10}$, $\mathrm{Ma} = 43{,}000$, $\varepsilon^* = 1.72$, $\Delta_r T = 0.05$. The pattern is fairly characteristic of flow in a rotating crucible when thermocapillary convection is included in the model. The strong eddy positioned halfway between the axis and the crucible wall results from buoyant convection, augmented by crucible rotation. It is too strong to permit a tip vortex from forming on the rotating crystal. Closer to the crucible wall we find the thermocapillary eddy. Although it is generated at the top surface of the melt, the Taylor-Proudman effect gives it depth.

The right side of Figure 6, shows the result of applying a 2-kG axial field. Since the same contour spacing for the Stokes stream function is used in both parts of the figure, the calming effect of the magnetic field is quite evident.

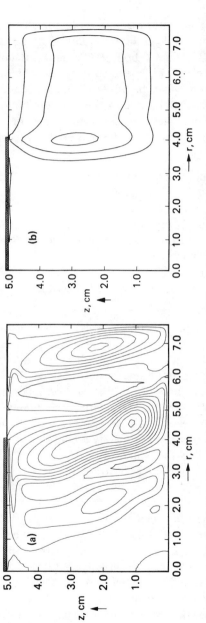

Figure 6 (Langlois & Lee 1983a) Meridional circulations in silicon growth. (*a*) No applied magnetic field. (*b*) Applied axial field of 2 kG. The contour spacing is 0.5 cm^3 s^{-1}.

Crystal Growth in Space

The advent of orbiting laboratories makes it possible to process materials in a low-gravity environment. Even in such vehicles, gravity does not vanish completely: There is a residue of about 10^{-4} times Earth normal because of gravity gradient, vehicle rotation, and solar wind. Moreover, the "microgravity" can vary somewhat because of attitude control maneuvers, movements of the crew, etc. The range of this "g jitter" is from about 10^{-5} to 10^{-3} times Earth normal. Nevertheless, even in the worst case, gravity is greatly reduced. Four implications for materials processing have been noted (Carruthers 1977):

1. Containerless handling of liquids becomes possible. Since the container is often a serious source of contamination, this development is of major significance.
2. Buoyant convection is greatly reduced. As a consequence, however, thermocapillary convection takes on added import (Schwabe 1981).
3. Sedimentation is eliminated.
4. Hydrostatic pressure, which influences phase equilibria, is reduced essentially to zero unless the experimenter wishes it otherwise.

One semiquantitative numerical simulation of Czochralski flow in microgravity has been published (Langlois 1980). It is unlikely that the bulk-flow approximation is valid under such circumstances. However, the material studied (aluminum oxide) has a rather high surface tension, which tends to suppress large-amplitude surface waves. The results should therefore give some idea about what happens, if we make due allowance for a smooth geometrical distortion of the entire melt volume.

The simulation suggested that, for Al_2O_3 growth in microgravity, thermocapillary convection is not merely a major effect—compared with buoyant convection, it is the whole story. Turning off the microgravity changed the stream function values by less than one part in a thousand. On the other hand, turning off thermocapillarity produced a flow field that was too weak by a factor of 2000 and a temperature field that was qualitatively wrong. The latter included regions of intense supercooling because of radiation from the free surface uncompensated by convective upwelling of warmer fluid. This suggests that, for hot melts, thermocapillary convection might actually be beneficial in microgravity: Without it, rather ingenious radiation shadowing would be needed to prevent flaring.

Two of the aforementioned reviews (Hurle 1977, Pimputkar & Ostrach 1981) treat crystal growth in space quite extensively. In view of the recent emergence of the space shuttle as a working vehicle, anything that could be

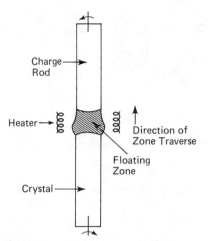

Figure 7   Schematic depiction of floating zone melting.

added now would probably be obsolete before the present review appears in print.[3]

OTHER METHODS OF GROWING CRYSTALS FROM THE MELT

When large crystals are unnecessary, other procedures may have advantages over Czochralski growth. Without attempting an extensive treatment, we conclude this review by summarizing two of the more important methods.

Zone Melting

In this technique, a heating element is gradually passed over a polycrystalline solid, melting it, and leaving a single crystal after the hot zone passes. This can be done in a boat or crucible, but nearly containerless growth can be achieved in a terrestrial environment by using the variant called *floating-zone melting*, wherein the hot zone is passed over a rod supported only at its ends (Figure 7). The crystal and nutrient rods are

---

[3] The above was written in November 1983. The typescript is now on my desk for proofreading, and the first Spacelab mission has just ended. Preliminary reports indicate that one of the experiments, under the direction of L. Napolitano, illustrates the predominance of thermocapillary convection in microgravity. Motion of tracer particles in fluid bridging a rather wide (5 cm) gap between plates can be switched on and off at will by controlling the relative temperature of the plates.

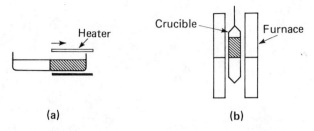

(a)   (b)

Figure 8   Schematic depiction of normal-freezing techniques. (*a*) Chalmers. (*b*) Bridgman-Stockbarger.

sometimes differentially rotated, for the same reasons that the crystal and crucible may be rotated in Czochralski growth.

In terrestrial applications of floating-zone melting, the gravitational force limits the size of the float zone. The smaller geometry tends to make thermocapillary convection more important, and this behavior has been intensively studied (Schwabe 1981). Digital simulation of melt flow in this technique has not been carried out as extensively as for Czochralski growth, but some work has been done (Kobayashi & Wilcox 1982).

Normal Freezing (Other Than Czochralski)

In zone melting, only a small portion of the charge is molten at any time. When the entire charge is initially molten and then is directionally solidified, the process is called *normal freezing*. Czochralski growth is an example, but there are others (Figure 8). With the Bridgman technique, a crucible of melt is lowered through the furnace. In the closely related Stockbarger technique, heat is extracted from the bottom of the system without relative motion of crucible and furnace. The same idea can be used with a horizontal rather than vertical orientation. In the Chalmers technique, sometimes called horizontal Bridgman, the melt is contained in a boat that is gradually withdrawn from a furnace (or, equivalently, the furnace is moved away from the boat). All of these techniques involve melt convection, and this too has been the subject of some study (Azouni 1981, Pimputkar & Ostrach 1981, Chang & Brown 1983, Crochet et al. 1983b, 1984).

Literature Cited

Azouni, M. A. 1981. Time-dependent natural convection in crystal growth systems. *PhysicoChem. Hydrodyn.* 2:295–309
Balasubramaniam, R., Ostrach, S. 1984. Fluid motion in the Czochralski method of crystal growth. *PhysicoChem. Hydrodyn.* 5:3–18
Brandle, C. D. 1977. Simulations of fluid flow in $Gd_3Ga_5O_{12}$ melts. *J. Cryst. Growth* 42:400–4

Burggraaf, P. S. 1982. GaAs bulk-crystal growth technology. *Semicond. Int.* June: 44–68

Carruthers, J. R. 1967. Radial solute segregation in Czochralski growth. *J. Electrochem. Soc.* 114:959–62

Carruthers, J. R. 1975. Crystal growth from the melt. In *Treatise on Solid State Chemistry*, ed. N. B. Hannay, 5:325–406. New York: Plenum

Carruthers, J. R. 1977. Crystal growth in a low gravity environment. *J. Cryst. Growth* 42:379–85

Carruthers, J. R. 1979. Dynamics of crystal growth. In *Crystal Growth: A Tutorial Approach*, ed. W. Bardsley, D. T. J. Hurle, J. B. Mullin, pp. 157–88. Amsterdam: North-Holland

Carruthers, J. R., Nassau, K. 1968. Nonmixing cells due to crucible rotation during Czochralski crystal growth. *J. Appl. Phys.* 39:5205–14

Chandrasekhar, S. 1961. *Hydrodynamic and Hydromagnetic Stability.* Oxford: Clarendon. 652 pp.

Chang, C. J., Brown, R. A. 1983. Radial segregation induced by natural convection and melt/solid interface shape in vertical Bridgman growth. *J. Cryst. Growth* 63: 343–64

Chedzey, H. A., Hurle, D. T. J. 1966. Avoidance of growth-striae in semiconductor and metal crystals grown by zone-melting techniques. *Nature* 210:933–34

Crochet, M. J., Wouters, P. J., Geyling, F. T., Jordan, A. S. 1983a. Finite-element simulation of Czochralski bulk flow. *J. Cryst. Growth.* 65: 153–65

Crochet, M. J., Geyling, F. T., Van Schaftingen, J. J. 1983b. Numerical simulation of the horizontal Bridgman growth of gallium arsenide crystal. *J. Cryst. Growth.* 65:166–72

Crochet, M. J., Geyling, F. T., Van Schaftingen, J. J. 1984. Finite element methods for calculating the growth of semi-conductor crystals. Presented at Int. Symp. Finite Elem. Methods Flow Probl., 5th, Austin

Golub, G. H., Langlois, W. E. 1979. Direct solution of the equation for the Stokes stream function. *Comput. Methods Appl. Mech. Eng.* 19:391–99

Goss, A. J., Adlington, R. E. 1959. Effects of seed rotation on silicon crystals. *Marconi Rev.* 22:18–36

Hoshikawa, K., Kohda, H., Hirata, H., Nakanishi, H. 1980. Low oxygen content Czochralski silicon crystal growth. *Jpn. J. Appl. Phys.* 19:L33–36

Hoshikawa, K., Hirata, H., Nakanishi, H., Ikuta, K. 1981. Control of oxygen concen-

tration in CZ silicon growth. *Proc. Int. Symp. Silicon Mater., Sci., Technol., 4th, Minneapolis. Volume on Semiconductor Silicon*, pp. 101–12. Princeton, N.J.: Electrochem. Soc.

Hurle, D. T. J. 1977. Hydrodynamics in crystal growth. In *Crystal Growth and Materials*, ed. E. Kaldis, H. J. Scheel, pp. 550–69. Amsterdam: North-Holland

Hurle, D. T. J., Jakeman, E. 1981. Introduction to the techniques of crystal growth. *PhysicoChem. Hydrodyn.* 2:237–44

Jones, A. D. W. 1983a. An experimental model of the flow in Czochralski growth. *J. Cryst. Growth* 61: 235–44

Jones, A. D. W. 1983b. Spoke patterns. *J. Cryst. Growth.* 63:70–76

Jones, A. D. W. 1984. Hydrodynamics of Czochralski growth—A review of the effects of rotation and buoyancy forces. In *Progress in Crystal Growth and Characterization.* Vol. 10. In press

Kim, K. M. 1982. Suppression of thermal convection by transverse magnetic fields. *J. Electrochem. Soc.* 129:427–29

Kim, K. M., Schwuttke, G. H., Smetana, P. 1981. Apparatus for Czochralski silicon crystal growth through axial magnetic field fluid flow damping. *IBM Tech. Discl. Bull.* 24:3376–77

Kim, K. M., Kran, A., Smetana, P., Schwuttke, G. H. 1983. Computer simulation and controlled growth of large diameter Czochralski silicon crystals. *J. Electrochem. Soc.* 130:1156–60

Kobayashi, N. 1978. Computational simulation of the melt flow during Czochralski growth. *J. Crystal. Growth* 52:425–34

Kobayashi, N. 1980. Computer simulation of heat, mass and fluid flows in a melt during Czochralski crystal growth. *Comput. Methods Appl. Mech. Eng.* 23:24–33

Kobayashi, N. 1981. Heat transfer in Czochralski crystal growth. In *Preparation and Properties of Solid State Materials*, ed. W. R. Wilcox, 6:119–253. New York/Basel: Marcel Dekker

Kobayashi, N., Arizumi, T. 1975. Computational analysis of the flow in a crucible. *J. Cryst. Growth* 30:177–84

Kobayashi, N., Wilcox, W. R. 1982. Computational studies of convection due to rotation in a cylindrical floating zone. *J. Cryst. Growth* 59:616–24

Kyrazis, D. T. 1977. A numerical modeling of the fluid dynamic stability of Hagen-Poiseuille flow. *Rep. UCRL-52289*, Lawrence Livermore Lab., Livermore, Calif.

Landau, L. D., Lifshitz, E. M. 1959. *Fluid Mechanics.* London/Paris/Frankfurt: Pergamon. 536 pp.

214 LANGLOIS

Langlois, W. E. 1977a. Vorticity-streamfunction computation of incompressible fluid flow with an almost-flat free surface. *Appl. Math. Model.* 1 : 196–98

Langlois, W. E. 1977b. Digital simulation of Czochralski bulk flow in a parameter range appropriate for liquid semiconductors. *J. Cryst. Growth* 42 : 386–99

Langlois, W. E. 1979a. Effect of the buoyancy parameter on Czochralski bulk flow in garnet growth. *J. Cryst. Growth* 46 : 743–46

Langlois, W. E. 1979b. Crucible with convex-inward bottom for oxide crystal growth. *IBM Tech. Discl. Bull.* 22 : 326

Langlois, W. E. 1980. Digital simulation of Czochralski bulk flow in microgravity. *J. Cryst. Growth* 48 : 25–28

Langlois, W. E. 1981a. Convection in Czochralski growth melts. *PhysicoChem. Hydrodyn.* 2 : 245–61

Langlois, W. E. 1981b. Conservative differencing procedures for rotationally symmetric flow with swirl. *Comput. Methods Appl. Mech. Eng.* 25 : 315–33

Langlois, W. E. 1982. A parameter sensitivity study for Czochralski bulk flow of silicon. *J. Cryst. Growth* 56 : 15–19

Langlois, W. E. 1983. Czochralski bulk flow of silicon at large melt aspect ratio. *J. Cryst. Growth* 63 : 67–69

Langlois, W. E., Lee, K. J. 1983a. Digital simulation of magnetic Czochralski flow under various laboratory conditions for silicon growth. *IBM J. Res. Dev.* 27 : 281–84

Langlois, W. E., Lee, K. J. 1983b. Czochralski crystal growth in an axial magnetic field: effect of Joule heating. *J. Cryst. Growth* 62 : 481–86

Langlois, W. E., Shir, C. C. 1977. Digital simulation of flow patterns in the Czochralski crystal pulling process. *Comput. Methods Appl. Mech. Eng.* 12 : 145–52

Langlois, W. E., Walker, J. S. 1982. Czochralski crystal growth in an axial magnetic field. *Proc. Conf. Comput. Asymptotic Methods Boundary Intern. Layers, 2nd, Dublin*, pp. 209–304. Dublin : Boole

Lee, K. J., Langlois, W. E., Kim, K. M. 1984. Digital simulation of oxygen transfer and oxygen segregation in magnetic Czochralski growth of silicon. *Physico-Chem. Hydrodyn.* 5 : 135–41

Lighthill, M. J. 1953. Theoretical considerations on free convection in tubes. *Q. J. Mech. Appl. Math.* 6 : 398–439

Metz, E. P. A., Miller, R. C., Mazelski, R. 1962. A technique for pulling single crystals of volatile materials. *J. Appl. Phys.* 33 : 2016–17

Mihelčič, M., Schroeck-Pauli, C., Wingerath, K., Wenzl, H., Uelhoff, W., van der Hart, A.

1981. Numerical simulation of forced convection in the classical Czochralski method, in ACRT and CACRT. *J. Cryst. Growth* 53 : 337–54

Mihelčič, M., Schröck-Pauli, C., Wingerath, K., Wenzl, H., Uelhoff, W., van der Hart, A. 1982. Numerical simulation of free and forced convection in the classical Czochralski method and in CACRT. *J. Cryst. Growth* 57 : 300–17

Miller, D. C. 1982. The role of fluid flow phenomena in the Czochralski growth of oxides. In *Materials Processing in the Reduced Gravity Environment of Space*, ed. G. E. Rindone, pp. 373–87. Amsterdam : Elsevier

Mueller-Krumbhaar, H., Scheel, H. J. 1974. Controlled crystal pulling with accelerated crucible rotation. *IBM Tech. Discl. Bull.* 17 : 903–4

Patrick, W. J., Westdorp, W. A. 1977. Control of oxygen in silicon crystals. *US Patent No.* 4,040,895

Pimputkar, S. M., Ostrach, S. 1981. Convective effects in crystals grown from the melt. *J. Cryst. Growth* 55 : 614–46

Robertson, D. S. 1966. A study of the flow patterns in liquids using a model Czochralski crystal growing system. *Br. J. Appl. Phys.* 17 : 1047–50

Scheel, H. J. 1972. Accelerated crucible rotation: a novel stirring technique in high-temperature solution growth. *J. Cryst. Growth* 13/14 : 560–65

Scheel, H. J., Mueller-Krumbhaar, H. 1980. Crystal pulling using ACRT. *J. Cryst. Growth* 49 : 291–96

Scheel, H. J., Schulz-Dubois, E. O. 1971a. Flux growth of large crystals by accelerated crucible-rotation technique. *J. Cryst. Growth* 8 : 304–6

Scheel, H. J., Schulz-Dubois, E. O. 1971b. Crystal rotation in crystal growth from melt. *IBM Tech. Discl. Bull.* 14 : 1631

Scheel, H. J., Schmid, H., Schulz-Dubois, E. O. 1971. Crystal growth with accelerated crucible rotation. *IBM Tech. Discl. Bull.* 3 : 1571

Schulz-Dubois, E. O. 1972. Accelerated crucible rotation: hydrodynamics and stirring effect. *J. Cryst. Growth* 12 : 81–87

Schwabe, D. 1981. Marangoni effects in crystal growth melts. *PhysicoChem. Hydrodyn.* 2 : 263–80

Shiroki, K. 1977. Simulations of Czochralski growth on crystal rotation rate influence in fixed crucibles. *J. Cryst. Growth* 40 : 129–38

Suzuki, T., Isawa, N., Okubo, Y., Hoshi, K. 1981. CZ silicon crystal grown in a transverse magnetic field. *Proc. Int. Symp. Silicon Mater., Sci., Technol., 4th, Minneapolis. Volume on Semiconductor Silicon*, pp. 90–100. Princeton, N.J.: Electrochem. Soc.

Turovskii, B. M., Mil'vidskii, M. G. 1962. Modeling the mixing of a melt during the growing of crystals according to the Czochralski method. *Sov. Phys.-Crystallogr.* 6:606–8

Utech, H. P., Flemings, M. C. 1966. Eliminating of solute banding in indium antinomide crystals by growth in a magnetic field. *J. Appl. Phys.* 37:2021–24

Wilson, L. O. 1978. A new look at the Burton, Prim, and Slichter model of segregation during crystal growth from the melt. *J. Cryst. Growth* 44:371–76

Witt, A. F., Herman, C. J., Gatos, H. C. 1970. Czochralski-type crystal growth in transverse magnetic fields. *J. Mater. Sci.* 5:822–24

Ann. Rev. Fluid Mech. 1985. 17:217–37

SOUND TRANSMISSION IN THE OCEAN

Robert C. Spindel

Department of Ocean Engineering, Woods Hole Oceanographic Institution, Woods Hole, Massachusetts 02543

1. INTRODUCTION

The ocean is a fluid waveguide, bounded by the air-sea surface above and the topography of the bottom below, through which sound propagates according to the scalar wave equation

$$D^2 p = c^2 \nabla^2 p, \tag{1}$$

where p is the acoustic pressure perturbation, c is the speed of sound, and the operator $D = \partial/\partial t + \mathbf{u} \cdot \nabla$ represents the material derivative in the presence of a fluid velocity field $\mathbf{u}$. The sound speed and $\mathbf{u}$ are functions of all spatial coordinates and time, but simplifying assumptions are usually made. Sound-speed variations in range are weak compared with vertical variations; to first order they are neglected. The velocity field of the fluid is generally taken to be zero because it is much less than the sound propagation speed. Furthermore, for a large class of problems of practical interest, the temporal scale of oceanic variation is much greater than the duration of an acoustic transmission, so that variations with t can be neglected. Thus the classic problem in sound transmission in the ocean consists of solving the simplified wave equation

$$\frac{\partial^2 p}{\partial t^2} = c^2 \nabla^2 p \tag{2}$$

for a variety of sound speed versus depth profiles, $c(z)$, and boundary conditions. During the period between 1912 (when the Titanic disaster stimulated work on acoustic echo-ranging systems for iceberg detection) and the Second World War, the predominant activity in ocean acoustics was the development of hydroacoustic transducers for transmitting and

217

0066–4189/85/0115–0217$02.00

receiving sound, mostly for echo- and depth-ranging systems (Lasky 1974). Acoustic modeling requirements were not very sophisticated and were satisfied by solutions to this range- and time-invariant form of the wave equation. Much of the research in modern ocean acoustics has to do with the more complicated propagation picture that emerges when less restrictive, more realistic conditions are allowed.

During and following World War II, the complexion of underwater acoustics changed. Submarine and surface sonar systems were designed to operate at long ranges and with high power. Their performance was compromised by random volume inhomogeneities, by surface and bottom reflections, and by more subtle forms of range variation in the (mostly) depth-dependent sound-speed profile. Submarine warfare led to the development of passive listening, detection and localization systems that required more detailed knowledge of the refractive-index variations in the ocean. Toward the end of the War, the existence of a ubiquitous deep sound channel was discovered (Ewing & Worzel 1948). Efficient use of the channel, which allowed acoustic propagation to thousands of kilometers, called for models that could handle range-varying sound-speed profiles. The military sector was not alone in its requirement for a more complete understanding of ocean acoustic properties. Oil and mineral exploration in coastal waters used acoustic prospecting techniques that depended upon an understanding of the interaction of sound with the sea floor. During this period, there were systematic measurements of the sound-speed environment of various ocean basins (Brown 1954, Jones & Von Winkle 1965), as well as of the ambient-noise background (Wenz 1962). Extensive measurements of bottom roughness were made in a variety of oceans in order to characterize scattering regimes (MacKenzie 1960, Clay 1966, Hampton 1974). Models of acoustic scattering from the time-varying surface of the sea were developed for incorporation into wave-equation boundary conditions (Eckart 1953, Fortuin 1970). Other aspects of the acoustic environment were also investigated. Viscous and shear absorption and attenuation due to dissolved salts were measured (Schulkin & Marsh 1962, Urick 1963, Thorp 1967, Mellen & Browning 1977, Schulkin & Marsh 1978), and acoustic scattering from fish, bubbles, and turbulence was studied (Andreeva 1964, Mintzer 1953, Chernov 1960, Tatarski 1961).

At the present time, the acoustic effects of the sea surface, bottom, and volume still are incompletely known and constitute the limiting factors in our understanding of sound transmission in the ocean. Even when detailed knowledge of the boundary and volume exists, it cannot easily be incorporated into appropriate boundary conditions for the solution of (1). Thus, much of the present research in ocean acoustics is focused on understanding the effects of physical ocean processes on sound transmission, as well as numerical and analytic methods of modeling acoustic propagation.

2. THE OCEAN ACOUSTIC ENVIRONMENT

As sound propagates through the ocean waveguide, it is subject to a wide variety of physical processes that alter the refractive index of the medium. There is the background sound-speed profile, which is treated largely as a deterministic feature and which, by analogy with the atmosphere, represents the climatology of the ocean. It is the depth-dependent profile $c(z)$ in (2) that has a vertical scale consistent with mean ocean depths (order 5 km) and horizontal variations limited only by the size of ocean basins, which are thousands of kilometers in extent. It is temporally variable, especially in the upper several hundred meters, on seasonal time scales. These latter changes are largely due to the influence of atmospheric inputs.

Superposed on the deterministic, climatological sound-speed profile is a smaller mesoscale structure that, by definition, comprises spatial scales of tens to hundreds of kilometers and time scales of days to several months. The class of phenomena comprising this category includes the complex rings and eddies spawned by the Gulf Stream and other current systems, oceanic fronts formed by the interaction of differing water masses, and major boundary currents. During the past several decades the magnitude of the contribution of the mesoscale to total ocean transport has become more evident (Wunsch 1981). It is conjectured that as much as 90% of the kinetic energy in the ocean is contained in variable mesoscale eddy fields rather than in the mean ocean circulation. Mesoscale variability, again by analogy with the atmosphere, is referred to as the ocean weather. The counterparts of high- and low-pressure centers are the anomalously warm and cold eddies of the mesoscale. Their spatial scales are roughly equivalent, but time constants in the atmosphere are measured in days; in the ocean, they are measured in months. The mesoscale can be treated deterministically, if sufficiently dense sampling of the sound-speed field has been accomplished, or as a stochastic perturbation of the climatological mean.

There is an ever finer and less energetic scale of perturbation due to oceanic mixing processes, small-scale circulation, turbulence, and instability phenomena. These are generally treated as random processes. They have vertical spatial scales of tens of meters to centimeters, but they can range to kilometers in the horizontal. The size of these disturbances is comparable to the size of an acoustic wavelength. Their temporal scale ranges from fractions of seconds up to inertial periods (hours). It spans the period over which many acoustic measurements are taken. Examples of processes comprising this group are internal waves, fine structure, microstructure, and turbulence.

The ocean acoustic field is almost completely determined by the sound speed, or index-of-refraction field, which in turn depends primarily on temperature, pressure, and salinity. The functional relationship between these parameters is expressed empirically and is given by a recent formula (Del Grosso 1974), simplified to first-order terms:

$$c = 1402.392 + 5.011T + 0.156P + 1.329S, \tag{3}$$

where c is in m s^{-1}, T in °C, P in kg cm^{-2}, and S in parts per thousand. Sound speed depends strongly on T and P and only weakly on S, since salinity variations are small. Over the world's oceans, sound speeds rarely exceed the limits $1430 < c < 1530$ m s^{-1}.

At mid-latitudes, $c(z)$ typically has, at a depth of about 1000 m, a value about 2 to 3% less than its values at the surface and bottom. This result is due primarily to the opposing effects of temperature, which dominates near the warmer surface and causes a rapid decrease in $c(z)$ as it decreases with depth, and hydrostatic pressure, whose increase with depth eventually results in an increasing $c(z)$. The transition results in a sound-speed minimum that focuses acoustic energy. In equatorial regions, the depth of the sound-speed minimum is nearer 1300 m; it gets shallower with increasing latitude, and in polar regions the minimum is at the surface. Representative mid-latitude Atlantic sound-speed profiles are shown in Figure 1.

A qualitative picture of oceanic sound propagation can be obtained by solving (2) in the geometrical-optics limit, where the sound field can be described completely in terms of rays whose characteristics depend only on c as a function of depth. Differential parametric equations for the rays can be derived from the eikonal equation or directly from Fermat's principle. In the two-dimensional form applicable to a horizontally isotropic ocean, the ray trajectories are described by a generalized form of Snell's law,

$$\frac{d\theta}{dx} = -\frac{\cos^2 \theta}{c}\frac{dc}{dz}, \tag{4}$$

where θ is the angle of the ray with the horizontal coordinate x. This equation may be integrated from any initial starting point and angle. The minus sign indicates that rays tend to bend downward in regions of positive sound-speed gradient and upward in regions of negative gradient. Thus, above the depth of the sound-speed minimum, rays curve downward, and below the minimum they curve upward, thereby exhibiting ducted, or channeled, propagation. Acoustic signals therefore cycle through the water column and can propagate to thousands of kilometers in range without ever interacting with the surface or bottom. The spatial wavelength of a complete cycle consisting of one upward and one downward loop is a function of the sound-speed profile. This scale length for energy refocusing is called a *convergence zone* and is typically 50 km.

The paths of wholly refracted (eigen-) rays connecting a source and receiver at 1300 m depth and 320 km range are shown in Figure 2. Many rays from the source arrive at the receiver from different directions and at different times, a pattern that gives rise to multipath propagation. The characteristic sound channel is clearly evident. Convergence-zone loops, regions devoid of rays (called *shadow zones*), and regions of partial focusing (called *caustics*), formed by the intersection of adjacent rays, are also present. The sequence of ray arrivals is also shown in the figure, from which it can be seen that most rays are resolved in time. They transit with unique average speeds that are functions of the ray geometry.

The ultimate range to which wholly refracted sound propagates is primarily a function of acoustic frequency f. In the absence of boundary

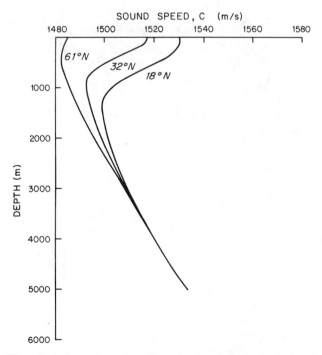

Figure 1   Climatological sound-speed profiles at various latitudes in the North Atlantic during the spring. The depth of the sound-speed minimum decreases with increasing latitude.

losses and chemical attenuation, absorptive losses are roughly proportional to f^2. Geometric spreading loss is proportional to R^{-2} for nonfocused, omnidirectional radiation and to R^{-1} for channeled propagation. Practical considerations (involving achievable source intensities, receiver sensitivities, and noise) limit ranges to hundreds of meters for $50 < f < 500$ kHz; to kilometers for $5 < f < 50$ kHz; to tens of kilometers for $1 < f < 5$ kHz; and to hundreds of kilometers for $f < 1$ kHz. For frequencies below 100 Hz, the range of transmission is limited by boundary interactions and not by fluid properties.

The time-varying ocean surface acts as a nearly perfect reflector, scattering in space almost all sound energy incident upon it and, because of its temporal variability, dispersing it in frequency. The ocean bottom has highly variable acoustic characteristics spanning the range from nearly

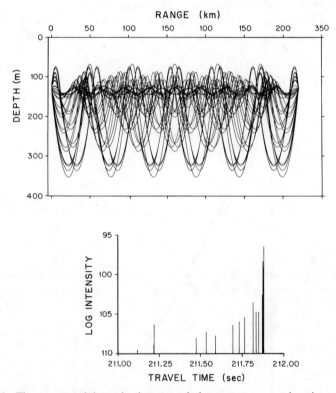

Figure 2   The upper panel shows the eigenray paths between a source and receiver at 1300 m depth separated by 320 km in a typical North Atlantic sound-speed environment. The lower panel shows the sequence of ray arrival times. The height of the arrival is proportional to the logarithm of the intensity.

perfect reflection to almost total attenuation. Seafloor roughness results in significant scattering.

These are the major features underlying ocean sound propagation. The large body of literature and research that has accumulated deals mostly with systematic measurements of acoustic propagation and oceanic parameters and with methods of solving (1) and (2) under various sound-speed and boundary conditions. A great deal of attention has been given both to establishing the conditions of validity for ray-theory solutions and to modifying ray theories to enable them to work in situations where the approximations of geometrical optics are not strictly valid. In general, ray-theory solutions are simpler than wave or normal-mode solutions, and it is often easier to interpret the ray picture. But ray theory is limited in applicability to high frequencies and cannot handle diffraction, caustics, and some scattering and focusing effects. It also breaks down near turning points, regions where rays reverse their up-down sense of direction. On the other hand, wave theories give complete solutions applicable to all conditions of frequency and sound-speed gradient. Unfortunately, unless simplifying assumptions are made, full solutions of the wave equation, including the effects of the surface and bottom boundaries (as well as range-dependent variations in the sound-speed profile), are difficult to obtain either numerically or analytically.

The effects of meso-, fine-, and microscale oceanic variations on acoustic signals, the development of efficient propagation models for range- and depth-varying propagation, and the inverse application of acoustics to measure ocean processes are three major topics in current ocean acoustics research. The meso-, fine-, and microscales are important because the performance of acoustic instruments and systems is ultimately limited by the intensity of these processes and by the spatial and temporal coherence scales they impose on the sound-speed field. A need has arisen for improved acoustic models, computationally efficient forms of the wave equation, that allow the calculation of the acoustic field in severely range- and depth-varying environments, such as those of the fronts and eddies of the mesoscale. Finally, present substantial understanding of the effects of ocean variability on acoustic transmissions has created the possibility of measuring ocean processes by extracting information from the acoustic field itself, since the propagating acoustic wave continually stores information about the environment through which it passes. Inverse methods can be applied to appropriate acoustic transmissions to deduce features of the ocean interior. These topics are reviewed in what follows.

There are many other active research areas, including shallow-water acoustics (Ingenito et al. 1978, Jensen & Kuperman 1983); very-high-frequency ($f > 50$ kHz) (Anderson & Zahuranec 1977) and very-low-frequency ($f < 10$ Hz) propagation; under-ice sound transmission and scattering from the frozen sea surface (Diachok 1980, Mikhalevsky 1981); sub-ocean-bottom (seismic) acoustics (Kuperman & Jensen 1980); acoustic diffraction at bathymetric discontinuities such as seamounts (Jordan & Medwin 1981), islands, and the shoreline; and various surface and bottom scattering effects. Some recent summaries of this research are available (Urick 1982, Clay & Medwin 1977).

3. ANALYTIC MODELS

Numerical models of acoustic propagation fall into two categories. First, there are the range-independent models, which assume a cylindrically symmetric ocean with a sound-speed profile that is a function of depth only. Surface and bottom boundaries are generally considered to be parallel planes. These models are fairly well developed, but they are inadequate for describing the acoustic field in a realistic, range-varying ocean. Second, there are range-dependent models, which allow for two- or three-dimensional variations in the sound-speed profile and can accommodate nonparallel boundaries. They are in a more elementary stage of evolution than the range-independent models (DiNapoli & Deavenport 1979), but they are crucial to understanding the effects of ocean variability on sound transmission.

The Parabolic Equation

The parabolic approximation to the wave equation was introduced into underwater acoustics by F. D. Tappert in 1972 (Spofford 1973, Hardin & Tappert 1973, Tappert 1977), although it had found prior extensive applications in optics. Its importance to sound propagation in the ocean is that it provides a computationally efficient means of solving for the acoustic field in an environment where the sound-speed field varies in range as well as depth. With this model, the effects of both large- and small-scale oceanic inhomogeneities can be examined. Baer (1980, 1981) and Lawrence (1983) used the parabolic approximation to model the effects of a Gulf Stream ring and a Tasman Sea eddy, respectively; these are both mesoscale disturbances. At the other end of the scale, Flatte & Tappert (1975) used it to model the effect of internal waves, which have vertical scales of meters.

The nature of the approximation is made clear by assuming a harmonic pressure disturbance in (2) to obtain the elliptic Helmholtz equation

$$\frac{\partial^2 p}{\partial r^2} + \frac{2a}{r}\frac{\partial p}{\partial r} + \frac{\partial p}{\partial z} + k_0^2\mu^2 p = 0, \tag{5}$$

where $k_0 = \omega/c_0$ and $\mu = c_0/c(r, z)$ are the acoustic wave number and index of refraction defined with respect to a reference sound speed c_0. For $a = 0$,

r and z are Cartesian coordinates; for $a = 1/2$, they are cylindrical coordinates; and for $a = 1$, they are spherical coordinates. The parabolic method consists of first representing the pressure disturbance in (5) as

$$p = (k_0 r)^{-a} \psi(r, z) \exp(i k_0 r) \tag{6}$$

to obtain

$$\frac{\partial^2 \psi}{\partial r^2} + 2 i k_0 \frac{\partial \psi}{\partial r} + \frac{\partial^2 \psi}{\partial z^2} + k_0^2 [(1 - \mu^2) + a(1 - a)(k_0 r)^{-2}] \psi = 0. \tag{7}$$

The approximation consists of neglecting the term $\partial^2 \psi / \partial r^2$, assumed small compared with $2 i k_0 \partial \psi / \partial r$. The resulting equation (in Cartesian coordinates) is

$$\frac{\partial^2 \psi}{\partial z^2} + 2 i k_0 \frac{\partial \psi}{\partial x} + k_0^2 [1 - \mu^2] \psi = 0. \tag{8}$$

The conditions for validity of this approximation are that the scale of variation of the medium be larger than the acoustic wavelength, that propagation paths remain close to the horizontal (wholly refracted paths rarely exceed $\theta = \pm 15°$), and that the backscattered field be small compared with the outgoing field.

The advantages of (8) over (7) are primarily numerical, since (7) is elliptic and must be solved in the entire region simultaneously. On the other hand, (8) can be marched out in range from some initial field at starting location x_0. Thus a range-dependent environment can be easily handled. A typical technique is to Fourier transform (8) in depth, march one step in range, and then transform back. Since two Fourier transforms are required for each range increment, the calculation can be time consuming for long ranges, high frequencies, or steep sound-speed gradients where many depth and range points must be computed. Surface and bottom interactions can also be difficult to handle. Thus the method is most appropriate for long-range, deep-ocean sound propagation, where low acoustic frequencies are used.

A fast and accurate algorithm using a split-step technique and the Fast Fourier transform has received widespread attention (Hardin & Tappert 1973, DiNapoli & Deavenport 1979). It was used to produce the Figure 3 parabolic equation and geometrical optics ray trace comparison for a shallow acoustic source. The ray solution predicts no acoustic field between convergence zones, whereas the parabolic equation gives a more realistic picture.

Unfortunately, the parabolic equation introduces errors in the phase velocities of propagating normal modes that can become severe even when the intrinsic assumptions underlying the approximation have been satisfied (McDaniel 1975a,b). There have been numerous attempts to resolve this difficulty; it still constitutes an active area of current research (Perkins & Baer 1982). Also, the parabolic equation gives no transit-time information. Hence it cannot be used when acoustic travel times or the arrival structure of wide-band pulses are needed, as in the case of ocean-bottom seismics or position triangulation. Although it is possible to use the parabolic equation for wide-band signals by repeatedly applying the model at discrete frequencies spanning the bandwidth of the source and then coherently

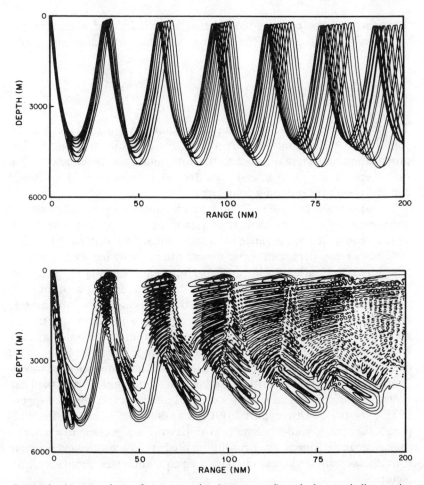

Figure 3 A comparison of ray acoustics (upper panel) and the parabolic-equation approximation (lower panel) for the acoustic field generated by a shallow sound source. The lower panel is contoured in acoustic intensity and shows the effects of diffraction. (Courtesy of the Office of Naval Research.)

summing the results at desired ranges, the computational burden of this procedure has prevented widespread application. However, the recent availability of high-speed, dedicated microcomputers may soon make wide-band modeling with the parabolic equation practical. In spite of these difficulties, the ability of the parabolic equation to handle range-varying propagation environments makes it a uniquely effective tool in modern ocean acoustics.

Range-Dependent Ray Solutions

In order to overcome the two present major shortcomings of the parabolic method—its inability to supply travel-time information and its compu-tational cost for high frequencies ($f > 500$ Hz), steep angles, and high gradients—there have been a number of recent developments in ray theory. In particular, range dependence is accommodated by allowing the sound speed to vary smoothly, and analytically, with range (Cornyn 1973); by segmenting the region between vertical sound-speed profiles into triangular sections and forcing a particular connection between them (Bucker 1971); or by tracing rays through discrete range intervals over which the environment is assumed constant (Weinberg & Zabalgogeazcoa 1977). The latter method is usually rapid, but difficulties arise at the edges of the range intervals, where discontinuities in the sound profile cause spurious effects. The analytic variation method leads to closed-form expressions for the sound speed but not for the rays. Although none of these methods is entirely satisfactory, the technique of segmenting into triangular sections has found substantial use in spite of the fact that establishing the connection between sections can lead to substantial errors unless done with care.

Path Integrals, Modes, Finite Elements

There are a number of other methods that have been used to obtain range-dependent solutions to the wave equation, though as yet they have limited application. Dashen (1979) has used Feynman path integrals to solve the parabolic equation, and Flatte (1983) has used this method to calculate the moments of the acoustic field in the presence of stochastic internal-wave fluctuations. A range-dependent normal-mode solution has been developed by Kanabis (1972). It appears to be able to handle substantial sound-speed gradients, but only limited comparisons with other models have been made. It is based on mode coupling between range sections that are themselves considered range independent in sound speed. Finite-element solutions have been proposed (Kalinowski 1979), but they are relatively undeveloped at the present time. They allow inclusion of bottom interactions in a fairly easily formulated representation, as well as arbitrary sound-speed variations, but they can become computationally unmanageable because of the large number of unknown parameters that arise from the discretization process.

4. MESOSCALE AND INTERNAL-WAVE EFFECTS

The mesoscale falls between the larger scale of the general (ocean basin and gyre) circulation and the smaller effects of internal waves, fine-structure, and microstructure. The perturbations these phenomena impose on the mean climatological sound-speed profile decrease in intensity from a maximum of about 2% (30 m s^{-1}) in the case of strong fronts. Mesoscale effects include planetary (or Rossby) waves, which are solutions of the linearized quasi-geostrophic equation of fluid dynamics in the presence of the variable Coriolis parameter. Planetary waves are the basis for the general class of phenomena known as eddies. They have spatial scales of 100 km in the horizontal, extend from the ocean surface (where they are most intense) to the bottom, and have temporal scales of 100 days. They are spatially and temporally much larger than the acoustic wavelength and acoustic travel times, respectively. Their effect on the acoustic field is treated deterministically except in the case of (proposed) long-term (>1 year) transmissions. Eddy fields have been observed in virtually all of the world's oceans. A review of eddy research is provided by Robinson (1983).

The ocean internal-wave field results from instabilities in the density stratification of the ocean on a rotating globe. A vertically displaced particle of water is restored to its equilibrium depth by a buoyancy force that is balanced by gravitational acceleration according to Newton's law. The displaced particle tends to oscillate with a frequency

$$n(z) = \left[\frac{g}{\rho}\frac{\partial\rho}{\partial z}\right]^{1/2}, \tag{9}$$

where ρ is the potential density and g is the gravitational constant. A solution of the dynamic equations of fluid motion driven by the Coriolis force is a set of horizontally traveling waves spanning the frequency range from $n(z)$ to the inertial frequency $\omega_i = (2\pi \sin L)/12$ h, where L is the latitude. At 30°N the inertial frequency is 1 cycle/day. The buoyancy frequency decreases approximately exponentially with depth. It has a peak of about 10 cycles/hour at the surface and an e^{-1} depth of about 1.5 km. The vertical scale of internal waves is about 10 m; the horizontal scale is hundreds of meters to kilometers. Thus internal waves have time scales comparable with acoustic transit times and vertical spatial scales comparable with the acoustic wavelength. Their acoustic effects are treated statistically.

Eddies and Fronts

Sound-speed contours through a Gulf Stream ring are shown in Figure 4. The eddy appears as a lenslike perturbation of the background sound-speed field. This is a cold-core eddy; the low-sound-speed interior is due to entrained cool water. It is most intense in the upper 1500 m where it raises the depth of the sound channel axis by about 500 m. The eddy will affect shallow cycling rays more than axial ones, will distort the multipath arrival pattern, and will change the energy distribution of the acoustic field.

These effects have been studied using geometric-optics ray theory (Vastano & Owens 1973, Weinberg & Zabalgogeazcoa 1977, Weinberg & Clark 1980, Henrick et al. 1980). The parabolic-equation method also has been applied to study eddy effects, especially changes in acoustic intensity (Baer 1980, 1981, Lawrence 1983). These authors examine ray-path changes due to vertical refraction, horizontal refraction, and eddy motion. They find that eddies significantly affect all characteristics of the acoustic field: Convergence zones can shift by tens of kilometers; regions of prior intense ensonification can become shadow zones; the multipath arrival sequence

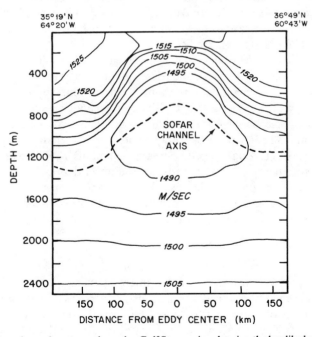

Figure 4   Sound-speed contours through a Gulf Stream ring showing the lenslike behavior of such eddies. (From Gemmel & Khedouri 1974.)

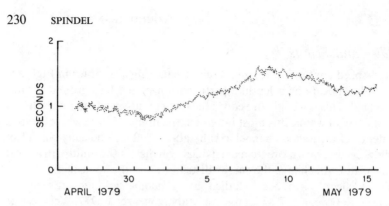

Figure 5   Travel-time variations along an acoustic path through a moving Gulf Stream meander. Only the last two seconds of total travel time are shown. The peak-to-peak variation is almost 0.6 s and arises from a north-south shift in the meander of about 300 km.

can change dramatically; and rays can be horizontally refracted by as much as a degree.

There have been a limited number of experiments to confirm these modeling results. In the Tasman Sea, Nysen et al. (1978) examined transmissions into, out of, and through warm-core eddies. Mosley (1979) conducted similar tests in an Atlantic cold-core eddy, while Beckerle et al. (1980) made observations in Atlantic warm-core eddies. For the most part, observed effects are well predicted.

The effects of frontal systems are equally strong. Temperature discontinuities across front boundaries can cause sound-speed changes of 20–30 m s^{-1}. An example of the shift in arrival time of an acoustic pulse transmitted through a moving Gulf Stream meander is shown in Figure 5 (Spindel & Spiesberger 1981). Variations in intensity of the received signal were as much as 10 dB. The abrupt rise of the sound channel in going from south to north through the Gulf Stream (or similar) front also causes a shift in acoustic energy distribution toward shallower depths.

Most studies of mesoscale effects have dealt with strong frontal systems and eddies. There is a lack of experience, either numerical or experimental, with less energetic, more typical eddies.

Internal Waves

Internal waves are considered to be the limiting factor in ocean transmissions at most frequencies of practical interest (50 Hz to 20 kHz). They cause amplitude and phase variations of acoustic-ray paths that are analogous to scintillations of optical paths caused by refractive-index variations in the atmosphere. They also limit the stability of acoustic paths, both temporally and spatially. Below 100 Hz, long acoustic wavelengths (> 30 m) are less affected by internal waves, and above 20 kHz, where wavelengths are less than a few centimeters, it is likely that fine-structure effects are more important.

The internal-wave field is characterized by a stationary, homogeneous, and horizontally isotropic frequency–wave number spectrum (Garrett & Munk 1975) that gives rise to variations in the refractive index of the ocean $c/c_0 \simeq 10^{-4}$. Horizontal scales are typically a factor of ten to a hundred greater than the (order) 10 m vertical scales, and fluctuations are most intense near the surface, decreasing rapidly with depth [approximately as $n^3(z)$].

A recent summary of acoustic–internal wave effects is provided by Flatte (1983). This work is based on approximately a decade's effort to employ the techniques of wave propagation in a random medium to the problem of acoustic–internal wave interaction. An analytic framework is constructed to analyze the moments of the received waveform and the statistics of acoustic intensity and phase along single acoustic paths. The character of acoustic fluctuations is expressed entirely by two parameters related to the strength and size of the time- and space-dependent internal-wave inhomogeneities. They are used to construct a strength–diffraction diagram that defines various forms of internal wave–acoustic scattering. These range from a region of minimal scattering, which corresponds to the geometric-optics limit, to fully saturated scattering, where amplitude and phase statistics along a ray obey random-walk (Rayleigh) statistics. In the intervening regions there are varying degrees of scattering (and randomness). Comparisons of this theory with data have been made, and most observations that fall in the region of validity of the theory have been corroborated. Almost all comparisons have been to time-dependent signal properties; comparisons with spatial characteristics remain to be done.

The theory has several limitations in spite of its successes. It is probably not adequate at frequencies below 100 Hz, nor is it valid near caustics or in regions of steep vertical sound-speed gradients. Also, it is a theory of single paths, requiring significant extrapolation for the common case of non-resolved multipath reception.

5. INVERSE TECHNIQUES

Inverse techniques are a broad class of methods used to estimate parameters in systems that can be modeled by a set of equations. They have found wide application in geophysics (Backus & Gilbert 1967), medicine (Bates et al. 1983), and other fields. The general idea is to solve the set

$$\mathbf{d} = F(\mathbf{p}) + \varepsilon \qquad (10)$$

for the parameters **p**, where **d** is a vector of data, ε is measurement noise, and F is a known operator. This expression is a statement of the forward problem. The methods of inversion generally are based on the theory of linear optimal estimation (Liebelt 1967), but nonlinear techniques can also be used. Inversion algorithms can be applied to various measures of the ocean acoustic field, such as intensity, phase, or travel time. They have been used to estimate properties of the ocean bottom (Newton 1981), to predict the ocean sound-speed profile (DeSanto 1984), and to reconstruct three-dimensional images of the interior sound-speed field (Munk & Wunsch 1979). In the latter case, inverse methods are used to infer the sound-speed field, which in turn is used to deduce physical ocean properties.

The use of inverse methods implies that a satisfactory solution to the forward problem can be obtained, i.e. a model can be constructed to account for observed properties of the field. The success that present acoustic and geoacoustic models have achieved in this respect has been one of the accomplishments of the past decade.

Bottom Acoustics

At frequencies below several hundred hertz, sound waves penetrate the ocean bottom and interact with sub-bottom unconsolidated marine sediments and other geologic structures. In general, marine sediments have the properties of porous, fluid-filled media that support both shear and compressional waves. Energy is reflected, refracted, and absorbed. The characteristics of bottom scattering are important inputs into numerical propagation models.

Inverse techniques in which bottom parameters are obtained from measurements of the reflected acoustic field have been considered by a number of authors. A summary of methods is provided by Schwetlick (1983). A method proposed by Bleistein & Cohen (1979) uses a relationship between the scattered field of an impulse source and the Fourier transform of the characteristic function of the scatterers to reconstruct an image of the scatterers. It is applicable to cases where sub-bottom scatterers can be treated as localized anomalies. For cases where the inhomogeneities are small and slowly varying, an inverse technique based on an integral equation for the impedance (ρc, the product of density and sound speed) of the bottom is derived.

In many instances, the sub-bottom can be modeled as a series of sediment layers, each characterized by a unique sound-speed and density profile. There is a rich literature dealing with wave propagation in such layered media (Brekhovskikh 1960). Thomson (1984) proposes a method for reconstructing the impedance profile when this model applies. It is based on work by Candel et al. (1980) using an impulsive source but allowing for some (realistic) bandwidth and aperture limits. The acoustic field is decomposed into upgoing and downgoing waves within the bottom, and the reflection coefficient is obtained in the form of a nonlinear Fourier transform, which is inverted to yield the impedance-versus-depth profile of the bottom. A wide-band impulse source is assumed, and the response of the bottom is obtained via forward modeling. The data consist of the amplitude and phase of the reflected field as a function of frequency. If two impulses are used, the ρc impedance product can be separated into density and sound-speed profiles.

A second method employs narrow-band harmonic sources (Frisk et al. 1980, Stickler 1983). In this case, variations of the incident angle produce a set of data with angle as the independent variable, rather than frequency as is the case with impulsive sources. Again, the impedance profile can be separated into ρ and c profiles if an additional measure is introduced. Performing the experiment at two discrete frequencies satisfies this requirement.

These and other methods have been tested against synthetic data with encouraging results. The difficulty in applying inverse techniques to actual ocean measurements rests with requirements imposed on the experimental parameters. Source and receiver geometries must be well controlled, and acoustic source signatures must be precisely known and of sufficient bandwidth to resolve sub-bottom features. Few comparisons with field data have been made.

Acoustic Tomography

Inverse techniques have been applied to the measurement of mesoscale variations by using acoustic travel time as a measure of sound-speed fluctuations. The ocean waveguide is probed by acoustic pulses launched from the periphery of an ocean volume. The transit times through the interior are altered by mesoscale sound-speed changes, which are in turn estimated by inverse methods. The procedure is similar to that employed in computerized medical tomography, in which an image is reconstructed from projections (e.g. X-rays). In ocean tomography, an image of the sound-speed field is reconstructed from vertical slices between multiple acoustic sources and receivers.

The method is particularly attractive for a number of reasons. First, images are obtained from periphery measurements; it is not necessary to physically enter or disturb the interior. Second, it has the potential to measure fairly large ocean areas with relatively few instruments, since the information grows quadratically with the number of instruments, rather than linearly as with conventional point measurements. Third, a consequence of multipath propagation is that a single source-receiver pair samples the ocean vertically in a manner that would otherwise require a large number of vertically distributed instruments. Finally, the integrating properties of acoustic propagation produce average quantities that cannot be obtained with point measurements.

The tomographic measurement uses the travel times τ_i along i ray paths (eigenrays) to invert for c, where the travel times are given by

$$\tau_i = \int_{\text{path}_i} [c(x, y, z, t)]^{-1} ds, \tag{11}$$

where ds is an incremental distance along the path. The problem is usually linearized by assuming a reference state c_0 with perturbation δc (the climatological mean and additive perturbations); this results in an expression for the travel-time variations,

$$\delta\tau_i = \int_{\text{path}_i} c_0^{-2} \delta c \, ds. \tag{12}$$

The method has been tested successfully with synthetic data by Munk & Wunsch (1979). It has also been tested with field data obtained from a feasibility experiment conducted in the southern North Atlantic (Ocean Tomography Group 1982, Cornuelle 1983). In the latter case, the dynamics of a mesoscale eddy were observed for several months via three-dimensional sound-speed images produced by tomographic inversions. As in the case of sub-bottom sound-speed profile inversions, controlled experimental parameters are required (Spindel 1980, Spindel et al. 1982). Figure 6 is a sample of the inverted travel-time data. It is a sound-speed contour map for one day and at one depth during the experiment, and it shows a low-sound-speed, cold-core eddy in the center of the tomographic array. A series of maps such as this one, at various depths and sequenced in time, track the evolution of this mesoscale feature.

6. CONCLUDING REMARKS

This article has summarized only enough of the fundamentals of sound transmission in the ocean to give the nonexpert reader a basic understanding of the effects of the ocean environment on underwater acoustic propagation. Three prominent areas of current research—analytic modeling, mesoscale and internal-wave acoustic effects, and acoustic inverse techniques—are reviewed in greater detail. However, there are many other topics of intense interest that have not been covered: acoustic scattering from the air-water interface (or from the water-ice interface in Northern latitudes) and from bubbles, fish, and the ocean bottom; the transmission of

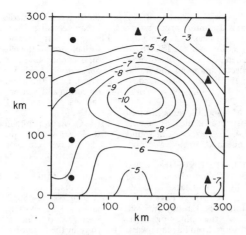

Figure 6 Contour map of sound speed for one day and at one depth (700 m), obtained from an acoustic tomography experiment. Sound transmitters were located at the positions marked by heavy dots; receivers are marked by triangles. The contours are in m s^{-1} with respect to a reference speed of 1510 m s^{-1}. The experiment resulted in a sequence of maps like this at various depths for a period of several months; these maps showed the evolution of a cold, low-sound-speed eddy.

very-low-frequency (< 10 Hz) signals, and characteristics of propagation at very high (> 50 kHz) frequencies; the spatial and temporal coherence of acoustic signals, and acoustics in shallow waters. The interested reader can pursue these subjects in such publications as the *Journal of the Acoustical Society of America* or in texts by Clay & Medwin (1977) and Tolstoy & Clay (1966).

ACKNOWLEDGMENT

I gratefully acknowledge the support of the Office of Naval Research under contract numbers N00014-84-C-0199 and N00014-82-C-0152. This is contribution number 5644 of the Woods Hole Oceanographic Institution.

Literature Cited

Anderson, N. R., Zahuranec, B. J. 1977. *Oceanic Sound Scattering Predictions.* New York: Plenum. 859 pp.

Andreeva, I. B. 1964. Scattering of sound by air bladders of fish in deep sound-scattering ocean layers. *Sov. Phys. Acoust.* 10: 17–20

Backus, G. E., Gilbert, J. F. 1967. Numerical applications of a formalism for geophysical inverse problems. *Geophys. J. R. Astron. Soc.* 13: 247–76

Baer, R. N. 1980. Calculations of sound propagation through an eddy. *J. Acoust. Soc. Am.* 67: 1180–85

Baer, R. N. 1981. Propagation through a three-dimensional eddy including the effects on an array. *J. Acoust. Soc. Am.* 69: 70–75

Bates, R. H. T., Gander, K. L., Peters, T. M. 1983. Overview of computerized tomography with emphasis on future developments. *Proc. IEEE* 71: 356–72

Beckerle, J. C., Baxter, L., Porter, R. P., Spindel, R. C. 1980. Sound propagation through eddies southeast of the Gulf Stream. *J. Acoust. Soc. Am.* 68:1750–67

Bleistein, N., Cohen, J. K. 1979. Inverse methods for reflector mapping and sound speed profiling. In *Ocean Acoustics*, ed. J. A. DeSanto, pp. 225–42. New York: Springer-Verlag. 285 pp.

Brekhovshikh, L. M. 1960. *Waves in Layered Media.* New York: Academic. 561 pp.

Brown, R. K. 1954. Measurement of sound velocity in the ocean. *J. Acoust. Soc. Am.* 26:64–72

Bucker, H. P. 1971. The RAVE (ray-wave) method. *Proc. NATO Conf. Geom. Acoust.*, ed. B. W. Conolly, R. H. Clark, 1:32–36. La Spezia, Italy: SACLANTCEN

Candel, S. M., deFillipi, F., Launay, A. 1980. Determination of the inhomogeneous structure of a medium from its plane-wave reflection response. Parts I and II. *J. Sound Vib.* 68:571–95

Chernov, L. A. 1960. *Wave Propagation in a Random Medium.* New York: McGraw-Hill. 168 pp.

Clay, C. S. 1966. Coherent reflection of sound from the ocean bottom. *J. Geophys. Res.* 71:2037–46

Clay, C. S., Medwin, H. 1977. *Acoustical Oceanography: Principles and Applications.* New York: Wiley. 544 pp.

Cornuelle, B. 1983. *Inverse methods and results from the 1981 acoustic tomography experiment.* PhD thesis. MIT/Woods Hole Oceanogr. Inst., Cambridge/Woods Hole, Mass. 359 pp.

Cornyn, J. J. 1973. GRASS: a digital-computer ray-tracing and transmission-loss-prediction system. *NRL Rep. 7621,* Nav. Res. Lab., Washington DC

Dashen, R. 1979. Path integrals for waves in random media. *J. Math. Phys.* 20:894–920

Del Grosso, V. A. 1974. New equation for the speed of sound in natural waters (with comparison to other equations). *J. Acoust. Soc. Am.* 56:1084–91

DeSanto, J. A. 1984. Oceanic sound speed profile inversion. *IEEE J. Oceanic Eng.* OE-9:12–17

Diachok, O. 1980. Arctic hydroacoustics. *Cold Reg. Sci. Technol.* 2:185–201

DiNapoli, F. R., Deavenport, R. L. 1979. Numerical models of underwater acoustic propagation. In *Ocean Acoustics*, ed. J. A. DeSanto, pp. 79–157. New York: Springer-Verlag. 285 pp.

Eckart, C. 1953. The scattering of sound from the sea surface. *J. Acoust. Soc. Am.* 25:566–70

Ewing, M., Worzel, J. L. 1948. Long range sound transmission. *Geol. Soc. Am. Mem.* 27. v+35 pp.

Flatte, S. M. 1983. Wave propagation through random media: contributions from ocean acoustics. *Proc. IEEE* 71:1267–94

Flatte, S. M., Tappert, F. D. 1975. Calculation of the effect of internal waves on oceanic sound transmission. *J. Acoust. Soc. Am.* 58:1151–59

Fortuin, L. 1970. Survey of the literature on reflection and scattering of sound waves at the sea surface. *J. Acoust. Soc. Am.* 47:1209–28

Frisk, G. V., Oppenheim, A. V., Martinez, D. R. 1980. A technique for measuring the plane-wave reflection coefficient of the ocean bottom. *J. Acoust. Soc. Am.* 68:602–12

Garrett, C., Munk, W. H. 1975. Space-time scales of internal waves: a progress report. *J. Geophys. Res.* 80:291–97

Gemmel, W., Khedouri, E. 1974. A note on sound ray tracing through a Gulf Stream eddy in the Sargasso Sea. *Nav. Oceanogr. Off. Tech. Rep. 6150-21-74,* Nav. Oceanogr. Off., Washington DC. 21 pp.

Hampton, L., ed. 1974. *Physics of Sound in Marine Sediments.* New York: Plenum. 567 pp.

Hardin, R. H., Tappert, F. D. 1973. Applications of the split-step Fourier method to the numerical solution of nonlinear and variable coefficient wave equations. *SIAM Rev.* 15:423–34

Henrick, R. F., Jacobson, M. J., Seigmann, W. L. 1980. General effects of currents and sound-speed variations on short-range acoustic transmissions in cyclonic eddies. *J. Acoust. Soc. Am.* 67:121–34

Ingenito, F., Ferris, R. H., Kuperman, W. A., Wolf, S. N. 1978. Shallow water acoustics: summary report. *NRL Rep. 8179,* Nav. Res. Lab., Washington DC

Jensen, F. B., Kuperman, W. A. 1983. Optimum frequency of propagation in shallow water environments. *J. Acoust. Soc. Am.* 73:813–19

Jones, L. M., Von Winkle, W. A. 1965. Sound velocity profiles in an area south of Bermuda. *Underwater Sound Lab. Rep. 632,* Nav. Underwater Syst. Cent., New London, Conn.

Jordan, E. A., Medwin, H. 1981. Scale model studies of scattering and diffraction at a seamount. *J. Acoust. Soc. Am.* 69:S59

Kalinowski, A. J. 1979. Application of the finite element method to acoustic propagation in the ocean. *NUSC Tech. Rep. S891,* Nav. Underwater Syst. Cent., New London, Conn.

Kanabis, W. B. 1972. Computer programs to calculate normal mode propagation and applications to analysis of explosive sound data in the Bifi range. *NUSC Tech. Rep. 4319,* Nav. Underwater Syst. Cent., New London, Conn. 37 pp.

Kuperman, W. A., Jensen, F. B., eds. 1980.

Bottom-Interacting Ocean Acoustics. New York: Plenum. 717 pp.

Lasky, M. 1974. A historical review of underwater acoustic technology 1916–1939. *J. Underwater Acoust.* 24:597–621

Lawrence, M. W. 1983. Modeling of acoustic propagation across warm core eddies. *J. Acoust. Soc. Am.* 73:474–85

Liebelt, P. B. 1967. *An Introduction to Optimal Estimation.* New York: Addison-Wesley. 273 pp.

MacKenzie, K. V. 1960. Reflection of sound from coastal bottoms. *J. Acoust. Soc. Am.* 32:221–31

McDaniel, S. T. 1975a. Propagation of a normal mode in the parabolic approximation. *J. Acoust. Soc. Am.* 57:307–11

McDaniel, S. T. 1975b. Parabolic approximations for underwater sound propagation. *J. Acoust. Soc. Am.* 58:1178–85

Mellen, R. H., Browning, D. G. 1977. Variability of low-frequency sound absorption in the ocean: ph dependence. *J. Acoust. Soc. Am.* 61:704–6

Mikhalevsky, P. N. 1981. Characteristics of cw signals propagated under the ice in the Arctic. *J. Acoust. Soc. Am.* 70:1717–22

Mintzer, D. 1953. Wave propagation in a randomly inhomogeneous medium. *J. Acoust. Soc. Am.* 25:922–27

Mosley, W. 1979. Summary of Freddex developments prior to at-sea phase. *Freddex Rep. No. 1*, Nav. Res. Lab., Washington DC. 70 pp.

Munk, W., Wunsch, C. 1979. Ocean acoustic tomography: a scheme for large-scale monitoring. *Deep-Sea Res.* 26A:123–61

Newton, R. G. 1981. Inversion of reflection data for layered media: a review of exact methods. *Geophys. J. R. Astron. Soc.* 65:191–215

Nysen, P. A., Scully-Power, P., Browning, D. G. 1978. Sound propagation through an East Australian current eddy. *J. Acoust. Soc. Am.* 63:1381–88

Ocean Tomography Group. 1982. A demonstration of ocean acoustic tomography. *Nature* 299:121–25

Perkins, J. S., Baer, R. W. 1982. An approximation to the three-dimensional parabolic-equation method for acoustic propagation. *J. Acoust. Soc. Am.* 72:515–22

Robinson, A. R., ed. 1983. *Eddies in Marine Science.* New York: Springer-Verlag. 609 pp.

Schulkin, M., Marsh, H. 1962. Absorption of sound in sea water. *J. Acoust. Soc. Am.* 34:864–65

Schulkin, M., Marsh, H. W. 1978. Low-frequency sound-absorption in the ocean. *J. Acoust. Soc. Am.* 63:43–48

Schwetlick, H. 1983. Inverse methods in the reconstruction of acoustical impedance profiles. *J. Acoust. Soc. Am.* 73:1179–86

Spindel, R. C. 1980. Multipath processing for ocean acoustic tomography. *Proc. IEEE EASCON 80*, pp. 165–70

Spindel, R. C., Spiesberger, J. L. 1981. Multipath variability due to the Gulf Stream. *J. Acoust. Soc. Am.* 69:982–88

Spindel, R. C., Worcester, P. F., Webb, D. C., Boutin, P. R., Peal, K. R., Bradley, A. M. 1982. Instrumentation for ocean acoustic tomography. *Proc. IEEE OCEANS 82*, pp. 92–99

Spofford, C. W. 1973. A synopsis of the AESD workshop or acoustic-propagation modelling by non-ray-tracing techniques. *AESD Tech. Note TN 73-05*, Off. Nav. Res., Washington DC

Stickler, D. C. 1983. Inverse scattering in a stratified medium. *J. Acoust. Soc. Am.* 74:994–1005

Tappert, F. D. 1977. The parabolic equation method. In *Wave Propagation and Underwater Acoustics*, ed. J. B. Keller, J. S. Papadakis, pp. 224–27. Berlin: Springer-Verlag

Tatarski, V. I. 1961. *Wave Propagation in a Turbulent Medium.* New York: McGraw-Hill. 285 pp.

Thomson, D. J. 1984. An inverse method for reconstructing the density and sound speed profiles of a layered ocean bottom. *IEEE J. Oceanic Eng.* OE-9:18–25

Thorp, W. H. 1967. Analytic description of the low-frequency attenuation coefficient. *J. Acoust. Soc. Am.* 42:270

Tolstoy, I., Clay, C. S. 1966. *Ocean Acoustics: Theory and Experiment in Underwater Sound.* New York: McGraw-Hill. 293 pp.

Urick, R. J. 1963. Low-frequency sound attenuation in the deep ocean. *J. Acoust. Soc. Am.* 35:1413–22

Urick, R. J. 1982. *Sound Propagation in the Sea.* Los Altos, Calif: Peninsula

Vastano, A. C., Owens, G. E. 1973. On the acoustic characteristics of a Gulf Stream cyclonic ring. *J. Phys. Oceanogr.* 3:470–78

Weinberg, N. L., Clark, J. G. 1980. Horizontal acoustic refraction through ocean mesoscale eddies and fronts. *J. Acoust. Soc. Am.* 68:703–6

Weinberg, N. L., Zabalgogeazcoa, X. 1977. Coherent ray propagation through a Gulf Stream ring. *J. Acoust. Soc. Am.* 62:888–94

Wenz, G. M. 1962. Acoustic ambient noise in the ocean: spectra and sources. *J. Acoust. Soc. Am.* 34:1936–56

Wunsch, C. 1981. Low frequency variability of the sea. In *Evolution of Physical Oceanography*, ed. B. A. Warren, C. Wunsch, pp. 342–77. Cambridge, Mass: MIT Press. 623 pp.

Ann. Rev. Fluid Mech. 1985. 17: 239–66

FLUID MODELING OF POLLUTANT TRANSPORT AND DIFFUSION IN STABLY STRATIFIED FLOWS OVER COMPLEX TERRAIN[1]

William H. Snyder[2]

Meteorology and Assessment Division, Environmental Sciences Research Laboratory, US Environmental Protection Agency, Research Triangle Park, North Carolina 27711

1. INTRODUCTION

Investigations of pollutant transport and dispersion in the atmosphere over complex relief are critical for the protection of air quality, because industrial enterprises and other sources of air pollution frequently locate within complex terrain. The ability to predict ground-level concentrations of air pollutants released from sources in or near complex terrain is required in order to determine the environmental impact of existing sources, to evaluate alternative new source locations, designs, and controls, and to estimate the effects of possible modifications to existing sources. Mathematical models that reliably predict concentrations when plumes are affected by complex terrain are not yet available. Field studies are very expensive and time consuming, and their results are not generally transferable to other sites.

Wind-tunnel studies on dispersion of effluents from industrial plants located in complex terrain have been conducted for over 40 years. Usually these studies were designed to answer specific questions, such as the

---

[1] The US Government has the right to retain a nonexclusive royalty-free license in and to any copyright covering this paper.
[2] On assignment from the National Oceanic and Atmospheric Administration, US Department of Commerce.

239

suitability of a particular plant location or the stack height necessary to avoid downwash; results have not been generally applicable to other sites. In the early wind-tunnel studies, care was taken to ensure a uniform wind profile and low turbulence intensity in the flow approaching the model. More recent studies have shown that the effects of shear and turbulence intensity (and scales) in the approach flow can strongly affect the basic flow structure over a model (e.g. the turbulence can effect changes in the locations of separation and reattachment points on the lee sides of hills). Therefore, in the past 15 years serious attempts have been made to simulate the atmospheric boundary layer, including stratification, and to conduct "generic studies" in which idealized terrain features have been used to obtain basic physical understanding. These studies are opposed to "engineering case studies" in which specific questions are answered with regard to particular installations. Generic studies are most useful when combined with theoretical modeling, because theory and experiments tend to direct each other.

   This paper summarizes the results of recent stratified towing-tank studies designed to (a) obtain basic understanding of flow and diffusion in complex terrain, (b) provide guidance on locating sources in complex terrain, and (c) provide "rules-of-thumb" for estimating concentrations when a source is located in complex terrain. Many (neutral) wind-tunnel studies have been conducted with the same purpose; these are summarized by Snyder (1984). The present review, however, concentrates on the dramatic and intriguing effects of the stratification.

2. SIMILARITY CRITERIA

Many factors affect the dispersion of pollutants in the atmosphere, including thermal effects, topography, rotation of the Earth, and buoyancy and moisture content of the effluent. Fluid model studies are desirable mostly because essential variables can be controlled, conditions can be reproduced, and the time and expense are much less than in full-scale studies. It is not generally possible, however, for all the factors influencing atmospheric dispersion to be included in a model. Normally, the similarity criteria are conflicting in some sense; it may be necessary to model one physical process at the expense of another. Put another way, certain nondimensional parameters in the prototype must be duplicated in the model for correct modeling. Almost invariably, duplication of all non-dimensional parameters is impractical or impossible. Therefore, decisions must be made as to which parameters are dominant; the less important ones must be ignored.

2.1   Basic Parameters

The basic nondimensional parameters for modeling atmospheric processes of different scales are generally agreed upon (Cermak et al. 1966, Snyder 1972, Cermak 1976, Snyder 1981) and require only brief mention here.

The Rossby number represents the ratio of advective or local accelerations to Coriolis accelerations. Snyder (1981) concluded from a review of the literature that the Rossby number need be considered when modeling prototype flows with length scales greater than about 5 km under neutral or stable conditions in relatively flat terrain. In modeling flows in complex terrain, we may expect local accelerations to be much more significant than in flat terrain; therefore, prototypes with length scales significantly larger than 5 km may be modeled while ignoring the Rossby number.

Three additional parameters are generally dismissed through use of the concept of Reynolds-number independence. They are (a) the Reynolds number itself, which is a ratio of inertial to viscous forces; (b) the Peclet number, which is the product of the Reynolds number and Prandtl number; and (c) the Reynolds-Schmidt product. These three parameters represent ratios of inertial properties of the flow to molecular properties of the medium. The basic concept of Reynolds-number independence is that if the flow is of sufficiently large Reynolds number, the main structure of the turbulence will be almost totally responsible for the transport of momentum and contaminant (heat or mass). Molecular properties (viscosity and diffusivities) contribute very little to the bulk transfer of momentum, mass, or contaminant; the main consequence of the molecular properties is to smooth out the very small-scale discontinuities of velocity, concentration, or temperature (i.e. the main effect is confined to setting the high wave-number cutoff of the respective spectra). However, molecular properties can be important in the flow very close to a smooth surface (Snyder 1981).

Current practice indicates that sufficiently large Reynolds numbers (sufficiently large to assure Reynolds-number independence) are attainable, at least for sharp-edged geometrical structures in ordinary meteorological wind tunnels. However, more work needs to be done to ascertain critical Reynolds numbers for more streamlined surfaces.

The square of the Froude number represents the ratio of inertial forces to buoyancy forces. It is closely related to the Richardson number, commonly used in the meteorological literature to describe the basic stability of the atmospheric boundary layer. However, a different interpretation of the Froude number is quite useful in considering stably stratified flow over hilly terrain. For example, when the flow approaching an isolated hill of height h has a uniform velocity profile and a linear density gradient, the appropriate

Froude number is

$$Fr = [U^2/(gh\Delta\rho/\rho)]^{1/2}, \tag{1}$$

where $\Delta\rho$ is the density difference between the base and the top of the hill. The square of the Froude number is the ratio of kinetic energy to potential energy, i.e. the ratio of the kinetic energy of a fluid parcel in the approach flow to the change in potential energy of a fluid element as it is raised from the base to the top of the hill. If the Froude number is much less than unity (very strong stratification), there is insufficient kinetic energy in the approach flow to raise fluid parcels from the base to the top of the hill. For a two-dimensional hill perpendicular to the wind, this would result in upstream blocking (Long 1972). For a three-dimensional hill, the fluid, rather than being blocked, can go round the sides. The Froude number is, therefore, probably the most important individual parameter to be matched when simulating strongly stable flow and diffusion in complex terrain.

2.2 *Boundary Conditions*

Because of the specification of zero velocity at solid boundaries, geometrical similarity (nondistorted models) is required. However, minute details need not be reproduced. Objects about the size of the surface-roughness length need not be reproduced in geometrical form; however, an equivalent aerodynamic roughness must be established. Over-roughening may be required to satisfy a roughness Reynolds-number criterion (Snyder 1981).

The approach flow must, in general, simulate the atmospheric boundary layer. Adequate simulations of the neutral atmospheric boundary layer are possible through a variety of techniques. Simulations of diabatic boundary layers have been accomplished using wind tunnels with heated and cooled floors; however, present technology allows only relatively small deviations from neutrality (mildly stable or mildly unstable conditions). The major problem with wind-tunnel simulation of strongly stable boundary layers is that in order to obtain small Froude numbers (e.g. Fr $\simeq$ 0.2, very strong stability), wind speeds must be reduced to the order of a few centimeters per second. Wind tunnels are generally extremely difficult to operate at low speeds (< 1 m s^{-1}) because they are designed to operate efficiently at their maximum speeds. At these low speeds, the screens and honeycombs designed to reduce turbulence, swirl, and external disturbances are completely ineffective; seemingly minor temperature variations across the test section or leakage through holes or seams in the tunnel sidewalls will easily induce secondary flows that are difficult to cope with. Also, measurement of the flow structure is exceedingly difficult with conventional instrumentation. Finally, because of the very small flow speeds, Reynolds numbers are also quite small, bringing into play the question of the Reynolds-number independence of the flow.

A saltwater-stratified towing tank provides an alternative method of obtaining the required strongly stable conditions without sacrificing the Reynolds-number independence condition. Because the kinematic viscosity of water is a factor of 15 smaller than that of air, a factor of 15 in the Reynolds number may be gained by using water as the medium (assuming the same model size and flow speeds).

Other sacrifices must be made, however, in using a towing tank. Because the model is towed through a quiescent medium, the effective approach flow has a uniform velocity profile and zero turbulence intensity. This approach flow does not simulate the nighttime stable atmospheric boundary layer, which is generally very shallow (< 100 m in depth), weakly turbulent, and possesses strong shear (generally a near-linear increase in speed with height). Frequently, however, pollutant sources discharge effluents at much higher elevations, i.e. *above* the shallow stable boundary layer, where the plume may be transported long distances with little or no dispersion [for example, see the cover photograph to the *Fourth Symposium on Turbulence, Diffusion and Air Pollution* (American Meteorological Society 1979)]. The only dispersion of the effluent (before the plume encounters an obstacle) occurs because of turbulence created *at the release point* (the stack). Because in very stable conditions the diffusion of chimney plumes is largely determined by initial mixing within the plume (Weil 1983), towing-tank studies can provide reasonable simulations of such conditions.

2.3   Dividing-Streamline Height

In Equation (1), a Froude number was defined that may be used to describe the basic structure of strongly stratified flows over hills. For three-dimensional hills, this structure may be usefully envisioned as composed of two layers: a lower layer of essentially horizontal flow in which plumes from upwind sources impinge directly on the hill surface, and an upper layer in which they pass over the hilltop. This structure was suggested by the theoretical arguments of Drazin (1961) and demonstrated through towing-tank experiments by Riley et al. (1976), Brighton (1978), and Hunt & Snyder (1980). Hunt & Snyder performed saltwater-stratified towing-tank studies using a simple bell-shaped hill with a uniform velocity profile and a linear density gradient in the flow approaching the hill. They showed that under these conditions the depth of the lower layer is predicted as

$$H_D/h = 1 - \text{Fr}, \tag{2}$$

where H_D is the dividing-streamline height, h is the hill height, and Fr is the Froude number as defined in Equation (1).

This definition of Froude number, however, is limited in its applicability to uniform wind profiles and linear density gradients. In order to apply the dividing-streamline concept to atmospheric flows, Snyder et al. (1984) derived a more general integral formula for predicting the dividing-streamline height for arbitrary shapes of wind profiles and stable density gradients. This integral formula was, in fact, suggested much earlier in a note by Sheppard (1956) based on kinetic/potential energy exchange arguments:

$$(1/2)\rho U_\infty^2(H_D) = g \int_{H_D}^{h} (h-z)(-d\rho/dz)\, dz, \qquad (3)$$

where ρU_∞^2 is to be evaluated far upstream at the elevation H_D, and $d\rho/dz$ is the vertical potential density gradient (also far upstream). This formula must be solved iteratively for the unknown H_D. It reduces to the simpler formula [Equation (2)] using the boundary conditions applicable to that case. [It also reduces to another simple formula derived and verified by Snyder et al. (1980) for determining whether an elevated (step) inversion will surmount a hill.]

This dividing-streamline height normalized by the hill height, H_D/h, is suggested here as a more general similarity parameter to be used in place of the Froude number for strongly stratified flows. The utility of this parameter is immediately evident: Plumes released below H_D must impact on the hill surface and/or be transported around the sides with the essentially horizontal flow of the lower layer; plumes released above H_D may pass over the hilltop and, in many respects, behave as if they were released at an elevation $H_s - H_D$ upwind of a hill of height $h - H_D$ (i.e. as if a ground plane were inserted at the dividing-streamline height H_D).

The parameter H_D/h is useful only if it is positive. [Negative results could be obtained, for example, from Equation (2) if Fr > 1.] Therefore, this parameter constitutes our definition of "strongly stratified flows" ($0 \leq H_D/h \leq 1$) wherein a lower layer of essentially horizontal flow exists. For weakly stratified flows, we revert to the Froude number defined in Equation (1). In that case, Fr > 1 and a lower layer of essentially horizontal flow does not exist.

The strongly stratified flow results described in the latter part of this paper are couched primarily in terms of the dividing-streamline height.

3. TOWING-TANK OPERATIONS

Maxworthy & Browand (1975) described various types of facilities designed to produce stratified flows. The Environmental Protection Agency (EPA) towing tank (1.2 m × 2.4 m × 25 m) uses a two-pump system and a mixing valve to fill the tank with a predesignated shape of stable density profile; it is not limited to the linear density profiles of the two-tank filling method. Typically, a 2.5-cm layer of freshwater is first fed into the tank. A weakly concentrated solution of saltwater is then added through a distribution piping system on the floor. This heavier solution flows in under and lifts the freshwater. This process continues, with each new mixture being slightly heavier than the previous mixture, until the tank is filled. A typical fill consists of a linear density profile of 43 distinct layers; the top layer is freshwater, and the bottom layer is saturated saltwater. At the end of the filling operation, the individual layers are visibly distinct; overnight, molecular diffusion tends to smooth the steps in the density profile into a continuous gradient such that individual layers are no longer discernible by eye or by measurement.

The density profiles tend to erode toward neutral at the surface and bottom of the tank because of the towing operations and simple molecular diffusion, although the constant linear gradient is maintained in the central portion. Because of its physical size and the expense, it is not practical to refill the whole tank every few days. Instead, a floating skimmer is used to remove the nonlinear (neutral) layer at the top; the original level is then restored by introducing an additional layer of saturated brine at the bottom. Figure 1 shows density profiles measured about two weeks apart,

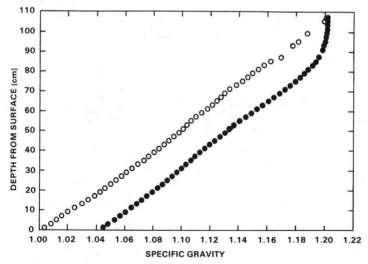

Figure 1   Density profiles measured in stratified towing tank: ○, after initial fill; ●, after 36 tows and 5 partial refills (totaling approximately 22 cm).

during which time 36 tows and 5 skims and refills (totaling about 22 cm in depth) were made. The linearity and shape were well maintained, but an increasingly deep layer of nearly constant density fluid accumulated at the bottom. This gradual change in the bottom boundary condition does not appear to significantly affect the study results (Snyder et al. 1984).

As a matter of convenience, the hill models are mounted on a flat square baseplate, and this apparatus is turned upside down, suspended from the carriage to the point at which the baseplate is submerged a few millimeters below the water surface; the apparatus is then towed the length of the tank. However, in discussion of the flow structure and plume behavior, the results are described as if the model were right side up.

Maxworthy & Browand (1975) provided a good description of instrumentation and techniques used for flow visualization and flow measurements; therefore, only the concentration measurement system is described here.

The effluent used both for flow visualization and as a quantitative tracer for concentration measurements is a blue food dye diluted with sufficient saltwater to produce a plume that is neutrally buoyant at the release height (typically, 1 part dye to 15 parts saltwater). This mixture is emitted from a bent-over "stack," isokinetically to form stream tubes and nonisokinetically to obtain a prescribed plume shape.

During the tow, samples are drawn through ports fixed on the hill surface or on rakes downstream. A vacuum system, shown schematically in Figure 2, is used to withdraw approximately 50 cm^3 of sample through each of the sampling ports into individual test tubes. The concentrations of dye in these samples are analyzed using a colorimeter. Standards of known concen-

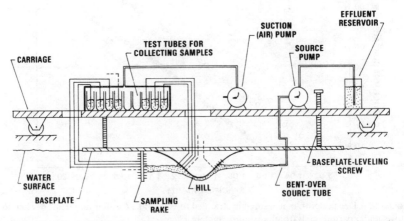

Figure 2 Schematic diagram showing system for collecting dye samples in towing tank.

tration are analyzed to establish a calibration curve for the colorimeter, and a numerical scheme is used to interpolate between calibration points. [Snyder & Hunt (1984) supply additional details.] This technique permits the measurement of dilutions in the range of 5 to 5000; using a larger concentration of dye in the effluent permits measurements of dilutions to perhaps 50,000.

One problem with the dye is that it leaves a residual in the tank, so that the background normally increases with every tow. The dye is controlled by the addition of a chlorine bleaching agent. The water is pH-balanced and chlorinated before being placed in the tank. The bleaching process is rather slow; a period of an hour or so is required to bleach out any dye in the tank. A drop of sodium thiosulfate in the sample test tubes neutralizes the chlorine in the sample and prevents bleaching (and hence deterioration) of the samples collected. The dye neutralizes the chlorine during the bleaching process, so that the tank may require periodic rechlorination by injecting bleach through a towed rake. In addition, the skimming operation removes contaminated water, and the refill operation restores clean and chlorinated water to the surface.

4. THREE-DIMENSIONAL HILLS

4.1 *Flow Structure*

Plume behavior in complex terrain is dramatically altered by the addition of stable stratification: Plume growth is severely inhibited, possibly reduced to zero (Britter et al. 1983); if plumes are released below the dividing-streamline height (see Section 2.3), they may be forced to impact on the upwind surface of the hill and travel round the sides (Hunt & Snyder 1980); if released not far above the dividing-streamline height, the plume centerlines may approach very close to the hill surface, resulting in large surface concentrations (Snyder & Hunt 1984). Under appropriate conditions, the plume may be rolled up into a free-stream rotor above the hilltop or possibly into an attached rotor on the lee side of the hill (Castro et al. 1983). Vortex shedding in horizontal planes is also possible in the lower-layer flow. Examples of each of these phenomena are presented in the following discussion.

The dividing-streamline or two-layer concept is useful because the transport and diffusion in each layer may be analyzed independently of one another using different but well-established techniques for each layer. For example, because the flow in the lower layer is approximately horizontal, the velocity field outside the wake may be treated as a two-dimensional potential flow about a cylinder. The cylinder shape is defined by the contour of the hill at the plume height (Figure 3). Given this mean flow field and the fact that stable stratification limits vertical diffusion, the calculation of dispersion from a point source at height H_s (below H_D) is the same as that from a *line* source near a cylinder (Hunt et al. 1979, Weil et al. 1981).

In the upper layer, buoyancy and inertial forces control the flow as it passes over the hill. Rowe et al. (1982) and Bass et al. (1981) have suggested that this upper-layer flow is approximately potential flow. This is not strictly correct, because the stratification above H_D has important effects on the vertical convergence and horizontal divergence of the streamlines (as well as on the diffusion). A better approximation, alluded to in Section 2.3 and demonstrated in Section 4.3, is to treat plumes in the upper layer as if a ground plane were inserted at the dividing-streamline height. By definition, the dividing-streamline height of the upper-layer flow is zero (because H_D defines the boundary between the upper and lower layers). Therefore, the Froude number of the upper-layer flow is unity, and the flow must be treated as if Fr = 1.

Figure 4 illustrates the horizontal flow field in the lower layer. In this case, the hill was a model of Cinder Cone Butte in southwestern Idaho

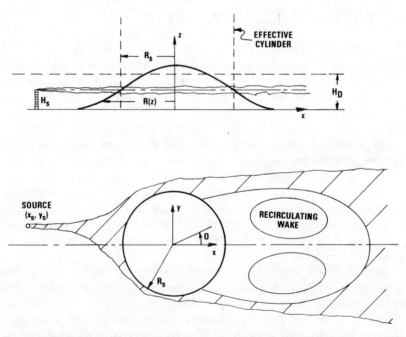

Figure 3 Simplification of plume impingement on three-dimensional hill under very stable stratification: (*top*) side view of plume impinging on hill at low Froude number; (*bottom*) section through hill, plume, and wake at $z = H_s$.

(further discussed in Section 5). The stratification was linear, the dividing-streamline height was $H_D/h = 0.8$, and the dye streamers were released at $H_s/h = 0.3$. Notice that the pattern in the lee of the hill (Figure 4, *bottom*) looks very much like the familiar Kármán vortex street observed downwind of two-dimensional cylinders in neutral flow.

Figure 5 is a view of a free-stream rotor as well as vortex roll-up on the front surface of a hill. In this case, the stratification was linear, the dividing-streamline height was $H_D/h = 0.6$, and the hill was a triangular ridge with base length equal to height and crosswind width equal to four times the height. Castro et al. (1983) have studied the lee-wave structure as a function of the stratification and the crosswind aspect ratio. They have shown that the ratio of the crosswind width of the ridge to its height has a negligible effect on the dividing-streamline height.

The conditions for the existence of a vortex on the upwind slope of a hill are not clear. Brighton (1978) calculated vertical deflections of streamlines in stratified flow round a hill; he demonstrated that both vertical shear in the flow approaching the hill and the hill slope have significant (but opposing) influences. Experiments by Snyder et al. (1984) in a stratified

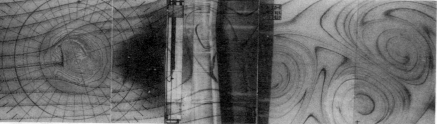

Figure 4   Flow in horizontal planes around Cinder Cone Butte : (*top*) oblique view with H_D = 0.8h, release height $z = 0.3h$; (*bottom*) top view showing Kármán vortex street downstream with $H_D = 0.8h$, release height $z = 0.6h$. Building roof shows in background of bottom center photograph.

shear flow (wind tunnel) tended to confirm the strong and opposing influences of the shear and hill slope; however, more work needs to be done to ascertain the conditions favorable to such vortex formation. Their results, however, suggested that Equation (3) is a good indicator of dividing-streamline heights, even in strong shear flows, for the vast majority of real hills (i.e. maximum slopes less than 26°).

Snyder et al. (1980, 1984) also analyzed an extensive range of laboratory observations and measurements of stably stratified flow over a variety of shapes and orientations of hills with different upwind density and velocity profiles. With a few understandable exceptions (ones that would not generally restrict application to the atmosphere), they reported that the integral formula [Equation (3)] is valid for predicting the dividing-streamline height for a wide range of stable density and velocity profiles and a wide range of three-dimensional hill shapes and orientations.

Some discussion of separation on the lee slopes of hills is in order. Assuming the fluid is infinite in its vertical extent, the fundamental lee-wave mode will have a wavelength of $\lambda = 2\pi U/N$ (N is the Brunt-Väisälä frequency $[=(g/\rho)\,d\rho/dz)]^{1/2}$) independent of hill shape. When a hill is of moderate slope and when this wavelength is comparable with the overall length of the hill, large-amplitude lee waves are to be expected. The pressure field of the wave acts to suppress separation on the lee face, although it may promote separation downstream, possibly forming a rotor. If the wavelength λ is long compared with the overall length of the hill (weak stratification), we may expect ordinary boundary-layer separation on the lee face, much as in neutral-flow separation. If λ is short compared with the hill length (strong stratification), we may expect separation to be lee-wave controlled, i.e. separation will occur at the first trough of the fundamental lee wave. But all of this depends on the hill slope as well. If the hill slope is small enough, separation will never occur; if it is large enough, separation will always occur near the crest. In very rough terms, for the moderately

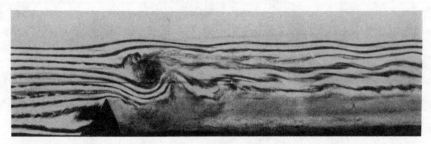

Figure 5   Centerplane streamers over triangular ridge with $H_D/h = 0.6$. Cross-stream width of ridge is $4h$.

sloped hills described here, separation on the lee face is controlled by the lee-wave field for $Fr < 0.8$, it is suppressed for $0.8 < Fr < 1.2$, and it is boundary-layer controlled for $Fr > 1.2$. Additional information is provided by Hunt & Snyder (1980), Snyder et al. (1980), and Castro et al. (1983) (and references therein); however, more information is clearly needed to establish the locations and types of separations that occur as functions of hill slope and stratification.

4.2 Plumes in the Lower Layer

The main characteristic of plumes emitted upwind of a hill but below H_D is that they impinge on the hill surface, split, and travel round the sides of the hill (Figure 6). Upwind, the plumes are largely constrained to move in horizontal planes, and vertical diffusion is severely limited. They are frequently rolled up within an upwind vortex as they impinge on the hill surface. This behavior does not appear to be regular, steady, or predictable, although the "diameter" of the vortices (i.e. vertical excursions of fluid parcels) appears to be limited to approximately $Fr \cdot h$. The vertical plume

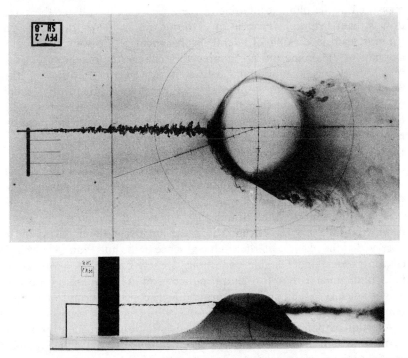

Figure 6   Top and side views of plume impingement on hill with $H_D/h = 0.80$ and $H_s/h = 0.78$.

dimensions increase suddenly and substantially at the upwind stagnation point, whether due to the vortex roll-up or to the vertical divergence of streamlines. These vertical widths are roughly maintained as the plumes are swept round the hill surface. The plumes lose elevation in traveling round the sides of the hill. Plumes that hug the hill surface leave it at the point where the flow separates (generally 100 to 110° from the upstream stagnation point, much as a streamline separates from the surface of a two-dimensional cylinder). Plumes emitted close enough to the stagnation line tend to be entrained into the wake region and rather rapidly regain their upstream elevation while mixing through the depth of the jump. Beyond that point, these entrained plumes tend to be vigorously mixed horizontally across the wake, leading to small wake concentrations. Whether or not they are entrained, plumes are generally affected by vortex shedding or low-frequency oscillations of the wake. These wake oscillations appear to induce oscillations in the plume upwind of the hill, causing it to waft from one side of the hill to the other.

Snyder & Hunt (1984) showed that under these conditions ($H_s < H_D$), the maximum surface concentration was essentially equal to the concentration measured at the plume centerline in the absence of the hill (at the same downstream distance). The location of the maximum concentration was on the upstream face. A small lateral displacement of the source from the stagnation streamline did not appreciably change the magnitude of the maximum concentration, but instead it moved the location of the maximum concentration to the side of the hill. Consequently, small oscillations in wind direction may be expected to result in a covering of the hill with the maximum concentration for short periods and to significantly reduce the average concentration. Finally, a slightly larger displacement of the source (i.e. a distance comparable with the plume width in the absence of the hill) caused the plume to miss the hill entirely, indicating a very strong sensitivity of surface concentration to wind direction (for further discussion, see Section 5).

It is important to understand that when a plume impinges on a downwind hill, surface concentrations can be essentially equal to those observed at the *plume centerline* in the absence of the hill. Plume centerline concentrations under stable conditions are typically 3 to 4 orders of magnitude greater than for neutral plumes and are even more orders of magnitude larger than maximum ground-level concentrations from stable plumes over flat terrain.

A terrain amplification factor for neutral conditions is frequently defined as the ratio of the maximum surface concentration observed in the presence of a hill to the maximum that would be observed from an identical source in flat terrain. Use of this terrain amplification factor in stable conditions would result in extremely large numbers that would be difficult to interpret. Therefore, a different terrain amplification factor is defined for stable flows: the ratio of the maximum observed surface concentration in the presence of a hill to that observed at the plume centerline at the same downwind distance in the absence of the hill.

4.3 Plumes in the Upper Layer

Plumes emitted above H_D, of course, are transported over the hilltop; however, if the release height is close to the dividing-streamline height, they spread broadly but thinly to cover the entire hill surface above H_D (Figure 7). Unlike plumes released at or below H_D, plume material reaches the hill surface only by *diffusion* perpendicular to the plume centerline. Plume meander as observed at or below H_D is absent. As H_s is increased relative to H_D, the point of first contact of the plume with the hill surface moves toward the hilltop; further increases in H_s move the contact point to the lee side of the hill (Figure 8). These features, in combination with the steadiness in the flow and thus the plume direction, have resulted in some of the largest

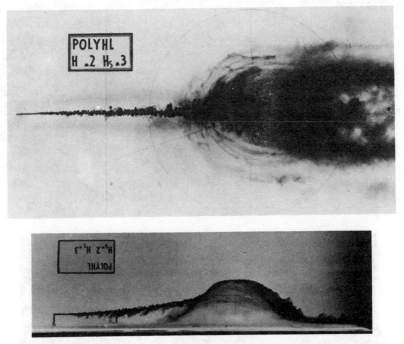

Figure 7 Top and side views of plume spreading over top of hill with $H_D = 0.2h$ and $H_s = 0.3h$.

surface concentrations observed under any conditions. With still further increases in H_s, the plume moves off the hill surface, and surface concentrations diminish rapidly.

As discussed in Section 4.1, some mathematical models treat the upper layer as potential flow; this is not strictly correct, because potential flow does not allow for the effects of the stratification. Also, as discussed in Section 4.1, the upper-layer flow may be treated as flow over a smaller hill with a Froude number of one ($H_D = 0$). We now examine these two approximations in some detail.

Figure 9 shows a comparison of hill centerline concentrations from sources of three different heights, measured (in one case) in the neutral and uniform approach flow of a wind tunnel and (in another case) in the strongly stratified (Fr = 1 or $H_D = 0$) towing tank (Snyder & Hunt 1984). These concentrations have been normalized by the maximum concentrations found at the centers of the plumes in the absence of the hill (to compensate for the differing diffusivities in the wind tunnel and towing tank).

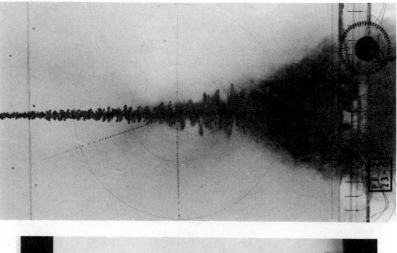

Figure 8 Lee-side contact of plume with hill with $H_D/h = 0.6$ and $H_s/h = 1.0$.

Pursuant to the discussion at the end of Section 4.1, we may expect boundary-layer separation on the lee side of the hill in the neutral wind-tunnel flow, but total suppression of separation due to stratification in the towing tank. The point of flow separation on the lee side of the hill under neutral conditions is clearly evident in Figure 9; a sharp decrease in concentration was observed at $x/h \approx 0.5$ ($z/h \approx 0.9$) for stack heights of $0.13h$ and $0.39h$. In contrast, at $Fr = 1$ the flow swept the plumes continuously down the lee side of the hill (no separation), so that changes in concentration along the hill surface were relatively smooth. Because of the

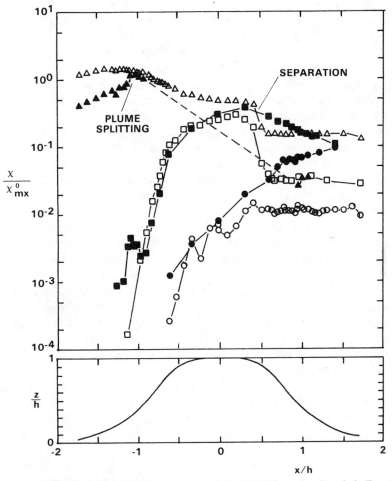

Figure 9   Hill centerline concentrations: open symbols, neutral flow; closed symbols, $Fr = 1$. Values of H_s/h: $\triangle$, 0.13; $\square$, 0.39; $\bigcirc$, 0.55.

separation, lee-side concentrations between the neutral and stratified cases differed by up to factors of ten. Upwind of the separation point, the effect of the stratification appeared to be small, with concentrations differing by much less than a factor of two. A tentative conclusion here is that for rough estimates, potential flow may be a fair approximation for the upper-layer flow in strongly stratified conditions *if* the maximum surface concentration is found to occur well upwind of the separation point in neutral flow or if separation does not occur in neutral flow. Implicit here is the assumption that appropriate diffusivities are used in calculating the plume spread.

It is interesting to note that the maximum surface concentration decreased by a factor of three when the source height was increased by the same factor (from $0.13h$ to $0.39h$). Theoretical arguments suggest that over level ground this decrease would have been H_2^2/H_1^2, or a factor of nine.

We have previously alluded to the idea that the upper-layer flow may be treated as flow over a smaller hill with a Froude number of unity, i.e. as if a ground plane were inserted at the dividing-streamline surface. It is known from laboratory studies that streamline trajectories are not contained within horizontal planes, even when well within the lower layer. On the other hand, concentration measurements suggest that a flat-surface approximation may yield reasonable estimates. From a practical view-point, mathematical models are vastly simplified if such an assumption yields reasonable estimates of surface concentration. Snyder & Lawson (1984) conducted a series of towing-tank studies to examine surface concentration patterns resulting from upwind plumes released in the upper layer and, more specifically, to test the adequacy of the assumption of a flat dividing-streamline surface. They attempted to answer the latter question not from a detailed analysis of the shape of such a surface, but from the more practical comparison of surface concentration patterns.

In the Snyder & Lawson experiments, linear density gradients were established in the tank, and the hill was towed at a speed U such that the Froude number Fr was 0.5 $(H_D/h = 0.5)$. Effluent was released at a height of $0.6h$, and the resulting hill-surface concentration patterns were measured. A second tow was conducted in which the entire model (hill, baseplate, and stack as a unit) was raised out of the water to the point where the water surface was precisely at the dividing-streamline height, i.e. the water surface was at half the hill height. The model was towed at the same speed as in the full-immersion tows, so that the Froude number of this now half-height hill was unity and all streamlines passed over the hill top. The flat water surface thus forced a flat dividing-streamline surface. The resulting surface-concentration patterns were then compared with the full-immersion patterns to ascertain the effects of a flat dividing-streamline surface.

Figure 10 shows a comparison of the concentration distributions resulting from the full- and half-immersion tows. The plumes were spread broadly to cover essentially the entire surface above half the hill height. (The dotted circle on the figure marks half the hill height.) The contours on the windward side of the hill are roughly circular, although somewhat elongated in the streamwise direction. In the half-immersion case, no concentrations were observed below half the hill height, of course, because that portion of the model was out of the water. In the full-immersion case, the plume "hugged" the lee side of the hill and was swept to elevations considerably lower than half the hill height.

Figure 11 is a scatter diagram comparing, on a point-by-point basis, the

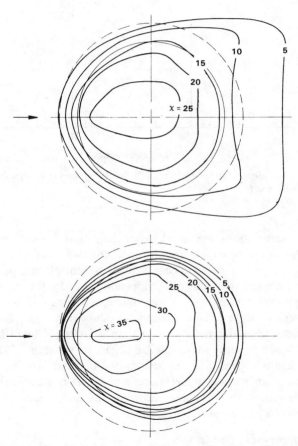

Figure 10 Concentration distributions measured on the hill surface with $H_D/h = 0.5$ and $H_s/h = 0.6$: (*top*) fully submerged ; (*bottom*) half submerged. Dotted circle indicates half the hill height.

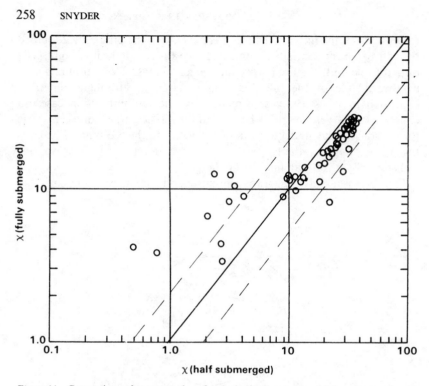

Figure 11   Comparison of concentrations for the half submerged and fully submerged cases with $H_D/h = 0.5$ and $H_s/h = 0.6$.

surface concentrations measured in the half- and full-immersion cases. Within the region of large concentrations, the two cases compared quite favorably; the half-immersion case yielded concentrations approximately 10 to 20% larger than the full-immersion case. In the region of low concentration, however, quite large differences occurred (worst case: a factor of ten). However, a close examination shows that in all cases in which the concentrations differed by more than a factor of two, the sampler locations were very close to half the hill height, i.e. either at $0.505h$ or $0.51h$.

These results suggest that the assumption of a flat dividing-streamline surface in a mathematical model is a reasonable approximation to make, at least with regard to predicting the locations and values of maximum concentrations and areas of coverage.

4.4   Further Discussion

Figure 12 presents an overview of the maximum surface concentrations measured in $H_s/h \times$ Fr space for the bell-shaped hill of Snyder & Hunt

(1984). Overlaid on this graph are the dividing-streamline height ($H_D/h = 1$ $-$ Fr), the boundary layer, and somewhat speculative isoconcentration lines. These lines were drawn to fit the data as measured, but a generous allowance was made for scatter in order that they might also satisfy our physical intuition. The graph suggests that the largest concentrations occur when the source release is near the dividing streamline (the solid line in the graph) and that they decrease rapidly with distance to the right of this line (larger stack heights or Froude numbers). This rapid decrease is due to the fact that as the stack height or Froude number is increased, the contact point moves upward over the hilltop and then down the lee side; further increases in $H_s - H_D$, much above the thickness of the plume in the absence of the hill (Σ_z^0), result in the plume lifting off the surface with no contact at all. Note that if $H_s - H_D \approx \Sigma_z^0$, a significant surface concentration can arise because of the downward deflection of the streamlines onto the hill. If the source height is less than the dividing-streamline height (left of the line), the measurements suggest that the maximum surface concentrations are

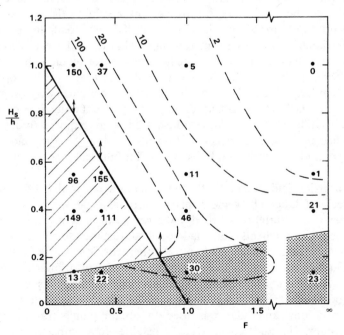

Figure 12 Isoconcentration contours (dashed lines) in $H_s/h \times$ Fr space. Solid line represents $H_s/h = 1 -$ Fr. Numbers represent concentrations measured at those points. The stippled area denotes the surface boundary layer, while the striped area represents roughly uniform concentration. The arrows indicate Σ_z^0/h, the depth of the plume at the location of the hill.

roughly uniform in this region. However, visual observations during the tows suggested that the lower the stack height, the more prominent was the plume meander (hence, lower concentrations), although the actual surface concentration seemed to depend on the precise point at which the plume entered the vortex on the upwind slope. The size and elevation of these vortices appeared to vary from one tow to the next (indeed, even during a single tow) and were not investigated in detail.

For the very small stack heights, the plumes dispersed relatively quickly in the surface boundary layer over the baseplate, so that concentrations were essentially independent of the hill Froude number. Recall that the hill Froude number is a bulk parameter that characterizes the overall flow over the hill; the Froude number characterizing the boundary-layer flow is much larger because of the smaller scale of the boundary layer ($\delta \ll h$) and the reduced density gradient caused by the mixing.

5. COMPARISONS WITH FIELD EXPERIMENTS

While much has been learned about flow and dispersion under stable conditions, the applicability of towing-tank results to the atmosphere is sometimes questioned. The height of the dividing streamline is directly proportional to wind speed; in a towing tank, the wind speed is constant. The unsteadiness of the wind in the field, however, can allow the plume to impact on the front surface at one moment and sweep over the top at the next. Similarly, a meandering wind will result in a plume being swept from one side of the hill to the other, resulting in significant reductions of surface concentration. Carefully designed field studies are required to determine the effects of these aspects of atmospheric flows and to aid in the interpretation of towing-tank results so that full-scale predictions can be made.

In a major field study recently completed, Lavery et al. (1982) collected six weeks of data on flow and dispersion around a 100-m-high isolated hill (Cinder Cone Butte) in a broad, flat river basin in southwestern Idaho. Detailed measurements were made of wind, turbulence, and temperature profiles in the approach flow and at other positions on the hill. Sulfur hexafluoride (as a tracer) and smoke (for flow visualization) were released from a mobile crane that allowed flexibility in positioning the source (height and location). One hundred samplers on the hill collected data on surface concentrations, and lidar was used to obtain plume trajectories and dimensions. The study concentrated on stable plume impaction.

One particular hour from that field study was selected for simulation in the towing tank (Snyder & Lawson 1981). That hour was 0500 to 0600, 24 October 1980 (Case 206), which may be characterized as very stable, i.e.

light winds and strong stable temperature gradients. Measurements made during the towing-tank experiments included ground-level concentrations under various stabilities and wind directions, vertical distributions of concentration at selected points, plume distributions in the absence of the hill, and visual observations of plume characteristics and trajectories.

This series of tows showed that the surface-concentration distributions were extremely sensitive to changes in wind direction. For example, Figure 13 shows that the distribution shifted from the north side of the hill to the south side with a shift of only 5° in wind direction. Comparisons of individual distributions with field results showed very much larger maximum surface concentrations and much narrower distributions in the

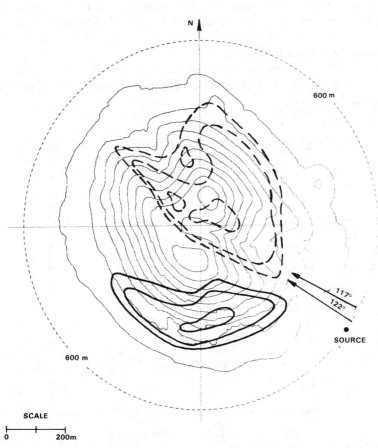

Figure 13   Concentration distributions measured during individual tows with $H_s/h = 0.31$ and $H_D/h = 0.38$; wind direction: —— 117°, – – – 122°.

model results. To account for this variability in the winds measured during the hour, a matrix of 18 tows (three wind directions × six wind speeds) was conducted, and the concentration patterns were superimposed. A scatter plot of superimposed model concentrations versus field concentrations (Figure 14) shows a marked improvement over the previous single-tow comparisons. The largest model concentrations were within a factor of two of the highest field values.

It is interesting to note that, whereas the *location* of the maximum shifted dramatically with small shifts in wind direction, the *value* of the maximum changed very little with changes in wind direction or wind speed. Maximum surface concentrations approached those at the plume centerline in the absence of the hill during individual tows, but because of the extreme sensitivity of the location to wind direction the plume was "smeared" broadly across the hill surface as the wind direction changed through only a few degrees. Therefore, short-term averages (≈ 5 min) in the field may be expected to approach plume centerline concentrations; longer-term averages (≈ 1 h) may be expected to be reduced by factors of from five to ten (or more, depending upon the magnitudes of the fluctuations in wind speed and direction).

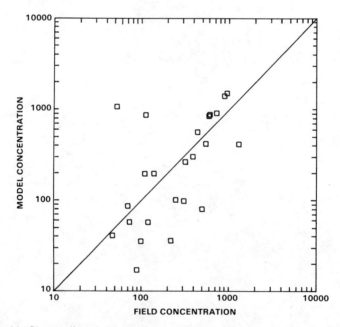

Figure 14 Scatter diagram comparing superposition of concentration distributions measured over Cinder Cone Butte model with field distributions.

6. TWO-DIMENSIONAL HILLS: LIMITATIONS OF TOWING TANK

In our previous discussions of the dividing-streamline concept, we implicitly assumed that the portion of the flow with insufficient kinetic energy to surmount the hilltop was able to pass around the sides of the hill. However, if the model is two-dimensional (spans the width of the tank), the fluid with insufficient kinetic energy is blocked or trapped upstream. Because of continuity, the blocked fluid must conserve its volume. Therefore, as the distance between the model and the upstream endwall of the tank decreases, the blocked fluid is "squashed," i.e. because the length of the blocked region must decrease with distance along the tow, the depth of the fluid that was initially blocked must increase. Eventually, all of the fluid upstream of the model will spill over the top. The fixed end on the tank, in effect, provides a uniform approach-flow velocity profile at a distance upstream of the model that varies as the experiment progresses.

Snyder et al. (1984) addressed this problem through a series of towing-tank studies. They used a series of "infinite" ridges of quite different cross-sectional shape to test the validity of the "steady-state" assumption of flow upwind of two-dimensional ridges under strongly stratified conditions.

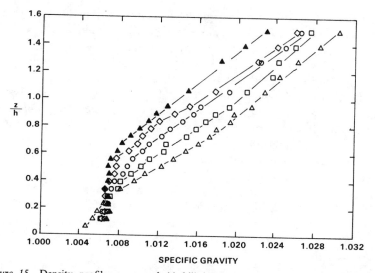

Figure 15 Density profiles measured 11 hill heights upstream of a two-dimensional triangular ridge: △, initial density profile. Sampling interval centered at $x = 1.3$ m (□), 6.7 m (○), 11.8 m (◇), and 17.7 m (▲) from start of tow.

Three ridge shapes were used: a vertical fence, a triangular ridge with a slope of 63°, and a witch of Agnesi $[1/(1 + x^2)]$ with a maximum slope of 39°. Results are shown here only for the triangular ridge. A density sampling rake was positioned 11 hill heights upwind of the ridge and towed with the ridge. Density samples were collected just prior to the start of the tow, during the first and last meter of the tow, and at the one-third and two-thirds points of the tow (the total length of the tow was approximately 20 m). The experiments were conducted at a Froude number of 0.5 (based on the tow speed and the undisturbed density profile), so that the calculated depth of blocked fluid (the dividing-streamline height) was $0.5h$.

Figure 15 shows the density profiles collected during these tows. The initial near-linear density profile was continuously modified during the tow at a position 11 hill heights upstream, and the results show that upstream conditions did not reach steady state. The profiles tended toward neutral at elevations below half the hill height.

The conclusion from this series of studies was that steady-state conditions are not established in strongly stratified flows (i.e. $Fr < 1$) over two-dimensional ridges in finite-length towing tanks. The "approach flow" velocity and density profiles changed continuously during the tow. Thus, these experiments have no analog in the real atmosphere.

7. CONCLUSIONS

The effects of stratification on plume behavior in complex terrain are dramatic and fascinating. Depending on the release height and the Froude number, plumes may be observed to impact on hills, to sweep narrowly over the tops of hills, or to be entrained into rotors or vortices of numerous types. The stratification may either enhance or inhibit separation on the lee sides of hills. A stratified towing tank is a useful tool to enhance basic physical understanding of such phenomena and to provide practical guidance or rules of thumb for locating sources and estimating likely impacts of such sources in complex terrain.

The Froude number is quite likely the most important single parameter to be matched when simulating strongly stable flow and diffusion in complex terrain. A closely related but more general parameter, the dividing-streamline height, provides for a more useful interpretation of strongly stratified flows. The dividing-streamline forms the boundary between a lower layer of essentially horizontal flow and an upper layer that passes over the hilltop.

Plumes released in the lower layer impact on the hill surface; the resulting surface concentrations essentially equal those observed at the center of the plume in the absence of the hill.

For practical purposes, a plume released in the upper layer can be treated as a release from a shorter stack upwind of a shorter hill, i.e. as if a ground plane were inserted at the dividing-streamline height. A comparison of field and laboratory observations of concentration patterns on a hill surface under strongly stratified conditions showed very good correspondence when wind speed and direction variability were accounted for in the model experiments. Concentration distributions on the hill surface were found to be extremely sensitive to slight changes in wind direction.

Finally, strongly stratified towing-tank experiments on flows over two-dimensional ridges were found to have no analog in the real atmosphere because of the unsteadiness created by the finite length of the tank.

Literature Cited

American Meteorological Society. 1979. *Preprints Volume, Symposium on Turbulence, Diffusion and Air Pollution, 4th,* Reno, Nev. 676 pp.

Bass, A., Strimaitis, D. G., Egan, B. A. 1981. Potential flow model for Gaussian plume interaction with simple terrain features. *Rep. to Environ. Prot. Agency under Contract No. 68-02-2759,* Research Triangle Park, N.C. 201 pp.

Brighton, P. W. M. 1978. Strongly stratified flow past three-dimensional obstacles. *Q. J. R. Meteorol. Soc.* 104:289–307

Britter, R. E., Hunt, J. C. R., Marsh, G. L., Snyder, W. H. 1983. The effects of stable stratification on turbulent diffusion and the decay of grid turbulence. *J. Fluid Mech.* 127:27–44

Castro, I. P., Snyder, W. H., Marsh, G. L. 1983. Stratified flow over three dimensional ridges. *J. Fluid Mech.* 135:261–82

Cermak, J. E. 1976. Aerodynamics of buildings. *Ann. Rev. Fluid Mech.* 8:75–106

Cermak, J. E., Sandborn, V. A., Plate, E. J., Binder, G. H., Chuang, H., et al. 1966. Simulation of atmospheric motion by wind tunnel flows. *Rep. No. CER66JEC-VAS-EJP-GHB-HC-RNM-SI17,* Fluid Dyn. Diffus. Lab., Colo. State Univ., Fort Collins. 102 pp.

Drazin, P. G. 1961. On the steady flow of a fluid of variable density past an obstacle. *Tellus* 13:239–51

Hunt, J. C. R., Puttock, J. S., Snyder, W. H. 1979. Turbulent diffusion from a point source in stratified and neutral flows around a three-dimensional hill. Part 1. Diffusion equation analysis. *Atmos. Environ.* 13:1227–39

Hunt, J. C. R., Snyder, W. H. 1980. Experiments on stably and neutrally stratified flow over a model three-dimensional hill.

J. Fluid Mech. 96:671–704

Lavery, T. F., Bass, A., Strimaitis, D. G., Venkatram, A., Greene, B. R., et al. 1982. *EPA Complex Terrain Model Development: First Milestone Report—1981, EPA-600/3-82-036,* Environ. Prot. Agency, Research Triangle Park, N.C. 304 pp.

Long, R. R. 1972. Finite amplitude disturbances in the flow of inviscid rotating and stratified fluids over obstacles. *Ann. Rev. Fluid Mech.* 4:69–92

Maxworthy, T., Browand, F. K. 1975. Experiments in rotating and stratified flows: oceanographic application. *Ann. Rev. Fluid Mech.* 7:273–305

Riley, J. J., Liu, H. T., Geller, E. W. 1976. A numerical and experimental study of stably stratified flow around complex terrain. *Rep. No. EPA-600/4-76-021,* Environ. Prot. Agency, Research Triangle Park, N.C. 41 pp.

Rowe, R. D., Benjamin, S. F., Chung, K. P., Havlena, J. J., Lee, C. Z. 1982. Field studies of stable air flow over and around a ridge. *Atmos. Environ.* 16:643–53

Sheppard, P. A. 1956. Airflow over mountains. *Q. J. R. Meteorol. Soc.* 82:528–29

Snyder, W. H. 1972. Similarity criteria for the application of fluid models to the study of air pollution meteorology. *Boundary-Layer Meteorol.* 3:113–34

Snyder, W. H. 1981. Guideline for fluid modeling of atmospheric diffusion. *Rep. No. EPA-600/8-81-009,* Environ. Prot. Agency, Research Triangle Park, N.C. 200 pp.

Snyder, W. H. 1984. Terrain aerodynamics and plume dispersion: a perspective view gained from fluid modeling studies. *Proc. Symp. Tibetan Plateau Mt. Meteorol,* Beijing. In press

Snyder, W. H., Britter, R. E., Hunt, J. C. R.

1980. A fluid modeling study of the flow structure and plume impingement on a three-dimensional hill in stably stratified flow. *Proc. Int. Conf. Wind Eng., 5th*, ed. J. E. Cermak, 1:319–29. New York: Pergamon

Snyder, W. H., Hunt, J. C. R. 1984. Turbulent diffusion from a point source in stratified and neutral flows around a three-dimensional hill. Part II. Laboratory measurements of surface concentrations. *Atmos. Environ.* In press

Snyder, W. H., Lawson, R. E. Jr. 1981. Laboratory simulation of stable plume dispersion over Cinder Cone Butte: comparison with field data. See Lavery et al. 1982, pp. 250–304

Snyder, W. H., Lawson, R. E. Jr. 1984. Stable plume dispersion over an isolated hill: releases above the dividing-streamline height. In *EPA Complex Terrain Model Development: Fourth Milestone Report—1984*, Environ. Prot. Agency, Research Triangle Park, N.C. In press

Snyder, W. H., Thompson, R. S., Eskridge, R. E., Lawson, R. E. Jr., Castro, I. P., et al. 1984. The structure of strongly stratified flow over hills: Dividing-streamline concept. *J. Fluid Mech.* In press

Weil, J. C. 1983. Application of advances in planetary boundary layer understanding to diffusion modeling. *Preprints Vol., Symp. Turbul. Diffus., 6th, Boston* pp. 42–46. Boston: Am. Meteorol. Soc.

Weil, J. C., Traugott, S. C., Wong, D. K. 1981. Stack plume interaction and flow characteristics for a notched ridge. *Rep. No. PPRP-61*, Md. Power Plant Siting Program, Martin Marietta Corp., Baltimore. 92 pp.

Ann. Rev. Fluid. Mech. 1985. 17 : 267–87

MATHEMATICAL MODELING FOR PLANAR, STEADY, SUBSONIC COMBUSTION WAVES

D. R. Kassoy

Department of Mechanical Engineering, University of Colorado, Boulder, Colorado 80309

Introduction

The propagation of steady combustion waves in a premixed gas has occupied the imagination of scientists since flames were first observed. Fundamental ideas about the possible modes of propagation probably evolved initially from the classical theory of one-dimensional, steady compressible flow with heat addition. The class of discontinuous waves characterized by pressure and density decrease (deflagrations) has a continuous spectrum of propagation speeds, from the ultrasubsonic to an $O(1)$ value associated with a sonic speed behind the wave. Structural properties of deflagrations and the physical mechanisms controlling their propagation were considered much later. The most enduring and prolific studies are concerned with the steady, isobaric, low-Mach-number flame, which propagates by conductive preheating and forward radical diffusion. Less evident, but of considerable importance, are the structural investigations of subsonic but $O(1)$ Mach number combustion waves, found initially in the descriptions of idealized planar detonations. Here, the reaction zone behind the thin shock wave propagates like a convecting thermal explosion, basically unaffected by restrictive transport-property effects. Transitional combustion waves, with propagation Mach numbers intermediate to those mentioned previously, have been modeled only recently. Clarke (1983a) finds that an increase in the Mach number causes the influence of transport effects on the flame structure to wane. The entire reaction process is increasingly dominated by a convective-reactive balance.

267

0066–4189/85/0115–0267$02.00

The popularity of the planar, premixed, steady deflagration model is a reflection of its conceptual simplicity rather than a measure of how frequently it is observed in the natural environment. In fact, the experimentalist must work with some finesse to produce a flow field with the appropriate characteristics. Nonetheless, the planar model is a marvelous paradigm that can be used to test mathematical formulation and solution techniques, such as asymptotic and numerical methods. In addition, novel physical effects, like complex chemistry, can be incorporated easily into the model to facilitate the investigation of new phenomena. Of course, the mathematical modeling of this conceptually simple process has not been without controversy. One can recall the technical discussions of the relative importance of radical diffusion compared with conduction in low-speed flame theory, the diversity of views on the importance of transport processes in higher-speed combustion waves, and the extensive consideration given to the cold-boundary difficulty in both cases to recognize that this subject has had a lively period of gestation and maturation.

This article begins with a brief description of selected paradigm models for low-, transitional-, and high-speed deflagrations. A comparison is made of discontinuous combustion-wave modeling and fully structural theories. This includes an exploration of the relationship between the doubly infinite model of a flame and that in a half-space. The physical viability and mathematical content of these models is next examined. It is concluded that the half-space model, traditionally associated with the burner-attached flame, provides the only physically accurate and mathematically well-defined representation of a steady-state event. A unified formulation is then given for the half-space subsonic combustion-wave problem. Activation-energy asymptotics are used as the basis of the mathematical analysis. This is followed by a description of the changes in wave structure with propagation speed. Transport effects, which have a profound effect on low-speed flames, are shown to be of reduced importance when the propagation Mach number increases slightly above the value associated with flame separation in the burner-attached model. High-speed combustion waves propagate like convecting thermal explosions, entirely independent of conduction and diffusion.

Paradigm Models of Steady Combustion Waves

The classical theory of one-dimensional, steady compressible flow with heat addition can be used to describe the properties of a discontinuous combustion wave separating two exact equilibrium end states (e.g. Williams 1965, Ahlborn & Liese 1981). Given the initial thermodynamic state of the unburned material and the total available chemical heat release, one can use conservation of mass, momentum, and energy and the state equation to construct the Hugoniot line shown schematically in Figure 1. The Rayleigh line is constructed in terms of the initially unknown propagation (input) Mach number. If, for example, the final equilibrium state pressure is prescribed, then the appropriate intersection of the Rayleigh and Hugoniot lines can be located. The intersection defines the propagation Mach number and other burned-state values. An example is shown in Figure 1 (line R_B) for a deflagration, characterized as a subsonic propagating wave across which the pressure and density decrease, and the temperature and gas speed increase. In general, the propagation Mach number can range from an ultrasubsonic value, associated with very small pressure changes (line R_A) to the so-called $O(1)$ Chapman-Jouguet value (line R_C) associated with a sonic velocity behind the wave.

In a real combustion system it is difficult to prescribe enough information a priori to use this classical theory as a unique predictor of jump conditions and/or wave speed. For example, in most experimental configurations associated with planar flames, the pressure change cannot be specified arbitrarily. If, on the other hand, the propagation (input) Mach number is prescribed, as in a burner experiment, the required initial equilibrium state cannot be maintained. In fact, the mathematical requirement of exact equilibrium in the upstream unburned gas is impossible to meet in physical

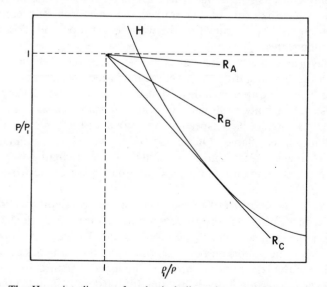

Figure 1   The Hugoniot diagram for classical discontinuous deflagration waves. The Hugoniot line *H* is constructed for a specified heat addition. Each Rayleigh line *R* corresponds to a specific mass flow rate.

systems, because reactive mixtures obeying the Arrhenius kinetic law have finite reaction rates at all nonzero temperatures.

In order to develop better insight into the nature of propagating combustion waves, one must find structural solutions connecting properly defined end states. Rate-dependent processes, absent from classical jump theory, must be incorporated into the modeling.

The modern theory for an isobaric low-Mach-number flame, driven by heat release from a high-activation-energy reaction, was developed by Bush & Fendell (1970). They describe the structure of a freely propagating flame in a doubly infinite field between two exact equilibrium states, a flow which represents the most natural extension of classical discontinuous combustion-wave theory. The Arrhenius formula is altered to produce an artificial reaction rate that vanishes identically at the upstream ambient temperature in order to ensure an upstream equilibrium state. Activation-energy asymptotics are used to show that the entire flame is dominated by transport-property effects. The thickness of the inert preheating zone is determined by a balance of convective and conductive processes. In contrast, the subsequent thinner reaction zone, where the temperature is basically the adiabatic flame value, is characterized by a competition between chemical heat release and conduction. While the magnitude of the small propagation Mach number is determined by asymptotic scaling arguments alone, its specific value depends on a constant of proportionality, an eigenvalue that can be determined uniquely from the continuous solution providing a transition between exact up- and downstream equilibrium states.

In physical systems, reactions occur at finite rates at any nonzero temperature. The resulting weak heat release generates a spatial tempera-ture gradient in the approaching cold mixture. Johnson (1963) showed that the existence of finite upstream gradients, no matter how small, implies that a large range of flame speeds are possible for the general activation-energy low-speed flame. Of course, this possibility was known by Hirschfelder et al. (1953), who considered the flame originating from a finite location where gradients are present (the half-space or burner-attached flame).

Clarke & McIntosh (1980) and Buckmaster & Ludford (1982), for example, have formulated the modern version of the burner-attached flame, based on high-activation-energy considerations. The gas mixture appears at a weakly nonequilibrium point. Normal Arrhenius kinetics governs the evolution of the reaction downstream. The flame structure dominated by transport effects is qualitatively like that for the doubly infinite model, and the same magnitude for the propagation Mach number is found. There is, however, significant heat loss to the origin, and the maximum flame temperature may be less than the adiabatic value. The propagation Mach number, related directly to the input mass flow rate, can vary significantly within the magnitude compatible with the theory. These changes affect the gradients at the origin, the location of the reaction zone relative to the origin, and the details of the structure within each zone. The restrictive character of the doubly infinite flame propagation Mach number disappears because finite gradients at the origin, regardless of size, can adjust to changes in the input mass flux. These models are representative of laboratory flat flame burner experiments.

When the propagation Mach number of the half-space flame reaches a sufficiently large critical value, although still very small relative to unity, the gradients at the origin become very small relative to those in the inert preheat zone. The active part of the flame, located far from the origin, is similar to that found in the doubly infinite model. In fact, to a first approximation, the propagation Mach numbers of both flames are identical. Further increases in the propagation speed cannot be accommodated by standard theory in which transport properties dominate.

Clarke (1983a) develops the high-activation-energy theory for slightly faster half-space flames. When the propagation Mach number is slightly larger than that permitted in classical theory, the usual flame structure, consisting of an inert zone followed by a reaction zone, is separated from the origin by a thicker region in which a balance of convection, conduction, and weak reaction prevails. A further increase in the Mach number causes the appearance of a convective-reactive zone adjacent to the origin that precedes the structure just mentioned. For sufficiently large values of the Mach number, but still very small relative to unity, the classical structure disappears downstream because the fuel is consumed in the dominant upstream convective-reaction zone. Here each fluid particle's energy increases purely as a result of chemical heat addition. An ignition process occurs like that in an adiabatic thermal explosion (Semenov 1928); this process has been discussed in contemporary terms by Kassoy (1975). In this nonconducting, nondiffusing system with small Mach number, there is no interaction between neighboring fluid particles. Compressibility effects are negligible because the Mach number is still very small.

The structural features of flames that propagate at $O(1)$ subsonic Mach numbers are embedded in the early descriptions of one-dimensional idealized detonation waves (Fickett & Davis 1979). Adamson (1960), among others, used perturbation methods to show that the physics in the reaction zone behind the leading shock wave is described by a balance of convection, reaction, and compressibility effects. As long as the ratio of the intermolecular collision time to the fastest reaction time is small (a physically plausible constraint), transport effects are negligible. High-activation-energy asymptotics are used by Bush & Fendell (1971) to develop the modern description of a Chapman-Jouguet detonation. The structural features of the reaction zone are qualitatively similar to those found earlier. Insofar as the flame itself is concerned, these models are of the half-space variety. The initial conditions pertain to those found at the back end of a fully structured shock, where gradients are small but finite. As a result, a large range of propagation speeds is possible. Related models of flames have been considered more recently by Clarke (1983b) and Kassoy & Clarke (1984), who show in a most explicit way that high-speed flames should be thought of as thermal explosions convecting in a compressible flow. Gradients at the origin of these finite subsonic Mach number flows are exceptionally small but finite. Once again a large range of propagation Mach numbers are possible, because even these tiny gradients can adjust to changes in input conditions. The major structural components of the high-speed flame consist of a reaction induction zone, followed by a much shorter region in which the reaction rate increases dramatically. The former is characterized by a convective-reactive-compressibility balance, where the temperature deviates only slightly from the initial value. The subsequent singular growth of the deviation defines a critical point known as the ignition delay location. There the reaction rate becomes large, and a rapidly varying gasdynamic process occurs. A balance of convective, reactive, and compressibility effects exists here as well.

The doubly infinite space model has also been used to describe flames that propagate at $O(1)$ subsonic Mach numbers. For example, Kapila et al. (1983) describe a high-activation-energy, compressible-flow model in which heat addition from reaction is vanishingly small relative to the internal energy of the unburned gases. An ignition temperature, thought of as a gas mixture property, is defined to ensure that the unburned mixture is inert until conduction preheating raises the mixture temperature to the ignition value. The propagation Mach number, a function of the ignition temperature, is determined uniquely by the single transitional solution that is compatible with the initial and final equilibrium states. It is not known whether the ignition temperature can be identified as a true material property, independent of an experimental apparatus.

The Kapila et al. (1983) model shows that the high-speed flame is controlled by a balance of fluid convection, chemical heat release, and compressibility effects. Transport property effects are essentially negligible. In contrast, Stewart & Ludford (1983) have developed a doubly infinite space model for a high-speed reaction zone in which the length scale of the entire flame is a modest multiple of the mean free path and in which heat release occurs in a thinner, embedded reaction zone dominated by transport effects. The reaction zone is located where the temperature is close to an ignition temperature, which is thought of as a material property of the unburned gas mixture. The flame speed is a unique function of the ignition temperature, independent of the detailed reaction process. Implicit in this model is the requirement that the characteristic reaction time be much shorter than the average intermolecular collision time, a condition difficult to justify on kinetic grounds. The structural features of this model differ significantly from those of Kapila et al. (1983) and Kassoy & Clarke (1984). In particular, the extremely short region of active chemical heat release is to be contrasted with an extended zone of compressible reactive flow in the latter works.

Historical reviews of these matters can be found in Williams (1965), Margolis (1980), Buckmaster & Ludford (1982), Clarke (1983a, b), Kassoy & Clarke (1984), and Clavin (1984), for example.

Critique of the Doubly Infinite Plane Model

In the previous section it was noted that the theory of discontinuous combustion waves shows that the propagation speed of deflagrations ranges from nearly zero to the Chapman-Jouguet value. A particular value can be determined for a specified initial thermodynamic state, heat addition, and pressure change across the wave. When the last quantity is exceedingly small, waves with the properties of low-speed flames are found. The propagation Mach number is proportional to the square root of the pressure drop. In contrast, the structural theory for such a flame produces a unique propagation speed that is dependent on the characteristic reaction rate. It is an eigenvalue compatible with the continuous solution between two precise equilibrium states. The small pressure drop is calculated from a decoupled momentum equation subsequent to the combustion calculation. Both theories require that the upstream gas be in an equilibrium state, a condition that cannot be met precisely in a real system.

The uniqueness of the propagation Mach number is a bit less clear when one considers higher-speed doubly infinite flames. Then the Mach number depends on the assumed value of an ignition temperature that is supposed to be an innate material property. It would appear that this amounts to replacing one unknown, the propagation Mach number, by another, the ignition temperature, that is not defined a priori.

The difficulties with the doubly infinite model of a freely propagating steady flame arise from the artificial mathematical conditions imposed on the describing system to ensure that equilibrium prevails in the upstream state. Only in this way can a well-posed, steady-state mathematical problem be defined. Unfortunately, the model is not physically viable for real chemical processes. Zeldovich (1980) has recognized that, in fact, the freely propagating flame is not amenable to steady-state analysis. Rather, the properly defined unsteady problem involves a hot deflagration with a short characteristic reaction time propagating into a cold mixture with a much longer characteristic reaction time. The short time reflects the large reaction rate of high-temperature combustion, while the latter is related to the thermal explosion time of a cool reactive gas mixture. In the conceptual sense, the flame propagates in a quasi-static manner, with conditions at a given instant determined by the momentary properties of the slowly reacting unburned mixture. Multiple-time-scale analysis based on high-activation-energy asymptotics can certainly be developed for this problem. In fact, the preliminary model of Zeldovich (1980) for an unsteady low-speed flame provides a starting point for a physically attractive formulation of such problems.

It seems reasonable to conclude that steady-state flames are studied most profitably in the context of a half-space model. This has the virtue of replicating real experimental conditions. In addition, the mathematical model is based on reaction kinetics governed by normal Arrhenius kinetics and requires no consideration of an ignition temperature. If appropriate consideration is given to determining the finite gradients at the origin, then one can describe a unified theory for subsonic combustion waves with propagation Mach numbers up to the Chapman-Jouguet value. A summary of this theory, drawn from a variety of sources, is given in succeeding sections.

The Mathematical Model

Planar, subsonic combustion waves are described by the steady, one-dimensional equations for a compressible, viscous, heat-conducting, diffusing, reacting gas. The general system (Williams 1965) can be simplified without diminishing the appropriate representation of important physical processes. In terms of material properties, it is assumed that the species specific heats and binary diffusion coefficients are equal and are dependent on the local thermodynamic state. The former condition simplifies the energy equation, while the latter permits Fick's law to be derived from the general equation for the diffusion velocity. The species molecular weights are assumed equal in order to use a perfect-gas equation of state for the mixture. A global decomposition reaction $A \rightarrow B$ is used to represent the chemistry. The reaction rate is described by the Arrhenius law.

The nondimensional equations for mass and momentum conservation (integrated once), for conservation of reactant and energy, and for state can be written, respectively, as

$$\hat{\rho}\hat{u} = 1, \qquad (4/3)(\hat{\mu}\hat{u}' - \sigma_1) = \hat{p} + \hat{u} - 1, \tag{1}$$

$$\hat{y}' = \frac{\mathrm{Le}_1}{\mathrm{Pr}_1}(\hat{\rho}\hat{D}\hat{y}')' - \left\{\frac{B\,e^{-1/\varepsilon\hat{T}_M}(\mu_1/p_1)}{\gamma M_1^2}\right\}\hat{\rho}\hat{y}\,e^{-1/\varepsilon[(1/\hat{T})-(1/\hat{T}_M)]}, \tag{2}$$

$$\hat{C}_p\hat{T}' = \frac{1}{\mathrm{Pr}_1}(\hat{\lambda}\hat{T}')' + \left\{\frac{B\,e^{-1/\varepsilon\hat{T}_M}(\mu_1/p_1)}{\gamma M_1^2}\right\}\hat{\rho}\hat{y}h\,e^{-1/\varepsilon[(1/\hat{T})-(1/\hat{T}_M)]}$$

$$+ (\gamma-1)M_1^2(\hat{u}\hat{p}' + (4/3)\hat{\mu}\hat{u}'^2), \tag{3}$$

$$\hat{\rho}\hat{T} = 1 + \gamma M_1^2\hat{p} \tag{4}$$

where $\hat{\mu}$, $\hat{D}$, $\hat{C}_p$, and $\hat{\lambda}$ are specified functions of $\hat{T}$ and $\hat{p}$ at most.

The nondimensional variables are defined with respect to analogous reference values at the origin (subscript "1"), with the exception of the pressure $\hat{p} = (p - p_1)/\rho_1 u_1^2$. The fuel mass fraction is represented by $\hat{y}$. Primes denote spatial derivatives with respect to $\hat{x} = x/x_R$, where the reference length is given by $x_R = \mu_1/\rho_1 u_1$. The latter quantity is a modest multiple of the mean free path when the Mach number [see (5), below] is a reasonable fraction of unity, and much larger for smaller Mach numbers.

In (2) and (3) the initial Lewis (Le_1) and Prandtl (Pr_1) numbers have standard definitions, and γ is the ratio of initial specific heats. The quantity $\hat{T}_M$ represents the maximum nondimensional temperature in the entire combustion wave and must be found in the course of analysis. In addition, the equations

$$\varepsilon = RT_1/E, \qquad M_1^2 = u_1^2/\gamma RT_1, \qquad h = \Delta H/C_{p1}T_1, \qquad \sigma_1 = u_1' \tag{5}$$

represent the small-activation-energy parameter, the initial Mach number, the heat-release parameter (the ratio of the total heat release ΔH to initial enthalpy), and the initial (unknown) strain rate, respectively. The quantity B in (2) and (3) represents the preexponential factor in the Arrhenius law.

The parameter in braces in (2) and (3),

$$B\,e^{-1/\varepsilon\hat{T}_M}(\mu_1/p_1)/\gamma M_1^2, \tag{6}$$

is proportional to the ratio of the intermolecular collision time (μ_1/p_1) to the fastest possible reaction time $B^{-1}\exp(1/\varepsilon\hat{T}_M)$, divided by the local input Mach number squared. The time ratio itself must be small in all practical problems because substantial reaction progress requires many molecular collisions.

At the origin, the specified conditions are

$$\hat{x} = 0; \qquad \hat{T} = \hat{u} = 1, \qquad \hat{y} - (\mathrm{Le}_1/\mathrm{Pr}_1)\hat{y}' = 1, \tag{7}$$

where the latter condition reflects the initial balance of convection and diffusion of fuel. All the variables have finite gradients at the origin. Far downstream it is assumed the system relaxes to a final equilibrium state where

$$\hat{x} \to \infty; \qquad \hat{p} = \hat{p}_\infty, \qquad \hat{y} = 0, \tag{8}$$

and all derivatives vanish.

One can integrate the fuel and energy equations across the entire domain $0 \le \hat{x} < \infty$ and employ (7) and (8) to show that

$$\int_1^{\hat{T}_\infty} \hat{C}_p \, d\hat{T} = h - \text{Pr}_1^{-1} \hat{T}'(0) - (\gamma - 1) M_1^2 [(1/2)(\hat{u}_\infty^2 - 1) + (4/3)\sigma_1]. \tag{9}$$

The first term on the right-hand side represents the total chemical heat release, while the second term describes the heat loss at the origin. The last term characterizes the effects of compressibility and dissipation. When $M_1^2 \ll 1$ and $\hat{C}_p = \text{Pr}_1 = 1$, then

$$\hat{T}_M = \hat{T}_\infty = 1 + h - \hat{T}'(0), \tag{10}$$

where $1 + h$ is the classical adiabatic flame temperature.

Solutions to (1)–(4) are to be found in the limit $\varepsilon \to 0$ for $M_1 = M_1(\varepsilon) < 1$. The input Mach number M_1 is assumed to be specified. The combustion-wave structure is sought both for $M_1 = o(1)$ and $M_1 = O(1)$ when $\varepsilon \to 0$. Additional details can be found in Kassoy & Clarke (1984).

The Low-Speed Flame

The half-space model for a low-speed flame has been described in modern terms by Carrier et al. (1978), Clarke & McIntosh (1980), and Buckmaster & Ludford (1982), who cite many additional references. In this section, a unified summary of that work is given in terms of the model in the previous section. For simplicity of description and notation, it is assumed that $\text{Le}_1 = \text{Pr}_1 = \hat{C}_p = 1$.

Consider first a region in which $1 \le \hat{T} < \hat{T}_M$ and the reaction-time parameter defined in (6) is at most algebraically large relative to ε. The latter condition implies that $M_1^2(\varepsilon)$ is basically exponentially small. In the limit $\varepsilon \to 0$, Equations (1)–(4) reduce to the inert preheat equations, $\hat{u} = \hat{\rho}^{-1} = \hat{T}$, and

$$\hat{y}' = (\hat{\rho}\hat{D}\hat{y}')', \qquad 1 = \hat{y}(0) - \hat{y}'(0), \tag{11}$$

$$\hat{T}' = (\hat{\lambda}\hat{T}')', \qquad \hat{T}(0) = 1, \tag{12}$$

as long as the concentration and temperature gradients are larger than the neglected reaction terms. Solutions can be written in terms of the

Lagrangian variable

$$\xi = \int_0^{\hat{x}} \hat{\rho} \, d\hat{x} \tag{13}$$

when the model transport properties $\hat{D} = \hat{T}^2$, $\hat{\lambda} = \hat{T}$ are used. One finds

$$\hat{T} = 1 + \hat{T}'(0)(e^{\xi} - 1), \tag{14}$$

which approaches $\hat{T}_M$ [defined in (10)] smoothly when

$$\xi \to \xi_M = \ln\left[1 + ((\hat{T}_M - 1)/\hat{T}'(0))\right] \tag{15}$$

and

$$\hat{y} = 1 - e^{(\xi - \xi_M)}, \tag{16}$$

which vanishes at ξ_M. The temperature gradient $\hat{T}'(0)$, representing heat loss to the origin, must be found in the course of analysis.

The limiting process used to derive (11) and (12) fails when $\hat{T}_M - \hat{T} = O(\varepsilon)$, $\hat{y} = O(\varepsilon)$, and $\hat{x}_M - \hat{x} = O(\varepsilon)$, where $\hat{x}_M = \xi_M + T'(0)\left[\exp(\xi_M) - 1 - \xi_M\right]$ is found from the inverse of (13). This implies that the appropriate reaction-zone variables are

$$\hat{T} = \hat{T}_M - \varepsilon\theta, \qquad \hat{y} = \varepsilon Y, \qquad \hat{x} = \hat{x}_M - \varepsilon\hat{s}. \tag{17}$$

These can be used in (1)–(4) to show that in the limit $\varepsilon \to 0$, a fundamental balance between transport and reaction effects exists when the reaction-time parameter defined in (6) is $O(\varepsilon^{-2})$. It follows that the input or propagation Mach number can be defined by

$$M_1^2(\varepsilon) = \varepsilon^2 B \, e^{-1/\varepsilon\hat{T}_M} (\mu_1/p_1)/\gamma\Omega, \qquad \Omega = O(1), \tag{18}$$

where Ω, a specified constant of proportionality, has a limited range of values (described below). The reaction-zone equations take the form

$$\theta''(\hat{s}) = (\Omega/\hat{T}_M^2)h \, e^{-\theta/\hat{T}_M}, \qquad Y''(\hat{s}) = (\Omega/\hat{T}_M^2)Y \, e^{-\theta/\hat{T}_M}. \tag{19}$$

Solutions must match with those from the inert preheat zone,

$$\theta(\hat{s} \to +\infty) \sim \hat{s}\hat{T}'(0)/\hat{T}_M, \qquad Y(\hat{s} \to +\infty) \sim \hat{s}\hat{T}'(0)/h\hat{T}_M, \tag{20}$$

and satisfy the downstream boundary conditions

$$\theta(\hat{s} \to -\infty) = Y(\hat{s} \to -\infty) = 0. \tag{21}$$

A study of the first integral of (19)–(21) can be used to demonstrate that a continuous solution is found only if

$$\hat{T}'(0) = (2\Omega)^{1/2}\hat{T}_M^2 \, e^{-\xi_M} \tag{22}$$

This equation can be combined with (10) and (15) to show that

$$\hat{T}_M = h^{1/2}/(2\Omega)^{1/4}, \qquad \hat{T}'(0) = 1 + h - h^{1/2}(2\Omega)^{-1/4}, \tag{23}$$

where h, defined in (5), is the specified heat-release parameter. The values of $\hat{T}_M$ and $\hat{T}'(0)$ depend upon the allowable range of Ω.

Two interesting extremes are apparent. When $\Omega \to \Omega_{max} = h^2/2$, then $\hat{T}_M \to 1$, $\hat{T}'(0) \to h$, and $\xi_M \sim x_M \to 0$. In effect, the flame cannot be sustained because of excessive heat loss to the origin. No substantial temperature rise or reactant consumption is observed. This extinction limit corresponds to a minimized value of M_1^2 in (18).

At the other extreme is the limiting form $\Omega \to \Omega_{min} = h^2/2(1+h)^4$, for which $\hat{T}'(0) \to 0$, $\hat{y}'(0) \to 0$, $\hat{T}_M \to 1 + h$, and $\xi_M \sim \hat{x}_M \to \infty$. In this case M_1^2 is maximized, and the flame moves far away from the origin. The initial gradients vanish identically. As a result, Ω_{min} is nothing more than the eigenvalue obtained in the doubly infinite flame model. This demonstrates in a transparent manner that the classical model for a freely propagating flame is embedded in the more physically appealing model of a burner-attached flame. It should always be viewed in that context!

The traditional low-speed theory fails when $\Omega \to \Omega_{min}$ because the convection and transport terms in (2) and (3) become as small as the neglected reaction term. The solutions in (14) and (16) can be used in (2) and (3) to show that a nonuniformity occurs when $\hat{x} = O(1)$ if

$$\hat{T}'(0) = O(\varepsilon^{-2} e^{-1/\varepsilon[1-(1/\hat{T}_M)]}), \qquad \hat{T}_M = 1 + h. \tag{24}$$

This implies that when $\Omega \to \Omega_{min}$, a revised asymptotic analysis is required to describe reaction initiation in a region adjacent to the burner where there is an inherent balance of convection, transport, and reaction effects. The usual low-speed flame structure appears farther downstream.

THE LOW-SPEED SEPARATED FLAME   Small perturbations from the initial state in the reaction-initiation zone, $\hat{x} = O(1)$, can be described by the transformations $(\hat{T}, \hat{y}) = 1 + \beta(\varepsilon)(g, -c)$, where the parameter

$$\beta(\varepsilon) = \Omega\varepsilon^{-2} e^{-1/\varepsilon[1-(1/\hat{T}_M)]}, \qquad \Omega \to \Omega_{min}^+, \qquad \hat{T}_M = 1 + h \tag{25}$$

is chosen to produce a distinguished limit of (1)–(4) with a balance of convection, transport, and reaction effects. One finds the basic approximations, $\hat{u} = \hat{\rho}^{-1} = 1$ and the energy equation

$$g' = g'' + h, \qquad g(0) = 0, \tag{26}$$

as long as $\beta g/\varepsilon = o(1)$. The temperature solution can be written as

$$\hat{T} \sim 1 + \beta(\varepsilon)[D(e^{\hat{x}} - 1) + h\hat{x}] + \cdots, \tag{27}$$

where D is an integration constant to be determined. The exponential term arises from the homogeneous solution to (26) and represents a temperature rise due to heat conduction from the adjacent downstream zone. Weak chemical heat addition in (26) is the source of the algebraic term. The faster exponential growth causes a nonuniformity when $\beta e^x = O(\varepsilon)$. This implies the existence of a subsequent region defined by $\hat{x} = \tilde{t}(\varepsilon) + \tilde{x}$ and $T = 1 + \varepsilon \tilde{T}$, where

$$\tilde{t}(\varepsilon) = \varepsilon^{-1}[1 - (1/\hat{T}_M)] - 3\ln(1/\varepsilon) - \ln \Omega \qquad (28)$$

represents a translational shift of the coordinate system. The resulting energy equation $\tilde{T}' = \tilde{T}''$ displays an inert balance of convection and conduction. It has the solution $\tilde{T} = D \exp(\tilde{x})$, which represents a continuation of the exponential growth in (27). Here again there is a nonuniformity when $\varepsilon \tilde{T} \sim \varepsilon \exp(\tilde{x}) = O(1)$. This implies that when $\tilde{x} = \ln(1/\varepsilon) + O(1)$, there will be an $O(1)$ variation in $\hat{T}$. Similar results can be obtained for $\hat{y}$.

The shifted inert preheat zone exists in a region $x^* = O(1)$ defined by $\tilde{x} = \ln(1/\varepsilon) + x^*$. The limiting forms of (1)–(4) are identical to those in (11) and (12). However, the initial conditions shown in the latter pair are replaced by the matching conditions $y(x^* \to -\infty) \sim 1$, $T(x^* \to -\infty) \sim 1$. It should be emphasized that this mathematical system in the x^* region is identical to that derived in the inert zone of the doubly infinite low-speed flame model. The solutions describe the smooth increase of temperature toward $\hat{T}_M = 1 + h$ at $x^* = 0$ and the decrease of fuel concentration to zero at the same location. One also finds the integration constant $D = h$. In the subsequent thin reaction zone where $x^* = \varepsilon X$, the describing equations are identical in form to those in (19), with $\hat{s}$ replaced by $-X$. The upstream matching conditions are given by $\theta(X \to -\infty) \sim -hX/\hat{T}_M$ and $Y(X \to -\infty) \sim -X/\hat{T}_M$. It follows that in the lowest-order approximation, the parameter $\Omega = \Omega_{\min}$. This is expected because the mathematical problem posed is the same as that in the doubly infinite flame model.

This brief description of the separated low-speed flame shows that the traditional thin reaction zone is located at a nondimensional stand-off distance

$$d(\varepsilon) = \frac{1}{\varepsilon}\left(1 - \frac{1}{\hat{T}_M}\right) - 2\ln\frac{1}{\varepsilon} + \ln\frac{1}{\Omega_{\min}} \qquad (29)$$

found from (28) and the translational transformations relating $\hat{x}$, $\tilde{x}$, and x^*. It is notable that the traditional structure is preceded by a region in which weak chemical reactions, represented by the h term in (26), are initiated. Corrections to $\Omega_{\min}$ can be obtained by using higher-order analysis.

Transitional-Speed Flames

Clarke (1983a) recognizes that there are profound changes in the structure and propagation physics of a flame as M_1^2, defined in (18), increases beyond the special value associated with $\Omega = \Omega_{\min}$. When $O(\varepsilon^2) \ll \Omega < \Omega_{\min}$, the reaction process is initiated in a region adjacent to the origin in which a balance of convection and reaction exists. The small temperature disturbance, similar to that in (27), grows algebraically as a result of weak local heat release because the exponential term is suppressed. Farther downstream there is a region in which a balance of convective, reactive, and transport effects prevails. There the temperature perturbation has the full character of that in (27). The growth rate is dominated by the exponential term, reflecting more profound conductive heating from a subsequent classical inert preheating region, beyond which one finds the usual thin reaction zone containing important transport effects. Reactant consumption in the upstream zones is exponentially small relative to the limit $\varepsilon \to 0$, so that the major heat release takes place in the reaction zone.

When $\Omega \to O(\varepsilon^2)$ from above, reaction zones controlled by transport effects cannot exist. In this case, fuel consumption increases in upstream zones dominated by convection-reaction balances. The fuel concentration reaching the reaction zone decreases, causing the latter to broaden. An increasingly large fraction of the total heat release occurs in the upstream zones, and proportionally less is generated in the reaction zone. Eventually the inert preheat zone–reaction zone structure disappears because the heat release in the latter zone vanishes. The technical details of these complex variations can be found in Clarke (1983a).

In order to consider the structure for larger propagation Mach numbers, the results in (25)–(27) are examined when exponential growth is suppressed in the initial region. The parameter β, defined in (25), is considered now with $\Omega = \Omega(\varepsilon) = o(1)$. In this case, $g = h\hat{x}$ because $D = 0$. Since $g'' = 0$ in (26), this means that the initial region exhibits a basic balance of convection and weak reaction. The nonuniformity associated with temperature perturbation growth occurs when $\beta g = O(\varepsilon)$ or $\hat{x} = O(\varepsilon/\beta)$. Equation (27) implies that $\hat{T} = 1 + O(\varepsilon)$ in that circumstance. This means that there is a subsequent zone defined by $\hat{x} = (\varepsilon/\beta)s$, $(\hat{T}, \hat{y}) = 1 + \varepsilon(\phi, -w)$, where the limiting form of the energy equation and the appropriate matching condition are

$$\phi'(s) = he^\phi, \qquad \phi(s \to 0) \sim hs, \tag{30}$$

where h is the constant defined in (5). The solution to this convection-reaction equation can be used to write the full temperature as

$$\hat{T} \sim 1 + \varepsilon \ln\left(\frac{1}{1-sh}\right). \tag{31}$$

One observes that a singularity exists when the induction distance $s = s_i = h^{-1}$ is reached. This solution behavior is identical to that observed in time-dependent thermal-explosion processes, described in contemporary terms by Kassoy (1975). There it is shown that the nonuniformity in (31) can be resolved efficaciously by employing the nonlinear scaling transformation

$$s = s_i - H(\hat{T}; \varepsilon) e^{-1/\varepsilon[1-(1/\hat{T})]} \tag{32}$$

in the next rapid-reaction region, where $O(1)$ variations in $\hat{T}$ and $\hat{y}$ take place. If (32) is used in (3), an equation for H can be obtained in the form

$$1 = \frac{d}{d\hat{T}}\left[\beta e^{1/\varepsilon[1-(1/\hat{T})]} \lambda\left(\frac{H}{\hat{T}^2} - \varepsilon \frac{dH}{d\hat{T}}\right)^{-1}\right] + \hat{\rho}\hat{y}h\left(\frac{H}{\hat{T}^2} - \varepsilon \frac{dH}{d\hat{T}}\right), \tag{33}$$

where smaller compressibility and dissipation terms have been ignored. The term on the left-hand side represents convection, while conduction is described by the first term on the right-hand side and chemical heat release by the second. In order to understand the influence of conduction on the process described by (33), conditions on $\beta(\varepsilon)$ are sought that suppress this effect. It follows from (33) that the conduction term will be vanishingly small in the limit $\varepsilon \to 0$ if

$$\beta e^{1/\varepsilon[1-(1/\hat{T})]}/\varepsilon H(T) = o(1) \tag{34}$$

for the range $1 < \hat{T} < \hat{T}_M = 1 + h$. The crucial condition occurs for $\hat{T} \to \hat{T}_M$ because the exponential is maximized at this value. Given the definition of β in (25) with $\Omega = \Omega(\varepsilon)$, the critical condition is

$$\lim_{\hat{T} \to \hat{T}_M} \Omega = o(\varepsilon^3 H(\hat{T} \to \hat{T}_M)). \tag{35}$$

The behavior of the H function is ascertained by combining (33) with the analogous fuel-concentration equation and $\hat{\rho} = 1/\hat{T}$ to find

$$h\hat{y} = \hat{T}_M - \hat{T}, \qquad H = \hat{T}^3/(\hat{T}_M - \hat{T}). \tag{36}$$

The limit process used to find this result fails when $\hat{y} = O(\varepsilon)$, $H = O(\varepsilon^{-1})$, which occurs when $\hat{T}_M - \hat{T} = O(\varepsilon)$. Then from (35), the condition

$$\Omega(\varepsilon) = o(\varepsilon^2) \tag{37}$$

guarantees that transport effects are negligible in the reaction zone described by the limiting form of (33). It follows from the definition of the propagation Mach number in the first part of (18) that when

$$O(B e^{-1/\varepsilon \hat{T}_M} (\mu_1/p_1)) \ll M_1^2 \ll O(1), \tag{38}$$

transport property effects are essentially negligible in the entire reaction process. These small-Mach-number flames are basically convecting thermal explosions because the speed at which the reaction propagates into unburned material depends only on the reaction rate itself.

It is most interesting to note that when $\Omega = O(\varepsilon^2)$, all three terms in the energy equation in (33) are $O(1)$ when $\hat{T}_M - \hat{T} = O(\varepsilon)$, $\hat{y} = O(\varepsilon)$, $H = O(\varepsilon^{-1})$, and $s_i - s = O(\varepsilon^{-1} \exp[-\varepsilon^{-1}(1 - \hat{T}_M^{-1})])$. This means that just beyond the rapid-reaction region, there is a relaxation zone defined by $\hat{T} = \hat{T}_M - \varepsilon\Phi$, $\hat{y} = \varepsilon m$, $\hat{x} = \hat{x}_i + \zeta$, $\hat{x}_i = (\varepsilon/\beta)s_i$, and $\Omega = \hat{\Omega}\varepsilon^2$, where, for example, the energy equation (3) has the limiting form

$$-\Phi'(\zeta) = -\hat{T}_M \Phi'' + \hat{\Omega}h\hat{T}_M^{-1}m\,e^{-\Phi/\hat{T}_M^2}. \tag{39}$$

In this region the final bits of fuel are consumed in a process governed by a balance of convection, conduction, and reaction effects. In contrast, the upstream regions are structurally similar to that for $\Omega = o(\varepsilon^2)$, dominated by a convection-reaction balance. This point emphasizes that when $\Omega = O(\varepsilon^2)$ one observes the last of the transport-property effects at the downstream end of the rapid-reaction zone.

The Subsonic High-Speed Flame

The theory for transitional-speed flames fails when $M_1 = O(1)$ in the limit $\varepsilon \to 0$, because the previously ignored compressibility effects alter the temperature distribution in the combustion wave. In particular the $\hat{u}\hat{p}'$ factor in (3) acts like a sink for thermal energy, which reappears as flow kinetic energy. A mathematical model of this compressible flow with distributed heat addition, including its fundamental physical properties, is described by Clarke (1983b) and Kassoy & Clarke (1984). The description here is extracted from elements of the latter paper.

The reaction process is initiated in $\hat{x} = O(1)$ where there is a balance of convection, conduction, weak reaction, and compressibility effects in the energy equation. In dimensional terms the length scale is a modest multiple of the mean free path, the shortest length scale for which a continuum theory is valid. The nondimensional temperature variation is given by

$$\hat{T} \sim 1 + \alpha(\varepsilon)[R\hat{x} + Q(1 - \exp(-\lambda_2^2 \hat{x}))], \qquad \alpha(\varepsilon) = \frac{B\,e^{-1/\varepsilon}(\mu_1/p_1)}{\gamma M_1^2}, \tag{40}$$

where $\alpha(\varepsilon) = \beta(\varepsilon)$ if (18) and (25) are combined formally for $M_1^2 = O(1)$. In

(40), $R = (1 - \gamma M_1^2)(1 - M_1^2)^{-1} h$, Q, and λ_2^2 are functions of γ, M_1^2, Pr_1, h, and initial gradients of temperature and speed. It should be noted that the very small magnitude of $\alpha(\varepsilon)$ in (40) is proportional to the ratio of the collision time, μ_1/p_1, to the much longer reaction time at the initial (cold) temperature $B^{-1} e^{1/\varepsilon}$. The temperature increases like $\hat{x}$, as long as $R > 0$, because of weak heat release. It follows that an accelerating combustion process evolves downstream of the origin only when $M_1^2 < \gamma^{-1}$. When $\gamma^{-1} \leq M_1^2 < 1$, the temperature actually declines as a result of the thermal-sink effect of the compressibility term in (3). When $R > 0$, one finds that the magnitudes of the fuel consumption, speed increase, and pressure drop are $O(\alpha)$ and that the algebraic growth of the temperature perturbation in (40) persists there as well.

The nonuniformity in (40) for $\hat{x} \to \infty$ can be used to show that there is a subsequent larger region, defined by the scaling $\hat{x} = (\varepsilon/\alpha)\bar{x}$, where a balance of convection, reaction, and compressibility effects determines the temperature distribution

$$\hat{T} \sim 1 + \varepsilon \ln (1 - R\bar{x})^{-1}. \tag{41}$$

The variation is reminiscent of that for transitional-speed flames in (31) and emphasizes the convecting thermal-explosion character of these high-speed combustion waves. When the induction-delay distance $\bar{x} = \bar{x}_i = R^{-1} > 0$ is reached, the temperature perturbation becomes unbounded, and one can expect a significant and rapid increase in temperature and a decrease in fuel concentration. The value of the induction distance is minimized for $M_1^2 \ll 1$, corresponding to the result from (31), and becomes very large when $M_1^2 \to (\gamma^{-1})$ from below because the temperature rise is vitiated by the compressibility sink. In dimensional terms, $x_i \leq O(10\,\mathrm{m})$ for typical values of the physical and kinetic parameters involved.

The nonuniformity in (40) can be resolved by employing a transformation like that in (32), with s replaced by $\bar{x}$. As a result, one can derive an energy equation from (3) that contains in part the terms in (33), but that has in addition (on the right-hand side) compressibility and dissipation effects described by

$$(\gamma - 1)M_1^2 \left[\hat{u} \frac{d\hat{p}}{d\hat{T}} + (4/3)\alpha\, e^{1/\varepsilon[1 - (1/\hat{T})]} \left(\frac{H}{\hat{T}^2} - \varepsilon \frac{dH}{d\hat{T}} \right)^{-1} \left(\frac{d\hat{u}}{d\hat{T}} \right)^2 \right].$$

Arguments identical to those used following Equation (33) can be invoked here to show that transport effects, including dissipation, are negligible compared with convection, reaction, and compressibility as long as the collision-time magnitude μ_1/p_1 is short compared with the fastest possible reaction time $B^{-1} \exp(1/\varepsilon\hat{T}_M)$, where $\hat{T}_M$ must be found from the solution in this rapid-reaction zone. Generally, $\hat{T}_M < 1 + h$.

Kassoy & Clarke (1984) have shown that the required solution can be constructed from the scaled energy, fuel species, mass, and momentum conservation equations and the equation of state. These can be written, respectively, as

$$H = [1 + (\gamma - 1)(d\hat{u}^2/d\hat{T})M_1^2/2](\hat{T}^2\hat{u}/h\hat{y}), \tag{42}$$

$$h(1 - \hat{y}) = \hat{T} - 1 + (\gamma - 1)(\hat{u}^2 - 1)M_1^2/2, \tag{43}$$

$$\hat{\rho}\hat{u} = 1, \quad \hat{u} = 1 - \hat{p}, \quad \hat{\rho}\hat{T} = 1 + \gamma M_1^2\hat{\rho}, \tag{44}$$

where $\hat{T}$ is the independent variable.

In contrast to the analogous solutions for the transitional-speed flames where $M_1^2 \ll 1$, this system describes a fully compressible flow with distributed heat addition. For a given value of M_1^2 there are four classes of solutions to consider, the properties of which depend on the value of h relative to a critical value

$$h_c = (1 - \gamma M_1^2)((3\gamma - 1) - \gamma(3 - \gamma)M_1^2)/8\gamma^2M_1^2. \tag{45}$$

When $0 < h < h_c$, $\hat{T}$ and $\hat{u}$ increase monotonically until $\hat{y} = 0$. The maximum temperature obtained is less than

$$\hat{T}_c = 1 + (1 - \gamma M_1^2)^2/4\gamma M_1^2.$$

When the total heat addition is critical ($h = h_c$), the monotonic temperature increase ends at $\hat{y} = 0$ with $\hat{T} = \hat{T}_c$. At this point the local Mach number has increased to $M_c^2 = \gamma^{-1}$. This is a critical value beyond which the compressibility thermal-sink effect causes the flow temperature to decline. As a result, when $h_c < h < h_{CJ} = (1 - M_1^2)^2/2(\gamma + 1)M_1^2$, the temperature in the rapid-reaction zone actually increases to the maximum possible value $\hat{T}_c$ and then decreases downstream until $\hat{y} = 0$. The Chapman-Jouguet case $h = h_{CJ}$ exhibits the maximum allowable flow acceleration to $M = 1$ when $\hat{y} = 0$ and $T = \hat{T}_{CJ} = 4\gamma(\gamma + 1)^{-2}\hat{T}_c$. Numerical examples given in Kassoy & Clarke (1984) show that the maximum value of heat addition for these steady flows (h_{CJ}) is extremely limited unless $M_1 \lesssim 0.2$ for $\gamma = 1.4$. For larger M_1 values, $\hat{T}_c \lesssim 3$, which means that these reaction zones are fairly cool. It should be noted that the value of $\hat{T}_M$ is less than T_c for $0 < h < h_c$ and equal to $\hat{T}_c$ for all larger h. The spatial variation $\hat{T}(\bar{x})$ is obtained from (32), with $\bar{x}$ replacing s.

The rapid-reaction zone solutions described by (42)–(44) are not uniformly valid when $\hat{y} = O(\varepsilon)$. As a result, there are additional downstream relaxation zones where the final bits of fuel are consumed and the far-field equilibrium state is approached.

Conclusions

A unified theory has been described for the steady-state burner-attached flame evolving downstream from an initial nonequilibrium point. The mathematical model is well posed when finite gradients are permitted at the origin, and it is unencumbered by artificial reaction rates and ambiguously defined ignition temperatures.

Traditional, transport-property-dominated flames exist for values of the effective propagation Mach numbers in (18) when $\Omega_{min} < \Omega < \Omega_{max}$. Extinction occurs when $\Omega \to \Omega_{max}$ because too much heat is lost to the burner to sustain a high-temperature reaction zone and $T_M \to 1$. When $\Omega \to \Omega_{min}$, one finds that the burner heat transfer vanishes, that $\hat{T}_M = 1 + h$, and that the distance to the flame is unbounded. The latter result represents a failure of traditional low-speed flame theory.

A more precise theory for $\Omega \to \Omega_{min}$ shows that the flame is located at a finite distance d from the burner [given in (29)], which is large compared with the transport-property-dominated flame thickness itself. The shifted flame is described by a rational, doubly infinite field model. Classical, cold-boundary difficulties do not arise, because the inert preheat zone evolves naturally from an upstream induction zone in which there is a balance of convection, conduction, and weak reaction effects. There, the existence of small but finite gradients, measured by $\beta(\varepsilon)$ in (25), allows the ill-posedness of the traditional doubly infinite formulation to be overcome.

As Ω is reduced below Ω_{min}, increasingly large amounts of fuel are consumed in convection-reaction-dominated zones upstream of the more traditional reaction-diffusion zone of the classical flame. In fact, when $\Omega = O(\varepsilon^2)$, $O(1)$ amounts of fuel are burned in the upstream thermal-explosion-like region. The remaining $O(\varepsilon)$ quantity is consumed in a thin convection-conduction-reaction zone described by (39). In effect, the classical flame structure has vanished.

Finally, when $M_1 = O(1)$ in the limit $\varepsilon \to 0$, the combustion wave is described in terms of a fully compressible flow with distributed heat addition. With the exception of a very thin initiation zone adjacent to the origin, transport effects play no role in these flames. The induction-delay zone and the subsequent rapid-reaction region are dominated by a balance of convection, reaction, and compressibility. Each moving fluid particle experiences heat release due to chemical reaction, as well as energy interchange due to compressibility effects. Under appropriate circumstances $h_c < h \le h_{CJ}$, this interaction can generate nonmonotonic temperature variations in the rapid-reaction zone. The temperature decrease reduces the reaction rate and enlarges the zone of significant heat release.

It should be emphasized that the characteristic dimensional length scales of the combustion wave change significantly as the value of M_1 increases. For example, the order of magnitude of the reaction-diffusion zone of a classical flame is obtained from $\Delta = [(\mu_1/p_1)B^{-1} \exp(1/\varepsilon\hat{T}_M)C_1^2]^{1/2}$, where C_1 is the speed of sound at the origin and $\hat{T}_M = 1 + h$. In comparison, the length of the preceding inert convection-conduction zone is Δ/ε. When $\Omega \to \Omega_{min}^+$, the flame stand-off distance is given by Δ/ε^2.

If the propagation Mach number is increased into the transitional regime, the vigorous reaction zone is separated from the burner by the distance $\Omega^{-1/2}\varepsilon^2 \exp(\varepsilon^{-1}(1 - \hat{T}_M^{-1}))\Delta$ for $\Omega(\varepsilon) = o(1)$. The reaction-zone thickness itself is $\Omega^{-1/2}\varepsilon^2 \exp(\varepsilon^{-1}(\hat{T}^{-1} - \hat{T}_M^{-1}))\Delta$ for $1 < \hat{T} < \hat{T}_M = 1 + h$. The dimension of this region is far larger than that for the classical reaction zone, and even thicker reaction regions are found when $M_1 = O(1)$. In this respect, it is noted that for a common set of initial thermodynamic conditions, high-speed reactions zones are relatively thick compared with those in transport-dominated flames. If, however, the high-speed flame is initiated at a relatively high temperature, the dimensions are reduced substantially.

ACKNOWLEDGMENTS

This work was supported by the Army Research Office under Contract DAAG29-82-K-0069. A special word of appreciation is extended to Prof. J. F. Clarke, who knew the meaning of transition.

Literature Cited

Adamson, T. C. 1960. On the structure of plane detonation waves. *Phys. Fluids* 3: 706–14

Ahlborn, B., Liese, W. 1981. Heat-flux induced wave fronts. *Phys. Fluids* 24: 1955–66

Buckmaster, J. D., Ludford, G. S. S. 1982. *Theory of Laminar Flames*. Cambridge: Cambridge Univ. Press. 266 pp.

Bush, W. B., Fendell, F. E. 1970. Asymptotic analysis of laminar flame propagation for general Lewis numbers. *Combust. Sci. Technol.* 1: 421–28

Bush, W. B., Fendell, F. E. 1971. Asymptotic analysis of the structure of a steady planar detonation. *Combust. Sci. Technol.* 2: 271–85

Carrier, G. F., Fendell, F. E., Bush, W. B. 1978. Stoichiometry and flameholder effects on a one-dimensional flame. *Combust. Sci. Technol.* 18: 33–46

Clarke, J. F. 1983a. On changes in the structure of steady plane flames as their speed increases. *Combust. Flame* 50: 125–38

Clarke, J. F. 1983b. Combustion in plane steady compressible flow. *J. Fluid Mech.* 136: 139–61

Clarke, J. F., McIntosh, A. C. 1980. The influence of a flameholder on a plane flame, including its static stability. *Proc. R. Soc. London Ser. A* 372: 367–92

Clavin, P. 1984. Introduction to modern laminar flame theory. *Proc. Int. Colloq. Dyn. Explos. and React. Syst.*, 9th. In press

Fickett, W., Davis, W. C. 1979. *Detonation*. Berkeley: Univ. Calif. Press. 485 pp.

Hirschfelder, J. O., Curtiss, C. F., Campbell, D. E. 1953. The theory of flames and detonations. *Symp. (Int.) Combust., 4th*, pp. 190–211. Baltimore: Williams & Wilkins

Johnson, W. E. 1963. On a first-order boundary value problem from laminar flame theory. *Arch. Ration. Mech. Anal.* 13: 46–54

Kapila, A. K., Matkowsky, B. J., van Harten,

A. 1983. An asymptotic theory of deflagrations and detonations. I. The steady solutions. *SIAM J. Appl. Math.* 43:491–519

Kassoy, D. R. 1975. Perturbation methods for mathematical models of explosion and ignition phenomena. *Q. J. Mech. Appl. Math.* 28:63–74

Kassoy, D. R., Clarke, J. F. 1984. The structure of a steady high speed deflagration with a finite origin. *J. Fluid Mech.* In press

Margolis, S. B. 1980. Bifurcation phenomena in burner-stabilized premixed flames.

Combust. Sci. Technol. 22:143–69

Semenov, N. N. 1928. Zur Theorie der Verbrennungsprozesses. *Z. Phys.* 48:571–82

Stewart, D. S., Ludford, G. S. S. 1983. Fast deflagration waves. *J. Méc. Théor. Appl.* 3:463–87

Williams, F. A. 1965. *Combustion Theory.* Reading, Mass: Addison-Wesley. 447 pp.

Zeldovich, Y. 1980. Flame propagation in a substance reacting at initial temperature. *Combust. Flame* 39:219–24

Ann. Rev. Fluid Mech. 1985. 17 : 289–320

FLUID MECHANICS OF COMPOUND MULTIPHASE DROPS AND BUBBLES

Robert E. Johnson

Department of Theoretical and Applied Mechanics, University of Illinois, Urbana, Illinois 61801-2983

S. S. Sadhal

Department of Mechanical Engineering, University of Southern California, Los Angeles, California 90089-1453

INTRODUCTION

This review deals with the fluid mechanics of compound multiphase drops and bubbles. While the term multiphase has most often been used in reference to systems with two or more phases (i.e. traditional multiphase flow), here we use it in reference to drops and bubbles that are compound liquid-liquid or gas-liquid drops surrounded by a distinct third fluid. Examples of compound liquid-liquid drops occurring in three-phase emulsions and of single gas-liquid drops generated under controlled laboratory conditions are shown in Figures 1 and 2. Fluid-mechanical studies of compound drops are for the most part still in the early stages of development. Lately, however, there has been a general surge of interest in understanding the behavior of such drops as a result of their occurrence in a variety of engineering systems. In this review, after a brief survey of the fields in which multiphase drops are of interest, we discuss the state of the art as we see it. While this subject covers various thermophysical and physico-chemical aspects, we focus on the fluid mechanics.

In the case of liquid-liquid compound drops it appears that this subject was first studied by Chambers & Kopac (1937) and Kopac & Chambers (1937) in connection with investigations of the coalescence of living cells

289

0066–4189/85/0115–0289$02.00

and oil drops. Kopac & Chambers interpreted the observations of cell-oil coalescence using a simple analysis of the coalescence of two immiscible drops. Their analysis, however, was far from complete. Various aspects of this problem have remained of interest in cell biology (Gershfeld & Good 1967), but the primary interest in this problem presently lies in more traditional aspects of chemical engineering. The first thorough theoretical analysis of static two-fluid drop configurations was completed by Torza & Mason (1970) in an investigation of three-phase interactions. Their paper included a substantial experimental part that examined two-fluid drops in shear and electric fields, and they also studied the dynamics of coalescence, i.e. engulfing. Their theoretical analysis, however, was restricted to static drops in the absence of gravity. One of the few theoretical studies on the translation of two-fluid drop configurations has been that of Johnson & Sadhal (1983). Their analysis, however, was limited to small Reynolds numbers and capillary numbers and to the case when one of the fluids comprising the compound drop is present in a thin layer. Other purely theoretical studies of translation include that of Rushton & Davies (1983), who considered a compound drop consisting of concentric spheres. Other

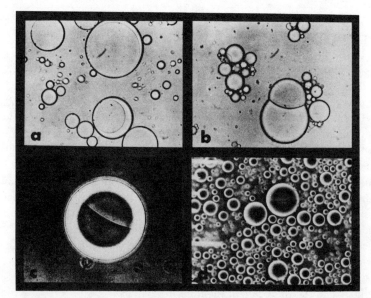

Figure 1   Compound drops in three-phase emulsions: (*a*) Silicone oil dibutylphthalate drops in distilled water +0.1 vol% Dowfax 9N15, (*b*) distilled water/castor oil drops in Silicone oil, (*c,d*) compound drops of two incompatible polymer solutions having a common solvent (toluene) in water (+emulsifier). Reproduced, with permission, from Torza & Mason (1970).

analyses of liquid-liquid compound drops have been associated with drop oscillations and stability (Patzer & Homsy 1975, Saffren et al. 1982, Landman & Greenspan 1982, Landman 1983, 1984a,b), and less well-known areas where two-fluid drops are of interest include the fabrication of inertial confinement fusion (ICF) target pellets, the break-up of compound liquid jets (Hertz & Hermanrud 1983), and the augmentation of evaporative cooling of drops (Mori et al. 1981).

The related configuration of a compound gas-liquid drop has recently been receiving considerable attention as a result of an increased interest in direct-contact heat- and mass-exchangers. Specific applications that involve gas-liquid compound drops range from direct-contact condensation or vaporization, a rather popular idea, to novel methods of blood oxygenation (Li 1971a,b, Li & Asher 1973, Ollis et al. 1972). A typical problem of concern in direct-contact heat transfer is that of a liquid drop vaporizing in a immiscible, less volatile liquid medium. The compound drop is generally rising or sinking as a result of buoyancy while the gaseous portion of the drop is growing from the attached liquid. The reverse situation involving a condensing vapor bubble is also a problem of general interest. Although there is a fairly large number of experimental studies concerned with translating gas-liquid compound drops in connection with heat and mass transfer (Mercier et al. 1974, Hayakawa & Shigeta 1974,

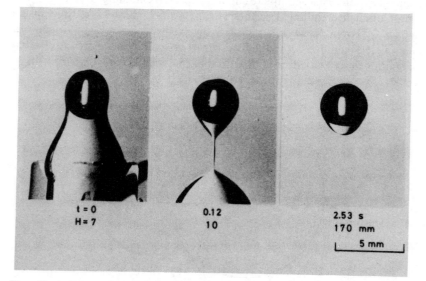

Figure 2 A compound gas-liquid drop (R113-86% aqueous glycerol) translating in aqueous glycerol. Reproduced, with permission, from Mori (1978).

Selecki & Gradon 1976, Tochitani et al. 1977a,b, Mori 1978), there is a striking absence of rigorous fluid-mechanical analysis. The exceptions to this are the previously mentioned studies of translating and oscillating liquid-liquid drops, which are also applicable here for gas-liquid drops.

Gas-liquid compound drops are also found as transient configurations during vapor explosions of drops near the superheat limit (Shepherd & Sturtevant 1982, Lasheras et al. 1980, Moore 1959). Such a phenomenon frequently occurs during the vaporization of multicomponent fuel drops. During the transient stages of spontaneous vaporization of superheated drops, one observes compound gas-liquid drops where, in a time of the order of a microsecond, a gas bubble grows from a drop. The very short time scale involved in vapor explosions makes this case fundamentally different from the situation involved in direct-contact heat exchangers.

STATIC COMPOUND DROPS

As already mentioned, Torza & Mason (1970) completed the first thorough analysis of static two-fluid drop configurations. Their work was based on the assumption that the equilibrium configuration is solely determined by the surface tensions σ_{ij} ($i \neq j \neq k = 1, 2, 3$) of the three possible fluid-fluid interfaces (Figure 3). The effects of gravity, fluid motion, interparticle forces, etc., were not accounted for in their analysis. Three basic equilibrium configurations exist: complete engulfing, partial engulfing, and nonengulfing. Complete engulfing describes a drop in which phase 1 is entirely inside phase 3. Torza & Mason adopted the term *2-singlet* (2s) to denote a completely engulfed two-fluid drop, since two interfaces are present. Partial engulfing describes a drop similar to that shown in Figure 3, where a single

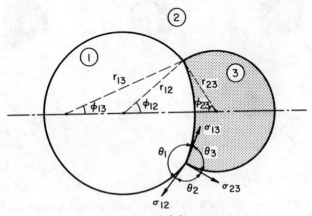

Figure 3   Compound-drop geometry.

compound drop consists of three fluid-fluid interfaces that intersect at a three-fluid contact line. Following Torza & Mason, we use the term *3-singlet* (3s) to describe such a drop. Lastly, the nonengulfing configuration is that for which the 13 interface will not form spontaneously. For consistency these are referred to as a pair of *1-singlets* (1s). In this case, phase 1 and phase 3 are completely engulfed by the bulk phase denoted as phase 2.

The occurrence of a particular configuration was shown by Torza & Mason (1970) to be determined by the values of three spreading coefficients S_1, S_2, and S_3 defined by

$$S_i = \sigma_{jk} - (\sigma_{ij} + \sigma_{ik}), \qquad (i \neq j \neq k = 1, 2, 3).$$

Adopting a convention whereby phase 2 is the bulk phase and phase 1 is the phase for which $\sigma_{12} \geqslant \sigma_{23}$, we have $S_1 < 0$, and the spreading coefficient criterion presented by Torza & Mason is

complete engulfing (2s): $S_1 < 0, S_2 < 0, S_3 > 0,$

partial engulfing (3s): $S_1 < 0, S_2 < 0, S_3 < 0,$

nonengulfing (1s pair): $S_1 < 0, S_2 > 0, S_3 < 0.$

Although a particular configuration type is determined by the surface tensions alone, we shall see later that a detailed description of the resulting configuration also depends on the phase 1 and phase 3 drop volumes. The only deficiency of the paper by Torza & Mason (1970) is that the proof of the above criterion appears only in a PhD thesis on which the paper is based (Torza 1970), and careful examination of that proof reveals that it is not quite rigorous. Nonetheless, a more careful proof shows that the results are indeed correct, although we shall see that the criterion can actually be simplified somewhat. More recently, Mori (1978) reexamined the spreading-coefficient criterion using a rather crude order-of-magnitude analysis and rederived the results originally found by Torza & Mason.

In the absence of any fluid motion and buoyancy, the jump in the normal stress at each interface is constant and equal to the pressure difference. Consequently, if we assume that the interfacial tensions are constant, then the three interfaces of the compound two-fluid drop will be spherical with radii denoted by r_{12}, r_{23}, and r_{13}. In this case the configuration of a compound drop is governed by the following equations: (a) the prescribed phase 1 and phase 3 volumes V_1 and V_3,

$$V_1 \equiv \frac{4}{3}\pi b_1^3 = \frac{\pi}{3} r_{12}^3 [2 + \cos\phi_{12}(3 - \cos^2\phi_{12})]$$

$$+ \frac{\pi}{3} r_{13}^3 [2 - \cos\phi_{13}(3 - \cos^2\phi_{13})], \tag{1}$$

$$V_3 \equiv \frac{4}{3}\pi b_3^3 = \frac{\pi}{3}r_{23}^3[2+\cos\ \phi_{23}(3-\cos^2\ \phi_{23})]$$

$$-\frac{\pi}{3}r_{13}^3[2-\cos\ \phi_{13}(3-\cos^2\ \phi_{13})], \tag{2}$$

where ϕ_{12}, ϕ_{23}, and ϕ_{13} give the positions of the three-fluid contact line relative to each interface, and b_1 and b_3 are the radii of the two drops in the nonengulfed configuration (1s pair); (b) the force balance at the contact line,

$$\sigma_{23}\sin\ \theta_2-\sigma_{13}\sin\ \theta_1 = 0, \tag{3}$$

$$\sigma_{12}+\sigma_{23}\cos\ \theta_2+\sigma_{13}\cos\ \theta_1 = 0, \tag{4}$$

where θ_1, θ_2 are the contact angles at the three-fluid contact line (see Figure 3); and (c) the geometrical constraints,

$$r_{12}\sin\ \phi_{12} = r_{13}\sin\ \phi_{13} = r_{23}\sin\ \phi_{23}, \tag{5,6}$$

giving the distance from the symmetry axis to the contact line, and

$$\theta_1 = \pi+\phi_{13}-\phi_{12}, \tag{7}$$

$$\theta_2 = \phi_{12}+\phi_{23}, \tag{8}$$

a consequence of the fact that we have three intersecting spherical surfaces ($\theta_3 = 2\pi-\theta_1-\theta_2$). If we now express the force balance at the contact line along directions parallel to the symmetry axis and perpendicular to it, Equations (3) and (4) can be rewritten as

$$\sigma_{12}\sin\ \phi_{12}-\sigma_{23}\sin\ \phi_{23}-\sigma_{13}\sin\ \phi_{13} = 0, \tag{9}$$

$$\sigma_{12}\cos\ \phi_{12}+\sigma_{23}\cos\ \phi_{23}-\sigma_{13}\cos\ \phi_{13} = 0. \tag{10}$$

With the use of (5) and (6), it can be seen that (9) is equivalent to the normal-stress condition

$$\frac{\sigma_{12}}{r_{12}} = \frac{\sigma_{23}}{r_{23}}+\frac{\sigma_{13}}{r_{13}}. \tag{11}$$

Note that in the force balance at the three-fluid contact line we have neglected the effects of a line tension, since in all practical cases of interest it is extremely small compared with the forces due to the surface tensions.

Lastly, if the curvature of the 13 interface in Figure 3 is reversed, i.e. if the origin of the radius r_{13} were to the right of the 13 interface instead of to the left as shown, then the correct equations for the volumes [Equations (1) and (2)] are given by interchanging the subscripts 1 and 3. Similarly, the subscripts 1 and 3 would also have to be interchanged in Equations (7) and (8). This situation, however, is not possible if $r_{12} < r_{23}$, since we have already assumed that phase 1 is such that $\sigma_{12} \geqslant \sigma_{23}$ and therefore that

$p_1 - p_2 > p_3 - p_2$ giving $p_1 > p_3$, which requires the curvature of the 13 interface to be as shown in Figure 3.

If the interfacial tensions σ_{ij} and the volumes V_1 and V_3 are specified, then Equations (1)–(8) are eight equations for the eight unknowns r_{12}, r_{13}, r_{23}, ϕ_{12}, ϕ_{13}, ϕ_{23}, θ_1, and θ_2. The criterion for a 3-singlet, however, is easily deduced from Equations (3) and (4), which give

$$S_2 = \sigma_{13} - \sigma_{12} - \sigma_{23} = -2\sigma_{23} \sin \tfrac{1}{2}\theta_2 \cos \tfrac{1}{2}\theta_3 / \sin \tfrac{1}{2}\theta_1, \tag{12}$$

$$S_3 = \sigma_{12} - \sigma_{13} - \sigma_{23} = -2\sigma_{23} \cos \tfrac{1}{2}\theta_2 \sin \tfrac{1}{2}\theta_3 / \sin \tfrac{1}{2}\theta_1. \tag{13}$$

Since $\theta_3 = \pi - \phi_{13} - \phi_{23}$, where ϕ_{13} and ϕ_{23} are between 0 and π for a 3-singlet, and since $\theta_3 \geqslant 0$ for a physically realizable situation (note Figure 3), we must have $0 \leqslant \theta_3 \leqslant \pi$. Furthermore, in order for the force balance (3) to be satisfied, clearly we must have $0 \leqslant \theta_k \leqslant \pi$ ($k = 1, 2$). Consequently, Equations (12) and (13) show that the requirement for a 3-singlet is that $S_2 \leqslant 0$ and $S_3 \leqslant 0$. Recall that by convention we have $S_1 < 0$, since we assumed that $\sigma_{12} \geqslant \sigma_{23}$. From (12) and (13) we can also evaluate S_2 and S_3 for the limiting cases when the drops just touch (i.e. nonengulfing) and when phase 1 is entirely inside phase 3 and the two drops just touch (i.e. complete engulfing). For the nonengulfing limit, $\phi_{12} = \phi_{13} = \phi_{23} = 0$; therefore, from (7), (8), (12), and (13) we have $S_2 = 0$ and $S_3 = -2\sigma_{23}$. Similarly, for the complete engulfing limit, $\phi_{12} = \phi_{13} = \pi$ and $\phi_{23} = 0$, giving $S_2 = -2\sigma_{23}$ and $S_3 = 0$. Since $S_2 = 0$ and $S_3 = 0$ correspond to the limiting cases, a true 3-singlet, i.e. a drop for which there are three distinct interfaces, actually requires that $S_2 < 0$ and $S_3 < 0$.

All that remains to complete the spreading-coefficient criterion is to show that the nonengulfing and complete-engulfing configurations are minimum-surface-energy configurations for $S_2 \geqslant 0$, $S_3 < 0$ and $S_3 \geqslant 0$, $S_2 < 0$, respectively. In the first case this is done by considering the difference in the surface energy between a pair of 1-singlets and a 3-singlet having an infinitesimal 13 interface, i.e. a 3-singlet with ϕ_{12}, ϕ_{13}, and $\phi_{23} \ll 1$ [see Figure 4 (top)]. For these configurations the surface energy is given by

$$E^{(1)} = \sigma_{12}A_{12}^{(1)} + \sigma_{23}A_{23}^{(1)}, \quad \text{(1s pair)}, \tag{14}$$

$$E^{(3)} = \sigma_{12}A_{12}^{(3)} + \sigma_{23}A_{23}^{(3)} + \sigma_{13}A_{13}^{(3)}, \quad \text{(3s)}, \tag{15}$$

where $A_{ij}^{(k)}$ denotes the area of the ij interface for the configuration denoted by k. With b_1 and b_3 as the radii of the two 1-singlets, we then have

$$A_{12}^{(1)} = 4\pi b_1^2, \qquad A_{23}^{(1)} = 4\pi b_3^2, \tag{16}$$

$$A_{12}^{(3)} = 2\pi r_{12}^2(1 + \cos \phi_{12}) \simeq 4\pi r_{12}^2(1 - \tfrac{1}{4}\phi_{12}^2 + \ldots),$$

$$A_{23}^{(3)} = 2\pi r_{23}^2(1 + \cos \phi_{23}) \simeq 4\pi r_{23}^2(1 - \tfrac{1}{4}\phi_{23}^2 + \ldots), \tag{17}$$

$$A_{13}^{(3)} = 2\pi r_{13}^2(1 - \cos \phi_{13}) \simeq \pi r_{13}^2\phi_{13}^2 + \ldots.$$

From the drop-volume equations evaluated for small ϕ_{ij} we obtain $b_1^3 \simeq r_{12}^3[1 + O(\phi_{ij}^4)]$ and $b_3^3 \simeq r_{23}^3[1 + O(\phi_{ij}^4)]$, and for small ϕ_{ij} Equations (5) and (6) give $r_{12}\phi_{12} \simeq r_{13}\phi_{13} \simeq r_{23}\phi_{23}$. Consequently, from (14) and (15) we find

$$E^{(3)} - E^{(1)} \simeq (\sigma_{13} - \sigma_{12} - \sigma_{23})\pi r_{12}^2\phi_{12}^2 = S_2\pi r_{12}^2\phi_{12}^2, \tag{18}$$

and therefore $E^{(1)} \leqslant E^{(3)}$ for $S_2 \geqslant 0$, and a 1s pair is the lower surface-energy configuration whenever $S_2 \geqslant 0$. Note that $S_2 \geqslant 0$ implies that $\sigma_{13} \geqslant \sigma_{12} + \sigma_{23}$, which gives $\sigma_{12} - \sigma_{13} \leqslant -\sigma_{23}$ leading to $S_3 \leqslant -2\sigma_{23}$. We therefore have shown that a pair of 1-singlets (nonengulfed configration) is the minimum-energy configuration whenever $S_2 \geqslant 0$ and $S_3 \leqslant -2\sigma_{23}$.

In the same manner we can consider the energy difference between a completely engulfed configuration (phase 1 entirely inside phase 3) and a 3-singlet having an infinitesimal 13 interface [Figure 4 (*bottom*)]. In this case $\phi_{12}, \phi_{13} \simeq \pi$ and $\phi_{23} \simeq 0$. The surface energy of the engulfed configuration is given by

$$E^{(2)} = \sigma_{13}A_{13}^{(2)} + \sigma_{23}A_{23}^{(2)}, \tag{19}$$

and the 3s configuration has a surface energy given by (15). Here the surface

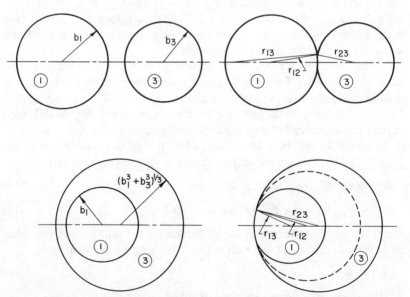

Figure 4   Compound-drop configurations used for the determination of the minimum surface energy configurations. (*Top*) A pair of 1-singlets and a 3-singlet configuration near the limit of a 1s pair; (*bottom*) a 2-singlet and a 3-singlet configuration near the limit of a 2-singlet.

areas are given by

$$A_{12}^{(2)} = 4\pi b_1^2, \qquad A_{23}^{(2)} = 4\pi(b_1^3 + b_3^3)^{2/3},$$

$$A_{12}^{(3)} = \pi r_{12}^2 \delta_{12}^2 + \dots, \tag{20}$$

$$A_{23}^{(3)} = 4\pi r_{23}^2(1 - \tfrac{1}{4}\phi_{23}^2 + \dots),$$

$$A_{13}^{(3)} = 4\pi r_{13}^2(1 - \tfrac{1}{4}\delta_{13}^2 + \dots),$$

where $\phi_{12} \simeq \pi - \delta_{12}$ and $\phi_{13} \simeq \pi - \delta_{13}$ ($\delta_{12}, \delta_{13} \ll 1$). Lastly, the volume equations (1) and (2) give $b_3^3 \simeq r_{23}^3 - r_{13}^3 + O(\delta_{ij}^4)$ and $b_1^3 \simeq r_{13}^3 + O(\delta_{ij}^4)$, and we have from (5) and (6) that $r_{12}\delta_{12} \simeq r_{13}\delta_{13} \simeq r_{23}\phi_{23}$. We then find that

$$E^{(3)} - E^{(2)} \simeq (\sigma_{12} - \sigma_{13} - \sigma_{23})\pi r_{12}^2 \delta_{12}^2 = S_3 \pi r_{12}^2 \delta_{12}^2, \tag{21}$$

and therefore the completely engulfed configuration has a lower surface energy and is the preferred configuration when $S_3 \geqslant 0$. Furthermore, if $S_3 \geqslant 0$ we must also have $S_2 \leqslant -2\sigma_{23}$.

There also exists the possibility of a completely engulfed configuration in which the two phases are interchanged, namely, phase 3 entirely inside phase 1. However, minimum-surface-energy considerations analogous to those discussed above readily show that a 3-singlet is the lower energy configuration if $S_1 < 0$, and this condition is assumed at the outset. In addition, note that the present development includes fluid lenses at a plane interface as a limiting case in which one of the drops becomes infinitely large (e.g. $r_{12} \to \infty$).

Finally, if we nondimensionalize the spreading coefficients by σ_{23} and note that $S_3/\sigma_{23} = -S_2/\sigma_{23} - 2$, the spreading-coefficient criterion can be most simply expressed as

complete engulfing (2s): $\qquad S_2/\sigma_{23} \leqslant -2,$

partial engulfing (3s): $\qquad -2 < S_2/\sigma_{23} < 0, \tag{22}$

nonengulfing (1s pair): $\qquad S_2/\sigma_{23} \geqslant 0,$

where by convention $S_1 \leqslant 0$. This criterion is depicted schematically in Figure 5 along with the data given by Torza & Mason (1970). The agreement is generally good, except for a few cases where the theory predicts a 2s configuration but observations indicate a 3s configuration. Such discrepancies are likely to be due to gravity forces or the effects of fluid motion.

As noted in the Introduction and depicted in Figure 6, a problem of general interest in direct-contact heat transfer is that of a vapor bubble growing by vaporization from an attached liquid drop while rising or falling due to buoyancy in a relatively hot bulk fluid. Conversely, in direct-contact condensation the related problem is that of a vapor bubble rising and condensing in a cold bulk fluid. These phenomena have received a considerable amount of attention in recent years (Sideman & Taitel 1964, Sideman & Isenberg 1967, Isenberg & Sideman 1970, Tochitani et al. 1977a,b, Mori 1978, Higeta et al. 1979, Pinder 1980, Simpson et al. 1973, 1974). The analysis of the heat-transfer process in these problems generally requires knowledge of the surface areas of the three fluid-fluid interfaces or, equivalently, the contact-line positions ϕ_{ij}. Although viscous effects associated with the drop motion and the effects of gravity are often important in these problems and are discussed later, the simple static compound-drop theory frequently does an adequate job in predicting 3-singlet drop configurations. In particular, one would expect the static theory to be a good approximation for cases having small capillary numbers, i.e. when the viscous forces are small compared with surface-tension forces. In this regard, Mori (1978) obtained good agreement when he compared the observed configurations of translating two-phase drops with the predicted ones by using the simple static spreading-coefficient theory. However, it is surprising that although the static-drop theory has successfully predicted configuration types (i.e. 2s versus 3s, etc.) for translating drops, no further comparisons have been made. For example, in

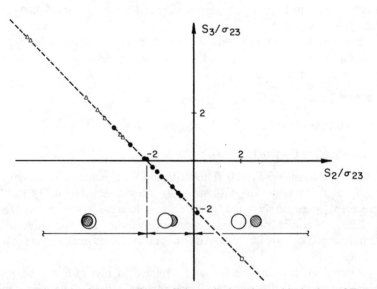

Figure 5 Schematic of the spreading coefficient criterion along with the data of Torza & Mason (1970). △, observed 2s configuration; ● observed 3s configuration; □, observed 1s pair. Dashed line is $S_3/\sigma_{23} = -S_2/\sigma_{23} - 2$.

the literature it is common to find data for open angle (corresponding to $\pi - \phi_{12}$) plotted against vapor ratio or vaporization ratio, but comparisons of these data with the predictions for open angle from the static-drop theory have apparently been overlooked, and researchers have often been concerned with complicated empirical correlations instead. The vapor ratio is defined as the ratio of vapor volume to total compound-drop volume $V_1/(V_1 + V_3)$, and the vaporization ratio is generally defined as the ratio of vapor mass to the total drop mass. In an attempt to demonstrate that the static theory is useful, we have compared data for open angle from two experiments with that predicted by the static theory. (Note that we actually plot the data versus ϕ_{12} instead of $\pi - \phi_{12}$.) One difficulty in making this

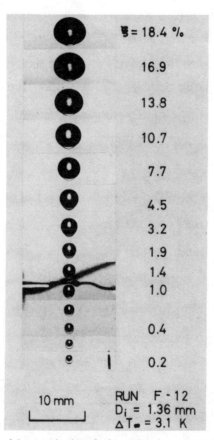

$\xi = 18.4\ \%$

16.9

13.8

10.7

7.7

4.5

3.2

1.9

1.4

1.0

0.4

0.2

10 mm

RUN F - 12
D_i = 1.36 mm
ΔT_∞ = 3.1 K

Figure 6   A sequence of the vaporization of a furan drop in aqueous glycerol, where ξ is the vaporization ratio, D_i the initial drop diameter, and ΔT_∞ the temperature difference. Reproduced, with permission, from Tochitani et al. (1977a).

comparison is that all three values of the surface tension are often not presented. We have therefore made sensible estimates for these when necessary. From Equations (1), (2), and (5)–(8) we can write the vapor ratio as

$$\xi_v = V_1/(V_1 + V_3) = f(\phi_{12}, \theta_1)/f(\phi_{12}, -\theta_2), \tag{23}$$

where

$$f(\phi_{12}, \alpha) = 2 + \cos \phi_{12}(3 - \cos^2 \phi_{12})$$
$$- \frac{\sin^3 \phi_{12}}{\sin^3 (\phi_{12} + \alpha)} \{2 + \cos (\phi_{12} + \alpha)[3 - \cos^2 (\phi_{12} + \alpha)]\}. \tag{24}$$

Typically, the surface tensions would be given, and therefore the contact angles θ_1, θ_2 needed in (23) are obtained from (3) and (4) as

$$\cos \theta_1 = \tfrac{1}{2}(\sigma_{12}/\sigma_{13})\{(\sigma_{23}/\sigma_{12})^2 - (\sigma_{13}/\sigma_{12})^2 - 1\},$$
$$\cos \theta_2 = -\tfrac{1}{2}(\sigma_{12}/\sigma_{23})\{(\sigma_{23}/\sigma_{12})^2 - (\sigma_{13}/\sigma_{12})^2 + 1\}. \tag{25}$$

In Figure 7 we have plotted ϕ_{12} versus ξ_v (vapor ratio) from (23), along

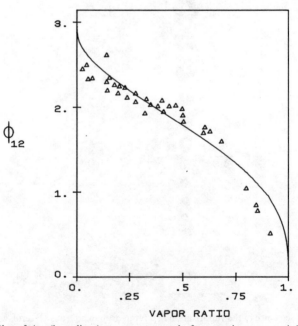

Figure 7 Plot of ϕ_{12} (in radians) versus vapor ratio from static compound-drop theory, compared with the data of Pinder (1980) for translating compound drops.

MULTIPHASE DROPS 301

with the data given by Pinder (1980) for a liquid drop comprised of a cyclopentane/hexane mixture (phase 3) vaporizing in water (phase 2). In this case, although σ_{13} (vapor-cyclopentane/hexane interface) is given as 19.5 dyn cm^{-1}, we were forced to estimate σ_{23} (water-cyclopentane/hexane) as approximately 50 dyn cm^{-1} and σ_{12} (vapor-water) as 66 dyn cm^{-1}. The comparison is actually quite reasonable, and we feel that the present theoretical result has a sounder basis than the empirical result proposed by Pinder. In fact, valid concerns about the foundations of Pinder's empirical formulation have been expressed by Higeta & Mori (1981). In Figure 8 we have also compared the static-drop theory with the data obtained by Tochitani et al. (1977a) for a pentane drop vaporizing in aqueous glycerol. Here σ_{13} (vapor-pentane interface) is given as 18 dyn cm^{-1}, and we estimate the surface tensions σ_{23} (pentane-glycerol) and σ_{12} (vapor-glycerol) as 45 and 61 dyn cm^{-1}, respectively. In this case the data are plotted against vaporization ratio ξ, which is related to vapor ratio by $\xi = \rho_1\xi_v/[\rho_3(1-\xi_v) + \rho_1\xi_v]$, where ρ_1 is the vapor density and ρ_3 is the density of the liquid

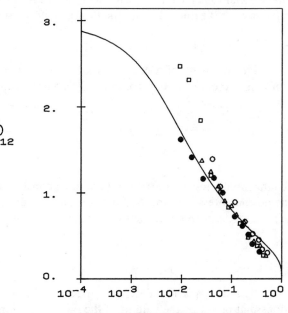

Figure 8 Plot of ϕ_{12} (in radians) versus vaporization ratio from static compound-drop theory, compared with the data of Tochitani et al. (1977a) for translating compound drops. Data correspond to different initial drop diameters: $\bigcirc D_i = 0.80$ mm, $\triangle D_i = 0.98$ mm, $\square D_i = 1.20$ mm, $\bullet D_i = 1.42$ mm.

comprising the drop. This choice has the advantage of expanding the scale for vapor ratios ξ_v near unity, e.g. for $\rho_3/\rho_1 = 250$ and $\xi_v = 0.99$ we have $\xi = 0.284$. Again the agreement in Figure 8 is quite reasonable.

TRANSLATING COMPOUND DROPS

As already noted, fluid motion and the related viscous effects can significantly alter compound-drop configurations. Unfortunately, the level of complexity of such problems has seriously slowed analytical progress. For steady translation of a compound drop, the general problem consists of solving the Navier-Stokes equation for each of the three phases with boundary conditions at the three interfaces coupling the flow regions together. In particular, for phase i ($i = 1, 2, 3$) we have

$$\rho_i(\mathbf{u}^{(i)} \cdot \nabla)\mathbf{u}^{(i)} = -\nabla p^{(i)} + \mu_i \nabla^2 \mathbf{u}^{(i)} + \mathbf{G},$$
$$\nabla \cdot \mathbf{u}^{(i)} = 0,$$
(26)

where μ_i is the dynamic viscosity of phase i and $\mathbf{G}$ is the body force (gravity). The boundary conditions for a translating drop with respect to the body frame of reference are the following: (a) the flow is uniform at infinity,

$$\mathbf{u}^{(2)} \to \mathbf{U} \quad \text{as} \quad |\mathbf{x}| \to \infty;$$
(27)

(b) the normal component of velocity vanishes (as a result of mass conservation) and the tangential component is continuous at the ij interface,

$$u_n^{(i)} = u_n^{(j)} = 0,$$
(28)

$$u_s^{(i)} = u_s^{(j)};$$
(29)

and (c) the tangential stress is continuous (assuming constant σ_{ij}), and the jump in the normal stress is proportional to the surface tension times the sum of the principal curvatures of the surface κ_{ij} at the ij interface,

$$\tau_{ns}^{(i)} = \tau_{ns}^{(j)},$$
(30)

$$\tau_{nn}^{(i)} - \tau_{nn}^{(j)} = \sigma_{ij}\kappa_{ij} \quad \text{(no sum on } i \text{ and } j\text{)}.$$
(31)

In addition to the above equations describing the fluid motion, we also have two equations that prescribe the volume of the two phases comprising the drop and two equations for force equilibrium at the contact line, i.e. Equations (3) and (4).

The nonlinearity of the governing equations and the fact that the drop shape (i.e. the shape of the three interfaces) is unknown make the solution to this problem extremely difficult. However, when the Reynolds number and capillary number are small and phase 3 is in a relatively thin layer surrounding phase 1, the problem is amenable to asymptotic analysis. This is partly because the nonlinear inertial terms may be neglected, and partly because the interfaces can be treated as perturbations from the undeformed spherical configuration. This limiting case has been examined by Johnson & Sadhal (1983) and Sadhal & Johnson (1983), and it represents a first step toward identifying the relevant physical parameters that control the configurations of translating compound drops. An additional primary restriction of this work is that the compound drop must be nearly spherical, i.e. the radii of the 13 and 23 interfaces are both required to be small perturbations of the radius of the 12 interface (Figure 9). This of course requires that the surface tensions satisfy $\sigma_{12} \approx \sigma_{23} + \sigma_{13}$. Fortunately, many experiments meet this requirement, as is borne out by the fact that compound drops are frequently observed to be nearly spherical.

The problem considered by Johnson & Sadhal (1983) is solved using a perturbation expansion in terms of the small thinness parameter ε

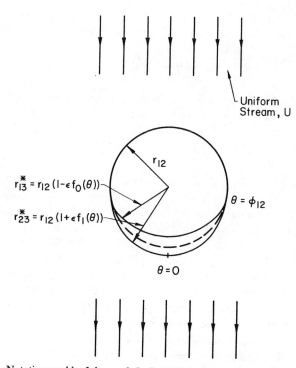

Figure 9   Notation used by Johnson & Sadhal (1983) for translating compound drops.

$= t_{max}/r_{12}$, where t_{max} is the maximum thickness of phase 3. Fluid motion in the thin layer or fluid film (phase 3) is driven by the shear stresses at the 13 and 23 interfaces, and the analysis of the flow in the film is similar to that of classical lubrication theory. In other words, shear stresses are balanced by the induced pressure field in the film. As is always the case, the drop shape is determined by the normal-stress boundary condition. For a static drop this condition requires the interfaces to be spherical. However, flow-induced deviations from sphericity are considered here. The results that appear in the original paper by Johnson & Sadhal are not in their most useful form for comparison with the results of the previous section or with experiments; consequently, we recast some of the key results of their work in terms of the present development and indicate a few new results.

In the formulation by Johnson & Sadhal (1983), the radii of the interfaces of phase 3 are taken relative to the origin of r_{12} (as shown in Figure 9) and are expressed as

$$r^*_{13} = r_{12}[1 - \varepsilon f_0(\theta)],$$
$$r^*_{23} = r_{12}[1 + \varepsilon f_1(\theta)],$$
(32)

where the film profile functions f_0 and f_1 are found to be related to the nondimensional film thickness $t(\theta)$ (nondimensionalized by t_{max}) by

$$f_0(\theta) = \frac{\sigma_{23}}{\sigma_{13} + \sigma_{23}} t(\theta), \qquad f_1(\theta) = \frac{\sigma_{13}}{\sigma_{13} + \sigma_{23}} t(\theta).$$
(33)

The normal-stress boundary conditions ultimately lead to the following integro-differential equation for the thickness $t(\theta) = f_0(\theta) + f_1(\theta)$:

$$\frac{\partial^2 t}{\partial \theta^2} + \cot \theta \, \frac{\partial t}{\partial \theta} + 2t = -\beta \left\{ p(0) + \int_0^\theta \frac{\tau_{23} + \tau_{13}}{t(\theta')} \, d\theta' \right\}.$$
(34)

Here $p(0)$ is the film pressure at $\theta = 0$ (nondimensionalized by $\mu_2 U/r_{12}$), τ_{23} and τ_{13} are known shear stresses exerted on phase 3 by phase 2 and phase 1, respectively (nondimensionalized by $\mu_2 U/r_{12}$), and $\beta = (\mu_2 U/\sigma_{23})(1 + \sigma_{23}/\sigma_{13})/\varepsilon^2$ is a measure of the magnitude of the viscous effects. Since the film thickness is a maximum at $\theta = 0$, the boundary conditions for (34) are (a) $t = 1$ and $\partial t/\partial \theta = 0$ at $\theta = 0$ (by symmetry), and (b) $t = 0$ at $\theta = \phi_{12}$ (since the film vanishes at the contact line). Note that three boundary conditions are required because $p(0)$ is unknown and is determined as part of the solution. For fixed β, different values of $p(0)$ correspond to solutions having different contact-line positions ϕ_{12}.

Imagining for a moment that both β and ϕ_{12} are known, we can solve Equation (34) to find the thickness $t(\theta)$ and hence the drop shape. However,

ϕ_{12} is actually unknown and must also be determined as part of the solution. The additional constraint needed to fully determine the drop configuration comes from the conditions at the contact line. If the surface tensions are prescribed, then the contact angles θ_1 and θ_2 are determined by (3) and (4), and we must require that $\varepsilon df_1/d\theta = \tan\theta_2$ and $\varepsilon df_0/d\theta = \tan\theta_1$ at $\theta = \phi_{12}$. Consequently, the condition that determines ϕ_{12} is

$$\varepsilon \frac{dt}{d\theta}\bigg|_{\theta=\phi_{12}} = \tan\theta_1 + \tan\theta_2. \tag{35}$$

Note that since $\varepsilon = t_{max}/r_{12}$ is assumed small, Equation (35) implies that the contact angles θ_1 and θ_2 are close to π for such a case.

It is instructive to compare the formulation here with that of the preceding section for static drops. If $U = 0$, we have $\beta = 0$, leading to $t(\theta) = \frac{1}{2}\beta p(0)(\cos\theta/\cos\phi_{12}-1)$, which is the thickness corresponding to spherical 13 and 23 interfaces. Note that $\beta p(0)$ does not vanish when $U = 0$ because the product of β and $p(0)$ is actually independent of U. Furthermore, since $t = 1$ at $\theta = 0$, we find that $\frac{1}{2}\beta p(0) = \cos\phi_{12}/(1-\cos\phi_{12})$, which gives $t(\theta) = (\cos\theta-\cos\phi_{12})/(1-\cos\phi_{12})$; therefore we obtain from the contact-angle condition (35)

$$\varepsilon\frac{dt}{d\theta}\bigg|_{\theta=\phi_{12}} = -\varepsilon\frac{\sin\phi_{12}}{1-\cos\phi_{12}}$$
$$= \tan\theta_1 + \tan\theta_2 \simeq -\varepsilon(\theta_1^{(1)}+\theta_2^{(1)}), \tag{36}$$

where we have written $\theta_k \simeq \pi - \varepsilon\theta_k^{(1)}$ $(k = 1, 2)$. Consequently, the volume V_3 is given by

$$V_3 \simeq 2\pi r_{12}^3\varepsilon \int_0^{\phi_{12}} t(\theta)\sin\theta\,d\theta$$
$$= 2\pi r_{12}^3\varepsilon(\tfrac{1}{2}+\tfrac{1}{2}\cos^2\phi_{12}-\cos\phi_{12})/(1-\cos\phi_{12}), \tag{37}$$

and since $V_1 \simeq \frac{4}{3}\pi r_{12}^3$, we obtain from (36) and (37)

$$\xi_v = \frac{V_1}{V_1+V_3} \simeq 1-\tfrac{3}{2}\varepsilon(\theta_1^{(1)}+\theta_2^{(1)})$$
$$\times (1/\sin\phi_{12}-\cot\phi_{12}-\tfrac{1}{2}\sin\phi_{12}). \tag{38}$$

Note that this is the same result as that obtained from (23) in the previous section with $\theta_k \simeq \pi - \varepsilon\theta_k^{(1)}$ $(k = 1, 2)$.

The effect of translation on compound-drop configurations can be examined by considering the film volume V_3 for various values of β. From

(35) and (37) we have

$$\frac{V_3}{2\pi r_{12}^3} = \varepsilon \int_0^{\phi_{12}} t(\theta) \sin \theta \, d\theta = -(\tan \theta_1 + \tan \theta_2) F(\phi_{12}, \beta),\tag{39}$$

where

$$F(\phi_{12}, \beta) = -\int_0^{\phi_{12}} t(\theta) \sin \theta \, d\theta \left| \frac{dt}{d\theta} \right|_{\theta = \phi_{12}},\tag{40}$$

and $t(\theta)$ is determined by (34). For the static case, $F(\phi_{12}, 0) = (\frac{1}{2} + \frac{1}{2}\cos^2 \phi_{12} - \cos \phi_{12})/\sin \phi_{12}$. In Figure 10 we have plotted the angle ϕ_{12} versus the volume factor $F(\phi_{12}, \beta)$ for a few values of β. As β increases, the driving force on the film increases, and we notice that the extent of the film decreases (i.e. ϕ_{12} decreases) for fixed values of the volume factor F. These results, however, cannot generally be used to compare with experiments directly. The difficulty is due to the fact that $\beta = (\mu_2 U/\sigma_{23})(1 + \sigma_{23}/\sigma_{13})/\varepsilon^2$ is typically not known because t_{max} or, equivalently, ε is not generally measured. More often, only the velocity, fluid viscosities, and surface

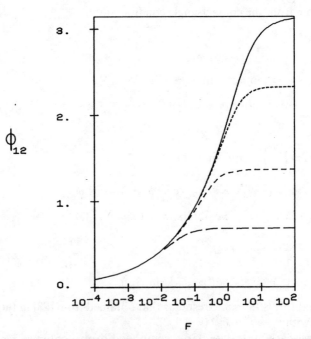

Figure 10 Plot of ϕ_{12} versus the volume factor F for a few values of the parameter β, which measures the magnitude of the viscous effects. ——— $\beta = 0$, ---- $\beta = 0.067$, — — — $\beta = 0.667$, ——— $\beta = 6.667$.

tensions would be measured, and frequently not all of these are presented in the literature. In any event, if all of these were available the capillary numbers would be known, and we would have available the parameter

$$\beta^* = \beta\varepsilon^2 = \frac{\mu_2 U}{\sigma_{23}}\left(1 + \frac{\sigma_{23}}{\sigma_{13}}\right), \tag{41}$$

which, after substituting for ε^2 from (35), can be written as

$$\beta^* = (\tan \theta_1 + \tan \theta_2)^2 C(\phi_{12}, \beta), \tag{42}$$

where

$$C(\phi_{12}, \beta) = \beta\left/\left[\frac{dt}{d\theta}\bigg|_{\theta=\phi_{12}}\right]^2\right..$$

Since t is a function of ϕ_{12} and β, and ϕ_{12} is a function of F and β (Figure 10), it is useful to plot the capillary-number factor C appearing in (42) versus F for various values of β. This is done in Figure 11. The way in which Figures 10 and 11 might be used is that for prescribed contact angles

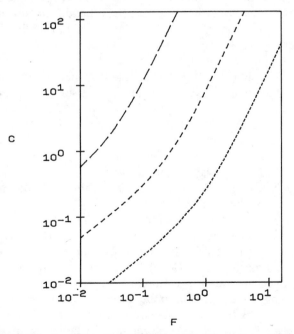

Figure 11 The capillary number parameter C versus the volume factor F for a few values of the parameter β, which measures the magnitude of the viscous effects. $----$ $\beta = 0.067$, $---$ $\beta = 0.667$, $\underline{\qquad}$ $\beta = 6.667$.

θ_1 and θ_2, capillary number β^* (i.e. factor C), and volume V_3 (i.e. factor F), Figure 11 could be used to obtain β, and then Figure 10 yields ϕ_{12} or the open angle. At present, however, sufficient data are simply not available for a reasonable comparison between theory and experiment.

In addition to predicting compound-drop configurations, Johnson & Sadhal (1983) present an expression for the drag force on the compound drop as a function of ϕ_{12}, μ_1/μ_2, and β. To leading order this result is independent of β and is given by

$$\text{Drag} = 4\pi\mu_2 U r_{12} \left\{ \frac{\mu_2}{4\pi(\mu_1 + \mu_2)} [2\phi_{12} + \sin \phi_{12} - \sin 2\phi_{12} - \tfrac{1}{3} \sin 3\phi_{12}] \right.$$

$$\left. + \frac{\mu_2 + \tfrac{3}{2}\mu_1}{\mu_1 + \mu_2} \right\}. \tag{43}$$

For fixed viscosity ratio μ_1/μ_2, Equation (43) monotonically increases from the Hadamard-Rybczynski result at $\phi_{12} = 0$ to the Stokes drag on a solid sphere at $\phi_{12} = \pi$. The higher-order corrections to (43) that involve β are quite complex and are omitted here.

Using a fundamentally different approach, Mori (1978) attempted to account for the effects of flow-induced deformation on translating gas-liquid drop configurations by considering a highly simplified model in which phase 3 (liquid) was assumed to extend beyond $\theta = \phi_{12}$ in an extremely thin film that terminated at $\theta = \phi_{12}^*$. Mori assumed that there was no fluid circulation within the film and that the surface tension of the film varied with the film thickness. Balancing the surface-tension gradient with the shear stress and assuming the flow could be approximated by Stokes flow past a solid sphere, Mori obtained a relation between capillary number and the extent of the thin film, i.e. ϕ_{12} and ϕ_{12}^*. Furthermore, a relation between ϕ_{12} and capillary number was deduced by assuming that the film reached all the way around to the front of the bubble ($\phi_{12}^* \to \pi$). Unfortunately, comparisons with his experiments were inconclusive, and the existence of such films was not verified.

Other papers concerned with the analysis of translating compound drops are few and far between. However, for the case of completely engulfed compound drops, an additional noteworthy study is the brief communication by Rushton & Davies (1983). They considered the low-Reynolds-number buoyancy-driven translation of a completely engulfed compound drop, assuming that phase 3 was present as a uniform layer about phase 1, i.e. the 23 and 13 interfaces were concentric spheres. (Note that our convention requires phase 1 inside phase 3 for complete engulfing, and therefore our notation has 1 and 3 interchanged from that used by Rushton & Davies.) The principal result of their work was the following expression for the drag force on the compound drop:

$$\text{Drag} = 4\pi\mu_2 U r_{23} \left\{ \frac{\mu_{12} + 6\mu_{32}^2 G(R) + \mu_{32}(2 + 3\mu_{12})F(R)}{\mu_{12} + 4\mu_{32}^2 G(R) + 2\mu_{32}(1 + \mu_{12})F(R)} \right\}, \tag{44}$$

where $\mu_{12} = \mu_1/\mu_2$, $\mu_{32} = \mu_3/\mu_2$, $R = r_{13}/r_{23}$, and

$$F(R) = \frac{(1+R)(2R^2 + R + 2)}{(1-R)(4R^2 + 7R + 4)}, \qquad G(R) = \frac{1 - R^5}{(1-R)^3(4R^2 + 7R + 4)}. \tag{45}$$

The drag is found to be always less than or equal to the Stokes drag on a solid sphere. One limit of interest is that of a thin film, i.e. $R = r_{13}/r_{23} \rightarrow 1$, in which case (44) tends to the Stokes drag on a solid sphere, in agreement with Equation (43) for $\phi_{12} = \pi$. A point not discussed by Rushton & Davies, however, is whether the concentric configuration is actually an equilibrium configuration. Rushton & Davies did not examine the force on the engulfed phase, i.e. phase 1. In fact, this concentric configuration will be an equilibrium configuration only in very special cases when the body force on phase 1 due to gravity is such as to exactly balance the viscous forces driving it off center. Recent studies by S. S. Sadhal & H. N. Oguz (unpublished work) have shown that for the eccentric 2s drop in translation, the engulfed fluid cannot be in equilibrium in the absence of a body force on it. They find that with the body force present, the equilibrium configuration, when it exists, is very sensitive to small changes in the buoyant force on the engulfed fluid. For $\rho_3 < \rho_1$ and $(\rho_1 V_1 + \rho_3 V_3) < \rho_2(V_1 + V_3)$, the drop rises with phase 1 off-center toward the rear. Another equilibrium state is found with phase 1 off-center toward the front, but this state is unstable. Similarly, for $\rho_1 < \rho_3$ and $\rho_2(V_1 + V_3) < (\rho_1 V_1 + \rho_3 V_3)$, the drop sinks with phase 1 in possible stable equilibrium toward the rear. However, for many of these cases if $\mu_3 \gg \mu_2$, a stable state of equilibrium may be found with phase 1 on either side of the center. It must be noted that the above density requirements do not guarantee an equilibrium state but are the necessary conditions. The total drag force for equilibrium cases appears to change very little with changes in eccentricity. For the nonequilibrium cases, the total drag force has a very strong dependence on the relative velocity of the phase 1 sphere.

Outside the low-Reynolds-number regime there has not been any rigorous analysis of translating compound drops. With modern numerical techniques, however, such problems can surely be addressed. Our own efforts, in fact, are presently leaning in this direction. In addition, the complicating effects of drop-drop interactions in droplet clouds and trains and the effect of bubble growth on gas-liquid drops during translation are all areas in need of further study because of their importance in applications. We should note that there have been a number of studies concerned with the growth of translating compound gas-liquid drops, including Sideman & Isenberg (1967), Isenberg & Sideman (1970), Sideman & Taitel (1964), Sideman & Hirsch (1965), Tochitani et al. (1977b), Selecki & Gradon (1976), and Maron Moalem & Miloh (1976), to mention only a few. These papers, however, are not generally concerned with the fluid mechanics of compound drops or the issue of drop configurations; instead, they are primarily concerned with the analysis of the heat-transfer process. They typically use simple fluid-mechanical models as an input to the heat-transfer problem. For example, Isenberg & Sideman (1970) assume that the flow field is that of a translating and radially growing sphere in potential flow. Selecki & Gradon (1976) decouple, in an ad hoc way, the analysis of droplet translation from the radial-growth problem, which they assume is governed by the Rayleigh equation. Maron Moalem & Miloh (1976) consider the heat transfer of a completely engulfed drop and include an approximate solution for the motion of the inner drop relative to the outer. Their fluid-mechanical analysis, however, involves an ad hoc combination of inviscid-flow theory and Stokes-flow theory. While they assume that the drag on the inner drop in the absence of boundaries is governed by the Hadamard-Rybczynski formula valid for Stokes flow, they account for the effect of the outer drop's boundary on the inner drop by using potential-flow theory. For an extensive review of heat-transfer-related problems involving compound drops, the reader is referred to Sideman & Moalem-Maron (1982) and Sideman (1966).

On a slightly different note, Shankar et al. (1981) and Shankar & Subramanian (1983) considered completely engulfed bubbles in connection with a study of the removal of bubbles from molten-glass drops in the near absence of gravity. They examined the motion of a completely engulfed bubble due to fluid motion in the surrounding drop that was generated by surface-tension gradients. The surface-tension gradients were due to a prescribed temperature field on the surface of the drop. Attention was limited to small Reynolds numbers and small Marangoni numbers. The bulk phase (phase 2) is not present in this problem, since the compound drop is assumed to be in a vacuum. Materials processing in space laboratories was the primary application for this problem. The principal result of their work is an expression for the bubble migration velocity, which is found as a function of the ratio of the bubble and drop radii, the distance between the centers of the drop and bubble, and the prescribed temperature field. Analytical results were presented for the limiting case of small bubbles.

Experimental studies of the fluid mechanics of translating compound drops include papers by Simpson et al. (1973, 1974), Mercier et al. (1974),

Hayakawa & Shigeta (1974), Mori et al. (1977b), Tochitani et al. (1977a), and Higeta et al. (1979), as well as the papers already referred to by Torza & Mason (1970), Mori (1978), and Pinder (1980). In addition, references to numerous studies slanted more heavily toward apsects of heat transfer can be found in the review by Sideman & Moalem-Maron (1982). The experimental findings generally show that at small Reynolds numbers the rise speeds of gas-liquid drops fall between the Stokes result for a solid sphere and that for a bubble. Such a result has always been attributed to the retarding effect of the liquid portion of the drop and is consistent with the predictions of Equation (43). However, direct comparisons have not yet been made. The data by Tochitani et al. (1977a) for vaporizing drops indicate that as the vaporization ratio ζ increases, the drag often decreases from the Stokes drag on a solid sphere. This is consistent with the fact that as ζ increases, ϕ_{12} decreases [e.g. Equation (40)], and hence the drag would decrease [Equation (45)], departing further from the Stokes result. An interesting observation by Higeta et al. (1979) was that the angle ϕ_{12} for a condensing n-pentane compound drop rising in glycerol is generally larger than that for a vaporizing n-pentane drop. Higeta et al. suggested that the condensing case is different from the evaporating case, because pentane vapor condensing on the glycerol-vapor interface, which is later driven around to the liquid phase at the rear of the drop, acts as a contaminant altering the surface tension of the glycerol-vapor interface. However, in order for ϕ_{12} to increase it would be necessary for the surface tension of the glycerol-vapor interface to increase, and this seems unlikely. We believe than an explanation of this observation requires a careful examination of the effects of the condensation and evaporation processes. At larger Reynolds numbers, Mercier et al. (1974) found that the presence of liquid attached to a bubble can significantly affect the bubble's configuration and therefore significantly alter the drag and rise speed. In contrast, Simpson et al. (1974) found reasonably good agreement between predictions for drag from single-phase bubble theory with his observations of gas-liquid compound drops. An interesting feature ar large Reynolds number is that the generally unsteady translation attributed to vortex shedding induces a sloshing motion in the liquid portion of the compound drop, which may in turn influence the overall unsteady motion of the drop. Torza & Mason (1970) have made a number of other interesting observations of compound drops, including their behavior in shear flow and electric fields and the mechanism of engulfing. The engulfing process was suggested to consist of two processes: penetration and spreading. Some of the excellent photographs of the engulfing phenomena taken by Torza & Mason are shown in Figure 12. A thorough analysis of the engulfing process has yet to be completed.

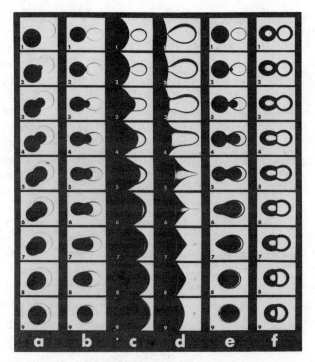

Figure 12 Photographs of the engulfing process for (1) distilled water/Ucon oil (LB-285) compound drops in Silicone oil (sequences *a–d*), (2) distilled water/Ucon oil (50-HB-55) drops in Silicone oil (sequence *e*), and (3) distilled water (+10% vol Tween 20)/castor oil drops in Silicone oil (sequence *f*). The drops were brought together using a shear flow in sequence (*a*) and a horizontal electrical field in sequences (*b–f*). In all sequences, phase 1 is dark. Reproduced, with permission, from Torza & Mason (1970).

STABILITY OF COMPOUND DROPS

The natural modes of oscillation and the stability of compound drops are areas that have attracted a fair amount of attention. Although the oscillations of single-phase drops were examined by Lamb as early as 1932 (Lamb 1932), the treatment of compound drops did not appear until recently. Miller & Scriven (1968) investigated theoretically the case of a liquid drop in another liquid separated by a viscoelastic surface phase. The first work in this area treating the case where the drop consists of two distinct bulk phases appears to be that of Patzer & Homsy (1975), who studied the rupture of thin spherical fluid films. They considered phases 1 and 2 to be identical fluids separated by a concentric spherical shell of phase-3 liquid. The basic state is the situation in which there is no motion at all. Then, by applying a small disturbance to this state, they carried out a linear stability analysis. Two limiting cases of dynamical tangential stress at an interface were examined. The first one is the case of a contaminant-free interface that has continuity of velocity and tangential stress. The second case corresponds to interfaces contaminated by surfactants, resulting in the no-slip condition. In addition, these interfaces have continuity of normal velocity and total normal stress. In the normal-stress condition, the surface tension as well as the disjoining pressure was incorporated into the fluid shell. The solution was obtained by expanding the displacements of the interfaces, the velocities, and the vorticities resulting from the disturbance in terms of spherical harmonics. These variables may be written as

$$f^k = F_n^k(r)Y_l^m(\theta, \phi)e^{ct} \tag{46}$$

where f^k is a generalized variable corresponding to the disturbance. Through an exchange-of-stabilities argument, it was shown that the exponent c is real for the no-slip case. For the case of mobile interfaces, c was assumed to be real. The onset of instability, therefore, takes place through the mode $c = 0$.

With the expansion (46) the stability problem reduced to a set of eight homogeneous algebraic equations, with the status of the stability given by the sign of the corresponding determinant. The stability curve for the lowest possible unstable mode is

$$S = A/4\pi\sigma r_{13}^2 = [0.903(\delta_c/r_{13})]^4, \tag{47}$$

where A is the Hamaker constant, σ is the interfacial tension, and δ_c is the critical thickness of the film. At this critical thickness the disturbances begin to grow and lead to the rupture of the film. With decreasing Sheludko number S, the film stability increases. Interestingly, this result is independent of the viscosity ratio $\mu_3/\mu_1 = \mu_3/\mu_2$. Also, the type of tangential boundary condition imposed (no-slip or mobile) seems to have little effect on the critical thickness. This is owing to the relatively high significance of the disjoining pressure in the normal-stress condition.

Recently, Landman (1984a) examined the stability in the absence of the disjoining pressure. The rupture of the film does not take place in this case, and the drop regains its spherical shape from a disturbed state. She discusses this behavior in considerable depth. In her analysis the phases 1 and 2 are different fluids, and hence there are two different interfacial tensions participating in the dynamics. The dynamical effects of the external fluid (phase 2), however, are considered negligible.

By carrying out the spherical-harmonics expansion for small disturbances and neglecting the inertial effects, Landman obtained two real and negative eigenvalues describing an aperiodic decay to spherical shape. Of the two decay rates, the lower one corresponds to the two interfaces being out of .phase, i.e. having opposite displacements. The higher decay rate corresponds to in-phase displacements. This mode is considerably weaker, and therefore it is likely that only the out-of-phase mode would be observed in a laboratory.

The inviscid limit of compound-drop oscillations has been examined both theoretically and experimentally by Saffren et al. (1982). Their work was motivated by an interest in containerless materials processing and the fabrication of fusion target pellets. The analysis was relatively straightforward, and for each mode they found a high- and low-frequency oscillation. The higher-frequency oscillation (which they termed the "bubble" mode) corresponded to both interfaces oscillating in phase, and the lower-frequency mode (which they called the "sloshing" mode) corresponded to an out-of-phase oscillation. These were both demonstrated by experiments in a neutral buoyancy tank. The expected results for the limiting cases corresponding to a simple drop (no engulfed phase, $r_{13} \to 0$), a rigid host or bulk fluid ($\rho_2 \to \infty$), a rigid core or engulfed phase ($\rho_1 \to \infty$), a thin shell, and a thick shell were all verified. An interesting experimental observation was that core centering occurred during oscillations for drops that were not initially concentric. Compound drops that were prepared such that their surfaces were nonconcentric would become concentric within a few cycles of oscillation. Saffren et al. suggested that a higher order of approximation is necessary to determine the centering force, since to the order of the approximations used in their work any position of the core or engulfed drop is neutrally stable. Experimental observation of the core-centering phenomenon has also been reported by Lee et al. (1982).

Another series of interesting investigations, in the limit of very viscous interior fluid, has been pursued by Landman & Greenspan (1982) and Landman (1983). The work began with the study of drop crenations (surface rippling) and spiculations for coated drops due to the addition of fluid to the outer shell or to the removal of fluid from the bulk phase of the drop. The primary application was aimed at interpreting the distortion of biological cells where the mass supply was envisioned to come from the bulk fluid. The best-known example of this phenomenon in cell biology is that of red blood cells, which develop crenate margins in hypertonic solutions. In their analysis, Landman & Greenspan treat the coating like a compressible surface fluid. The discussion of this work seems appropriate here because the subsequent development (Landman 1983), in which the coating is a bulk fluid, follows naturally. In both cases, the dynamical properties of the continuous phase are not taken into consideration.

Landman & Greenspan (1982) first obtained the solution to the basic state, which consists of simply supplying mass to the surface phase without any disturbance. The supply rate decays exponentially in time, so that the surface density $\Gamma(t)$ is given by

$$\Gamma(t) = \Gamma(0)[1 + C - Ce^{-\lambda t}]. \tag{48}$$

This corresponds to total mass increase by a factor $(1 + C)$.

They then considered small disturbances of this basic state. This is effected through defining the surface of the drop by the function

$$r_{12} = a\{1 + \varepsilon\zeta^*(\theta, \phi, t)\}, \tag{49}$$

where a is the radius of the drop in the basic state. The disturbance ζ^* is expanded in terms of spherical harmonics, viz.

$$\zeta^* = \text{Re} \sum_{n=2}^{\infty} \sum_{m=0}^{\infty} Y_{mn}(t)e^{im\phi}P_n^{(m)} (\cos \theta), \tag{50}$$

along with pressure, velocity components, surface density, and surface tension in similar forms. The bulk-fluid momentum and continuity equations were solved with the boundary conditions given at the surface defined by (49). These boundary conditions include the balance of the normal and tangential stresses, together with the continuity of tangential velocity at the surface. The normal velocity is, of course, given by $\partial r_{12}/\partial t$. In satisfying the stress boundary conditions, Landman & Greenspan included the effects of surface tension as well as the surface shear and the surface dilatation forces. The surface tension was taken to be a linear function of the surface density, and the surface viscosity coefficients were taken to be uniform. The authors also made the assumption that $\sigma_0 a/\mu v \ll 1$, where σ_0 is the surface tension in the basic state. This assumption allowed the exclusion of the inertial terms in the bulk-phase momentum equations. The complete development led to a second-order linear differential equation for the amplitudes $Y_{mn}(t)$, which turns out to be independent of m. For sufficiently large initial deposition rates of the surface fluid, the solution set displays an initial growth period up to a maximum followed by an exponential decay. Alternatively, for sufficiently large surface viscosities or a small enough bulk viscosity, similar characteristics for the amplitudes are found. For such behavior of $Y_{mn}(t)$ the encapsulated drop, at early times, takes the shape of a crenated sphere. The crenations decrease in size and number until at longer times the drop becomes spherical.

In the case when phase 3 is a bulk fluid, fluid is added to the shell at an exponentially decaying rate, as before. The basic state again consists of simple growth of the shell with no disturbance to the drop. Disturbances of the type given by (49) and (50) were assumed for each of the interfaces 13 and

23. The oscillations were therefore described by two sets of amplitudes. By satisfying the continuity of velocities and stresses at the interfaces, a system of two linear first-order differential equations was obtained for the two sets of amplitudes. The momentum equations were solved in the "very viscous" limit, so that the inertia terms were negligible.

As in the previous case, the deposition of fluid into the shell gives rise to crenations, which attenuate in time and disappear. However, the bulk-fluid shell theory leads to fewer crenations than in the case of the surface-fluid shell. Landman (1983) attributed this to the fact that the surface fluid is described by surface viscosities, surface compressibility, and one coefficient of surface tension, while the bulk-fluid shell has a bulk viscosity, no compressibility, and two surface-tension coefficients. She inferred from this behavior that a simple relation of the type

surface viscosity = (bulk viscosity) × (thickness)

does not seem to exist.

In another recent study, Landman (1984b) developed a theoretical model for a red blood cell. In this treatment the cell is considered to be a viscous Newtonian fluid coated by a viscoelastic solid membrane. Here also, crenations develop by the addition of more surface material. The results of the analysis show that in this case the crenations are sustained for approximately hundredfold longer time periods than in the case of a viscous fluid membrane.

With regard to stability Plesset & Sadhal (1982) have used diffusion theory to examine the thermodynamic stability of a vapor bubble that is entirely engulfed by a liquid-gas solution. The compound drop is assumed to be in a bulk fluid in which the gas is insoluble. The analysis shows that although two equilibrium states may exist corresponding to a small and large bubble radius, only the larger bubble radius is found to be stable. In a related paper on this subject, Mori et al. (1977a) presented a thermodynamic stability criterion based on Gibbs thermodynamic potential and found good agreement with experiments using Freon liquid-vapor compound drops in glycerine. The work of Plesset & Sadhal provided a clearer physical interpretation of the stability analysis and experiments presented by Mori et al.

COMPOUND DROPS IN VAPOR EXPLOSIONS

Recently a number of very interesting observations have been made by Shepherd & Sturtevant (1982) of compound gas-liquid drops during the rapid evaporation of single drops of butane at the superheat limit, i.e. a vapor explosion. Interest in this phenomenon stems from accidental industrial vapor explosions and natural events involving interactions of molten lava with water. This subject is also of interest in the area of vaporizing multicomponent fuel drops (see, for example, Lasheras et al. 1980, Law 1978). Shepherd & Sturtevant observed that on a microsecond ' time scale a single bubble grows within a superheated drop. During the rapid evaporation a fundamental dynamic instability on the surface of the bubble was observed for the first time. The authors believe that the instability is driven by the evaporative process, and they found that the vaporization rate was two orders of magnitude greater than that which would be calculated by classical theory. The bubble growth rate, however, is found to be only marginally higher than what classical theory would predict. The instability mechanism was identified as being analogous to the Landau instability of laminar flames. Other interesting results include the observation that at later stages of evaporation, when the bubble has grown large enough to contact the surface of the drop (3s configuration), the bubble surface protrudes into the bulk fluid and a toroidal wave pattern is present on the protruding surface. This was believed to be the result of a vapor jet generated at the evaporating liquid-vapor interface that impinges on the bubble surface. Some, if not all, of these features can be seen in Figure 13.

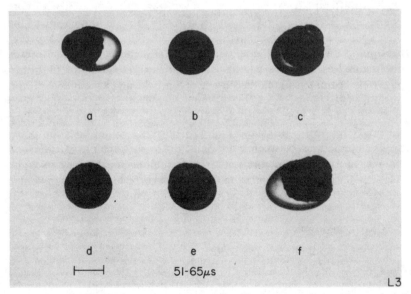

Figure 13 Fully developed two-phase drops produced by a vapor explosion. The roughened evaporating surface and toroidal wave pattern present on the protruding bubble surface can be seen. The dark region is the gas phase. Reproduced, with permission, from Shepherd & Sturtevant (1982).

The precursor to the formation of a two-phase drop and a potential vapor explosion is the nucleation of a bubble in a superheated liquid-liquid emulsion. Although the mechanics of the nucleation process is somewhat outside the scope of the present review, we briefly note that this phenomenon has been studied by Moore (1959), Apfel (1971) Jarvis et al. (1975), and Avedisian & Andres (1978). These studies collectively have examined bubble nucleation entirely inside a drop, as well as nucleation at the interface between a drop and the bulk fluid. The size of a bubble that is in unstable equilibrium is determined by calculating the extremum of the reversible, isothermal work required to form a bubble. A bubble of the critical size will grow if the temperature is increased slightly or if the liquid pressure is decreased slightly. These studies have also determined the rate of nucleation using standard thermodynamic considerations analogous to those used in homogeneous nucleation theory.

CONCLUDING REMARKS

We have attempted in this review to introduce the reader to the relatively new subject of compound drops. We have clearly avoided a detailed discussion on the effect of surfactants on drop/bubble motion, and we have made no attempt to cover the heat/mass transfer associated with compound drops. It is our hope that this review will provide stimulation so that others will become interested in this subject, and that, in addition, experts in heat/mass transfer and bioengineering will benefit from the present discussion. Researchers studying the fluid mechanics of compound drops have only begun to scratch the surface of this field, and a great deal of work remains to be done. Some of the noteworthy areas that remain to be explored in greater detail include translation of compound drops at large Reynolds numbers, translation of growing and collapsing compound drops, interactions between compound drops in three-phase emulsions, stability of translating compound drops, engulfing, and vapor explosions. As the analysis of compound drops matures, it is hoped that a closer contact will develop between theoreticians and experimentalists.

ACKNOWLEDGMENTS

The authors would like to thank R. P. Smet for his assistance in preparing the figures, the National Science Foundation for their generous support during the last few years (NSF MEA 81-07564, MEA 83-11495, MEA 83-51432), and the USC Faculty Research and Innovation Fund. Thanks are also due to those individuals who provided the excellent photographs and to Vivian Sprau, who did a splendid job typing the manuscript.

Literature Cited

Apfel, R. E. 1971. Vapor nucleation at a liquid-liquid interface. *J. Chem. Phys.* 54:62–63

Avedisian, C. T., Andres, R. P. 1978. Bubble nucleation in superheated liquid-liquid emulsions. *J. Colloid Interface Sci.* 64:438–53

Chambers, R., Kopac, M. J. 1937. The coalescence of living cells with oil drops. I. Arbacia eggs immersed in sea water. *J. Cell. Comp. Physiol.* 9:331–43

Gershfeld, N. L., Good, R. J. 1967. Line tension and the penetration of a cell membrane by an oil drop. *J. Theor. Biol.* 17:246–51

Hayakawa, T., Shigeta, M. 1974. Terminal velocity of two-phase droplet. *J. Chem. Eng. Jpn.* 7:140–42

Hertz, C. H., Hermanrud, B. 1983. A liquid compound jet. *J. Fluid Mech.* 131:271–87

Higeta, K., Mori, Y. H. 1981. Surface-area prediction for two phase drop in an immiscible liquid (letter to the editor). *Can. J. Chem. Eng.* 59:410–11

Higeta, K., Mori, Y. H., Komotori, K. 1979. Condensation of a single vapor bubble rising in another immiscible liquid. In *Heat Transfer—San Diego 1979*, ed. R. W. Lyczkowski, *AIChE Symp. Ser.* 75:256–65

Isenberg, J., Sideman, S. 1970. Direct contact heat transfer with change of phase: bubble condensation in immiscible liquids. *Int. J. Heat Mass Transfer* 13:997–1011

Jarvis, T. J., Donohue, M. D., Katz, J. L. 1975. Bubble nucleation mechanisms of liquid droplets superheated in other liquids. *J. Colloid Interface Sci.* 50:359–68

Johnson, R. E., Sadhal, S. S. 1983. Stokes flow past bubbles and drops partially coated with thin films. Part 2. Thin films with internal circulation—a perturbation solution. *J. Fluid Mech.* 132:295–318

Kopac, M. J., Chambers, R. 1937. The coalescence of living cells with oil drops. II. Arbacia eggs immersed in acid or alkaline calcium solutions. *J. Cell. Comp. Physiol.* 9:345–61

Lamb, H. 1932. *Hydrodynamics*. Cambridge: Cambridge Univ. Press. 738 pp. 6th ed. Reprinted, 1945, by Dover (New York)

Landman, K. A. 1983. On the crenation of a compound liquid droplet. *Stud. Appl. Math.* 69:51–74

Landman, K. A. 1984a. Stability of a viscous compound fluid drop. *AIChE J.* In press

Landman, K. A. 1984b. A continuum model for a red blood cell transformation: sphere to crenated sphere. *J. Theor. Biol.* In press

Landman, K. A., Greenspan, H. P. 1982. On the crenation of coated droplets. *Stud.*

Appl. Math. 66:189–216

Lasheras, J. C., Fernandez-Pello, A. C., Dryer, F. L. 1980. Experimental observations on the disruptive combustion of free droplets of multicomponent fuels. *Combust. Sci. Technol.* 22:195–209

Law, C. K. 1978. Internal boiling and superheating in vaporizing multicomponent droplets. *AIChE J.* 24:626–32

Lee, M. C., Feng, I., Elleman, D. D., Wang, T. G., Young, A. T. 1982. Generation of a strong core-centering force in a submillimeter compound droplet system. *Proc. Int. Colloq. Drops Bubbles, 2nd, Nov. 19–21, 1981*, ed. D. H. Le Croisette. *Jet Propul. Lab. Publ. 82-7*

Li, N. N. 1971a. Separation of hydrocarbons by liquid membrane permeation. *Ind. Eng. Chem. Process Des. Dev.* 10:214–21

Li, N. N. 1971b. Permeation through liquid surfactant membranes. *AIChE J.* 17:459–63

Li, N. N., Asher, W. J. 1973. Blood oxygenation by liquid membrane permeation. In *Chemical Engineering in Medicine, Adv. Chem. Ser.* 118:1–14

Maron Moalem, D. M., Miloh, T. 1976. Theoretical analysis of heat and mass transfer through eccentric spherical fluid shells at large Peclet number. *Appl. Sci. Res.* 32:395–414

Mercier, J. L., da Cunha, F. M., Teixeira, J. C., Scofield, M. P. 1974. Influence of enveloping water layer on the rise of air bubbles in Newtonian fluids. *J. Appl. Mech.* 96:29–34

Miller, C. A., Scriven, L. E. 1968. The oscillations of a fluid droplet immersed in another fluid. *J. Fluid Mech.* 32:417–35

Moore, G. R. 1959. Vaporization of superheated drops in liquids. *AIChE J.* 5:458–66

Mori, Y. H. 1978. Configurations of gas-liquid two-phase bubbles in immiscible liquid media. *Int. J. Multiphase Flow* 4:383–96

Mori, Y. H., Hijikata, K., Nagatani, T. 1977a. Fundamental study of bubble dissolution in liquid. *Int. J. Heat Mass Transfer* 20:41–50

Mori, Y. H., Komotori, K., Higeta, K., Inada, J. 1977b. Rising behavior of air bubbles in superposed liquid layers. *Can. J. Chem. Eng.* 55:9–12

Mori, Y. H., Nagai, K., Funaba, H., Komotori, K. 1981. Cooling of freely falling liquid drops with a shell of an immiscible volatile liquid. *J. Heat Transfer* 103:508–13

Ollis, D. F., Thompson, J. B., Wolynic, E. T.

1972. Catalytic liquid membrane reactor. I. Concept and preliminary experiments in acetaldehyde synthesis. *AIChE J.* 18:457–58

Patzer, J. F. II, Homsy, G. M. 1975. Hydrodynamic stability of thin spherically concentric fluid shells. *J. Colloid Interface Sci.* 51:499–508

Pinder, K. L. 1980. Surface-area prediction for two phase drops in an immiscible liquid. *Can. J. Chem. Eng.* 58:318–24

Plesset, M. S., Sadhal, S. S. 1982. On the stability of gas bubbles in liquid-gas solutions. *Appl. Sci. Res.* 38:133–41

Rushton, E., Davies, G. A. 1983. Settling of encapsulated droplets at low Reynolds numbers. *Int. J. Multiphase Flow* 9:337–42

Sadhal, S. S., Johnson, R. E. 1983. Stokes flow past bubbles and drops partially coated with thin films. Part 1. Stagnant cap of surfactant film—exact solution. *J. Fluid Mech.* 126:237–50

Saffren, M., Elleman, D. D., Rhim, W. K. 1982. Normal modes of a compound drop. *Proc. Int. Colloq. Drops Bubbles, 2nd, Nov. 19–21, 1981, Monterey, Calif.*, pp. 7–14, ed. D. H. Le Croisette. *Jet Propul. Lab. Publ.* 82-7

Selecki, A., Gradon, L. 1976. Equation of motion of an expanding vapour drop in an immiscible liquid medium. *Int. J. Heat Mass Transfer* 19:925–29

Shankar, N., Subramanian, R. S. 1983. The slow axisymmetric thermocapillary migration of an eccentrically placed bubble inside a drop in zero gravity. *J. Colloid Interface Sci.* 94:258–75

Shankar, N., Cole, R., Subramanian, R. S. 1981. Thermocapillary migration of a fluid droplet inside a drop in a space laboratory. *Int. J. Multiphase Flow* 7:581–94

Shepherd, J. E., Sturtevant, B. 1982. Rapid evaporation at the superheat limit. *J. Fluid Mech.* 121:379–402

Sideman, S. 1966. Direct contact heat trans-fer between immiscible liquids. *Adv. Chem. Eng.* 6:207–86

Sideman, S., Hirsch, G. 1965. Direct contact heat transfer with change of phase: condensation of single vapor bubbles in an immiscible liquid medium. Preliminary studies. *AIChE J.* 11:1019–25

Sideman, S., Isenberg, J. 1967. Direct contact heat transfer with change of phase: bubble growth in three-phase systems. *Desalination* 2:207–14

Sideman, S., Moalem-Maron, D. 1982. Direct contact condensation. *Adv. Heat Transfer* 15:227–81

Sideman, S., Taitel, Y. 1964. Direct-contact heat transfer with change of phase. Evaporation of drops in an immiscible liquid medium. *Int. J. Heat Mass Transfer* 7:1273–89

Simpson, H. C., Beggs, G. C., Nazir, M. 1973. Evaporation of butane drops in brine. *Int. Symp. Fresh Water from the Sea, 4th,* 3:409–20

Simpson, H. C., Beggs, G. C., Nazir, M. 1974. Evaporation of a droplet of one liquid rising through a second immiscible liquid. A new theory of the heat transfer process. *Proc. Int. Heat Transfer Conf., 5th, Tokyo,* 5:58–63

Tochitani, Y., Mori, Y. H., Komotori, K. 1977a. Vaporization of single drops in an immiscible liquid. Part I. Forms and motions of vaporizing drops. *Wärme Stoffübertrag.* 10:51–59

Tochitani, Y., Nakagawa, T., Mori, Y. H., Komotori, K. 1977b. Vaporization of single liquid drops in an immiscible liquid. Part II. Heat transfer characteristics. *Wärme Stoffübertrag.* 10:71–79

Torza, S. 1970. Interfacial phenomena in shear and electric fields. PhD thesis. McGill Univ., Montreal

Torza, S., Mason, S. G. 1970. Three-phase interactions in shear and electrical fields. *J. Colloid Interface Sci.* 33:67–83

Ann. Rev. Fluid Mech. 1985. 17 : 321–58

THE RESPONSE OF TURBULENT BOUNDARY LAYERS TO SUDDEN PERTURBATIONS

A. J. Smits

Department of Mechanical and Aerospace Engineering, Princeton University, Princeton, New Jersey 08544

D. H. Wood

Department of Mechanical Engineering, University of Newcastle, Newcastle, New South Wales 2308, Australia

1. INTRODUCTION

This review deals with the behavior of turbulent boundary layers subjected to sudden perturbations. Such flows are frequently encountered in practice. For example, when offshore breezes encounter a coastline, the sudden change in surface roughness (often accompanied by a similar change in surface heat flux) can have a significant effect on the flow. In the flow over an airfoil or turbine blade, rapid changes in pressure gradients and surface curvature may occur, and blowing or suction may be applied at some point along the airfoil as a form of boundary-layer control. Possibly the most dramatic example of a suddenly applied perturbation is the extremely rapid variation in pressure and temperature experienced in a shock-wave/boundary-layer interaction.

Apart from practical reasons, the study of the behavior of a perturbed boundary layer is also of considerable fundamental interest. For instance, Clauser (1956) applied Maxwell's concept of a "black box" to describe turbulent boundary layers, and he suggested that we may improve our physical understanding by observing their response to different outside

0066–4189/85/0115–0321$02.00

influences (see Figure 1). Later work by Coles (1962), Townsend (1965), Bradshaw & Ferriss (1968a), and the surveys by Tani (1968) and Schofield (1973) also represent important contributions to the subject. In addition, the recent reviews by Bradshaw (1973, 1975, 1976) and Hefner et al. (1983) cover some of the material considered here, as do the proceedings of the 1968 Stanford conference (Kline et al. 1969) and the 1980–81 Stanford conferences (Kline et al. 1981). In this review, we discuss this work to see what we have learned regarding the contents of Clauser's "black box."

Our understanding of turbulent shear layers may be judged in some sense by examining our ability to predict their behavior. On the most fundamental level, this prediction requires the solution of the complete time-dependent Navier-Stokes equations. Such an approach is clearly beyond our current analytical and computational capacity, and it seems likely that this situation will continue into the foreseeable future. As an approximate technique for solving the equations of motion, large-eddy simulation looks very promising (Reynolds 1976), but at this stage it still must be regarded as a research tool, and its application is far from routine. On a more practical engineering level, we generally deal with the time-averaged equations coupled with some model for the Reynolds-stress behavior. For relatively simple flows, such as a two-dimensional zero-pressure-gradient boundary layer on an adiabatic flat plate, this engineering approach appears to be fairly reliable (see, for instance, Kline et al. 1969, Coles & Hirst 1969). Nevertheless, it is not clear how far these calculation methods can be extended to the complex flows that often occur outside the laboratory. The addition of a moderate pressure gradient, for example, does not seem to present any severe problems. By way of contrast, a small amount of wall curvature can usually expose many shortcomings in the turbulence model, and for these flows existing techniques generally require additional modeling.

We can therefore identify three levels of complexity in describing the perturbation response of a turbulent boundary layer. As a matter of definition, we label flows "simple" if the perturbation is so weak, or so gradual, that the response exhibits a form of local similarity or "self-preservation" (that is, if the flow is adequately described using local length and velocity scales). By inference, simple flows are those flows that can be predicted satisfactorily using existing techniques. On a higher level of complexity, we can identify flows subjected to sudden, severe perturbations. Here we usually require additional modeling; either the perturbation is large (such as a sufficiently severe change in surface roughness) or it produces extra physical effects (such as the presence of surface curvature). Even though these flows may violate the thin-shear-layer approximation in the immediate region of the perturbation, they are generally recognizable as

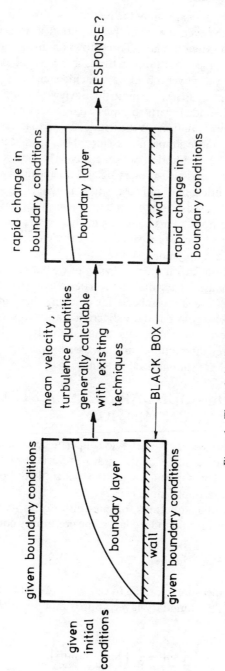

Figure 1 The turbulent boundary layer as a black box.

boundary layers for practically all of the recovery region. The third level of complexity arises, however, when the perturbation is so severe that the boundary-layer approximation fails. These flows are usually characterized by a change of species [see, for example, Bradshaw & Wong (1972), where a boundary layer separates over a backward-facing step].

Largely for reasons of space, we limit this review to investigations that study the effect of sudden perturbations that essentially preserve the boundary-layer approximation. We exclude flows that result from the interaction of two or more simple flows, since these have been reviewed by Bradshaw (1975, 1976). In addition, we generally only consider flows where the initial boundary layer is self-preserving and the mean flow is nominally two-dimensional. Our primary purpose is to highlight common features of the boundary-layer response and to evaluate the implications for engineering calculation methods. This assessment has only recently become possible; until a few years ago, sufficient turbulence measurements of suitable quality were simply not available.

We begin in Section 2 with a brief review of turbulent-boundary-layer characteristics, emphasizing the distribution of time and length scales. This review suggests a classification scheme for different perturbations, and in Sections 3, 4, and 5 we consider these different classes in turn. Flows with more than one perturbation acting simultaneously, including shock-wave/boundary-layer interactions, are discussed separately in Section 6, and the conclusions are summarized in Section 7.

2. REVIEW OF BOUNDARY-LAYER STRUCTURE AND CALCULATION METHODS

Before we proceed to discuss the behavior of perturbed boundary layers, it is useful to summarize some of the more important characteristics of undisturbed layers. This summary will serve to identify the general features of the perturbation response more clearly.

We begin with the equations of motion. For a two-dimensional incompressible turbulent boundary layer, the continuity equation is

$$\frac{\partial U}{\partial x} + \frac{\partial V}{\partial y} = 0, \tag{1}$$

where x and U are measured parallel to the wall in Figure 1 and y and V at right angles to it. The boundary-layer approximation to the momentum equation is given by

$$U \frac{\partial U}{\partial x} + V \frac{\partial U}{\partial y} = -\frac{1}{\rho} \frac{dP}{dx} + \frac{1}{\rho} \frac{\partial}{\partial y} \left(\mu \frac{\partial U}{\partial y} - \rho \overline{uv} \right). \tag{2}$$

Capital letters signify mean quantities, and lower-case letters denote fluctuating quantities—Reynolds averaging has been used throughout. As usual, P denotes the pressure, ρ is the density, and μ is the viscosity. All current engineering calculation methods (for incompressible flows) begin with (1) and (2). To solve these equations, however, some "closure" model must be introduced to evaluate the unknown turbulent shear stress $\tau \equiv -\rho\overline{uv}$. Reynolds (1976) classifies these models into four categories according to their level of complexity, and we consider these categories in turn. In "zero-equation" models, an eddy viscosity or mixing length relates τ directly to $\partial U/\partial y$. It is widely suggested, though far from proven, that such models are inaccurate in severely perturbed flows. In "one-equation" models, an additional equation is solved to obtain τ. For example, the well-known method of Bradshaw et al. (1967; hereinafter BFA) uses the equation for the transport of turbulent energy:

$$U\frac{\partial k}{\partial x} + V\frac{\partial k}{\partial y} = -\overline{uv}\frac{\partial U}{\partial y} + \frac{\partial}{\partial y}(\overline{pv}+\tfrac{1}{2}\overline{q^2v}) + \varepsilon, \tag{3}$$

$$\underset{\text{Advection}}{} \qquad \underset{\text{Production}}{} \qquad \underset{\text{Diffusion}}{} \quad \underset{\text{Dissipation}}{}$$

where $2k = \overline{q^2} = \overline{u^2}+\overline{v^2}+\overline{w^2}$. If we assume that the diffusion, the dissipation length scale $L_\varepsilon = (-\overline{uv})^{3/2}/\varepsilon$, and the stress ratio $a_1 = \overline{uv}/k$ are all universal functions of y/δ, Equation (3) becomes an equation for the shear stress. In "two-equation" models, as exemplified by the well-known k-ε model, Equation (3) is used, but the dissipation is calculated using a separate equation (see, for example, Rodi 1981). "Stress-equation" models make up the final category. These models use a separate transport equation for each component of the Reynolds stress, in addition to an equation for the dissipation.

As the number of model equations increases, the computational costs increase as well. In addition, the physical understanding of individual terms in each equation decreases. As Rubesin et al. (1977) pointed out, greater complexity does not necessarily lead to greater accuracy. Whatever the complexity of the model, however, it is probably most useful to discuss the boundary-layer behavior in terms of structure parameters and length scales. These parameters usually form the core of any turbulence model because they presumably have a more predictable behavior than the dimensional turbulence quantities themselves. Examples include the stress ratio a_1, the dissipation length scale L_ε, and the turbulent Prandtl number $Pr_t [= (\overline{v\theta}\ \partial U/\partial y)/(\overline{uv}\ \partial T/\partial y)]$, which is often used to calculate the transport of heat.

A useful qualitative observation is that a boundary layer may be split up into two fairly distinct regions: a small inner region characterized by relatively large velocity gradients, and an outer region where the velocity gradients are much lower. This distinction is made primarily on the basis of empirically observed velocity and length scales. In the inner layer, for example, the velocity scale u_τ is defined using the kinematic shear stress at the wall ($u_\tau \equiv \sqrt{\tau_w/\rho_w}$), and the length scale is v/u_τ, where v is the kinematic viscosity. These are often called wall variables. In the outer layer, the velocity scale for the velocity defect ($U_e - U$) is also u_τ, but the appropriate length scale is the boundary-layer thickness δ. (U_e is the local velocity of the flow external to the boundary layer.) In the region where the inner and outer layers overlap, dimensional arguments alone will give the well-known "log law" for the mean-velocity profile (Millikan 1938), i.e.

$$u^+ = \frac{1}{K} \ln y^+ + C, \tag{4}$$

where $u^+ = U/u_\tau$, $y^+ = yu_\tau/v$, and K and C are constants. By differentiating (4), we see that the velocity gradient in the logarithmic region varies inversely with the distance from the wall. In practice, the log law extends approximately over the region $y^+ \geqslant 30$ and $y/\delta \leqslant 0.2$.

For a rough wall, the origin for y in (4) is shifted by an amount that depends on the roughness function, as discussed in Section 3. For a compressible boundary layer (with heat transfer), the logarithmic variation of velocity is still observed if the velocity is scaled appropriately to take account of the density variation (Van Driest 1951). When there is heat transfer (with negligible buoyancy), the appropriate temperature scale is θ_τ, where $\theta_\tau = Q_w/(\rho C_p u_\tau)$ and Q_w is the local wall heat flux. The dimensional arguments for the overlap region lead to a log law for the mean-temperature profile; that is,

$$\frac{T_w - T}{\theta_\tau} = \frac{1}{K_\theta} \ln y^+ + C_\theta, \tag{5}$$

where T_w is the wall temperature and K_θ and C_θ are empirical constants.

In the wall region of an unperturbed boundary layer, Equation (3) reduces to a balance between the local processes of production and dissipation of turbulent energy. This state is often called "local energy equilibrium" and represents a special case of self-preservation. Local equilibrium, with the added assumption that $L_\varepsilon = Ky$, was used by Townsend (1961) to derive the log law. In fact, the log law appears to be less sensitive to departures from equilibrium than this analysis suggests. Since the equation for the shear stress τ does not have a corresponding local equilibrium form, it is often easier to assess the response of the wall region to a perturbation by using (3). In general, however, it is the behavior of the structural parameters relating directly to τ that are the most interesting, since τ is the unknown in (2).

Usually, the initial response to a perturbation is confined to a region that eventually spreads across the whole boundary layer. Only after this has occurred can the relaxation process be completed. Hence, it is important to distinguish between the spreading rate of the perturbation and the response time of the flow parameters. Bradshaw (1973) suggested that a suitable response time for the stress-containing eddies is the ratio of the turbulence energy k to its rate of production. A corresponding relaxation length X is U times this response time. Taking a typical value of a_1 as 0.3 gives $X \simeq U/(0.3 \ \partial U/\partial y)$. In the middle of the layer, using a 1/7th power law, $U/(\partial U/\partial y) \simeq 3$ and therefore $X \approx 10\delta_0$. A similar argument for the relaxation length of a passive contaminant field also gives $X \approx 10\delta_0$, and it is generally true that the outer layer responds very slowly to a disturbance. As the wall is approached, however, X tends to zero; that is, the flow adjusts very quickly in the inner region. This estimate for X remains useful in perturbed flows, even though these simple arguments no longer strictly apply. In more general terms, the distinction between the wall layer and the outer layer remains important after a strong perturbation.

The relaxation length also depends on the quantity measured. This occurs largely because of the increasing sensitivity of the turbulence terms as they increase in order. For example, the shear stress may take longer than the mean velocity to relax, partly because τ becomes self-preserving at roughly the same rate as $\partial U/\partial y$. Unless otherwise stated, all estimates of X used here were taken from the mean-velocity results; even then, it appears that $10\delta_0$ is an underestimate.

The only analytical method applicable to perturbed boundary layers is rapid-distortion theory [see Hunt (1977) for a recent review]. The theory neglects all the triple-product terms and usually the dissipation as well. It is valid only for "weak" turbulence where the length l over which the disturbance is applied is short, in the sense that $l/U \ll \Lambda/k^{1/2}$ (where Λ, the integral scale of the turbulence, is roughly equal to δ in the outer layer). Although the theory may be capable of describing the initial response to a sudden perturbation, it cannot in principle describe the whole relaxation process.

Since time and length scales vary with distance from the wall, the nature of the response to a change in boundary conditions will depend on where the disturbance is applied. Typical "distributions" of initial perturbations are shown qualitatively in Figure 2. The vertical scale is arbitrary for all three cases considered. A step change at the surface obviously localizes the disturbance and so is the simplest possible change in boundary conditions. A more diffuse effect is produced by a rapid change in pressure gradient,

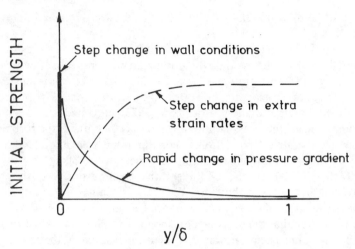

Figure 2   Distribution of initial perturbation strength. The vertical scale is arbitrary.

since its major influence is felt near the wall and its effect decreases away from the wall (see Section 4). When extra strain rates are imposed by a rapid change in boundary conditions, such as in a change in surface curvature, the initial effect varies roughly as the inverse of $\partial U/\partial y$, and therefore it is felt most strongly in the outer layer.[1] Hence, Figure 2 gives a convenient indication of the complexity of the response to a change in boundary conditions and provides a basis for the organization of the discussion that follows.

3.   DISTURBANCES INITIALLY AFFECTING THE WALL REGION

Here we consider step changes or impulses in surface roughness, wall heat flux or wall temperature, and blowing or suction. Step changes can usually occur in two ways. For instance, surface roughness can change from smooth to rough ($S \to R$) or rough to smooth ($R \to S$). When two step changes of opposite character occur successively, we have an "impulse." True impulsive changes are difficult to produce experimentally, but, for example, the perturbation produced by a short strip of roughness is a

---

[1] It is also possible to produce a localized disturbance in the outer layer by using a small two-dimensional obstacle. Such disturbances are of fundamental interest and have a direct application to drag reduction, but since the available data were recently reviewed extensively by both Hefner et al. (1983) and Bushnell (1984), they are not considered here. In this connection, the recent paper by Bandyopadhyay (1984) should also be mentioned.

reasonable approximation to an impulse in roughness (S → R → S) if the length of the strip is small compared with $10\delta_0$.

In all these cases, the initial disturbance is located precisely at the wall, and the outward propagation of the disturbance is characterized by the formation of an internal layer. The height of this layer, δ_i, marks the outward extent of the flow that is influenced by the new boundary condition, and the rate at which δ_i grows may be found, without great accuracy, by superimposing successive mean-velocity or temperature profiles. The growth rate appears to be typical of a thin shear layer. For example, Wood (1981, 1982) found that for step changes in surface roughness, δ_i increased approximately as $x^{0.8}$, where x was the distance downstream of the step. Although alternative correlations were given by Jackson (1976), Schofield (1981), and Deaves (1981), it is clear that the growth rate is slow. Similar results were found for the thermal internal layer formed by a step change in heat flux (Antonia et al. 1977).

We begin the detailed discussion of individual cases by considering step changes in surface roughness (shown schematically in Figure 3). Upstream of the step, the boundary layer is assumed to be fully developed with a logarithmic region described by

$$\frac{U}{u_\tau} = \frac{1}{K} \ln (y/z_0), \tag{6}$$

where z_0 is the "roughness length." For a smooth wall, comparison with (4) gives

$$z_0 = (v/u_\tau) \exp [-KC]. \tag{7}$$

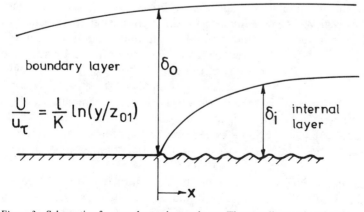

Figure 3   Schematic of a step change in roughness. The coordinate y is normal to x.

Downstream of the step, the layer slowly adjusts to the change in surface roughness. Sufficiently far downstream, the mean velocity in the inner region is again given by Equation (6) if the roughness length appropriate to the new surface is used. For a rough surface,

$$z_0 = (v/u_\tau) \exp \left[- K(C + \Delta U/u_\tau) \right], \tag{8}$$

where the roughness function $\Delta U/u_\tau$ measures the departure of the fully developed mean-velocity profile from the smooth-wall distribution. [For further details see Perry et al. (1969), who also discuss the difficulty of locating the origin for the velocity profile on a rough wall.]

The size of the step change can be measured by either $M^* \equiv z_{01}/z_{02}$ (where the subscripts 1 and 2 refer to the upstream and downstream values, respectively) or $M = \ln M^*$, which is the usual measure. Alternatively, the log law can be fitted to the mean-velocity measurements and the value of $(C + \Delta U/u_\tau)$ found at varying distances from the step. This method gives an indication of the relaxation distance X. For example, as the strength of the step decreases $M^* \to 1$, it follows that $M \to 0$, and $\Delta U/u_\tau$ tends to the fully developed value at all downstream locations. The values of M used in this review are taken from Wood (1981, 1982).

Measurements of the nearly self-preserving response to a small change in roughness were given by Antonia & Wood (1975). Large step changes, however, are intrinsically more interesting, and the most extensive measurements of (zero pressure gradient) boundary-layer response to large step changes in roughness are those of Antonia & Luxton (1971, 1972). Immediately after a smooth-to-rough step ($M = -5.0$), they observed a rapid rise in the wall shear stress; within the internal layer the shear stress varied nearly linearly from τ_w to the undisturbed value at δ_i. The advection and diffusion became significant within the internal layer even before it had penetrated the outer layer, indicating the breakdown of local equilibrium in the wall region. As a consequence, the dissipation length parameter L_ε was considerably reduced from its equilibrium value of Ky. The relaxation length X was found to be relatively small, as self-preservation appeared to be re-established within $20\delta_0$ of the step. In contrast, a large rough-to-smooth step ($M = 5.2$) decreased the wall shear stress and increased L_ε within the internal layer. Recovery was far from complete by the last measurement station at $16\delta_0$, demonstrating that the internal layer grew less rapidly than in the S → R case. Since the magnitude of M was similar in both cases, the results demonstrate the essential nonlinearity of boundary-layer response to large perturbations.

Additional experiments include those of Bradley (1968) and Schofield (1975). Bradley measured S → R and R → S steps in the atmosphere ($M = -4.8$ and 5.3, respectively). His mean-velocity and surface-shear results were qualitatively similar to those of Antonia & Luxton. Unfortunately, he did not take any turbulence measurements. Schofield obtained mean velocities for a R → S step in an adverse pressure gradient flow. The results implied a rapid return to self-preservation after the step.

The only supersonic boundary-layer measurements for a step change in roughness are those of Fenter & Lyons (1958) and Berg (1977). The latter investigation is by far the more complete and includes mean and fluctuating measurements for several S → R steps and a single R → S step. The free-stream Mach number was approximately 6, and the wall was near-adiabatic. The relaxation distance for the mean flow was found to be $10\delta_0$ for S → R steps and $14\delta_0$ for R → S steps, whereas the relaxation distance for the fluctuating properties was approximately one-and-a-half times as long. No triple products were measured, but it seems that the general behavior of the results is entirely similar to that observed in subsonic flows.

The only measurements available for a nearly impulsive change in surface roughness are those of Andreopoulos & Wood (1982). They measured the response of a smooth-wall incompressible boundary layer to a perturbation produced by a sandpaper strip of length $\sim 3\,\delta_0$. At the end of the strip, τ_w was nearly three times the undisturbed value. The response was considerably more complicated than a simple superposition of step changes would suggest. The relaxation was very slow, and the recovery was still incomplete at the last measuring point, located $55\delta_0$ downstream of the strip. Figure 4a shows the wall shear-stress distribution, and Figure 4b gives some of the shear-stress results from that experiment. The roughness caused a sharp peak in τ, which moved outward and slowly decayed downstream.

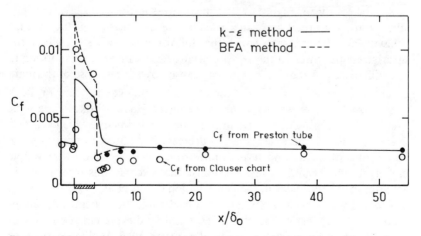

Figure 4a Measured and calculated wall shear stress for "impulse" in surface roughness (Andreopoulos & Wood 1982).

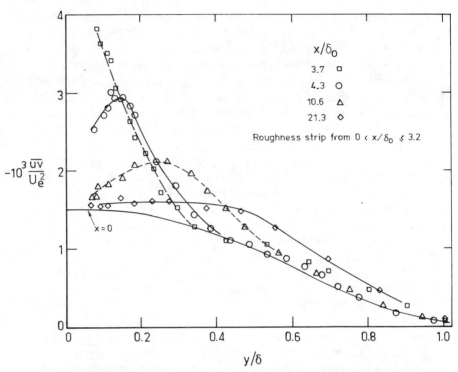

Figure 4b Shear-stress profiles for "impulse" in surface roughness (Andreopoulos & Wood 1982). Curves for visual aid only.

As can be seen from Figure 4*b*, the shear stress in the outer layer has not fallen to the self-preserving value, even by the last measuring station. This behavior may be contrasted with that following a $S \to R$ step, where the shear stress increases to the self-preserving level without overshooting it. The difference in behavior is probably caused by the response of the main generation term for τ, $\overline{v^2} \, \partial U / \partial y$, and the triple-product term $\partial \overline{u^2 v} / \partial y$. Both these terms result from the nonlinearity of the Navier-Stokes equations. The first term is always positive in roughness experiments, which suggests that it is easier to increase τ than to decrease it. Indirect evidence for the importance of the second term comes from the observation that simple models for the boundary-layer response that exclude this term result in equations that are linear in the perturbation (Smits et al. 1979b).

These experiments have stimulated considerable theoretical and computational work. Most of the early theoretical studies assumed either self-preservation in the modified region (Townsend 1965, Mulhearn 1977) or particular relationships for the mean velocity and shear stress (Panofsky &

Townsend 1964, Deaves 1981). The major practical shortcoming of the self-preserving assumption is the inability to account for the upstream influence that is particularly noticeable after R → S steps (see Wood 1982). Later work includes the calculations by Wood (1978) and Andreopoulos & Wood (1982), who used the BFA method. Figure 4a shows the wall stress predicted by both the BFA and the k-ε model. Comparison with the experimental shear-stress distribution (not shown here) also suggests that the BFA model is more accurate.

Most computational work has attempted to reproduce Bradley's measurements, perhaps because of the practical importance of step changes in meteorological flows. The most sophisticated attempt to date is that of Rao et al. (1974), who used a stress-equation model. Most of the significant discrepancies between these calculations and Bradley's S → R results were attributed by Wood (1978) to a poor choice of z_0 for the smooth surface. Wood's calculations used both the original BFA method with the turbulent diffusion ignored and the modified BFA method (Bradshaw & Unsworth 1976), which accounts for nonlocal effects on L_ε. The results support Townsend's (1976, p. 311) conclusion that the diffusion is not important in the wall region after single steps, and they indicate that the major nonlocal effect is the advection of turbulent kinetic energy or, by implication, the mean transport of shear stress. As a practical example of the usefulness of models for step changes in roughness, the Engineering Sciences Data Unit of London (ESDU 1982) used Deaves' (1981) analysis to incorporate the effects of roughness changes on building wind-load estimation.

Next we consider step changes in wall heat flux. In these experiments, a thermal internal layer of height δ_T, analogous to δ_i, can be found from the mean-temperature profiles. Here, we only treat flows where buoyancy effects are negligible and temperature acts as as a passive scalar. Furthermore, the usual models for the turbulent Prandtl number reduce to functions of y only [see Reynolds (1975) and Browne & Antonia (1979) for examples]. Under these assumptions, the equation for the mean temperature becomes linear, and therefore if Pr_t is an adequate description of the turbulent transfer of heat, the mean temperature field will respond linearly to a perturbation in wall heat flux.

If the wall temperature differs from the free-stream value upstream of the step, we have a logarithmic variation of mean temperature described by Equation (5). This can be written in a form similar to Equation (6) by defining a "thermal roughness" parameter

$$z_0 = (v/u_\tau) \exp\left[-K_\theta C_\theta\right]. \tag{9}$$

By analogy with the behavior downstream of a change in roughness, the temperature profile downstream of a change in heat flux may be written in logarithmic form by introducing $\Delta T/\theta_\tau$, so that (9) becomes

$$z_0 = (v/u_\tau) \exp\left[-K_\theta(C_\theta + \Delta T/\theta_\tau)\right]. \tag{10}$$

In all the experiments performed to date, the flow was at a uniform temperature either before the step or well downstream of the step. When the temperature is uniform throughout the flow, the temperature log law is obviously meaningless, and therefore there is no thermal equivalent of M or M^*. Step changes in roughness and heat flux can still be compared, however, on the basis of a step size defined by $(C + \Delta U/u_\tau)$ or $(C_\theta + \Delta T/\theta_\tau)$. Those experiments for which information on $\Delta U/u_\tau$ or $\Delta T/\theta_\tau$ was available are compared in Figure 5. The step sizes were calculated using K_θ and C_θ values given in the original papers. These constants unfortunately show considerable experimental scatter, mainly because it is difficult to measure the wall heat flux accurately. Nevertheless, Figure 5 indicates that the data for steps in roughness and heat flux span a similar range of step size.

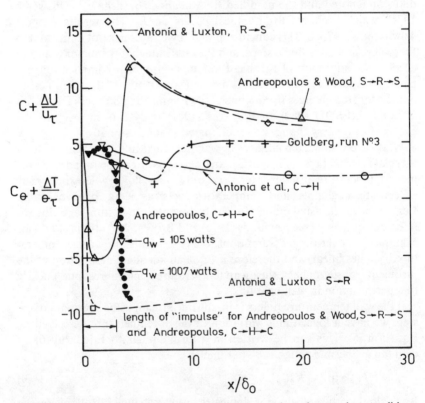

Figure 5 Strength of perturbation versus x/δ_0 for changes in surface roughness, wall heat flux, and pressure gradient.

The first detailed measurements of a step change from a cold to heated wall (C → H) were made by Johnson (1957, 1959), who pointed out the long relaxation length of the temperature field. Blom (1970) and Fulachier (1972) provided further measurements, but the work of Antonia et al. (1977) is undoubtedly the most detailed and complete. Close to the step, the latter authors observed a breakdown in local thermal equilibrium near the wall; that is, the production and destruction of the mean-square temperature fluctuation $\overline{\theta^2}$ were not equal. This imbalance is associated with the relatively high value of $(C + \Delta T/\theta_\tau)$ near the step (see Figure 5) and the increase in the thermal dissipation length scale $L_\theta \equiv (\overline{v\theta})^2 [(-\overline{uv})^{1/2}\varepsilon_\theta]^{-1}$ above the equilibrium value $K_\theta y$. Here ε_θ is the dissipation of $\overline{\theta^2}$. To within experimental uncertainty, they found a collapse of $\overline{v\theta}/u_\tau\theta_\tau$ when plotted against y/δ_T. However, the ratio $a_{1\theta} \equiv \overline{v\theta}/(\overline{\theta^2})^{1/2}(-\overline{uv})^{1/2}$, which is analogous to a_1, did not become constant in the internal layer until some distance downstream ($x/\delta_0 \geqslant 6$). The turbulent Prandtl number Pr_t varied significantly across the flow at all x.

The response to a sudden decrease in wall heat flux (H → C) has not been measured in comparable detail, perhaps because the general level of the mean and fluctuating temperature, and hence the experimental accuracy, decreases with x. For this reason, the measurements by Charnay et al. (1979) and Subramanian & Antonia (1981) were limited to $x/\delta_0 \leqslant 14$. The former found a significant breakdown in thermal equilibrium at $x/\delta_0 \simeq 7$ caused by high levels of advection and diffusion of $\overline{\theta^2}$ in the wall region. The latter found that δ_T grew less rapidly than in the C → H case.

The only measurements available for a nearly impulsive change in heat flux are those by Andreopoulos (1983), who measured the thermal field developing within a fully developed boundary layer as it passed over a heated wall of length $\sim 3\delta_0$. The second step (H → C) was not quite equal and opposite the first step (C → H) because of a small longitudinal heat transfer to the downstream "cold" surface. The second internal layer grew less rapidly than the first by an amount that is unlikely to be caused by the small longitudinal heat transfer alone, and this observation is consistent with the results of the single-step experiments.

Unfortunately, none of the H → C cases are the direct inverse of any of the C → H experiments, and the interpretation of Andreopoulos's results is complicated by the short length of the heated surface. Nevertheless, the evidence suggests that the mean field does not respond in a linear manner. This can only be caused by the nonlinear behavior of $\overline{v\theta}$, which in turn is inconsistent with the usual models for Pr_t.

Perhaps the only true impulsive change in wall conditions is provided by a thin line source of heat (Shlien & Corrsin 1976) or passive contaminant

(Poreh & Cermak 1964). The linearity of the governing equations for the mean temperature suggests that this problem can be considered as a $C \rightarrow H$ step followed by a $H \rightarrow C$ step. Furthermore, the outward extent of the thermal field will be marked by the coincident internal layers for the (coincident) steps. The downstream development of the internal layers cannot be inferred accurately from either set of data, although the similarity plot of Poreh & Cermak (their Figure 5) suggests that the internal layers grow as $x^{0.8}$ (that is, at nearly the same rate as δ_i after a step change in roughness).

Theoretical or computational analyses of the flows just considered include Townsend's (1965) analysis based on the assumption of self-preservation and Kays' (1966) semiempirical expression for the Reynolds analogy factor. Antonia et al. (1977) found this latter expression to be in reasonable agreement with their results. Bradshaw & Ferriss (1968b) extended the BFA method to calculate the behavior of a passive contaminant and achieved reasonable agreement with the measurements of Poreh & Cermak (1964). The same results were also computed with reasonable success by Wassel & Catton (1977), who used an eddy-viscosity model. Browne & Antonia (1979) compared four $C \rightarrow H$ experiments with calculations of the Bradshaw & Unsworth (1976) method, which assumes a constant Pr_t, and the Patankar & Spalding (1970) method, which does not. As would be expected, the largest discrepancies between measurement and calculation occurred close to the step in all cases. Launder & Samaraweera (1979) developed a two-equation turbulence model with equations for $\overline{u\theta}$ and $\overline{v\theta}$ and obtained reasonable agreement with the results of Antonia et al. A similar model was used by El Tahry et al. (1981) to successfully reproduce the measurements of Poreh & Cermak.

The final class of perturbed flows to be considered in this section is those flows with a sudden change in wall blowing or suction. A large amount of relevant data are available, but these flows are difficult to set up experimentally, and much of the available data appear to be influenced by departures from two-dimensionality, inadequate skin-friction measurements, low-Reynolds-number effects, and effects of surface roughness. In addition, comprehensive turbulence measurements are simply not available. Nevertheless, two rather careful experiments (Squire et al. 1977, Simpson 1971) deserve mention. Squire et al. studied a step increase in blowing at Mach 1.8, where F changes from zero to 0.0045, and a step decrease in blowing at Mach 3.6, where F changes from 0.0029 to zero. Here, $F \equiv \rho_w V_w / \rho_e U_e$, where $\rho_w V_w$ is the transpired mass flux. We have not included these results in Figure 5 because the usual log law is fundamentally altered by blowing or suction. Squire et al. found good agreement between the results from the step increase and calculations according to the BFA method and an eddy-viscosity formulation. The comparison for the step decrease was not quite so good. Simpson (1971) studied the effect of a change in wall suction for an incompressible flow. He also made detailed measurements following a step increase in wall blowing, although the results were confined to mean-velocity measurements. Over the step, V_w/U_e increased from zero to 0.008. As in the R → S case, the wall shear stress decreased rapidly downstream of the step. Self-preservation was restored at around $30\delta_0$ in all cases. Simpson found that the BFA method gave reasonable estimates for δ_i, which he called the penetration point. It is possible that the success of the calculation methods merely indicates that the perturbation is too weak to significantly alter the turbulence structure. It is also possible that a larger change in suction or blowing would destroy the thin shear-layer behavior of the response. Furthermore, large amounts of suction can induce "relaminarization" (Narasimha & Sreenivasan 1979), and this phenomenon is outside the scope of the present review.

   In summary, the qualitative response of a boundary layer to a perturbation initially confined to the wall is quite clearly described in terms of an internal layer slowly propagating into a relatively undisturbed external layer. The quantitative response depends on the magnitude and direction of the step size. For example, S → R and C → H steps relax faster than R → S and H → C steps of comparable magnitude. In other words, large steps lead to a highly nonlinear response. In all cases, local equilibrium breaks down near the wall and the advection becomes important. It also seems likely that diffusion becomes significant after impulsive perturbations. In addition, the dissipative length scale near the wall increases above the equilibrium value for R → S and C → H and decreases below this value for S → R steps. The nonlinearity of the response in the outer layer appears to result from the behavior of the generation term and the triple-product term in the equation for the shear stress. After a R → S step, or a S → R → S impulse, the "excess" shear-stress levels in the outer layer decay slowly to reach the self-preserving level. In contrast, after a S → R step (and presumably a R → S → R impulse), the shear stress in the outer layer increases relatively quickly toward the self-preserving level.

4.   CHANGES IN PRESSURE GRADIENT

To the boundary-layer approximation, $\partial P/\partial y = 0$ for the layer on a flat plate, and a change in dP/dx is felt equally across the layer. However, if (1) and (2) are used to form an equation for $\partial U/\partial y$, it can be seen that in the outer layer where $\mu\, \partial^2 U/\partial y^2$ is small, $\partial U/\partial y$ cannot immediately respond to a change in dP/dx (Bradshaw & Ferriss 1965). Since it is $\partial U/\partial y$ and not P

that appears in the equations for τ and k, it is clear that in the outer layer the turbulence cannot respond immediately to a change in pressure gradient; the total head is approximately conserved, and the mean-velocity profile is shifted without altering $\partial U/\partial y$ (Stratford 1959, Townsend 1962, Bradshaw & Galea 1967). In contrast, very near the wall, the mean-momentum equation (2) indicates a balance between pressure gradient and viscous forces; that is, dP/dx is balanced by $v\partial^2 U/\partial y^2$, and so the inner-layer velocity gradient and turbulent stresses are affected immediately by a change in pressure gradient.

Thus the distribution of the initial disturbance shown in Figure 2 refers to the effect on the turbulence and not the mean velocity. The region of disturbed turbulence should propagate outward from the wall as an internal layer of sorts, but it is clear that this layer does not contain all the disturbed flow (in contrast to the internal layers described in Section 3). Whether this outward propagation can be clearly observed depends largely on the downstream behavior of the outer layer.

There are some new, important effects produced by rapid changes in pressure gradient. For instance, positive $\partial U/\partial x$ in a two-dimensional incompressible flow is associated with negative $\partial V/\partial y$, and Townsend (1961) suggested that negative $\partial V/\partial y$ tends to flatten the large eddies and reduce their contribution to the Reynolds stress. By implication, positive $\partial V/\partial y$ would tend to increase the Reynolds stress independently of any change in $\partial U/\partial y$. In other words, $\partial V/\partial y$ may have a greater effect in the outer layer (where $\partial U/\partial y$ is smallest) than would be suggested by an order-of-magnitude analysis of (3). Bradshaw (1973) did not find the experimental evidence conclusive, however, and the effects of $\partial V/\partial y$ in boundary layers have generally been considered rather weak. In contrast, recent computational work by Hanjalić & Launder (1980) and Galmes et al. (1983) has suggested that these effects may be more important than previously thought; the question obviously requires resolution.

In a general sense, $\partial V/\partial y$ is an example of an extra strain rate, and most extra strain rates exert a significant influence on the boundary-layer behavior. For example, it is clear that in compressible flows pressure gradients give rise to extra strain rates that are very important (Bradshaw 1974). These extra strain rates are discussed more fully in Section 5; here, we confine ourselves to a discussion of incompressible flows.

Step changes in pressure gradient can be produced by the sudden application or removal of either a favorable or an adverse pressure gradient. It must be noted that (a) a boundary layer negotiating the removal of a favorable pressure gradient has not been investigated thus far, and (b) the imposition of a large, favorable pressure gradient may cause the initially turbulent boundary layer to revert to a laminar-like state. Because of this "relaminarization," and the usually concomitant low-Reynolds-number effects, we do not discuss these flows here. They were reviewed extensively by Narasimha & Sreenivasan (1979) and were also considered at the 1980/81 Stanford conference (Kline et al. 1981). Finally, (c) there appear to be no experimental results available for an impulsive application of pressure gradient, although some analytical work was performed by Fernholz (1966) and Bradshaw & Ferriss (1968a).

Mean-velocity and turbulence measurements following the removal of an adverse pressure gradient were made by Bradshaw & Ferriss (1965) and Goldberg (1966). The wall shear stress increased after the step by around 70% in the former study and 340% in the latter. Bradshaw & Ferriss found no appreciable change to the log-law constants, while Goldberg did; the values of $(C + \Delta U/u_\tau)$ for his run #3 are shown in Figure 5. (The "step" was assumed to be at $x = 1.6$ inches.) In both experiments, the mixing length increased in the wall region. Since the advection also increased in this region, it is clear that local equilibrium broke down close to the step.

The sudden application of a relatively weak adverse pressure gradient was investigated by Bradshaw (1967a). The wall shear stress decreased by about 40% over a distance of $50\delta_0$. The effect on structure parameters such as a_1 and L_ε appeared to be rather weak, and the diffusion and advection remained approximately equal (and small) throughout the boundary layer. The computations presented for this flow at the 1968 Stanford conference were generally more accurate than those for the flow studied by Bradshaw & Ferriss (1965). Samuel & Joubert (1974) studied the response of the boundary layer to the application of a stronger pressure gradient that was increasingly adverse (both dP/dx and d^2P/dx^2 were positive). Here, the wall shear stress decreased by a factor of four over a distance of about $80\delta_0$, and the parameter a_1 fell by about 15% over this distance. The log law, however, was maintained throughout the flow. The predictions for this flow are of rather mixed quality (see Cantwell & Kline 1981). Rather strangely, the three methods that gave the best results include two eddy-viscosity models and one Reynolds-stress model.

When an even larger step change in adverse pressure gradient is produced, the flow generally separates at some distance downstream. The most detailed investigation of such a flow is undoubtedly that of Simpson et al. (1977, 1981). A similar, more recent experiment was performed by Cutler & Johnston (1984). In both sets of experiments, the upstream boundary layer developed in a slightly favorable pressure gradient before being suddenly decelerated. The log law was maintained until quite close to separation in both cases, but the parameter $a_1' = -\overline{uv}/(\overline{u^2} + \overline{v^2})$ was reduced. For example, a_1' fell by about 60% prior to separation in the Cutler & Johnson experiment. Nevertheless, a comparison between Simpson et al.'s data and the BFA method gave reasonably good agreement up to the separation point.

In summary, for an attached boundary layer, the response to changes in pressure gradient is not as spectacular as that to changes in surface roughness. This is largely because it is difficult to impose a change in pressure gradient that is both large and rapid without inducing separation (compare, for instance, Goldberg's results and the roughness step results shown in Figure 5). Although the log law appears to be unaffected, except perhaps for the strongest possible perturbations, structure parameters such as a_1 and the mixing length can be significantly altered; for a step from zero to adverse pressure gradient, these parameters are reduced, and for a step from adverse to zero pressure gradient they increase.

5.   DISTURBANCES INITIALLY AFFECTING THE OUTER LAYER

There exists a class of perturbations that is characterized by the application of strain rates additional to the "simple" shear $\partial U/\partial y$. These strain rates are generally introduced by changing the flow geometry. For example, longitudinal streamline curvature introduces the extra strain rate $e = \partial V/\partial x$, streamline divergence or convergence is associated with the strain rate $\partial W/\partial z$, and compression or dilatation is characterized by the summation of extra strain rates represented by the divergence of the velocity field. Since the strain rate $\partial U/\partial y$ continually decreases away from the wall, extra strain rates appear to have their strongest relative effect in the outer region.

Once again, we can distinguish three different perturbations. In the first case, an undisturbed boundary layer experiences the sudden application of an extra strain rate; in the second case, a boundary layer, disturbed by an extra strain rate over some considerable distance, experiences the sudden removal of this strain rate; and in the third case, an undisturbed boundary layer experiences an impulsive application of an extra strain rate.

Bradshaw (1973) demonstrated that simple analogies can often be drawn between the effects of different extra strain rates, and therefore these phenomena can usually be treated as a single class. It must be pointed out that Bradshaw was generally concerned with extra strain rates small compared with the principal strain rate $\partial U/\partial y$. When this ratio is no longer small, the qualitative analogy between the effect of different extra strain rates may break down.

The nature of the response depends on the strength of the perturbation, as well as the distance over which it is applied. For step changes in extra strain rate, a convenient measure of the strength is given by $e/(\partial U/\partial y)$. This ratio will vary from 0.01 or 0.02 for "weak" extra strain rates to 0.1 or 0.2 for "strong" extra strain rates. When the extra strain rate is applied over a time comparable with the large-eddy lifetime, the slow response of the large eddies suggests that a better measure of the perturbation strength is given by the integral of the extra strain rate over the time it acts, i.e. $I = \int e \, dt$ (Smits et al. 1979c). For an "impulse" of curvature, I is simply given by the total turning angle, whereas for dilation we have $I = 1/\gamma \ln p_2/p_1$, where p_2/p_1 is the pressure ratio (Hayakawa et al. 1982). Although these numbers are small, parameters such as the dissipation length scale appear to be affected by a factor $F = [1 + \alpha e/(\partial U/\partial y)]$, or $F = (1 + \alpha I)$, where $\alpha = O(10)$. The effect on skin friction or Stanton number is similar in magnitude, and it seems that extra strain rates generally have an influence an order of magnitude larger than that suggested by the mean-motion equations. For example, Thomann (1968) showed that when a supersonic boundary layer encounters a curved wall where the ratio of boundary-layer thickness to radius of curvature is $\delta/R = 0.02$, the Stanton number decreases by approximately 20% for convex curvature and increases by about the same amount for concave curvature. The longitudinal pressure gradient was zero in both cases.

The response depends on the sign of the extra strain rate. For example, concave curvature is "destabilizing" and tends to increase turbulent activity, whereas convex curvature is "stabilizing" and suppresses turbulent mixing. Similarly, divergence and compression are destabilizing, while convergence and dilatation are stabilizing. The qualitative explanations for these observations are quite straightforward, but unfortunately these arguments considerably underestimate the observed consequences (see Bradshaw 1973).

The three extra strain rates considered here are longitudinal curvature, lateral divergence, and bulk dilatation. The first two can occur in compressible or incompressible flow; the third is confined to compressible flows. A discussion of the behavior under the action of more than one strain rate is deferred until Section 6.

A good example of a relatively weak step change from a flat wall to one with longitudinal curvature is given by the experiment of Muck (1982). [See also Muck et al. (1984) and Hoffmann et al. (1984).] The curvature was either concave or convex and persisted for approximately $50\delta_0$ downstream of the point where the curvature was first applied. The radius of curvature was kept constant, and the curvature parameter δ/R varied from about 0.01 to 0.012 for the convex side, and from 0.01 to 0.02 for the concave side. There was a short region of fairly strong pressure gradient near the start of the curvature. The overall effects due to the change in pressure were rather small, however, and the effects of curvature could be clearly distinguished.

This work represents the culmination of previous work at Imperial College by Meroney & Bradshaw (1975) and Hoffmann & Bradshaw (1978), who used the same apparatus, and it may be regarded as the most complete set of data for perturbations in mild curvature. Similar, but less complete, experiments on convex and concave walls were performed by Ramaprian & Shivaprasad (1978) using $\delta/R \simeq 0.013$.

The Imperial College work shows that

> the effects of convex and concave curvature on boundary layers are totally different, even qualitatively: convex curvature tends to attenuate the pre-existing turbulence, apparently without producing large changes in statistical average eddy shape, while concave curvature results in the quasi-inviscid generation of longitudinal ("Taylor-Görtler") vortices, together with significant changes in the turbulence structure, induced either directly by the curvature or indirectly by the vortices.
>
> From the point of view of calculation methods, the implication is that although stabilizing and destabilizing curvature are qualitatively connected by a common dimensional analysis, the quantitative differences are such that the one cannot be regarded as a useful guide to the treatment of the other. Specifically, rates of change of turbulence structural parameters with curvature parameter are likely to be discontinuous at zero curvature, and in particular the response time of a turbulent boundary layer to convex curvature, implying mere attenuation, is very much less than the response time to concave curvature, implying re-organization of the eddy structure. [Muck et al. 1984]

The application of convex curvature has only a very small influence on a_1, whereas a similar step in concave curvature eventually increases a_1 by about 25%. The dissipation length scale is strongly affected in both flows; in the middle of the layer, it decreases to half its original value on the convex wall and just about doubles on the concave wall. This trend is reflected in the mean-velocity profiles on the concave wall by the appearance of a "dip" below the log law, implying an increase in the near-wall length scale above its equilibrium value of Ky.

Perhaps the most surprising aspect of the application of concave curvature is the appearance of steady longitudinal roll-cells. These roll-cells seem to be a general phenomenon associated with concave curvature, although their span-wise spacing appears to be dictated by upstream disturbances. They cause span-wise variations in all boundary-layer properties, and the magnitude of these variations depends on the strength of the curvature. These roll-cells can dominate the downstream behavior if they are strong enough, and they obviously pose a problem for conventional calculation methods.

As examples of a stronger step change in curvature, we consider the measurements of So & Mellor (1972, 1973, 1975), Gillis et al. (1980), and Gillis & Johnston (1983). The experiments of So & Mellor provided almost the first detailed turbulence measurements in boundary layers perturbed by surface curvature. The measurements were performed on both convex and concave walls, and the parameter δ/R was kept constant at approximately 0.08 and 0.12, respectively, over a distance of approximately $50\delta_0$. The work of Gillis et al. was confined to a study of convex curvature. Two experiments were performed, one with δ_0/R approximately 0.1 and another with $\delta_0/R = 0.05$. The curvature persisted for about $20\delta_0$ and $40\delta_0$, respectively, and the total turning angle in both cases was 90°. Thereafter, the curvature was suddenly removed, and the boundary layer was allowed to relax on a flat plate with zero pressure gradient.

Gillis & Johnston went to considerable trouble to eliminate the pressure gradients near the beginning and end of the curved section. In the So & Mellor experiments, however, there was a short region of intense pressure gradient near the start of the curvature (favorable for the convex experiment, adverse for the concave experiment). These pressure gradients did not seem to have any significant effect on the boundary-layer behavior, except perhaps for a short region just downstream of the beginning of curvature; both So & Mellor and Gillis & Johnston observed an essentially similar response.

These experiments show that the sudden application of strong convex curvature results in rather dramatic changes to the turbulence structure. Within a few boundary-layer thicknesses the shear stress for $y/\delta \geqslant 0.5$ collapses to virtually zero. In the region near the wall, the shear stress decreases less dramatically, but here the most interesting aspect is a change of scale; $-\overline{uv}/u_\tau^2$ after a sustained region of convex curvature scales with y/R rather than y/δ. These observations led Gillis & Johnston to propose a two-layer model for the perturbed boundary layer, consisting of an inner "active" layer where the behavior of the turbulence was more or less "normal" and an outer layer containing nearly isotropic "debris" from the thick, turbulent boundary layer upstream of curvature. The absence of shear stress in the outer layer means there is no production mechanism, and consequently the turbulence there decays as it moves downstream.

The sudden removal of strong convex curvature is followed by a very slow recovery of the boundary layer. Gillis & Johnston found that the skin-friction coefficient, for example, showed no sign of recovery for about $40\delta_0$ downstream, although this recovery behavior appears to scale with x/R rather than x/δ_0. The slow response seems to be a direct result of the structural changes caused by the upstream convex curvature; the almost total absence of shear stress in the outer layer suggests a destruction of the flow "history," and so the recovery is dominated by the slow growth of the inner, active layer. This interpretation is supported by the fast recovery of the parameter a_1 — once the convex curvature is removed, the newly created turbulence retains little memory of the upstream history (Smits et al. 1979c).

The response to the sudden application of strong concave curvature seems to be slower than the corresponding step in convex curvature. This observation is in accord with the results for smaller steps. The major differences between a strong step versus a weak step seem to depend crucially on the steadiness of the longitudinal roll-cells that are formed. In the experiment by So & Mellor (1975), for instance, spanwise variations were weak, but this observation does not exclude the possibility that randomly moving roll-cells were present. We would intuitively expect these roll-cells to have a greater influence if they were fixed more firmly in place, as may occur when upstream disturbances are strong enough.

An experiment that realizes the impulse condition reasonably closely was performed by Smits et al. (1979c), who studied the response of a boundary layer to short regions of both convex and concave curvature of length $6\delta_0$ and $12\delta_0$, respectively. The curvature was strong, and δ_0/R was approximately 0.2 and 0.1 for the convex and concave curvatures, respectively. The total turning angle was 30°. One additional case for a 20° concave bend was also tested.

At the exit of the bend on the convex side, the behavior is very similar to that observed by Gillis & Johnston. The shear stress and the stress ratio a_1 have collapsed in the outer half of the boundary layer, and there is an implied decrease in length scale. The downstream recovery is marked by a rapid relaxation of a_1 and a slower relaxation of the shear stress itself. The recovery in τ, shown in Figure 6a, is similar to that following an impulse of roughness, shown in Figure 4b. In both cases the shear stress in the outer layer overshoots the self-preserving level before reducing to it. The curved-wall results were generally well predicted by Gibson et al. (1981) using a k-ε model, although the overshoot τ was overestimated. The return of the shear stress and the skin friction to their self-preserving value is considerably quicker than that seen by Gillis & Johnston. The implication is that a prolonged region of convex curvature is more effective in destroying the memory of the outer layer, providing further evidence for the slow adjustment of the large eddies.

At the exit of the bend on the concave side, the boundary layer showed the by now expected large increases in all the turbulent stresses. In addition, a_1 increased by about 80% regardless of the turning angle, and there was a marked increase in the length scale. A set of steady longitudinal vortices was observed to form, and these continued almost unchanged in strength and spacing to the end of the test section ($\simeq 70\delta_0$ long). Perhaps the most spectacular feature of the relaxation process was that the decay of the initially high turbulent intensity was not monotonic; the Reynolds stresses in the outer layer eventually collapsed to well below the level at entry to the bend. This collapse was caused by the great decrease in $\partial U/\partial y$ that resulted from the intense shear-stress gradients within and just downstream of the bend (see Figure 6b). Once the production mechanism had been so effectively reduced, the turbulent stresses simply decayed; this result is a rather surprising consequence of an impulse in destabilizing curvature.

One more recent investigation (Baskaran 1983, Smits et al. 1981) on the effect of longitudinal curvature should be mentioned. The study consisted of two related experiments. In the first, the boundary layer developed on a flat plate before entering a short region of concave curvature, followed immediately by a prolonged region of convex curvature. The configuration resembled a circular hill of height-to-chord ratio 0.162 placed on a flat plate, with small concave blending regions fore and aft. The flow eventually separated in the adverse pressure gradient region on the lee side. The measurements showed that the effects of concave curvature, including the appearance of longitudinal roll-cells, quickly decay, and the subsequent behavior is dominated by the influence of pressure gradient and convex curvature. In view of the strong curvature effects and the alternating pressure gradients, it was perhaps surprising that the method of Bradshaw & Unsworth (1976) was in excellent agreement with the skin-friction distribution almost to the point of separation.

In the second experiment, the boundary layer developed on a sym-

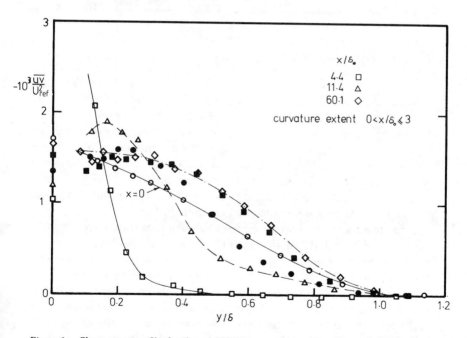

Figure 6a   Shear-stress profiles for "impulse" in convex curvature [turning angle 30° (Smits et al. 1979c)]. Curves for visual aid only.

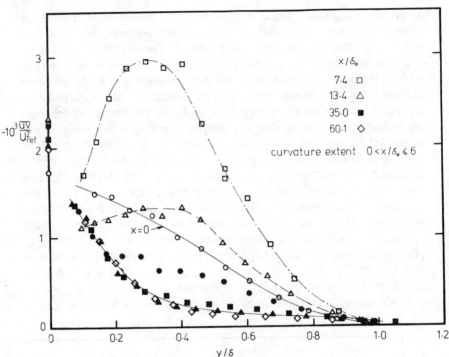

Figure 6b Shear-stress profiles for "impulse" in concave curvature [taken at a crest in skin friction, with turning angle of 30° (Smits et al. 1979c)]. Curves for visual aid only.

metrical circular-arc airfoil, and its radius of curvature and thickness-to-chord ratio were the same as the convex hill of the first experiment. The results suggest that the boundary layer on the airfoil behaves very similarly to the active, "internal" layer observed on the hill and acts as a model for this layer. Baskaran (1983) proposed that both these layers are dominated by the effects of pressure gradient rather than curvature, and he thereby provided an explanation for the observation that separation occurred at almost the same position in the two experiments.

Step changes in extra strain rates other than curvature have not been documented in such detail. For example, Smits et al. (1979a) reviewed the work on the effects of lateral divergence and concluded that, up until that time, there seemed to be no experiment designed to investigate these effects in the absence of other special effects like three-dimensionality or transverse curvature. Smits et al. studied the boundary layer on an axisymmetric "cylinder-flare" body, in which the boundary layer initially grows on a cylinder in axial flow and then passes to a conically diverging flare with an included angle of 40°. The flow experiences a step change in extra strain rates. In the blending region, both concave curvature and lateral divergence are present; the interaction between these two destabilizing extra strain rates is discussed in Section 6. Unfortunately, this experiment gives little information regarding a step change in divergence; the presence of concave curvature obscures the response to divergence alone, and only the near-asymptotic state under the continued application of divergence is relatively unambiguous.

The final extra strain rate considered in this section is bulk dilatation. By its nature, dilatation is confined to compressible flows, and in practice it is seen most clearly in flat-plate supersonic boundary layers subjected to longitudinal pressure gradients. The presence of the pressure gradient means that no direct evidence for the effect of dilatation exists, but we probably know enough about the effects of pressure gradients to extract the effects of dilatation (as long as shocks are not present). Notable experiments include those of Zwarts (1970), Peake et al. (1971), and Lewis et al. (1972). Bradshaw (1974) used these results to make the first deductions regarding the effects of dilatation, but unfortunately these experiments do not include turbulence measurements. This situation has continued until recently. Dussauge & Gaviglio (1981) used rapid-distortion theory to confirm the significance of bulk dilatation as an extra strain rate, but their experimental work on a sudden expansion was complicated by the presence of longitudinal curvature; this study is discussed in the next section, where we examine flows with multiple perturbations. Rubesin et al. (1977) have shown that at least two of the current transport equation models for compressible flow will reproduce bulk compression effects, at least qualitatively. The evidence suggests that dilatation and divergence act in a similar way. Physically, this can be expected, since both effects tend to reduce the cross-sectional area of fluid elements in a streamwise plane and thereby increase the spanwise component of vorticity (Bradshaw 1973).

6.   FLOWS WITH MULTIPLE PERTURBATIONS

Thus far, we have treated flows experiencing the application of a single kind of perturbation, either as a step change or as an impulse. In many practical configurations, however, more than one perturbation may occur. For example, a step change in wall curvature is usually accompanied by a region of nonzero pressure gradient. In a subsonic flow, the effects of a change in pressure gradient have been reasonably well documented, and in most cases we can make satisfactory allowance for these effects. In a supersonic flow, however, the application of an adverse pressure gradient is accompanied by the effects of the compressive extra strain rate div U. Another, more complex example is given by a shock-wave/boundary-layer interaction.

Here, the flow first experiences a shock wave, followed by the simultaneous effects of adverse pressure gradient, dilatation, and possibly longitudinal curvature.

In this section, we consider such complex flows. Until recently, very few well-documented experiments were available for study. Many workers were apparently deterred by the experimental difficulties as well as the difficulties associated with interpreting the results. This situation seems to be changing somewhat, and as we shall see, recent studies have revealed some interesting observations.

It would be naive to expect that the combined influence of different disturbances is given by a simple summation of their separate effects, and the available evidence bears this out. For example, Smits et al. (1979a) studied the incompressible flow over an axisymmetric cylinder-flare body in which the boundary layer developed in axial flow over a circular cylinder before diverging over a conical flare. The behavior far downstream of the cylinder-cone transition was typical of a diverging boundary layer (see Section 5). Within the transition region and for a short distance downstream, however, the flow experienced the combined effects of concave curvature and streamline divergence. Both these effects were destabilizing, yet the combined effect was considerably less than might be expected from each effect acting separately. The behavior also differed qualitatively; a concave-curvature impulse is known to produce longitudinal roll-cells through a Taylor–Görtler-like instability (Smits et al. 1979c), yet in the cylinder-flare experiment, no such roll-cells were detected. It appears that the amplification of longitudinal vorticity by concave curvature and the amplification of spanwise vorticity by streamline divergence interact nonlinearly in such a way as to prevent the formation of these roll-cells, at least in the mean.

Similar interactive effects between different strain rates were observed by Smits & Joubert (1982), who studied the flow over two bodies of revolution. For each body of revolution, the longitudinal curvature was always convex, but the lateral divergence experienced over the forebody changed to convergence over the afterbody. It was observed that the combined effects of convergence and convex curvature (both stabilizing effects) produced an overall stabilizing effect of about the expected magnitude. In contrast, the combined effects of divergence and convex curvature (the first destabilizing, the second stabilizing) appeared to produce a strong stabilizing influence that was considerably greater than that expected from the effect of convex curvature alone.

Similar examples can be drawn from two-dimensional supersonic flows. For instance, in supersonic flow, adverse pressure gradients are often produced by curving the test wall (see, for example, Sturek & Danberg

1972a,b, Laderman 1980). On a plane wall, we observe the combined effects of pressure gradient and compressive extra strain rate (as discussed in Section 5). With a curved wall, however, we have the additional effect of concave curvature present. Both extra strain rates are destabilizing, and we might expect some interesting interactive effects. Recent experiments include those by Chou & Childs (1983) and the study reported by Taylor & Smits (1984) and Jayaram & Smits (1985). The latter investigations are by far the most complete. A nominally zero pressure-gradient layer was subjected to a short region of concave curvature, followed by a flat recovery section. The total turning angle was $8°$. Measurements were made for two cases, $\delta_0/R \approx 0.1$ and $\delta_0/R \approx 0.02$, for which the corresponding lengths of curved wall were $1.4\delta_0$ and $7\delta_0$. In both cases, the shock wave formed outside the boundary layer, and the flow inside the layer experienced a near-isentropic compression. The data include mean-flow profiles as well as turbulence properties measured with normal and inclined hot wires.

The results showed a surprising similarity to those observed by Smits et al. (1979a) in the subsonic cylinder-flare experiment. For example, the mean-velocity profiles showed a characteristic dip below the log law, indicating an increase in the length scale. In addition, there was no evidence for the presence of longitudinal roll-cells. It seems possible, therefore, that the qualitative similarity between compression and divergence (both reduce the cross-sectional area of fluid elements in the streamwise plane and thereby amplify cross-stream vorticity) continues to hold even in the presence of another extra strain rate.

For both values of δ_0/R, the maximum shear stress increased by a factor of five, while $\overline{u^2}$ increased by a factor of four. Measurements of $\overline{v^2}$ and $\overline{w^2}$ were not made, but the $\overline{u^2}$ results suggest an increase in a_1 of about 20%, similar to that observed in the cylinder-flare experiment. The two curved-wall cases, however, showed a markedly different relaxation behavior. For $\delta_0/R = 0.1$, τ and $\overline{u^2}$ remained high, but for $\delta_0/R = 0.02$, τ decreased considerably while $\overline{u^2}$ did not, indicating that recovery of the turbulence structure may depend critically on the detailed nature of the disturbance.

These experiments on concavely curved walls may be contrasted with the work of Dussauge & Gaviglio (1981) (see also Dussauge 1981), who studied the behavior of a boundary layer as it passes through an expansion produced by a sudden deflection of the wall at a sharp edge. Here the flow experienced a short region of favorable pressure gradient, as well as the stabilizing effects of convex curvature and bulk dilatation. Dussauge & Gaviglio measured the longitudinal turbulence intensity and the static temperature fluctuation intensity, and they found that the interaction reduced $\overline{u^2}$ to half its upstream value. The temperature/velocity correlation coefficient R_{uT} remained constant at a value of 0.8, giving strong support to Morkovin's (1962) "strong Reynolds analogy." Furthermore, a rapid distortion analysis demonstrated reasonable agreement with the experimental results. Specifically, the dilatation appeared to be responsible for about two thirds of the observed reduction in $\overline{u^2}$, with the rapid part of the pressure-strain term contributing the remaining third. The choice for the pressure-strain term model was not crucial; three different models all gave about the same result (see also Hayakawa et al. 1983a).

The final class of flows we review are those that occur when a shock wave interacts with a boundary layer. The shock wave may be produced by an external shock-wave generator, in which case it generally interacts with a flat-plate boundary layer, or it may be produced by a compression corner where the flow is turned through some angle. In both instances, the flow tends toward separation as the shock strength increases.

The first case is usually tested in an axisymmetric configuration to avoid three-dimensional effects. The investigations by Kussoy & Horstman (1975) and Rose (1973) are two good examples. In these experiments, the boundary layer first experiences a shock wave, followed by a short region of continued compression. It is unlikely that transverse curvature had any significant effect in either experiment, and the response of the boundary layer is therefore attributable to the combined effect of the shock wave, the subsequent adverse pressure gradient, and the associated compressive extra strain rate.

In the second case, the interaction is produced by a compression corner, and the boundary layer experiences the additional effect of concave curvature. Two complete sets of data exist: one from Poitiers (Lee 1979, Ardonceau et al. 1979, Ardonceau 1984), the other from Princeton (Settles et al. 1979, Hayakawa et al. 1982, 1983b, Muck & Smits 1983, 1984). Both investigations studied the effect of shock strength, and the flows range from attached to separated. The measurements include extensive turbulence data.

Qualitatively it seems that all shock-wave/boundary-layer interactions produce similar results, regardless of how the interaction is generated. Immediately on passing the shock wave, the turbulent stresses increase sharply. Downstream, under the action of further compression and perhaps streamline curvature, the amplification continues for some distance before the flow begins to relax. The detailed behavior depends on the position within the boundary layer (as expected from the discussion in Section 3). Near the wall, the amplification is rather steep and the relaxation follows quickly. Further out in the boundary layer, the amplification is much slower and continues for a longer distance downstream (Hayakawa et al. 1983b). The level of the amplification depends on shock strength, although not all components of the Reynolds stress are affected equally. When the interaction is weak (that is, when the flow is far from separation), the ratio $\overline{uv}/\overline{u^2}$ typically increases through the shock. This behavior was predicted by Debieve et al. (1982) using a discontinuity model in the Reynolds-stress transport equation. (The analysis also agreed well with their limited experimental results.) When the interaction is strong, however, the ratio $\overline{uv}/\overline{u^2}$ decreases through the shock.

To understand the influence of the shock wave more precisely, Anyiwo & Bushnell (1982) and Zang et al. (1982) analyzed the effect of a plane shock wave on an incident turbulent field. They found that several turbulence amplification or generation mechanisms exist, including (a) "direct" amplification as a consequence of the jump condition, (b) "generation" of turbulence by incident acoustic and entropy fluctuations, (c) "focusing" caused by distortions of the shock front, and (d) direct conversion of mean-flow energy into turbulence by shock oscillation. They concluded that the first- and last-named mechanisms were probably more important than the other two.

None of these amplification mechanisms appear to be significant if there is no separation. The overall pressure rise and turning angle seem to be more important. Compare, for example, the $8°$ corner study by Hayakawa et al. (1982) and the curved-wall study by Taylor & Smits (1984) and Jayaram & Smits (1985) ($\delta_0/R = 0.1$). The incoming boundary layer, the overall pressure rise, and the total turning angle were the same in each experiment. In the curved-wall experiment, the shock formed outside the boundary layer; in the compression corner, the shock penetrated almost to the wall. Yet the results showed that the overall turbulence amplification was virtually identical in the two experiments.

When the shock is strong, however, it obviously has a major influence on the flow. Firstly, a separated region appears. Secondly, shock oscillation is observed. Zang et al. (1982) showed this to be a powerful mechanism for turbulence amplification, but Muck & Smits (1984) proposed that if the shock movement is essentially random, then the mean-flow energy is transferred more to the normal stresses than to the organized motions contributing to the shear stresses. This suggestion explains the behavior of $\overline{uv}/\overline{u^2}$ in strong interactions and implies that the unsteadiness of the shock wave generates significant "inactive" motions, such as those discussed by Townsend (1961) and Bradshaw (1967b). Thirdly, three-dimensional effects are produced. Specifically, surface flow visualization by Settles et al. (1979) and turbulence measurements by Ardonceau (1984) suggest that longitudinal roll-cells may be formed. These features may be associated with the concave streamline curvature near separation and reattachment.

There have been many attempts to calculate these very complicated flows, albeit with mixed success (see, for example, Adamson & Messiter 1980). We have already mentioned the work of Dussauge & Gaviglio (1981) and Debieve et al. (1982). Their work indicates that analyses based on the concepts of rapid or discontinuous distortion may find some application in supersonic flow; their results look very promising. More traditional approaches have not been so successful. For example, Horstman et al. (1977) and Visbal & Knight (1983) calculated the Princeton compression corner flows using simple algebraic eddy-viscosity models. Both calculations performed rather poorly, perhaps because the destabilizing extra strain rates were neglected. Unfortunately, it is difficult to see how multiple perturbations, such as those experienced in a shock-wave/boundary-layer interaction can be modeled. We know that these effects do not necessarily add linearly; what we do not know a priori is how a particular nonlinear interaction may proceed. Further work is urgently required.

7.   CONCLUDING REMARKS

Because an undisturbed boundary layer consists of a wall region and an outer flow with different time scales, the initial distribution of a perturbation, as shown by Figure 2, has a significant effect on the boundary-layer response.

The simplest possible perturbation is localized at the wall. The most common examples are step changes in roughness or wall heat flux. The boundary layer responds to these perturbations by forming an internal layer that contains all the flow affected by the new boundary condition. For large changes in roughness, local equilibrium breaks down close to the step as the advection of turbulent energy becomes important. The evidence suggests that the diffusion is also important after an impulsive change in roughness. The outer-layer response is particularly sensitive to the sign of the step. For example, if the downstream surface is rough, self-preservation is quickly re-established, whereas if the downstream surface is smooth, the "excess" shear stress propagates across the layer (as shown in Figure 4b) and decays slowly toward self-preservation. This nonlinearity was attributed partly to the behavior of the triple-product terms. The response to changes in wall heat flux also appears to be nonlinear. A proper demonstration of this result would largely invalidate the use of a turbulent Prandtl number to describe the turbulent transfer of heat and, by implication, the use of an eddy diffusivity for turbulent dispersion calculations.

Changes in the pressure gradient, which predominantly influence the wall region initially, are among the most difficult perturbations to achieve, so this area remains one of the least explored. Local equilibrium breaks down for relatively large changes, and an internal layer of sorts propagates outward. None of the experiments reviewed were able to achieve the spectacular changes found after large changes in the wall condition.

Rapid changes in extra strain rates are predominantly felt in the outer layer, but the effects are large even for small values of the extra strain rate. Again this response is nonlinear. For example, the imposition of convex curvature causes the turbulence to collapse in the outer layer, while experiments on concave curvature show the presence of longitudinal roll-cells. In addition, there appears to be an upper limit on the curvature parameter δ_0/R, beyond which little extra change is observed in the convex measurements.

A further demonstration of the nonlinearity of the response is given by flows subjected to multiple perturbations. The combined influence of different disturbances is rarely given by a simple summation of their separate effects. It is therefore insufficient to study perturbations in isolation; the nature of their interaction must also be examined.

We have not seriously attempted to judge the performance of calculation methods for these perturbations. The small proportion of our references that were considered at the two Stanford conferences (and the smaller proportion that were actually computed) demonstrates the lack of any systematic attack on perturbed boundary layers. Before any major assessment can be done, particularly in reference to the levels of complexity in turbulence models considered in Section 2, problems such as the correct specification of boundary conditions must be resolved. As shown in Section 3, this is particularly important for the calculation of step changes in roughness.

It appears that such a systematic attack may well prove useful. For example, Figures 4b and 6a show striking similarities in the outer-layer behavior of the shear stress following impulses of roughness and convex curvature. This similarity also extends to the triple products (Andreopoulos & Wood 1982). Other turbulence parameters, such as a_1, behave differently. In the simpler impulse of roughness, a_1 does not alter significantly from its self-preserving value, while the changes following an impulse of curvature are significant. This gradation in response suggests that turbulence models, whose closure forms and constants have been optimized for self-preserving flows, may be successively tested against increasingly complex perturbations to investigate the extra modeling required. We regard this possibility as the most exciting conclusion of this review.

ACKNOWLEDGMENTS

We thank Y. Andreopoulos, who helped with the initial preparation of Section 3 and suggested the use of Figure 5, and D. M. Bushnell, R. A. Antonia, H. H. Fernholz, and J.-P. Dussauge for their helpful comments.

We are also indebted to L. S. Handelman for her editorial help and to L. A. Marchesano for her patient and careful typing. One of us (AJS) was supported by NASA under Grant NAGW-240 (monitored by R. A. Graves and G. Hicks) and AFOSR under Grant 84-0061 (monitored by M. Francis) while preparing this review.

Literature Cited[a]

Adamson, T. C. Jr., Messiter, A. F. 1980. Analysis of two-dimensional interactions between shock waves and boundary layers. *Ann. Rev. Fluid Mech.* 12:103–38

Andreopoulos, J. 1983. The response of a turbulent boundary layer to a double step-change in a wall heat flux. *ASME J. Heat Transfer* 105:841–45

Andreopoulos, J., Wood, D. H. 1982. The response of a turbulent boundary layer to a short length of surface roughness. *J. Fluid Mech.* 118:143–64

Antonia, R. A., Luxton, R. E. 1971. The response of a boundary layer to a step change in surface roughness. Part 1. Smooth to rough. *J. Fluid Mech.* 48:721–61

Antonia, R. A., Luxton, R. E. 1972. The response of a boundary layer to a step change in surface roughness. Part 2. Rough to smooth. *J. Fluid Mech.* 53:737–57

Antonia, R. A., Wood, D. H. 1975. Calculation of a turbulent boundary layer downstream of a small step change in surface roughness. *Aeronaut. Q.* 26:202–10

Antonia, R. A., Danh, H. Q., Prabhu, A. 1977. Response of a turbulent boundary layer to a step change in surface heat flux. *J. Fluid Mech.* 80:153–77

Anyiwo, J. C., Bushnell, D. M. 1982. Turbulence amplification in shock-wave boundary layer interaction. *AIAA J.* 20:893–99

Ardonceau, P. L. 1984. A detailed study of the turbulent flow in a supersonic shock-wave boundary layer interaction. *AIAA J.* In press

Ardonceau, P. L., Lee, D. H., Alziary de Roquefort, T., Goethals, R. 1979. Turbulence behavior in shock wave/boundary layer interaction. *AGARD CP-271, Pap.* 8

[a] Experiments marked [S1:XXXX], [S2:XXXX], or [FF:XXXX] were considered at the 1968 Stanford Conference (Kline et al. 1969), the 1980–81 Stanford Conferences (Kline et al. 1981), and by Fernholz & Finley (1977, 1981), respectively. In each case, XXXX is the appropriate catalog number.

Bandyopadhyay, P. R. 1984. Mean flow in boundary layers disturbed to alter skin friction. *ASME J. Fluids Eng.* In press

Baskaran, V. 1983. *Turbulent flow over a curved hill.* PhD thesis. Univ Melbourne, Austral.

Berg, D. E. 1977. *Surface roughness effects on the hypersonic turbulent boundary layer.* PhD thesis. Calif. Inst. Technol., Pasadena (Univ. Microfilms 77-17260) [FF:7703]

Blom, J. 1970. Experimental determination of the turbulent Prandtl number in a developing temperature boundary layer. *Int. Heat Transfer Conf., 4th, Paris, Versailles,* Vol. 2, Pap. FC 2.2

Bradley, E. F. 1968. A micrometeorological study of velocity profiles and surface drag in the region modified by a change in surface roughness. *Q. J. R. Meteorol. Soc.* 94:361–79

Bradshaw, P. 1967a. The response of a constant-pressure turbulent boundary layer to the sudden application of an adverse pressure gradient. *ARC R&M 3575,* Aeronaut. Res. Counc., Engl. [S1:3300]

Bradshaw, P. 1967b. "Inactive" motion and pressure fluctuation in turbulent boundary layers. *J. Fluid Mech.* 30:241–58

Bradshaw, P. 1973. Effects of streamline curvature on turbulent flow. *AGARDograph 169.* NATO

Bradshaw, P. 1974. The effect of mean compression or dilatation on the turbulence structure of supersonic boundary layers. *J. Fluid Mech.* 63:449–64

Bradshaw, P. 1975. Complex turbulent flows. *ASME J. Fluids Eng.* 97:146–54

Bradshaw, P. 1976. Complex turbulent flows. *Proc. Int. Congr. Appl. Math., 14th, Delft,* ed. W. T. Koiter, pp. 101–11. Delft: North-Holland

Bradshaw, P., Ferriss, D. H. 1965. The response of a retarded equilibrium turbulent boundary layer to the sudden removal of pressure gradient. *NPL Aeronaut. Rep. 1145,* Natl. Phys. Lab., Teddington, Engl. [S1:2400]

Bradshaw, P., Ferriss, D. H. 1968a. The effect of initial conditions on the development of turbulent boundary layers. *Aeronaut. Res. Counc. CP-986*, Engl.

Bradshaw, P., Ferriss, D. H. 1968b. Calculation of boundary layer development using the turbulent energy equation. IV. Heat transfer with small temperature differences. *NPL Rep. 1271*, Natl. Phys. Lab., Teddington, Engl.

Bradshaw, P., Galea, P. V. 1967. Step-induced separation of a turbulent boundary layer in incompressible flow. *J. Fluid Mech.* 27:111–30

Bradshaw, P., Unsworth, K. 1976. Computation of complex turbulent flows. *Proc. Lockheed-Georgia Viscous Flow Symp.*, pp. 447–98. Lockheed-Georgia Co.

Bradshaw, P., Wong, F. Y. F. 1972. The reattachment and relaxation of a turbulent shear layer. *J. Fluid Mech.* 52:113–35

Bradshaw, P., Ferriss, D. H., Atwell, N. P. 1967. Calculation of boundary layer development using the turbulent energy equation. *J. Fluid Mech.* 28:593–616

Browne, L. W. B., Antonia, R. A. 1979. Calculation of a turbulent boundary layer downstream of a step change in surface temperature. *ASME J. Heat Transfer* 101:144–50

Bushnell, D. M. 1984. Body-turbulence interaction. *AIAA Pap. 84-1527*

Cantwell, B. J., Kline, S. J., eds. 1981. *Proc. 1980–81 AFOSR-HTTM-Stanford Conf. Complex Turbul. Flow: Comparison of Computation and Experiment*, Vol. 3, Thermosci. Div., Stanford Univ., Calif.

Charnay, G., Schon, J. P., Alcaraz, E., Mathieu, J. 1979. Thermal characteristics of a turbulent boundary layer with inversion of the wall heat flux. In *Turbulent Shear Flows 1*, ed. L. J. S. Bradbury et al., pp. 104–8. Berlin: Springer-Verlag

Chou, J. H., Childs, M. E. 1983. An experimental study of surface curvature effects on a supersonic turbulent boundary layer. *AIAA Pap. 83-1672*

Clauser, F. H. 1956. The turbulent boundary layer. *Adv. Appl. Mech.* 4:1–51

Coles, D. E. 1962. The turbulent boundary layer in a compressible fluid. *Rand Corp. Rep. R-403-PR*

Coles, D. E., Hirst, E. A., eds. 1969. *Proc. Comput. Turbul. Boundary Layers, 1968 AFOSR-IFP-Stanford Conf.*, Vol. 2, Thermosci. Div., Stanford Univ., Calif.

Cutler, A. D., Johnston, J. P. 1984. An experimental investigation on the effect of initial shear-stress distribution on a separating turbulent boundary layer. *AIAA Pap. 84-1583*

Deaves, D. M. 1981. Computation of wind flow over changes in surface roughness. *J. Wind Eng. Ind. Aerodyn.* 7:65–94

Debieve, J.-F., Gouin, H., Gaviglio, J. 1982. Evolution of the Reynolds stress tensor in a shock-wave-turbulence interaction. *Indian J. Technol.* 20:90–97

Dussauge, J.-P. 1981. *Evolution de transferts turbulents dans une détente rapide, en écoulement supersonique.* Thèse de Doctorat d'Etat. Univ. Aix Marseille II, Fr.

Dussauge, J.-P., Gaviglio, J. 1981. Bulk dilation effects on Reynolds stress in the rapid expansion of a turbulent boundary layer at supersonic speed. *Proc. Symp. Turbul. Shear Flows, Davis, Calif.*, 2:33–38 [S2:8632]

El Tahry, S., Gosman, A. D., Launder, B. E. 1981. On the two- and three-dimensional dispersal of a passive scalar in a turbulent boundary layer. *Int. J. Heat Mass Transfer* 24:35–46

ESDU. 1982. Strong winds in the lower atmospheric boundary layer. Part 1. Mean-hourly wind speed. *Data Item 82026*, Eng. Sci. Data Unit, London, Engl.

Fenter, F. W., Lyons, W. C. 1958. An experimental investigation of the effects of several types of surface roughness on turbulent boundary layer characteristics at supersonic speeds. *Rep. DRL-411*, Def. Res. Lab., Univ. Tex., Austin. [FF:5804]

Fernholz, H. H. 1966. Eine grenzschicht-theoretische Untersuchung optimaler Unterschalldifusoren. *Ing.-Arch.* 35:192–201

Fernholz, H. H., Finley, P. J. 1977. A critical compilation of compressible turbulent boundary layer data. *AGARDograph 223.* NATO

Fernholz, H. H., Finley, P. J. 1981. A further compilation of compressible boundary layer data with a survey of turbulence data. *AGARDograph 263.* NATO

Fulachier, L. 1972. *Contribution à l'étude des analogies des champs dynamique et thermique dans une couche limite turbuleuse. Effet de l'aspiration.* Thèse Docteur des Sciences. Univ. Provence, Fr.

Galmes, J. M., Dussauge, J. P., DeKeyser, I. 1983. Supersonic turbulent boundary layers subjected to a pressure gradient: calculation with a k-ε model. *J. Méc. Théor. Appl.* 2:539–58

Gibson, M. M., Jones, W. P., Younis, B. A. 1981. Calculation of turbulent boundary layers on curved surfaces. *Phys. Fluids* 24:386–95

Gillis, J. C., Johnston, J. P. 1983. Turbulent boundary-layer flow and structure on a convex wall and its redevelopment on a flat wall. *J. Fluid Mech.* 135:123–53 [S2:0233]

Gillis, J. C., Johnston, J. P., Kays, W. M., Moffatt, R. J. 1980. Turbulent boundary layer on a convex, curved surface.

356   SMITS & WOOD

Rep. HMT-31, Thermosci. Div., Stanford Univ., Calif. [S2:0233]

Goldberg, P. 1966. Upstream history and apparent stress in turbulent boundary layers. *MIT Gas Turbine Lab. Rep. 85*, Mass. Inst. Technol., Cambridge

Hanjalić, K., Launder, B. E. 1980. Sensitizing the dissipation equation to irrotational strains. *ASME J. Fluids Eng.* 102:34–40

Hayakawa, K., Smits, A. J., Bogdonoff, S. M. 1982. Hot-wire investigation of an unseparated shock-wave/turbulent boundary layer interaction. *AIAA J.* 22:579–85

Hayakawa, K., Smits, A. J., Bogdonoff, S. M. 1983a. Turbulence measurements in a compressible reattaching shear layer. *AIAA J.* 22:889–95

Hayakawa, K., Muck, K. C., Smits, A. J., Bogdonoff, S. M. 1983b. The evolution of turbulence in shock-wave/boundary layer interactions. *Proc. Australasian Fluid Mech. Conf., 8th, Newcastle, Austral.*

Hefner, J. N., Anders, J. B., Bushnell, D. M. 1983. Alteration of outer flow structures for turbulent drag reduction. *AIAA Pap. 83-0293*

Hoffmann, P. H., Bradshaw, P. 1978. Turbulent boundary layers on surfaces of mild longitudinal curvature. *IC Aero Rep. 78-04*, Imperial Coll., London, Engl. [S2:0231/0232]

Hoffmann, P. H., Muck, K. C., Bradshaw, P. 1984. The effect of concave surface curvature on turbulent boundary layers. *J. Fluid Mech.* In press

Horstman, C. C., Settles, G. S., Vas, I. E., Bogdonoff, S. M. 1977. Reynolds number effects on shock-wave turbulent boundary-layer interactions. *AIAA J.* 15:1152–58

Hunt, J. C. R. 1977. A review of the theory of rapidly distorted turbulent flows and its applications. *Proc. Bienn. Fluid Dyn. Symp., 13th, Warsaw, Pol.*, pp. 121–52

Jackson, N. A. 1976. The propagation of modified flow downstream of a change in roughness. *Q. J. R. Meteorol. Soc.* 102:924–33

Jayaram, M., Smits, A. J. 1985. *The distortion of a supersonic turbulent boundary layer by bulk compression and surface curvature.* To be presented at AIAA Aerosp. Sci. Meet., 23rd, Reno, Nev.

Johnson, D. S. 1957. Velocity, temperature and heat-transfer measurements in a turbulent boundary layer downstream of a stepwise discontinuity in wall temperature. *ASME J. Appl. Mech.* 24:2–8

Johnson, D. S. 1959. Velocity and temperature fluctuation measurements in a turbulent boundary layer downstream of a stepwise discontinuity in wall temperature. *ASME J. Appl. Mech.* 26:325–36

Kays, W. M. 1966. *Convective Heat and Mass*

Transfer, p. 244. New York: McGraw-Hill. 387 pp.

Kline, S. J., Morkovin, M. V., Sovran, G., Cockrell, D. G., eds. 1969. *Proc. Comput. Turbul. Boundary Layers, 1968 AFOSR-IFP-Stanford Conf.*, Vol. 1, Thermosci. Div., Stanford Univ., Calif.

Kline, S. J., Cantwell, B. J., Lilley, G. M., eds. 1981. *Proc. 1980–81 AFOSR-HTTM-Stanford Conf. Complex Turbul. Flow: Comparison of Computation and Experiment*, Vol. 1, Thermosci. Div., Stanford Univ., Calif.

Kussoy, M. I., Horstman, C. C. 1975. An experimental documentation of a hypersonic shock-wave turbulent boundary layer interaction flow—with and without separation. *NASA TMX 62412* [S2:8651; FF:7501]

Laderman, A. J. 1980. Adverse pressure gradient effects on supersonic boundary-layer turbulence. *AIAA J.* 18:1186–95 [FF:7803]

Launder, B. E., Samaraweera, D. S. 1979. Application of a second moment turbulence closure to heat and mass transport. Part 1. Two-dimensional flow. *Int. J. Heat Mass Transfer* 22:1631–43

Lee, D. H. 1979. *Étude de l'évolution de la turbulence dans une interaction onde de choc-couche limite.* Thèse. Univ. Poitiers, Fr.

Lewis, J. E., Gran, R. L., Kubota, T. 1972. An experiment on the adiabatic compressible turbulent boundary layer in adverse and favorable pressure gradients. *J. Fluid Mech.* 51:657–72 [S2:8402; FF:7201]

Meroney, R. M., Bradshaw, P. 1975. Turbulent boundary layer growth over a longitudinally curved surface. *AIAA J.* 13:1448–53

Millikan, C. B. 1938. A critical discussion of turbulent flows in channels and circular tubes. *Proc. Congr. Appl. Mech., 5th*, pp. 386–92

Morkovin, M. V. 1962. Effects of compressibility on turbulent flows. In *Mécanique de la Turbulence*, ed. A. Favre, pp. 367–80. Paris: CNRS

Muck, K. C. 1982. *Turbulent boundary layers on mildly curved surfaces.* PhD thesis. Univ. London, Engl.

Muck, K. C., Smits, A. J. 1983. The behavior of a compressible turbulent boundary layer under incipient separation conditions. *Proc. Turbul. Shear Flow Conf., 4th, Karlsruhe*, pp. 215–20

Muck, K. C., Smits, A. J. 1984. Behavior of a turbulent boundary layer subjected to a shock induced separation. *AIAA Pap. 84-0097*

Muck, K. C., Hoffman, P. H., Bradshaw, P. 1984. The effect of convex surface curvature on turbulent boundary layers. *J. Fluid Mech.* In press

Mulhearn, P. J. 1977. Relations between surface fluxes and ocean profiles of velocity, temperature and concentration, downwind of a change in surface roughness. *Q. J. R. Meteorol. Soc.* 103:785–802

Narasimha, R., Sreenivasan, K. R. 1979. Relaminarization of fluid flows. *Adv. Appl. Mech.* 19:221–309

Panofsky, H. A., Townsend, A. A. 1964. Change of terrain roughness and the wind profile. *Q. J. R. Meteorol. Soc.* 90:147–55

Patankar, S. V., Spalding, D. B. 1970. *Heat and Mass Transfer in Boundary Layers.* London: Intertext Books. 225 pp. 2nd ed.

Peake, D. J., Brakman, G., Romeskie, J. M. 1971. Comparisons between some high Reynolds number turbulent boundary layer experiments at Mach 4 and various recent calculation procedures. *AGARD CP-93-71, Pap. 11* [S2:8401; FF:7102]

Perry, A. E., Schofield, W. H., Joubert, P. N. 1969. Rough wall turbulent boundary layers. *J. Fluid Mech.* 22:285–304

Poreh, M., Cermak, J. E. 1964. Study of diffusion from a line source in a turbulent boundary layer. *Int. J. Heat Mass Transfer* 7:1083–95

Ramaprian, B. R., Shivaprasad, B. G. 1978. The structure of turbulent boundary layers along mildly curved surfaces. *J. Fluid Mech.* 85:273–303

Rao, K. S., Wyngaard, J. C., Cote, O. R. 1974. The structure of the two dimensional internal boundary layer over a sudden change in surface roughness. *J. Atmos. Sci.* 31:738–46

Reynolds, A. J. 1975. The prediction of turbulent Prandtl and Schmidt numbers. *Int. J. Heat Mass Transfer* 18:1055–69

Reynolds, W. C. 1976. Computation of turbulent flows. *Ann. Rev. Fluid Mech.* 8:183–208

Rodi, W. 1981. Examples of turbulence models for incompressible flows. *AIAA J.* 20:872–79

Rose, W. C. 1973. The behavior of a compressible turbulent boundary layer in a shock-wave induced adverse pressure gradient. *NASA TN D-7092* [FF:7306]

Rubesin, M. W., Crisalli, A. J., Horstman, C. C., Acharya, M. 1977. A critique of some recent second order closure models for compressible boundary layers. *AIAA Pap.* 77-128

Samuel, A. E., Joubert, P. N. 1974. A boundary layer developing in an increasingly adverse pressure gradient. *J. Fluid Mech.* 66:481–505 [S2:0141]

Schofield, W. H. 1973. The effects of sudden discontinuities on turbulent boundary layer development. *Rep. ARL/ME-139*,

Aeronaut. Res. Lab., Melbourne, Austral.

Schofield, W. H. 1975. Measurements in adverse-pressure-gradient turbulent boundary layer with a step change in surface roughness. *J. Fluid Mech.* 70:573–93

Schofield, W. H. 1981. Turbulent shear flow over a step change in surface roughness. *ASME J. Fluids Eng.* 103:344–51

Settles, G. S., Fitzpatrick, T. J., Bogdonoff, S. M. 1979. Detailed study of attached and separated compression corner flowfields in high Reynolds number supersonic flow. *AIAA J.* 17:579–85 [S2:8631]

Shlien, D. J., Corrsin, S. 1976. Dispersion measurements in a turbulent boundary layer. *Int. J. Heat Mass Transfer* 19:285–95

Simpson, R. L. 1971. The effect of a discontinuity in wall blowing on the turbulent incompressible boundary layer. *Int. J. Heat Mass Transfer* 14:2083–97

Simpson, R. L., Strickland, J. H., Barr, P. W. 1977. Features of a separating turbulent boundary layer in the vicinity of separation. *J. Fluid Mech.* 79:553–94

Simpson, R. L., Chew, Y.-T., Shivaprasad, B. G. 1981. The structure of a separating turbulent boundary layer. Parts 1 and 2. *J. Fluid Mech.* 113:23–51, 53–73 [S2:0431]

Smits, A. J., Joubert, P. N. 1982. Turbulent boundary layers on bodies of revolution. *J. Ship Res.* 26:135–47

Smits, A. J., Eaton, J. A., Bradshaw, P. 1979a. The response of a turbulent boundary layer to lateral divergence. *J. Fluid Mech.* 94:243–68

Smits, A. J., Young, S. T. B., Bradshaw, P. 1979b. The effect of short regions of high surface curvature on turbulent boundary layers. *IC Aero TN 78-104*, Imperial Coll., London, Engl.

Smits, A. J., Young, S. T. B., Bradshaw, P. 1979c. The effect of short regions of high surface curvature on turbulent boundary layers. *J. Fluid Mech.* 94:209–42 [S2:0235]

Smits, A. J., Baskaran, V., Joubert, P. N. 1981. Measurements in a turbulent boundary layer flow over a two-dimensional hill. *Proc. Symp. Turbul. Shear Flows, 3rd, Davis, Calif.*

So, R. M. C., Mellor, G. L. 1972. An experimental investigation of turbulent boundary layers along curved surfaces. *NASA CR-1940*

So, R. M. C., Mellor, G. L. 1973. Experiment on convex curvature effects in turbulent boundary layers. *J. Fluid Mech.* 60:43–62

So, R. M. C., Mellor, G. L. 1975. Experiment on turbulent boundary layers on a concave wall. *Aeronaut. Q.* 16:25–40

Squire, L. C., Thomas, G. D., Marriott, P. G. 1977. Compressible turbulent boundary

358   SMITS & WOOD

layers with injection. *AIAA J.* 15:425–27

Stratford, B. S. 1959. The prediction of separation of the turbulent boundary layer. *J. Fluid Mech.* 5:1–16

Sturek, W. B., Danberg, J. E. 1972a. Supersonic turbulent boundary layer in adverse pressure gradient. Part I: The experiment. *AIAA J.* 10:475–80 [FF:7101]

Sturek, W. B., Danberg, J. E. 1972b. Supersonic turbulent boundary layer in adverse pressure gradient. Part II: Data analysis. *AIAA J.* 10:630–35 [FF:7101]

Subramanian, C. S., Antonia, R. A. 1981. Response of a turbulent boundary layer to a sudden decrease in wall flux. *Int. J. Heat Mass Transfer* 24:1641–47

Tani, I. 1968. Review of some experimental results on the response of a turbulent boundary layer to sudden perturbations. In Kline et al. 1969, pp. 483–94

Taylor, M., Smits, A. J. 1984. The effects of a short region of concave curvature on a supersonic turbulent boundary layer. *AIAA Pap. 84-0169*

Thomann, H. 1968. Effect of streamwise wall curvature on heat transfer in a turbulent boundary layer. *J. Fluid Mech.* 33:281–92 [FF:6800]

Townsend, A. A. 1961. Equilibrium layers and wall turbulence. *J. Fluid Mech.* 11:97–120

Townsend, A. A. 1962. The behavior of a turbulent boundary layer near separation. *J. Fluid Mech.* 12:536–54

Townsend, A. A. 1965. The response of a turbulent boundary layer to abrupt changes in surface conditions. *J. Fluid Mech.* 22:799–822

Townsend, A. A. 1976. *The Structure of Turbulent Shear Flow.* Cambridge: Cambridge Univ. Press. 429 pp. 2nd ed.

Van Driest, E. R. 1951. Turbulent boundary layer in compressible fluids. *J. Aeronaut. Sci.* 18:145–60

Visbal, M., Knight, D. 1983. Evaluation of the Baldwin-Lomax turbulence model for two-dimensional shock wave boundary layer interactions. *AIAA Pap. 83-1697*

Wassel, A. T., Catton, I. 1977. Diffusion from a line source in a neutral or stably stratified atmospheric surface layer. *Int. J. Heat Mass Transfer* 20:383–91

Wood, D. H. 1978. Calculation of the neutral wind profile following a large step change in surface roughness. *Q. J. R. Meteorol. Soc.* 104:383–92

Wood, D. H. 1981. The growth of the internal layer following a step change in surface roughness. *Rep. TN-FM 57*, Dept. Mech. Eng., Univ. Newcastle, Austral.

Wood, D. H. 1982. Internal boundary layer growth following a step change in surface roughness. *Boundary-Layer Meteorol.* 22:241–44

Zang, T. A., Hussaini, M. Y., Bushnell, D. M. 1982. Numerical computations of turbulence amplification in shock wave interactions. *AIAA Pap. 82-0293*

Zwarts, F. 1970. *Compressible turbulent boundary layers.* PhD thesis. McGill Univ., Montreal, Can. [S2:8411; FF:7007]

Ann. Rev. Fluid Mech. 1985. 17 : 359–409
Copyright © 1985 by Annual Reviews Inc. All rights reserved

MODELING EQUATORIAL OCEAN CIRCULATION

Julian P. McCreary, Jr.

Nova University Oceanographic Center, Dania, Florida 33004

1. INTRODUCTION

The field of equatorial oceanography has grown tremendously in the past decade. One reason for this growth has been the discovery of a rich variety of phenomena in the region. Another is the realization that sea-surface temperature (SST) anomalies in the equatorial oceans have a significant effect on global climate. Not surprisingly, a large number of models have been developed to study equatorial dynamics.

1.1 Observations

Since the focus of this paper is on the modeling of equatorial circulation, observational studies are not reviewed extensively here. Rather, this subsection briefly surveys some observations that have provided the stimulus for the development of equatorial theory. Two recent reviews discuss observations in greater detail. McCreary et al. (1981) is a collection of short articles on the subject, and Knox & Anderson (1984) is a comprehensive review of phenomena from all three tropical oceans.

EQUATORIAL CURRENTS   The Equatorial Undercurrent is a geostrophically balanced, subsurface, eastward jet that lies in the pycnocline just beneath the surface mixed layer. Since its rediscovery in the early 1950s, it has been detected on almost every cruise to the equatorial Atlantic and Pacific Oceans; the sole exception occurred during the 1982/83 El Niño, when the flow in the upper ocean in the central Pacific was westward for a period of about one month (Firing et al. 1983). Philander (1973) reviews early observations of the Undercurrent, while Firing et al. (1981) report a recent set of observations of the Pacific Equatorial Undercurrent taken for a period of time lasting longer than one year.

On several occasions an ageostrophic equatorial undercurrent, distinct

359

0066–4189/85/0115–0359$02.00

from the deeper one, has been observed in the thick surface layer of the western Pacific (Hisard et al. 1970, Donguy et al. 1984). On these occasions the surface-layer flow is not at all depth-independent; in the upper part of the layer there is westward drift driven by the prevailing trade winds, whereas in the lower part the flow is eastward. These currents respond rapidly to changes in the wind. For example, Hisard et al. reported that in April 1967 the trade winds suddenly reversed for a period of time. After eight days of westerly winds the sense of the surface-layer velocity shear was completely reversed, with eastward flow at the surface and westward flow at depth. The deeper, geostrophically balanced undercurrent was not significantly affected by the wind change, indicating that the dynamics of the two undercurrents are considerably different.

Two other equatorial currents are the subsurface countercurrents. These currents are geostrophically balanced, eastward flows located nearly symmetrically about the equator at depths somewhat greater than that of the Equatorial Undercurrent. At some locations they are entirely separate from the Undercurrent, but at others they are contiguous with it. They are appreciable flows, attaining speeds and volume transports that are typically 20 cm s^{-1} and 5–10 sverdrups, respectively. Tsuchiya (1972, 1975) noted their existence in the Pacific Ocean, and Cochrane et al. (1979) found them in the Atlantic Ocean as well. McPhaden (1984) briefly reviews these papers and discusses other observations of subsurface countercurrents.

EQUATORIALLY TRAPPED AND UNSTABLE WAVES Variability in the equatorial oceans is often interpreted as being due to linear, equatorially trapped waves. For example, Wunsch & Gill (1976) pointed out that sealevel variability at periods of about 10 days appears to be caused by resonant, equatorial gravity waves. Duing & Hallock (1979) and Hallock (1979) suggested that meanders of the Atlantic Equatorial Undercurrent with a period of about 16 days are wind-generated, Rossby-gravity waves. Knox & Halpern (1982) reported a pulse of currents and sea level that propagated eastward across the Pacific Ocean with a speed approaching 3 m s^{-1}, and they suggested that the event was an equatorial Kelvin wave.

Luyten & Swallow (1976) first reported the existence of deep currents throughout the water column in the western Indian Ocean. Similar deep currents have since been observed in the Pacific and Atlantic Oceans as well. These currents are characterized by a broad spectrum of vertical wavelengths, have time scales of one month and longer, and are trapped within a few degrees of the equator. Linear, equatorial wave theory has been used with some success to interpret various aspects of these observations (see, for example, Eriksen 1981). Particularly interesting is the presence of vertically propagating signals in the data. Weisberg & Horigan (1981) in the

Atlantic Ocean and O'Neill (1982) in the Indian Ocean interpreted variability at periods of 1–2 months to be due to vertically propagating Rossby-gravity waves. Luyten & Roemmich (1982) noted upward phase propagation of current variability at the semiannual period in the western Indian Ocean and suggested that this signal was likely to be a Rossby wave. Lukas & Firing (1984) presented evidence for a vertically propagating Rossby wave at the annual period in the Pacific Ocean.

Legeckis (1977) and Legeckis et al. (1983) reported the presence of cusplike waves located a few degrees north of the equator in the eastern and central Pacific Ocean. The waves were visibly apparent in satellite images of SST. They propagated westward and had a wavelength of about 1000 km and a period of about one month. There is no evidence of a spectral peak in the wind field with a similar wave number and frequency. It is likely, then, that this signal is an unstable wave, as suggested by Philander (1978b).

REMOTE FORCING   A remarkable property of the equatorial oceans is that wind fluctuations in one part of the basin can quickly affect the state of the ocean at locations thousands of kilometers away. A well-known example of a remotely forced event is the appearance of warm SST anomalies in the eastern Pacific during El Niño. Another example is the annual cycle of ocean circulation in the eastern Atlantic.

Bjerknes (1966, 1969) first noted that remote winds at Canton Island were highly correlated with SST off Peru at the time scales associated with El Niño. It is now clear that El Niño is one aspect of changes in ocean circulation thoughout the Pacific Ocean and of changes in the atmosphere on a global scale. Rasmussen & Carpenter (1982) and Wyrtki (1977, 1979, 1984) document the changes in SST, sea level, and wind stress that occur in the tropical Pacific during El Niño events, while Horel & Wallace (1981) summarize the global changes that occur in the atmosphere. The collection of papers in Witte (1983) discusses events that took place during the 1982/83 El Niño.

During the Northern Hemisphere summer, SST in the eastern Atlantic cools by about 5°C along the equator and the west coast of Africa. This cooling is particularly interesting because the annual cycle of the local winds is weak in the region. Moore et al. (1978) first suggested that the annual cooling might be remotely forced by the strong annual cycle of the equatorial zonal wind field in the western Atlantic. In support of this idea, Servain et al. (1982) demonstrated that SST in the eastern Atlantic is highly correlated with winds from the western Atlantic, but it is much less correlated with the local winds. Picaut (1983) discussed evidence that the upwelling signal along the coast of Africa is a remotely forced, propagating wave.

1.2   Theories

This paper reviews the models that have been used to study phenomena like those discussed above. The purpose of the paper is twofold: to present the basic principles of equatorial dynamics as succinctly as possible, and to survey recent developments in the field. It is hoped that those readers who are not familiar with the subject will find this paper to be a useful introduction, and that readers who are already familiar with the topic will find it to be a convenient reference source. Section 2 organizes the various models according to dynamical sophistication. Sections 3–5 discuss unforced solutions, solutions forced by switched-on winds, and solutions forced by periodic winds, respectively. Finally, Section 6 outlines the directions of current research in the study of El Niño, a subject of considerable interest at the present time. As we shall see, linear models are extremely useful for identifying important processes involved in equatorial dynamics. For this reason, some linear solutions are written down explicitly and discussed at length. Linear solutions are compared with nonlinear ones whenever possible.

A number of other review articles of equatorial models already exist to supplement this one. O'Brien (1979), Leetmaa et al. (1981), Cane & Sarachik (1983), and Knox & Anderson (1984) all present useful introductions to the subject. McCreary (1980) discusses the mathematical techniques used for finding analytic solutions to a variety of equatorial problems, and O'Brien et al. (1981) review models of El Niño. Gill (1983a) devotes a chapter of his book to the equatorial dynamics of the atmosphere and the ocean. Finally, the collection of articles in Nihoul (1985) is a good introduction to the new field of coupled ocean-atmosphere models.

2.   EQUATIONS OF MOTION

A wide variety. of ocean models have been used to study equatorial circulation. They range in dynamical complexity from simple, linear, surface-layer models to sophisticated, nonlinear, continuously stratified models. The sophisticated models are remarkably successful at producing realistic solutions. The simple models have proven to be invaluable tools for isolating important physical processes at the equator, and they have also produced quite realistic solutions. It is fair to say that our understanding of equatorial dynamics has progressed through a careful comparison of the solutions to models of all types. This section describes commonly used models and organizes them, as much as is possible, according to increasing sophistication.

2.1   Surface-Layer Models

The density structure of the equatorial ocean often consists of a well-mixed surface layer of constant density above a sharp pycnocline. Such a structure is almost always present in the western Pacific and Atlantic Oceans, where the mixed layer is 100–150 m deep. A number of equatorial models consider the ocean circulation in just this surface layer alone. Their underlying assumption is that the surface layer is decoupled from the deep ocean by the sharp pycnocline.

CONSTANT-THICKNESS MODELS   Some of the earliest equatorial models assume the surface layer has a constant thickness H. A simple set of equations describing the flow in such a layer is the linear set

$$u_t - fv + p_x = (vu_z)_z + v_h\nabla^2 u,$$

$$v_t + fu + p_y = (vv_z)_z + v_h\nabla^2 v, \tag{1}$$

$$u_x + v_y + w_z = 0, \qquad p_z = 0,$$

where u, v, and w are zonal, meridional, and vertical velocities, respectively; p is the pressure; f is the Coriolis parameter; and v_h and v are coefficients of horizontal and vertical eddy viscosity, respectively. Boundary conditions at the ocean surface are

$$vu_z = \tau^x, \qquad vv_z = \tau^y, \qquad w = 0 \quad \text{at} \quad z = 0, \tag{2}$$

and at the bottom of the layer are

$$u_z = v_z = w = 0 \quad \text{at} \quad z = -H, \tag{3}$$

where τ^x and τ^y are zonal and meridional components of surface wind stress. The bottom conditions ensure that the deep ocean does not influence the flow in the surface layer in any way.

It is possible to solve for the depth-averaged flow $(\bar{u}, \bar{v})$ independently from the shear flow $(u' = u - \bar{u}, v' = v - \bar{v})$. Integrating (1) over the water column and dividing by H gives

$$\bar{u}_t - f\bar{v} + p_x = \tau^x/H + v_h\nabla^2\bar{u},$$

$$\bar{v}_t + f\bar{u} + p_y = \tau^y/H + v_h\nabla^2\bar{v}, \tag{4}$$

$$\bar{u}_x + \bar{v}_y = 0.$$

The difference between Equations (4) and (1) gives the set that describes the shear flow,

$$u'_t - fv' = -\tau^x/H + (vu'_z)_z + v_h\nabla^2 u',$$

$$v'_t + fu' = -\tau^y/H + (vv'_z)_z + v_h\nabla^2 v', \tag{5}$$

$$u'_x + v'_y + w_z = 0,$$

and Equations (5) are solved subject to boundary conditions (2) and (3) with u and v replaced by u' and v'. Note that the two flow components are entirely separate from each other; that is, there are no terms in (4) and (5) that couple the components together. This convenient separation is possible only because the system neglects bottom stress and nonlinear terms.

The interesting and instructive property of this system is that the dynamics of each component are very different. Equations (4) are just the equations describing a barotropic ocean, and Equations (5) do not involve any pressure gradients. Each component responds very differently to the wind. For example, in response to a switched-on wind, the depth-averaged flow adjusts to a nonlocal Sverdrup balance by radiating barotropic waves, whereas the shear flow rapidly adjusts to a completely local balance (see the discussion in Section 4.1).

Various sets of equations similar to (1) have been studied. Stommel (1960) found steady-state solutions to (5) without horizontal mixing (see the discussion of Figure 2). Gill (1971) and McKee (1973) later included horizontal mixing and found solutions in a parameter range where horizontal mixing was a dominant process. Charney (1960) and Charney & Spiegel (1971) adopted no-slip boundary conditions (that is, $u = v = 0$ at $z = -H$), thereby allowing the deep ocean to have a strong influence (unrealistically strong) on the surface layer. Both of these latter studies also added nonlinear terms to Equations (1), but otherwise they simplified the system by reducing it to two dimensions in y and z, retaining only the effects of p_x by specifying it externally to be τ^x/H.

These models are useful in spite of their dynamical simplicity. One reason is that they are able to simulate features of equatorial flows. As noted in the Introduction, very strong shear flows do exist in the constant-density surface mixed layer near the equator. These currents respond rapidly to the wind, and they often fluctuate independently from currents in the deeper ocean. Another reason is that, as we shall see, shear flows similar to (5) also appear as parts of more sophisticated reduced-gravity and continuously stratified models.

REDUCED-GRAVITY MODELS   A prominent feature of the equatorial ocean is that the sharp, near-surface pycnocline can move vertically a considerable distance in a short period of time (of the order of 100 m or more in only a few months). Constant-thickness models are not capable of describing this movement. Reduced-gravity models are a direct extension of the constant-thickness models that do allow the thickness of the layer to vary. They assume that the ocean consists of a thin surface layer of density ρ overlying an infinitely deep lower layer of density $\rho + \Delta\rho$. The depth of the interface between the layers simulates the movement of the ocean pycnocline.

An example of a linear reduced-gravity system is just Equations (1)–(3), except that in (3)

$$w = -h_t \quad \text{at} \quad z = -H, \tag{6}$$

where h is the instantaneous thickness of the surface layer. A final equation, relating h to pressure, is

$$h = H + p/g', \tag{7}$$

where $g' = (\Delta\rho/\rho)g$; this relation follows from the assumption that there are no pressure gradients in the deep ocean. It is only in (7), through its dependence on $\Delta\rho$, that the deep ocean influences the layer.

Again it is possible to solve for the depth-averaged flow in the layer separately from the shear flow. The depth-independent equations are

$$\begin{aligned}
u_t - fv + p_x &= \tau^x/H + v_h\nabla^2 u, \\
v_t + fu + p_y &= \tau^y/H + v_h\nabla^2 v, \\
h_t/H + u_x + v_y &= 0,
\end{aligned} \tag{8}$$

where overbars have been dropped for convenience. The shear-flow equations are Equations (5) with the replacement

$$u'_x + v'_y + w_z = h_t/H. \tag{9}$$

The shear-flow component of reduced-gravity models is almost always ignored, and Equations (8) by themselves are referred to as the reduced-gravity equations. This neglect amounts to assuming that v is infinite in the surface layer. [For example, (34) is a solution to (5) for a switched-on wind. Note that u' and v' in (34) vanish as v tends to infinity.]

The modification (9) does not affect the shear flow, u' and v', at all; only w is changed by an additional part that varies linearly from zero at the ocean surface to $-h_t$ at $z = -H$. The presence of the term h_t/H in (8), however, changes the dynamics of the depth-independent component significantly. As a result, Equations (8) are essentially the same as Equations (20) describing one of the baroclinic modes of a continuously stratified ocean, where the separation constant c_n is given by $(g'H)^{1/2}$.

Several studies include nonlinear terms in reduced-gravity models. Often nonlinear terms involving only depth-independent quantities are added to the depth-independent equations (8). These terms generally do not affect solutions very much, primarily because $\bar{u}$ and $\bar{v}$ are typically not sufficiently large. Cane (1979a,b) developed a nonlinear, reduced-gravity system that included all the nonlinear terms. His model is a nonlinear version of the reduced-gravity system described above, except that it is formulated with only two degrees of freedom in the vertical. (The surface layer is divided into two sublayers. The upper sublayer has a constant thickness and is directly acted upon by the wind.) The nonlinearities affect the solutions significantly in this model, because they involve both the shear and depth-averaged components of the flow field (see the discussion in Section 4.1).

THERMODYNAMIC, MIXED-LAYER MODELS   Large and persistent SST anomalies often occur in the tropical oceans, a well-known example being the El Niño phenomenon in the Pacific Ocean. The models discussed above are not capable of studying SST anomalies, since the temperature of the layer is fixed. Several reduced-gravity models have recently been developed to overcome this deficiency. They include active thermodynamics, thereby allowing the temperature of the layer to vary. They also allow various processes (the terms, w_e and $\bar{w}$, defined below) to change the mass of the layer.

The thermodynamic model of Anderson & McCreary (1984) is a nonlinear extension of Equations (8) that also includes an additional equation for the heat content of the layer. The equations governing the mass and heat content of the layer are

$$h_t + (uh)_x + (vh)_y = w_e - \bar{w},$$

$$(hT)_t + (uhT)_x + (vhT)_y = \frac{Q_s}{\rho c_p} + w_e T_e - \bar{w} T, \tag{10}$$

where w_e is an entrainment velocity given by

$$h\Delta T w_e = 2\delta - h\frac{Q_s}{\rho c_p}, \tag{11}$$

T is the temperature of the layer, T_e is the constant temperature of the deep ocean, and $\Delta T = T - T_e$. The system involves three thermodynamic processes: Q_s, δ, and $\bar{w}$, where Q_s is a heat flux into the layer and is given by

$$Q_s = -\rho c_p \gamma (T - T^*), \tag{12}$$

δ is the rate of potential energy increase due to mechanical stirring by the wind, and $\bar{w}$ is a slow upwelling velocity in the deep ocean that is presumably driven by the background thermohaline circulation. All three processes are required in order for the layer to maintain itself: Q_s is necessary to keep the surface layer warmer than the deep ocean, δ always acts to increase the thickness of the layer by forcing entrainment of fluid from the deep ocean, and $\bar{w}$ prevents the layer thickness from increasing indefinitely by continually removing mass from the layer.

Schopf & Cane (1983) developed a thermodynamic, reduced-gravity model involving two active layers, each of which is described by a set of equations similar to (8) and (10). (The model is actually a bit more complicated than this, because the temperature in the second layer is assumed to vary linearly with depth; this extra degree of freedom requires an additional equation to determine the temperature at the base of the second layer.) Their system allows mass and heat to mix between the layers in a manner similar to the Anderson & McCreary model. The equations for the mass and heat content in the upper layer of their model are

$$h_t + (uh)_x + (vh)_y = w_e,$$

$$(hT)_t + (uhT)_x + (vhT)_y = \frac{Q_s}{\rho c_p} + w_e\{\theta(w_e)T_e + \theta(-w_e)T\}, \tag{13}$$

where w_e is determined by the expression

$$h\Delta T w_e \theta(w_e) = 2\delta - h\frac{Q_s}{\rho c_p} - 2\varepsilon_0 h, \tag{14}$$

where ε_0 is a dissipation term for potential energy, T_e is now the temperature at the top of the second layer and is not constant, and $\theta(\xi)$ is a step function.

It is instructive to contrast the mixing processes involved in these two models. The way each of them develops and maintains SST anomalies depends crucially on the parameterization of these processes. Some of the similarities and differences are evident from a direct comparison of Equations (10) and (11) with (13) and (14). For example, during entrainment ($w_e > 0$) the two systems are quite similar, with $\bar{w}$ and $\varepsilon_0 h$ playing analogous roles. During detrainment, however, they differ considerably because of the presence of step functions in (13) and (14). In addition, the two models have quite different states of rest, states in which the right-hand sides of (10) and (13) are set equal to zero.

2.2   Linear, Continuously Stratified Models

Layer models are limited because equatorial currents are not entirely confined to the surface mixed layer(s) of the ocean. A logical extension of Equations (1) is to include the effects of continuous stratification and to solve for the flow field throughout the water column. The following set of linear equations is such an extension. Under the conditions specified below, it is possible to find analytic solutions to this set. Primarily for this reason, the solutions of this system provide a foundation for much of our understanding of equatorial dynamics.

In a state of no motion, the model ocean has a stably stratified background density structure $\rho_b(z)$ and an associated Väisälä frequency

$N_b(z)$. A set of equations linearized about this background state is

$$u_t - fv + p_x = F - ru + (vu_z)_z + v_h \nabla^2 u,$$

$$v_t + fu + p_y = G - rv + (vv_z)_z + v_h \nabla^2 v,$$

$$u_x + v_y + w_z = 0, \qquad p_z = -\rho g, \tag{15}$$

$$\rho_t - g^{-1} N_b^2 w = (\kappa \rho_z)_z - \varepsilon \rho,$$

where the terms with the coefficients r and ε are Rayleigh friction and Newtonian cooling, respectively. Surface and bottom boundary conditions are again (2) and (3), with conditions (3) now being applied at the ocean bottom at $z = -D$. In addition, since the system includes the vertical mixing of heat, two conditions involving ρ must be applied at the ocean surface and bottom. In many calculations, wind stress is assumed to enter the ocean as a body force, F and G, with the separable form

$$F = \tau^x(x, y, t)Z(z), \qquad G = \tau^y(x, y, t)Z(z), \tag{16}$$

where $\int_{-D}^0 Z(z)\, dz = 1$. In this event, either there is no vertical mixing of momentum in the system, or a no-stress surface condition is adopted. A comprehensive discussion of similar sets of equations can be found in Veronis (1973) and in Moore & Philander (1978).

VERTICAL MODES   Without vertical mixing it is always possible to represent solutions to (16) as expansions of the vertical normal modes of the system. With vertical mixing this representation is possible only in special cases. One of these cases is considered by Moore (1980) and is discussed later in this section. Another occurs when v and κ are assumed to be inversely proportional to N_b^2, the form of mixing of heat is modified to be $(\kappa \rho)_{zz}$, and the boundary conditions on ρ are that $\kappa \rho = 0$ at the ocean surface and bottom (McCreary 1980, 1981). The following discussion assumes that these latter conditions hold.

The vertical modes of the system are the eigenfunctions $\psi_n(z)$, which satisfy the differential equation

$$\psi_{nz} = -\frac{N_b^2}{c_n^2} \int_{-D}^z \psi_n \, dz, \tag{17}$$

subject to the boundary condition

$$\frac{1}{c_n^2} \int_{-D}^0 \psi_n \, dz = 0, \tag{18}$$

and they are usually normalized so that $\psi_n(0) = 1$. It is convenient to order them so that the eigenvalues c_n decrease monotonically. The $n = 0$

eigenfunction is unique in that $c_0 = \infty$; this eigenfunction is the barotropic mode of the system. (If a free-surface condition is adopted, rather than $w = 0$, then $c_0 = \sqrt{gD}$.) The remaining eigenfunctions form an infinite set of barotropic modes.

Solutions to (15) are then given by

$$
u = \sum_{n=0}^{\infty} u_n \psi_n, \qquad v = \sum_{n=0}^{\infty} v_n \psi_n, \qquad p = \sum_{n=0}^{\infty} p_n \psi_n,
$$
$$
w = \sum_{n=0}^{\infty} w_n \int_{-D}^{z} \psi_n \, dz, \qquad \rho = \sum_{n=0}^{\infty} \rho_n \psi_{nz}, \tag{19}
$$

where the expansion coefficients are functions only of x, y, and t. The equations governing the expansion coefficients are

$$
u_{nt} - f v_n + p_{nx} = F_n - (\varepsilon + A/c_n^2) u_n + v_h \nabla^2 u_n,
$$
$$
v_{nt} + f u_n + p_{ny} = G_n - (\varepsilon + A/c_n^2) v_n + v_h \nabla^2 v_n, \tag{20}
$$
$$
p_{nt}/c_n^2 + u_{nx} + v_{ny} = -(\varepsilon + A/c_n^2) p_n/c_n^2,
$$

and also

$$
w_n = (\partial_t - \varepsilon - A/c_n^2) p_n/c_n^2, \qquad \rho_n = -p_n/g. \tag{21}
$$

For convenience, these equations assume unit Prandtl numbers, that is, $v = \kappa = A/N_b^2$ and $r = \varepsilon$. When the wind stress enters the ocean as a body force, the forcing of each mode is

$$
F_n = \tau^x(x, y, t) Z_n \bigg/ \int_{-D}^{0} \psi_n^2 \, dz, \qquad G_n = \tau^y(x, y, t) Z_n \bigg/ \int_{-D}^{0} \psi_n^2 \, dz, \tag{22}
$$

where $Z_n = \int_{-D}^{0} Z(z) \psi_n \, dz$. When the wind enters the ocean through the surface stress conditions, the only change is that $Z_n = 1$ in (22).

Vertical mixing appears in (20) in the form of a simple drag with a drag coefficient A/c_n^2, and therefore it affects each mode like the terms proportional to ε. The difference between the two types of mixing is that A/c_n^2 increases with mode number (eventually increasing like n^2), whereas ε does not. Consequently, vertical mixing typically does not influence the low-order modes much, but eventually it dominates the dynamics of the high-order modes (see the discussion of Figures 4a and 4b). Rayleigh friction and Newtonian cooling, on the other hand, affect all modes equally.

Models differ in the type of mixing that is present. Lighthill (1969), who first used and popularized this approach to study equatorial circulation, found solutions to the inviscid form of (15). Many of the models discussed in the next sections are also inviscid. McCreary (1981, 1984) included vertical mixing and in the latter model added horizontal mixing with the simple form $v_h v_{nxx}$. Models that solve (20) numerically include all the horizontal mixing terms (McCreary et al. 1984, Gent et al. 1983). The Gent et al. (1983) model used Rayleigh friction and Newtonian cooling rather than vertical mixing.

A limitation of this approach is that ρ must be fixed at the ocean surface. As a result, solutions necessarily cannot generate a responsive SST field. To avoid this difficulty, Rothstein (1983) developed a procedure to replace the surface condition on ρ with one on heat flux. First, he modified the surface condition on ρ from $\kappa\rho = 0$ to $\kappa\rho = \kappa\rho_s$, where $\rho_s(x, y, t)$ is an arbitrary density distribution. This change adds the forcing term $-(g\kappa\rho_s/c_n^2)/\int_{-D}^{0}\psi_n^2\ dz$ to the third equation of (20). Then he chose ρ_s in such a way that the heat-flux condition, $(\kappa\rho)_z = 0$, was approximately satisfied. By iterating the procedure, the surface condition of the approximate solution converged to the desired one.

A SURFACE MIXED LAYER   It is possible to incorporate a constant-thickness surface mixed layer in (15) simply by assuming that $N_b(z) = 0$ for $z > -H$ (McCreary 1981). The eigenfunction problem, as defined in (17) and (18), is still well posed, and so the eigenfunctions as well as the solutions (19) remain well behaved. The effect on the eigenfunctions is only that $\psi_{nz} = 0$ in the layer. Therefore, in the mixed layer u, v, and p are depth independent, w varies linearly, and ρ is identically zero.

Moore (1980) considered an interesting special case of Equations (15) that included a mixed layer like that defined in the preceding paragraph in which $v \neq 0$, but assumed that $v = 0$ in the deeper ocean. Boundary conditions at the bottom of the layer are that stress vanishes and w is continuous across $z = -H$. Thus, his model ocean consists of a viscid, constant-density surface layer overlying an inviscid, stratified, deep ocean.

Moore demonstrated that, as in the linear constant-thickness and reduced-gravity models discussed above, the response of this system can be separated into two components. The first is just the surface-layer shear flow described by (5). The second is the flow described by the inviscid form of (15), in which the wind stress enters the ocean as a body force spread uniformly throughout the layer. The important implication of this model, as for the constant-thickness and reduced-gravity models, is that a shear flow can exist in the surface layer that responds independently from the rest of the flow field.

MERIDIONAL MODES   It is also possible to represent solutions to (15) as expansions in meridional, rather than vertical, modes. This approach, however, suffers from several mathematical difficulties. One of them is that the equations of motion must be Fourier transformed in both space and time, and it is generally not possible to invert the transforms of the solutions with analytic methods. Another is that it is not easy to impose surface and bottom boundary conditions of the form (2) and (3), because these conditions couple together individual meridional modes (McPhaden 1981). No doubt as a result of these difficulties, only three oceanographic studies have used the approach (Wunsch 1977, Philander 1978a, McPhaden 1981), and each assumes a simple forcing of the form $e^{ikx-i\sigma t}$ for which it is not necessary to invert any transforms. Because of its limited applications, this approach is not discussed further in this paper.

2.3   Nonlinear, Continuously Stratified Models

There have been a number of studies of fully nonlinear, continuously stratified ocean models. Equations of motion are

$$u_t + uu_x + vu_y + wu_z - fv + p_x = (vu_z)_z + v_h\nabla^2 u,$$

$$v_t + uv_x + vv_y + wv_z + fu + p_y = (vv_z)_z + v_h\nabla^2 v,$$

$$u_x + v_y + w_z = 0, \qquad p_z = -\rho g, \tag{23}$$

$$\rho_t + u\rho_x + v\rho_y + w\rho_z = (\kappa\rho_z)_z + \kappa_h\nabla^2\rho,$$

and they are solved subject to boundary conditions (2) and (3) with two additional conditions on ρ. The obvious advantage of these equations is that they provide a way of studying the effects of nonlinearities on equatorial circulation. Their disadvantage is that solutions must be found numerically. As a result, it is sometimes difficult to identify the important processes at work in them.

3.   UNFORCED SOLUTIONS

Equatorially trapped waves are a complete set of solutions to the unforced version of (20) that have the form $e^{ikx-i\sigma t}$. They are an essential part of the response of virtually all stratified models. Their importance is apparent in that solutions to (20) forced by the wind can be represented entirely as packets of waves [as in the solutions (36) and (39)]. In addition, they are visible in the response of nonlinear models as well. This section first defines these waves and then discusses how they are affected by mixing, boundaries, currents, and nonlinearities.

Many of the quantities involved in the following discussion and in the rest of this paper depend in some way on the eigenvalues c_n and should be labeled with the subscript n. For notational simplicity this subscript is deleted, unless confusion might result from its absence.

3.1 Equatorially Trapped Waves

A useful approximation, one that is almost universally adopted in equatorial ocean models, is the equatorial β-plane approximation, which represents the Coriolis parameter by βy. With this approximation, free solutions to (20) can be conveniently expressed in terms of a discrete set of parabolic cylinder functions $\phi_{\mu_l}(\eta)$. These eigenfunctions satisfy the equation

$$(\phi_{\mu_l})_{\eta\eta} - \eta^2 \phi_{\mu_l} = -(2\mu_l + 1)\phi_{\mu_l}, \tag{24}$$

where $\eta = (\beta/c)^{1/2} y \equiv \alpha_0 y$. The length scale α_0^{-1} is the equatorial Rossby radius of deformation associated with a particular vertical mode. The eigenvalues are fixed by requiring that each ϕ_{μ_l} vanish at high latitudes. In this case, the eigenvalues are

$$\mu_l = l, \tag{25}$$

and the parabolic cylinder functions are the Hermite functions $\phi_l(\eta)$. It is useful to normalize them so that $\int_{-\infty}^{\infty} \phi_l^2 \, d\eta = 1$.

Inviscid equatorially trapped Rossby and gravity waves are

$$u_{lj} = A_{lj} \frac{c\alpha_0}{i\omega} \Phi_{lj}^- \, e^{ik_{lj}x - i\sigma t},$$

$$v_{lj} = A_{lj} \phi_l(\eta) \, e^{ik_{lj}x - i\sigma t}, \tag{26}$$

$$p_{lj} = A_{lj} \frac{c^2\alpha_0}{i\omega} \Phi_{lj}^+ \, e^{ik_{lj}x - i\sigma t},$$

where

$$\Phi_{lj}^{\pm}(\eta) = \frac{\left(\dfrac{l+1}{2}\right)^{1/2} \phi_{l+1}(\eta)}{ck_{lj}/\omega - 1} \pm \frac{\left(\dfrac{l}{2}\right)^{1/2} \phi_{l-1}(\eta)}{ck_{lj}/\omega + 1} \tag{27}$$

and $\omega = \sigma$. There are two wave numbers designated by the index j, associated with each value of l. They are

$$k_{l(\frac{1}{2})} = -\frac{\beta}{2\omega} \left\{ 1 \mp \left[1 - 4\frac{\omega^2}{\beta^2} \left(\alpha_l^2 - \frac{\omega^2}{c^2} \right) \right]^{1/2} \right\}, \tag{28}$$

where $\alpha_l^2 = \alpha_0^2(2l+1)$. With this choice, k_{l1} (k_{l2}) corresponds to a wave with westward (eastward) group velocity. When $\sigma^2/(\beta c)$ is greater than $l+\frac{1}{2} + \sqrt{l(l+1)}$ or less than $l+\frac{1}{2} - \sqrt{l(l+1)}$, k_{lj} is a real number. The higher-frequency waves are gravity waves, and the lower-frequency ones are Rossby waves. Intermediate-frequency waves have complex wave numbers and so decay either to the east or the west. Waves for which u_{lj} and p_{lj} are symmetric (antisymmetric) about the equator are referred to as symmetric (antisymmetric) waves; with this definition, waves for l odd (even) are symmetric (antisymmetric).

Two additional free solutions to (20), designated here by the index $l = -1$, have a zero meridional velocity field. The equatorially trapped Kelvin wave is

$$u_{-12} = p_{-12}/c = A_{-12}\phi_0(\eta) \, e^{ik-12x-i\sigma t}, \qquad v_{-12} \equiv 0, \qquad (29)$$

and has the dispersion relation

$$k_{-12} = \omega/c. \qquad (30)$$

The other wave, labeled the anti-Kelvin wave by Cane & Sarachik (1979), has a meridional structure $e^{\frac{1}{2}\eta^2}$ and a dispersion relation $k_{-11} = -\omega/c$. Because this wave grows indefinitely for large η, it is not a physically realistic solution in an unbounded ocean.

When $l = 0$, the radical in (28) is a perfect square, and the eastward-propagating wave has the dispersion relation

$$k_{02} = \omega/c - \beta/\omega. \qquad (31)$$

This wave is usually called the Rossby-gravity wave, since it has characteristics associated with both kinds of waves (see Figure 1); it is also

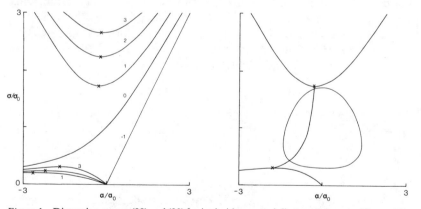

Figure 1   Dispersion curves (28) and (30) for inviscid, equatorially trapped waves. The curves are plotted in dimensionless coordinates, with $\alpha_0 = (\beta/c)^{1/2}$ and $\sigma_0 = (\beta c)^{1/2}$. The left panel shows the curves for the Kelvin wave ($l = -1$), for the Rossby-gravity wave ($l = 0$), and for the $l = 1, 2$, and 3 Rossby and gravity waves when the values of k_{lj} are real. Some of the curves are labeled with their corresponding values of l. The right panel shows the curves for the $l = 1$ Rossby and gravity waves, and it also includes the curves in the region where the values of k_{1j} are complex; the thin curve indicates the imaginary part of k_{1j}. The crosses indicate resonance points for the Rossby and gravity waves.

sometimes referred to as the Yanai wave. The westward-propagating wave has the dispersion relation $k_{01} = -\omega/c$. This root is extraneous because it is not a possible solution to (20). Note that the term involving ϕ_{l-1} in (27) is not well defined, since both the numerator and denominator vanish for $l = 0$.

Figure 1 plots the dispersion relations (28), (30), and (31). The left panel shows the curves for the Kelvin and Rossby-gravity waves, and those for $l = 1, 2,$ and 3 Rossby and gravity waves. Note that the Rossby-gravity curve resembles that for a Rossby wave when $k \ll 0$ and that for a gravity wave when $k \gg 0$. The right panel shows the dispersion curves for $l = 1$ Rossby and gravity waves, and also for the waves with complex values of k_{11} and k_{12}. Similar complex roots also occur for waves associated with other values of l.

It is possible to define a set of equatorially trapped waves even when the system includes vertical mixing, Rayleigh drag, and Newtonian cooling. According to (20) the waves are just those discussed above, except that $\omega = \sigma - iA/c^2 - i\varepsilon$. The only effect is that the wave numbers (28) and (30) now always have an imaginary part, and the waves decay in the direction of their group velocity. The waves still exist when $\sigma \to 0$, in which case they are purely damped waves.

It is not as easy to define a set of waves when there is horizontal mixing. McCreary (1984) included horizontal mixing of the form $\nu_h v_{nxx}$ in (20). In this case, the waves have the same horizontal structure as in (26) and (29); however, the dispersion relation for Rossby and gravity waves is quartic, rather than quadratic, so that there are four roots k_{lj} for each value of l. For more general forms of horizontal mixing, it is no longer possible to express the waves in terms of parabolic-cylinder functions.

3.2   The Long-Wavelength Approximation

At low frequencies the two roots in (28) are approximately given by $k_{l2} = -\beta/\omega$ and

$$k_{l1} = -\frac{\omega}{c}(2l+1). \tag{32}$$

The Rossby waves with eastward group velocity have short wavelengths (at the annual frequency $-2\pi/k_{l2} = 55$ km), and so they are not significantly excited by geophysically realistic (large-scale) winds. They are only important near western ocean boundaries, where they superpose to create narrow, intense western boundary currents. The waves with westward group velocity have long wavelengths [at the annual frequency and with $c = 100$ cm s^{-1}, $-2\pi/k_{l1} = 10,000/(2l+1)$ km] and are always an important part of the response of the tropical ocean. Note also that the long waves are nondispersive, a mathematically convenient relation.

A number of models take advantage of these properties by making the long-wavelength approximation. This approximation requires the zonal flow field to be in geostrophic balance; that is, it neglects the time-dependent and friction terms in the second of Equations (20). The effect is that the dispersion relations (28) are replaced by the single root (32), and the short waves are filtered completely out of the system.

3.3 Boundary Effects

NORTHERN AND SOUTHERN BOUNDARIES In the eastern Atlantic the coast of Africa near $5°N$ is oriented in an east-west direction and thus provides a northern boundary for equatorial flow. How are equatorially trapped waves affected by such a boundary? Hickie (1977) and Philander (1977) discuss waves in an ocean with a northern boundary located at $y = y_N > 0$. Cane & Sarachik (1979) consider the effects on the waves when the basin also has a southern boundary at $y = y_S < 0$. At these boundaries meridional flow must vanish, and so Equation (24) must be solved subject to the boundary condition $\phi_{\mu_l} = 0$ at $\eta = \alpha_0 y_N, \alpha_0 y_S$. As a result, the eigenvalues are no longer integers (exceptions are the special cases $y_N = 0$, $y_S = -\infty$ and, equivalently, $y_N = \infty$, $y_S = 0$ for which $u_l = 2l+1$), and the horizontal structures of the waves change.

Another effect is that the two waves disallowed in the unbounded ocean are now possible (Cane & Sarachik 1979). Since $\mu_0 \neq 0$, it follows that the $l = 0$ Rossby wave is no longer extraneous. Provided the basin has both a northern and southern boundary, the anti-Kelvin wave is also well defined everywhere in the basin. Neither of the waves, however, is really an equatorially trapped wave. Each owes its existence to the presence of the boundaries and has its largest amplitude there. This property is particularly evident for a low-frequency $l = 0$ Rossby wave when the ocean has a single northern (or southern) boundary that is located more than an equatorial Rossby radius (α_0^{-1}) from the equator. In this case, the $l = 0$ Rossby wave has all the properties of a coastally trapped Kelvin wave (Cane & Sarachik 1979).

EASTERN AND WESTERN BOUNDARIES Equatorially trapped waves reflect from eastern and western boundaries of the ocean, and these reflected waves can significantly affect the tropical ocean. Moore (1968) showed how to find the reflected waves when the boundaries are simple, meridionally oriented barriers. His method chooses the amplitudes of the reflected waves to ensure that $u = 0$ everywhere at the boundary. At an eastern boundary a chain of reflected waves is required that leads toward increasing values of l, whereas at a western boundary the chain leads toward decreasing values. Physically, the reason for this difference is that energy can only propagate poleward (equatorward) along an eastern (western) boundary. [See Moore

& Philander (1978) for a recent comprehensive discussion of this approach.]

To illustrate Moore's method, consider the reflection of an equatorial Kelvin wave [Equation (29)] from an eastern boundary at $x = 0$. The reflected waves are the Rossby or gravity waves [Equation (26)] with westward group velocity ($j = 1$), since reflected energy must be carried away from the boundary. One term of the u-field of the $l = 1$ wave has the meridional structure ϕ_0. By choosing the amplitude of the wave to be $A_{11} = (i\omega/c\alpha_0)\sqrt{2}(ck_{11}/\omega + 1)A_{-12}$, this part exactly cancels the u-field of the Kelvin wave at the boundary. The other part of the $l = 1$ wave has the structure ϕ_2, and it is now necessary to eliminate this u-field by using the $l = 3$ wave. It follows that an infinite set of waves is required to reflect the Kelvin wave, and according to (27) their amplitudes satisfy the recursion relations

$$A_{l+2,1} = [(ck_{l+2,1}/\omega + 1)/(ck_{l1}/\omega - 1)]\,[(l+1)^{1/2}/(l+2)^{1/2}]A_{l1}$$

for $l = 1, 3, 5, \ldots$. For sufficiently large values of l the wave numbers k_{l1} of the reflected waves are all complex (as in the right panel of Figure 1), and so this portion of the reflected wave packet remains trapped to the coast. Moore proved that these trapped waves sum to create a β-plane, coastal Kelvin wave that carries energy poleward at sufficiently high latitudes.

Moore's method can be applied directly to problems that are periodic in time (Cane & Sarachik 1981, McCreary 1981, 1984). It is also possible to use this approach to find the reflection of wave packets that are not periodic in time. The wave packets are first Fourier-transformed in time, then Moore's method is used to find the reflected waves associated with each Fourier component, and finally the inverse transform of the solution is evaluated. Anderson & Rowlands (1976) used this procedure to find the reflection of a step-function Kelvin wave [with the zonal structure $\theta(ct - x)$] from an eastern boundary. The resulting transforms were very complicated and had to be inverted numerically. McCreary (1976, 1980) and Cane & Sarachik (1977) found the reflection of various wave packets from both eastern and western boundaries, but they adopted the long-wavelength approximation. In this case, the inverse transforms can be easily found analytically.

Cane & Sarachik (1977, 1979) pointed out that at sufficiently low frequencies it is not necessary to use Moore's method to find the reflection of waves from a western boundary. In this limit, the key simplification is that reflected Rossby waves (with the dispersion relation $k_{l2} = -\beta/\omega$) are approximately nondivergent, and so their contribution to the reflected wave field can be represented conveniently in terms of a stream function ψ, rather than as an expansion in parabolic cylinder functions. One property of this nondivergent contribution is that it generates no net transport across the boundary [since $\int_{-\infty}^{\infty} \psi_y \, dy = \psi(\infty) - \psi(-\infty) = 0$]. Consequently, any mass flux into the boundary must be balanced by an outgoing flux associated only with the reflected equatorial Kelvin wave. This condition allows the amplitude of the Kelvin wave to be determined independently from the nondivergent contribution.

Several papers consider reflections of waves from sloping boundaries, i.e. boundaries described by the curve $y = \alpha^{-1}x$. Moore et al. (1981) discuss the reflection of $l = 1$ Rossby waves from a sloping western boundary. With $\alpha = 0$, the only reflected waves are the symmetric $l = 1, j = 2$ Rossby wave and an equatorial Kelvin wave. With $\alpha \neq 0$, the antisymmetric Rossby-gravity wave is also a part of the reflection, and for some frequencies it is the dominant part. Cane & Gent (1984) also study the reflection of various Rossby waves from a sloping western boundary, but they restrict their analysis to low-frequency waves in order to take advantage of the simplifications discussed in the preceding paragraph. They find that when the boundary slopes, the amplitude of reflected Kelvin waves varies considerably with frequency and vanishes entirely for some frequencies. Clarke (1983) considers the reflection of Kelvin waves from a sloping eastern boundary, but his solutions are valid only off the equator.

ISLANDS   Yoon (1981), Rowlands (1982), and Cane & duPenhoat (1982) have studied how islands affect equatorial waves. The islands in these studies have various simple forms. They all are meridionally oriented barriers (with a finite meridional extent L) that either are infinitesimally thin (a line island) or have a small zonal extent. Provided that an island is centered near the equator, waves are either transmitted or reflected depending largely on the size of L relative to α_0^{-1}: when $L < \alpha_0^{-1}$ the waves are not much affected by an island; when $L > 2\alpha_0^{-1}$ the islands might as well be barriers of infinite extent, and there is very little transmission past the island. Some solutions are for systems of islands representing various low-latitude island chains (like the Gilbert and Galapagos Islands in the Pacific Ocean, and the Maldives in the Indian Ocean). Individual islands in the chains each satisfy the criterion $L < \alpha_0^{-1}$. As a result, equatorial waves readily propagate through these island chains with little reflection.

3.4   *Interactions With Currents*

Several papers have considered the interaction of equatorially trapped waves with a background zonal current $U(y)$. The models are all reduced-gravity models, like (8), except that they also include interaction terms with $U(y)$. The presence of the current affects the waves in three different ways. First, it can simply act to modify their dispersion relations and to alter their meridional structures. Second, some waves can become unstable and extract energy from the current by barotropic instability. It is possible to derive criteria that must hold if the currents can be unstable. One of them is that $\mathrm{Re}(\sigma)/k = U(y)$ somewhere in the flow, where $\mathrm{Re}(\sigma)$ denotes the real part of σ. Finally, waves can also be absorbed near critical layers where $\sigma/k = U$, thereby contributing some of their energy to the current. At the present time there are no studies that explicitly treat the critical-layer absorption of equatorially trapped waves [although Schopf et al. (1981) discuss other effects of critical layers on their solutions], but there are several papers that discuss the other types of interactions.

In complementary studies, McPhaden & Knox (1979) and Philander (1979) considered the modifications of equatorially trapped waves caused by equatorial currents. Both studies avoided the possibility of unstable waves and critical-layer absorption by restricting their discussion to waves for which $\sigma/k > U(y)$ everywhere. McPhaden & Knox discussed effects of both eastward and westward currents on gravity and Kelvin waves, but they did not consider Rossby waves. The Kelvin waves are hardly affected at all by the currents, but gravity waves are visibly distorted. Philander also included Rossby waves in his discussion, but he restricted $U(y)$ to be a narrow eastward current like the Equatorial Undercurrent. The phase speeds of Rossby waves are significantly reduced by the current. Because the current is narrow, however, the decrease is much smaller than that for a current without shear (that is, for a constant U); in particular, the phase speeds never become eastward.

Philander (1978b) studied the effect of the surface currents in the tropical Pacific and Atlantic Oceans on equatorially trapped Rossby waves. He chose for $U(y)$ a profile with westward flow from 3°S to 3°N and eastward flow from 3°N to 10°N in order to simulate the South Equatorial Current and the North Equatorial Countercurrent. Because of the westward equatorial current, critical layers existed for the westward propagating Rossby waves, and some of them were unstable. Philander limited his discussion to a description of the most unstable waves and suggested that this instability may account for the unstable waves reported by Legeckis (1977) and Legeckis et al. (1983) in the Pacific.

Philander (1976) studied a reduced-gravity model with two active surface layers and included different background zonal currents $U_1(y)$ and $U_2(y)$ in the upper and lower layers, respectively. The upper current $U_1(y)$ was similar to that described in the previous paragraph, and $U_2(y)$ resembled the Equatorial Undercurrent. As a result of the presence of westward flow in $U_1(y)$, the system possessed a barotropically unstable wave. In addition, because the background currents involved vertical shear, baroclinic instability was also possible. Philander concluded, however, that baroclinic instability is not effective near the equator, and that the only tropical current that could be baroclinically unstable is the North Equatorial Current north of 10°N.

3.5   *Nonlinear Effects*

There are a few studies that consider the effect of nonlinearities on equatorially trapped waves. They all use reduced-gravity models like (8), except that they include some or all of the nonlinear terms. Boyd (1980a) demonstrated that nonlinearities can cause an equatorial Kelvin wave to break, but he concluded that the breaking of Kelvin waves is not likely to be important in realistic ocean basins. Boyd (1980b) derived an expression for the structure of Rossby solitons, and Kindle (1981, 1983) was able to generate them in a numerical model. Finally, in a series of papers, Ripa (1982, 1983a,b) considered resonant interactions of triads of equatorially trapped waves.

4.   SOLUTIONS FOR SWITCHED-ON WINDS

This section discusses the response of various equatorial models to a wind field that switches on at some initial time and thereafter remains steady. [See Weisberg & Tang (1983) for a discussion of the response of a reduced-gravity model to growing and propagating wind fields.] For convenience the wind is assumed to have the separable form

$$\tau^x = \tau^y = \tau_0 X(x) Y(y) \theta(t), \tag{33}$$

where $X(x)$ and $Y(y)$ are arbitrary functions that are usually zonally and meridionally bounded, and $\theta(t)$ is a step function in time. Both adjustments to equilibrium and the steady solutions themselves are discussed.

4.1   *Solutions to Surface-Layer Models*

The solution to (5) without horizontal mixing, subject to boundary conditions (2) and (3), is

$$u' = \frac{2}{H} \sum_{n=1}^{\infty} \left\{ \frac{f\tau^y + vm^2\tau^x}{v^2 m^4 + f^2} + \mathrm{Re}\left[\frac{\tau^x - i\tau^y}{-vm^2 + if} e^{ift} e^{-vm^2 t} \right] \right\}$$

$$\times \cos mz \; \theta(t),$$

$$v' = \frac{2}{H} \sum_{n=1}^{\infty} \left\{ \frac{-f\tau^x + vm^2\tau^y}{v^2 m^4 + f^2} + \mathrm{Re}\left[\frac{\tau^y + i\tau^x}{-vm^2 + if} e^{ift} e^{-vm^2 t} \right] \right\}$$

$$\times \cos mz \; \theta(t), \tag{34}$$

$$w = - \int_{-H}^{z} (u'_x + v'_y) \, dz,$$

where $m = n\pi/H$ (D. W. Moore, private communication). The time, speed, and width scales of the solution are apparent in (34). The flow field adjusts to equilibrium in a time scale $H^2/(v\pi^2)$, reaches speeds of the order of $2\tau_0 H/(v\pi^2)$, and has a width scale measured by $v\pi^2/(\beta H^2)$. For the parameter values of Figure 2, these scales are 6 days, 50 cm s^{-1} and 100 km, respectively. The solution can respond rapidly to the wind because it is not necessary to move isopycnals in the constant-density layer. Another property of (34) is that it is a local balance, that is, it is directly proportional to the strength of the wind $\tau_0 X(x) Y(y)$.

Recall that (34) is not the total response of the layer. For a constant-thickness model there is also a depth-independent contribution to the flow that responds like a barotropic mode. For reduced-gravity and continuously stratified models, this component adjusts much more slowly to the wind because it involves the displacement of isopycnals; this adjustment is discussed later in this section.

Figures 2a and 2b, adapted from Stommel (1960), show the steady response of (34) for both easterly and southerly winds, respectively. Parameter values are $v = 20$ cm^2 s^{-1}, $\beta = 2 \times 10^{-13}$ cm^{-1} s^{-1}, and $H = 100$ m, and the amplitude of the wind is 0.5 dyn cm^{-2}. The solutions have a number of features that were first discussed heuristically by Cromwell (1953). For easterly winds there is Ekman divergence at the equator forcing equatorial upwelling and a compensating return flow at depth. There is a surface drift at the equator in the direction of the wind. For southerly winds there is a northward surface drift across the equator. This drift causes a divergence and upwelling of fluid somewhat south of the equator, and a

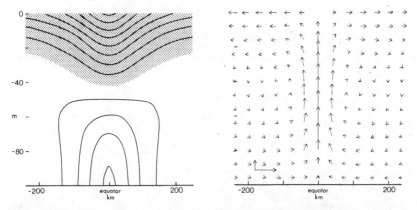

Figure 2a Vertical sections of zonal velocity (*left*) and meridional circulation (*right*) for a constant-thickness model forced by an easterly wind with an amplitude of 0.5 dyn cm^{-2}. The contour interval is 10 cm s^{-1}, there is no zero contour line, and the shaded regions indicate westward flow. Calibration arrows are 0.01 cm s^{-1} and 100 cm s^{-1} in the vertical and horizontal directions, respectively. The solution demonstrates that a strong, ageostrophic shear flow can exist in the equatorial mixed layer. After Stommel (1960).

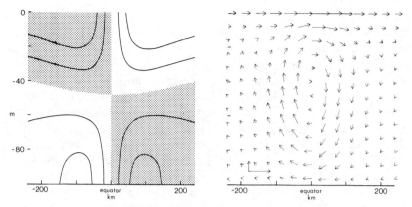

Figure 2b   As in Figure 2*a*, except that the wind is southerly. After Stommel (1960).

convergence and downwelling of fluid somewhat north of the equator. A limitation of these solutions is that all the currents are ageostrophic. Therefore the model does not provide an adequate explanation of the permanent, geostrophically balanced Equatorial Undercurrent. It does, however, provide a picture of the current structures that can rapidly develop in the surface layer in response to fluctuating winds.

Solutions to surface-layer models are sensitive to the choice of bottom boundary conditions. For example, with no-slip bottom conditions linear solutions do not have an undercurrent, but nonlinear solutions do (Charney 1960, Charney & Spiegel 1971). This result suggests, and has been used to argue, that nonlinearities are necessary for the existence of the Equatorial Undercurrent. A better conclusion is that the no-slip conditions are too stringent.

Solutions are also sensitive to the strength of nonlinearities (Charney 1960, Charney & Spiegel 1971). Cane (1979b) contrasts linear and nonlinear solutions in detail. His linear solutions have structures very similar to those in Figure 2 (except that they have only two degrees of freedom in the vertical), and it is useful to refer to this figure when discussing the changes caused by nonlinearities (also see Figure 5). For easterly winds, the nonlinear terms weaken or eliminate the westward surface drift and strengthen the eastward undercurrent. For westerly winds, they strengthen the eastward surface flow and weaken or eliminate the westward undercurrent. For southerly winds, the surface eastward current north of the equator shifts poleward and strengthens, and the subsurface westward flow shifts equatorward until it overlaps the equator.

Two important nonlinear terms are vu_y and wu_z, i.e. the advection of the zonal current by the meridional circulation. If these terms are sufficiently small, they act only to shift the position of the linear zonal currents and to distort their shape. [Robinson (1966) pointed out this property explicitly by treating these two terms, and others, as perturbations to a system of equations like Charney's (1960) system.] The two terms in Cane's solutions appear to act in this way. To illustrate this point, it is again useful to refer to the linear flow fields in Figure 2. For easterly winds, the convergence of fluid at depth advects eastward momentum toward the equator, thereby intensifying the undercurrent. The equatorial upwelling advects the undercurrent upward, thereby weakening the westward surface drift. For westerly winds, the direction of flow of all the currents in Figure 2a is reversed. In this case, surface convergence of eastward momentum intensifies the surface drift, and equatorial downwelling advects eastward flow downward, thereby weakening or eliminating the undercurrent. For southerly winds, the surface eastward flow is advected farther to the north and is concentrated and intensified. The deep westward flow north of the equator is advected south until it appears on the equator.

4.2 Solutions to Linear, Continuously Stratified Models

ZONALLY INDEPENDENT SOLUTIONS: THE YOSHIDA JET   Moore & Philander (1978) and Cane & Sarachik (1976) discussed inviscid solutions to (20) when the wind is zonally independent [i.e. $X(x) = 1$ in (33)], and McCreary (1980) found a similar solution when the system included vertical mixing. Provided that the zonal flow is assumed to be in geostrophic balance (the long-wavelength approximation), the solution for zonal winds with vertical mixing is

$$u_n = (fv_n + F_n)\frac{c^2}{A}(1 - e^{-A/c^2 t})\theta(t),$$

$$v_n = -\frac{1}{\alpha_0 c}\sum_{l=1}^{\infty}\frac{[\eta F]_l}{2l+1}\phi_l(\eta)\theta(t), \tag{35}$$

$$p_n = c^2\alpha_0 v_{n\eta}\frac{c^2}{A}(1 - e^{-A/c^2 t})\theta(t).$$

Far from the equator, Equations (35) describe a meridional Ekman drift, $fv_n + F_n = 0$. Near the equator where the Coriolis force is not sufficient to balance the wind, the meridional drift becomes a swift zonal jet. Note that in the absence of vertical mixing (the limit $A \to 0$), the jet continues to accelerate indefinitely. This flow is one of the distinguishing characteristics of equatorial dynamics and is referred to as the Yoshida jet. As we shall see, a similar equatorial jet still exists even when the wind stress is zonally bounded.

If the assumption of zonal geostrophy is not made, Equations (35) also involve oscillations at a discrete set of frequencies $(\beta c)^{1/2}(2l+1)$ (Moore & Philander 1978). These oscillations are equatorial inertial oscillations, and they are very similar to inertial oscillations at midlatitudes. Their periods, however, are much longer; for example, with $c = 100$ cm s^{-1} the period is $2\pi/(\beta c)^{1/2} = 16$ days. They appear in numerical models as well, and to weaken their effect the wind stress is usually turned on gradually, rather than abruptly as in (33).

ZONALLY BOUNDED SOLUTIONS: ADJUSTMENT TO EQUILIBRIUM   It is mathematically difficult to find solutions to (20) when the wind stress field is zonally bounded. The solution proceeds by Fourier transforming the equations, finding the solution for each Fourier component, and finally inverting the transforms. As in the Anderson & Rowlands (1976) study, it is necessary to resort to numerical methods to invert the transforms. When the long-wavelength approximation is adopted, however, it is possible to invert the transforms analytically (Cane & Sarachik 1976, 1977, McCreary 1976, 1980). In this case, the inviscid solution in response to a zonal wind of the form (33) is

$$
u_n = \tau_{0n} \frac{1}{\beta} Y_{yy} \int_{+\infty}^{x} X \, dx' \; \theta(t) + \left\{ \frac{\tau_{0n}}{2c} Y_0 \phi_0 \int_{-\infty}^{\infty} X \, dx' \; \theta(t) \right\}
$$

$$
- \sum_{l=1}^{\infty} A_l \Phi_{l1}^{-}(\eta) \frac{c\alpha_0}{\beta/\alpha_l^2} \int_{+\infty}^{x} X\left(x' + \frac{\beta}{\alpha_l^2} t\right) dx' \; \theta(t)
$$

$$
+ A_{-1} \phi_0 \int_{+\infty}^{x} X(x'-ct) \, dx' \; \theta(t),
$$

$$
v_n = -\tau_{0n} \frac{1}{\beta} Y_y X \theta(t) + \sum_{l=1}^{\infty} A_l \phi_l(\eta) X\left(x + \frac{\beta}{\alpha_l^2} t\right) \theta(t), \tag{36}
$$

$$
p_n = \tau_{0n}(Y - y Y_y) \int_{+\infty}^{x} X \, dx' \; \theta(t) + \left\{ \frac{\tau_{0n}}{2} Y_0 \phi_0 \int_{-\infty}^{\infty} X \, dx' \; \theta(t) \right\}
$$

$$
- \sum_{l=1}^{\infty} A_l \Phi_{l1}^{+}(\eta) \frac{c^2 \alpha_0}{\beta/\alpha_l^2} \int_{+\infty}^{x} X\left(x' + \frac{\beta}{\alpha_l^2} t\right) dx' \; \theta(t)
$$

$$
+ c A_{-1} \phi_0 \int_{+\infty}^{x} X(x'-ct) \, dx' \; \theta(t),
$$

where $[\eta Y]_l$, $[Y_n]_l$, and Y_l are the Hermite-expansion coefficients of ηY, Y_n, and Y, respectively, $\tau_{0n} = \tau_0 / \int_{-D}^{0} \psi_n^2 \, dz$, and

$$
A_l = \tau_{0n} \frac{\alpha_0}{\beta} \left(\frac{[\eta Y]_l}{2l+1} - [Y_n]_l \right), \qquad A_{-1} = \frac{\tau_{0n}}{2c} Y_0, \tag{37}
$$

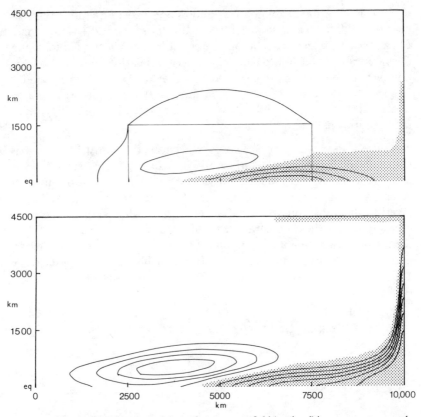

Figure 3   The time development of the surface pressure field (sea level) in response to a patch of easterly wind that is switched on initially. The response is shown at 1 month (*upper left*), 3 months (*lower left*), 13 months (*upper right*), and 5 yr (*lower right*). The horizontal structure of the wind is indicated in the upper left panel, and the maximum value of the wind stress is 0.5 dyn cm^{-2}. The contour interval is 10 cm, there is no zero contour line, and the shaded regions indicate a drop in sea level. The solution illustrates the adjustment of a baroclinic mode to Sverdrup balance. When the wind change is reversed, aspects of the solution compare favorably with sea-level observations during El Niño events. After McCreary & Anderson (1984).

are the amplitudes of packets of equatorially trapped Rossby and Kelvin waves (McCreary 1980). As in (35), there are no inertial oscillations because the long-wavelength approximation is adopted.

It is useful to trace the development of (36) in time. Initially the solutions behave just like the inviscid version of (35). For example, it is possible to show that at small times both u_n and p_n grow linearly in time. At larger times a packet of equatorially trapped waves radiates from the patch. After the passage of these waves, the solution is left in an equilibrium state very different from that in (35).

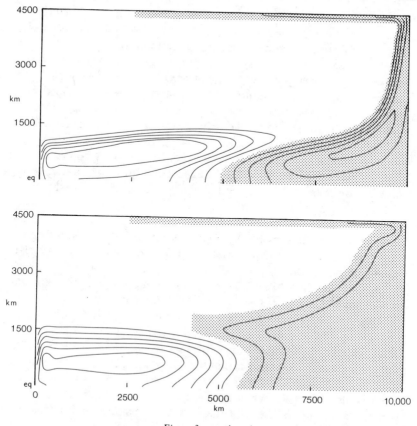

Figure 3—continued

Except for the terms in braces, the steady part of (36) describes a baroclinic mode in a state of Sverdrup balance, that is, a solution to the steady, inviscid version of Equations (20). (The concept of Sverdrup balance is usually used to describe depth-averaged properties of ocean circulation. The usage is extended in equatorial dynamics to apply also to circulation associated with individual baroclinic modes.) The terms in braces describe a steady, zonally independent equatorial jet, which is the analogue of the growing, inviscid Yoshida jet in (35). The jet does not accelerate indefinitely because the radiation of equatorially trapped waves allows the system to develop a zonal pressure gradient at the equator to balance the wind there. In a bounded basin the equatorial jet is not possible, since there can be no flow through the boundaries, and the steady solution is just Sverdrup flow. Finally, note that the solution is nonlocal in that it extends well beyond the zonal limits of the wind, the equatorial jet being an obvious example.

Figure 3, taken from McCreary & Anderson (1984), illustrates the role of equatorially trapped waves in the adjustment to Sverdrup balance. The figure shows the response of the surface pressure field when Equations (20) are forced by a patch of zonal wind symmetric about the equator. After one month, equatorial Kelvin waves have propagated into the eastern ocean, but the slower Rossby waves have not yet progressed very far into the western ocean. After three months the equatorial jet is well developed in the central and eastern ocean, as evidenced by the presence of strong meridional pressure gradients there. In the central ocean a zonal pressure gradient develops to balance the easterly wind. In the far eastern ocean the equatorial Kelvin waves have reflected from the eastern boundary as a packet of Rossby waves, and coastal Kelvin waves are also visible, carrying the reflection poleward along the boundary. The reflected Rossby waves begin to eliminate the equatorial jet, as evidenced by the absence of meridional pressure gradients near the eastern boundary. After 13 months the equatorial ocean is essentially in equilibrium with the wind. A measure of the spin-up time of the equatorial ocean, T, is the time it takes a Kelvin wave to propagate across the basin and a reflected $l = 1$ Rossby wave to return (Cane & Sarachik 1977). For this solution, $d = 10,000$ km and $c = 125$ cm s^{-1}, so that $T = 4(d/c) = 1$ yr. After five years the state of Sverdrup balance is well established in the tropical ocean. Rossby waves, propagating slowly across the ocean, are apparent only well off the equator.

The presence of vertical mixing in the model can change the equilibrium response markedly. One effect is that the propagating radiation field in (36) decays as it propagates with an e-folding time scale of c^2/A. The steady part of (36) is also changed. When mixing is weak, the steady solution still resembles Sverdrup flow, except that now currents decay away from the region of the wind. For example, to the east of the wind patch the equatorial jet decays with an e-folding scale of c^3/A. This effect is very important because it means that reflected waves from boundaries can no longer completely cancel the strong equatorial jet; as a result, the addition of mixing actually can *strengthen* equatorial currents (McCreary 1981). When mixing is strong, steady solutions do not resemble Sverdrup flow at all; instead, they have a steady balance that resembles (35).

STEADY SOLUTIONS   Figure 4a, taken from McCreary (1981), and Figure 4b show the equatorial circulation produced by (19), (20), and (21) when the ocean is forced by a patch of easterly and southerly winds, respectively, each with an amplitude of 0.5 dyn cm^{-2}. The sums in (19) are truncated after $n = 100$, and this number of modes is more than sufficient to ensure that the solution is well converged. The background density field has a strong pycnocline just beneath a surface mixed layer of constant density; the thickness of the layer is 75 m in Figure 4a and 50 m in Figure 4b. There is vertical mixing in the system of the form $v = A/N_b^2$, and the value of A is set

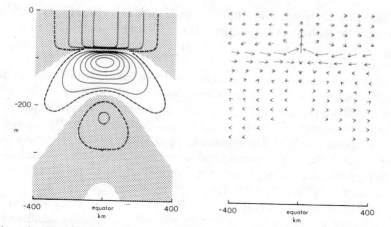

Figure 4a   Vertical sections of zonal velocity (*left*) and meridional circulation (*right*) at the center of the ocean basin for a linear, continuously stratified model forced by an easterly wind. The positions of eastern and western boundaries and the horizontal structure of the wind are similar to those in Figure 3, and the amplitude of the wind stress is 0.5 dyn cm^{-2}. The contour interval is 20 cm s^{-1}, the dashed contours are ± 10 cm s^{-1}, there is no zero contour line, and the shaded regions indicate westward flow. Calibration arrows are 0.005 cm s^{-1} and 10 cm s^{-1} in the vertical and horizontal directions, respectively. There is a geostrophically balanced Equatorial Undercurrent located in the pycnocline. After McCreary (1981).

so that v has a minimum value of 0.55 cm^2 s^{-1}. As in Figures 2*a* and 2*b*, the flow fields have realistic features. For easterly winds there is Ekman divergence, equatorial upwelling, and a compensating return flow at, or slightly above, the level of the core of the undercurrent. The undercurrent is located in the pycnocline and is in geostrophic balance, in good agreement

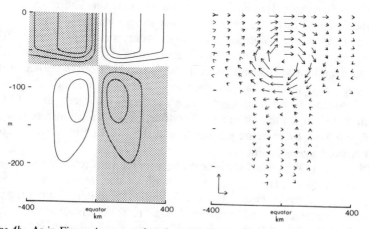

Figure 4b   As in Figure 4*a*, except that the ocean is forced by a southerly wind, and the contour interval is 5 cm s^{-1}. Similar to Rothstein (1983).

with the observations. For southerly winds there is northward drift across the equator with upwelling (downwelling) somewhat south (north) of the equator. There is a surface eastward current north of the equator, and a subsurface one to the south.

It is useful to discuss the dynamics of these solutions in terms of the steady balances of individual modes. Recall that the drag coefficients associated with vertical mixing, A/c^2, increase rapidly with mode number. For the low-order modes A/c^2 is small, and these modes nearly adjust to Sverdrup balance. They develop pressure gradients that tend to balance the wind, thereby limiting the strength of the equatorial currents. For the high-order modes A/c^2 is large, and these modes adjust to the steady two-dimensional balance in (35). They add up to produce the meridional circulation patterns in Figures 4a and 4b. The equatorial jet is strong only for intermediate modes (neither boundary waves nor damping significantly weakens it), and these modes add up to generate the Equatorial Undercurrent. The change in the character of the steady balances of individual modes is necessary in order to produce realistic equatorial flow fields. Solutions in which all the modes are either in Sverdrup balance or in two-dimensional balance are unrealistic (McCreary 1981).

As noted in Section 2.2, McPhaden (1981) found steady solutions to (15) by representing them as expansions in meridional modes. He assumed a steady zonal wind with the horizontal structure e^{ikx} and included in (15) only mixing of heat and momentum of the form $(vu_z)_z$, $(vv_z)_z$, and $\varepsilon\rho$. These restrictions were all necessary in order to be able to represent solutions as expansions in meridional modes. Solutions developed an undercurrent, similar to that in Figure 4a.

An advantage of McPhaden's approach is that it shows clearly how midlatitude Ekman dynamics change near the equator. There is a vertical-structure equation associated with each meridional mode that has six independent wave solutions. For high-order meridional modes (that contribute to the solution well off the equator), four of the solutions produce midlatitude Ekman layers at the top and bottom of the ocean, and the other two produce a geostrophically balanced, midlatitude Rossby wave. As the meridional mode number decreases, these six solutions gradually change their character. The four Ekman solutions are increasingly modified by pressure gradients, and the two Rossby waves become increasingly ageostrophic.

Rothstein (1983) found solutions that were relevant to the circulation in the eastern Pacific. He drove his model with an idealized representation of the southeast trade winds; the winds shifted gradually from primarily easterly in the central and western oceans to southerly near the eastern boundary. Well away from the eastern boundary, the solution had a strong

Equatorial Undercurrent, much like that in Figure 4a. Near the eastern boundary, the Undercurrent weakened and moved south of the equator. At the eastern boundary, part of the Equatorial Undercurrent joined with a coastal undercurrent south of the equator. In response to equatorial and coastal upwelling, SST was cool along the equator in the central and eastern oceans and along the southern coast.

As mentioned in the Introduction, subsurface eastward flow is not entirely confined to the region of the Equatorial Undercurrent; subsurface countercurrents also exist somewhat off the equator and at greater depths (Tsuchiya 1972, 1975, Cochrane et al. 1979). McPhaden (1984) pointed out that deep eastward currents, similar to these subsurface countercurrents, also exist in several equatorial models (McPhaden 1981, McCreary 1981, Philander & Pacanowski 1980); for example, they are visible on either side of the equator in Figures 4a and 5a. McPhaden found a steady solution and discussed the dynamics of its deep eastward currents in detail.

4.3   Solutions to Nonlinear, Continuously Stratified Models

Solutions to nonlinear models like (23) are still sensitive to the strength of vertical and horizontal mixing (Semtner 1981). Semtner & Holland (1980) minimized the effect of mixing in their model by choosing $v = \kappa = 1.5$ $cm^2 s^{-1}$ and by adopting a biharmonic form of horizontal mixing. Their solutions rapidly became unstable, with the Equatorial Undercurrent being barotropically unstable and the westward surface currents somewhat off the equator being baroclinically unstable (Philander 1976). Cox (1980) used somewhat larger values for mixing parameters. His solutions were still unstable, but the eddy activity was considerably less. Philander and coworkers typically use strong mixing in their models; for example, the solutions in Figures 5a and 5b have $v = 10\,cm^2 s^{-1}, \kappa = 1\,cm^2 s^{-1}, v_h = 10^7$ $cm^2 s^{-1}$, and $\kappa_h = 2 \times 10^7\ cm^2\ s^{-1}$. With these choices their solutions remain stable, and the effect of the nonlinear terms in their models is primarily to distort the linear solutions in recognizable ways. As noted by Semtner (1981), the observed Equatorial Undercurrent is more stable than the one produced in the Semtner & Holland model, suggesting that the stronger mixing used in other models is more realistic. He concluded that additional studies were needed to explore further the sensitivity of numerical models to the strength and form of mixing. One such study is that of Pacanowski & Philander (1981), which examines the use of vertical-mixing coefficients that depend on Richardson number.

Figure 5a, taken from Philander & Pacanowski (1980), and Figure 5b, taken from Philander & Delecluse (1983), contrast steady linear and

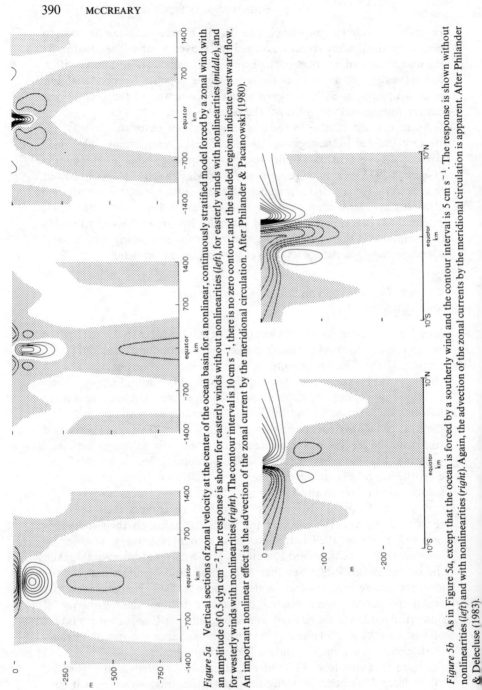

Figure 5a Vertical sections of zonal velocity at the center of the ocean basin for a nonlinear, continuously stratified model forced by a zonal wind with an amplitude of 0.5 dyn cm^{-2}. The response is shown for easterly winds without nonlinearities (*left*), for easterly winds with nonlinearities (*middle*), and for westerly winds with nonlinearities (*right*). The contour interval is 10 cm s^{-1}, there is no zero contour, and the shaded regions indicate westward flow. An important nonlinear effect is the advection of the zonal current by the meridional circulation. After Philander & Pacanowski (1980).

Figure 5b As in Figure 5a, except that the ocean is forced by a southerly wind and the contour interval is 5 cm s^{-1}. The response is shown without nonlinearities (*left*) and with nonlinearities (*right*). Again, the advection of the zonal currents by the meridional circulation is apparent. After Philander & Delecluse (1983).

nonlinear solutions in response to zonal and meridional winds, respectively, each with an amplitude of 0.5 dyn cm^{-2}. The initial density field of the models has a sharp near-surface pycnocline with a maximum density gradient at $z = -100$ m, and density is roughly constant for $z > -50$ m. The flow fields of both linear solutions are similar to those in Figures 4a and 4b. There is, however, considerable vertical shear in the surface layer ($z > -50$ m), so that these solutions have characteristics of those in both Figures 2 and 4. Just as for the surface-layer models, it is visibly apparent that important nonlinear effects are the advection of the zonal flow field by the meridional and vertical currents.

5.   SOLUTIONS FOR PERIODIC WINDS

This section discusses the response of various equatorial models to periodic winds. For most of the solutions, the wind is again assumed to have a simple separable form

$$\tau^x = \tau^y = \tau_0 X(x) Y(y) e^{-i\sigma t}. \tag{38}$$

The first part of this section describes some important properties of solutions to linear, continuously stratified models. The concepts of equatorial resonances, focal points, and beams are discussed. The second part describes several solutions to periodically forced models of various types.

5.1   Solutions to Linear, Continuously Stratified Models

The solution to (20) without horizontal mixing and when $X(x)$ is a zonally bounded function is

$$u_n = \sum_{l=0}^{\infty} \sum_{j=1}^{2} \frac{c\alpha_0}{i\omega} (A_{lj} + R_{lj})\Phi_{lj}^{-} \, e^{ik_{lj}x - i\sigma t} + (A_{-12} + R_{-12})\phi_0 \, e^{ik_{-12}x - i\sigma t},$$

$$v_n = \sum_{l=0}^{\infty} \sum_{j=1}^{2} (A_{lj} + R_{lj})\phi_l \, e^{ik_{lj}x - i\sigma t}, \tag{39}$$

$$p_n = \sum_{l=0}^{\infty} \sum_{j=1}^{2} \frac{c^2\alpha_0}{i\omega} (A_{lj} + R_{lj})\Phi_{lj}^{+} \, e^{ik_{lj}x - i\sigma t} + c(A_{-12} + R_{-12})\phi_0 \, e^{ik_{-12}x - i\sigma t}$$

(McCreary 1980, 1981). The amplitudes are

$$A_{lj} = \frac{\tau_{0n}\alpha_0}{i} \frac{(1/c)[\eta Y]_l - (k_{lj}/\omega)[Y_\eta]_l}{k_{lj} - k_{lj'}} \int_{L_j}^{x} e^{-ik_{lj}x'} \, dx',$$

$$A_{-12} = \frac{\tau_{0n}}{2c} Y_0 \int_{-\infty}^{x} e^{-ik_{lj}x'} \, dx', \tag{40}$$

where $j \neq j'$, $L_1 = -L_2 = \infty$, and $\omega = \sigma - iA/c^2 - i\varepsilon$. The quantities R_{lj} are the amplitudes of waves that are reflected from eastern and western boundaries. The solution (39) is clearly a complicated superposition of all possible equatorially trapped waves.

EQUATORIAL RESONANCES   In some circumstances the contribution of a single wave dominates all the others in (39). One such circumstance occurs at equatorial resonances. The amplitude of the Rossby and gravity waves in (40) contains the term $(k_{lj} - k_{lj'})^{-1}$, and this term is large near resonance points where $\mathrm{Re}(k_{lj}) = \mathrm{Re}(k_{lj'})$. Damping in the system prevents the term from ever becoming infinitely large by ensuring that the k_{lj}'s always have an imaginary part. Several resonance points are indicated by crosses in Figure 1.

There has been only one theoretical study of equatorial resonances. Wunsch & Gill (1976) found a solution somewhat like (39). They pointed out that properties of sea-level observations at various islands near the equator were consistent with the presence of resonant equatorially trapped gravity waves. Luther (1980), in a detailed study of island sea-level data, found more evidence of resonant equatorially trapped gravity waves.

RAY THEORY   Usually a considerable number of waves contribute to the solution (39), and it is not useful to isolate the contribution of a single wave. It is useful, however, to separate out the contributions of less complex pieces of the solution. The solution, (39) and (19), is a double sum of vertical modes (designated by the index n) and of the various types of waves associated with each mode (designated by the index l). Thus, the separation can proceed conveniently in two different ways. One way considers the response of an individual baroclinic mode, i.e. it specifies a value for n and carries out the summation over all values of l. The other considers only the response of waves of a particular type, i.e. it specifies a value for l and carries out the summation over n. It is possible to use ray theory to predict where energy associated with pieces defined in these two ways will go.

Focal points   Several studies have considered the response of a single vertical mode to low-frequency forcing by a zonal wind (Schopf et al. 1981, Cane & Sarachik 1981, Cane & Moore 1981). In each of them, equatorial Kelvin waves generated by the wind field over the interior ocean reflect from the eastern boundary as a packet of Rossby waves [the waves with amplitudes R_{l1} in (39)]. Energy associated with this packet does not propagate entirely westward, but tends to focus back to the equator. This focusing of energy leaves "shadow zones" in the ocean interior off the equator, i.e. regions where energy associated with reflected Rossby waves does not appear. Figure 6, taken from Schopf et al. (1981), nicely illustrates these features.

Schopf et al. (1981) use ray theory to explain these properties as follows. At any locality, assume that solutions to (20) have the form $A(x, y, t) \exp[ikx + i\phi(y) - i\sigma t]$. Then the dispersion relation for long-wavelength Rossby waves is

$$\sigma = \frac{-\beta k}{\beta^2 y^2/c^2 + l^2},$$

(41)

where $l \equiv \phi_y$ is a local meridional wave number that varies with y. (In their discussion, Schopf et al. did not make the long-wavelength approximation.) The slope of ray paths is the ratio of zonal and meridional group velocities, so that

$$\frac{dy}{dx} = \frac{\sigma_l}{\sigma_k} = -2\frac{\sigma}{\beta}\sqrt{-\frac{\beta k}{\sigma} - \frac{\beta^2 y^2}{c^2}}.$$

(42)

Equation (42) has the general solution

$$y = \left(-\frac{c^2 k}{\sigma \beta}\right)^{1/2} \cos\left[2\frac{\sigma}{c}x + \theta_0\right],$$

(43)

where θ_0 is an integration constant.

Now let the eastern ocean boundary be a meridional barrier located at $x = 0$. Exact and numerical solutions indicate that phase is very nearly constant along the coast for the reflected packet of Rossby waves, so that to a good approximation Equation (43) satisfies the boundary condition

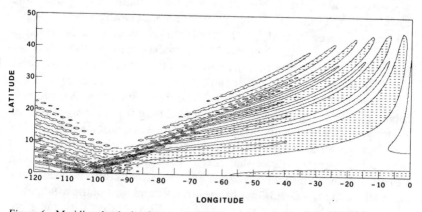

Figure 6 Meridional velocity field produced by equatorially trapped Rossby waves with $\sigma = 2\pi \, \text{yr}^{-1}$ and $c = 300 \, \text{cm s}^{-1}$ when phase is fixed to be constant along the eastern boundary [that is, $\theta_0 = 0$ in (43)]. The horizontal structure of the flow, rather than its amplitude, is important here. Energy focuses back onto the equator at a distance given approximately by (44). There are shadow zones where Rossby-wave energy does not appear. After Schopf et al. (1981).

$l \equiv \phi_y = 0$ at $x = 0$. According to (41), when $l = 0$ the value of y is $(c^2 k/\sigma\beta)^{1/2}$, and so it follows that the integration constant is $\theta_0 = 0$. With this choice, it is evident that all ray paths converge to the equator at the positions

$$x_{fi} = -\frac{\pi c}{4\sigma}(2i + 1), \tag{44}$$

$i = 0, 1, 2, \ldots$. When the exact dispersion relation is used instead of (41), these positions are only approximate focal points. Schopf et al. also found the ray paths when the eastern boundary was sloping. In that case θ_0 was a function of y, and the focal points were shifted off the equator.

Beams Ray theory can also be used to predict where energy associated with waves of a given type will go. In this case, one looks for approximate WKB solutions of the form $\exp[ikx + i\phi(z) - i\sigma t]$, and thereby defines a local vertical wave number $m \equiv \phi_z$. The dispersion relations for Kelvin, Rossby-gravity, and long-wavelength Rossby waves are just Equations (30), (31), and (32), with the replacement $c_n = N_b/|m|$. As a result, slopes of ray paths are

$$\frac{dz}{dx} = \frac{\sigma_m}{\sigma_k} = (2l + 1)\frac{\sigma}{N_b}\frac{|m|}{m}, \qquad -\frac{\sigma}{N_b}\frac{|m|}{m}, \qquad -\frac{\sigma}{N_b}\frac{|m|}{m}, \tag{45}$$

for long-wavelength Rossby waves, Rossby-gravity waves, and Kelvin waves, respectively. Thus, when phase propagates upward ($m > 0$), energy propagates downward to the west for long-wavelength Rossby waves and downward to the east for Rossby-gravity and Kelvin waves.

Figure 7a, taken from McCreary (1984), shows the response of the ocean when it is forced by a patch of zonal wind oscillating at the annual period. For convenience, in order that ray paths are straight lines, the background Väisälä frequency has a constant value $N_b = 0.0045 \text{ s}^{-1}$ beneath a surface mixed layer with a thickness of 75 m. The figure shows three sections of zonal velocity along the equator, taken at a time when the wind is most westerly. The ocean is unbounded, has only an eastern boundary, and has both boundaries in the upper, middle, and lower panels, respectively. All three panels clearly indicate that energy propagates into the deep ocean along ray paths. Without boundaries, energy descends into the deep ocean primarily along two beams that propagate east and west of the wind patch. To the east of the patch, the response is a Kelvin beam with the slope $-\sigma/N_b$, and to the west it is primarily an $l = 1$ Rossby beam with the slope $3\sigma/N_b$. When there is an eastern boundary, the Kelvin beam reflects from the eastern boundary as a collection of Rossby beams; the $l = 1, 3$, and 5 Rossby beams are clearly visible in the middle panel. With both boundaries, the response in the deep ocean is complicated because of the multiple reflection of the equatorial beams between the boundaries.

It is interesting that the contributions of individual vertical modes to the bounded basin solutions in Figure 7a all exhibit focal points at the positions predicted by (44), as in Figure 6. One might expect that near these focal points a particular baroclinic mode would stand out above all the others in the complete solutions. Curiously, there is no evidence at all of these focal points in Figure 7a. This result suggests that it is not always a useful exercise to isolate the contributions of individual baroclinic modes from the complete response.

Figure 7b, also taken from McCreary (1984), shows the response of the equatorial ocean when it is forced by a patch of meridional wind oscillating at a period of one month. As in Figure 7a, the background Väisälä frequency is constant beneath a surface mixed layer. The figure shows the meridional velocity field at a time when the winds are most southerly. The propagation of a beam of Rossby-gravity waves is evident in the figure. Even though the basin has both an eastern and western boundary, the response is considerably simpler than that in the lower panel of Figure 7a. The reason is that at a period of one month, no Rossby waves are available for the reflection from the eastern boundary. As a result, the beam reflects entirely poleward via β-plane Kelvin waves (Moore 1968; see the discussion in Section 3.3).

The solutions in Figures 7a and 7b assume that N_b is constant in the deep ocean. When there is a sharp pycnocline, the WKB approximation is not valid for low-order vertical modes, and some energy can reflect off the pycnocline rather than propagate through it (Philander 1978a). Rothstein et al. (1984) studied in detail the reflection of beams from pycnoclines of various sorts, and they found that there is surprisingly little reflection from the pycnocline. For realistic background density structures, nearly all the energy passed through the pycnocline along ray paths, just as predicted by ray theory.

5.2   Other Solutions

Wunsch (1977) studied the response of an inviscid, linear, continuously stratified model like (15), except that forcing in the model was conveniently represented as a surface distribution of vertical velocity (presumably driven by the wind). The forcing had the form of a westward-propagating sinusoidal wave with a period of one year, and the ocean basin was assumed to be infinitely deep and horizontally unbounded (although at the end of his paper, Wunsch reported effects introduced by a western ocean boundary). The solutions, in qualitative agreement with the observations, had a rich vertical structure that was narrowly confined to the equator.

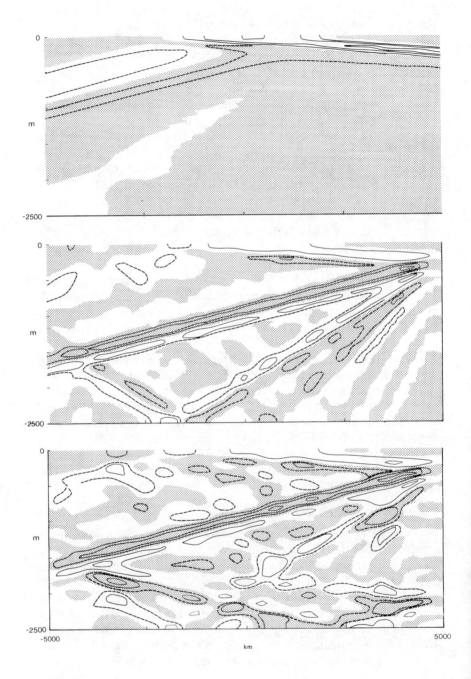

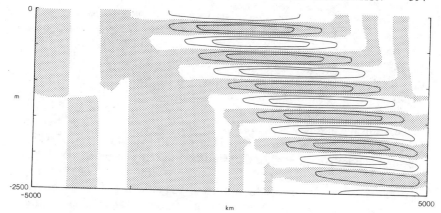

Figure 7b   Similar to Figure 7a, except that a section of meridional velocity is shown when the ocean is forced by a meridional wind oscillating at the monthly frequency. The basin has both an eastern and western boundary. The contour interval is 5 cm s^{-1}, there is no zero contour, and the shaded region indicates southward flow. A beam of Rossby-gravity waves propagates into the deep ocean with the slope $-\sigma/N_b$. There are no Rossby waves available to reflect this beam from the eastern boundary, and so it reflects entirely poleward as a beam of coastal Kelvin waves. After McCreary (1984).

In apparent contrast to the solutions in Figure 7a, Wunsch's solutions produced a great deal of energy in the deep ocean, even though the ocean basin lacked an eastern boundary. In fact, there is no contradiction between the two solutions. They differ because the forcing in Wunsch's model is not limited zonally. In Wunsch's solutions, energy still propagates into the deep ocean along ray paths that descend into the deep ocean at shallow angles. Therefore, all the energy present in the deep ocean in his solutions was

←————————————————————————————

Figure 7a   Vertical sections of zonal velocity along the equator when a linear, continuously stratified model is forced by a zonal wind oscillating at the annual frequency. The response is shown for an unbounded ocean (*top*), for an ocean with an eastern boundary (*middle*), and for an ocean with both an eastern and western boundary (*bottom*) at a time when the winds are most westerly. The wind stress is confined to the region $-2500 < x < 2500$ km, is essentially independent of y, and reaches a maximum amplitude of 0.125 dyn cm^{-2} in the center of the basin. The contour interval is 20 cm s^{-1}, there is no zero contour line, and shaded regions indicate westward flow. Dashed contours are ± 10 cm s^{-1} and are included only if they help to identify prominent features of the flow field. A strong beam of Kelvin waves, propagating into the deep ocean with the slope $-\sigma/N_b$, is evident in all the panels. When the ocean has an eastern boundary, the Kelvin beam reflects as a collection of Rossby beams, propagating into the deep ocean with the slopes $(2l+1)\sigma/N_b$. The flow pattern in the lower panel is complicated because of the multiple reflection of beams between both boundaries. After McCreary (1984).

necessarily generated by a remote part of the forcing. (With a western boundary the source of energy in the west was absent. The deep response was thus due entirely to remote forcing in the east.) In the solutions of Figure 7a, the zonal extent of the basin is effectively made much longer by the reflection of beams from basin boundaries.

Busalacchi & Picaut (1983) forced a reduced-gravity model (a single-baroclinic-mode model) with a realistic representation of the annual winds in the tropical Atlantic. They found that the annual sea-level response along the coast of Africa is most strongly influenced by zonal winds west of $10°W$. The alongshore component of the wind also affects coastal sea level, but because that component is weak, it does not affect the coastal ocean nearly as much as the remote winds do. According to (44) and with Busalacchi & Picaut's value of c (125 cm s^{-1}), a focal point should (just barely) exist in the Atlantic ($x_{f_0} \simeq -5000$ km). No focal point is apparent in their solution, however, most likely because the eastern boundary of their basin is a realistic representation of the actual (sloping) coast of Africa.

In a complementary study to that of Busalacchi & Picaut, McCreary et al. (1984) forced a linear, continuously stratified model with an idealized representation of the annual variation of the equatorial trade winds in the western Atlantic. As in the McCreary (1984) study, waves associated with several different modes superposed to form beams that propagated energy vertically as well as horizontally. Along the equator the response was predominantly a beam of equatorial Kelvin waves and a lowest-order ($l = 1$) Rossby beam. The coast of Africa at $5°N$ provided a northern boundary to the basin. Along this coast the response was primarily a beam of coastal Kelvin waves ($l = 0$ Rossby waves). The solution compared favorably with several observations from the Gulf of Guinea. In particular, it exhibited vertical propagation of phase along the coast of Africa at $5°N$, in good agreement with observations discussed by Picaut (1983).

In the tropical Indian Ocean there is a strong semiannual cycle to the zonal wind, with strong westerlies occurring during the transition periods between the Northeast and Southwest Monsoons. To investigate the effect of these winds on the ocean, Gent et al. (1983) forced a linear, continuously stratified model with a realistic representation of the semiannual cycle of the winds. Their ocean basin had eastern and western boundaries located at the longitudes $43°E$ and $97°E$, but it lacked northern and southern boundaries. Upward phase propagation existed in their solution and compared favorably with observations from the western Indian Ocean discussed by Luyten & Roemmich (1982). In contrast to the solutions in Figure 7a, equatorial beams were not apparent in their solutions; instead, focal points for several of the modes were visible in the complete solution. The discrepancy between this solution and others that exhibit a beamlike response is unresolved at the present time. The authors suggested that the difference is due to the complexity of the horizontal structure of the realistic wind field that drove their model, and that beams will only be apparent when $X(x)$ in (38) has a simple form. A more likely possibility is that the significant mixing in their model affected its response considerably.

Philander & Pacanowski (1981) forced a nonlinear, continuously stratified model with idealized zonal wind fields at various frequencies. The nonlinear terms affect the strong surface currents considerably. One obvious effect was that eastward currents are enhanced over westward currents. This strengthening is probably due to the advection of the near-surface zonal currents by the meridional and vertical flow field, in a manner like that discussed in Section 4 for the steady solutions. In contrast, nonlinearities did not affect the weaker deep currents very much. As in several of the linear models, the deep currents appeared to be dominated by a wind-driven Kelvin beam and an $l = 1$ Rossby beam (McCreary 1984).

Cox (1980) forced a nonlinear, continuously stratified model with a realistic representation of the annual cycle of the Pacific trades. At the time of the year when the surface currents were most intense, they became unstable and began to meander with a wavelength of 1000 km and a period of 1.1 months. The region of unstable currents then acted as a forcing region for waves propagating into the deep ocean. Cox noted the presence of Rossby-gravity waves near a depth of 2000 m at a position 20° east of the forcing region, and he used ray theory to explain how this energy propagated into the deep ocean. Thus, the Cox model generated a Rossby-gravity beam similar to the one shown in Figure 7b.

6. MODELS OF EL NIÑO

Early models of El Niño studied the response of the ocean to a relaxation of an idealized representation of the equatorial trade winds (Godfrey 1975, Hurlburt et al. 1976, McCreary 1976, 1977, O'Brien et al. 1981, Philander 1981). The response in Figure 3 is representative of these solutions, particularly of the solution of McCreary (1977). With the direction of the wind and of the response reversed, the figure shows the time development of surface pressure (sea level) in response to a relaxation of the equatorial trade winds confined to the central Pacific. Shortly after the winds relax, sea level rises rapidly in the eastern ocean (equivalent to a deepening of the pycnocline there) and drops in the western ocean, in association with a massive transfer of warm surface water from the western to the eastern ocean. This large-scale response is very similar to changes in sea level that have been observed to take place during El Niño events (Wyrtki 1977, 1979, 1984).

Current research generally falls into three categories. One type of effort finds solutions to ocean models that are forced by realistic, rather than idealized, winds. The advantage of the use of realistic winds is that solutions can be more closely compared with observations. Another type of research investigates the causes of SST anomalies during El Niño events. These studies use ocean models that include thermodynamics as well as dynamics. A third type studies how the ocean and the atmosphere interact during El Niño, by using coupled ocean-atmosphere models. This section summarizes a few papers that illustrate the direction of current research.

6.1 *Solutions for Realistic Winds*

Busalacchi & O'Brien (1981) forced a linear, reduced-gravity model with shipboard estimates of winds from the tropical Pacific during the period 1961–70. The correlation between time series of model layer thickness in the eastern ocean, h_e, and observed sea level from the eastern Pacific was remarkably high, suggesting that the simple model was correctly representing basic equatorial dynamics. The authors demonstrated that the variability in h_e was due to changes in the remote equatorial winds and involved a large-scale response, much like that in Figure 3. In particular, they showed that a westerly wind anomaly in the western equatorial Pacific was responsible for the onset of the 1965 and 1969 El Niño events.

Busalacchi et al. (1983a,b) extended the above calculations to include winds from the period 1971–78. They compared model layer thickness h with observed sea-level variability from the eastern, western, and central Pacific, and the comparisons were good in all three regions. In the eastern ocean, the strong 1972 and 1976 El Niño events were reproduced. As in the 1963 and 1969 events, they were initiated by a collapse of the trade winds in the western Pacific. In the western Pacific, the 1963, 1969, 1972, and 1976 El Niño events were followed by a large decrease in h and also a drop in observed sea level. The decrease in h was the result of the radiation of Rossby waves from a region of weakened trade winds in the central Pacific, again similar to the response in Figure 3. In the central Pacific, h was influenced both by Kelvin waves and by $l = 1$ Rossby waves.

6.2 *Thermodynamic Models*

Schopf & Harrison (1983) used a slightly modified version of the Schopf & Cane (1983) thermodynamic, mixed-layer model to study the development of SST anomalies during the onset of El Niño. In a series of three numerical experiments, they demonstrated that the development was sensitive to the initial SST field. In the first experiment, the ocean was initially in a state of rest, with uniformly warm SST everywhere in the basin. In the second, the ocean was initially in balance with mean southerly winds and had cool SST near the eastern boundary south of the equator. In the third, the ocean was in balance with mean easterly winds, with cool SST along the equator in the eastern and central ocean. In all three experiments, westerly winds were switched on initially in the western part of the basin. As in Figure 3, the radiation of Kelvin waves from the western Pacific and their subsequent reflection from the eastern boundary was the basic dynamical process occurring in the model. Warm anomalies occurred only in the latter two cases, which initially had regions of cool SST. The warming was caused predominantly by advective effects associated with the passage of wave fronts, rather than by mixed-layer processes, a result supported by the studies of Philander (1981) and of Gill (1983b).

Schopf (1983) discussed other solutions to the Schopf & Harrison model. In this study, air temperature was assumed to react rapidly to SST, rather than to be a fixed quantity. The effect was to decrease γ in (12) by a factor of 10, thereby increasing the e-folding decay time of SST signals from 50 to 500 days. Solutions developed SST anomalies that extended over a much larger area than those in the Schopf & Harrison study, in better agreement with observations.

The 1982 El Niño event differed from the more common events in several ways. One was that in 1982, warm SST anomalies appeared first along the equator in late summer and spread east, whereas during typical events warm SST anomalies appear first at the coast of South America and spread west. Harrison & Schopf (1984) argued that this difference was due only to the fact that the SST field at the time of the onset (summer) was different in 1982. During the spring, SST is cool only near the coast of South America, whereas in the summer a cool tongue exists along the equator as well. They contrasted two solutions to the Schopf & Harrison model that had different initial SST fields resembling the springtime and summertime situations in the Pacific. Model SST anomalies developed differently for the two cases, in a manner consistent with the observed differences.

Zebiak & Cane (1983) used a thermodynamic, mixed-layer model to follow the development of SST anomalies throughout an El Niño event. They first forced their model with a realistic representation of the seasonal cycle of the Pacific trade winds, and they then altered the wind field to include the wind anomalies associated with El Niño discussed by Rasmussen & Carpenter (1982). They considered the relative importance of various thermodynamic processes throughout the event and concluded that no single process (like horizontal advection) dominated at all times.

6.3   Coupled Ocean-Atmosphere Models

McCreary (1983) and McCreary & Anderson (1984) developed simple, coupled ocean-atmosphere models to study El Niño and the Southern

Oscillation. In both studies, the model ocean was a linear, reduced-gravity system like (8). Thermodynamics in the ocean was highly parameterized, with SST being either warm or cool depending on whether the thickness of the layer h was greater than or less than an externally specified thickness h_c. In the former study, the model atmosphere consisted of two patches of zonal wind stress τ_w and τ_h, where τ_w was a region of strengthened equatorial trade winds in the central ocean, and τ_h was a region of strengthened extra-equatorial trade winds in the eastern ocean. The two wind patches interacted with the ocean according to the ideas of Bjerknes (1966, 1969): when the eastern ocean was cool ($h < h_c$) τ_w switched on simulating an enhanced Walker circulation, and when the eastern ocean was warm ($h > h_c$) τ_h switched on simulating an enhanced Hadley circulation. In the latter study, the model atmosphere consisted of two wind patches, τ_w and τ_s, where τ_w was the same as above and τ_s was an idealized version of the annual cycle of the Pacific trade winds.

The interesting property of these two models is that, for reasonable choices of model parameters, they both oscillate at the long time scales associated with the Southern Oscillation. The dynamics of the oscillations is, however, quite different between the two systems. In the former, it is the time it takes Rossby waves generated by τ_h to cross the ocean that sets the oscillation period. In the latter, the coupled system has two states of equilibrium: one with τ_w switched on, and the other with τ_w switched off. The wind field τ_s acts as a "trigger" that switches the system from one near-equilibrium state to another.

The above models are limited in that both ocean thermodynamics and the model atmosphere are highly parameterized. Anderson & McCreary (1984) studied a more sophisticated system that included both active ocean thermodynamics and a dynamical, rather than empirical, atmosphere. The model ocean was the nonlinear, reduced-gravity, thermodynamic model described in Section 2 [see the discussion of Equations (10) and (11)]. The equations of motion of the model atmosphere were

$$ru_a - fv_a + p_{ax} = 0,$$

$$rv_a + fu_a + p_{ay} = 0,$$ (46)

$$\varepsilon p_a + c^2(u_{ax} + v_{ay}) = Q.$$

These equations describe the response of a single, baroclinic mode of the atmosphere to a source of latent heat Q. Despite its simplicity, this model atmosphere has been successfully used in studies of the tropical wind field associated with El Niño (Gill 1980, Zebiak 1982). Interactions between the ocean and the atmosphere were specified as follows. Wind stress driving the ocean was given by

$$\tau^x = \rho_a C_d' u_a, \qquad \tau^y = \rho_a C_d' v_a, \tag{47}$$

and the release of latent heat was related to the temperature of the ocean by

$$Q = -Q_0 \frac{T - T_c}{\bar{T}(0) - T_c} \theta(T - T_c), \tag{48}$$

where $\theta(T - T_c)$ is a step function, and $\bar{T}(0)$ is roughly the maximum possible value of T. According to (48), there is release of latent heat to the atmosphere only when the ocean becomes warmer than T_c.

Solutions were found numerically for both cyclic and bounded ocean basins. In both cases, instabilities grew to a finite amplitude and began to propagate slowly eastward at speeds of the order of 10 cm s^{-1}, and these disturbances existed for a wide range of parameters. Interestingly, they did not depend on the presence of the advection terms in the ocean; the response of the system was not appreciably changed when these terms were neglected. The bounded-basin solutions compared favorably with observations of El Niño and the Southern Oscillation in several respects; in particular, they oscillated at long time scales. They failed, however, to simulate the rapid onset of El Niño events.

The studies of Lau (1981) and of Philander et al. (1984) provide some insight into the cause of these disturbances. They both considered coupled systems consisting of a linear, reduced-gravity ocean model like (8) and an atmospheric model like (46). (Lau's model atmosphere also included time-derivative terms.) Both τ^x and τ^y were specified according to (47), and Q was taken to be directly proportional to the layer-thickness anomaly $h - H$. Lau greatly simplified the system by considering only coupling between equatorially trapped Kelvin waves in the ocean and atmosphere, and he found a dispersion relation for the coupled waves. One of the roots always had a phase speed very close to that of an atmospheric Kelvin wave. The other was an oceanic Kelvin wave for weak coupling. As coupling increased, however, the phase speed of the wave slowed to zero and then became positive imaginary, which indicated that it was now a growing instability. Philander et al. solved their system numerically without resorting to Lau's simplification. They also found a rapidly growing instability.

The instability in these two models develops in the following way. Suppose initially that $h - H$ is weakly positive in a patch centered on the equator, so that a Q exists to drive the atmosphere. This Q drives westerlies to the west of the patch and easterlies to the east of it. Provided the coupling is sufficiently strong, the effect of this convergent wind field on the ocean is to pile up more mass in the patch, thereby increasing $h-H$ and Q once again. There is no limit on the size of $h-H$ in the model, and so the instability grows indefinitely. The disturbance in the Anderson & McCreary model is caused by a similar mechanism. It does not grow indefinitely because Q is related to T, rather than $h-H$, and ocean thermodynamics prevents T from ever increasing much beyond $\bar{T}(0)$.

7. SUMMARY AND CONCLUSIONS

Equatorial models range in dynamical complexity from linear, constant-thickness, surface-layer models to fully nonlinear, continuously stratified, numerical models. The simpler models have proven to be very useful for identifying the important processes involved in equatorial dynamics. In particular, under certain conditions it is possible to find analytic solutions to the linear, continuously stratified system [(15)], and these solutions provide a basis for understanding much of equatorial dynamics. A thorough understanding of equatorial dynamics is only possible through a careful comparison of solutions to models of all types.

Equatorially trapped waves are a complete set of functions of the unforced version of (20). Without mixing they are the familiar set of Rossby and gravity waves, the Rossby-gravity wave and the Kelvin wave with the dispersion curves shown in Figure 1. With mixing the waves are damped in the direction of their group velocity. The meridional structures of the waves are modified by the presence of northern and southern boundaries, and two extra boundary-trapped waves are possible in this case. Waves reflect from eastern and western boundaries and from islands. Background zonal currents distort the meridional structure of the waves and can cause them to be unstable. There are, as yet, no studies of the critical-layer absorption of equatorially trapped waves.

The response to a switched-on wind differs considerably depending on the type of model and on the zonal structure of the wind. The shear flow in the surface layer of the ocean rapidly adjusts to equilibrium with the wind in a few days. In response to a zonally independent zonal wind, the inviscid solution to (20) is an accelerating equatorial current (the Yoshida jet). If the wind is zonally bounded, the inviscid solution adjusts to a state of Sverdrup balance plus a steady equatorial jet. Waves reflected from ocean boundaries act to eliminate this jet. The addition of mixing can strengthen equatorial currents by preventing reflected waves from completely canceling the jet.

Provided that mixing is sufficiently strong, solutions to continuously stratified models forced by steady winds compare favorably with observations. They produce quite realistic mixed-layer flows, undercurrents, and subsurface countercurrents. A comparison of steady, nonlinear and linear solutions suggests that the effect of nonlinearities is primarily to distort the linear solutions in recognizable ways. For example, an important nonlinear effect is the advection of the zonal currents in the direction of the meridional circulation. For smaller values of mixing parameters, nonlinear solutions do not reach a steady state, but rather they are unstable. If mixing is weak, eddy activity in the solutions is unrealistically large.

The response of a linear, continuously stratified ocean to periodic winds is a complex superposition of all possible equatorially trapped waves. It is not usually useful to isolate the contribution of any single wave; the exception is the case of equatorial resonance. Instead, ray theory is a useful technique for determining where energy associated with packets of waves will go. Under suitable conditions [$\theta_0 = 0$ in (43)], Rossby waves associated with a single vertical mode tend to focus their energy back on the equator. In contrast, waves associated with a single type of wave superpose to form beams that carry energy both vertically and horizontally at slopes given by (45). The propagation of beams into the deep ocean is evident in several linear and nonlinear solutions.

Early models of El Niño considered the response of the ocean to an idealized relaxation of the equatorial trade winds. Current research generally falls into three, not entirely exclusive, categories. One type of research forces ocean models with a realistic representation of the Pacific wind field. One model of this type indicated that a weakening of the trade winds in the far western Pacific initiated several El Niño events. Another demonstrated that the drop in sea level at western Pacific islands during El Niño was caused by the radiation of Rossby waves from the central Pacific. A second type of research uses thermodynamic models to study the generation and decay of SST anomalies during El Niño. Several models indicated that the advection of temperature is important during the onset of El Niño, but that other processes are important later on in the event. A third type of research uses coupled ocean-atmosphere models to investigate the ocean's role in setting the long time scales associated with El Niño and the Southern Oscillation. Several models do oscillate at long time scales, all of them for different dynamical reasons.

There is room for development in all the branches of equatorial modeling discussed in this review. However, significant advances in equatorial modeling have usually followed the discovery of an unexpected and intriguing phenomenon. A good example is the discovery of deep equatorial currents. There was an immediate need to develop models to explain how energy gets into the deep equatorial ocean. A recent example is the 1982/83 El Niño event. This event was unusual in many ways and has stimulated the development of both atmospheric and ocean models. A safe prediction is that the next few years will see a rapid growth in the field of tropical ocean-atmosphere interaction.

ACKNOWLEDGMENTS

The writing of this paper was sponsored by the National Science Foundation under grant No. OCE 79-21785 through PEQUOD and under grant No. ATM 82-05491. Necessary computations were performed on the CRAY-1 computer at the National Center for Atmospheric Research. NCAR is supported by the National Science Foundation.

Much of the discussion in Section 4.1, concerning the linear solution (34) and its relevance to surface-layer equatorial flows, is an outgrowth of discussions I had several years ago with Dennis Moore and Mike McPhaden. I am indebted to David Anderson, Neill Cooper, Pijush Kundu, Mike McPhaden, and Lew Rothstein, who suggested several improvements of an earlier version of this review. Finally, the efforts of Kevin Kohler and Kathy Maxson are greatly appreciated; without their assistance, the preparation of the manuscript in a timely manner would not have been possible.

Literature Cited

Anderson, D. L. T., McCreary, J. P. 1984. Slowly propagating disturbances in a coupled, ocean-atmosphere model. Submitted for publication

Anderson, D. L. T., Rowlands, P. B. 1976. The role of inertia-gravity and planetary waves in the response of a tropical ocean to the incidence of an equatorial Kelvin wave on a meridional boundary. *J. Mar. Res.* 34:295–312

Bjerknes, J. 1966. A possible response of the atmospheric Hadley circulation to equatorial anomalies of ocean temperature. *Tellus* 18:820–29

Bjerknes, J. 1969. Atmospheric teleconnections from the equatorial Pacific. *Mon. Weather Rev.* 97:163–72

Boyd, J. P. 1980a. The non-linear equatorial Kelvin wave. *J. Phys. Oceanogr.* 10:1–11

Boyd, J. P. 1980b. Equatorial solitary waves. Part I: Rossby solutions. *J. Phys. Oceanogr.* 10:1699–1718

Busalacchi, A. J., O'Brien, J. J. 1981. Interannual variability of the equatorial Pacific in the 1960's. *J. Geophys. Res.* 86:10901–7

Busalacchi, A. J., Picaut, J. 1983. Seasonal variability from a model of the tropical Atlantic. *J. Phys. Oceanogr.* 13:1564–88

Busalacchi, A. J., Takeuchi, K., O'Brien, J. J.

1983a. Interannual variability of the equatorial Pacific—revisited. *J. Geophys. Res.* 88:7551–62

Busalacchi, A. J., Takeuchi, K., O'Brien, J. J. 1983b. On the interannual wind-driven response of the tropical Pacific Ocean. In *Hydrodynamics of the Equatorial Ocean*, pp. 155–95. Amsterdam: Elsevier

Cane, M. A. 1979a. The response of an equatorial ocean to simple wind stress patterns: I. Model formulation and analytic results. *J. Mar. Res.* 37:233–52

Cane, M. A. 1979b. The response of an equatorial ocean to simple wind stress patterns: II. Numerical results. *J. Mar. Res.* 37:355–98

Cane, M. A., duPenhoat, Y. 1982. The effect of islands on low frequency equatorial motions. *J. Mar. Res.* 40:937–62

Cane, M. A., Gent, P. R. 1984. Reflections of low-frequency equatorial waves at arbitrary western boundaries. *J. Mar. Res.* In press

Cane, M. A., Moore, D. W. 1981. A note on low-frequency equatorial basin modes. *J. Phys. Oceanogr.* 11:1578–85

Cane, M. A., Sarachik, E. S. 1976. Forced baroclinic ocean motions: I. The linear equatorial unbounded case. *J. Mar. Res.* 34:629–65

Cane, M. A., Sarachik, E. S. 1977. Forced baroclinic ocean motions: II. The linear bounded case. *J. Mar. Res.* 35:395–432

Cane, M. A., Sarachik, E. S. 1979. Forced baroclinic ocean motions: III. The linear equatorial basin case. *J. Mar. Res.* 37:355–98

Cane, M. A., Sarachik, E. S. 1981. The response of a linear baroclinic equatorial ocean to periodic forcing. *J. Mar. Res.* 39:651–93

Cane, M. A., Sarachik, E. S. 1983. Equatorial oceanography. *Rev. Geophys. Space Phys.* 21:1137–48

Charney, J. G. 1960. Non-linear theory of a wind-driven homogeneous layer near the equator. *Deep-Sea Res.* 6:303–10

Charney, J. G., Speigel, S. L. 1971. The structure of wind-driven currents in homogeneous oceans. *J. Phys. Oceanogr.* 1:149–60

Clarke, A. J. 1983. The reflection of equatorial waves from oceanic boundaries. *J. Phys. Oceanogr.* 13:1193–1207

Cochrane, J. D., Kelly, F. J., Olling, C. R. 1979. Subthermocline countercurrents in the western equatorial Atlantic Ocean. *J. Phys. Oceanogr.* 9:724–38

Cox, M. D. 1980. Generation and decay of 30-day waves in a numerical model of the Pacific. *J. Phys. Oceanogr.* 10:1168–86

Cromwell, T. 1953. Circulation in a meridional plane in the central equatorial Pacific. *J. Mar. Res.* 12:196–213

Donguy, J., Eldin, G., Meyers, G., Morliere, A., Rebert, J. P. 1984. A thermal interpretation of an equatorial change in circulation in the western Pacific. *Trop. Ocean-Atmos. Newslett. No. 24*, pp. 9–10

Duing, W., Hallock, Z. 1979. Equatorial waves in the upper central Atlantic. *Deep-Sea Res.* 26(Suppl. 2):161–78

Eriksen, C. C. 1981. Deep currents and their interpretation as equatorial waves in the western Indian Ocean. *J. Phys. Oceanogr.* 11:48–70

Firing, E., Fenander, C., Miller, J. 1981. Profiling current meter measurements from the NORPAX Hawaii to Tahiti Shuttle experiment. *Data Rep. 39, HIG-81-2*, Hawaii Inst. Geophys., Honolulu. 146 pp.

Firing, E., Lukas, R., Sadler, J., Wyrtki, K. 1983. Equatorial Undercurrent disappears during 1982–1983 El Niño. *Science* 222:1121–23

Gent, P. R., O'Neill, K., Cane, M. A. 1983. A model of the semiannual oscillation in the equatorial Indian Ocean. *J. Phys. Oceanogr.* 13:2148–60

Gill, A. E. 1971. The Equatorial Current in a homogeneous ocean. *Deep-Sea Res.* 18:421–31

Gill, A. E. 1980. Some simple solutions for heat-induced tropical circulation. *Q. J. R. Meteorol. Soc.* 106:447–62

Gill, A. E. 1983a. *Atmosphere-Ocean Dynamics.* New York: Academic. 661 pp.

Gill, A. E. 1983b. An estimation of sea-level and surface-current anomalies during the 1972 El Niño and consequent thermal effects. *J. Phys. Oceanogr.* 13:586–606

Godfrey, J. S. 1975. On ocean spin-down. I: A linear experiment. *J. Phys. Oceanogr.* 5:399–409

Hallock, Z. 1979. On wind-excited, equatorially trapped waves in the presence of mean currents. *Deep-Sea Res.* 26(Suppl. 2):261–84

Harrison, D. E., Schopf, P. S. 1984. Kelvin-wave-induced anomalous advection and the onset of SST warming in El Niño events. *Mon. Weather Rev.* 112:923–33

Hickie, B. P. B. 1977. The effects of coastal geometry on equatorially trapped planetary waves. Part 1: Free oscillations in the Gulf of Guinea. Unpublished manuscript

Hisard, P., Merle, J., Voituriez, B. 1970. Equatorial Undercurrent at 170°E in March and April 1967. *J. Mar. Res.* 28:281–303

Horel, J. D., Wallace, J. M. 1981. Planetary scale atmospheric phenomena associated with the Southern Oscillation. *Mon. Weather Rev.* 109:813–29

Hurlburt, H., Kindle, J., O'Brien, J. J. 1976. A numerical simulation of the onset of El Niño. *J. Phys. Oceanogr.* 6:621–31

Kindle, J. C. 1981. On equatorial Rossby solitons. See McCreary et al. 1981, pp. 353–61

Kindle, J. C. 1983. On the generation of Rossby solitons during El Niño. In *Hydrodynamics of the Equatorial Ocean*, pp. 353–68. Amsterdam: Elsevier

Knox, A., Anderson, D. L. T. 1984. Recent advances in the study of the low-latitude ocean circulation. *Prog. Oceanogr.* In press

Knox, R., Halpern, D. 1982. Long-range Kelvin wave propagation of transport variations in Pacific Ocean equatorial currents. *J. Mar. Res.* 40(Suppl):329–39

Lau, K. M. 1981. Oscillations in a simple equatorial climate system. *J. Atmos. Sci.* 38:248–61

Leetmaa, A., McCreary, J. P., Moore, D. W. 1981. Equatorial currents: observations and theories. In *Evolution of Physical Oceanography*, pp. 184–96. Cambridge, Mass: MIT Press

Legeckis, R. 1977. Long waves in the eastern equatorial Pacific Ocean: a view from a geostationary satellite. *Science* 197:1179–81

Legeckis, R., Pichel, W., Nesterczuk, G. 1983. Equatorial long waves in geostationary satellite observations and in a multi-channel sea surface temperature analysis. *Bull. Am. Meteorol. Soc.* 64:133–39

Lighthill, M. J. 1969. Dynamic response of the Indian Ocean to the onset of the Southwest Monsoon. *Philos. Trans. R. Soc. London Ser. A* 265:45–93

Lukas, R., Firing, E. 1984. The annual Rossby wave in the central equatorial Pacific Ocean. Submitted for publication

Luther, D. S. 1980. *Observations of long period waves in the tropical oceans and atmosphere.* PhD thesis. Mass. Inst. Technol./Woods Hole Oceanogr. Inst., Cambridge/Woods Hole, Mass. (WHOI-8-17)

Luyten, J. R., Roemmich, D. H. 1982. Equatorial currents at semiannual period in the Indian Ocean. *J. Phys. Oceanogr.* 12:406–13

Luyten, J. R., Swallow, J. C. 1976. Equatorial undercurrents. *Deep-Sea Res.* 23:999–1001

McCreary, J. P. 1976. Eastern tropical ocean response to changing wind systems—with application to El Niño. *J. Phys. Oceanogr.* 6:632–45

McCreary, J. P. 1977. *Eastern ocean response to changing wind systems.* PhD thesis. Univ. Calif., San Diego

McCreary, J. P. 1980. Modelling wind-driven ocean circulation. *Tech. Rep. HIG-80-3*, Hawaii Inst. Geophys., Honolulu. 64 pp.

McCreary, J. P. 1981. A linear, stratified ocean model of the Equatorial Undercurrent. *Philos. Trans. R. Soc. London Ser. A.* 298:603–35

McCreary, J. P. 1983. A model of tropical ocean-atmosphere interaction. *Mon. Weather Rev.* 111:370–87

McCreary, J. P. 1984. Equatorial beams. *J. Mar. Res.* 42:395–430

McCreary, J. P., Anderson, D. L. T. 1984. A simple model of El Niño and the Southern Oscillation. *Mon. Weather Rev.* 112:934–46

McCreary, J. P., Moore, D. W., Witte, J. M., eds. 1981. *Recent Progress in Equatorial Oceanography: A Report of the Final Meeting of SCOR Working Group 47 in Venice, Italy.* Fort Lauderdale, Fla: Nova Univ./NYIT Press. 466 pp.

McCreary, J. P., Picaut, J., Moore, D. W. 1984. Effects of remote annual forcing in the eastern tropical Atlantic Ocean. *J. Mar. Res.* 42:45–81

McKee, W. D. 1973. The wind-driven equatorial circulation in a homogeneous ocean. *Deep-Sea Res.* 20:889–99

McPhaden, M. J. 1981. Continuously stratified models of the steady-state equatorial ocean. *J. Phys. Oceanogr.* 11:337–54

McPhaden, M. J. 1984. On the dynamics of Equatorial Subsurface Countercurrents. *J.*

Phys. Oceanogr. In press

McPhaden, M. J., Knox, R. A. 1979. Equatorial Kelvin and inertio-gravity waves in zonal shear flow. *J. Phys. Oceanogr.* 9:263–77

Moore, D. W. 1968. *Planetary-gravity waves in an equatorial ocean.* PhD thesis. Harvard Univ., Cambridge, Mass.

Moore, D. W. 1980. Modelling the tropical ocean circulation. In *Expert Lectures in Ocean Circulation Modelling*, ed. J. J. O'Brien (Rep. to French Govt.)

Moore, D. W., Philander, S. G. H. 1978. Modelling of the tropical ocean circulation. In *The Sea*, 6:319–61. New York: Wiley-Interscience

Moore, D. W., Hisard, P., McCreary, J. P., Merle, J., O'Brien, J. J., et al. 1978. Equatorial adjustment in the eastern Atlantic. *Geophys. Res. Lett.* 5:637–40

Moore, D. W., Brasington, J., Spielvogel, L. 1981. Reflection of equatorial planetary waves from an inclined coast. See McCreary et al. 1981, pp. 363–64

Nihoul, J. 1985. Coupled atmosphere-ocean models. *Proc. Int. Liège Colloq. Ocean Hydrodyn., 16th.* Amsterdam: Elsevier. In press

O'Brien, J. J. 1979. Equatorial oceanography. *Rev. Geophys. Space Phys.* 17(7): 1569–75

O'Brien, J. J., Busalacchi, A. J., Kindle, J. 1981. Ocean models of El Niño. In *Resource Management and Environment Uncertainty*, pp. 159–212. New York: Wiley-Interscience

O'Neill, K. 1982. Observations of vertically propagating equatorially trapped waves in the deep western Indian Ocean. *Tech. Rep. WHOI-82-11*, Woods Hole Oceanogr. Inst., Woods Hole, Mass. 162 pp.

Pacanowski, R. C., Philander, S. G. H. 1981. Parameterization of vertical mixing in numerical models of tropical oceans. *J. Phys. Oceanogr.* 11:1443–51

Philander, S. G. H. 1973. Equatorial Undercurrent: measurements and theories. *Rev. Geophys. Space Phys.* 11:513–70

Philander, S. G. H. 1976. Instabilities of zonal equatorial currents. *J. Geophys. Res.* 81:3725–35

Philander, S. G. H. 1977. The effects of coastal geometry on equatorial waves. *J. Mar. Res.* 35:509–23

Philander, S. G. H. 1978a. Forced ocean waves. *Rev. Geophys. Space Phys.* 16:15–46

Philander, S. G. H. 1978b. Instabilities of zonal equatorial currents, 2. *J. Geophys. Res.* 83:3679–82

Philander, S. G. H. 1979. Equatorial waves in the presence of the Equatorial Undercurrent. *J. Phys. Oceanogr.* 9:254–62

Philander, S. G. H. 1981. The response of the

equatorial ocean to a relaxation of the trade winds. *J. Phys. Oceanogr.* 11:176–89

Philander, S. G. H., Delecluse, P. 1983. Coastal currents in low latitudes. *Deep-Sea Res.* 30:887–902

Philander, S. G. H., Pacanowski, R. C. 1980. The generation of equatorial currents. *J. Geophys. Res.* 85(C2):1123–36

Philander, S. G. H., Pacanowski, R. C. 1981. Response of equatorial oceans to periodic forcing. *J. Geophys. Res.* 86(C3):1903–16

Philander, S. G. H., Yamagata, T., Pacanowski, R. C. 1984. Unstable air-sea interactions in the tropics. *J. Atmos. Sci.* 41:604–13

Picaut, J. 1983. Propagation of the seasonal upwelling in the eastern Atlantic. *J. Phys. Oceanogr.* 13:18–37

Rasmussen, E. M., Carpenter, T. H. 1982. Variations in tropical sea surface temperature and surface wind fields associated with the Southern Oscillation/El Niño. *Mon. Weather Rev.* 110:354–84

Ripa, P. 1982. Nonlinear wave-wave interactions in a one-layer reduced-gravity model on the equatorial beta-plane. *J. Phys. Oceanogr.* 12:97–111

Ripa, P. 1983a. Weak interactions of equatorial waves in a one layer model. Part I: General properties. *J. Phys. Oceanogr.* 13:1208–26

Ripa, P. 1983b. Weak interactions of equatorial waves in a one layer model. Part II: Applications. *J. Phys. Oceanogr.* 13:1227–40

Robinson, A. R. 1966. An investigation into the wind as the cause of the Equatorial Undercurrent. *J. Mar. Res.* 24:179–204

Rothstein, L. M. 1983. *A model of the equatorial sea surface temperature field and associated circulation dynamics.* PhD thesis. Univ. Hawaii, Honolulu

Rothstein, L. M., Moore, D. W., McCreary, J. P. 1984. Interior reflections of a periodically forced equatorial Kelvin wave. Submitted for publication

Rowlands, P. B. 1982. The flow of equatorial Kelvin waves and the Equatorial Undercurrent around islands. *J. Mar. Res.* 40:915–34

Schopf, P. S. 1983. On equatorial waves and El Niño. II: Effects of air-sea thermal coupling. *J. Phys. Oceanogr.* 13:1878–93

Schopf, P. S., Cane, M. A. 1983. On equatorial dynamics, mixed layer physics and sea surface temperature. *J. Phys. Oceanogr.* 13:917–35

Schopf, P. S., Harrison, D. E. 1983. On equatorial waves and El Niño. I: Influence of initial states on wave-induced currents and warming. *J. Phys. Oceanogr.* 13:936–48

Schopf, P. S., Anderson, D. L. T., Smith, R. 1981. Beta-dispersion of low-frequency Rossby waves. *Dyn. Atmos. Oceans* 5:187–214

Semtner, A. J. 1981. Recent models of the Equatorial Undercurrent. See McCreary et al. 1981, pp. 127–34

Semtner, A. J., Holland, W. R. 1980. Numerical simulation of equatorial ocean circulation. Part I: A basic case in turbulent equilibrium. *J. Phys. Oceanogr.* 10:667–93

Servain, J., Picaut, J., Merle, J. 1982. Evidence of remote forcing in the equatorial Atlantic Ocean. *J. Phys. Oceanogr.* 12:457–63

Stommel, H. 1960. Wind-drift near the equator. *Deep-Sea Res.* 6:298–302

Tsuchiya, M. 1972. A subsurface North Equatorial Countercurrent in the eastern equatorial Pacific Ocean. *J. Geophys. Res.* 77:5981–86

Tsuchiya, M. 1975. Subsurface countercurrents in the eastern equatorial Pacific Ocean. *J. Mar. Res.* 33:145–75

Veronis, G. 1973. Large scale ocean circulation. In *Advances in Applied Mechanics*, 13:1–92. New York: Academic

Weisberg, R. H., Horigan, A. M. 1981. Low-frequency variability in the equatorial Atlantic. *J. Phys. Oceanogr.* 11:913–20

Weisberg, R. H., Tang, T. Y. 1983. Equatorial ocean response to growing and moving wind systems with application to the Atlantic. *J. Mar. Res.* 41:461–86

Witte, J. M., ed. 1983. *Papers from 1982/83 El Niño/Southern Oscillation Workshop held at AOML/NOAA, Miami, Florida, November 3–4, 1983.* Reg. No. 4: Gov. Print. Off.

Wunsch, C. 1977. Response of an equatorial ocean to a periodic monsoon. *J. Phys. Oceanogr.* 7:497–511

Wunsch, C., Gill, A. E. 1976. Observations of equatorially trapped waves in Pacific sea level variations. *Deep-Sea Res.* 23:371–90

Wyrtki, K. 1977. Sea level during the 1972 El Niño. *J. Phys. Oceanogr.* 7:779–87

Wyrtki, K. 1979. The response of sea surface topography to the 1976 El Niño. *J. Phys. Oceanogr.* 9:1223–31

Wyrtki, K. 1984. Monthly maps of sea level in the Pacific during the El Niño of 1982 and 1983. In *Time Series of Ocean Measurements, IOC Tech. Ser. No. XX,* Vol. 2. In press

Yoon, J.-H. 1981. Effects of islands on equatorial waves. *J. Geophys. Res.* 86:10913–20

Zebiak, S. E. 1982. A simple atmospheric model of relevance to El Niño. *J. Atmos. Sci.* 39:2017–27

Zebiak, S. E., Cane, M. A. 1983. Modelling of sea surface temperature during El Niño. See Witte 1983, pp. 223–29

Ann. Rev. Fluid Mech. 1985. 17:411–45

THE KUTTA CONDITION
IN UNSTEADY FLOW

David G. Crighton

Department of Applied Mathematical Studies, University of Leeds, Leeds LS2 9JT, England

INTRODUCTION

In several papers published in the first decade of this century, Kutta and Joukowsky independently proposed that the lift on an airfoil at incidence in a steady unseparated flow is given by potential-flow theory with the unique value of the circulation that removes the inverse-square-root velocity singularity at the trailing edge. This proposal—tantamount to saying (cf. Batchelor 1967) that in the unsteady start-up phase the action of viscosity is such that, in the ultimate steady motion, viscosity can be explicitly ignored but implicitly incorporated in a single edge condition—is known as the *Kutta-Joukowsky hypothesis*. Subsequently the name *"Kutta condition"* (no doubt largely for brevity) has come to be used to connote the removal of a velocity singularity at some distinguished point on a body in *unsteady* flow.[1]

The condition has recently been applied to unsteadiness in a variety of mean configurations. These include trailing-edge flows with the same and with different flows on the two sides of the body upstream of the edge, attached leading-edge flows, and grossly separated flows past bluff bodies. Imposition of a Kutta condition on unsteady perturbations to one of these mean flows has a variety of physical ramifications. It represents the mechanism by which both the lift is changed and the amplitude and directivity of a sound field are modified. It is the analytical step that in many cases describes the conversion—almost total—of acoustic energy in an incident sound wave to energy of vortical motion on a shear layer; on

---

[1] Tani (1979) argues convincingly for the attribution of the trailing-edge condition to Kutta alone, as implied in the title of the present paper.

0066–4189/85/0115–0411$02.00

occasion, it also describes the reverse process—sound generation by vorticity in an edge interaction. It is arguably the key to the receptivity problem of the spatial hydrodynamic instability of shear flows, describing how external perturbations can couple locally to rotational perturbations, which then amplify freely as spatial instabilities; in this sense, it is the key to understanding how large-scale coherent features of turbulent flows evolve in response to forcing fields small in amplitude but coherent in phase. It seems, further, to be the phase-locking criterion for the operation of self-sustained oscillations in shear-layer feedback cycles, such as occur in many flows in nature and in engineering, and in some musical instruments.

There is both direct and indirect evidence for the validity of the Kutta condition in restricted regions of the multiparameter space needed to describe time-dependent perturbations of high-Reynolds-number separated flows. Recent years have seen a much better understanding of the nature of the Kutta condition, and in a few cases advances in "local interaction" theory for boundary layers have put the Kutta condition on a strong theoretical foundation—essentially by showing that, within restricted parameter ranges, only those "outer" potential flows that satisfy a Kutta condition are, in general, compatible (in the matched asymptotic expansion sense) with an acceptable multilayered "inner" viscous structure.

This review first describes the nature of and basis for a Kutta condition in unsteady flow. This is followed by a brief summary of the many applications of the Kutta condition and the fundamental role that it now appears to play. Finally, we discuss its possible further applications and those aspects calling for theoretical or experimental elucidation.

NATURE OF THE KUTTA CONDITION

To fix ideas, it is best to begin by considering a linearized unsteady problem in the absence of mean flow—the diffraction of a plane acoustic wave by a semi-infinite rigid plate with a sharp edge. If the problem is taken as inviscid, the least singular solution for the diffracted potential $\phi(\mathbf{x})$ $\exp(-i\omega t)$ (in response to some prescribed incident wave) has $\phi = O(|\mathbf{x}|^{1/2})$ near the edge, so that for the pressure and velocity we have $p = O(|\mathbf{x}|^{1/2})$ and $\mathbf{u} = O(|\mathbf{x}|^{-1/2})$. All other solutions differ from this minimally singular one by a linear combination of eigenfunctions

$$\phi_E^{(n)} = H_{n/2}^{(1)}(k_0 r) \cos \frac{n\theta}{2}, \quad (n = 0, 1, 2, \ldots), \tag{1}$$

where the plate occupies $(y = 0, x > 0)$, and k_0 is the acoustic wave number at frequency ω. One has to ask whether ϕ itself is in some sense an acceptable solution in some region of parameter space, despite its mildly singular velocities; whether in some other region of parameter space one or more of the $\phi_E^{(n)}$ might not permissibly be added (and then how would their as yet arbitrary coefficients be determined); or whether none of the solutions is acceptable and the problem as posed is meaningless. (It may seem pedantic to consider at such length a problem where the answer is familiar, but precisely the same points occur in more complicated problems of recent interest, and the correct answer—sometimes out of several candidates—is not yet established.)

Attempts have often been made in the past (and even more commonly in other wave fields such as electromagnetics and elasticity) to argue the relevance of one particular choice ϕ without enlarging the model problem to include other effects. Thus, for example, we can exclude all the $\phi_E^{(n)}$ by requiring that the energy in an arbitrarily small region around the edge should be finite, or that the energy flux out of a small surface enclosing the edge must vanish. These arguments work only for the simplest problems, however, and the only general way to settle the issue is to enlarge the context by including one or more effects so far neglected (nonlinearity, diffusion, finite-scale smooth-edge geometry, etc.), to treat the *general* solution for ϕ as merely an "outer" solution, and to determine the coefficients of the $\phi_E^{(n)}$ by matching to an acceptable (usually less singular) "inner" solution on a new small length scale associated with the newly introduced mechanism.

Suppose, first, that the plate remains rigid but is of finite thickness $2a$, with square edges, and that $\varepsilon = k_0 a \ll 1$. The diffraction problem for such a body cannot be solved exactly, but it can be treated by matched asymptotic expansions (Van Dyke 1975) in the limit $\varepsilon \to 0$. For the outer leading-order field we have the solution

$$\phi + \sum_{n=1}^{N} C_n \phi_E^{(n)},$$

with $\mathbf{x}$ scaled on k_0^{-1}, and for the inner we have to solve Laplace's equation, with $\mathbf{x}$ scaled on a.

The least singular forced inner solution has $\phi = O(|\mathbf{x}|^{3/4})$, corresponding to the square edges, for $|\mathbf{x}| \ll a$, but for $|\mathbf{x}| \gg a$ it has precisely the $|\mathbf{x}|^{1/2}$ form required for matching with the minimally singular outer solution (Crighton & Leppington 1973). Thus the singularity of the outer solution is weakened by appeal to finite inner scale, but the problem is not solved yet, for all the singular outer $\phi_E^{(n)}$ can be similarly matched to an inner field that is less singular. The situation is only completely resolved if one introduces (in linear theory) a mechanism that is believed always to lead to *finite* velocities—rounded-edge geometry, or viscosity. Alblas (1957) solved the half-plane diffraction problem with inclusion of viscous effects (with kinematic viscosity v); this poses a far more difficult Wiener-Hopf problem than the classical inviscid Sommerfeld problem. It can be shown from Alblas' work that if $\mathbf{x}$ is scaled on k_0^{-1} and $\varepsilon = k_0\delta = k_0(v/\omega)^{1/2} \to 0$, the field reduces at leading order to the irrotational minimally singular $\phi(\mathbf{x})$; while if $\mathbf{x}$ is scaled on the viscous length δ, the motion is rotational, satisfying the no-slip condition and with finite velocities everywhere within the Stokes layer around the plate and the edge.

It is in this sense that questions of nonuniqueness arising from the presence of eigensolutions and their associated edge singularities are to be resolved. When the matching of an acceptable inner solution to a leading-order outer field demands simply that the least singular outer solution be taken, we may loosely say that a Kutta condition is to be imposed on the outer solution at the edge. In the case of diffraction with no mean flow, this means that the pressure behaves like $|\mathbf{x}|^{1/2}$ near the edge, leaving the velocities singular like $|\mathbf{x}|^{-1/2}$. In the presence of mean flow (where the term "Kutta condition" is more commonly used), a less singular solution is usually possible—one in which *all velocities are finite*, though velocity gradients are usually infinite. Naturally, finiteness of the pressure or velocity can only be demanded in the leading-order outer solution; higher approximations, as $\varepsilon \to 0$ with the outer variables held fixed, usually involve increasingly strong singularities, which can nevertheless be matched at higher order to acceptable inner solutions.

Now in applications we are usually concerned with unsteady perturbations to a mean flow. On the outer scale (lengths made dimensionless with a geometric reference length), the mean flow can often be taken as a potential flow. If the potential flow remains attached to the body (an airfoil or turbomachinery guide vane, for example), the linearized unsteady flow will also be irrotational, but it may have discontinuities across a downstream wake of vanishing thickness. These unsteady discontinuities correspond to neutrally stable unsteady vorticity convected at the mean flow speed. They constitute an eigenfunction for the problem, and the velocity field they generate has $|\mathbf{x}|^{-1/2}$ singularities at the trailing edge from which the vorticity is shed. A Kutta condition may be satisfied here by adding an appropriate multiple of this eigensolution to cancel the velocity singularity in the forced solution (for some incident gust or sound wave) that has no wake discontinuities. This procedure has for more than 60 years been adopted for unsteady-airfoil problems; it determines the wake vorticity, the unsteady blade loading, and the radiated sound field.

If the mean flows on either side of the trailing edge are attached, but different in magnitude, there is a discontinuity in the outer *steady* flow across a thin downstream wake or shear layer. The aeroengine exhaust is the commonest example of this situation, though by no means the only one.

The shear layer in its vortex-sheet form is unstable to perturbations of any wavelength or frequency. For perturbations to mean shear layers created at an edge or separation point, it has now been conclusively established in numerous investigations over the past 20 years that *spatial-instability theory* provides the correct model; that is, we take time dependence $\exp(-i\omega t)$ with *real* ω and allow perturbations to amplify exponentially in the downstream direction while requiring boundedness in the transverse direction. There is good evidence that the finite behavior that in practice ensues downstream does not invalidate, even for elliptic equations, the predictions of initial exponential growth in spatial instability. Spatial-instability ideas were first applied to the mean vortex sheet between two streams by Helmholtz. When the vortex sheet is shed from a trailing edge, the coupling between instabilities and the upstream splitter plate leads again to velocity singularities at the edge, which can again be used to cancel those similar singularities induced when the plate–vortex sheet system is forced by some external perturbation. Thus the Kutta condition here is the criterion for solution of the "receptivity problem," to use the term apparently coined by Morkovin (cf. Morkovin & Paranjape 1971), i.e. the determination of the amplitude of an instability wave, given the amplitude of the external perturbation. This was first shown by Crighton & Leppington (1974) for the trailing-edge problem. After a period of initial exponential growth, the instabilities roll up and amalgamate (under the influence of nonlinearity and shear-layer divergence) to form the large-scale coherent turbulent structures that have been so intensively studied recently—and the Kutta condition then controls the initial formation of these structures and may be the key to active stabilizing control of the flow.

Until recently it was not thought possible to apply a Kutta condition to perturbations of an attached flow at a leading edge, and weak $|x|^{-1/2}$ singularities in velocity and pressure were accepted as inevitable. There are, however, two downstream instability mechanisms whose eigenfunctions may be used to apply an unsteady leading-edge Kutta condition. If the mean flow in which the edge is placed is a wake or jet flow, its inherent instability will couple with the edge to produce a singularity there that could cancel an externally induced singularity. This is the situation envisaged by Goldstein (1981), who shows, however, that there are several issues at stake here, and that the solution satisfying the leading-edge Kutta condition is not necessarily the correct one. In the case of a stable incident flow, there are boundary layers on the plate surfaces, and these can support Tollmien-Schlichting waves—though the growth rates of these will be very small in the vicinity of the edge where the boundary layers are thin. Howe (1981a,b) argues that the effect of these waves is to produce displacement-thickness fluctuations antisymmetric about the plane $y = 0$ of the plate, so that the rigid wall condition $\partial\phi/\partial y = 0$ is replaced by $\partial\phi/\partial y = \pm V \exp(i\kappa x)$ on $y = 0\pm$, $x > 0$, where κ may be taken as real (and equal to about $\omega/0.6\, U_\infty$, where ω is the disturbance frequency and U_∞ the free-stream speed). The constant V can be linearly related to the amplitude of some external forcing by imposing a leading-edge Kutta condition.

Consider finally the case of grossly separated mean flow—as, say, in the high-Reynolds-number flow past a cylinder. It is now fairly well accepted that the correct "outer" representation of the potential flow is provided by the Kirchhoff (1869) free-streamline theory (see Sychev 1972, Smith 1977, 1979, 1982, Messiter 1983). The surface streamline leaves the body a short distance downstream of the location (the Brillouin point) at which a hypothetical separating streamline would have the same curvature as the bluff body from which it detaches, and a curved vortex sheet separates the outer potential flow from an essentially stagnant wake. Helmholtz instabilities can develop on the separation shear layer, and again—perhaps surprisingly in view of the geometry—$|x|^{-1/2}$ velocity singularities are produced at the mean position of separation, both by an externally forced motion and by the coupling between the Helmholtz instability and the surface from which it springs. If a Kutta condition is imposed at separation (Goldstein 1984), the receptivity problem is solved for grossly separated flows over smooth surfaces.

Most flows of current interest are covered by the above examples, to which we return in a moment. However, grossly separated flows at sharp leading or trailing edges (as in the case of a sharp-edged airfoil at large angle of attack) are not covered so far, nor are perturbations to supersonic flows [though see Morgan (1974) for the possibilities at a trailing edge in supersonic flow, as well as Miles (1959) and Ashley & Landahl (1965, pp. 153, 256)].

LOCAL-INTERACTION THEORY

Steady Trailing-Edge Flows

The Kutta condition relates to the behavior of outer perturbations to a high-Reynolds-number flow as an edge or point of separation is approached. "Local interaction," or "triple-deck," theory for boundary layers now provides a detailed analytical and computational understanding of how the boundary layer accommodates the abrupt changes in conditions at a trailing edge while remaining attached upstream of the edge and evolving into the Goldstein near-wake form[2] a little distance downstream of it. A

---

[2] The Goldstein near-wake form is named in honor of Sydney Goldstein, not to be confused with Marvin Goldstein, whose work is fully described and cited in this paper.

similar viscous and nonlinear structure has now been found to describe a
great number of situations in which a boundary layer experiences an abrupt
(on the outer scale) change of conditions, even to the extent of describing the
structure at separation from a bluff body and matching it to a Kirchhoff
free-streamline flow. As a result, it is generally accepted that preservation of
a multilayered structure whose streamwise extent is always $O(\text{Re}^{-3/8} L)$
(where $\text{Re} = U_\infty L/v$, and L is a geometric scale) is *necessary* in any unsteady
motion that one would like to treat as a small perturbation of a known
mean flow of attached or free-streamline form. If the parameters (amplitude, frequency, etc.) of the perturbation do not allow the basic multilayered
structure of the mean flow to survive, separation will be provoked in the
unsteady flow at a time-dependent point well upstream of the trailing-edge
or steady-separation point, and no division into mean and perturbation
components will be possible. In the basic steady triple deck for a flat-plate
trailing edge, the pressure perturbation is a fraction $O(\text{Re}^{-1/4})$ of the
dynamic head $\rho_\infty U_\infty^2$, and it is required of further perturbations induced by
any other external cause (steady incidence, or an unsteady sound wave, for
example) not only that they have the appropriate scales to preserve the
integrity of the triple-deck, but that they also have a pressure variation of at
most $O(\text{Re}^{-1/4})$. Pressure rises beyond this cannot be accommodated at the
trailing edge without flow separation.

Reviews of local-interaction theory have been provided by several authors
(Stewartson 1974, 1981, Smith 1982, Messiter 1983) with almost exclusive
emphasis on steady flows. Only the briefest description is in order here.
Upstream of a trailing edge the incoming Blasius boundary layer has
thickness $\text{Re}^{-1/2}$, while downstream the Goldstein near-wake thickness is
$\text{Re}^{-1/2}x^{1/3}$ in terms of outer variables scaled on L. Transition between the
two takes place over the asymptotically short length $\text{Re}^{-3/8}$ [essentially as
found by Lighthill (1953) for the extent of upstream influence in supersonic
shock-wave–boundary-layer interaction]. Shortening of the x-scale makes
the main deck $(\text{Re}^{-1/2})$ inviscid, though rotational, and the viscous
boundary-layer equations hold only in a lower deck, or sublayer, of
transverse thickness $\text{Re}^{-5/8}$. In the sublayer the pressure is a function of x
only, but not a *known* function as it is for the thicker incoming boundary
layer. Instead, the flow in the sublayer and the perturbations in the upper
deck are coupled. In the upper deck, streamlines come from the irrotational
flow outside the upstream boundary layer. The upper-deck flow is therefore
irrotational and must then have the same transverse scale $(\text{Re}^{-3/8}L)$ as the
streamwise extent of the whole triple deck. Ahead of the edge the wall
boundary layer decelerates the flow, the displacement thickness increases,
and the pressure increases; downstream, the removal of the wall leads to
fluid acceleration, an initial convergence of streamlines, and a decrease in

pressure, so that a favorable pressure gradient is created in the triple deck. Changes in the pressure and flow direction in the upper deck are related, as in thin-airfoil theory, by a Hilbert transform pair for subsonic flow or by a local linear form for supersonic flow where the wave equation applies. These changes are functions of x only, and they are transmitted through the main deck to the lower deck where the Prandtl boundary-layer equations apply, driven by the upper-deck pressure and with a matching to the main deck that involves the flow-direction function. Thus, in place of inhomogeneous boundary-layer equations driven by a known external pressure, we have homogeneous boundary-layer equations with coupling in x to perturbations in the upper deck. The coupling is *local* on the geometric scale, but *nonlocal* for subsonic flow on the triple-deck scale $O(\mathrm{Re}^{-3/8}L)$.

The coupled problem has been satisfactorily solved for both subsonic and supersonic flow; the sublayer problem is analytically tractable if linearized about a constant mean shear, but the full problem needs to be solved numerically—which appears to show that a unique solution exists to the trailing-edge triple-deck problem. Solutions to the triple-deck problem are not completely smooth. There remains, for example, a discontinuity in pressure gradient $\partial p/\partial x$ at $x = 0$ that is resolved in an interior (twin-deck region) of size $\mathrm{Re}^{-1/2} \times \mathrm{Re}^{-1/2}$. An infinite number of smaller inner regions may be necessary to make the solutions arbitrarily smooth; in the smallest of these, of size $\mathrm{Re}^{-3/4} \times \mathrm{Re}^{-3/4}$, the full steady Navier-Stokes equations apply. If these higher-order nonuniformities are ignored, and only the leading-order triple-deck effect included, we have a first correction to the Blasius drag coefficient in the form

$$C_\mathrm{D} \sim 1.328\ \mathrm{Re}^{-1/2} + 2.66\ \mathrm{Re}^{-7/8}, \tag{2}$$

and this gives astonishingly good agreement with both full Navier-Stokes calculations and with experiments down to $\mathrm{Re} = 10$ and even lower, where the triple-deck length is a large fraction of the total plate length. The steady triple-deck structure is contained within the more complicated structures shown in Figures 1, 2, and 3.

A great number of steady perturbations to this basic aligned flat-plate problem have now been examined [incidence, camber, surface bumps and dents for both external and internal flows, corners, suction and blowing, etc.; see Stewartson (1981), Smith (1982), and Messiter (1983) for very full reviews]. The sizes and scales of these perturbations may seem "bizarre" (Smith 1982, p. 224)—for example, the effect of a bump of height $O(\mathrm{Re}^{-5/8}L)$ and streamwise scale $O(\mathrm{Re}^{-3/8}L)$ may not seem to be of general significance. However, the accepted belief is that disturbances that "conform" to the triple-deck scales [in particular, the streamwise extent $O(\mathrm{Re}^{-3/8})$ and viscous lower-deck thickness $O(\mathrm{Re}^{-5/8})$] and that provoke adverse pressure rises of no more than $O(\mathrm{Re}^{-1/4})$ are critical and canonical, in the sense that weaker disturbances are negligible while stronger disturbances will cause flow separation on a large scale. In the case of weaker disturbances, one can simply take a limit of the canonical triple-deck problem until some other scale intrudes; thus, for example, the problem for $O(\mathrm{Re}^{-3/8})$ bump lengths covers all lengths $O(\mathrm{Re}^{-\beta})$, where $0 < \beta < 3/4$.

Unsteady Trailing-Edge Flows

The same attitude applies to unsteady flows. Two cases have so far been examined in detail. Brown & Daniels (1975) (in a paper with the abbreviated running title "The Kutta Condition for an Oscillating Aerofoil") examine the oscillating airfoil, pitching or plunging, in a uniform incompressible stream. In addition to $\mathrm{Re} = U_\infty L/\nu$, there are nondimensional parameters $S = \omega L/U_\infty$ (where ω is the radian frequency) and an incidence amplitude α. To order S in relation to Re, it is argued that unless $S \gg 1$, the flow everywhere in the vicinity of the trailing edge will be quasi-steady, with time entering only parametrically and the flow at each instant given by the steady-incidence solutions of Brown & Stewartson (1970). The distinctive change occurs when time derivatives are brought into play in the viscous nonlinear lower deck, and this requires that the lower-deck thickness $\mathrm{Re}^{-5/8}L$ and the Stokes-layer thickness $(\nu/\omega)^{1/2}$ be comparable and thus that $S = O(\mathrm{Re}^{+1/4})$. Taking then the high-frequency limit of the potential-flow solution, with the vortex-wake strength adjusted to satisfy an unsteady Kutta condition, it is found that

$$\frac{p - p_\infty}{\rho_\infty U_\infty^2} \sim \alpha S^2 \left(\frac{x}{L}\right)^{1/2} \exp(-i\omega t) \tag{3}$$

for the surface pressure, and on the triple-deck scale ($x = \mathrm{Re}^{-3/8}L$) the right hand side is $O(\mathrm{Re}^{-1/4})$ if $\alpha = O(\mathrm{Re}^{-9/16})$. If α exceeds $O(\mathrm{Re}^{-9/16})$, an adverse pressure gradient will be created near the edge over part of the cycle that exceeds the inherent favorable pressure gradient supplied by the triple deck, and the flow will separate ahead of the trailing edge. If S exceeds $O(\mathrm{Re}^{1/4})$, the oscillations may be too rapid to preserve the triple-deck structure (though the linearized lower-deck problem was solved by Brown & Daniels without inconsistency in the further limit $S/\mathrm{Re}^{1/4} \to \infty$).

Given these scalings, Brown & Daniels identify the flow structure illustrated in Figure 1 below. The principal difference between this and the steady triple-deck lies in the appearance of "foredecks." In general, for $O(1) \le S \le O(\mathrm{Re}^{1/4})$, there is a transition from time-dependent perturbations to the Blasius boundary layer 2a to the quasi-steady main deck 4b through an upper foredeck of size $LS^{-1} \times L\,\mathrm{Re}^{-1/2}$, and from the

Stokes-layer solutions in 2b to a quasi-steady lower deck 4c through a lower foredeck of size $LS^{-3/2} \times L\,\mathrm{Re}^{-5/8}$, as shown by Brown & Cheng (1981). Thus, with the understanding that not all of the foredeck regions involve new solutions, one can regard the foredeck now as comprising a first and second deck of streamwise scale LS^{-1} and $LS^{-3/2}$, respectively, with each comprising upper and lower decks [as in Figure 1 of Brown & Cheng (1981)]. In the critical case $S = O(\mathrm{Re}^{1/4})$ the lower foredeck is redundant, and the Stokes layer merges directly with the lower element of the triple deck, as is necessary if the lower deck is to be genuinely unsteady.

Brown & Daniels solve for the flows in most of the regions shown, while leaving a formidable numerical problem unsolved for the lower deck. Here, dependence remains on two $O(1)$ parameters, α_0 and S_0, and the nonlinear unsteady boundary-layer equations are to be solved with a Hilbert transform relation between the driving pressure and the displacement function that appears in the outer boundary condition. The authors show, however, how the system plausibly admits matching to the upstream surface pressure and to the downstream Goldstein wake. This wake has a centerline whose oscillating displacement has (at large downstream distances on the lower-deck scale) just the right $x^{3/2}\exp(-i\omega t)$ form for matching with the vortex-sheet wake satisfying the unsteady Kutta condition. Upstream of the trailing edge a linearized treatment of the lower-deck problem can be carried out, taking the mean flow as one of constant shear (an inappropriate representation downstream.) A Wiener-Hopf problem is rendered relatively simple by the further approximation of very

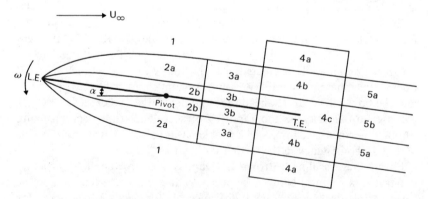

Figure 1   Multilayer structure in the neighborhood of a flat plate, oscillating about a pivot with angular frequency ω and incidence amplitude α in a stream U_∞. 1, potential flow; 2, perturbed Blasius boundary layer and inner Stokes layer; 3, foredeck; 4, triple deck; 5, modified Goldstein wake. (Reproduced, with permission, from Brown & Daniels 1975.)

high frequency ($S_0 \to \infty$), and the analytical solution confirms the matching features anticipated for the full nonlinear lower-deck problem.

This appears then to confirm the validity of the Kutta condition for $S = O(\mathrm{Re}^{1/4})$ and $\alpha = O(\mathrm{Re}^{-9/16})$. The full range of smaller values of S ($0 < S \ll \mathrm{Re}^{1/4}$) has been examined by Brown & Cheng (1981), with, of course, the constraint $\alpha S^2 = O(\mathrm{Re}^{-1/16})$ required for the triple-deck pressure. The foredeck structure now splits, as mentioned earlier, but Brown & Cheng show nevertheless that all of the matchings can be carried through that permit the use of computed solutions for the triple deck on a plate at steady incidence α. Here it is found, from the numerical work of Chow & Melnik (1977), that the flow on the suction side separates just at the trailing edge when $\mathrm{Re}^{1/16}\lambda^{-9/8}\alpha = 0.47$, where $\lambda = 0.3321$ is the Blasius skin-friction factor. [The limiting value had earlier been estimated by Brown & Stewartson (1970) to lie in the range 0.33 to 0.41 and to mark incipient "trailing-edge stall"; in the steady case $S = 0$, and conformity with the triple deck requires $\alpha = O(\mathrm{Re}^{-1/16})$.] Brown & Cheng are thus able to describe the satisfaction of a Kutta condition on the pitching or plunging airfoil (and, indeed, the derivation of the first viscous correction to the lift and moment) over a wide range of frequency and amplitude, right up to the separation condition, for $S \le O(\mathrm{Re}^{1/4})$. Comparison of these results with experiment follows in the next section.

Orszag & Crow (1970) first discussed the coupling between spatial instabilities on a vortex-sheet shear layer and the edge of the splitter plate from which the sheet emanates. At frequency ω the coupling effect decays algebraically downstream, and for $\omega x/U_\infty \gtrsim 1$ the field is dominated by a freely developing Helmholtz instability with eigenfunction

$$\exp\left\{\frac{\omega}{U_\infty}(x+i|y|)(1+i)-i\omega t\right\}.$$

It is assumed that there is uniform incompressible flow $(U_\infty, 0)$ above the plate ($x < 0, y = 0$) and stagnant fluid below, and that there is no external forcing field. Then if the pressure is required to decay upstream (it does so like $|x|^{-1/2}$), there is one solution, it is exponentially unbounded as $x \to +\infty$, and the vortex-sheet deflection as $x \to 0+$ is like $x^{1/2}$ $\exp(-i\omega t)$. Correspondingly, the pressure and the velocities in the moving stream have $|x|^{-1/2}$ singularities, but they are finite in the stagnant fluid; this is the "no-Kutta-condition" solution.

Orszag & Crow argued in favor of a "rectified Kutta condition," claiming that vortex shedding in the steady state would be just sufficient to ensure that the vortex sheet would at no time in the cycle bend down into the static fluid. This can be achieved by the superposition of a time-independent parabolic deflection of the sheet into the moving stream, of magnitude equal to the amplitude of the parabolic periodic displacement $x^{1/2}$ $\exp(-i\omega t)$. A parabolic displacement sustains no pressure jump, and it is therefore a candidate for a mean vortex-sheet location. When the superposition is made, there remain singularities at the vertex of the parabola in the moving stream; but it is argued that these are not really present in the flow domain and have been spuriously transferred there, in a familiar way, by the linearization.

In addition, Orszag & Crow obtained a third solution, also exponentially unbounded downstream, but now satisfying a "full Kutta condition," with $\eta(x) \sim x^{3/2} \exp(-i\omega t)$ and all pressures and velocities finite at the edge. This is achieved (in the absence of external forcing) at the expense of the upstream behavior, with the potentials and pressures growing upstream like $|x|^{1/2}$ and the velocities decaying like $|x|^{-1/2}$. Largely because of the upstream behavior, Orszag & Crow favored the rectified Kutta condition solution and regarded the full-Kutta-condition solution as "indefensible."

Extensions can be made to the cases in which there is a nonzero flow below the plate as well, and in which the flow is laterally confined by rigid walls. These induce no qualitative change to the upstream, edge, or downstream behavior. However, unless the mean flow on one side is very small, the argument in favor of the rectified Kutta condition is completely lost.

As a global solution, that satisfying the full Kutta condition looks unacceptable [though its edge behavior is similar to that prevailing when the no-Kutta-condition eigensolution is used to cancel an edge singularity induced by some external cause, like a gust or incident sound wave, and that is really all that matters in Daniels' (1978) examination of the viscous and nonlinear edge structure]. Attempts have been made [Crighton (1972); and Davis (1975) for the case of identical flow on the two sides] to match the full-Kutta-condition solution to an acoustic outer field; however, the correct interpretation of these solutions is that they contain an *incoming* acoustic field, and that there is no purely radiating solution satisfying the full Kutta condition (whether the upper and lower streams are equal or not), in agreement with Howe (1976).

If an edge singularity is induced by external forcing, the full Kutta condition can be satisfied by addition of an appropriate multiple of the no-Kutta-condition eigensolution, and then the scattered field from the plate/vortex-sheet combination has satisfactory behavior everywhere at infinity (decaying algebraically upstream if incompressible, radiating outward everywhere if compressible). A field of this kind was first constructed by Crighton & Leppington (1974) for excitation due to a line source in the static fluid. This amounts to the first solution of a receptivity problem for realistic geometry involving an edge from which the vortex sheet springs. It is to be observed that the full-Kutta-condition solution of Crighton & Leppington is causal [i.e. it could be reached as the long-time limit of a causal initial-value problem, the essential condition for which is that, for time dependence $\exp(-i\omega t)$, the fields should be analytic for all complex ω in Im $\omega > 0$]. However, the eigensolution is also causal, and therefore the full Kutta condition implies causality (for this problem) but not necessarily vice versa. Previous studies (Jones & Morgan 1972) of the response of a *doubly infinite* vortex sheet to excitation have shown the bounded solution (consisting simply of incident, reflected, and transmitted acoustic waves, with some local refinements) to be noncausal, and the addition of an appropriate freely developing spatial Helmholtz instability eigenfunction was mandated to comply with causality. If no edge is present, causality seems the only criterion to deal with the receptivity problem, and indeed it was used in this way by Tam (1971) and in several subsequent papers to explain the fine-scale wave fronts observed close to the thin shear layers immediately downstream of a jet nozzle. A splitter-plate edge is, however, invariably present, and even if it is far from the source of excitation, it appears that the satisfaction of a Kutta condition at the edge is of overriding importance and guarantees causality.

Daniels (1978) has examined the inner problem with the Orszag & Crow full-Kutta-condition eigensolution as the outer solution; the upstream behavior of this solution should have little bearing on the edge behavior and should certainly allow the question of whether $\eta \sim x^{3/2}$ and/or $\eta \sim x^{1/2}$ is compatible with an acceptable attached multilayered edge structure. Daniels assumes (realistically, from the experimental point of view) that only a finite length L of the splitter plate upstream of the edge is exposed to viscous effects, so that we can define Re $= U_\infty L/\nu$ and order the frequency parameter $S = \omega L/U_\infty$ and amplitude h/L in terms of Re. Here, h is the vortex-sheet deflection when $x = O(U_\infty/\omega)$, and U_∞ is the mainstream velocity above the plate. If $\eta \sim x^{3/2}$ as $x \to 0+$, the scalings $h/L \sim \text{Re}^{-7/16}$ and $S \sim \text{Re}^{1/4}$ lead to a match of the Stokes layer with the inner deck of the triple deck and ensure a genuine unsteady flow there, and they make the pressure $O(\text{Re}^{-1/4})$ in the triple deck.

Figure 2, taken from Daniels (1978), shows the edge structure with the above scalings. If $1 \ll S \ll \text{Re}^{1/4}$, the type of analysis used by Brown & Cheng (1981) presumably applies. The inner regions 13 and 14 and the Navier-Stokes region 15 in Figure 2 are needed only to resolve higher-order nonuniformities (principally pressure-gradient discontinuity and infinite slope of the dividing streamline) that survive the triple deck. The displacement boundary layer, 17, serves to reduce to zero the slip velocity generated by the steady motion induced below the plate by the $O(1)$ mainstream above. In fact, as pointed out by Rienstra (1979), the trailing-edge flow in

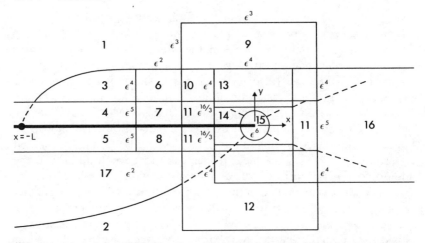

Figure 2   Multilayer structure near the trailing edge of a splitter plate separating uniform mean flow U_∞ in $y > 0$ from static fluid in $y < 0$. Viscous action starts upstream at $x = -L$ on the plate, and ε is written for $\mathrm{Re}^{-1/8}$. 1, potential flow ($y > 0$); 2, potential flow ($y < 0$); 3, Blasius boundary layer; 4, 5, Stokes layers; 6, 7, foredeck ($y > 0$); 8, foredeck ($y < 0$); 9–11, triple deck ($y > 0$); 11, 12, triple deck ($y < 0$); 13, 14, inner region; 15, Navier-Stokes region; 16, mixing layer; 17, displacement boundary layer. (Reproduced, with permission, from Daniels 1978.)

the absence of any external perturbation does not have any triple-deck structure here; a weak flow is induced below the plate, and the mainstream streamlines suffer a slight upward deflection in passing over the trailing edge, without the infinite transverse-velocity singularity that is removed by the triple deck when the mainstream velocities are the same on the two sides of the plate. A triple-deck structure is then evoked by any perturbation, but it may not be possible now for the triple deck to sustain a pressure change of $O(\mathrm{Re}^{-1/4})$, even though this may be a natural first choice.

Daniels (1978) obtains solutions to leading order in many of the regions of Figure 2 and gives plausible arguments for the overall consistency of the structure in terms of matching requirements. As in Brown & Daniels (1975), an analytic solution can be given (it is in fact precisely the solution of Brown & Daniels) for the linearized lower-deck response upstream of the edge; and the second-order terms in the behavior of this solution as $x \to \pm \infty$ (scaled on lower-deck thickness) can be identified as corresponding to a multiple of the no-Kutta-condition eigensolution. The behavior of the dividing streamline (separating fluid that has come from $-\infty$ above the plate from the rest) can also be found by expansion of the lower-deck solution as $x \to 0+$ with $y/x^{1/3} = O(1)$ (again with x, y scaled on lower-deck thickness). It is found that

$$y \simeq 0.895[x/a(t)]^{1/3}, \tag{4}$$

where $\lambda a(t)$ is the scaled skin friction just upstream of the edge ($\lambda = 0.3321$ and $a = 1$ in steady flow). The derivation requires that $a(t)$ be positive and so assumes that separation has not taken place upstream of the edge; it shows, at the same time, that the dividing streamline is at all times deflected upward into the moving stream. If the linearized solution for the lower deck is used to obtain $a(t)$, it is found that $a(t) = 1 + \delta \cos \omega t$, where $0 < \delta \ll 1$ and δ is proportional to the scaled amplitude h and to the lower-deck velocity gradient just upstream of the edge. This shows that the streamline behavior is indeed of the rectified Kutta condition kind, as envisaged by Orszag & Crow (1970), but the upward deflection is associated with viscous and nonlinear effects on a scale $x \sim \mathrm{Re}^{-3/8}L$ and is not something to be enforced on an outer solution scaled on L. The singularity in streamline deflection is something to be resolved at still smaller scales—presumably on the $\mathrm{Re}^{-3/4}L$ Navier-Stokes scale, if not earlier.

Daniels (1978) also examines the no-Kutta-condition outer solution, for which $p - p_\infty \sim (-x/L)^{-1/2} \exp(-i\omega t)$ as $x \to 0-$. The scaling $S \sim \mathrm{Re}^{1/4}$ is again needed, and to make the scaled pressure $O(\mathrm{Re}^{-1/4})$ on the triple-deck scale requires a scaled vortex-sheet deflection $h = O(\mathrm{Re}^{-1/16})$. For the lower deck the unsteady boundary-layer equations are again to be solved with the Hilbert transform relation between pressure and flow displacement; if this problem is again linearized about a uniform shear, a solution can be found that has $(-x)^{-1/2}$ decay as $x \to -\infty$ (in lower-deck variables), as demanded for matching with the no-Kutta-condition pressure. However, the pressure then tends to a finite nonzero periodic value as $x \to 0-$, and as a result there is an $O(\mathrm{Re}^{-1/4})$ pressure discontinuity on the $\mathrm{Re}^{-3/8}L$ scale. Such a large jump cannot be resolved at finer scales; the only jump that can be accommodated is one of $O(\mathrm{Re}^{-1/2})$, and this can (perhaps) be smoothed in the innermost $\mathrm{Re}^{-3/4} \times \mathrm{Re}^{-3/4}$ region in which the full Navier-Stokes equations apply. Therefore, we must take h smaller ($h = \mathrm{Re}^{-13/16} h_0$) in order to make $(p - p_\infty)/\rho_\infty U_\infty^2 = O(\mathrm{Re}^{-1/2})$ for $|x| = O(\mathrm{Re}^{-3/8}L)$. This makes the lower-deck equations *linear*, and the exact Wiener-Hopf solution already obtained applies in all detail, except for modification to satisfy the new matching requirement. Daniels examines the asymptotic solutions to the full Navier-Stokes equations as $x/(\mathrm{Re}^{-3/4}L) \to \pm\infty$, showing that there appears to be satisfactory matching to the triple deck and that the $O(\mathrm{Re}^{-1/2})$ pressure jump appears to be smoothed. Outside the Navier-Stokes region, but just upstream of the edge on the triple-deck scale, he finds the scaled skin friction in the form

$$\frac{\partial \bar{u}}{\partial \bar{y}}(\bar{x}, 0, t) \sim 1 + (-\lambda^{1/2}x/\mathrm{Re}^{-3/4}L)^{-4/3} 3^{-1/6}$$

$$\times 2^{13/4}\pi^{-2}(\tfrac{1}{3}!)^2 h_0 S^{3/4} \sin \omega t + \cdots, \tag{5}$$

which indicates that separation will take place upstream of the edge at

$$\frac{x}{L} \sim -\mathrm{Re}^{-3/4}\left(\frac{(\frac{1}{3}!)^2 2^{13/4}}{3^{1/6}\pi^2}\right)^{3/4} \frac{h_0^{3/4}S^{9/16}}{\lambda^{1/2}}. \tag{6}$$

At this stage it appears plausible that the no-Kutta-condition solution is consistent with an acceptable inner structure for small amplitudes $[O(\mathrm{Re}^{-13/16})]$, though a full numerical study of the Navier-Stokes region is really needed. Daniels then argues against this conclusion, as follows. Variations in the *steady* flow in the triple deck give rise to a pressure variation of order $\mathrm{Re}^{-1/4}\,(-x/\mathrm{Re}^{-3/8}L)^{2/3}$, at least when there is an $O(1)$ flow on both sides of the plate. For the Navier-Stokes solution one then has to require matching with this leading-order $(-x)^{2/3}$ term as well as matching of the next term, of order unity, with the unsteady finite value produced in the triple deck. Daniels argues that it is unlikely that the first two terms in the asymptotic behavior of the Navier-Stokes solution can be essentially arbitrarily chosen to match two independent triple-deck terms in this way, and he also claims that the argument extends to the Orszag & Crow case of one flow at zero velocity.

If the Navier-Stokes region does turn out to have the required properties, the unsteadiness of the no-Kutta-condition solution leads to just a small perturbation of the mean (viscous and nonlinear) flow produced by the interaction between the stream and the stagnant fluid. For $\mathrm{Re}^{-3/8}L \ll x \ll L$, the equation of the dividing streamline is found to be

$$\frac{y}{L} = \mathrm{Re}^{-1/2}0.895\left(\frac{x}{L}\right)^{1/3} + \mathrm{Re}^{-11/16}h_0 S^{1/2}4i\left(\frac{1+i}{\pi}\right)^{1/2}\left(\frac{x}{L}\right)^{1/2}e^{-i\omega t} + \cdots, \tag{7}$$

where the second term gives the required match with the no-Kutta-condition outer flow and is a small perturbation of the mean streamline deflection, of almost parabolic shape, given by the first term. Again, as envisaged by Orszag & Crow (1970), the dividing streamline at no time bends down into the static field.

Further work on the singular outer solutions and their induced viscous structure is much needed. There are often situations in which a flow appears to develop spontaneous oscillations in the absence of any obvious source of external excitation, and in these cases it is natural to represent the flow by an eigenfunction of, say, the Orszag & Crow kind. There are no full-Kutta-condition eigenfunctions that have satisfactory behavior at large distances, so that one has to ask whether the no-Kutta-condition eigensolutions are acceptable—with the implication that if they are not, as conjectured by Daniels (1978), then all spontaneous shear-layer oscillations are in fact *forced oscillations* (with the forcing then presumably associated with upstream feedback from some downstream obstacle).

Leading-Edge Flows

Unsteady Kutta conditions have been applied to leading-edge flows by Goldstein (1981) and Howe (1981a,b); the former assumes that the oncoming flow is rotational and capable of sustaining spatial instabilities, while the latter relies on Tollmien-Schlichting waves in the boundary layer to provide the eigensolution that can relieve an edge singularity induced by external forcing. For steady attached leading-edge flows there is no structure of the triple-deck kind. The full Navier-Stokes equations apply, and while there are coordinate expansions for the edge flow, these have only local validity and cannot be related to the fully developed downstream boundary layer. Time-dependent perturbations have only recently been considered (Goldstein 1983), and the validity of an unsteady Kutta condition is a completely open matter, though there is certainly indirect evidence for its validity in some flows. This is discussed in the section dealing with applications.

Separated Flows

The most recent use of a new form of Kutta condition has been with grossly separated flows around bluff bodies (Goldstein 1984). The mean structure that is to be perturbed was first proposed by Sychev (1972) and greatly strengthened by Smith (1977, 1979, 1982). For the leading-order outer flow, the particular Kirchhoff (1869) free-streamline flow is chosen that separates from the body at what Goldstein (in recognition of Brillouin's work on such flows; see Birkhoff 1950, p. 50) calls the *Brillouin point*. Here the separating streamline and body have the same curvature; the streamline would have infinite curvature if it were to separate elsewhere, while if there were earlier separation, the streamline would cut the body surface and later separation would lead to an infinite adverse pressure gradient. At leading order the wake behind the separating streamline is stagnant, at the upstream pressure p_∞. On the body scale we have, as the separation point $x = 0$ is approached (x is a local surface coordinate; $x = 0$ corresponds to $\theta = 55°$ from the forward stagnation point for flow past a circular cylinder),

$$S(x) \sim S_0 x^2 + S_1 x^{5/2} + \cdots \tag{8}$$

for the streamline shape as $x \to 0+$, with S_0 determined by the body curvature and S_1 by details of the Kirchhoff flow. For the surface pressure, we have

$$p_0(x, 0) \sim \tfrac{5}{2} S_1 |x|^{3/2} \quad \text{as } x \to 0-, \tag{9}$$

$$p_0(x, 0) = 0 \quad \text{for } x > 0. \tag{10}$$

The first viscous corrections to the free-streamline flow have been obtained for the circular cylinder; the expansion for the outer flow takes the form

$$p = p_0 + \mathrm{Re}^{-1/16} p_1 + \cdots \tag{11}$$

and is nonuniform in a triple-deck region whose structure is essentially that of the trailing-edge problem. Just upstream of the triple deck, the pressure has the trailing-edge form

$$p \sim -\mathrm{Re}^{-1/16} b \lambda^{9/8} |x|^{1/2} \tag{12}$$

as $x \to 0-$, while downstream, viscous forces are concentrated in a detached shear layer surrounding the curve

$$y \sim \mathrm{Re}^{-1/16} (\tfrac{2}{3} b \lambda^{9/8}) x^{3/2} \tag{13}$$

as $x \to 0+$ (on the body scale). Here, b is an unknown numerical constant (necessarily positive). The consistency of the triple-deck inner breakaway with an outer smooth separation requires that b be uniquely determined by the solution of the boundary-layer equations in the lower deck, with Hilbert-transform coupling to the upper deck and other matching conditions (together with other global effects related to the closing up of the wake, which have not yet been settled). Smith (1977) shows rather firmly that there *is* a unique b ($b \approx 0.44$).

Generally, the surface pressure gradient is favorable upstream of the Brillouin point, and separation is caused by an abrupt adverse pressure gradient induced in the triple deck by the $O(\mathrm{Re}^{-1/16})$ correction to the outer flow. The separation is smooth on the body scale, but it is mildly singular on the triple-deck scale ($y \propto x^{3/2}$); at still smaller scale, the singularity is removed. A composite representation of the separation streamline is

$$y = S(x) \sim S_B x^{3/2} + S_0 x^2 + S_1 x^{5/2} + \cdots, \tag{14}$$

where $S_B = O(\mathrm{Re}^{-1/16})$, and the effect of the first term is to displace the separation point downstream from its Brillouin location by an amount $O(\mathrm{Re}^{-1/16})$ on the body scale. The general scheme proposed by Sychev and Smith is illustrated in Figure 3.

Goldstein (1984) examines spatial instabilities on the separating streamline, motivated by the experimental results of Ahuja et al. (1983). The latter authors showed that gross separation from an airfoil can be strongly influenced by small-amplitude sound, even when the boundary layer ahead of separation is fully turbulent. The size of the separation bubble can be drastically reduced by a sound field. This effect is ascribed by Goldstein to enhanced entrainment, across the separation shear layer and into the stagnant wake, caused by the orderly wavelike structures prominent on the shear layer under acoustic excitation. The scale of such waves is small in comparison with the streamwise body dimension, and Goldstein therefore first constructs an outer solution, valid away from the separation region, that describes a freely amplifying high-frequency Helmholtz instability. Effects of slow mean-flow divergence are important, and they are fully and simply incorporated.

Next the coupling with the surface is examined, and a solution is found that matches the instability downstream. The eigensolution constructed in this way is very similar to the Orszag & Crow (1970) no-Kutta-condition eigensolution for the splitter-plate trailing edge. This is rather surprising, because although the geometry appears flat near the separation point, the pressure remains finite and the streamline shape behaves like $x^{3/2}$. Nonetheless, Goldstein shows that the inner solution, on scales small compared with the wake width but large compared with triple-deck scales, has singular velocities, infinite like $|x|^{-1/2}$.

Finally, Goldstein constructs a forced solution that satisfies the appropriate conditions both upstream on the rigid surface and downstream on the separation vortex sheet. No specific form is needed for the excitation; all that matters is the pressure gradient that would be induced at the Brillouin point by the forcing in the absence of separation. The forced solution also has $|x|^{-1/2}$ velocity singularities. The appropriate linear combination of forced solution and eigensolution, together with the application of a Kutta condition, removes the singularity; in addition, it solves the receptivity problem by linearly relating the instability-wave amplitude to the pressure gradient induced externally at the separation

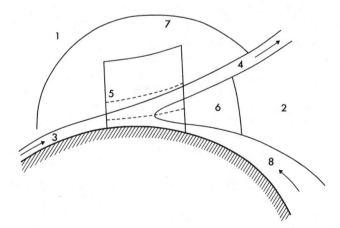

Figure 3   Multilayer structure at separation from a bluff body. 1, potential flow; 2, near-eddy region; 3, boundary layer; 4, separated shear layer; 5, triple deck; 6, 7, inviscid adjustment zones; 8, reversed-flow boundary layer. (Reproduced, with permission, from Smith 1979.)

point [assumed *not* to be a Brillouin point by the inclusion of the $O(\mathrm{Re}^{-1/16})$ term in the expression for $S(x)$].

To justify this procedure, Goldstein considers the cases in which the Kutta condition is not imposed, arguing then that the singularity (still present, but at higher order, with Kutta condition imposed) emphasizes spatial gradients near separation. Thus one would expect a nonlinear but quasi-steady free-streamline flow to replace the linearized unsteady flow as the singularity is approached, and Goldstein does indeed write down a free-streamline solution that satisfies the inviscid quasi-steady equations and separates from a moving point $x_s(t)$. He then shows that this solution, with *finite* velocities everywhere, matches the inner (singular) limit of the linearized solution, and so in effect shows that the inverse-square-root velocity singularity of the linearized solution can be accounted for by the motion of the separation point. The multiple of the scaled streamwise pressure gradient $P'_0(0)$, produced at separation by the external excitation, that determines the instability-wave amplitude in the Kutta condition case is replaced by $[P'_0(0) - x_s]$ when the separation point moves. If x_s is known somehow, the nonlinearization just described enables the receptivity problem to be solved, and if $|x_s| \ll P'_0(0)$, then the upshot of applying the nonlinearization is the *same* as that of application of the Kutta condition.

Suppose now that $S \ll \mathrm{Re}^{1/4}$, where S is an appropriate Strouhal number; this is the condition for the entire triple deck at separation to be quasi-steady (with the proviso that unsteady foredecks may be needed, as in Brown & Cheng 1981). Then Goldstein is able to relate the separation problem in a moving frame attached to the separation point to the steady problem analyzed by Smith (1977, 1979) to show that $|x_s|$ is indeed small compared with $P'_0(0)$, and therefore that the Kutta condition is validated for this quasi-steady case. As the frequency increases, the appropriate scaled x_s increases and the Kutta condition seems likely to fail. The fully unsteady trailing-edge flow was examined by Brown & Daniels (1975) and Daniels (1978) for frequencies so high that $S \sim \mathrm{Re}^{1/4}$, and the Kutta condition was shown to be still generally valid. If Goldstein's (1984) expression for x_s in terms of $P'_0(0)$ is extrapolated to $S \sim \mathrm{Re}^{1/4}$ (which is not valid, but may be indicative), it is found that x_s and $P'_0(0)$ have equal and opposite values; in this case, the Kutta condition for separation at a smooth surface would definitely fail.

Goldstein gives much more detail on this problem, and he particularly emphasizes that although the $O(\mathrm{Re}^{-1/16})$ term displacing the separation location from the Brillouin point is formally small, it is likely to be dominant on the appropriate unsteady inviscid scale at all Reynolds numbers for which the flow could possibly remain laminar. He also refers to the recent work of Cheng & Smith (1982) showing that on appropriately slender bodies the separation can take place an $O(1)$ distance downstream of the Brillouin point, as is the case for *turbulent* separation from a bluff body (according to Sychev & Sychev 1980). Much more remains to be done on unsteady separation, and also on turbulent separation and multilayered structures for turbulent trailing-edge flows. The results of Goldstein (1984) do, however, contain specific features, such as dependence on Reynolds number and frequency, that now require experimental study.

EXPERIMENTS

Unsteady trailing-edge flow for oscillating airfoils has been studied in numerous experiments. In some cases I believe the authors have misconceived the Kutta-condition issue; in others, only flow visualizations have been presented (with some, but by no means all, of the pertinent parameter values), from which it is not possible to derive an unambiguous interpretation. Moreover, in no single case has an experiment been planned and conducted in the context of triple-deck ideas (however loosely applied). Consequently this section refers only to a couple of experiments where a rational mechanical explanation seems either definitely possible or impossible. Indirect experimental evidence for or against the Kutta condition is described in the following section.

Nozzle Flows

For the splitter-plate trailing edge separating a low-speed stream from static fluid, Bechert & Pfizenmaier (1971) measured fluctuating and mean velocities in a circular jet by hot-wire anemometry, and pressure fluctuations near the jet in the static fluid by using a small microphone. The hot-wire and pressure measurements involved distances from the edges as small as $0.5\theta_e$ and $3\theta_e$, respectively, where θ_e is the exit boundary-layer momentum thickness. Larger distances at which measurements were taken were all, nonetheless, small compared with the hydrodynamic length U_J/ω, where U_J is the jet exit velocity and ω the frequency to which fluctuations were locked by small-amplitude loudspeaker excitation in the plenum chamber.

No mean deflection of the dividing streamline was seen on these scales, though no peak in unsteady velocity was seen near the edge either. The transverse pressure gradient was shown to vanish at the edge, but this does not discriminate between full Kutta condition with pressure $|x|^{5/2}$ in the still fluid and no Kutta condition with pressure $|x|^{3/2}$. If anything, the data support the full-Kutta-condition solution, though not convincingly; the difficulties have much more to do with the quantities measured than with the spatial resolution.

In a second series of experiments, Bechert & Pfizenmaier (1975) used a novel optical compensation measurement giving a resolution of 1–3 μm in the dividing-streamline location. Three Reynolds numbers (10^4, $5 \times 10^4, 10^5$) were used, based on jet diameter D and exit velocity U_J, as were a range of Strouhal numbers fD/U_J (with f the circular frequency) between 0.4 and 2.26. The exit-plane forcing levels were varied by 10 dB, and no significant nonlinearity in the jet response was found, even for the smallest x/D. Dependence of the dividing-streamline shape $h(x)$ on Reynolds number, at a given Strouhal number, was weak, especially at small x/D. Typical response curves are given in Figure 4 (which is Figure 10 of Bechert & Pfizenmaier 1975); here h^* is the streamline shape normalized by a length representing the forcing amplitude. The linearity of the response is seen, as is a "parabolic-shaped region near the nozzle edge" (Bechert & Pfizenmaier 1975, italics on p. 142). The authors were reluctant to interpret their results in terms of a full Kutta condition or the absence of a Kutta condition, and indeed they favored a linear combination of the two solutions, arguing that their configuration (with a circular jet and a finite boundary-layer thickness) did not in any case correspond to the plane vortex-sheet theory. I believe their configuration *is* appropriate; the circular geometry is irrelevant on any inner scale, and the finite boundary-layer effects are taken care of by the triple deck. It is not easy to decide on the effective streamwise length L to be used in $Re = U_J L/\nu$, but the dependence on this is weak, and any reasonable interpretation gives values for Re, $S = \omega L/U_J$, and a dimensionless forcing that are well inside the regions in which the full Kutta condition should apply.

Daniels (1978) argues rather convincingly that this is the case, and that Figure 4 shows a three-deck structure in which the outer behavior is like $x^{3/2}$, with a transition through the main deck to a behavior of the $x^{1/3}$ kind in the inner deck, as described earlier. The "parabolic" $x^{1/3}$ behavior occurs on the inner scales, not in the potential flow. As a qualification to this interpretation, however, it should be noted that if the unsteady flow is apparently linear at all scales, then the lower-deck behavior is expected to consist of small-amplitude oscillations about a *mean* curve $y \simeq 0.895x^{1/3}$ deflected into the moving stream, rather than the oscillation $x^{1/3}$ $\exp(-i\omega t)$ apparently seen here. These experiments come nearest, in my opinion, to addressing the Kutta condition through direct study of the trailing-edge flow; a repeat of them with the predictions of triple-deck theory to hand would be most valuable.

Oscillating Airfoils

There are many more reports of experiments on oscillating airfoils and on airfoils entering gusts or subject to acoustic excitation. No consistent pattern emerges; indeed, it rapidly becomes apparent that experiments

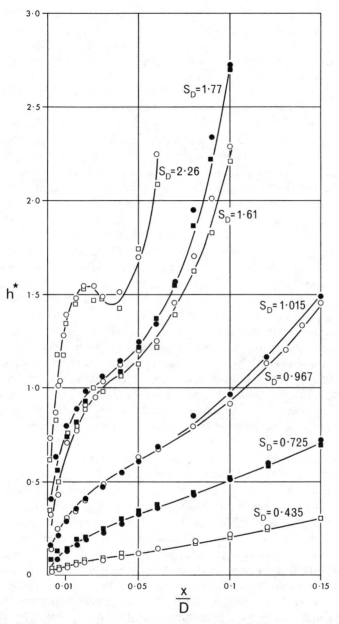

Figure 4 Optical compensation measurements of normalized dividing-streamline shape h^* as a function of downstream distance x/D; D is jet diameter, U_J jet exhaust speed, f frequency, $\mathrm{Re} = U_J D/\nu = 5 \times 10^4$, and $S_D = fD/U_J$ (with values indicated). The symbols $\bigcirc$ and $\bullet$ correspond to nozzle exit-plane forcing at 108 dB, and $\square$ and $\blacksquare$ to such forcing at 98 dB. (Reproduced, with permission, from Bechert & Pfizenmaier 1975.)

done in this area with whatever facilities happen to be available and without an underlying theoretical framework are bound to lead to the present totally unsatisfactory situation. Even in the most favorable conditions (of low frequency and low amplitude), the Kutta condition has not been adequately validated; measurements of the surface pressures near the edge may indicate a decrease toward zero of the loading, but the correlation between theory and experiment on the phase of the pressure near the edge—and indeed over much of the chord—is generally very poor. While the unsteady lift experienced may be reasonably predicted, the failure to correctly predict the phase is particularly serious for problems like the determination of flutter boundaries for a turbomachinery cascade.

In the visualization study of Ohashi & Ishikawa (1972), it is explicitly stated that the Kutta-Joukowsky condition was found to be valid at *all* test conditions, in the sense that the dividing streamline separated from the trailing edge at all times. In this study, airfoils were driven in a plunging mode, at Reynolds numbers around 10^4, over a reduced frequency range $\Omega = \omega L / U_\infty$ from about 1 to 20 and over a normalized amplitude range a/L $= 0.04$–0.45 (where a is the half-amplitude, and L the chord). Even the highest of these frequencies satisfies the condition $S \lesssim \mathrm{Re}^{1/4}$ of the Brown & Daniels (1975) theory, and so the applicability of the Kutta condition turns on the magnitude of the pressure rise produced. At the smallest oscillation amplitudes, the equivalent angle of incidence α can be said to satisfy the condition $\alpha = O(\mathrm{Re}^{-9/16})$ for the inviscid pressure rise to be accommodated in an attached trailing-edge flow, and one would expect the Kutta condition to hold. It is hard to believe that it holds for the higher oscillation amplitudes, where (as noted by the authors) the quasi-steady incidence may greatly exceed both the incidence (of 10–15°) for separation in steady flow and the incidence (around 2°) at which the Brown & Stewartson (1970) trailing-edge separation and stall would be expected. Some support for this view comes from Figure 8 of Ohashi & Ishikawa, where a curve Ω_2 is drawn in an $(\Omega, a/L)$ plane beyond which the abrupt shedding of large vortices of alternating sign replaces the more gentle and continuous shedding at lower Ω or a/L. The curve Ω_2 is not precisely defined, of course, but it may be fairly represented by the relation $\Omega^2 a/L = 3$, with the dependence on $\Omega^2 a/L$ being precisely that which arises in the inviscid unsteady pressure just outside the trailing region. (Recall that only a very narrow range of Reynolds number is involved in these experiments.) The observed change in vortex shedding is then consistent with the idea that for $\Omega^2 a/L \lesssim 3$ (and $\mathrm{Re} \sim 10^4$–10^5) the inviscid pressure supplied by the Kutta condition *can* be accommodated by the triple deck, but that for larger values periodic separation will occur upstream of the trailing edge.

In the experiments of Satyanarayana & Davis (1978), at a chord Reynolds number of 560,000, a symmetric airfoil was subjected to pitching oscillations of about 1° amplitude at frequency parameters $\Omega = \omega L / U_\infty$ between 0.1 and

2.4 [where the pressure measurements were taken at 40% chord and at four locations close to the edge (the closest at 97% chord)]. Agreement with the predictions of linear theory with full Kutta condition was good at all points, in both amplitude and phase, at $\Omega = 0.1$, and good in amplitude at $\Omega = 2.04$ but with significant phase discrepancies at the trailing edge. At $\Omega = 2.46$ there is an overprediction of the phase by about 70°, while the trailing-edge loading has a finite value comparable in amplitude and phase with that at 40% chord.

In these experiments, all the frequency parameters Ω (or S in our earlier notation) are less than $Re^{1/4}$ by at least a factor of 10, and if the Brown & Daniels high-frequency estimate of the inviscid pressure is used, this pressure seems safely to be tolerated by the triple deck at $Re \sim 10^6$. Thus, at first sight it appears surprising that the Kutta condition should fail in these circumstances. A possible explanation may, however, lie in the fact that all the values of Ω are so small that the high-frequency estimates are inappropriate and one should use the quasi-steady limit. There, as shown in Brown & Stewartson (1970), trailing-edge stall takes place at an incidence of about 2°—which is of at least the right order of magnitude to explain the Kutta-condition failure observed by Satyanarayana & Davis (1978). On the other hand, however, separation from the suction surface in genuinely "steady" flow is never observed at incidences as small as 1–2°, and in fact is not observed until 10–15° or so. A possible reconciliation of this dilemma may be provided by the recent work of Smith (1983), who argues that when the incidence first exceeds the rather small value of Brown & Stewartson, separation does indeed take place, but not in the catastrophic form called "stall." Instead, the flow reattaches just upstream of the trailing edge, and catastrophic stall is postponed to large incidences, perhaps as high as the 10–15° values often quoted as leading to separation.

More serious inconsistencies and discrepancies arise in most of the other published papers in this area; see the papers cited for further references. The paramount need is for experiments exploiting the scalings and balances of triple-deck theory, which has proved so successful (in comparison with experiment and full Navier-Stokes computations) in predicting the drag of airfoils and bluff bodies in separated flow. Indirect assessment of trailing-edge conditions is also sometimes possible—and then with spectacular results—as described in the next section.

APPLICATIONS AND CONSEQUENCES

Shear-Layer Receptivity

Differential mean flow past a splitter-plate trailing edge forms a downstream shear layer that (certainly for low-speed flows) sustains spatial instabilities over a range of low frequencies. It is natural in the first place to try to model the flow as inviscid, with a mean vortex sheet. If the sheet is doubly infinite, then appeal to causality is necessary (Jones & Morgan 1972) in order to determine the instability-wave amplitude at each frequency ω, given the nature of some external excitation (a point source, for example). This appeal is unnecessary if the presence of the splitter plate is taken into account and a Kutta condition applied, as in Crighton & Leppington (1974). The edge effect is dominant, even if the source and edge are remote—and the Kutta condition also guarantees causality but not vice versa.

For strictly incompressible flow, the Crighton-Leppington problem can be greatly simplified, and Bechert (1982, 1983) (using earlier work of his with U. Michel) has made these simplifications and studied the flow experimentally. If uniform flow U_∞ is separated from static fluid and effectively two-dimensional flow prevails, the tangential velocity u_2 in the stream is predicted to have the form

$$|u_2| = \frac{|\Delta p_{12}|}{\rho(U_\infty \omega l)^{1/2}} \frac{\pi^{1/2}}{2^{9/4}} \exp\left(\frac{\omega x}{U_\infty}\right), \tag{15}$$

where $|\Delta p_{12}|$ is the magnitude of the pressure difference across the splitter plate a distance l upstream from the edge. It is assumed here that $\omega x/U_\infty \gtrsim 1$, so that only the freely developing instability launched at the edge is significant. Figure 5, from Bechert (1983), shows the remarkable agreement obtained between this prediction and experiments (over a range of Strouhal numbers S_θ based on initial shear-layer thickness θ, controlled by upstream boundary-layer suction) in which Δp_{12} and u_2 were directly measured in a configuration corresponding reasonably well to that of the theoretical model. Further theory and experiment using a concentrated acoustic source excitation have now been carried out (D. W. Bechert, private communication), and the experiments confirm the predicted dependence of the instability-wave features on the source location. Results for a source very close to the edge are particularly simple, while those for a distant source show that the source is completely ineffective if located downstream of the edge near the shear layer, and that it is most effective in exciting an instability if located upstream.

Beyond the range of linear exponential growth, the vortical waves wind up into discrete vortices—vortex rings in the round-jet case—which comprise the large-scale orderly structures of the fully turbulent downstream flow. These structures are believed to be responsible for much of the entrainment into the flow and for much of the sound generation. Control of these processes by external excitation has been advocated and investigated for some 15 years [see, for example, Crighton (1981) for a review of the acoustic consequences of controlled excitation]. Direct control (by mechanical devices) of the nonlinear fully developed structures, with their random phases and azimuthal structure, seems quite impracticable. The obvious route to possible control of the coherent structures lies in the linear control provided by the Kutta condition that determines the receptivity of the initial shear layer.

As noted earlier, Goldstein (1984) has interpreted the control by Ahuja et al. (1983) of the separated-flow bubble on an airfoil as arising from enhanced entrainment associated with orderly instabilities on the separated shear layer. An attempt to verify Goldstein's receptivity formula for grossly separated flows—in appropriately simplified circumstances in the first place—along lines similar to those in Bechert's trailing-edge work would be of the greatest interest.

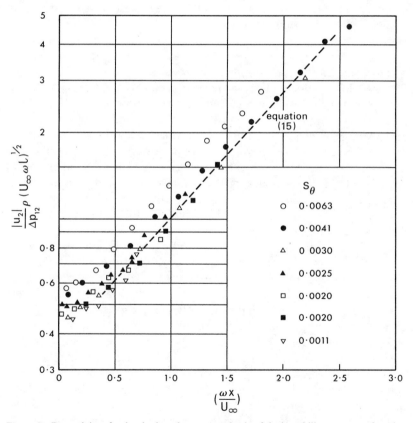

Figure 5   Receptivity of a simple shear layer; magnitude of the instability wave as a function of downstream distance. Broken line is the prediction of Equation (15). (Reproduced, with permission, from Bechert 1983.)

For leading-edge flows the Kutta condition again serves as the receptivity criterion if it is assumed (cf. Howe 1981a,b) that the boundary layer can support Tollmien-Schlichting waves, which provide an unsteady displacement-thickness condition on the outer flow. Then the problem for a leading edge in uniform flow with external excitation is very similar to that for a trailing edge with a downstream vortex wake—except that the energy-conversion processes are different (see the next section). The receptivity problem at the leading edge has recently been treated from a more fundamental point of view by Goldstein (1983). He shows that there is first a small edge region governed by the full Navier-Stokes equations, and that then the unsteady boundary-layer equations take over, these developing a double-deck structure with increasing x, with the upper deck involving a modified Blasius flow and the lower a Stokes shear wave. Because the equations at this level are parabolic, they have eigensolutions that need to be included, but these all decay with x, with the lowest order decaying most rapidly. These solutions to the unsteady boundary-layer equations are not asymptotic to solutions of the Navier-Stokes equations farther downstream, and the solutions needed there turn out to satisfy the Orr-Sommerfeld equation with slowly varying coefficients. There is an overlap domain in which it is shown, curiously, that the lowest-order boundary-layer eigensolution, decaying most rapidly in x, ultimately matches the amplifying Tollmien-Schlichting wave, while the other eigensolutions continue to decay. Once the amplitude of the lowest-order eigensolution is determined in terms of the external forcing (which needs a numerical solution of the leading-edge problem followed by a resolution into a particular solution plus eigenmodes), the receptivity problem is settled. It is not yet known whether or not this is equivalent, in some parameter ranges, to the Kutta condition of Howe.

If the upstream flow is itself inviscidly unstable (and the divided downstream flow may or may not be), there are more possibilities, as shown by Goldstein (1981) and discussed by Crighton (1981). There is a forced bounded solution (as $|\mathbf{x}| \to \infty$) with $|\mathbf{x}|^{-1/2}$ velocity singularities, and there is an eigenfunction with similar singularities and in general involving both an incident instability and a different one developing downstream of the edge. The general solution is singular at the edge and noncausal, contains an instability incident from upstream, and is exponentially unbounded downstream. Three obvious choices of the eigensolution constant give (a) a solution singular at the edge and bounded downstream, but noncausal and involving an incident instability; (b) a solution satisfying the Kutta condition at the edge, but unbounded downstream and (unlike the trailing-edge problem) noncausal and involving the incident instability; or (c) a solution that is causal and contains no instability incident from upstream, but is singular at the edge and unbounded downstream. Goldstein (1981) prefers the last of these, but there are difficulties in the interpretation of some of the exponential wave fields that lead me to think that the issue remains open and that solution (b) satisfying the leading-edge Kutta condition may be acceptable.

Energy-Conversion Mechanisms

At low Mach numbers, the trailing-edge Kutta condition represents a mechanism for sound absorption; incident acoustic energy is converted to vortical energy on the downstream shear layer. There it may be neutrally convected in uniform flow or amplified in an unstable flow—but that is irrelevant unless the vorticity can be scattered into sound by some downstream interaction; the energy-conversion process is a local one at the trailing edge.

The effect was first seen quantitatively in experiments on the reflection and transmission of sound across the exit of a subsonic jet nozzle; if W_T is the net acoustic power (incident minus reflected) propagating toward the nozzle along the jet pipe and W_R is the power radiated in the far field, then it is found that $W_R \ll W_T$ and that most of the incident power is given to the vortex wake. If we let $\Delta = 10 \log_{10} W_R/W_T$, so that $(-\Delta)$ is the attenuation in decibels, then typical results are shown in Figure 6, where Δ is plotted versus $k_0 R$ $(= \omega R/c_0$, with R the nozzle radius and c_0 the sound speed). The solid curve corresponds to an approximation

$$\frac{W_R}{W_T} = \frac{(1 + \frac{4}{3}M^2)}{4M}(k_0 R)^2 \tag{16}$$

given by Bechert (1980), where M is the jet Mach number. Bechert postulates that the low-frequency open-end reflection coefficient has the same value (-1) in the presence of the jet flow as it does in the absence of flow. Cargill (1982) shows this postulate to be equivalent to satisfying a Kutta condition, and he also shows that the M^2 terms above are incorrect, with the result

$$\frac{W_R}{W_T} = \frac{(1 + \frac{1}{3}M^2)}{4M(1 - M^2)^3}(k_0 R)^2. \tag{17}$$

This has the effect of raising the solid curve in Figure 6 uniformly in $k_0 R$ by 0.1, 0.9, 2.8, and 7.2 dB for the four respective values of M, improving further what is already impressive agreement between theory and experiment and indirectly confirming the applicability of the Kutta condition.

The trailing-edge interaction may actually lead to an enhancement of acoustic power at the expense of the mean flow—but that is a finite-Mach-number effect (as in Rienstra 1981), and at low Mach number there is always

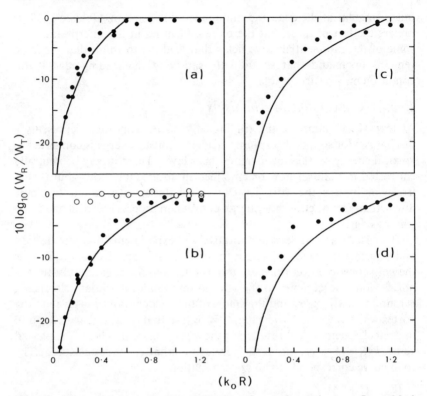

Figure 6   Absorption of sound in transmission through a nozzle carrying mean flow at Mach number M. W_R, W_T defined in text; k_0 is the acoustic wavenumber at frequency ω, R the nozzle radius. Solid line is the prediction of Equation (16). Symbols (●) denote measurements at (*a*) $M = 0.1$, (*b*) $M = 0.3$, (*c*) $M = 0.5$, and (*d*) $M = 0.7$. Symbols (○) in (*b*) indicate measured values at $M = 0$ and confirm energy conservation in absence of mean flow. (Reproduced, with permission, from Bechert 1980.)

a loss of acoustic power to vortical fluctuations. This conversion has been extensively studied by Howe (1979, 1980a,b), who has studied the reflection and transmission of waves through perforated screens in the presence of flow. The perforations may take the form of two-dimensional slits or circular apertures, and the Kutta condition is applied at all trailing edges (the upstream semicircular arcs in the case of circular holes with grazing mean flow). Bias flow normal to the screen has been considered as well as grazing flow. A sample result is the following simple expression for the power loss Δ per unit area, for waves normally incident on a screen with slit perforations (open area ratio σ), at low Strouhal number in grazing flow at

Mach number M:

$$\Delta = \frac{(\frac{1}{2}\pi\sigma)^2 + M^2}{(\frac{1}{2}\pi\sigma + M)^2}. \tag{18}$$

This function has a minimum value of $\frac{1}{2}$, which gives 3-dB attenuation, when $M = \frac{1}{2}\pi\sigma$; however, larger attenuations are possible if the sound makes repeated reflections from the perforated walls of a duct or enclosure. Many further applications can be found in references in the papers cited here; though these detailed studies are all recent, the underlying ideas can be found in patent applications for practical noise suppression devices going back to 1916!

At a leading edge the energy exchange is different, according to Howe's (1981a,b) displacement-thickness theory. Here it can be shown that antisymmetric displacement-thickness fluctuations on the two sides of a thin splitter plate lead, via the Kutta condition, to the transfer of energy from the mean flow to the acoustic field, with a sound wave incident from downstream extracting the maximum acoustic energy from the mean flow. The energy-transfer equations for leading and trailing edges in fact have precisely the same forms, except for overall sign; but whereas the energy of a sound wave incident on the trailing edge is needed to energize the wake, no such energy is required at the leading edge, where the wave perturbs a preexisting nonradiating mean vorticity field and causes the generation of acoustic energy additional to that incident on (and then reflected and diffracted by) the splitter plate.

Howe's first application of this idea was to the drive mechanism of the flue organ pipe. Here he ignored the singular behavior at the trailing edge where the driving jet separates, and instead he applied a leading-edge Kutta condition at the downstream leading edge of the mouth. This provided the displacement-thickness fluctuations downstream from the mouth that drive the organ-pipe resonator, and the leading-edge energy extraction from the mean flow is needed to overcome energy losses through various dissipative processes (diffusion, radiation, etc.). The claim for this view of the operating mechanism is strongly supported by impressive prediction of the threshold blowing velocities needed to energize the first four modes.

Application of a trailing-edge Kutta condition as well would require explicit treatment of the instability waves on the jet shear layers across the mouth of the instrument. Then we have the jet edge-tone configuration, if the rest of the resonator is ignored. This configuration has been much studied, but no rational theory to predict the frequencies of its operating stages has so far been published. The present author (Crighton 1984) has tried to remedy this situation by calculating the phase changes around the

various constituents of the edge-tone feedback cycle using linear hydrodynamic stability and boundary-value-problem techniques, and by assert-
ing that the frequency stages are determined by the application of a Kutta
condition at the upstream (trailing) edges of the duct from which the jet
emanates. Vortex shedding here is forced by the upstream feedback
pressure field generated when the jet interacts with the downstream
(leading) edge, where no Kutta condition is enforced. One prediction from
this work is the relation for the nth stage of operation, $h/\lambda = n + 7/24$, where
n is an integer, λ the sinuous instability wavelength at the operating stage
considered, and h the streamwise width of the mouth; this relation is to be
compared with a famous experimental correlation $h/\lambda = n + 1/4$ (where,
however, the mean value of the constants measured is actually 0.27 rather
than 0.25), and this similarity again represents encouraging indirect
evidence for the trailing-edge Kutta condition.

Finally, we should mention the deceptively difficult problem of the shear
layer across a slit in a plate with mean flow on one side only. Self-sustained
oscillations are often observed in practical cases involving flows over slits
and cutouts, and it is natural to seek a prediction of possible eigenfrequen-
cies using a linear vortex-sheet model. Several attempts have been made in
this direction; the most satisfactory are those of Howe (1981b), who derived
an expression for the vortex-sheet displacement containing two constants
of integration, and showed that if these were used to ensure that the vortex
sheet was attached to the plate at the upstream and downstream edges of
the slit, then there would be no net energy exchange between the mean and
unsteady flows and any initial disturbance would simply decay. The two
constants can be used to enforce a Kutta condition at the trailing
(upstream) edge, but at the downstream edge the vortex-sheet displacement
has an inverse-square-root singularity. There is, nevertheless, an extraction
of energy from the mean flow for frequencies within a finite band, or
equivalently a range of flow speeds U_∞ over which a wall-cavity resonance
at frequency ω can be excited (in the common case when the slit forms the
mouth of a cavity).

Although reasonable agreement with experiment was found, Howe
modified this theory by the inclusion of displacement-thickness fluctuations on the downstream plate, so as to mitigate the leading-edge
singularity and to model the large effect on the downstream boundary layer
of vorticity periodically ejected from the slit on either side of the plate. The
extra constant associated with the displacement-thickness fluctuations can
be used to reduce the vortex-sheet singularity to a simple jump discontinuity at the downstream edge. This leads to several modifications of
the earlier predictions, some of them slight, but others more serious. In
particular, at all but the lowest Strouhal numbers the displacement-

thickness fluctuations cause a volume flux through the slit whose magnitude is almost the same as that generated by the vortex-sheet motion, but of opposite sign. This effect would substantially change the radiated sound. Again, however, the comparisons with several sets of data (for the acoustic impedance of the slit and for the frequency ranges for self-sustained oscillations) are generally very convincing and offer strong evidence both for the almost ubiquitous application of the trailing-edge Kutta condition and the idea of displacement-thickness fluctuations controlled by a leading-edge Kutta condition.

CONCLUSIONS

Local-interaction theory provides a firm basis for understanding the applicability of an unsteady Kutta condition, whether in attached leading- or trailing-edge flows or in separated flows around bluff bodies. The steady-flow predictions of triple-deck theory have been remarkably well validated in comparison with experiment and Navier-Stokes calculations, even at very modest Reynolds numbers. Similar validation of predictions for unsteady flow has not yet been made, and it will remain difficult to understand the unsteady performance of airfoils until theory and experiment are brought together here.

New phenomena arising as consequences of the application of an unsteady Kutta condition continue to be reported, many of them in aeroacoustics and the general field of the interaction between acoustic fields and unsteady stable or unstable vortical waves. The leading-edge Kutta condition has already had important applications, and we can expect further applications and a deeper understanding of it in relation to the inner viscous and nonlinear flow structure near the edge and farther downstream. The Kutta condition for separated free-streamline flows is even more recent, and it may have important implications for the control of separation by active forcing.

Little has been said about supersonic flows, although their triple-deck structure is relatively well explored for steady flow. The edge behavior of inviscid perturbations is quite different from that in subsonic flow; for instance, the vortex-sheet deflection $h(x, t)$ in the Orszag & Crow (1970) type of trailing-edge problem has the form $h(x, t) \sim x \exp(-i\omega t)$ as $x \to 0+$, and no more smooth behavior is possible. This variation has not yet been related to any triple-deck structure, nor has any significant effect been reported as arising from the vortex-sheet instabilities launched by this edge behavior.

Finally, we must disappoint the pessimistic reader by saying that the notion of a Kutta condition and its many and varied consequences do not immediately collapse if the boundary-layer flow is fully turbulent. Many of the aeroacoustic experiments involve fully turbulent boundary layers, and yet convincing agreement is found with predictions based on the application of a Kutta condition to the outer potential flow. One of the most recent developments in triple-deck theory (see reviews cited) has involved the matching of a steady outer flow to a multilayered edge structure (or a separation-point structure) governed by mean turbulent equations (with some degree of "modeling"); the establishment of criteria for a Kutta condition on unsteady perturbations coupled to a turbulent mean state may not be too far off.

Literature Cited

Ahuja, K. K., Whipkey, R. R., Jones, G. S. 1983. Control of turbulent boundary layer flows by sound. *AIAA Pap. 83-0726, Aeroacoust. Conf., 8th, Atlanta, Ga.*

Alblas, J. B. 1957. On the diffraction of sound waves in a viscous medium. *Appl. Sci. Res. Sect. A* 6:237–62

Ashley, H., Landahl, M. T. 1965. *Aerodynamics of Wings and Bodies.* Reading, Mass: Addison-Wesley. 279 pp.

Batchelor, G. K. 1967. *An Introduction to Fluid Dynamics.* Cambridge: Cambridge Univ. Press. 615 pp.

Bechert, D. W. 1980. Sound absorption caused by vorticity shedding, demonstrated with a jet flow. *J. Sound Vib.* 70: 389–405

Bechert, D. W. 1982. Excited waves in shear layers. *DFVLR Forschungsber. 82-83.* 150 pp.

Bechert, D. W. 1983. A model of the excitation of large scale fluctuations in a shear layer. *AIAA Pap. 83-0724, Aeroacoust. Conf., 8th, Atlanta, Ga.*

Bechert, D., Pfizenmaier, E. 1971. On the Kutta condition at the nozzle discharge edge in a weakly unsteady nozzle flow. *DLR FB71-09.* Transl. as *R. Aircr. Establ. Libr. Transl. No. 1617* (From German)

Bechert, D., Pfizenmaier, E. 1975. Optical compensation measurements on the unsteady exit condition at a nozzle discharge edge. *J. Fluid Mech.* 71:123–44

Birkhoff, G. 1950. *Hydrodynamics.* Princeton, N.J.: Princeton Univ. Press. 186 pp.

Brown, S. N., Cheng, H. K. 1981. Correlated unsteady and steady laminar trailing-edge flows. *J. Fluid Mech.* 108:171–83

Brown, S. N., Daniels, P. G. 1975. On the viscous flow about the trailing edge of a rapidly oscillating plate. *J. Fluid Mech.* 67:743–61

Brown, S. N., Stewartson, K. 1970. Trailing-edge stall. *J. Fluid Mech.* 42:561–84

Cargill, A. M. 1982. Low-frequency sound radiation and generation due to the interaction of unsteady flow with a jet pipe. *J. Fluid Mech.* 121:59–105

Cheng, H. K., Smith, F. T. 1982. The influence of airfoil thickness and Reynolds number on separation. *Z. Angew. Math. Phys.* 33:151–80

Chow, R., Melnik, R. E. 1977. Numerical solutions of the triple-deck equations for laminar trailing-edge stall. *Proc. 5th Int. Conf. Numer. Methods Fluid Dyn., Lecture Notes in Physics* 59:135–44. Springer

Crighton, D. G. 1972. Radiation properties of the semi-infinite vortex sheet. *Proc. R. Soc. London Ser. A* 330:185–98

Crighton, D. G. 1981. Acoustics as a branch of fluid mechanics. *J. Fluid Mech.* 106: 261–98

Crighton, D. G. 1984. The operating stages of the jet edge-tone feedback cycle. *J. Fluid Mech.* In press

Crighton, D. G., Leppington, F. G. 1973. Singular perturbation methods in acoustics: diffraction by a plate of finite thickness. *Proc. R. Soc. London Ser. A* 335:313–39

Crighton, D. G., Leppington, F. G. 1974. Radiation properties of the semi-infinite vortex sheet: the initial value problem. *J. Fluid Mech.* 64:393–414

Daniels, P. G. 1978. On the unsteady Kutta condition. *Q. J. Mech. Appl. Math.* 31:49–75

Davis, S. S. 1975. Theory of discrete vortex noise. *AIAA J.* 13:375–80

Goldstein, M. E. 1981. The coupling between flow instabilities and incident disturbances at a leading edge. *J. Fluid Mech.* 104:217–46

Goldstein, M. E. 1983. The evolution of Tollmien-Schlichting waves near a leading edge. *J. Fluid Mech.* 127:59–81

Goldstein, M. E. 1984. Generation of instability waves in flows separating from smooth surfaces. *J. Fluid Mech.* 145 : 71–94

Howe, M. S. 1976. The influence of vortex shedding on the generation of sound by convected turbulence. *J. Fluid Mech.* 76 : 711–40

Howe, M. S. 1979. Attenuation of sound in a low Mach number nozzle flow. *J. Fluid Mech.* 91 : 209–29

Howe, M. S. 1980a. Aerodynamic sound generated by a slotted trailing edge. *Proc. R. Soc. London Ser. A* 373 : 235–52

Howe, M. S. 1980b. The influence of vortex shedding on the diffraction of sound by a perforated screen. *J. Fluid Mech.* 97 : 641–53

Howe, M. S. 1981a. The role of displacement thickness fluctuations in hydroacoustics, and the jet-drive mechanism of the flue organ pipe. *Proc. R. Soc. London Ser. A* 374 : 543–68

Howe, M. S. 1981b. On the theory of unsteady shearing flow over a slot. *Philos. Trans. R. Soc. London Ser. A* 303 : 151–80

Jones, D. S., Morgan, J. D. 1972. The instability of a vortex sheet on a subsonic stream under acoustic radiation. *Proc. Cambridge Philos. Soc.* 72 : 465–88

Kirchhoff, G. 1869. Zur Theorie freier Flüssigkeitsstrahlen. *Z. reine Angew. Math.* 70 : 289–98

Lighthill, M. J. 1953. On boundary layers and upstream influence: II. Supersonic flow without separation. *Proc. R. Soc. London Ser. A* 217 : 478–507

Messiter, A. F. 1983. Boundary-layer interaction theory. *J. Applied Mech., Trans. ASME* 50 : 1104–13

Miles, J. W. 1959. *The Potential Theory of Unsteady Supersonic Flow.* Cambridge : Cambridge Univ. Press. 220 pp.

Morgan, J. D. 1974. The interaction of sound with a semi-infinite vortex sheet. *Q. J. Mech. Appl. Math.* 27 : 465–87

Morkovin, M., Paranjape, S. V. 1971. Acoustic excitation of shear layers. *Z. Flugwiss.* 9 : 328–35

Ohashi, H., Ishikawa, N. 1972. Visualization study of flow near the trailing edge of an oscillating airfoil. *Bull. JSME* 15 : 840–47

Orszag, S. A., Crow, S. C. 1970. Instability of a vortex sheet leaving a semi-infinite plate.

Stud. Appl. Math. 49 : 167–81

Rienstra, S. W. 1979. *Edge influence on the response of shear layers to acoustic forcing.* PhD thesis. Tech. Univ., Eindhoven, Neth. 116 pp.

Rienstra, S. W. 1981. Sound diffraction at a trailing edge. *J. Fluid Mech.* 108 : 443–60

Satyanarayana, B., Davis, S. 1978. Experimental studies of unsteady trailing-edge configurations. *AIAA J.* 16 : 125–29

Smith, F. T. 1977. The laminar separation of an incompressible fluid streaming past a smooth surface. *Proc. R. Soc. London Ser. A* 356 : 433–63

Smith, F. T. 1979. Laminar flow of an incompressible fluid past a bluff body: the separation, reattachment, eddy properties and drag. *J. Fluid Mech.* 92 : 171–205

Smith, F. T. 1982. On the high Reynolds number theory of laminar flows. *IMA J. Appl. Math.* 28 : 207–81

Smith, F. T. 1983. Interacting-flow theory and trailing-edge separation—no stall. *J. Fluid Mech.* 131 : 219–49

Stewartson, K. 1974. Multistructured boundary layers on flat plates and related bodies. In *Advances in Applied Mechanics,* ed. C.-S. Yih, 14 : 145–239. New York : Academic

Stewartson, K. 1981. D'Alembert's paradox. *SIAM Rev.* 23 : 308–43

Sychev, V. V. 1972. On laminar separation. *Izv. Akad. Nauk SSSR, Mekh. Zhidk. Gaza* 3 : 47–59. Transl., 1974, in *Fluid Dyn.* 7 : 407–19 (From Russian)

Sychev, V. V., Sychev, Vik. V. 1980. On turbulent separation. Transl., 1981, as *NASA TM-76634* (From Russian)

Tam, C. K. W. 1971. Directional acoustic radiation from a supersonic jet generated by shear layer instability. *J. Fluid Mech.* 46 : 757–68

Tani, I. 1979. The wing section theory of Kutta and Zhukovski. In *Recent Developments in Theoretical and Experimental Fluid Mechanics,* ed. U. Müller, K. G. Roesner, B. Schmidt, pp. 511–16. Berlin : Springer-Verlag

Van Dyke, M. D. 1975. *Perturbation Methods in Fluid Mechanics.* Stanford, Calif : Parabolic. 271 pp.

Ann. Rev. Fluid Mech. 1985. 17 : 447–85

TURBULENT DIFFUSION FROM SOURCES IN COMPLEX FLOWS

J. C. R. Hunt

Department of Applied Mathematics and Theoretical Physics and Department of Engineering, University of Cambridge, Cambridge CB3 9EW, England

1. INTRODUCTION

The dispersion of matter and heat in turbulent flows is generally analyzed in different ways depending on whether the matter and heat are released from distributed sources, such as heat at the wall of a pipe, or whether (as in this review) they are released from a single source that is small compared with the scale of the flow. There are many examples of such types of dispersion in engineering fluid mechanics, such as the spreading of a flame in a highly turbulent engine flow, the dispersion of one or more species emitted from pipes into large chemical reactors, and the heat released from local overheating in nuclear reactor subassembly channels. There are also many examples where continuous or sudden sources of pollutant or heat are discharged into the atmosphere or into aqueous environments. Usually these discharges occur in complex flows with inhomogeneous turbulence, such as boundary layers over level surfaces, or in flows impinging on surfaces, such as hills in the atmosphere or underwater ridges in the oceans.

The aim of this review is primarily to summarize the current ideas and theories about the basic mechanisms for dispersion from localized sources in complex turbulent flows. A brief consideration of the examples already given indicates some of the characteristic features of such flows: in-homogeneity and unsteadiness of the turbulence, the shear and the convergence and divergence of the mean flow, the non-Gaussianity of the turbulence, recirculation of the mean flow, and the presence of surfaces.

In most practical dispersion problems, many of these effects occur simultaneously, but in the dispersion from small sources and in the vicinity

447

0066–4189/85/0115–0447$02.00

of the sources, one can discriminate both theoretically and experimentally between these effects and learn something about them. It is possible to make full use of the statistical theory of Taylor (1921) and its various generalizations to complex flow, first reviewed by Batchelor & Townsend (1956). Near the source, the displacements of fluid elements passing through the source are sensitive to the precise nature of the statistics of the velocity field, but farther away from the source the effects of many different processes make for statistical "blurring" and can enable certain simplifications to be made (Chatwin & Allen 1985).

The development of new instruments and methods of data analysis is greatly facilitating the measurement and analysis of the displacements of fluid elements in turbulent flows, which is why the statistical theory is likely to increase in importance.

This paper is a development of earlier reviews by Hunt et al. (1979a) and Hunt (1982), with a greater emphasis placed on physical explanation of the processes. In Section 2 the basic theory is summarized for the connection between the mean concentration and the statistics of displacements of fluid elements. In Section 3 these displacements are analyzed in converging and diverging flows and in shear flows, in cases where the turbulence statistics do not vary significantly across the "plume" (the "thin plume" approximation). In Section 4 the effects of nonhomogeneity of the turbulence across the plume are considered, as are those of diffusion near surfaces. We also discuss some approximate methods used to model these complex processes. In Section 5 we briefly summarize and label the effects by introducing "diffusion factors."

There is not enough space here to review the recent work on concentration fluctuations in simple and complex flows; however, we do briefly discuss the developing work on random flight models of turbulent diffusion from sources in complex flows [see Durbin (1983) for a recent review]. Many of the ideas and techniques described here were initially developed for studying dispersion in complex flows around buildings and hills in the atmospheric boundary layer. These applications have recently been reviewed by Wilson & Britter (1982) and by Egan (1984).

2.   ANALYSIS OF TURBULENT DIFFUSION IN TERMS OF DISPLACEMENT OF FLUID ELEMENTS

2.1   General Distribution of Sources

When a substance S (or heat) is released into a turbulent flow from a source, it is transported by the motion of the fluid elements and by diffusion of molecules (or diffusion of molecular motion in the case of heat), which can be expressed as an equation for the flow (in units of S crossing unit area per unit time in the ith direction) by

$$f_i^*(\mathbf{x}, t) = u_i^* C^* - D\ \partial C^*/\partial x_i, \tag{2.1}$$

where u_i^* is the ith component of the velocity of the fluid element at $(\mathbf{x}, t)$, C^* is the concentration of S within the fluid element per unit volume, and D is the molecular diffusivity of S. Since there is not, in *general*, any close relation between u_i^* and C^*, it is generally necessary to consider the history of the molecules of S as they move from the source to $(\mathbf{x}, t)$ in order to calculate the diffusion from the source. It is essential to distinguish carefully between how S is transported by fluid elements and how it is transported by molecular motion (Saffman 1960; see Figure 1).

In most environmental or industrial turbulent flows, the Peclet numbers based on the turbulent velocity scale u_0 and length scale L_0, $U_0 L_0/D$, or $U_0 L_0/K$ are large ($\gtrsim 10^2$). The dimensional arguments [on the basis of (2.1)] and detailed analysis show that matter or heat are transported *in the mean* largely by the mean and turbulent velocity field, i.e.

$$\overline{|u_i^* C^*|} \gg D\ \overline{\partial C^*/\partial x_i}.$$

However, this does not mean that molecular diffusion can generally be neglected in calculating turbulent diffusion; in some cases it is a controlling process—for example, in determining concentration fluctuations (Chatwin & Sullivan 1979) and chemical reactions in turbulent flows, in determining the flux and concentration near surfaces where the normal velocity decreases to zero, and in turbulent flows controlled by stabilizing buoyancy forces; in the latter case, the molecular diffusion of the denser and lighter

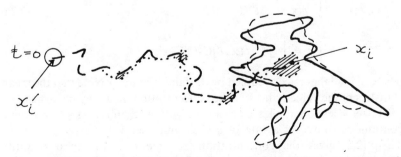

Figure 1 Schematic diagram showing displacements of marked fluid elements and molecules of S released into the flow at $t = 0$: ———— fluid element traveling from x_i' to x_i; ······· track of typical molecule; ⊂⊃ volume occupied by molecules of S originally at x_i'. "Boundaries" at time t indicate some large fraction of the original fluid elements (———) and molecules (———) released at time $t = 0$.

Figure 2 Schematic diagram showing how some fluid elements ($\rightarrow \rightarrow \rightarrow$) arriving at x_i come from a small region near x_i', and others ($- \times - \times -$) do not.

species between fluid elements controls the displacement of the fluid elements (see also Section 4).

We begin our discussion of turbulent dispersion from a source by neglecting molecular diffusion, since such a simplification is often a good approximation to the original problem (and it certainly helps in gaining an understanding of the problem). Under such an assumption, the concentration of S within any fluid element remains constant, $dC^*/dt = 0$, and so the concentration of S at any point ($\mathbf{x}$) is determined by whether or not fluid elements at $\mathbf{x}$ passed through the source. The two main objectives of a calculation are to predict the mean concentration C and to predict the mean square of the fluctuation concentration $\overline{c^2}$ and the flux $\overline{u_i^* C^*}$, where $C^* = C + c$ and $\bar{c} = 0$ by definition.

Let $p(\mathbf{x}, t; \mathbf{x}', t')\, d\mathbf{x}'$ be the probability of a fluid element being displaced by the velocity field $\mathbf{u}$ from within a distance $d\mathbf{x}'$ of the point $\mathbf{x}'$ (at time t') to the point $\mathbf{x}$ (at time t) in a statistical ensemble of turbulent flows. If a certain quantity M units of S is released into the flow, then the concentration C^* of a cloud of S at the point $\mathbf{x}'$ (time t') is $C_0(\mathbf{x}')$. By taking the product of C^* and C^{*2} at t' for each fluid element, and its probability of reaching x at time t, the means of C^* and C^{*2} are given by

$$C(\mathbf{x}, t; t') = \int_{\mathscr{D}} p(\mathbf{x}, t; \mathbf{x}', t') C_0(\mathbf{x}')\, d\mathbf{x}', \tag{2.2a}$$

$$\overline{C^{*2}} = C^2 + \overline{C^2} = \int_{\mathscr{D}} p(\mathbf{x}, t; \mathbf{x}', t') C_0^2(\mathbf{x}')\, d\mathbf{x}', \tag{2.2b}$$

where $\mathscr{D}$ is all the space occupied by the fluid (Figure 2). Note that the mean statistics of the velocity field or the concentration field may be unsteady.

Matter is less often released as a fixed volume than it is as a flow from a continuous source at a rate of $q(\mathbf{x}', t')$ units per volume per unit time. Then (2.2) gives the concentration at time t of the total quantity $C_0(\mathbf{x}', t')\, d\mathbf{x}' = q\, dt\, d\mathbf{x}'$ of substance released at time t'. Thus, the concentration at time t of all the substance released within $\mathscr{D}$ up to this time is

$$C(\mathbf{x}, t) = \int_{\mathscr{D}} \int_{-\infty}^{t} p(\mathbf{x}, t; \mathbf{x}', t') q(\mathbf{x}', t')\, dt'\, d\mathbf{x}'. \tag{2.2c}$$

Since the probability of finding an element somewhere is one, as is the probability that an element *came* from somewhere, $p(\mathbf{x}, t; \mathbf{x}', t')$ must satisfy the integral conditions

$$\int_{\mathscr{D}} p(\mathbf{x}, t; \mathbf{x}', t') \, d\mathbf{x} = \int_{\mathscr{D}} p(\mathbf{x}, t; \mathbf{x}', t') \, d\mathbf{x}' = 1. \tag{2.3}$$

Then, by integrating (2.2c), the important conservation conditions for C and $\overline{C^{*2}}$ can be derived for a finite release

$$\int C(\mathbf{x}, t) \, d\mathbf{x} = \int C_0(\mathbf{x}') \, d\mathbf{x}' = M_0, \tag{2.4a}$$

$$\int C^{*2}(\mathbf{x}, t) \, d\mathbf{x} = \int (C^2 + \overline{c^2}) \, d\mathbf{x} = \int C_0^2 \, d\mathbf{x}', \tag{2.4b}$$

and for a continuous source

$$\int_{\mathscr{D}} C(\mathbf{x}, t) \, d\mathbf{x} = \int_{-\infty}^{t} \int_{\mathscr{D}} q(\mathbf{x}', t') \, d\mathbf{x}' \, dt'. \tag{2.4c}$$

Equations (2.4a,c) merely state that the total amount of substance S in the fluid is conserved. Equation (2.4b) provides an upper limit for estimating $\overline{c^2}$ (Chatwin & Sullivan 1979).

2.2 Small Sources

It is useful to make mathematical idealizations using Dirac delta functions of sources that are small compared with typical dimensions of the flow.

For a release of a small volume of S at $\mathbf{x}^s$, we have

$$C_0(\mathbf{x}', t') = M_0 \delta(\mathbf{x}' - \mathbf{x}^s). \tag{2.5}$$

Substituting this into (2.2a) leads to

$$C(\mathbf{x}, t) = M_0 p(\mathbf{x}, t; \mathbf{x}^s, t'). \tag{2.6}$$

For a small continuous source at $\mathbf{x}^s$, releasing Q units per second, we have

$$q(\mathbf{x}', t') = Q(t') \delta(\mathbf{x}' - \mathbf{x}^s). \tag{2.7a}$$

Then from (2.2) the mean concentration can be calculated, even when there is no mean flow, such as in thermal convection or in near-wakes behind obstacles. In many cases S is released into a strong cross-flow (e.g. a chimney). If this is idealized as having a uniform constant velocity U_0 in the x_1 direction and as being much stronger than the velocity fluctuations u_1 in that direction (i.e. $U_0^2 \gg \overline{u_1^2}$), then all fluid elements take a time $U/(x - x')$ to move from x' to x, so that

$$p(\mathbf{x}, t; \mathbf{x}', t') = p_2(y, z, t; y', z', t')\delta(x - x' - U(t - t')), \tag{2.8}$$

where $p_2 \, dy' \, dz'$ is the probability that a particle moves from near y', z' at t' to y, z at t for any x or x'. Substituting (2.7), (2.8) into (2.2c) leads to

$$C(\mathbf{x}, t) = (Q(t^s)/U_0)p_2(y, z, t; y^s, z^s, t^s), \tag{2.9}$$

where $t^s = t - (x - x^s)/U_0$. Thus, for small finite volume or steady sources, Equations (2.6) and (2.9) show that the mean concentration is directly proportional to the probability distribution of particle displacements *from the source*.

Thus, the nth moments of the mean concentration distribution are related to the nth moments of the probability distribution of displacement and hence to the mean of the nth moments $\overline{X_i^n}$. For example, the mean position of a cloud of contaminant and the second moment of its distribution about the source (usually denoted by σ_α^2) are given by

$$\begin{Bmatrix} \overline{X_\alpha} = \\ \overline{X_\alpha^2} = \sigma_\alpha^2 \end{Bmatrix} = \frac{1}{M_0} \int \begin{Bmatrix} (x_\alpha - x_\alpha^s) \\ (x_\alpha - x_\alpha^s)^2 \end{Bmatrix} C(\mathbf{x}, t) \, d\mathbf{x} = \int \begin{Bmatrix} (x_\alpha - x_\alpha^s) \\ (x_\alpha - x_\alpha^s)^2 \end{Bmatrix} p(\mathbf{x}, t; \mathbf{x}^s, 0) \, d\mathbf{x}. \tag{2.10}$$

The spreading relative to the mean position $x_i^s + \overline{X}_i$ (as opposed to the source position x_i^s) is defined by

$$\Sigma_\alpha^2 = \overline{(X_\alpha - \overline{X}_\alpha)^2} = \overline{X_\alpha^2} - \overline{X}_\alpha^2 = \sigma_\alpha^2 - (\overline{X}_\alpha)^2. \tag{2.11}$$

An average width is given by $L = (\Sigma_1 \Sigma_2 \Sigma_3)^{1/3}$. In cases where the tails of the concentration distribution need to be known (e.g. plumes originating from ground-level releases, or explosive regions in inflammabale clouds), higher moments (e.g. $\overline{X_\alpha^3}$) are also important.

If a cloud of marked fluid elements or particles is released at some initial time t^s, then one wants to know the displacement of the instantaneous cloud center $\{X_i\}$ and some measure of the instantaneous dimensions of the cloud $\{\Sigma_i\}$ or $\{L\} = (\{\Sigma_1\} \{\Sigma_2\} \{\Sigma_3\})^{1/3}$. The random functions $\{X_i\}$ and $\{\Sigma_i\}$ are defined by

$$\{X_\alpha\} = \int (x_\alpha - x_\alpha^s)C^*(\mathbf{x}, t) \, d\mathbf{x} \bigg/ \int C^* \, d\mathbf{x},$$

$$\{\Sigma_\alpha\}^2 = \int (x_\alpha - x_\alpha^s - \{X_\alpha\})^2 C^*(\mathbf{x}, t) \, d\mathbf{x} \bigg/ \int C^* \, d\mathbf{x}. \tag{2.12}$$

Batchelor (1952) showed that

$$\overline{\{X_\alpha\}} = \overline{X_\alpha}, \qquad \overline{\{\Sigma_\alpha\}^2} + \overline{(\{X_\alpha\} - \overline{X}_\alpha)^2} = \Sigma_\alpha^2,$$

$$\Sigma_\alpha^2 = \tfrac{1}{2}\overline{(X_\alpha^{(a)} - X_\alpha^{(b)})^2},$$

where (a) and (b) refer to any pairs of particles in the cloud. These expressions imply that (a) the mean cloud center coincides with the mean position of individually related particles at some initial time t^s, (b) the spreading of the cloud relative to its own center plus the meandering of the cloud's center equals the spread of individually related particles, and (c) the spreading of the cloud relative to its own center is equal to one half the mean square *two-particle dispersion*. Eventually the cloud becomes much larger than the scale of turbulent motions, and particle displacements become decorrelated, so that

$$\overline{(\{X_\alpha\} - \overline{X}_\alpha)^2} \to 0 \quad \text{and} \quad \Sigma_\alpha^2 = \overline{\{\Sigma_\alpha\}^2}.$$

If we use the conservation condition (2.4), the second moments provide an estimate of the mean concentration C_σ in the mean cloud (cloud center $\overline{X}_i$), with representative volume $L^3 = \Sigma_x\Sigma_y\Sigma_z$, of $C_\sigma \sim M_0/L^3$. On the other hand, the mean concentration as averaged over the center of each cloud (cloud center $\{X_i\}$) is $C_\sigma \sim M_0/\{L\}^3$, where $\{L\}^3 = \{\Sigma_x\}\{\Sigma_y\}\{\Sigma_z\}$.

2.3 Equations for the Displacements of Fluid Elements

We have reviewed the general relations between the mean concentration distribution and the statistics of displacement of fluid elements and have indicated how the essential features of turbulent diffusion of S from many different kinds of concentration distributions in many different inhomogeneous unsteady source flows can be predicted by calculating the first and second moments of the displacements of fluid elements.

The kinematic equation that defines the displacement X of particles (or fluid elements) passing through the source P at x^s can be written in terms of the mean velocity U and the fluctuating velocity u at $X + x$ (at a given time t) as an Eulerian equation

$$\frac{dX}{dt} = (U + u)(X + x^s, t), \tag{2.13}$$

where $\bar{u} = 0$. In terms of the mean and fluctuating velocity of the particle at time t, the displacement X can be expressed as a Lagrangian equation

$$\frac{dX}{dt} = (V + v)\langle t \rangle, \tag{2.14}$$

454   HUNT

where $\bar{\mathbf{v}} = 0$; in general, $\mathbf{V}\langle t \rangle \neq \mathbf{U}(\mathbf{X}, t)$, as is shown in Section 4. (We shall take the time of release as $t = 0$ unless stated otherwise.) In the next section, we consider various simplified forms of this equation that also correspond to various kinds of practical problems. The main aim in each case is to calculate the first two moments of the components of the displacement $\mathbf{X}$, viz. $\overline{X_\alpha}$, $\overline{(X_\alpha - \bar{X}_\alpha)^2}$ for $\alpha = 1, 2, 3$.

3. "NARROW PLUMES" IN STEADY COMPLEX FLOWS

3.1 Equations for the Displacement in Narrow Plumes

In most complex turbulent flows, the mean velocity and the turbulence experienced by fluid elements change considerably as the elements travel from the source. In some cases the mean flow advects the fluid elements into regions with different turbulence; in other cases the turbulence itself transports the fluid elements into a region of different turbulence characteristics. In this section we concentrate on the former kind of problem, which occurs when sources of pollution are located upwind of hills or buildings in the atmosphere or, in general, of changes in surface roughness, or when heated turbulent flows impinge onto downwind obstacles, such as heat-exchanger tubes or turbine blades (Figure 3, Figure 4a).

The characteristic scales that determine the dispersion in these flows are

U_0 and u_0: typical mean and root mean square (rms) turbulent velocities,

L_0: typical integral scale of the turbulence,

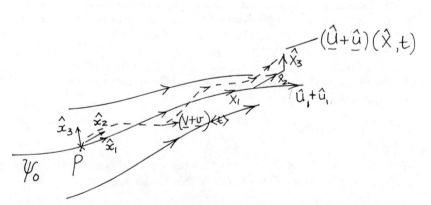

Figure 3 Definition sketch of mean streamlines ($\rightarrow\rightarrow$); the coordinate system ($\hat{x}$) is defined relative to the mean streamline ψ_0 through the source P, the Lagrangian $(\mathbf{V} + \mathbf{v})\langle t \rangle$ and Eulerian velocities $(\hat{\mathbf{U}} + \hat{\mathbf{u}})$, and the displacements $\hat{\mathbf{X}}$.

H: length scale over which the mean flow and turbulence change.

If we assume the mean velocity field $U(x)$ to be steady, a mean streamline ψ_0 can be drawn through the source P, and a coordinate system $(\hat{x}_i)$ can be constructed parallel and normal to this streamline, with its origin at the source. If over the time t of travel from the source P, the particles do not diffuse a distance far from the mean streamline compared with the scale H of the inhomogeneity of the mean flow, then the mean velocity of a particle at $\mathbf{X} + \mathbf{x}^s$ is given by its value on ψ_0, $\hat{U}_\alpha$, and its derivatives $\partial\hat{U}_\alpha/\partial x_j$ on ψ_0:

$$\hat{U}_\alpha(\hat{X}_1, \hat{X}_2, \hat{X}_3, t) = \hat{U}_\alpha(X_1, t) + \sum_j \hat{X}_j \frac{\partial \hat{U}_\alpha}{\partial x_j}(X_1, t) + \cdots. \tag{3.1}$$

By definition the normal components of mean velocity are zero on the mean streamline, so $\hat{U}_2 = \hat{U}_3 = 0$ on ψ_0.

Then, (2.13) and (3.1) reduce to three coupled equations for the three components of displacement:

$$d\hat{X}_1/dt = \hat{U}_1(\hat{X}_1, t) + \sum_{j=2,3} \hat{X}_j\, \partial U_1/\partial \hat{x}_j + \hat{u}_1(\hat{\mathbf{X}}, t), \tag{3.2}$$

$$d\hat{X}_\alpha/dt = \sum_{j=2,3} \hat{X}_j\, \partial U_\alpha/\partial \hat{x}_j + \hat{u}_\alpha(\hat{\mathbf{X}}, t). \tag{3.3}$$

If the turbulence is weak enough (see Section 4), the mean Lagrangian velocity $\mathbf{V}$ is equal to the mean velocity $\mathbf{U}(x, t)$, and then $\hat{u}_\alpha(\hat{\mathbf{X}}, t) = v_\alpha\langle t\rangle$.

3.2 Normal Diffusion in Straining Flows

Consider a symmetrical straining flow where $\partial\hat{U}_2/\partial\hat{x}_3 = \partial\hat{U}_3/\partial\hat{x}_2 = 0$, such as a flow over the centerline of a symmetric three-dimensional hill (above the surface layer), a flow in a contraction, or any two-dimensional flow. Then, Equation (3.3) for the diffusion normal to the mean streamline can be further simplified to

$$\frac{d\hat{X}_\alpha}{dt}\langle t\rangle = \hat{X}_\alpha\langle t\rangle\, \partial\hat{U}_\alpha/\partial x_\alpha\langle t\rangle + \hat{v}_\alpha\langle t\rangle, \qquad \alpha = 2, 3. \tag{3.4}$$

It follows that the solution for the random displacement of a fluid element passing through P (i.e. $\hat{X}_\alpha = 0$ when $t = 0$) is

$$\hat{X}_\alpha\langle t\rangle = \beta_\alpha^{-1} \int_0^t \beta_\alpha \hat{v}_\alpha\langle t'\rangle\, dt', \tag{3.5}$$

where the straining function at the position of the particle at time t is

$$\beta_\alpha\langle t\rangle = \exp\left(-\int_0^t (\partial\tilde{U}_\alpha/\partial x_\alpha)\langle t'\rangle\, dt'\right).$$

This is a random function if $\partial \hat{U}_\alpha / \partial \hat{x}_\alpha$ varies along the streamlines (since $\hat{X}_1 \langle t \rangle$ is random), as for example in a stagnation flow with a strong turbulence. But if the turbulence is weak enough, or if the strain rate only varies slowly over the turbulence scale, the random element of the strain function is negligible, i.e.

$$\left| \int_0^{\hat{x}_1} (\partial \hat{U}_\alpha / \partial x_\alpha v_1 / \hat{U}_1^2) \, d\hat{x}_1 \right| \ll \left| \int_0^{\hat{x}_1} (\partial \tilde{U}_\alpha / \partial x_\alpha / \hat{U}_1) \, d\hat{x}_1 \right|.$$

In practice, this condition is generally satisfied where $u_0/U_0 \lesssim 0.2$, and it allows the strain function to be written as a deterministic function of position along ψ_0, i.e.

$$\beta_\alpha \langle t \rangle = \beta_\alpha(\hat{x}_1),$$

where

$$t = \int_0^{\hat{x}_1} d\hat{x}_1' / \hat{U}_1(\hat{x}_1').$$

By taking the ensemble average of (3.4), we obtain the mean square displacement in terms of an integral over a product of the cross-correlation at times t', t'' and the strain function, i.e.

$$\overline{X_\alpha^2} \langle t \rangle = \beta_\alpha^{-2}(\hat{x}_1) \int_0^t \int_0^t \beta_\alpha \langle t' \rangle \beta_\alpha \langle t'' \rangle \rho_\alpha \langle t', t'' \rangle \, dt' \, dt'', \tag{3.6a}$$

where $\rho_\alpha \langle t', t'' \rangle = \overline{v_\alpha \langle t' \rangle v_\alpha \langle t'' \rangle}$. For later reference, it is useful to define the normalized correlation

$$R_\alpha = \rho_\alpha / (\overline{v_\alpha^2} \langle t' \rangle \overline{v_\alpha^2} \langle t'' \rangle)^{1/2}. \tag{3.6b}$$

No assumptions about the homogeneity or spatial scale of the turbulence have been made so far. However, to *calculate* (as opposed to measuring) ρ_α in complex flows, it is necessary to make further assumptions, which again are often found to be valid in practice.

3.3 Calculating the Turbulence Correlations

The change of the covariance of the fluctuating velocities at times t', t'' in a complex shear flow with converging, diverging, and varying streamlines depends on the relative magnitude of three time scales: the times of travel from the source t; the time for the mean flow and the turbulence to be distorted ($T_d \sim H/U_0$, where H is the length scale of the obstacle or contraction); and the Lagrangian time scale T_L, which essentially is the time scale over which the velocities of fluid elements are correlated.

If $t \lesssim T_L$ and $u_0 \ll U_0$, then any changes in the turbulence structure along the mean streamline are largely determined by the distortion of the turbulence by the velocity gradients of the mean flow, by the effects of any obstacles, and (in a stratified fluid) by local changes in the density gradient, provided that any of these effects occur within a distance L_0 of the mean streamline. These are linear processes in which any changes in the turbulence are determined by the initial vorticity and velocity of the turbulence near P. When $t \gtrsim T_L$, the advection and distortion of the turbulence vorticity by the turbulent velocity field, which is a chaotic decorrelating and nonlinear process, begin to be as large as by the linear or "rapid" processes. The linear analysis is often called "rapid distortion theory" (Townsend 1976, Deissler 1968, Hunt 1973).

Very close to the source, ρ_α reduces to the mean square velocity at P, i.e. when

$$t \ll T_d, \qquad t \ll T_L,$$

then

$$X_\alpha = \hat{v}_\alpha\langle 0\rangle t \quad \text{and} \quad \rho_\alpha = \overline{\hat{v}_\alpha^2}\langle 0\rangle = \overline{\hat{u}_\alpha^2}(\mathbf{x}^s, 0) \qquad (R_\alpha = 1). \tag{3.7a}$$

Over this short time scale, even in a uniform mean flow, the particles move in straight lines. [Thus the concentration distribution can easily be calculated from (2.9)—for example, for a continuous line source Q in a steady turbulent flow, we have

$$C = Q p_{\mathbf{u}}(\hat{u}_2, \hat{u}_3)/(\hat{x}_1^2/\hat{U}_1),$$

where $p_{\mathbf{u}}$ is the probability distribution of the two normal velocity components and

$$\hat{u}_2 = \hat{U}_1\hat{x}_2/\hat{x}_1, \qquad \hat{u}_3 = \hat{U}_1\hat{x}_3/\hat{x}_1. \tag{3.7b}$$

By a distance farther downwind such that $t \sim T_d$, the turbulence changes by a number of processes:

1. As the mean streamline enters a region of scale H where the mean velocity changes, its gradients lead to straining and rotation of the initial vorticity of the turbulence (e.g. region I in Figure 4a). Since the initial vorticity is weakly correlated with initial velocity (or not at all in isotropic turbulence), the changes in the turbulent velocity produced by straining are weakly correlated with the initial velocity. Their magnitude increases with (H/L_0) and with the total strain $[\hat{U}_1(\hat{x}_1)/U_0 - 1]$.

2. If the mean streamline moves to within a distance n of a surface or obstacle, the turbulence of scale $k^{-1} \gtrsim n$ is changed by the "blocking" of the eddies by the "image" eddies, which ensure that there is no normal fluctuating velocity across the surface. It is mainly the larger scales of the blocked normal fluctuations that contribute to large-scale fluctuations

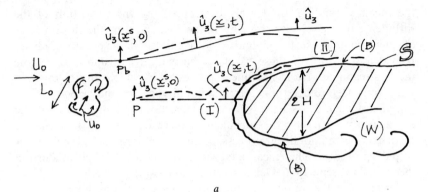

a

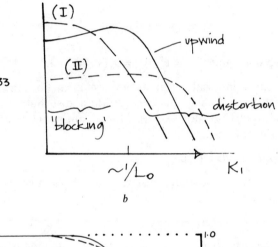

b

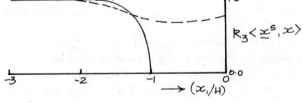

c

Figure 4   Turbulent flow around a bluff obstacle. (*a*) The scales of upwind turbulence, the regions of the flow, and the different characteristic trajectories of particles released in weakly or strongly straining flows. (*b*) Typical spectra of turbulent velocity components normal to the mean streamlines upwind and in regions I and II when $L_0 \simeq 2H$. (*c*) Normalized correlation $R_3\langle x^s, x\rangle$ between the normal velocity at the source position $x^{(s)}$ and along the mean streamline at x_1 at a later time t equal to the travel time. Flow around a circular cylinder of radius H; $L_0 \ll H$, $x^{(s)} = (-3H, 0)$ ———, $x^s = (-3H, H/2)$ ---; $L_0 \gg H$, any $x^{(s)}$ ⋯.

parallel to the surface. The smaller scales are not affected, so not only are the velocity fluctuations changed but also the correlation with the initial velocity is changed. The changes in the spectra of the normal velocity component $\hat{u}_3$, caused by the distortion and blocking effects in regions I and II of Figure 4a, are shown in Figure 4b. These are based on measurements by Bearman (1972), Britter et al. (1979), Huot et al. (1984), and others, and the theory of Hunt (1973) and Hunt & Graham (1978).

3. If the mean streamline is close enough to a surface or an obstacle, it can enter a new field of turbulence (region B)—for example, in the boundary layer or wake of an obstacle or in the internal layer downwind of a change in surface roughness. Generally this new field of turbulence is weakly correlated with the upwind turbulence, but the largest scales of the upwind turbulence persist and may control the diffusion in the new field of turbulence (Panofsky et al. 1982, Hunt 1982).

4. If there are density gradients in the upwind flow, the changes in the mean flow affect these density gradients and hence, if buoyancy forces are significant, the turbulence. This process has not been explored, but its effect is seen in the behavior of plumes in wind-tunnel and field studies (e.g. Hunt & Snyder 1980).

Computations plotted in Figure 4c show how $\hat{R}_{33}$ changes in weakly turbulent flow around a circular cylinder when $L_0 \ll H$ [using the theory of Hunt (1973, Section 6)]. Note that except close to the stagnation point when the source is on the centerline, R_3 is greater than about 0.5. This means that the trajectories of fluid elements are more irregular than in the upwind flow, but that they do not contain many rapid changes in direction. Typically, $(\overline{\hat{u}_3^2})^{1/2}$ varies by about a factor of 2 over this range.

If the incident turbulence is isotropic and its scale L_0 is small compared with H, useful expressions can be derived for small changes in $\overline{u_\alpha^2}$ by utilization of rapid distortion theory (Townsend 1976, p. 77). For example, in a steady two-dimensional flow where $\hat{U}_2 = 0$,

$$\overline{v_3^2} = \overline{u_3^2} = u_0^2 \{1 + \tfrac{4}{5}[(\hat{U}_1(t)/\hat{U}_1(0)) - 1] + \cdots\},$$

where

$$\hat{U}_1(t)/\hat{U}_1(0) - 1 \ll 1$$

and

$$\hat{R}_{33} = \hat{R}_{22} = 1 + O((t/T_\mathrm{d})^2). \tag{3.8a}$$

If the scale of the turbulence L_0 is much *larger* than the scale (H) of the inhomogeneities of the mean flow, the former is controlled by the blocking action at nearby boundaries [e.g. large-scale atmospheric turbulence impinging onto a building (Britter et al. 1979)]. In this case, $T_d \ll T_L$ and $R_3 = R_2 = 1$. Thus, the calculation of ρ_α reduces to the calculation of $\hat{U}_\alpha^2$. For flow around a free-standing obstacle, the changes of turbulent fluctuations are essentially quasi-steady fluctuations; the turbulence induces a small change $\delta\alpha$ in the direction of approach flow, where $\overline{(\delta\alpha)^2} = \overline{u_{30}^2}/U_0^2$. Here $\overline{u_{30}^2}$ is the mean square normal turbulent velocity upwind,

$$\overline{u_3^2} = (\partial \tilde{U}_3/\partial\alpha)^2 \overline{u_{30}^2}/U_0^2, \tag{3.8b}$$

where the derivative in the normal mean velocity with respect to α can be expressed in terms of the derivatives of the U_3, U_1 components and the angle θ between the mean streamline and the x_1-axis:

$$\partial \tilde{U}_3/\partial\alpha = (\partial U_3/\partial\alpha) \cos \theta - (\partial U_1/\partial\alpha) \sin \theta.$$

Figure 5c shows computations and measurements of $\overline{u_3^2}$ near a circular cylinder in homogeneous turbulence. If the mean flow is potential flow on the stagnation line, $\overline{u_3^2}$ increases by a factor of two, but by less if L_0 is comparable with H (Hunt & Mulhearn 1973).

In some kinds of boundary-layer flow where the turbulence scales are large compared with the scales of mean velocity gradients [e.g. in the presence of thermal convection or large-scale free-stream turbulence (Hunt 1984)], the "blocking" effect is also the dominant cause of distortion of the turbulence. Then turbulence over any surface undulations is more affected by the changes in the "blocking" or "image"-induced turbulence caused by the changes in surface slope than by distortion of the vorticity of the turbulence. However the simple quasi-steady analysis only leads to $\overline{u_3^2} = 0$ and $\overline{u_1^2} = (\hat{U}_1/U_0)^2 \overline{u_{10}^2}$. Higher-order analysis of the blocking process is necessary.

In stationary homogeneous turbulence, ρ_α has the special form

$$\rho_\alpha(t', t'') = \overline{u_\alpha^2} R_\alpha(t' - t''). \tag{3.9}$$

This is a commonly used approximation wherever the turbulence time scale T_L is *small* compared with the time scale T_D on which the turbulence of the marked fluid elements is changing. This can be justified for sources near surfaces in turbulent boundary layers (Hunt & Weber 1979); it is only an approximation in the wake of a source that generates its own small-scale turbulence [e.g. the classic heated cylinder (Hinze 1975, p. 505)] or far downwind of sources in inhomogeneous turbulence. This is an implicit assumption in many turbulence models, which assume small relaxation times for most turbulence mechanisms.

3.4   *Calculation of Dispersion*

Given the covariances of the turbulent velocity components, the mean square dispersion can be calculated for various flows and various scales.

1. For short travel times and small distortion, or if $R_\alpha = 1$, the expression (3.6a) reduces to an integral involving only the rms of the turbulent velocity and the strain function

$$\hat{\sigma}_\alpha \langle t \rangle = \overline{(\hat{X}_\alpha^2)^{1/2}} \langle t \rangle = \beta_\alpha^{-1} \langle t \rangle \int_0^t \beta_\alpha \langle t' \rangle \sqrt{\overline{v_\alpha^2}} \langle t' \rangle \, dt' \tag{3.10}$$

for

$$t \ll T_{\rm L}, \qquad t < T_{\rm D}.$$

2. For long travel times, or in very small-scale turbulence, it is useful to rewrite (3.6b) as

$$\hat{\sigma}_\alpha^2 = 2\beta_\alpha^{-2} \int_0^t \beta_\alpha \langle t' \rangle K_\alpha \langle t' \rangle \, dt', \tag{3.11a}$$

where

$$K_\alpha \langle t' \rangle = \beta_\alpha^{-1} \langle t' \rangle \int_0^{t'} \beta_\alpha \langle t'' \rangle \rho_\alpha \langle t', t'' \rangle \, dt''. \tag{3.11b}$$

Then if ρ_α satisfies (3.9), K_α reduces to the usual form of the turbulent diffusivity:

$$K_\alpha \langle t' \rangle = \overline{v_\alpha^2} \, T_{\rm L}^{(\alpha)}, \tag{3.11c}$$

where $T_{\rm L}^{(\alpha)} = \int_{-\infty}^t R_\alpha \langle t - t' \rangle \, dt'$ is the Lagrangian time scale of the α component. Equation (3.11a) reduces to Taylor's (1921) results for stationary turbulence, and Equation (3.11b) shows how K_α depends on the strain function when $T_{\rm L}$ is comparable with the distortion time scale $T_{\rm d}$.

3. In the limit of two-dimensional flows, where $\partial \hat{U}_3 / \partial \hat{x}_3 = -\partial \hat{U}_1 / \partial \hat{x}_1$ and $\hat{U}_2 = 0$, the strain functions β_3, β_2 reduce to simple forms. Since

$$\int \partial \hat{U}_3 / \partial \hat{x}_3 \, dt = -\int (\partial \hat{U}_1 / \partial \hat{x}_1 / \hat{U}_1) \, d\hat{x}_1,$$

it follows that

$$\beta_3 = \hat{U}_1(\hat{x}_1) / \hat{U}_1(0), \qquad \beta_2 = 1, \tag{3.12}$$

so (3.6a) reduces to

$$\hat{\sigma}_3^2 = \overline{\hat{X}_3^2} = \frac{1}{\hat{U}_1(\hat{x}_1)^2} \int_0^{\hat{x}_1} \int_0^{\hat{x}_1} \rho_3 \langle x_1', x_1'' \rangle \, d\hat{x}_1' \, d\hat{x}_1'' \tag{3.13a}$$

and

$$\hat{\sigma}_2^2 = \overline{\hat{X}_2^2} = \int_0^t \int_0^t \rho_2 \langle t', t'' \rangle \, dt' \, dt''. \tag{3.13b}$$

When $t \ll T_L$, or $R_\alpha = 1$, these reduce to

$$\hat{\sigma}_3 = \frac{1}{\hat{U}_1(\hat{x}_1)} \int_0^{\hat{x}_1} (\overline{\hat{u}_3^2})^{1/2}(\hat{x}_1') \, d\hat{x}_1', \qquad \hat{\sigma}_2 = \int_0^t (\overline{\hat{u}_2^2})^{1/2}(t') \, dt'. \tag{3.13c}$$

Thus the rms displacement normal to the mean streamline in the plane of the flow is directly proportional to the distance between the mean streamline and adjacent streamlines [i.e. $\hat{U}_1(0)/\hat{U}_1(x_1)$] and to a double integral of the covariance of the turbulent velocities of a fluid element at two distances $(\hat{x}_1', \hat{x}_1'')$ along the mean streamline.

The criterion for when the streamline convergence or divergence dominates over turbulent diffusion can be derived by differentiating (3.13a):

$$d\hat{\sigma}_3^2/d\hat{x}_1 = \left(-\frac{2d\hat{U}_1/d\hat{x}_1}{\hat{U}_1(\hat{x}_1)} \hat{\sigma}_3^2 + \frac{2}{\hat{U}_1(\hat{x}_1)^2} \int_0^{\hat{x}_1} \rho_3 \langle \hat{x}_1, \hat{x}_1' \rangle \, d\hat{x}_1' \right). \tag{3.14}$$

Thus, near the source where $t < T_L$, the criterion is that

$$\Sigma \sim \hat{\sigma}_3 \, d\hat{U}_1/d\hat{x}_1/(\overline{\hat{u}_3^2})^{1/2} \gtrsim 1,$$

and where $t > T_L$, it is

$$\Sigma \sim \hat{\sigma}_3^2 \, d\hat{U}_1/d\hat{x}_1/(\overline{\hat{u}_3^2} T_L) \gtrsim 1.$$

Thus the wider the plume, the greater is the relative effect of the mean straining.

Equations (3.13c) and (3.8) have been used by Hunt & Mulhearn (1973) to calculate $\hat{\sigma}_3^2$ and $\hat{\sigma}_2^2$ when $t \ll T_L$ and $u_0 \ll U_0$ downwind of sources in homogeneous turbulent flow. Figure 5a shows how the divergence of the streamline accelerates diffusion of particles from a source on the stagnation line, while diffusion from a source *off* the stagnation line is first accelerated by the divergence and then reduced by the convergence of the streamlines.

Figure 5b shows computations for the effects on $\overline{X_2^2}$ and $\overline{X_3^2}$ of reducing the scale L_0 of turbulence, so that it is much smaller than H. But it is assumed that the turbulence is weak enough that $t \lesssim T_L$ or

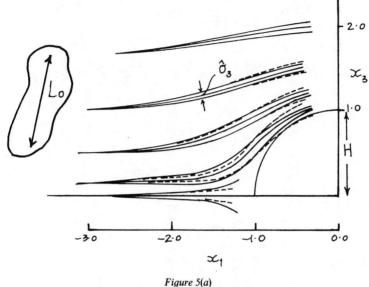

Figure 5(a)

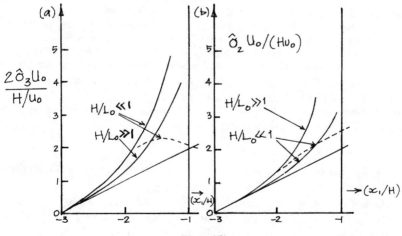

Figure 5(b)

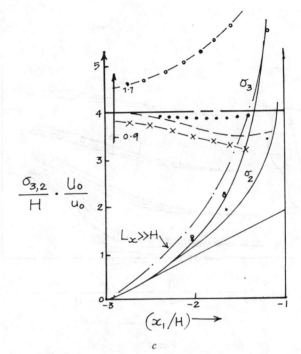

$$\frac{\sigma_{3,2}}{H} \cdot \frac{U_0}{u_0}$$

$$(x_1/H) \longrightarrow$$

c

Figure 5 Diffusion from a point source located upwind of a circular cylinder placed in weak homogeneous turbulent flows. (a) Calculation of mean square displacements normal to the mean streamlines in the plane of the flow $\hat{\sigma}_3$ (—) and perpendicular to it $\hat{\sigma}_2$ (---); large-scale turbulence $L_0 \gg H$. Source locations defined in terms of H. (From Hunt & Mulhearn 1973.) (b) Calculations of the effect of varying the integral scale of the turbulence on $\hat{\sigma}_3$ and $\hat{\sigma}_2$; —— source at (-3.00), --- source at $(-3.12, 0.5)$. (c) Comparisons between calculations and experimental measurements by R. E. Britter of turbulence and diffusion on the stagnation line of a circular cylinder (Hunt et al. 1979a). *Velocity:* ··· measurements of $(\overline{u_2^2})^{1/2}$; --- measurements of $(\overline{u_3^2})^{1/2}$; theory $(L_x/H) \gg 1$ $(\ll 1)$ for $(\overline{u_3^2})^{1/2}$ -O-O- (-×-×-). *Diffusion:* O ● measurements of σ_3, σ_2 from source at $(x_1/H) = -3$; —— approximate theory using measurements of $\overline{u_2^2}$, $\overline{u_3^2}$ [Equation (3.10)]; -·- asymptotic theory $(L_x/H) \gg 1$.

$T_L (\sim L_0/U_0) \ll D/U_0$. If we use rapid distortion theory, these calculations first require the evaluation of triple integrals over the wave-number spectrum of the upwind turbulence in order to obtain $\rho_\alpha \langle \dot{x}_1', \dot{x}_1'' \rangle$ (Hunt 1973, pp. 690–91) and then the evaluation of the double integrals of (3.13a,b) along the mean streamline. For a circular cylinder as L_0/H decreases, the compression of vortex lines parallel to $\hat{x}_1$ leads to normal turbulence $\overline{u_3^2}$ decreasing, to R_3 decreasing, and as a result of both, to normal diffusion $\overline{X_3^2}$ decreasing on the stagnation line. But the transverse component of turbulence $\overline{u_2^2}$ increases, while R_2 decreases, which leads to a small net change in $\overline{\hat{X}_2^2}$.

Figure 5c shows the results of measurements by R. E. Britter of diffusion normal to the stagnation line tracer gas released from a source upwind of a circular cylinder in approximately homogeneous grid turbulence, where $H/L_x = 0.32$. The measurements of $\hat{\sigma}_3$ and $\hat{\sigma}_2$ lie between the theoretical limits for $H/L_0 \gg 1$ and $H/L_0 \ll 1$ (Hunt et al. 1979a).

However, by assuming that $R_\alpha \simeq 1$, and by using the *measured* values of $\overline{u_3^2}$ and $\overline{u_2^2}$ and the calculated value of $\hat{U}_1$, we can calculate the diffusion quite accurately. In these straining flows, if the approximate values of $\hat{\sigma}_2^2$ and $\hat{\sigma}_3$ are required, then only the *initial* values of $\overline{u_2^2}$ and $\overline{u_3^2}$ need to be known, provided that the mean velocity variation of $\hat{U}_1$ is specified [because as (3.14) shows, $\hat{\sigma}_3^2$ can be dominated by the convergence and divergence of the streamlines]. In the experiments cited, the ratio of $\hat{\sigma}_3$ to the radius of the cylinder (H) increased to about 0.25, so in practice the narrow-plume approximation does not have to be very strictly satisfied.

4. For the special case of three-dimensional flows, which are axisymmetric about the x_1-axis (for example, flow around a hemisphere on a plane), we have

$$\partial\hat{U}_3/\partial\hat{x}_3 = -1/R\frac{\partial}{\partial\hat{x}_1}(\hat{U}_1 R), \qquad \partial\hat{U}_2/\partial\hat{x}_2 = 0,$$

where $R(\hat{x}_1)$ is the distance from the axis of symmetry of the streamline through the source, which is taken to be large compared with the plume thickness $\hat{\sigma}_3$. Then the strain functions have the simple forms

$$\hat{\beta}_3 = \hat{U}_1(\hat{x}_1)R(\hat{x}_1)/(\hat{U}_1(0)R(0)), \qquad \hat{\beta}_2 = 1.$$

Thus, $\hat{\beta}_3$ differs from the two-dimensional function by the increase in the radius along the mean streamline. In flow over an obstacle where the rate $R(\hat{x}_1)/R(0)$ typically varies by much more than $\hat{U}(\hat{x}_1)/\hat{U}_1(0)$, diffusion normal to the mean streamlines is more strongly reduced by *converging streamlines* than in two-dimensional flow. But when $\hat{U}_1(\hat{x}_1)$ decreases and $R(\hat{x}_1)$ increases in a stagnation region, the normal diffusion is less strongly amplified by diverging streamlines.

In fact, if the source is *on* the axis of symmetry (or the stagnation line of an obstacle), then $R \to 0$ and $\partial\hat{U}_3/\partial\hat{x}_3 = -\frac{1}{2}\partial\hat{U}_1/\partial\hat{x}_1$, so

$$\hat{\beta}_3 = \sqrt{\hat{U}_1(\hat{x}_1)/\hat{U}_1(0)}.$$

Thus the straining function β_3 is only equal to the square root of its two-dimensional value for the same change in mean velocity.

These axisymmetric flows are a special case of many three-dimensional

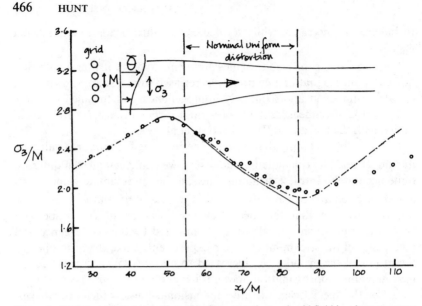

Figure 6   Variation of the width of a region between a hot and cold fluid in a homogeneous grid turbulence before, during, and after a region of converging streamlines. ○ experiments; —— self-preserving solution; – · – · – rapid distortion statistical calculations. (From Keffer et al. 1978.)

flows where the mean streamlines converge in one direction and diverge in another. In a wind-tunnel study of turbulent diffusion in such a flow, Keffer et al. (1978) showed how as flow enters a region of converging streamlines, the mean square displacements of particles from an upwind source can begin to shrink (see Figure 6). In this study, the authors actually measured the thickness of the thermal mixing layer between hot air above and cool air below the mean streamline. The statistical calculations, which agreed with the measurements, also suggested that if a converging flow continues long enough, the normal component of turbulence can be amplified sufficiently so that the mean square displacements can begin to increase again. [This occurs after a distance of order $\hat{U} \ln \Sigma/(d\hat{U}_1/d\hat{x}_1)$.] Keffer et al. (1978) also showed how in converging flows an assumption that the plume has a self-preserving structure also provides a simple, but specific, method of calculating the plume width.

3.5   Thin Clouds in Shear Flows

The thin-plume or thin-cloud equations [(3.2) and (3.3)] have been used to analyze the effects of a shear on the longitudinal spreading of clouds. If $\partial \hat{U}_1/\partial \hat{x}_3 > 0$, but all other mean gradients are zero, the equations for X_1 and X_3 are coupled; but X_3 can be calculated from (3.3) with $\partial \hat{U}_3/\partial \hat{x}_\alpha = 0$

and then substituted into (3.2). The main result from this analysis is that $\overline{(X_1 - \bar{X}_1)^2}$ increases in proportion to $T_L(\partial \hat{U}_1/\partial \hat{x}_3)^2 t^3 \overline{v_3^2}$ (Monin & Yaglom 1971). These results have been generalized to *unsteady* shear flows by Maxey (1979).

4. BROAD PLUMES IN HOMOGENEOUS TURBULENCE

4.1 *Diffusion Near the Source*

In most practical problems, one needs to know more about diffusion processes than can be gained from the "thin-plume" approximations in order to calculate the concentration distribution of the contaminant when it has diffused from the mean streamline a distance $\hat{\sigma}_\alpha$ comparable with the scale H of the inhomogeneities of the turbulence. The scale H may be of the order of or large compared with L_0, the scale of the turbulence. In such a case it is not possible to assume that the structure of turbulence is constant across the plume. In some flows the fluid elements take a much greater time than an integral time scale T_L to move from the mean streamline into the regions of different diffusivity; in others, they move rather rapidly into these regions and carry with them the "memory" of the turbulence near the source.

We begin by considering the latter situation, soon after the release time ($t = 0$), and we restrict the discussion to normal diffusion in a shear flow $U = (U_1(x_3), 0, 0)$ where the turbulence is only a function of x_3. (The results can be generalized to inhomogeneous turbulence in straining flows.)

The displacement of the particle X_3 can be expressed in terms of the normal turbulent velocity component as

$$\frac{dX_3}{dt} = v_{3s}\langle 0 \rangle + \int_0^t \frac{dv_3}{dt} \langle t' \rangle \, dt'$$

$$= v_{3s} + \int_0^t (\partial u_3/\partial t + U_1 \, \partial u_3/\partial x_1 + u_j \, \partial u_3/\partial x_j)(\mathbf{X}, t') \, dt'. \tag{4.1}$$

Therefore, the mean vertical velocity of particles soon after release from a source ($t \ll T_L$) is (Monin & Yaglom 1971, Hunt 1982)

$$d\bar{X}_3/dt = \left(\overline{u_1 \frac{\partial u_3}{\partial x_1}} + \overline{u_2 \frac{\partial u_3}{\partial x_2}} + \overline{u_3 \frac{\partial u_3}{\partial x_3}} \right) t + \cdots,$$

which reduces in an incompressible flow, homogeneous in x_1, x_2, to

$$d\bar{X}_3/dt = (\partial \overline{u_3^2}/\partial x_3)t + \cdots. \tag{4.2}$$

Thus, although the mean vertical velocity at any point is zero, particles *marked* or *released* at a point drift upward soon after their release. (This also means that these marked particles on average have drifted *downward* before passing through the source!) Note that half the contribution to this drift comes from the increase in $\overline{u_3^2}$ with x_3, and half from the correlation between u_1 and $\partial u_3/\partial x_1$ and between u_2 and $\partial u_3/\partial x_2$. This can be seen in upward thermals in convection or in downward eddies below a free surface where the entrainment into the eddies leads to such correlations (Figure 7a).

Willis & Deardorff (1976) provided a good test of Equation (4.2) by measuring particles in thermal convection in a laboratory tank. In such a flow, the gradient of vertical turbulence increases toward the surface $(\partial \overline{u_3^2}/\partial x_3/\alpha x_3^{-1/3})$, so the mean displacement $\bar{X}_3(t)$ can be greater for particles released from a source near the surface than for a source released higher up! (Figure 7b). Note that in the calculations of Section 3, this mean drift relative to the streamlines could be ignored because the turbulence was assumed to be locally homogeneous in directions normal to the mean streamline.

Equation (4.2) has been discussed in several recent papers, mainly in the context of developing random-flight models of turbulent diffusion (e.g. Legg & Raupach 1982, Ley & Thomson 1983, Janicke 1981, Wilson et al. 1981).

When one is interested in calculating how $\overline{X}_3$ and $\overline{X_3^2}$ vary with *distance* x_1 from the source, the mean velocity gradient and the correlations between u_1 and u_3 have to be considered. If $x_3 = 0$ on the mean streamline, then

$$dX_3/dt = (U_1(X_3) + u_1(\mathbf{X}, t)) \frac{dX_3}{dx_1} = u_3(\mathbf{X}, t).$$

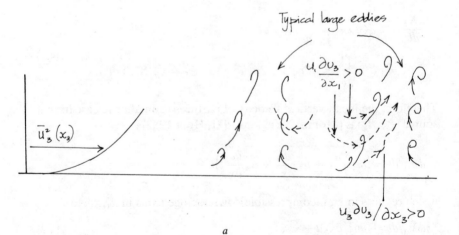

a

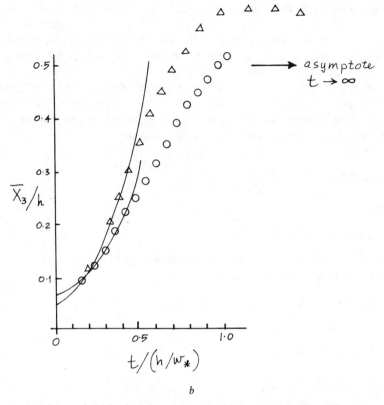

Figure 7 Mean trajectories in inhomogeneous turbulence. (*a*) Profile of mean-square turbulent vertical velocity, and typical trajectories into large-scale rising eddies. (*b*) Measurements of X_3, the mean vertical displacement from sources near the surface in thermal convection ($\triangle - x_3^s/h = 0.05$; $\bigcirc - x_3^s/h = 0.067$), compared with (4.2) (Willis & Deardorff 1976); $\partial \overline{u_3^2}/\partial x_3$ is calculated from the measurements in the same laboratory experiment, h is the depth of the layer, and w_* is the characteristic velocity of turbulence driven by buoyancy.

Expanding near the source yields

$$\frac{d\bar{X}_3}{dx_1} = -\frac{\overline{u_1 u_3}}{U_1^2} + \left(\frac{\partial \overline{u_3^2}}{\partial x_3} - \overline{u_3^2}\frac{dU_1/dx_3}{dU_1}\right)(x/U_1) + \cdots, \tag{4.3}$$

which shows that the mean vertical displacement at a given *distance* x_1 downwind is positive if $-\overline{u_1 u_3} > 0$. This is because in a rising eddy, which typically moves more slowly than the mean velocity, a fluid element is displaced upward a distance of about $x_1 u_3/(U_1 - |u_1|)$, which is more than if it were in a downward eddy [when the displacement would be about $x_1 u_3/(U_1 + |u_1|)$]. This phenomenon was first noted by Hinze & Van der

Hegge Zijnen (1951). But at a slightly later time, the rising eddies then move far enough into the faster mean flow (dU_1/dx_3) that their vertical displacement, *at given x*, is *less* than for the falling eddies. In homogeneous turbulence at large times, the latter situation continues to be the dominant trend (Hunt 1982).

The mean rate of increase of $\overline{X_3^2}$, or the square of the plume width, can be calculated from (4.1) by first expressing $X_3\langle t \rangle$ in terms of the initial velocity of a fluid element at a source v_{3s} and the change in velocity Δv_3 experienced by the fluid element as it moves away (Figure 8a):

$$X_3 = v_{3s}t + \int_0^t (t-t') \, d\Delta v_3/dt\langle t' \rangle \, dt', \tag{4.4}$$

where

$$v_3\langle t \rangle = v_{3s} + \Delta v_3 \langle t \rangle \quad \text{and} \quad v_{3s} = v_3\langle 0 \rangle = u_3(\mathbf{x}^s, 0).$$

Then, by expressing X_3 in terms of the displacements X_{3s} produced by v_{3s}, and ΔX_3 by the change in velocity, we have

$$\overline{X_3 dX_3/dt} = \overline{X_{3s}v_{3s}} + \overline{\Delta X_3 v_{3s}} + \overline{X_{3s}\Delta v_3} + \overline{\Delta X_3 \Delta v_3}$$

where

$$\overline{X_{3s}v_{3s}} = \overline{v_{3s}^2}\, t,$$

$$\overline{\Delta X_3 v_{3s} + X_{3s}\Delta v_3} = \int_0^t (2t-t')\overline{v_3\langle 0 \rangle \, dv_3/dt\langle t' \rangle} \, dt',$$

and

$$\overline{\Delta X_3 \Delta v_3} = \int_0^t \int_0^t (t-t') \, \overline{dv_3/dt\langle t' \rangle \, dv_3/dt\langle t'' \rangle} \, dt' \, dt''.$$

In inhomogeneous unsteady turbulence, the velocity correlations in these integrals are expressed in terms of the correlation functions and variance as follows:

$$\overline{v_3\langle t \rangle v_3\langle t' \rangle} = [\overline{v_3^2\langle t \rangle v_3^2\langle t' \rangle}]^{1/2} R_3\langle t, t' \rangle, \tag{4.5a}$$

from which it follows, by differentiation, that

$$\overline{v_3\langle 0 \rangle \frac{dv_3}{dt}\langle t \rangle}$$

$$= \sqrt{\overline{v_3^2}\langle 0 \rangle} \left[\frac{d}{dt} \sqrt{\overline{v_3^2}\langle t \rangle} \cdot R_3\langle 0, t \rangle + \sqrt{\overline{v_3^2}\langle t \rangle} \frac{\partial R_3}{\partial t}\langle 0, t \rangle \right]. \tag{4.5b}$$

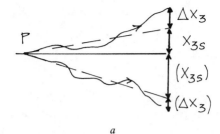

a

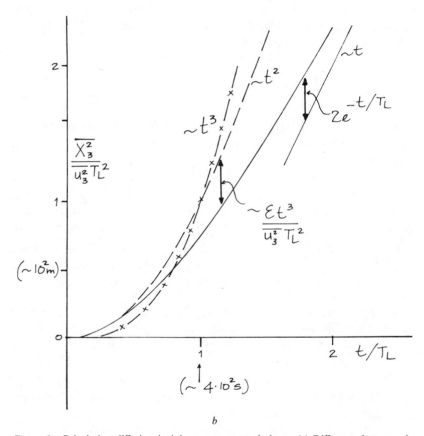

b

Figure 8   Calculating diffusion in inhomogeneous turbulence. (*a*) Difference between the displacements of the initial trajectory (— —) and the actual trajectory (———). (*b*) Mean square X_3^2 displacements in homogeneous (———) and inhomogeneous turbulence ($\partial \overline{u_3^3}/\partial x_3 > 0$) (–×–×–); diffusion caused by initial turbulence alone (— —). Typical magnitudes for plume width at given travel time in a convective boundary layer are indicated on the axes, following Lamb (1982).

Over short times ($\tau_L \ll t \ll T_L$), the derivative of R_3 is related to the "structure function" by

$$\overline{(v_3\langle t\rangle - v_3\langle 0\rangle)^2} = 2\overline{v_3^2}\langle 0\rangle(1 - R_3(0,t) + O(t^2))$$
$$= 2(\pi B\varepsilon)t, \tag{4.5c}$$

where $\tau_L (\sim R^{-1/4}T_L)$ is the Lagrangian microscale (Tennekes 1975), ε is the energy dissipation rate per unit mass in the turbulence, and B is a universal constant ($B \simeq 1/\pi$ if $\varepsilon \simeq \overline{v_3^2}/T_L$). This is not inconsistent with recent measurements by Hanna (1981), who found $B \simeq 0.2$. Note that the change in the mean square velocity of fluid elements over short times only depends on ε and the isotropic small-scale structure of small-scale turbulence, which is found in all kinds of turbulent flows.

Hence if $\tau_L \ll t \ll T_L$, it follows that

$$\overline{X_3^2} = \overline{v_3^2}\langle 0\rangle t^2 + \left(\frac{1}{2}\frac{d\overline{v_3^2}}{dt}\langle 0\rangle - \frac{1}{3}\pi B\varepsilon\right)t^3 + \cdots, \tag{4.6}$$

where at $t = 0$ the Lagrangian velocity and its derivative can be expressed in terms of local Eulerian derivatives as

$$\overline{v_3^2}\langle 0\rangle = \overline{u_3^2}(\mathbf{x}^s, 0), \qquad d\overline{v_3^2}/dt\langle 0\rangle = \left(\frac{\partial \overline{u_3^3}}{\partial x_3} + \frac{\partial \overline{u_3^2}}{\partial t}\right)(\mathbf{x}^s, 0). \tag{4.7}$$

Thus in stationary homogeneous turbulence (where $d\overline{v_3^2}/dt = 0$), the initial growth of a plume is reduced by the effect of the growing negative correlation between the initial velocity and the acceleration (i.e. $\overline{\Delta X_3 v_{3s}} + \overline{X_{3s}\Delta v_{3s}} < 0$), which only depends on the universal structure of the smaller scales of turbulence in the inertial subrange (a result essentially due to Novikov 1963).

In many flows near boundaries (especially shear-free flows) where the turbulence is inhomogeneous (e.g. thermal convection, free-surface flows, etc.), the marked particles encounter more vigorous turbulence as they diffuse upward, and so the net effect is to increase $\overline{X_3^2}$ in proportion to $\partial \overline{u_3^3}/\partial x_3$ (Hunt 1984). In turbulent thermal convection, $\partial \overline{u_3^3}/\partial x_3$ is approximately constant with height in the lower one fifth of the boundary layer. If the source is very close to the ground ($x_3 = 0$), where $\overline{u_3^2}(\mathbf{x}^s) \simeq 0$ and $T_L \simeq 0$, it follows that near the ground (van Dop et al. 1984)

$$\overline{X_3^2} \simeq \tfrac{1}{2}(\partial \overline{u_3^3}/\partial x_3)(\mathbf{x}^s)t^3, \tag{4.8a}$$

while near an elevated source

$$\overline{X_3^2} \simeq \overline{u_3^2}t^2. \tag{4.8b}$$

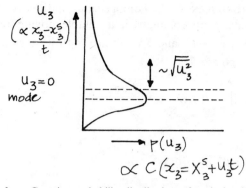

$$\propto C \left(x_3 = X_3^s + u_3 \frac{t}{} \right)$$

Figure 9   Effect of non-Gaussian probability distributions of vertical turbulent velocity near the source.

This "t^3" law can be derived by a similarity argument (Yaglom 1972) and has been observed in the laboratory and field measurements of Willis & Deardorff (1976) and Moninger et al. (1978) for ground-level sources. It goes some way toward explaining the surprising result (Lamb 1982) that in the atmosphere (Figure 8b) the vertical width of plumes from surface sources exceeds that of elevated sources. It is not clear why these formulae for diffusion near the source are valid for particles that move well above the source!

In Equations (2.9) and (3.7b) it was shown that the mean concentration distribution (say, in the x_3 direction near a source in a cross flow) is proportional to the probability distribution of u_3 at the source. So if $\overline{u_3^3} > 0$, the mode of the skewed probability distribution is negative, and therefore the height at which $C(x_3)$ is a maximum drops below the source, while the mean plume height $x_3^s + \overline{X}_3$ increases. This is a typical feature of convective flows where the buoyancy forces generate $\overline{u_3^3}$ ($d\overline{u_3^3}/dt$ increases with $g\overline{u_3^2\theta}/\overline{\theta} > 0$, where θ is the temperature). But in shear layers (e.g. wakes of hills or buildings) where $\partial \overline{u_3^2}/\partial x_3 > 0$, $\overline{u_3^3}$ is usually negative ($d\overline{u_3^3}/dt$ is of the order of $-\overline{u_3^2}\,\partial\overline{u_3^2}/\partial x_3$). In this case, both the heights x_3 of the concentration distribution move toward the region of large vertical turbulence (Figure 9; Lamb 1982, Hunt 1982).

4.2   Diffusion Over Longer Times

The rate of increase of the mean square displacement $\overline{X_3^2}$ of fluid elements normal to the mean streamline is given by integrating (4.1):

$$d\overline{X_3^2}/dt = 2\int_0^t \rho_3\langle t, t'\rangle\, dt. \tag{4.9a}$$

HOMOGENEOUS TURBULENCE   In homogeneous stationary turbulence (to which grid turbulence approximates in a suitable moving frame), $\rho_3\langle t, t'\rangle = \rho_3\langle\tau\rangle$ (where $\tau = t - t'$), and $d\overline{X_3^2}/dt$ grows to its asymptotic value of $2\overline{v_3^2}T_L$, so that when $t \gg T_L$,

$$\overline{X_3^2} \sim 2t\overline{v_3^2}T_L - \int_0^\infty \tau\rho_3\langle\tau\rangle \, d\tau, \tag{4.9b}$$

where

$$T_L = \int_{-\infty}^t \rho_3\langle t, t'\rangle \, dt'/\overline{v_3^2}\langle t\rangle \tag{4.9c}$$

in stationary conditions.

But in stably stratified turbulence, where buoyancy forces inhibit the unlimited growth of $\overline{X_3^2}$, it is commonly observed (though until recently, seldom discussed) that $d\overline{X_3^2}/dt$ increases downwind of the source and then decreases to nearly zero. Also, in any flow with upper and lower boundaries, we have $d\overline{X_3^2}/dt \to 0$. Since $\overline{X_3^2}$ must be positive, it follows that $T_L \simeq 0$ and $\int_0^\infty \tau\rho_3\langle\tau\rangle \, d\tau < 0$ in such a flow, which means that the autocorrelation function $\rho_3\langle\tau\rangle$ has to be negative over some of its range (Pasquill 1974, Csanady 1964, Pearson et al. 1983). In stably stratified flows, the marked fluid elements must change their density if they are to travel large vertical distances, so the distinctions in Section 2 between the molecules or particles of the contaminant and the marked fluid elements are important. The case of stable stratification is just one of many turbulent flows where the Lagrangian mean square velocities and time scales are neither theoretically nor experimentally known. There have been only a few measurements in homogeneous turbulent flows (e.g. Snyder & Lumley 1971) and almost none in inhomogeneous flows. One hopes that new experimental and data-analysis techniques and computer simulations of turbulent flows will soon remedy this situation.

SHEAR FLOWS   Nevertheless, useful heuristic concepts and models have been developed for inhomogeneous flows largely based on (4.9a). As shown in Section 4.1, the mean Lagrangian velocity, perpendicular to the mean streamlines (say $V_3\langle t\rangle$), may not be zero in such flows. Then it is useful to differentiate between the *fluctuating*, $v_3\langle t\rangle$, and mean Lagrangian velocities, defined for the shear flow of Section 4.1 by

$$u_3(\mathbf{X}, t) = V_3\langle t\rangle + v_3\langle t\rangle. \tag{4.10}$$

Thus, in inhomogeneous turbulence the Lagrangian autocorrelation function $\rho_3\langle t, t'\rangle$ has to be defined in terms of $v_3\langle t\rangle$, as in (3.6a), and T_L [defined by (3.11c)] is the time scale of the fluctuating Lagrangian velocity.

In some flows, $T_L\langle t\rangle$ is small compared with the time in which the particles travel a significant distance (H, the scale of the flow). An example is when particles are released near the surface of a turbulent boundary layer T_L (Hunt & Weber 1979). Even in this inhomogeneous flow, the limiting form of (4.9) when $t/T_L \gg 1$ can be applied:

$$d(X_3 - \bar{X}_3)^2/dt \simeq 2\overline{v_3^2}\langle t\rangle T_L\langle t\rangle,$$

from which it follows that

$$\overline{X_3^2} \simeq 2t\overline{v_3^2}\langle t\rangle T_L\langle t\rangle + \overline{X_3^2}. \tag{4.11}$$

Equation (4.11) shows how in such flows the mean square displacement of fluid elements across mean streamlines is produced by turbulent diffusion and mean drift. In the case of surface release in a turbulent boundary layer, these two components are about equal.

Both of these effects have been modeled by a number of approximate techniques. The most widely used of them is to assume that the mean flux F_i is proportional to the local mean concentration gradient:

$$F_i = K_{ij}\, \partial C/\partial x_j. \tag{4.12}$$

Then C is defined by the advective diffusion equation (Monin & Yaglom 1971)

$$\partial C/\partial t + U_k \frac{\partial C}{\partial x_k} = \frac{\partial}{\partial x_i}\left(K_{ij}\frac{\partial C}{\partial x_j}\right). \tag{4.13}$$

In order for this equation to model the growth of $\overline{X_3^2}$, K_{ij} must be a function of the travel time t, i.e. $K_{33} = \frac{1}{2}d\overline{X_3^2}/dt$, and it cannot therefore be simply a function of the local flow properties, an assumption that is convenient to make but that is often incorrect. In homogeneous stationary turbulence, when the Lagrangian autocorrelation is given by $\rho_3\langle \tau\rangle = \overline{u_3^2}\, e^{-\tau/T_L}$, since $\overline{v_3^2} = \overline{u_3^2}$ (Lumley 1962), it follows that

$$K_{33} = \overline{u_3^2}\, T_L(1 - e^{-t/T_L}), \tag{4.14}$$

which implies that K_{33} only becomes independent of travel time when t is large compared with T_L. [For diffusion from an elevated source (100 m) in the atmosphere, this corresponds to a downwind distance of about $\frac{1}{2}$ km, or for a source at half radius from the wall of a pipe, $2\frac{1}{2}$ radii down the pipe]. Another implication of the diffusion equation is that in homogeneous turbulence, the concentration distribution $C(x_3)$ is Gaussian. For example, in a steady cross flow, the concentration distribution is given by $C = Q\exp(-x_3^2/2\sigma_3^2)/(\sqrt{2\pi}\sigma_3 U_1)$, where Q is the source strength.

No assumption about the probability distribution of the turbulence is made in the derivation of $\overline{X_3^2}$ in terms of ρ_3, so only in certain turbulent flows can C be given by the Gaussian distribution.

Since the probability distribution of the velocity u_3 is approximately Gaussian in most shear layers, it follows from (2.9) and (3.7b) that $C(x_3)$ is Gaussian near the source. But (4.6) shows that dX_3/dt changes a little farther downwind ($t \sim T_L$) because of the small-scale accelerations and the large-scale inhomogeneities of the turbulence, which means that at this stage $p(x_3)$, and therefore $C(x_3)$, ceases to be Gaussian.

Much farther downwind in the idealized case of homogeneous stationary turbulence, when $t \gg T_L$, the displacements $X_3 = \int u_3 \, dt$ are equal to a sufficiently large sum of independent components that (by the central limit theorem) X_3 must have a Gaussian distribution.

In principle these limitations to (4.12) can be overcome by specifying the flux F_3 in terms of the whole space and time distribution of $\partial C/\partial x_3$—not a very practical or physically revealing procedure! However, "Gaussian plume modeling" (as it is called) forms the basis of most air-pollution dispersion calculations.

Recent wind-tunnel measurements of $C(x_3)$ have shown in detail how its distribution is not exactly Gaussian. On the basis of these and other measurements, Berkowicz & Prahm (1980) suggest that F_3 should be expressed in terms of a weighted integral of $\partial C/\partial x_3$ over space and time, but there are obvious problems in how to define this integral!

Diffusion in inhomogeneous shear-flow turbulence, where the turbulence varies with x_3, is usefully modeled by assuming the form for the variation of K_{33} with x_3. If in a large region there is a linear gradient of K_{33}, i.e.

$$K_{33} = K + x_3 K',$$

then (4.13) shows how the mean height and width of a plume are affected by the gradient

$$\overline{\partial X_3} \, dt = K' = \partial K_{33}/\partial x_3 \tag{4.15a}$$

and

$$d\overline{X_3^2}/dt = 2K + 2\overline{X_3} \, d\overline{X_3}/dt \tag{4.15b}$$

(Pasquill 1974).

Note that (4.15b) agrees with (4.11) when $t \gg T_L$ if $K = \overline{v_3^2} T_L$, which shows how K should be defined in terms of the fluctuating Lagrangian velocity. However, (4.15a) does not agree with the results *near* the source [Equations (4.2) and (4.6)]. But if the expression (4.14) is heuristically taken to apply to inhomogeneous turbulence, so that

$$K = \overline{u_3^2}(x_3) T_L(x_3)(1 - e^{-t/T_L(x_3)}), \tag{4.16}$$

then from (4.15a) when $t \ll T_L$, $d\overline{X}_3/dt = t \, \partial \overline{v_3^2}/\partial x_3$, which does agree with (4.2). However, because (4.16) contains no information about $\overline{u_3^3}$, this expression for $\overline{X_3^2}$ cannot lead to the correct expression for $\overline{X_3^2}$ at small times (i.e. 4.16).

For fluid elements released from an elevated source in the lower part of the boundary layer [where $\overline{v_3^2} \simeq \overline{u_3^2} \simeq$ constant and $T_L \simeq x_3/(3\sqrt{\overline{u_3^2}})$], we obtain from (4.15) and (4.16), for $t \ll T_L$,

$$d\overline{X}_3/dt = 0, \tag{4.17a}$$

and when

$$t \gg T_L^s, \qquad \overline{X}_3 \to \overline{X}_3(0) + K'[t - 2T_L^s], \tag{4.17b}$$

where T_L^s is the time scale at the source.

These equations (4.17a,b) agree well with the laboratory measurements of heat released from a wire in a turbulent boundary layer (Shlien & Corrsin 1976) and in a turbulent pipe flow (Frost 1981) (Figure 10). They also demonstrate how the effect of the *gradient* of diffusivity takes a time of about

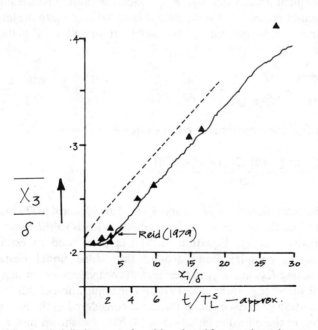

Figure 10 Mean vertical displacement of particles released from an elevated source $x_3^s = 0.2 \, \delta$ in a turbulent boundary layer of thickness δ. ▲, experimental measurements of mean height of the temperature distribution (Shlien & Corrsin 1976); —— numerical simulations [600 particles (Durbin & Hunt 1980, Reid 1979)]; – – – solution of diffusion equation ($K_{33} = 0.4 \, u_* x_3$).

2 Lagrangian time scales to affect the mean drift of the plume. The physical reason is clear: It takes that time before the fluid elements have moved upward or downward into turbulence with a significantly different time scale from that at the source, and then for the longer time scale eddies above the source to diffuse the plume upward more than the small time scale eddies below the source can diffuse the plume downward. This is different from the drift induced by the gradient of $\overline{u_3^2}$ at the source, which can act more quickly [as shown by Equation (4.2)].

An alternative method of modeling this process is by simulating random walks or random flights of fluid elements, subject to assumptions about the values of $\overline{v_3^2}$, $T_L(x_3)$, and other statistics of the turbulence. For recent reviews and developments, see Durbin (1983) and van Dop et al. (1984).

STRAINING AND SHEAR FLOWS   Now consider again diffusion in straining flows where the turbulence is, locally, approximately homogeneous and the probability distribution of $\hat{u}_3, \hat{u}_2$ is approximately Gaussian. Given these conditions, the results in (3.11) might be expected to be the same as those derived from the advection diffusion equation for the mean concentration C using turbulent diffusivities $\hat{K}_{22}, \hat{K}_{33}$, which vary along the streamline but are constant normal to the streamline at each value of $\hat{x}_1$. In the limit of the thin-plume approximations in a symmetric straining flow, Equation (4.13) becomes

$$\hat{U}_1 \frac{\partial C}{\partial \hat{x}_1} + \hat{x}_2 \frac{\partial \hat{U}_2}{\partial \hat{x}_2} \frac{\partial C}{\partial \hat{x}_2} + \hat{x}_3 \frac{\partial \hat{U}_3}{\partial \hat{x}_3} \frac{\partial C}{\partial \hat{x}_3} = \hat{K}_{22} \frac{\partial^2 C}{\partial \hat{x}_2^2} + \hat{K}_{33} \frac{\partial^2 C}{\partial \hat{x}_3^2}. \tag{4.18}$$

The solution for a continuous point source is

$$C = \frac{Q \exp\left[-(\hat{x}_3^2/(2\hat{\sigma}_3^2) + \hat{x}_2^2/(2\hat{\sigma}_2^2))\right]}{\hat{U}_1 2\pi\hat{\sigma}_3\hat{\sigma}_2}, \tag{4.19}$$

where the expressions for $\hat{\sigma}_3(\hat{K}_{33})$ and $\hat{\sigma}_2(\hat{K}_{22})$ are identical to (3.11a). Thus in this situation, the advective diffusion equation leads to the same result as the statistical theory. Equation (4.19) has been used to estimate the dispersion in complex flows around three-dimensional obstacles by approximating $\hat{K}_{22}$ and $\hat{K}_{33}$ to their undisturbed values, rather than to their distorted values (e.g. Hunt et al. 1979b). The application of this formula to the calculation of surface concentration is considered in the next section.

The use of the diffusion equation (4.13) has also shown how shear and many other kinds of complex mean flow and diffusivity variations affect dispersion far downstream of sources; this topic is covered in the review by Chatwin & Allen (1985) in this volume (see also Smith 1981).

4.3   Diffusion Onto a Surface

There are two main ways in which contaminant from a source reaches a solid surface in a turbulent flow: either it is directly advected by the combined action of the mean flow and turbulence onto the surface near a stagnation point or line (Figure 4a) or, where the the mean flow is parallel to the surface, it is advected by the turbulence alone (Figure 11). In both cases, wherever fluid elements reach the surface, either there is a local stagnation point on the surface or the local streamline is parallel to the surface. Locally (outside the sublayer) we have $u_n^* = \lambda n$, where n is the local normal. So the time for any fluid element to reach the surface from $n(\int_n^0 dn'/\lambda n')$ is infinite.

Therefore, molecular diffusion has to be considered. But as in many other aspects of turbulent flows, this mechanism does not necessarily depend on the actual *magnitude* of the fluid's molecular properties (in this situation, the diffusivity).

MEAN FLOW ONTO SURFACES   In the first case, we consider matter released from a source on the mean stagnation line. The mean concentration just upstream of the obstacle can be calculated from the results in Section 3 if the mean square displacement is small. For a steady line source upwind of a cylindrical obstacle, placed in a homogeneous Gaussian turbulence, it can be calculated from (4.19) as

$$C = \frac{\exp[-x_3^2/(2\sigma_3^2)]}{\sqrt{2\pi}\sigma_3 U_1}. \tag{4.20}$$

Since $\sigma_3 U_1 = [\int_0^{x_1} \int_0^{x_1} \rho_3 \langle x_1', x_1'' \rangle \, d\dot{x}_1' \, d\dot{x}_1'']^{1/2}$ is finite as $x_1 \to x_{1\,\text{stag}}$ (where $x_{1\,\text{stag}}$ is the value of x_1 at the stagnation point) for large or small-scale turbulence, it follows that $C(x_1, 0)$ is finite as $x_1 \to x_{1\,\text{stag}}$. Since for a steady source, there is an infinite time for the material to diffuse by molecular diffusion to the surface, the surface concentration C_{stag} is approximately equal to $C(x_1 \to x_{1\,\text{stag}}, 0)$ *for a line source*. The same result also holds *for a point source* upwind of a three-dimensional stagnation point.

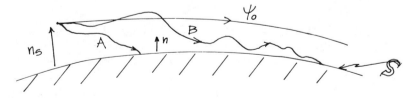

Figure 11   The different kinds of trajectories for particles reaching a surface in a turbulent boundary layer.

It is interesting to note that C_{stag} can theoretically be greater than $C_{stag}^{(0)}$, the concentration at the location of the stagnation point in the absence of the obstacle. For a line source near a circular cylinder with radius H, we have

$$C_{stag}/C_{stag}^{(0)} = \frac{-(|x_1^{(s)}| - H^2/|x_1^s|)}{(|x_1^s| - H)} \quad \text{for large-scale turbulence} \quad (4.21a)$$

$$= \left(\frac{|x_1^s| + H^2/|x_1^s| - 2H}{|x_1^s| - H} \right)^{1/2}$$

for very small-scale turbulence with a constant diffusivity.   (4.21b)

Note that this ratio is greater or less than unity, depending on the turbulence structure and the source position x_1^s (Hunt & Mulhearn 1973).

For a *point* source in a two-dimensional stagnation flow, the mean centerline concentration is

$$C \propto q/(\sigma_3 \sigma_2 U_1) \propto \frac{1}{\sigma_2}, \quad \text{since} \quad \sigma_3 U_1 = O(1). \quad (4.22)$$

But, from (3.13b), it follows that $\sigma_2 \to \infty$ as $x_1 \to x_{1stag}$, since the travel time $t \to \infty$. Thus, Equation (4.22) implies that $C \to 0$ as $x_1 \to x_{1stag}$. Consequently, the surface concentration in this case depends critically upon any streamwise transport by small-scale turbulence and molecular diffusion. In general, when the distance n from the surface is small enough that $n \, \partial U_n/\partial n \lesssim \sqrt{\overline{u_n^2}(n)}$, then the movement of the marked particles to the surface is determined by the small-scale eddies of the scale n and finally by molecular diffusion. Consequently, C_{stag} can be finite but is significantly less than $C_{stag}^{(0)}$. As the source moves further upwind, the effect of the extra growth of σ_2 near the stagnation point becomes relatively smaller. So $C_{stag}/C_{stag}^{(0)} \to 1$ as $|x_1^s|/H \to \infty$. If the molecular diffusivity is large enough that $D/n \gtrsim \sqrt{\overline{u_n^2}(n)}$, it (rather than the turbulence) determines the surface concentration. Chatwin's (1974) exact solutions for the diffusion of a puff of contaminant in laminar flow at a two-dimensional stagnation point could be used to estimate C_{stag}.

At an axisymmetric three-dimensional stagnation point, the product $\hat{\sigma}_2 \hat{\sigma}_3 \hat{U}_1$ is of order unity (Hunt et al. 1979b), so $C_{stag} \sim C_{stag}^{(0)}$.

Measurements have been made by R. E. Britter & J. S. Puttock at the University of Cambridge on C_{stag} and $C_{stag}^{(0)}$ on a circular cylinder and a sphere in grid turbulence for sources placed at certain distances upwind. They showed, for example, that $C_{stag}/C_{stag}^{(0)}$ has a value of 0.3 for the circular cylinder and 0.8 for the sphere with the source two radii upwind ($|x_1^s|/H \simeq 3$).

This ratio of $C_{stag}/C_{stag}^{(0)}$ has been a subject of considerable controversy in the study of air-pollution dispersion, with some proponents suggesting that the ratio should be much less than one ("because the obstacle spreads out the plume") and others suggesting that it should be 2.0 ("because the structure reflects the plume")! This controversy stimulated the extensive set of field and laboratory measurements of surface concentration on three-dimensional hills in stably stratified flows (Snyder 1985, Lavery et al. 1982). A striking feature of the results is that C_{stag} is usually within a factor of 2 of $C_{\text{stag}}^{(0)}$ at stagnation regions. [In strong stable conditions the flow is largely horizontal. There is little vertical motion or diffusion, so the diffusion from a point source resembles that of a line source; hence, Equation (4.21b) provides a useful estimate (Hunt et al. 1979b, Snyder & Hunt 1984).

MEAN FLOW PARALLEL TO THE SURFACE   The surface concentration of contaminant released from a source above the surface in a turbulent flow depends on how the marked fluid elements move to within a small distance of the surface (Figure 11). In the limit where $x_3^s \lesssim u_3^s T_L^s$, large-scale turbulent eddies with characteristic velocity and time scale T_L^s impinge downward on the surface, producing a local three-dimensional stagnation point. (A) The previous analysis of a *mean* three-dimensional stagnation flow suggests that the surface concentration within the eddy should therefore be about the same as if the surface was not present! This turns out to be a useful way of estimating the value C_{mx} and location x_{mx} of maximum surface concentrations from elevated sources in the atmospheric boundary layer in the presence of strong thermal convection (Venkatram 1980, Lamb 1982):

$$C_{\text{mx}} \sim Q/(U_0(u_3^s t_{\text{mx}})^2) \sim Q/(U_0(x_3^s)^2),$$

$$x_{1_{\text{mx}}} \sim U_0 t_{\text{mx}}, \qquad t_{\text{mx}} \sim x_3^s/u_3^s.$$

It also provides some physical basis for modeling diffusion in thermal convection by considering random updrafts and downdrafts (e.g. Misra 1982). Even in a neutral boundary layer, the same mechanism has been suggested as determining the maximum surface concentration (Pasquill 1974, p. 203), in which case the travel time to the maximum surface concentration would be about $x_3^s/(\overline{u_3^2})^{1/2}$, which is about $3T_L^s$. This result agrees to within $\pm 0.5T_L$ with laboratory and field measurements for elevated point sources.

Further downwind in these two kinds of boundary layers, any fluid elements reaching the surface will have traveled for many Lagrangian time scales and will have moved away from and toward the boundary several times. (B) So the surface acts as a distributed source of contaminant to increase the concentration downwind. The former process of direct impingement of eddies onto the surface (A) is poorly modeled by the diffusion equation, but the latter process (B) downwind is approximately modeled by this equation, with the diffusivity normal to the surface being a prescribed function of distance from the surface. [In this case, a useful reciprocal theorem shows that the surface concentration at x from an elevated source at x^s is the same as the concentration at the source position x^s from a ground-level source at x if the flow is reversed (Smith 1957).]

In strongly stable conditions in the atmospheric boundary layer, $K_{33}(x_3)$ is approximately *constant* with height above a thin surface layer. This is the only condition in which it is quite correct to assume that the effect of the surface on the mean concentration is the same as a mirror-image source below the surface at $n = -n_s$ (or $x_3 = -x_3^s$ on a plane), a basic element in the technique of "Gaussian plume modeling" of dispersion in the atmosphere introduced by O. G. Sutton and developed by F. Pasquill and F. A. Gifford [see Pasquill (1974) for details].

In complex turbulent flows around obstacles, the diffusion of the marked fluid elements to the surface is determined by the evolution of the upwind turbulence as it is distorted and blocked by the obstacle *and* by the local, small-scale turbulence generated on the surface of the obstacle. The interaction of these processes is poorly understood. However, estimates have been made of surface concentrations from sources placed upwind of cylindrical and hemispherical obstacles. These estimates have been made by calculating the distribution of concentration in the plume using the diffusion equation (with the diffusivity determined by upwind turbulence) and the "thin-plume" approximation, applying the boundary condition $\partial C/\partial n = 0$ on the surface. Except near the stagnation point, this process leads to surface concentrations that are about twice the value obtained by simply using the distribution from the thin-plume calculation; in other words, where the plume is advected *along* the surface, this calculation suggests that there is effectively an image plume below the surface (Hunt & Mulhearn 1973, Hunt et al. 1979b, Egan 1984). Thus, even in quite complex flows the image technique can be useful, but it must be used with caution. Further research is necessary to extend our very limited understanding of diffusion onto surfaces.

5. DISCUSSION

The circumstances in which different aspects of complex flows affect diffusion from point sources can be summarized succinctly in terms of "diffusion" factors:

1. Convergence/divergence of streamlines: $\hat{\sigma}_3 \, \partial \hat{U}_3/\partial \hat{x}_3/(\hat{U}_1 \, \partial \hat{\sigma}_3/\partial \hat{x}_1)$. This factor indicates whether the rate of streamline convergence or divergence is comparable with the diffusion by turbulence alone.

2. Mean shear across the plume: $(\hat{\sigma}_3 \, \partial \hat{U}_1/\partial \hat{x}_3/\hat{U}_1)$. This expression indicates the ratio of the deflection of a plume by shear (to the low-velocity side) and the plume width.

3. Inhomogeneity of the turbulence: $\hat{\sigma}_3 \overline{u_3^2} \; \partial \overline{u_3^2}/\partial \hat{x}_3$ and $K_{33}\hat{\sigma}_3 \; \partial K_{33}/\partial \hat{x}_3$. These factors indicate the ratio of the mean drift of the plume and the growth of its width caused by the gradients in the variance near the source and the diffusivity downstream (or large times).

4. Non-Gaussianity: $\overline{u_3^3}/(\overline{u_3^2})^{3/2}$ and $\hat{\sigma}_3 \; \partial \overline{u_3^3}/\partial x_3/(\overline{u_3^2})^{3/2}$. The former expression indicates the ratio of the difference between the mean vertical motion of the positions of the maximum concentration and the growth rate of the plume or cloud width near the source. The latter indicates the ratio of the additional growth rate of a plume or cloud associated with the gradients of turbulence and the growth rate contributed by the turbulence at the source.

5. Surface effects: these are significant when $n_s/\hat{\sigma}_3 \sim 1$, a condition occurring when the distance n from the mean streamline to the surface is comparable with the plume or cloud width.

There are other factors that are also important in complex flows, such as (a) the effects of the spatial or temporal reversing of the flow (Smith 1983) in recirculating separated flows and in wave motion, and (b) in stably stratified flows the effects on turbulence and diffusion of varying density gradients, a problem that has barely been looked at. An important conclusion is that most of these effects are controlled by different mechanisms and described by different mathematical models near the source ($t \lesssim T_L$) and far from the source ($t \gg T_L$).

ACKNOWLEDGMENTS

This review draws upon the work and ideas of many colleagues, at Cambridge and elsewhere, to whom I am grateful. I regret that many of their contributions have not been appropriately described or acknowledged here. Some of this work was done with Flow Analysis Associates of Ithaca, New York, on a project sponsored by the state of Maryland Department of Natural Resources, as well as with the Wave Propagation Laboratory of NOAA in Boulder, Colorado.

Literature Cited

Batchelor, G. K. 1952. Diffusion in a field of homogeneous turbulence. II. The relative motion of particles. *Proc. Cambridge Philos. Soc.* 48: 345–62

Batchelor, G. K., Townsend, A. A. 1956. Turbulent diffusion. In *Surveys in Mech-

anics*, ed. G. K. Batchelor, R. E. Davies, pp. 352–99

Bearman, P. W. 1972. Some measurements of the distortion of turbulence approaching a two-dimensional body. *J. Fluid Mech.* 53: 451–67

Berkowicz, R., Prahm, P. 1980. On the spectral turbulent diffusivity theory for homogeneous turbulence. *J. Fluid Mech.* 100:433–48

Britter, R. E., Hunt, J. C. R., Mumford, J. C. 1979. The distortion of turbulence by a circular cylinder. *J. Fluid Mech.* 92:269–301

Chatwin, P. C. 1974. The dispersion of contaminant released from instantaneous sources in laminar flow near stagnation points. *J. Fluid Mech.* 66:753–66

Chatwin, P. C., Allen, C. M. 1985. Mathematical models of dispersion in rivers and estuaries. *Ann. Rev. Fluid Mech.* 17:119–49

Chatwin, P. C., Sullivan, P. J. 1979. The relative diffusion of a cloud of passive contaminant in incompressible turbulent flow. *J. Fluid Mech.* 91:337–55

Csanady, G. T. 1964. Turbulent diffusion in a stratified flow. *J. Atmos. Sci.* 21:439–47

Deissler, R. G. 1968. Effects of combined two-dimensional shear and normal strain on weak locally homogeneous turbulence and heat transfer. *J. Math. Phys. (Cambridge, Mass.)* 47:320

Durbin, P. A. 1983. Stochastic differential equations and turbulent dispersion. *NASA Ref. Publ. No. 1103*

Durbin, P. A., Hunt, J. C. R. 1980. Dispersion from elevated sources in turbulent boundary layers. *J. Méc.* 19:679–95

Egan, B. A. 1984. Transport and diffusion in complex terrain. *Proc. Oholo Conf.* (To be published in *Boundary-Layer Meteorol.*)

Frost, S. 1981. *Temperature dispersion in turbulent pipe flow.* PhD dissertation. Univ. Cambridge, Engl.

Hanna, S. R. 1981. Lagrangian and Eulerian time scales in the daytime boundary layer. *J. Appl. Meteorol.* 20:242–49

Hinze, J. O. 1975. *Turbulence.* New York: McGraw-Hill. 790 pp. 2nd ed.

Hinze, J. O., Van der Hegge Zijnen, B. G. 1951. Local transfer of heat in anisotropic turbulence. *Inst. Mech. Eng. Gen. Discuss. Heat Transfer*, p. 188

Hunt, J. C. R. 1973. A theory of turbulent flow round two-dimensional bluff bodies. *J. Fluid Mech.* 61:625–706

Hunt, J. C. R. 1982. Diffusion in the stable boundary layer. In *Atmospheric Turbulence and Air Pollution Modelling*, ed. F. T. M. Nieuwstadt, H. van Dop, pp. 231–74. Dordrecht: Reidel

Hunt, J. C. R. 1984. Turbulence structure in thermal convection and shear-free boundary layers. *J. Fluid Mech.* 138:161–84

Hunt, J. C. R., Mulhearn, P. J. 1973. The dispersion of pollution by the wind near two-dimensional obstacles. *J. Fluid Mech.* 61:245–74

Hunt, J. C. R., Graham, J. M. R. 1978. Free-stream turbulence near plane boundaries. *J. Fluid Mech.* 83:209–35

Hunt, J. C. R., Snyder, W. H. 1980. Experiments on stably and neutrally stratified flow over a model three-dimensional hill. *J. Fluid Mech.* 96:671–704

Hunt, J. C. R., Weber, A. H. 1979. A Lagrangian statistical analysis of diffusion from a ground level source in a turbulent boundary layer. *Q. J. R. Meteorol. Soc.* 105:423–43

Hunt, J. C. R., Britter, R. E., Puttock, J. S. 1979a. Mathematical models of dispersion of air pollution around buildings and hills. *Proc. IMA Symp. Math. Modelling Turbul. Diffus. Environ.*, pp. 145–200. London: Academic

Hunt, J. C. R., Puttock, J. S., Snyder, W. H. 1979b. Turbulent diffusion from a point source in stratified and neutral flows around a three-dimensional hill. I. Diffusion equation analysis. *Atmos. Environ.* 13:1227–39

Huot, J. P., Rey, C., Arbey, H. 1984. Distortion of turbulence approaching a plate held normal to the flow. *Phys. Fluids* 27:541–43

Janicke, L. 1981. Particle simulation of inhomogeneous turbulent diffusion. *Proc. Int. Tech. Meet. NATO-CCMS, Palo Alto, Calif.*, pp. 527–36. New York: Plenum

Keffer, J. F., Kawall, J. A., Hunt, J. C. R., Maxey, M. R. 1978. The uniform distortion of thermal and velocity mixing layers. *J. Fluid Mech.* 86:465–90

Lamb, R. G. 1982. Diffusion in the convective boundary layer. In *Atmospheric Turbulence and Air Pollution Modelling*, ed. F. T. M. Nieuwstadt, H. van Dop, pp. 159–230. Dordrecht: Reidel

Lavery, T. F., Bass, A., Strimaitis, D. G., Venkatram, A., Greene, B. R., et al. 1982. *EPA Complex Terrain Modelling Program First Modelling Report 1981, US EPA Contract No. 68-02-3421*

Legg, B. J., Raupach, M. R. 1982. Markov chain simulation of particle dispersion in inhomogeneous flows: the mean drift velocity induced by a gradient in Eulerian velocity variance. *Boundary-Layer Meteorol.* 24:3–13

Ley, A. J., Thomson, D. J. 1983. A random walk model of dispersion in the diabatic surface layer. *Q. J. R. Meteorol. Soc.* 109:867–80

Lumley, J. L. 1962. The mathematical nature of the problem of relating Lagrangian and Eulerian statistical functions in turbulence. In *Mécanique de la Turbulence*, pp. 12–16. Paris: Editions CNRS

Maxey, M. R. 1979. *Aspects of unsteady turbulent shear flows.* PhD dissertation. Univ. Cambridge, Engl.

Misra, P. K. 1982. Dispersion of non-

TURBULENT DIFFUSION 485

buoyant particles inside a convective boundary layer. *Atmos. Environ.* 16:239–44

Monin, A. S., Yaglom, A. M. 1971. *Statistical Fluid Mechanics*, Vol. 1. Cambridge, Mass: MIT Press. 769 pp.

Moninger, W. R., Frisch, A. S., Campbell, W. C., Strauch, R. G. 1978. Doppler radar measurements of plume dispersal and dissipation rates in the boundary layer. *Proc. Conf. Radar Meteorol., 18th*, pp. 49–54. Boston: Am. Meteorol. Soc.

Novikov, E. A. 1963. Method of random forces in turbulence theory. *J. Exp. Theor. Phys.* 44:2159–68

Panofsky, H. A., Larko, D., Lipschutz, R., Stone, G., Bradley, E. F., et al. 1982. Spectra of velocity components over complex terrain. *Q. J. R. Meteorol. Soc.* 108:215–30

Pasquill, F. 1974. *Atmospheric Diffusion*. Chichester: Ellis Horwood (Wiley). 429 pp.

Pearson, H. J., Puttock, J. S., Hunt, J. C. R. 1983. A statistical model of particle motions and vertical diffusion in a homogeneous stratified turbulent flow. *J. Fluid Mech.* 129:219–49

Reid, J. D. 1979. Markov chain simulations of vertical dispersion in the neutral surface layer for surface and elevated releases. *Boundary-Layer Meteorol.* 16:3–22

Saffman, P. G. 1960. On the effect of the molecular diffusivity in turbulent diffusion. *J. Fluid Mech.* 8:273–83

Shlien, D. J., Corrsin, S. 1976. Dispersion measurements in a turbulent boundary layer. *Int. J. Heat Mass Transfer* 19:285–95

Smith, F. B. 1957. The diffusion of smoke from a continuous elevated point source into a turbulent atmosphere. *J. Fluid Mech.* 2:49–76

Smith, R. 1981. Effect of non-uniform currents and depth variations upon steady discharges in shallow water. *J. Fluid Mech.* 110:373–80

Smith, R. 1983. The contraction of contaminant distributions in reversing flows. *J. Fluid Mech.* 129:137–51

Snyder, W. H. 1985. Fluid modeling of pollutant transport and diffusion in stably stratified flows over complex terrain. *Ann. Rev. Fluid Mech.* 17:239–66

Snyder, W. H., Hunt, J. C. R. 1984. Turbulent diffusion from a point source in stratified and neutral flows around a three-dimensional hill. Part II. Laboratory measurements of surface concentrations. *Atmos. Environ.* In press

Snyder, W. H., Lumley, J. L. 1971. Some measurements of particle velocity autocorrelation functions in a turbulent flow. *J. Fluid Mech.* 48:41–71

Taylor, G. I. 1921. Diffusion by continuous movements. *Proc. London. Math. Soc.* 20:196–212

Tennekes, H. 1975. Eulerian and Lagrangian time microscales in isotropic turbulence. *J. Fluid Mech.* 67:561–67

Townsend, A. A. 1976. *The Structure of Turbulent Shear Flow*. Cambridge: Cambridge Univ. Press. 429 pp. 2nd ed.

van Dop, H., Nieuwstadt, F. T. M., Hunt, J. C. R. 1984. Random walk models for particle displacements in inhomogeneous unsteady turbulent flows. Submitted for publication

Venkatram, A. 1980. Dispersion from an elevated source in a convective boundary layer. *Atmos. Environ.* 14:1–10

Willis, G. E., Deardorff, J. W. 1976. A laboratory model of diffusion in the convective boundary layer. *Q. J. R. Meteorol. Soc.* 102:427–45

Wilson, D. J., Britter, R. E. 1982. Estimates of building surface concentrations from nearby point sources. *Atmos. Environ.* 16:2631–46

Wilson, J. D., Thurtell, G. W., Kidd, G. E. 1981. Numerical simulations of particle trajectories in inhomogeneous turbulence. II. Systems with variable turbulent velocity scale. *Boundary-Layer Meteorol.* 21:423–41

Yaglom, A. M. 1972. Turbulent diffusion in the surface layer of the atmosphere. *Izv. Acad. Sci. USSR Atmos. Oceanic Phys.* 8:333–40

Ann. Rev. Fluid Mech. 1985. 17 : 487–522

GRID GENERATION FOR FLUID MECHANICS COMPUTATIONS

Peter R. Eiseman

Department of Applied Physics and Nuclear Engineering,
Columbia University, New York, NY 10027

INTRODUCTION

Fluid mechanics is understood in a descriptive way through experimental observation, mathematical analysis, and numerical simulation. When the understanding is required for flows with complex internal structure and with complicated regional boundaries, insight is gained primarily from experiments and simulations. Numerical simulations are motivated by the prospect of economically obtaining a detailed flow-field description. In each instance, a governing system of flow equations is analytically formulated over the region and is solved in an approximate form on a computer. Various numerical methods have been devised for such approximations, and most depend upon some representation of the flow-field region by means of finitely many points and the interconnections between them. When each regional boundary is given by a sequence of connected points, the efficiency and accuracy of the various methods are enhanced, since boundary conditions can be applied without interpolation. Further enhancement comes when the connectivity pattern is regular. The most regular and thus preferable patterns are those that result from coordinate transformations.

With the application of transformations, regions with topological complexity are consistently treated even when points are in motion. During the course of simulation, motion can be advantageously used to adaptively resolve the significantly varying solution quantities. From a geometric viewpoint, the quantities determine a surface over the physical region. The resolution of the surface as it evolves then determines the adaptive movement.

0066–4189/85/0115–0487$02.00

Because of the typically required generality, most of the useful coordinate transformations are nonorthogonal. While many two-dimensional regions or surfaces can efficiently be given conformal coordinates to match a variety of geometric configurations, pointwise distributions cannot be arbitrarily specified on boundaries. Upon specification, conformality with its simple metric structure is no longer available. The next best structure comes from orthogonal coordinates: Availability is also limited, but not as severely as with conformal coordinates. In higher dimensions both are virtually eliminated because of their severely restricted range of applications.

Two primary categories for arbitrary coordinate generation have been developed. They are algebraic methods and partial differential equation methods. They include both fixed and moving grids, and control over the results is the main issue in the respective developments. Control can be applied precisely and adaptively and in a variety of ways. In this review, these methods are analyzed by examining their strengths and weaknesses. Before embarking upon the detailed methodology involved in coordinate generation, some general observations are first made concerning applications. This establishes the overall setting in terms of geometry, topology, and motion.

THE APPLICATION OF COORDINATE TRANSFORMATIONS

With one transformation, Cartesian coordinates on a rectangular region are typically mapped into coordinates that cover the physical region and match boundaries. When the transformation is applied to a discrete representation, the connectivity pattern is preserved. On a rectangular region, regularity arises naturally from uniform spacing in each Cartesian direction. The simplest pattern occurs when the points are connected only along coordinate directions. This is called a Cartesian grid. Under the mapping, the Cartesian grid goes into a corresponding physical-space grid, where boundaries are defined by discrete coordinate curves. With the transformation, the flow-field computations can be performed entirely on the fixed Cartesian grid, regardless of the geometry or motion of the boundaries. The Cartesian grid then represents the rectangular computational space.

The same computational space can be used to treat multiply connected physical regions that arise when there are a number of solid bodies in the field. The basic methods come from the identification of Cartesian-coordinate sections with either artificial or actual boundaries. Artificial boundaries are used to connect bodies with the physical boundary or with other bodies to essentially reduce the number of bodies. These connections are transmissive boundaries for the fluid and are called branch cuts, patches, or simply cuts. Actual boundaries are determined by the geometry of the physical bodies. These are prescribed in correspondence with either internal or boundary Cartesian sections. The internal sections are used alone as a slit or in a combination to remove a local region. In physical space, the mapped result is wrapped around the given body. With the slit, physical boundary data must be doubly specified so that its image is opened. With the local region, the boundary grid inherits the Cartesian corners or edges even if they do not physically exist. In the contrary sense, any Cartesian section may also map to a physical boundary with corners or edges. Barring irregularities in the physical boundaries, the Cartesian identifications tend to produce further irregularities that appear as coordinate singularities on the boundaries. While special numerical treatment can usually be applied at singularities, the accuracy of a simulation can suffer, particularly if significant properties of the fluid occur near the boundaries. The fundamental reason for the singularities comes directly from the consideration of a multiply connected physical region. The use of a single coordinate transformation and the associated computational space is the primary cause for the placement of the singularities directly on the boundaries.

To move the singularities off the boundaries and away from potentially large flow variations, more than one coordinate system is required. The immediate consequence is an assembly of computational spaces and identifications. Altogether, this may appear as a general rectilinear region covered by a Cartesian grid. With junctures between systems corresponding to the identifications, coordinate transformations are then applied to produce grids for each system, which upon assembly cover the physical region in a patchwise fashion.

On a local level, the physical-space grid retains the simple, regular connectivity pattern of each coordinate grid. The only places where the pattern is altered are those that correspond to the moved singularities and that consequently represent only a minor part of the entire region. On a global level, the grid connectivity is changed in a way that usually represents a higher degree of conformity with the actual topology of the region. This occurs because solid bodies in the field often get individual local coordinate grids wrapped around them. The local grids essentially provide a good structure for the modeling of nearby fluid-mechanical phenomena.

From a global perspective, the outer boundaries of the local systems are considered as a new collection of bodies that can be fewer in number if some of them are joined together. The physical-space grid is typically constructed from such collections with the introduction of further artificial transmissive boundaries, which altogether determine its global topology.

With the selection of grid topology, certain families of coordinate curves or surfaces are selected for enhancement in number on a purely local basis. The local body-oriented grids provide an immediate example: The number of coordinate curves or surfaces that wrap around each body can be adjusted separately. As a consequence, the simulation process can be optimized in the vicinity of each body. A similar optimization is not generally available for one global transformation, since local requirements cause global changes that would be wasteful and possibly conflicting. As in the case of a single transformation, the simulation process can be done entirely with respect to a fixed computational space, even when the physical grid is in motion. The movement is restricted only by the fixed choice of grid topology, which is a possible problem only when distinct bodies collapse upon each other.

Once a grid has been generated to match the physical boundaries and to have some conformity with the regional topology, simulation accuracy can be enhanced through grid movement. While maintaining the simple, regular connectivity pattern, points are pushed into positions that more accurately represent the desired solution to the governing system of fluid equations. Although the entire solution would theoretically be needed to drive the pointwise motion, the intrinsic character is often quite accurately detected by a small number of salient quantities. For example, in supersonic compressible flow, pressure is readily identified as a salient quantity, since shock waves are detected by its rapid variation. Altogether, the collection of salient quantities efficiently summarizes the solution data. On application, a feedback cycle is formed in which the solution is advanced by some computational algorithm and the grid is moved in response. Such cycles that couple movement with solution procedures are called *adaptive grid techniques*. When the motion is supplemented or replaced by arbitrary local changes in the number of points, the adaptive procedures lose the logically regular ordering of grids and inherit a data-management problem. When overlapping grids are employed, the problem becomes one of data transfer and its numerical consequences. The discussion herein, however, is restricted only to coordinate transformations and the resulting grids.

ALGEBRAIC GRID GENERATION

With coordinate transformations defined by algebraic formulas, the physical region can be given a continuous description; this is often done in an explicit manner. From the application of such transformations, arbitrarily large grids can be efficiently generated. The definition and the application form a process known as *algebraic grid generation*.

Classically, transformations have been globally defined by analytic functions of a complex variable and by direct shearing. The analytic functions produce conformal coordinates that are inherently nonsingular and over which the fluid-dynamic equations assume their simplest general form. In addition, potential-flow solutions are readily available and are often quite useful. The fundamental limitations, however, are a loss of control over boundary distributions and a practical restriction to two dimensions. For an overview of conformal mapping techniques, the reader is referred to the surveys by Moretti (1980) and Ives (1982).

The limitations of conformal coordinates are removed with the classical shearing transformations. The price is nonorthogonality, which typically adds complication to the fluid-dynamic equations. In a practical sense, the complications have a minor impact and may not even appear in the discrete form. As a consequence, the primary general development starts with shearing transformations. These and the closely related Hermite transformations are just global interpolations in one direction. The first and most fundamental objective is to insert the maximum possible amount of control over the grid in a given direction. This is provided by the multisurface transformation that includes both shearing and Hermite transformations as special cases. The same control is available in all directions by using Boolean sums of multisurface transformations. In the earlier shearing and Hermite cases, the transformations were often called transfinite in view of the fact that a generally infinite number of boundary points were interpolated when directions were combined. This is in comparison with the finite use of only vertex data. Our discussion in this section starts with shearings and proceeds to the multisurface method; we then develop the Boolean operations and conclude with a look at applications.

Shearing and Hermite Transformations

When an arbitrary positive function f over some domain is multiplied by a scale factor η, the result is either above or below the original function, depending upon whether or not the scale factor is greater than or less than unity. In particular, when η varies from 0 to 1, the product ηf varies from 0 to f. At each fixed point in the domain, this variance monotonically traces out a vertical line segment extending from the domain up to the function value. A coordinate system is defined by taking the line segments together with the curves or surfaces corresponding to ηf. Since this process represents a vertical shearing of f onto its domain, it is called a shearing transformation. In two dimensions with Cartesian coordinates (x, y), the transformation is given by

$$x = \xi,$$
$$y = \eta f(\xi),$$ (1)

and is illustrated in Figure 1 for the case of a nozzle geometry. In the displayed discrete form, a Cartesian (ξ, η) space grid is mapped onto the nozzle-conforming grid. The first equation determines the vertical lines with constant ξ values. Upon insertion into the second equation, it yields the scaled form with respect to the domain of f.

While the shearing as expressed in Equation (1) can be extended to include a general bottom boundary and arbitrary distributions in each variable, the basic process is done vertically. To perform a shearing in any direction, the transformation must be written in vector form. In this spirit, Equation (1) is rewritten for the position vector (x, y) as

$$(x, y) = (\xi, 0) + \eta[(\xi, f(\xi)) - (\xi, 0)],$$ (2)

where the bottom and top boundaries are explicitly given by $(\xi, 0)$ and $(\xi, f(\xi))$, respectively. With bottom and top boundaries replaced by arbitrary vector functions $\mathbf{P}_1(\xi)$ and $\mathbf{P}_2(\xi)$, the general shearing transformation is given by

$$\mathbf{P}(\xi, \eta) = \mathbf{P}_1(\xi) + \eta[\mathbf{P}_2(\xi) - \mathbf{P}_1(\xi)]$$ (3)

for the position vector $\mathbf{P}(\xi, \eta)$ with Cartesian components typically for n dimensions. The boundary coordinates are given by the vector $\boldsymbol{\xi}$ of length $n-1$. When $n = 2$, this becomes the curve parameterization ξ. For each fixed boundary coordinate ξ, the shearing is done along the line segment from $\mathbf{P}_1(\xi)$ to $\mathbf{P}_2(\xi)$ as η varies from 0 to 1. Shearing in arbitrary directions is readily illustrated with the choice $\mathbf{P}_1(\xi) = (0, 0)$ and $\mathbf{P}_2(\xi) = (\cos \xi, \sin \xi)$, which produces polar coordinates on the unit disk.

In two dimensions, Eiseman (1978) presented a thorough geometric analysis of shearing transformations and considered applications to

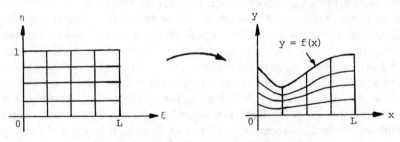

Figure 1   A shearing transformation for a nozzle.

cascades of airfoils. In a polar manner, a family of coordinate curves that looped about an airfoil were generated along lines from the airfoil to an outer boundary on which periodic alignment was enforced in a vertical direction matching top and bottom portions. With a periodic parameterization for the outer loop P_2, the parameterization for the airfoil P_1 was chosen to yield orthogonality at its surface. The purpose was to locally approximate boundary-layer coordinates. However, because of the linear pseudo-radial coordinate curves, the periodic matching was only possible for positions. The derivatives at the periodic juncture were discontinuous. In a more general context, the patching together of shearing transformations gives rise to slope discontinuities at the junctures. While a smoothing can be done as a postprocessing step, a direct algebraic formulation is nicer.

The most direct manner of producing smoothness is to use classical Hermite interpolation. To specify one derivative on each boundary in addition to locations, the cubic polynomials must be employed. In terms of the position vector, the cubic Hermite transformation is given by

$$P(\xi, \eta) = (1 - 3\eta^2 + 2\eta^3)P(\xi, 0) + \eta^2(3 - 2\eta)P(\xi, 1)$$

$$+ \eta(1 - \eta)^2 \frac{\partial P}{\partial \eta}(\xi, 0) + \eta^2(\eta - 1)\frac{\partial P}{\partial \eta}(\xi, 1), \tag{4}$$

where the coefficients are chosen so that evaluations of P or $\partial P/\partial \eta$ at the endpoints of η reduce to the given specifications on the right-hand side. For this to happen, each of these four possibilities produces the desired term with a coefficient of unity and the remaining three with coefficients of zero. Higher-order Hermite cases follow this same general pattern, first by allowing more derivatives at the boundaries to be specified. In the simple cubic case, a typical application is the specification of boundary orthogonality. In continuation, the most general Hermite form includes the same type of interpolation at a finite number of η values. Function values and derivatives up to various possibly distinct levels are interpolated as each point is taken in succession in going from boundary to boundary. The algebraic format is clear, but complicated, and thus is not given here. As a final remark on Hermite interpolation, we note that we could consider cases where the interpolation skips either function values or derivatives less than the highest interpolated order at some points. This is called a *defective Hermite interpolation*. For polynomials, it leads to the Hermite-Birkhoff interpolation problem, which is discussed in Prenter (1975, pp. 414–19).

The Multisurface Transformation

A set of constraints upon the grid arises in a natural way when there is a need to enhance accuracy and efficiency or to extend capabilities. Typical constraints come from the requirements for resolution, smoothness, topology, embedding, and uniformity. The shearing and Hermite transformations arose primarily to meet the requirements at the boundaries. These transformations are special cases of the general multisurface transformation that was developed by Eiseman (1979) to provide the necessary control over the grid at all locations and in a very precise sense. To illustrate the level of control and how constraints arise, an example of a grid from the multisurface transformation is displayed in Figure 2. The grid was designed for two flow-field simulations in which the effect of wind-tunnel walls could readily be examined. While the grid globally expands to an exact polar system for the far-field, local Cartesian grids were smoothly embedded about the airfoil. Taking the wind-tunnel walls to be Cartesian segments above and below the airfoil, we can apply boundary conditions directly for the wind-tunnel part, with concurrent grid alignment near the

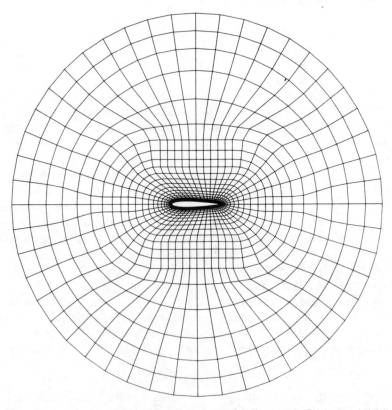

Figure 2   A grid to simulate the effect of wind-tunnel walls upon the flow over an airfoil.

walls. The separate simulation with far-field boundary conditions then produces flow-field data at the same grid points as the simulation with walls included. In addition to the embedding, the grid was clustered near the airfoil on which orthogonality was enforced.

THE CONSTRUCTION The multisurface transformation is designed for coordinate generation from a boundary $A(\xi)$ to a boundary $B(\xi)$ by means of an independent variable η and a smooth vector field established by interpolation of only first derivatives. From the vantage point of the general position vector $P(\xi, \eta)$, the interpolation may be interpreted as a deficient form of Hermite interpolation. For interpolation points $\eta_1 < \eta_2 < \cdots < \eta_{N-1}$, the vector field is given by

$$\frac{\partial P}{\partial \eta}(\xi, \eta) = \sum_{k=1}^{N-1} \phi_k(\eta) \frac{\partial P}{\partial \eta}(\xi, \eta_k), \tag{5}$$

where the functions ϕ_k vanish at all interpolation points except $\eta = \eta_k$ (where the value is unity). By integration from $A(\xi)$, the position vector is given by

$$P(\xi, \eta) = A(\xi) + \sum_{k=1}^{N-1} G_k(\eta) \frac{\partial P}{\partial \eta}(\xi, \eta_k),$$

where

$$G_k(\eta) = \int_{\eta_1}^{\eta} \phi_k(\zeta) \, d\zeta. \tag{6}$$

At the endpoint $\eta = \eta_{N-1}$, the final position should be located at $B(\xi)$. Upon setting

$$V_k(\xi) = G_k(\eta_{N-1}) \frac{\partial P}{\partial \eta}(\xi, \eta_k), \tag{7}$$

the endpoint condition becomes

$$B(\xi) = A(\xi) + \sum_{k=1}^{N-1} V_k(\xi). \tag{8}$$

Starting with $P_1(\xi) = A(\xi)$, successive vector positions are determined by the partial sums

$$P_i(\xi) = P_1(\xi) + \sum_{k=1}^{i-1} V_k(\xi), \tag{9}$$

which, in an inverse sense, yields

$$V_k(\xi) = P_{k+1}(\xi) - P_k(\xi). \tag{10}$$

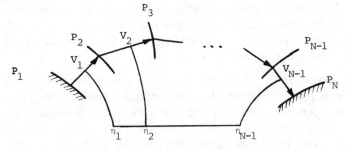

Figure 3   The construction of the multisurface transformation.

When the derivative in Equation (7) is expressed in terms of Equation (10) and is inserted into Equation (6), the position vector is given by

$$\mathbf{P}(\xi, \eta) = \mathbf{P}_1(\xi) + \sum_{k=1}^{N-1} \frac{G_k(\eta)}{G_k(\eta_{N-1})} [\mathbf{P}_{k+1}(\xi) - \mathbf{P}_k(\xi)], \tag{11}$$

where $\mathbf{P}_1(\xi) = \mathbf{A}(\xi)$ and $\mathbf{P}_N(\xi) = \mathbf{B}(\xi)$. The equation represents a coordinate transformation that is entirely constructed from the curves or surfaces $\mathbf{P}_1, \mathbf{P}_2, \ldots, \mathbf{P}_N$, each of which is expressed with respect to the same space of curvilinear variables ξ. For this reason, it is called the *multisurface transformation*. In correspondence with the interpolation points of Equation (5) and the vector directions of Equation (10), the constructive surfaces are assumed to be ordered in a monotone sequence that simply partitions the space between the boundaries $\mathbf{P}_1$ and $\mathbf{P}_N$. The intermediate surfaces $\mathbf{P}_2, \mathbf{P}_3, \ldots, \mathbf{P}_{N-1}$ and the choice of interpolants ϕ_k are used to control the grid so that various constraints can be satisfied as the grid is generated from boundary $\mathbf{P}_1$ to boundary $\mathbf{P}_N$. Because of the ratio of integrals G_k in the multisurface transformation, the normalization of the interpolants is not required. In Figure 3, a two-dimensional illustration is given.

INTERPOLANTS   When there is only one interpolation, a constant function is the simplest choice. The result is the shearing transformation, which assumes the general form of Equation (3) when the unit interval $0 \leqslant \eta \leqslant 1$ is used. The cubic Hermite transformation of Equation (4) is obtained with two intermediate surfaces and global quadratic polynomial interpolants for the points $\eta_1 = 0, \eta_2 = 1/2$, and $\eta_3 = 1$. The exact Hermite format appears when the two intermediate control surfaces $\mathbf{P}_2$ and $\mathbf{P}_3$ are traded for the boundary derivatives. From the pure Hermite formulation, however, there is virtually no indication of any limitation on the size of the magnitudes. One is simply required to use a trial-and-error approach. By contrast, a natural geometric limitation comes from the assumption that there is some space between P_2 and P_3. This prevents the interpolation of vector directions in Equation (5) that would experience an abrupt backward flip. The corresponding coordinate curves then do not backtrack and thereby cause the grid to fold upon itself. In summary, a direct geometric interpretation of Hermite derivatives is provided in a manner that can be simply used to avoid the folding problem.

Among the various possible interpolation functions for the multisurface transformation, those that vanish off local regions can be used to establish local controls. Such interpolants have been constructed with piecewise polynomials. As a result of the integration for G_k, the level of derivative continuity is one greater than that of the piecewise polynomals. Relative to neighboring points of interpolation, the range of nonzero values for each interpolant is as small as possible subject to the constraints of derivative continuity, of curvature generality, and of uniformity admission. The details are given by Eiseman (1982a,b). An illustration of the general interpolants corresponding to the first two levels of derivative continuity is displayed in Figure 4. The interpolation functions at or near the endpoints appear as truncated versions. The piecewise linear case of Figure 4a yields first-derivative continuity and is generally applicable in two dimensions but not in three. For three dimensions, the second-derivative continuity and curvature generality from the piecewise quadratics of Figure 4b are required. The essential features of each interpolant, however, are the same. Both provide local control at a very precise level. The basic control stems from a dependency upon only the closest constructive surfaces. To illustrate this fact, we examine the simple piecewise linear case in graphical form without the supporting mathematical theory. In Figure 5, a curve segment corresponding to the interval $\eta_k \leqslant \eta \leqslant \eta_{k+1}$ is displayed at a fixed value of ξ. The segment is determined by the three successive surfaces indicated by dots at the endpoints of the line segments. The curve leaves one line at $P(\xi, \eta_k)$ with matching tangent direction and enters the next at $P(\xi, \eta_{k+1})$ in the same manner. A general point within the interval is noted by $P(\xi, \eta)$. The

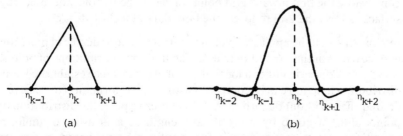

(a)   (b)

Figure 4 Local interpolation functions for the multisurface transformation. (a) Piecewise linear with continuity. (b) Piecewise quadratic with continuity up to first derivatives.

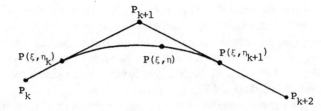

Figure 5   A curve segment determined by the piecewise linear interpolants.

curve segment is depicted without inflection points, and this is supported by the theory. With the local piecewise quadratics, the same pattern is continued by adding surfaces to each end of the construction to produce a dependency upon five consecutive surfaces. The corresponding segment naturally assumes a more arbitrary form. At or near the boundaries, each form is slightly modified to reflect the truncation of the interpolation functions.

With the freedom to choose intermediate control surfaces, the local dependency in η can be used to locally manipulate the grid without altering it elsewhere. Because of the algebraic formalism, the manipulation is done precisely and represents a high level of local control. An example with the local piecewise linear interpolants was given in Figure 2, where local Cartesian grids were smoothly embedded into a polarlike global grid about an airfoil so that the effect of wind-tunnel walls could be examined. Another useful embedding is the smooth and precise inclusion of boundary-layer coordinates within a global system. In addition to the many possible specifications for such local coordinate forms, a further application is to produce smoothness. When the boundaries have slope discontinuities, the classical shearing and Hermite transformations will propagate those discontinuities across the grid unless sufficient clustering is applied along the boundaries at the requisite locations. Without clustering, the local controls can be used to limit the propagation, even to the extent that it is gone before the first inward grid point. In effect, the use of smooth nearby surfaces allows the grid to forget the boundary at a close distance.

UNIFORMITY   In the application of the controls, a major degree of precision is achieved with the capability to establish uniform distributions of points, curves, or surfaces in either a local or a global sense. Once established, any distribution function can be composed with the transformation to achieve a desired effect without distortion. An obvious example is the distribution of points along a curve by using its arc length as a measure of uniform conditions. Uniform conditions for families of coordinate curves or surfaces, by contrast, are much more complex. While the direct and simple approach would be to take the arc length of each curve in the η variable, the result could still be a nonuniformly distributed family. This will happen if the curves deviate sufficiently from straight lines. No deviation occurs with the shearing transformation, and with this transformation the desired uniformity appears directly as a result of linearity in η along the line segments. Along the more arbitrarily constructed curves, the same uniformity can also be directly achieved by a projection onto line segments or, more generally, onto prescribed vector directions. With a given vector $\tau(\xi)$ to define the direction of measurement, the projected curve assumes the length

$$s_p(\xi, \eta) = [P(\xi, \eta) - P_1(\xi)] \cdot \tau(\xi) \tag{12}$$

starting from $P_1(\xi)$. A uniform family of curves is then obtained when the length s_p along τ is linear in η. In a graphic sense, the distance measurements can be viewed as if a sequence of yardsticks were laid across the region, with one for each fixed ξ. If we use the intuitive notion of yardsticks, the distinction between uniform and nonuniform can be visually detected in an almost analytic way. A global example is displayed in Figure 6.

To admit uniformity, the choice of interpolants for the multisurface transformation is restricted to satisfy the relationship

$$\sum_{k=1}^{N-1} \frac{\phi_k(\eta)}{\phi_k(\eta_k)} = 1. \tag{13}$$

Once satisfied, the chosen interpolants lead to conditions upon the placement of intermediate surfaces in order to achieve uniformity either globally or locally. The global polynomials and the local piecewise-linear cases generally satisfy the condition. The local piecewise quadratics have to be constructed to satisfy it. The respective development is given by Eiseman

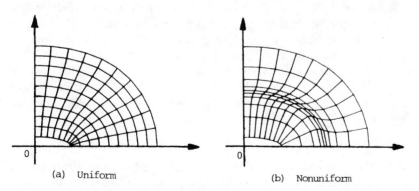

(a) Uniform   (b) Nonuniform

Figure 6 Grids about an ellipse.

(1979, 1982a,b). On application, the desired uniform distribution of curves or surfaces is prescribed in an a priori sense by precise, simple algebraic formulas. The resulting transformation then has the uniformity analytically built into its formulation. This formalism is particularly valuable in the construction of local coordinate forms.

Boolean Operations

The multisurface transformation represents the synthesis and extension of classical algebraic techniques for coordinate construction in one direction. The capability to combine directions comes with Boolean operations on projectors that correspond to transformations. Although the highest level of control is available with the projector for the general multisurface transformation, the essential properties of combining directions is more readily displayed with the projector for the classical shearing transformation.

To start, we first consider the collection (space) of all transformations $F(\xi, \eta)$ over the domain of ξ and η variables. When the two boundaries $P_1(\xi)$ and $P_2(\xi)$ are given in correspondence with $\eta = 0$ and $\eta = 1$, the collection contains the shearing transformation of Equation (3) and many other transformations that match the two boundaries. The selection of shearing is the same as a projection Q_η from the entire collection onto the shearing element. Analytically, the shearing projector Q_η is defined by

$$Q_\eta[F] = (1 - \eta)F(\xi, 0) + \eta F(\xi, 1), \tag{14}$$

where the form of Equation (3) is retrieved upon setting $P(\xi, \eta) = Q_\eta[F]$, $P_1(\xi) = F(\xi, 0)$, and $P_2(\xi) = F(\xi, 1)$. In a formal algebraic context, a projector is an operator whose double application produces the same result as its original application. The shearing operation is readily seen to satisfy this condition, which is symbolically stated as $Q_\eta^2 = Q_\eta$.

At the boundaries of the ξ variables, the shearing projector yields a transformation where the corresponding image boundaries are generated by straight line segments. Such lateral boundaries, however, are often too specialized. They simply arise from a construction in one direction that does not contain any data about lateral variations. To include the requisite data, other directions must be considered in a similar format. In two dimensions with $0 \leqslant \xi \leqslant 1$, a shearing projector for the only other direction is given by

$$Q_\xi[F] = (1 - \xi)F(0, \eta) + \xi F(1, \eta), \tag{15}$$

where the specifications for $F(0, \eta)$ and $F(1, \eta)$ define an arbitrary pair of lateral boundaries for Q_η. While the simple sum of projectors for the ξ and η directions would include data for all boundaries, the corners would be doubly counted. To correct this situation, a pure corner interpolation must be subtracted from the sum. Such an interpolation comes from the successive application of the two projectors. The result is called the *"tensor product"* interpolant and is given by

$$Q_\xi Q_\eta[\mathbf{F}] = (1-\xi)(1-\eta)\mathbf{F}(0,0) + \xi(1-\eta)\mathbf{F}(1,0)$$

$$+(1-\xi)\eta\mathbf{F}(0,1) + \xi\eta\mathbf{F}(1,1). \qquad (16)$$

The order of application is unimportant, since the commutative property $Q_\xi Q_\eta = Q_\eta Q_\xi$ is clearly satisfied. The sum with the corner adjustments is called the *"Boolean sum,"* is denoted by $\oplus$, and is defined by

$$Q_\xi \oplus Q_\eta = Q_\xi + Q_\eta - Q_\xi Q_\eta. \qquad (17)$$

Using the known commutativity of the product, it is an easy matter to check that the Boolean sum is a projector in a formal algebraic sense. The projected result $Q_\xi \oplus Q_\eta[\mathbf{F}]$ is also readily seen to match all four boundaries. With different projectors, the multidirectional matching of various properties follows when the commutative property holds. In the bidirectional cubic Hermite case, it follows from derivative commutativity at the corners. Within the general multisurface context, it comes from a well-defined control net of constructive surfaces.

Upon application of multidirectional matching, the interpolation can be executed in one direction at a time. With the Boolean sum of Equation (17) in the form $Q_\xi + Q_\eta(1-Q_\xi)$, the two-dimensional process is split into the steps

$$\mathbf{F}_1 = Q_\xi[\mathbf{F}]$$

and $\qquad (18)$

$$\mathbf{F}_2 = \mathbf{F}_1 + Q_\eta[\mathbf{F} - \mathbf{F}_1],$$

where $\mathbf{F}_2(\xi,\eta)$ is the position vector defining the transformation. With (ξ,η,ζ) coordinates for three dimensions, corresponding Q_ξ, Q_η, Q_ζ must be used in some fashion. When all bounding surfaces are specified, the transformation is given by the Boolean sum $Q_\xi \oplus Q_\eta \oplus Q_\zeta[\mathbf{F}]$, which, as before, is similarly split into steps as

$$\mathbf{F}_1 = Q_\xi[\mathbf{F}],$$

$$\mathbf{F}_2 = \mathbf{F}_1 + Q_\eta[\mathbf{F} - \mathbf{F}_1], \qquad (19)$$

$$\mathbf{F}_3 = \mathbf{F}_2 + Q_\zeta[\mathbf{F} - \mathbf{F}_2].$$

When only edges are specified, Boolean products and sums are used to define the transformation as $(Q_\xi Q_\eta \oplus Q_\xi Q_\zeta \oplus Q_\eta Q_\zeta)[\mathbf{F}]$. When only corners are specified, the transformation assumes the pure product form $Q_\xi Q_\eta Q_\zeta[F]$, which (with the shearing projector) yields standard trilinear interpolation. In continuation, the pattern of specifications can be done in a mixed sense. Interpreting products as intersections and sums as unions, transformations can be readily constructed for specifications of various assortments of bounding surfaces, edges, or corners.

With the interpolation of continuum quantities such as full curves and surfaces, the algebraic methods using Boolean sums have been called transfinite to reflect the generally infinite number of interpolation points. The theory of transfinite interpolation was developed as an outgrowth of the earlier blending-function techniques of Coons (1964) for the mathematical representation of surfaces required in computer-aided geometric design, as in Barnhill & Riesenfeld (1974). The basic development with Boolean operations was done by Gordon (1971). The use in coordinate generation is described by Gordon & Hall (1973) and Gordon & Theil (1982). The stepwise application is explicitly presented by Smith (1982) for the Hermite case of arbitrary degree.

Applications

Algebraic grid-generation techniques have been developed to meet a variety of problem constraints in a precise manner. The principal constructive elements are (*a*) the multisurface transformation for its control in one direction and (*b*) the Boolean operations for extending the control to all directions. Typical previously mentioned applications arose from constraints for geometry, resolution, smoothness, local embedding, and grid patching. Further applications arise in conjunction with other techniques. Through an algebraically defined system of coordinates, orthogonal trajectories can be computed with enhanced accuracy from the algebraic definition. The various orthogonal trajectory techniques for the production of orthogonal grids are reviewed by Eiseman (1982c). Nearly conformal transformations have been obtained by using a sequence of simple algebraic conformal mappings followed by a shearing. The conformal maps bring the geometry nearly onto a line or a circle. The shearing makes the match exact, but it destroys precise conformality. Such cases were examined by Caughey (1978). If we use the local controls within the multisurface transformation, the exact match can be accomplished with a rapid blending into either Cartesian or polar coordinates. The final transformation is then nearly conformal adjacent to the boundaries and exactly conformal elsewhere.

The automation required to systematically generate algebraic grids for a wide variety of applications was established in two dimensions with the software system developed by Eiseman (1982d). The eventual manipulation of the system with interactive graphics will streamline the process for many applications. By contrast, such automation is not available in three dimensions, where grid generation tends to be done on a case-by-case basis.

PARTIAL DIFFERENTIAL EQUATIONS METHODS

The techniques for the generation of arbitrary curvilinear coordinates come from either explicit or implicit definitions. Correspondingly, the coordinates are usually given either algebraically or as the solution to differential equations. In the latter category, a numerical solution procedure is required to find the grid-point locations from the defining system and the associated boundary conditions that must represent the physical geometry.

Hyperbolic Methods

When only one physical boundary is specified, hyperbolic partial differential equations may be used to obtain a grid by spatially marching from the given boundary. The remaining boundaries are determined by the solution and are geometrically unimportant in cases such as the external flow about a single object. With the requirement to prescribe only the object and the spacing along it, the hyperbolic methods proceed in marching steps that give the spacing in the outward direction. The inherent efficiency of the methods is due to the use of a single sweep through physical space. In two dimensions, orthogonal grids are obtained and are smooth provided that the boundary is free from slope discontinuities. A fundamental development was given by Stadius (1977), and one which was well suited to body concavity was presented by Steger & Sorenson (1980). The hyperbolic methods, however, are restricted in the amount of control that can be applied.

Elliptic Methods and Concavity Controls

The element of control over the grid has been more fully developed with the methods based upon elliptic partial differential equations. These we shall refer to as *elliptic methods*. As in the previous algebraic development, conformal mapping can be done in two dimensions with specifications of only boundary geometry. Typically, a pair of Laplace equations is solved subject to Cauchy-Riemann boundary conditions. The basic departure from conformal mapping comes with the specification of pointwise distributions along the boundaries. From this departure, some resolution requirements for the flow can be inserted. In comparison, the boundary distributions conflict with conformal conditions and produce nonorthogonal coordinates.

The earliest successful development was formally reported by Winslow (1967), who started with a Laplace system that determines a conformal mapping from the physical region into the space of curvilinear variables. Since the grid had to be generated in the physical space, the system had to be reformulated by an interchange of dependent and independent variables. With the reformulation, the physical boundaries are then represented by pointwise (Dirichlet) boundary conditions on the rectangular region of curvilinear variables. When the basic Laplace system determines a nonsingular mapping, the inverse mapping is biased toward conformal conditions. This tends to impose a global smoothness. The conflicting Dirichlet data produce nonorthogonality, which is most intense at the boundaries and gradually decays toward conformality as the interior is approached. From the discrete viewpoint of grid mappings, conformality is observed when small Cartesian squares are mapped approximately into squares of varying size.

The next major step in the development of elliptic methods was given by Thompson et al. (1974), who added periodic boundary conditions to produce branch cuts for various topological configurations and who also suggested that control over the grid could be accomplished by altering the original Laplace system. The alteration is to consider a pair of Poisson equations by including specifications for the right-hand sides. These are called *forcing terms* and are general functions of the curvilinear variables. The particular form to be used was established later by Thompson et al. (1977). Without forcing terms, Mastin & Thompson (1978) were able to show that the two-dimensional system analytically defined a nonsingular transformation. In higher dimensions or with forcing functions, nonsingularity is not generally assured. However, in two dimensions, the implication is that the basic system will reliably produce nonoverlapping grids, and that departures therefrom can be balanced against a solidly established reliability factor.

ONE DIMENSION   To examine the basic Poisson formulation in the simplest possible manner, our discussion starts in one dimension and then extends to higher dimensions. For a physical-space location x and a curvilinear variable ζ, the one-dimensional Laplacian is just the second derivative ζ_{xx}. The Laplace equation is given by its vanishing. An equally spaced grid in ζ then yields a corresponding equally spaced grid in x, since the analytic solution produces a linear relationship. To deviate from linearity and the associated uniform grid, the second derivative is forced to have prescribed positive and negative values. This gives the function $\zeta(x)$ respective upward and downward concavity in amounts that increase with the size of the force. The result is a Poisson formulation. The force as a control over concavity is employed for the purpose of clustering points in the physical space. Clustering is done at fixed locations in the curvilinear space, with

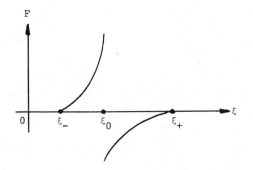

Figure 7   A one-dimensional forcing function for one cluster.

corresponding forcing terms each given as functions of ξ. The physical clustering locations x are then determined by the solution.

To illustrate the control, we consider a single cluster that starts at ξ_-, is centered at ξ_0, and ends at ξ_+. The associated forcing function $F(\xi)$ for the Poisson equation

$$\xi_{xx} = F(\xi) \tag{20}$$

is depicted in Figure 7. The solution is qualitatively displayed in Figure 8 and under the assumption of nonsingularity is a strictly monotone function. Upon approaching ξ_- from below, the solution is linear, since F is zero. At some physical point x_- corresponding to ξ_-, the solution starts to rise in the concave-upward direction with increasing intensity until ξ_0 is reached

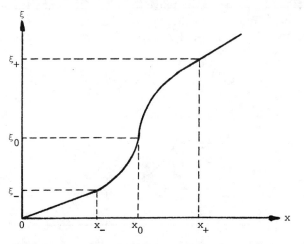

Figure 8   A one-dimensional Poisson solution for one cluster.

for some x_0. The forcing function F then changes sign to flip concavity to the downward direction. This decreases in intensity upon approaching x_+, which corresponds to the point ξ_+ where F reaches zero. Above ξ_+ and beyond x_+, the vanishing F causes the solution to return to linearity. With the uniform grid for ξ, the physical-space grid is most tightly clustered at the location x_0, where the solution has its greatest slope. On either side the slope steadily decreases toward constant conditions with increasing distance from x_0. The constant slopes on each side may be different, and if they are, the associated grid sections will be uniform but will have different spacing.

The requirements for equal spacing on each side are evident when the Poisson equation is transformed by using the chain rule to interchange the dependent and independent variables. The transformed equation, which is also needed for the numerical solution, is given by

$$x_{\xi\xi} = -x_\xi^3 F(\xi). \tag{21}$$

A direct integration, however, yields

$$\frac{1}{x_\xi^2} = c + 2 \int F(\xi) \, d\xi \tag{22}$$

for some constant c. A general plot of the right-hand side is presented in Figure 9. The constant of integration c gives the constant slope for the uniform grid before ξ_-. The uniform grid following ξ_+ will have different spacing if the second constant value is distinct from c. In the figure, a smaller value is depicted. This would correspond to a larger spacing Δx, since for constant $\Delta \xi$, the plotted values give $(\Delta \xi / \Delta x)^2$. To match the spacing, the values on each side must be the same, and the forcing function must then integrate to zero over the range from ξ_- to ξ_+. If the function is

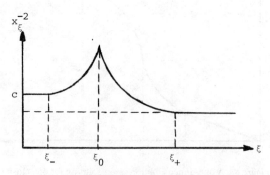

Figure 9   The first Poisson integral for one dimension.

antisymmetric about ξ_0, then the zero integral is assured, but ξ_- and ξ_+ must be at equal distances from ξ_0.

The inclusion of a cluster point, while preserving the original uniform grid elsewhere, is accomplished with a forcing function defined separately for the four successive segments determined by ξ_-, ξ_0, and ξ_+. Upon relaxing the requirement for exact preservation, the function can be defined to asymptotically decay on the two segments adjoining ξ_0. The form is very close to the previous case displayed in Figure 7. The ξ_- and ξ_+ can now be interpreted as the points where the decay has just become virtually complete. With the assumption of antisymmetry, the choice of simple exponential segments leads to the Poisson equation

$$\xi_{xx} = -\text{sgn}(\xi-\xi_0)ae^{-b|\xi-\xi_0|}, \tag{23}$$

which was suggested in two dimensions by Thompson et al. (1977). The sign function sgn is $+1, 0,$ and -1 corresponding to arguments that are positive, zero, and negative, respectively. The intensity of the cluster is controlled by the coefficient a; the size, by the rate of decay b. When periodic boundary conditions are imposed for the equation, the decay must be carried periodically through the grid a requisite number of times. This can be shortened by a sufficiently rapid decay. When multiple clusters are desired, the same considerations apply, and the right-hand side of Equation (23) is extended with a cluster summation index i for $a, b,$ and ξ_0.

TWO DIMENSIONS   The further extension to two dimensions assumes the same format, although the curvilinear locations for clustering are now both points and lines rather than just points. For curvilinear variables (ξ, η), the Poisson system assumes the form

$$\xi_{xx}+\xi_{yy} = P(\xi,\eta),$$
$$\eta_{xx}+\eta_{yy} = Q(\xi,\eta). \tag{24}$$

Clustering is accomplished relative to the uniform conditions defined by the pure Laplace system where the right-hand sides vanish. The forcing function P for the first equation controls the curves in physical space that correspond to constant values of ξ. As in the one-dimensional case from Equation (23), the function $P(\xi,\eta)$ is conveniently taken as a sum of exponential-type terms. To uniformly cluster curves of constant ξ about the curve for ξ_0, a term of the form

$$a \, \text{sgn}\,(\xi-\xi_0) \exp\,(-c|\xi-\xi_0|) \tag{25}$$

is included. To locally cluster curves of constant ξ toward the point with parameter value η_0 on a curve of fixed ξ_0, a term of the form

$$b \, \text{sgn}\,(\xi-\xi_0) \exp\,(-d[(\xi-\xi_0)^2+(\eta-\eta_0)^2]^{1/2}) \tag{26}$$

is also included. As earlier, a and b give the clustering intensity, while c and d give the decay rates to establish the range of clustering. Each controlling coefficient a, b, c, and d is positive and upon inclusion in $P(\xi, \eta)$ would be appropriately subscripted along with the desired point locations (ξ_0, η_0) and line locations ξ_0. Negative values for a and b would produce the opposite effect of repulsion. The forcing function $Q(\xi, \eta)$ for the curves with constant η is similarly structured by interchanging the above roles for ξ and η in our discussion of $P(\xi, \eta)$. The analytic solution to the generating system [Equation (24)] has derivative continuity to all orders except along the lines where clustering is imposed. There, continuity is obtained only up to first derivatives because of the flipping of concavity from the sign functions in Equations (25) and (26).

At the boundaries, only one direction of concavity is needed, and it is determined by a simple truncation of the general form upon translation to either side. When the pointwise distribution on a boundary is fixed as a result of Dirichlet conditions, the effect of forcing functions is to change intersection angles and spacings from the adjacent curve. For a boundary with constant ξ, the forcing function P changes the spacing, while Q alters the angles of the intersecting constant-η curves. The actual specification of the angles and spacings requires an iterative determination of P and Q. This was done by Steger & Sorenson (1979).

Rather than adjustments purely at the boundary, the spacing along the boundary may also be desired within the interior of the region. To propagate boundary distributions toward the interior, there must be a suitable definition of the forcing functions. As a simple illustrative example, consider the Cartesian coordinates on a rectangle with an x distribution $\xi = x^3$ and a uniform y distribution $\eta = y$. Upon substitution into the Poisson system of Equation (24), the forcing functions for these coordinates are found to be $P = 6\xi^{1/3}$ and $Q = 0$. Along with the distributions prescribed only on the boundaries, the Poisson system has been constructed from its solution. If the same boundary data were given to the Laplace system obtained by setting $P = 0$, then the x distribution would not be propagated across the rectangle. Instead, points would spread out in adjusting themselves to approximate conformal conditions in the interior.

In the general setting, the propagation of distributions from the boundaries is accomplished by an interpolation of forcing functions from the boundaries. The boundary forces are obtained from the Poisson system by using the pointwise distribution and the assumption of boundary orthogonality. The orthogonality assumption replaces the transverse derivative data in the system with the boundary curvature distribution. From the boundary forces, the interpolation is done between opposing faces in the (ξ, η) space. Once the Poisson system has been established with the interpolated forcing functions, the boundary orthogonality assumption is discarded. The solution then proceeds directly from the usual boundary data. The propagation of distributions by forcing-function interpolation is primarily due to Middlecoff & Thomas (1979) and Warsi & Thompson (1976).

In the propagation technique and in the motivating Cartesian example, the Poisson system is used to determine the forcing functions. With a view toward modifying existing coordinate systems, the same determination can be more arbitrarily applied to establish corresponding Poisson systems. Their modification then results in a corresponding modification of the coordinates. Clustering would then be done relative to the determined system.

With the forcing functions depending only upon the curvilinear variables, the application of clustering is relative to only curvilinear location. The internal points or curves about which clusters are desired cannot be given specified locations in the physical region. Rather, each cluster affects the others, and altogether the Poisson solution determines their respective locations. This can be an advantage, however, since possibly unimportant locations would not have to be specified and would thereby relieve the user of the effort. This determination can also be a disadvantage if the location is important. A potential remedy is suggested by Thompson (1982b), who considers a forcing-function dependence on physical rather than curvilinear variables.

To numerically solve the Poisson system with any of the above forcing functions, the dependent and independent variables must be interchanged by using the matrix inverse to the chain rule. The transformed Poisson system is given by

$$g_{22}x_{\xi\xi} - 2g_{12}x_{\xi\eta} + g_{11}x_{\eta\eta} = -g[x_\xi P - x_\eta Q],$$

$$g_{22}y_{\xi\xi} - 2g_{12}y_{\xi\eta} + g_{11}y_{\eta\eta} = -g[y_\xi P - y_\eta Q],$$

(27)

where

$$g_{11} = x_\xi^2 + y_\xi^2, \qquad g_{12} = x_\xi x_\eta + y_\xi y_\eta, \qquad g_{22} = x_\eta^2 + y_\eta^2$$

are metric coefficients and $g = \det(g_{ij})$ is the square of the Jacobian $x_\xi y_\eta - x_\eta y_\xi$. Geometrically, the coefficients come from the differential expansion of squared arc length along arbitrary curves, with measurement relative to the given coordinate system. On the rectilinear boundary for the curvilinear variables, the boundary data for various bodies are now explicitly inserted as Dirichlet boundary conditions. Branch cuts are directly given by periodic boundary conditions and, as in the case of clustering, have locations that are determined by the solution and that cannot be specified in advance.

The same Poisson format, the clustering, and the interchange of variables can be given in three dimensions. The format is simple: The clustering is for points, curves, and surfaces, and the interchange is more detailed. The entire process is much more complicated, is done on a case-by-case basis, and depends upon grid generation for two-dimensional surfaces.

ADAPTIVE GRID MOVEMENT

The numerical simulation of fluid-mechanical phenomena often involves solutions with rapid variations over short distances and at unpredictable locations. Moreover, a solution may even be multivalued. When the grid is predetermined by the physical region and the resolution requirements at expected locations, the accurate simulation can be jeopardized if the large solution variations should appear somewhere on a scale that is too small for the prescribed spacing of points. To model such phenomena, the grid must be adaptively moved into positions that most accurately represent the solution as it evolves.

The Consolidation of Adaptive Data—A Monitor Surface

The severe behavior in the solution is usually well monitored by a small number of salient physical quantities. Their use clearly obviates the need to create a monitoring quantity by means of a formal error estimate. Altogether, the salient quantities form a vector that can be evaluated at each point in physical space. Collectively, the evaluations determine a surface over the physical space of the same dimension as the space. The surface then simply consolidates the resolution requirements into a single object. When salient quantities have large variations at distinct or nonconflicting locations, their combined behavior can be adequately followed by using some monotone scalar function of the quantities. The effect of employing suitable scalar functions is to define a monitoring surface in a space of the lowest possible dimension. When all quantities are collapsed into one scalar function, the dimension is only one greater than the physical space. Once established, the evolving monitor surface becomes the object that must be accurately resolved by the appropriately adapting grid.

The methods for adaptive movement consider the surface either as a geometric entity or as a function from which derivative data are taken. With the function viewpoint, gradients are directly given by first derivatives, and curvature is roughly approximated by second derivatives. With the geometric viewpoint, gradients are implicitly detected by surface arc-length measurements, and the actual curvatures are employed directly. Moreover, the treatment of multivalued solutions can be done in a direct surface-oriented fashion.

The comparison between the two viewpoints is most readily seen when the surface is defined by a single scalar quantity u. In one dimension, with location x, the assumption of constant arc-length increments means that $(1 + u_x^2)^{1/2} dx$ is constant. Large gradients u_x increase the first factor and cause dx to shrink to maintain constancy. When the increase is substantial, the first factor is essentially the magnitude of the gradient itself. In the discrete form, the one-dimensional surface grid is uniform, and the projection down into physical space produces a corresponding grid that resolves gradients. If the surface changes direction on a scale that is too small for the uniform arc-length grid, then points must be moved to provide adequate resolution. The precise measure of direction changes is the curvature. In terms of the physical-space location x, it is given by $k = u_{xx}(1 + u_x^2)^{-3/2}$. From the surface viewed as a function, it is usually approximated by u_{xx}, which is accurate only when u_x is much less than unity in magnitude. In an application where u_x varies rapidly from zero to a large value, the estimate u_{xx} of k correspondingly varies from a highly accurate one to one that is much too large because of a missing large denominator. The consequence is that the desired resolution is pulled off-center toward the steep gradient region.

In higher dimensions, the use of arc length and curvature is required for the same purposes. The application of each, however, is less well defined, since there is no unique measure of a uniform distribution of surface points, nor is there a universal choice for curvature. The complicating feature comes from the combination of distinct directions. The measure of uniformity can be established in terms of equal cell volumes, equal arc lengths along curves, or some other desirable mapping property such as conformality. Directional properties are witnessed by noting that cells with fixed volumes can become elongated, while cells with fixed edge lengths can be collapsed. In contrast, conformality produces well-structured cells that can rapidly dilate in size. On balance, these uniformity measures are reasonable for a wide variety of situations, albeit each has distinct strengths and weaknesses. As with uniformity, there are also distinct choices for curvature clustering. Along any curve in the surface, curvature is split into a geodesic part and a normal part to separately describe curve and surface bending rates. The geodesic part gives rates only for the curve, while the normal part gives rates only for the surface. For adaptive movement, complications from the physical boundaries are treated with geodesic curvature, while complications from the solution are treated with normal curvature.

The basic surface geometry is given by the normal curvature, which is formally a measure of the rate at which surface-tangent planes vary as the curve is traversed. In two dimensions, both Gaussian and mean curvatures are defined in terms of the two principal normal curvatures, which are the maximum and minimum values, respectively, when all angular directions are considered. The Gaussian and mean curvatures are the product and the average, respectively. When the surface has a fold corresponding to a disturbance such as a straight shock front, the normal curvature has a minimum of zero along the front and some large maximum in the perpendicular direction. The Gaussian curvature vanishes as a result of the factor of zero. Consequently, it cannot be used for clustering to the front. Similarly, mean curvature vanishes on minimal surfaces and cannot be used for clustering on such surfaces (e.g. Scherk's minimal surface: $e^z \cos x = \cos y$). A more reasonable and easily applied approach is to use normal curvatures directly.

When the surface is viewed as a function and second derivatives are taken, a situation similar to the previously discussed one-dimensional case arises. As in one dimension, large gradients tend to shift clustering regions off-center. In addition, the sole use of derivatives along coordinate curves in physical space fails to detect the full bends in the surface. The neglected parts correspond to the sideways tilting motion of the tangent planes as the curve is traversed.

For a space of any dimension, the same function and geometric views of the surface can be taken and compared. To accurately represent the surface, the geometric view is required and is consistent with local Taylor series estimates on the surface. The function viewpoint approximates the geometric quantities in varying degrees of quality and is consistent with local Taylor series estimates from physical space. To continue in more detail, basic differential geometry is needed and is available in Laugwitz (1965). The application to adaptive movement is discussed further in Eiseman (1983a).

Underlying each viewpoint is the presumption that the surface is sufficiently smooth to yield reasonable approximations to the various derivatives and geometric quantities. With the representation in the discrete form of a grid, a numerical filter can usually be applied to remove spurious oscillations and to round off sharp discontinuities. When a filter is needed and the solution must be exactly preserved, the surface must be stored separately and then smoothed. In addition, the temporal evolution of the surface may also require a smoothing operation to prevent wiggles in time. The corresponding filter then requires surface data over a time period up to at most the new solution level.

Movement Strategies

With the salient solution properties consolidated into a smooth monitor surface, the creation of accurate grid representations produces general movement strategies that are not tied to a particular numerical solution procedure or physical problem. The generality is needed to address a wide variety of simulations in a nearly optimal way. Further optimization would be expected when special techniques are developed for special situations. The moving finite-element method described by Miller & Miller (1981) and the modified-equation technique proposed by Klopfer & McRae (1981) are both optimal in their respective measures, and they are restricted to distinct types of solution procedures. For a class of convection-diffusion problems, the transformation into simpler diffusion equations was simultaneously considered by Piva et al. (1982) and Ghia et al. (1983). Under this physical restriction, the movement is determined by the Poisson system of Equation (27) with velocity-dependent forces. While further such studies are available for specific situations, the discussion here is for general strategies that not only are simpler to implement in many situations but also address the more complicated simulations that typically arise in fluid mechanics.

The general adaptive grid simulations directly use the monitor surface to feed the solution properties back into the movement part of the processes. The primary objective of the feedback process is to maximize accuracy for the given number of points at the lowest possible cost for movement should that be a significant part of the simulation. Attempts to mathematically formulate the objective were made by Rheinboldt (1983) and Babuška et al. (1983b).

PROPORTIONALITY With the stated objective, the grid motion is established directly in the physical region or on the monitor surface by using weight functions. The typical application is to require that the product of the weight and some measure of cell size be constant or nearly so. An increase in the weight then causes the cell sizes to shrink and to produce a desired clustering of points. Balanced against the arbitrary application of clustering is the requirement that the grid be well structured, in the sense that deviations from orthogonality are moderate and that global smoothness, or equivalently uniformity, appears when clustering is absent or has moved elsewhere. When smoothness is explicitly included, the simplest and most useful form of the weight is given by the linear combination

$$w = 1 + \sum_{k=1}^{m} c_k M_k, \tag{28}$$

where the M_k are the magnitudes of attracting quantities and the corresponding c_k are the coefficients that determine the respective levels of significance. The smoothness is provided by the choice of unity for the first term, which replicates the measure of cell size under the intended multiplication.

In one dimension or for curves in higher dimensions, the cell size is the differential element of arc length ds that is taken either in the physical region or on the overlying monitor surface. The transformation to a curvilinear variable ξ is defined when the product $w\,ds$ is proportional to $d\xi$. The grid motion comes from the assumed constant value of $d\xi$, which forces ds to vary inversely with w. The result is called an equidistribution of the weight. When w is a function of s, the resolution is prescribed at fixed physical or surface locations. If we assume that ξ is in the unit interval, a direct integration yields the transformation

$$\xi = \frac{F(s)}{F(s_{\max})}, \tag{29}$$

where

$$F(s) = s + \sum_{k=1}^{m} c_k \int_{0}^{s} M_k(z)\, dz. \tag{30}$$

When all the coefficients c_k are set to 0, the transformation is linear and a uniform grid results. With the normalization by $F(s_{\max})$, the proportionality constant is explicitly given. Had differentiation been performed instead, the constant would then have been implicitly determined from the boundary conditions to the resulting differential equation $w\xi_{ss} - w_s\xi_s = 0$.

With the linearity of the weight carried over into $F(s)$, the coefficients can be given a precise interpretation by using the evaluation at $s_{\max}$. Dwyer (1983) observed that the total amount of a quantity M_k was represented by its total integral I_k, and that the ratio $r_k = c_k I_k / F(s_{\max})$ would then be the fractional amount. The prescription of fractions is more natural, since precise numbers of points can be assigned for each quantity once the total number of points is given. The conversion of the prescribed fractions r_k into weighting coefficients $c_k = (r_k/I_k)F(s_{\max})$ is done once $F(s_{\max})$ is determined. The determination comes from Equation (30) with an evaluation at $s_{\max}$ and a substitution for each c_k. Upon solution, we obtain $F(s_{\max}) = s_{\max}/(1 - r_1 - \cdots - r_m)$. Underlying the expression for c_k is the assumption that the total amount of each quantity I_k is not zero. Otherwise, there would be a division by 0. In practice, such a division is avoided by adding a very small positive quantity to each I_k.

While the integrals I_k represent the total amount of a given quantity M_k, it should be noted that the contribution to clustering must occur somewhere. At the very least, the associated locations are spread over some of the arc length. Moreover, contributions from distinct quantities may also occur at some of the same places. When the places for clustering are localized and are each the result of a single quantity, the total fraction of the points for the places can be precisely specified. This is accomplished by adding the appropriate local arc length to each integral, by removing that total added length from s_{max} to get an adjusted maximal arc length, and then by proceeding as before. When there are intervals where clustering comes from more than one quantity, the same type of adjustments are made by simply accumulating all contributions for the interval and removing the appropriate amounts from the previous totals. Such adjustments are only possible when the quantities vanish or almost vanish everywhere except on localized intervals.

A similar development leading up to such functions occurs when the weight is a function of only ξ. This means that clustering is done relative to curvilinear, rather than physical, locations and is thus free to move as in the elliptic method. The weight then appears in reciprocal form, and the roles of ξ and s are formally reversed. The consequent transformation gives s as a function of ξ. Since equal spacing is prescribed for ξ, the desired grid in s is evaluated without interpolation, in contrast to the transformation of Equation (29).

Although further generality would come with weights defined by arbitrary monotonic functions of the magnitudes M_k, the adaptive methods derived from proportionality statements typically use the linear form and thereby inherit the interpretation of coefficients in terms of fractions. White (1979, 1982) takes the monitor surface as the actual solution, chooses s as its arc length, and effectively has $w = 1$. The simulations are also performed with respect to s rather than the physical distance x. This permitted White (1982) to smoothly simulate multivalued behavior in the nonlinear wave equation $u_t + uu_x = 0$. Relative to the arc length of the actual solution, Ablow & Schecter (1978) consider curvature clustering with the curvature magnitude M_1 in Equation 28 for $m = 1$. For arbitrary curves bounding two-dimensional regions, Eiseman (1979) employs the same transformation in the equivalent geometric format of approximate circular normal images. In two dimensions, Dwyer et al. (1980) examine a unidirectional adaptation along a fixed family of coordinate curves in the physical region. By taking the curves one at a time, they develop the movement as a sequence of one-dimensional problems. Relative to the arc length s of each curve, gradient and curvature clustering are inserted with M_1 and M_2 for the estimated size of the first and second derivatives of the monitor surface. Also in physical space, Gnoffo (1982) studies gradient adaptation in two directions with first derivatives for M_1. The sequence of curves now extends through one coordinate family at a time in a cyclic fashion that is iterated until convergence. Returning to the geometric viewpoint, Ablow (1982) considers solution surface arc length along coordinate curves. The partial differential equations describing equal arc length are iteratively solved.

As a synthesis and extension of the previous methods, Eiseman (1983a) presented a general alternating-direction adaptive strategy that uses the linear weight of Equation (28) for arbitrary surfaces. The process is executed sequentially on coordinate curves that are grouped by direction. A cyclic alternation between an arbitrary number of directions is iterated until convergence. Relative to surface arc length, the resolution of basic geometry is done with normal curvature. The coefficient for the weight with normal curvature is determined by a maximum fraction of points and a factor that varies from 0 to almost unity as curve arc length goes from a minimal value to a large one. The dynamic shift of coefficient from curve to curve ensures that a large increase in arc length would more strongly pull points into the bending locations that otherwise might be missed. The remaining locations in such a circumstance are well represented by arc length. With the adaptation being done on a curve-by-curve basis, the use of normal curvature is particularly important, since no clustering is done for curves that may bend within the surface during the iterative process. Rather, clustering is done only for the surface, and (as a favorable by-product) coordinate curves are pulled into alignment with folds in the surface. The importance of alignment has been accented by Anderson & Rai (1982). The geodesic curvature is the measure of curve bending within the surface and is employed only for surface boundaries that can nontrivially arise from the physical region. To resolve surface boundaries, its magnitude is included in the weight of Equation (28) with a coefficient that decays in the inward direction to shift clustering toward surface geometry. With clustering relative to arc-length uniformity, the grid structure can become skewed in places. To prevent excessive skewness, the deviations from orthogonality can be used in the weight to pull toward their equidistribution. While the 1 in the expression for w gives the uniformity measure of arc length, the measure can be essentially changed to equal volumes with a weight term for the Jacobian. Further weighting choices are possible, and these result in a balancing problem between the relative importance of the desired properties.

When the same weight is applied to surface volume elements, the uniformity measure is directly given by equal volumes. With weighted volume elements on the surface, Eiseman (1983b) geometrically constructs a local molecule that is relaxed in a point iterative style. For a two-dimensional surface grid, the molecule about each point is determined by cells corresponding to the four quadrants surrounding it. A separate balance along each coordinate curve is performed in a mean-value sense. For a given curve, this balance is between the total weighted volume on either side of the transverse coordinate curve. In each quadrant, the volume is that of the triangle that contains the point. The weight is applied at the barycenter. The weight center for a given side is then the vector sum for two adjacent triangles. Upon projection to the given curve, a weight is established on the curve at some location that is nearly a third of the outward distance to the neighboring grid point. The mean value is then taken along the curve to determine a new curvilinear location. From the new curvilinear locations, the new surface location is determined by bilinear interpolation. With the bilinear communication, the molecule can also be created directly in the space of curvilinear variables by using the weighted surface volumes. The Cartesian format then provides some algebraic simplification. With the molecule in either location, the transformation is always nonsingular as a continuum, and this is a distinct advantage. However, since it is evaluated at only one point in each molecule, the nonsingularity can be destroyed for the grid. This can occur only when the four quadrilateral cells about a surface point form a patch with severe concavity. To correct this possible situation, the sides adjacent to a point of concavity can be extended to determine values that limit movement along the facing curve. The use of such limiters assures that the grid will never overlap during the adaptive movement. In the context of a general connectivity triangular mesh, a robustness similar to that from nonoverlap comes with restructuring strategies [as reported by Fritts & Boris (1979), who considered Lagrangian movement]. Full adaptive movement was developed by Erlebacher (1984) and Erlebacher & Eiseman (1984). The basic movement is in physical space, curvature attraction is done with an approximation to the mean curvature of the monitor surface, and the movement molecule is a surface-area-weighted average over the triangles about a point. Global movement is done in a point iterative sense. With such iteration on geometrically constructed local molecules, the movement is determined by an implicitly defined elliptic system of partial differential equations.

ELLIPTIC SYSTEMS   The Poisson system of Equation (24) is also directly applicable to adaptive movement. Mastin & Thompson (1983) considered the inclusion of monitor-surface derivatives in the forcing functions. Each derivative quantity is with respect to curvilinear variables and can be separately prescribed for each Poisson equation. The forcing function for each direction is a scaled first derivative of the quantity with respect to the same direction. To restrict the possibility of grid overlap, bounds were established for the scaling coefficients.

From a geometric view of the monitor surface, the Poisson system can be expressed with respect to the surface. The previous Laplacians are simply replaced by the Laplace-Beltrami operator, which is described by Warner (1970) in the context of Hodge theory. Proceeding directly from the vector expression for the inverse form of Equation (27), we may use the Gauss equations for second derivatives to suitably state the system. This was done by Warsi (1982). Another related development for surfaces was presented by Thomas (1982), who reasoned from a full three-dimensional Poisson system under the assumptions that curves entered the surface orthogonally and with no curvature. In each development, mean curvature appears as a source term. Although the surface grids were considered as parts of three-dimensional transformation techniques, they can be applied for adaptive movement. Conversely, the surface grids established for adaptive purposes can also be applied within the context of fixed higher-dimensional transformations.

THE VARIATIONAL SETTING A natural framework for elliptic methods is provided by a variational setting. The main distinction is that the parameters for grid control appear in global integrals, rather than in direct proportionality statements or explicitly in source terms. The previous controls were for direct elemental relationships and concavity. When a global linear combination of integrals is considered, the controlling coefficients are for the average quantities represented by integrals over the entire physical region rather than only parts of it. The grid is determined by taking the extremum of the integral combination, which produces the generating system in the form of Euler equations. On comparing the methods in the format of differential equations, we can observe the distinctive locations for control. The primary difference arises from the differential elements employed and the measure of uniformity.

In an isolated sense, the integral of a weighted volume is stationary over the physical region with physical differentials when the weighted volume is constant. The result is a proportionality statement for volumes. Upon returning to the integral, however, a change to curvilinear differentials indicates that the weight is applied to squared volumes rather than to simple weighted volumes. With the general linear weight of Equation (28), the leading term of unity would produce uniformity in the form of equal volumes. Other terms could shift it toward other measures such as conformality. Conformality would appear if M_1 is taken as the sum of the square residual of the Cauchy-Riemann equations. Given a large enough coefficient c_1 to dwarf the effect of the leading unit term, the integral would be essentially split into a Cauchy-Riemann part and the remaining weight without unity. Since the conformal part depends upon transformation derivatives, the stationary condition is now more complex than the previous proportionality statement. In terms of the metric coefficients, the integrand for the conformal part is $h_s = (g^{11} + g^{22})$, where the g^{ij} are the matrix inverse elements to g_{ij} in Equation (27). When the conformal part alone is stationary, the Laplace system from Winslow (1967) is obtained. This fact was first observed by Yanenko et al. (1978), who balanced the conformal part against a weighted volume $h_V = wJ^\alpha$ and a Lagrangian part $h_L = \| V_F - V_G \|^2$ (where V_F and V_G are fluid and grid velocities, respectively). With the integrals over the space-time volume element $dx\,dy\,dt$ conveniently denoted by I_s, I_V, and I_L, the total integral for the grid is given by $I_s + \lambda_V I_V + \lambda_L I_L$, where λ_V and λ_L are adjustable constants that with increased size bias toward weighted volume adaptivity and Lagrangian tracking, respectively.

The next development beyond Yanenko et al. (1978) was the inclusion of a separate orthogonality control, which in light of the conformality measure for uniformity might appear to be redundant. The main utility, however, is that orthogonality can be enhanced without the effect of dilating cell volumes, as would happen with conformality. Removing the Lagrangian measure I_L and the associated dt, Brackbill & Saltzman (1982) considered integrands with the cross metrics g_{12}^2 and $(g^{12})^2$, which lead to integrals I_0 and I'_0 and parameters λ_0 and λ'_0. If we drop the prime notation, then Euler equations are generated from $I_s + \lambda_0 I_0 + \lambda_V I_V$ with $\alpha = 1$. The conformal part I_s was called smoothness. Relative to smooth conditions from I_s, the level of adaptivity is adjusted with λ_V, and the grid structure is improved with increases in λ_0.

The metric forms in the integrals can be carried directly over to arbitrary surfaces as well as into higher dimensions. In the process with surfaces, an additional complication arises, since the chain rule must be employed to adjust to fixed local surface coordinates that enter by composition. In particular, the variational formulations can be done directly on the monitor surface.

SUMMARY

The need for grid generation arose from the discrete requirements for the numerical simulation in fluid mechanics with the geometric and topological complexity of physical regions and with the possibility of rapid solution variations at unpredictable locations. The general topological setting is examined to establish the overall formats for the desired grids without addressing the specific details required of the methods to meet the consequent variety of constraints.

The element of control over the grid is the most fundamental aspect needed to satisfy the constraints from topology, geometry, and solution variations. The most precise level of control is available from algebraic methods. This is established with the multisurface transformation in one direction and with the extension into multiple directions by means of Boolean operations. From a more relaxed level of precision, elliptic partial differential equation methods are developed from the Laplace system representing conformal conditions up to the Poisson system, where convexity controls are established and examined.

The general feedback cycle for adaptive movement is considered by using a monitor surface to consolidate the feedback data into a single object. With the natural objective of accurately representing this object, strategies for movement are considered. These cover methods from direct proportionality statements, Poisson systems, and variational formulations by taking the monitor surface either as a function over physical space or as a geometric entity. The basic control in the various methods is in the form of weight functions. With the placement in distinct spots in distinct formats, the consequent effects, although similar, are also distinct. How, when, and where control is applied give the separation between the methods and the important effects on the grid in the varying levels of response, directness, and precision.

ACKNOWLEDGMENT

This paper was written under US Air Force sponsorship with Grant No. AFOSR-82-0176A, monitored by John P. Thomas, Jr.

Literature Cited

Ablow, C. M., Schechter, S. 1978. Campylotropic coordinates. *J. Comput. Phys.* 27:351–62

Ablow, C. M. 1982. Equidistant mesh for gas dynamic calculations. See Thompson 1982a, pp. 859–64

Anderson, D. A., Rai, M. M. 1982. The use of solution adaptive grids in solving partial differential equations. See Thompson 1982a, pp. 317–38

Babuška, I., Chandra, J., Flaherty, J. E., eds. 1983a. *Adaptive Computational Methods for Partial Differential Equations*. Philadelphia: SIAM. 251 pp.

Babuška, I., Miller, A., Vogelius, M. 1983b. Adaptive methods and error estimation for elliptic problems of structural mechanics. See Babuška et al. 1983a, pp. 57–73

Barnhill, R. E., Riesenfeld, R. F., eds. 1974. *Computer Aided Geometric Design*. New York: Academic. 326 pp.

Brackbill, J. U., Saltzman, J. S. 1982. Adaptive zoning for singular problems in two dimensions. *J. Comput. Phys.* 46:342–68

Caughey, D. A. 1978. A systematic procedure for generating useful conformal mappings. *Int. J. Numer. Methods Eng.* 12:1651–57

Coons, S. A. 1964. Surfaces for computer-aided design of space forms. *Project MAC*, Design Div., Dept. Mech. Eng., Mass. Inst. Technol., Cambridge

Dwyer, H. A. 1983. *Grid adaption for problems with separation, cell Reynolds number, shock–boundary layer interaction, and accuracy.* Presented at AIAA Aerosp. Sci. Meet., 21st, Reno, Nev. *Pap. AIAA-83-0449*

Dwyer, H. A., Kee, R. J., Sanders, B. R. 1980. An adaptive grid method for problems in fluid mechanics and heat transfer. *AIAA J.* 18 : 205–12

Eiseman, P. R. 1978. A coordinate system for a viscous transonic cascade analysis. *J. Comput. Phys.* 26 : 307–38

Eiseman, P. R. 1979. A multi-surface method of coordinate generation. *J. Comput. Phys.* 33 : 118–50

Eiseman, P. R. 1982a. Coordinate generation with precise controls over mesh properties. *J. Comput. Phys.* 47 : 331–51

Eiseman, P. R. 1982b. High level continuity for coordinate generation with precise controls. *J. Comput. Phys.* 47 : 352–74

Eiseman, P. R. 1982c. Orthogonal grid generation. See Thompson 1982a, pp. 193–234

Eiseman, P. R. 1982d. Automatic algebraic coordinate generation. See Thompson 1982a, pp. 447–64

Eiseman, P. R. 1983a. Alternating direction adaptive grid generation. *Proc. AIAA Comput. Fluid Dyn. Conf., 6th, Danvers, Mass.,* pp. 339–48

Eiseman, P. R. 1983b. Adaptive grid generation by mean value relaxation. See Ghia & Ghia 1983, pp. 29–34

Erlebacher, G. 1984. *Solution adaptive triangular meshes with application to plasma equilibrium.* PhD thesis. Columbia Univ., New York, N.Y. 208 pp.

Erlebacher, G., Eiseman, P. R. 1984. *Adaptive triangular mesh generation.* Presented at AIAA Fluid Dyn., Plasmadyn., Lasers Conf., 17th, Snowmass, Colo. *Pap. AIAA-84-1607*

Fritts, M. J., Boris, J. P. 1979. The Lagrangian solution of transient problems in hydrodynamics using a triangular mesh. *J. Comput. Phys.* 31 : 173–215

Ghia, K. N., Ghia, U., eds. 1983. *Advances in Grid Generation,* Vol. 5. New York: ASME. 219 pp.

Ghia, K., Ghia, U., Shin, C. T. 1983. Adaptive grid generation for flows with local high gradient regions. See Ghia & Ghia 1983, pp. 35–48

Gnoffo, P. A. 1982. A vectorized, finite volume, adaptive-grid algorithm for Navier-Stokes. See Thompson 1982a, pp. 819–36

Gordon, W. J. 1971. Blending-function methods of bivariate and multivariate interpolation and approximation. *SIAM J. Numer. Anal.* 8 : 158–77

Gordon, W. J., Hall, C. A. 1973. Construction of curvilinear coordinate systems and applications to mesh generation. *Int. J. Numer. Methods Eng.* 7 : 461–77

Gordon, W. J., Thiel, L. C. 1982. Transfinite mappings and their application to grid generation. See Thompson 1982a, pp. 171–92

Ives, D. C. 1982. Conformal grid generation. See Thompson 1982a, pp. 107–36

Klopfer, G. H., McRae, D. S. 1981. The nonlinear modified equation approach to analyzing finite difference schemes. *Proc. AIAA Comput. Fluid Dyn. Conf., 5th, Palo Alto, Calif.,* pp. 317–33

Laugwitz, D. 1965. *Differential and Riemannian Geometry.* New York: Academic. 238 pp.

Mastin, C. W., Thompson, J. F. 1978. Elliptic systems and numerical transformations. *J. Math. Anal. Appl.* 62 : 52–62

Mastin, C. W., Thompson, J. F. 1983. *Adaptive grids generated by elliptic systems.* Presented at AIAA Aerosp. Sci. Meet., 21st, Reno, Nev. *Pap. AIAA-83-0451*

Middlecoff, J. F., Thomas, P. D. 1979. Direct control of the grid point distribution in meshes generated by elliptic equations. *Proc. AIAA Comput. Fluid Dyn. Conf., 4th, Williamsburg, Va.,* pp. 175–79

Miller, K., Miller, R. 1981. Moving finite elements I. *SIAM J. Numer. Anal.* 18 : 1019–32

Moretti, G. 1980. Grid generation using classical techniques. See Smith 1980, pp. 1–36

Piva, R., DiCarlo, A., Favini, B., Guj, G. 1982. Adaptive curvilinear grids for large Reynolds number viscous flows. In *Lecture Notes in Physics,* 170 : 414–19. Berlin/Heidelberg/New York: Springer-Verlag

Prenter, P. M. 1975. *Splines and Variational Methods.* New York: Wiley Interscience. 323 pp.

Rheinboldt, W. C. 1983. Feedback systems and adaptivity for numerical computations. See Babuška et al. 1983a, pp. 3–19

Smith, R. E., ed. 1980. *Numerical Grid Generation Techniques, NASA CP 2166.* 574 pp.

Smith, R. E. 1982. Algebraic grid generation. See Thompson 1982a, pp. 137–70

Stadius, G. 1977. Construction of orthogonal curvilinear meshes by solving initial value problems. *Numer. Math.* 28 : 25–48

Steger, J. L., Sorenson, R. 1979. Automatic mesh-point clustering near a boundary in grid generation with elliptic partial differential equations. *J. Comput. Phys.* 33 : 405–16

Steger, J. L., Sorenson, R. L. 1980. Use of hyperbolic partial differential equations to generate body fitted coordinates. See Smith 1980, pp. 463–78

Thomas, P. D. 1982. Numerical generation of composite three-dimensional grids by quasilinear elliptic systems. See Thompson 1982a, pp. 667–86

Thompson, J. F., ed. 1982a. *Numerical Grid Generation.* New York: North Holland. 909 pp.

Thompson, J. F. 1982b. Elliptic grid generation. See Thompson 1982a, pp. 79–106

Thompson, J. F., Thames, F. C., Mastin, C. W. 1974. Automatic numerical generation of body-fitted curvilinear coordinate system for field containing any number of arbitrary two-dimensional bodies. *J. Comput. Phys.* 15:299–319

Thompson, J. F., Thames, F. C., Mastin, C. W. 1977. TOMCAT—A code for numerical generation of boundary-fitted curvilinear coordinate systems on fields containing any number of arbitrary two-dimensional bodies. *J. Comput. Phys.* 24:274–302

Warner, F. W. 1971. *Foundations of Differentiable Manifolds and Lie Groups.* Glenview, Ill./London: Scott, Foresman. 270 pp.

Warsi, Z. U. A. 1982. Basic differential models for coordinate generation. See Thompson 1982a, pp. 41–78

Warsi, Z. U. A., Thompson, J. F. 1976. Machine solutions of partial differential equations in the numerically generated coordinate systems. *Rep. MSSU-EIRS-ASE-77-1,* Miss. State Univ., Miss. State

White, A. B. Jr. 1979. On selection of equidistributing meshes for two-point boundary-value problems. *SIAM J. Numer. Anal.* 16:472–502

White, A. B. Jr. 1982. On the numerical solution of initial boundary-value problems in one space dimension. *SIAM J. Numer. Anal.* 19:683–97

Winslow, A. M. 1967. Numerical solution of the quasilinear Poisson equation in a nonuniform triangle mesh. *J. Comput. Phys.* 2:149–72

Yanenko, N. N., Kovenya, V. M., Lisejkin, V. D., Fomin, V. M., Vorozhtsov, E. V. 1978. On some methods for the numerical simulation of flows with complex structure. In *Lecture Notes in Physics,* 90:565–78. Berlin/Heidelberg/New York: Springer-Verlag

Ann. Rev. Fluid Mech. 1985. 17 : 523–59

COMPUTING THREE-DIMENSIONAL INCOMPRESSIBLE FLOWS WITH VORTEX ELEMENTS[1]

A. Leonard

Computational Fluid Dynamics Branch, NASA Ames Research Center, Moffett Field, California 94035[2]

1. INTRODUCTION

This article is concerned with the numerical simulation of three-dimensional, unsteady vortical flows of an incompressible fluid at high Reynolds number. In many cases of interest, the fluid containing vorticity occupies only a small fraction of the total fluid volume, and for incompressible flows it is sufficient to follow only the evolution of the vorticity field. The velocity field can then be determined from the vorticity field and boundary conditions, when desired. In addition, from the theorems of Helmholtz and Kelvin, we know that for a uniform-density inviscid fluid, tubes of vorticity retain their identity and move as material entities.

These facts, established over a century ago, plus some recent theoretical results form the basis for the numerical schemes discussed in this review. Lagrangian vortex elements are convected at or near the local fluid velocity, and the vorticity vector associated with each element is strained by the local velocity gradient. Roughly speaking, these elements are sections of a vortex tube or filament and are required only where the vorticity is nonzero. Thus, rather complex nonlinear flows may be represented by a relatively small number of computational elements. As an extreme example, we cite the

[1] The US Government has the right to retain a nonexclusive royalty-free license in and to any copyright covering this paper.

[2] This article was written while the author was at the Graduate Aeronautical Laboratories, California Institute of Technology, Pasadena, California 91125.

523

two-dimensional flow represented by the dynamics of four point vortices, governed by eight nonlinear ordinary differential equations. This flow is known to be chaotic or turbulent-like (Aref 1983). The situation appears to be similar in three dimensions, except that a larger number of computational elements is required to obtain an accurate simulation of the inviscid equations of motion. (The two-dimensional point-vortex method has the luxury of satisfying the inviscid equations exactly regardless of the number of vortices used.) Experience has shown that some fairly complex three-dimensional flows may be simulated with a hundred or so vortex elements requiring only modest computer power. In this particular class of flows, the vorticity is assumed to be in the form of isolated vortex tubes with a uniform core structure within each tube. Theoretical studies during the last 10 to 15 years have pointed the way to accurate dynamical equations for the vortex elements in this situation. The simulation of flows of this type, using one computational filament per physical vortex tube, is described in Section 2.

For problems involving a relatively smooth continuum of vorticity, or variable core structure within a single tube of vorticity, a dense collection of vortex elements or filaments should be employed. Simulations of this type tend to push available computer resources to their limits. Schemes for these applications have been developed within the last five years and are considered in Section 3. In Section 4 we present a summary and some conclusions.

Earlier reviews by Saffman & Baker (1979) and Leonard (1980b) discuss vortex interactions in two and three dimensions and their computation. For a general overview of a wide range of vortex flows—natural and man-made—the book by Lugt (1983) is recommended. Most unsteady three-dimensional vortical flows are unstable and necessarily become turbulent or transitional. Highly developed Eulerian finite-difference and spectral methods are now available to tackle these flows. See Rogallo & Moin (1984) for a recent review of this rapidly developing field. The methods described in this paper are similar to nonlinear panel methods used to compute low-speed, three-dimensional steady wakes of aerodynamic bodies. Panel methods, however, are not discussed further in this paper. The reader should consult Belotserkovskii (1977) or Hoeijmakers (1983) for further information.

Helmholtz and Kelvin

We state here the relevant results from the works of Helmholtz and Kelvin of roughly a hundred years ago. The reader may wish to consult Truesdell (1954), Serrin (1959), or Batchelor (1967) for further details and original references.

Let $\mathbf{u}(\mathbf{x}, t)$ be the velocity field and $\boldsymbol{\omega}(\mathbf{x}, t) = \nabla \times \mathbf{u}$ be the vorticity field. The vorticity transport equation is the curl of the momentum equation and, for constant-density flow, is given by

$$\frac{D\boldsymbol{\omega}}{Dt} = \boldsymbol{\omega} \cdot \nabla\mathbf{u} + \nu\nabla^2\boldsymbol{\omega}, \tag{1}$$

where $D/Dt = \partial/\partial t + \mathbf{u} \cdot \nabla$ and ν is the kinematic viscosity. It is a simple matter to show (Batchelor 1967) that an infinitesimal line element $\delta\mathbf{l}$ satisfies the evolution equation

$$\frac{D\delta\mathbf{l}}{Dt} = \delta\mathbf{l} \cdot \nabla\mathbf{u}. \tag{2}$$

Comparing (1) and (2) we see that for inviscid motion, vortex lines move as material lines (Helmholtz).

The collection of all vortex lines that pierce a given surface patch S defines a three-dimensional vortex tube. The tube may be generated by moving the surface patch through space along the vortex lines. We define the strength or circulation Γ of this tube by a surface integral over the patch as follows:

$$\Gamma = \int_S \boldsymbol{\omega} \cdot d\mathbf{S}. \tag{3}$$

Because the vorticity field is solenoidal ($\nabla \cdot \boldsymbol{\omega} = 0$), the circulation is the same for all oriented surface patches that define a given vortex tube (Helmholtz). By Stokes' theorem, the circulation may also be expressed as a line integral over the curve λ bounding the patch S,

$$\Gamma = \int_\lambda \mathbf{u} \cdot d\lambda. \tag{4}$$

If we assume that λ moves as a material curve, then in general Γ is a function of time given by the integral of (1),

$$\frac{d\Gamma}{dt} = -\nu \int_\lambda \nabla \times \boldsymbol{\omega} \cdot d\lambda, \tag{5}$$

so that for an inviscid flow we have (Kelvin)

$$\frac{d\Gamma}{dt} = 0. \tag{6}$$

Thus a tube of vorticity retains its identity as it moves with the fluid.

The fluid velocity can be determined as follows. From the relations

$$\nabla \cdot \mathbf{u} = 0 \quad \text{and} \quad \boldsymbol{\omega} = \nabla \times \mathbf{u}, \tag{7}$$

we find that

$$\nabla^2 \mathbf{u} = -\nabla \times \boldsymbol{\omega}, \tag{8}$$

so that $\mathbf{u}$ may be expressed as

$$\mathbf{u}(\mathbf{x}) = -\frac{1}{4\pi} \int \frac{(\mathbf{x} - \mathbf{x}') \times \boldsymbol{\omega}(\mathbf{x}')d\mathbf{x}'}{|\mathbf{x} - \mathbf{x}'|^3} + \nabla\phi, \tag{9}$$

where ϕ is the potential associated with the homogeneous solution of (8) required to satisfy boundary conditions (for example, a free-stream velocity or flow tangency at a solid boundary). Equation (9) is known as the Biot-Savart law for the velocity field.

2. MOTION OF ISOLATED VORTEX TUBES

Thin-Filament Approximation

In many applications the vorticity is confined to a small number of isolated tubes of vorticity or vortex filaments, e.g. aircraft trailing vortices and vortex rings. We consider first the problem of determining equations of motion for these filaments. In reality the vorticity in each vortex is distributed over a finite core with a characteristic radius σ but, in analogy with the point-vortex representation of two-dimensional flows, it is tempting to idealize these flows as systems of space curves, each with zero cross-sectional area and constant circulation. For a single space curve C, the vorticity field then has the representation

$$\boldsymbol{\omega}(\mathbf{x}) = \Gamma \int_C \delta[\mathbf{x} - \mathbf{r}(s')] \frac{\partial \mathbf{r}}{\partial s'} \, ds', \tag{10}$$

where $\mathbf{r}(s)$ is the space curve of the filament parameterized by arc length s, and Γ is the filament circulation. Using (9) we find that the velocity field induced by this filament in an unbounded domain, with no interior boundaries, is then given by

$$\mathbf{u}(\mathbf{x}) = -\frac{\Gamma}{4\pi} \int_C \frac{[\mathbf{x} - \mathbf{r}(s')] \times \dfrac{\partial \mathbf{r}}{\partial s'} \, ds'}{|\mathbf{x} - \mathbf{r}|^3}. \tag{11}$$

As long as the field point $\mathbf{x}$ does not approach within a distance σ of any part of the curve, Equation (11) is a good approximation for the velocity field of the original finite-core vortex. As $\mathbf{x}$ approaches $\mathbf{r}(s)$ on the smooth space curve of the idealized vortex, $\mathbf{u}$ diverges as $1/|\mathbf{x} - \mathbf{r}(s)|$. A similar divergence is present in the two-dimensional point-vortex method, but the point vortex in two dimensions has a bounded self-induced velocity (equal to zero). On the other hand, the three-dimensional, zero cross-section line vortex has an infinite self-induced velocity anywhere its curvature is nonzero. If we attempt to evaluate $\mathbf{u}(\mathbf{r}(s))$ we obtain, for s' close to s,

$$\mathbf{u}(\mathbf{r}(s)) = \frac{\Gamma}{4\pi}\left[\frac{\partial \mathbf{r}}{\partial s} \times \frac{\partial^2 \mathbf{r}}{\partial s^2}\int \frac{ds'}{|s-s'|} + O(1)\right], \tag{12}$$

which results in a logarithmic divergence (Batchelor 1967).

Clearly we must take into account the structure of the vortex core. To start, we assume that the vorticity within the core is given by $\omega_0(r)$ in a locally toroidal coordinate system (r, θ, s), and we then study simple flows amenable to analysis—the motion of a ring vortex and linearized disturbances on a rectilinear vortex. These considerations should provide clues for the general case. The analysis, however, is subtle. For example, a straightforward application of the Biot-Savart law applied to the center of the filament $(r = 0)$ yields an incorrect result for the speed of a vortex ring. One problem is that the assumption $\omega(r, \theta) = \omega_0(r)$ does not satisfy the inviscid equations to sufficient accuracy. In toroidal coordinates, the vorticity transport equation for axisymmetric inviscid flow with $u_s = 0$ is

$$\frac{D}{Dt}\left(\frac{\omega(r, \theta)}{h_s}\right) = 0, \tag{13}$$

where $h_s = 1 + \kappa r \sin\theta$ is the metric coefficient for the s direction and κ is the local curvature. As a result, the equations of motion demand a slight adjustment in the assumed vorticity distribution. For example, a vortex ring of radius R with the assumed distribution $\omega_0(r) = \Gamma/\pi\sigma^2$ for $r < \sigma$, and $\omega_0(r) = 0$ for $r > \sigma$ (solid-body rotation), actually has the distribution $\omega(r, \theta) = \omega_0(r)(1 + \kappa r \sin\theta)$ within a boundary that is a slightly deformed circle (Norbury 1973).

Careful analyses (Fraenkel 1970, Saffman 1970) show that the speed of a vortex ring of radius R with $\sigma/R \ll 1$ is given by

$$U_R = \frac{\Gamma}{4\pi R}\left[\log\left(\frac{8R}{\sigma}\right) - \frac{1}{2} + \int_0^\sigma \chi^2(r)\frac{dr}{r} + \int_\sigma^\infty (\chi^2(r) - 1)\frac{dr}{r}\right], \tag{14}$$

where

$$\chi(r) = \frac{1}{\Gamma}\int_0^r 2\pi r' \omega_0(r')\, dr'$$

is the fraction of circulation within radius r. The last integral in the brackets of (14) is the required modification if the characteristic radius σ is smaller than the radius beyond which the vorticity is zero, e.g. for a Gaussian core, $\omega_0(r) = (\Gamma/\pi\sigma^2)\exp(-r^2/\sigma^2)$. For solid-body rotation we obtain Kelvin's result (Lamb 1932)

$$U_R = \frac{\Gamma}{4\pi R}\left[\log\left(\frac{8R}{\sigma}\right) - \frac{1}{4}\right]. \tag{15}$$

Kelvin (1880) also found the dispersion relation for long bending waves on a rectilinear vortex with a constant-vorticity core. These are helical disturbances proportional to $\exp(i\Omega t + im\theta + ikx)$, where m is the azimuthal mode in cylindrical coordinates and k is the axial wave number. His result for $m = \pm 1$, generalized to an arbitrary core distribution, is

$$\Omega = \pm \frac{\Gamma k^2}{4\pi}\left[\log\left(\frac{2}{k\sigma}\right) - \gamma + \int_0^\sigma \chi^2(r)\frac{dr}{r} + \int_\sigma^\infty (\chi^2(r)-1)\frac{dr}{r}\right], \tag{16}$$

where γ is Euler's constant.

Probably the first attempt to modify the Biot-Savart law (11) to account for finite core effects for general application was by Rosenhead (1930), who arrived at the expression

$$\mathbf{u}(\mathbf{x}) = -\frac{\Gamma}{4\pi}\int_C \frac{[\mathbf{x}-\mathbf{r}(s')]\times\dfrac{\partial\mathbf{r}}{\partial s'}\,ds'}{(|\mathbf{x}-\mathbf{r}|^2+\mu^2)^{3/2}}, \tag{17}$$

where $\mu = O(\sigma)$ but is otherwise unspecified.

Another procedure, suggested by Hama (1962, 1963), is to cut off the line integral of (12) for $|s'-s| < \delta\sigma$. Crow (1970), in his stability analysis of aircraft trailing vortices, showed that the proper choice of δ [$\log(2\delta) = 1/4$] would produce the asymptotically correct speed of a vortex ring with a core of constant vorticity [Equation (15)], as well as the correct dispersion relation for long waves on a rectilinear vortex [(16) for the case $\omega_0(r) = $ constant].

Moore (1972), preferring the computational aspects of Rosenhead's modification (17), chose $\mu^2 = \alpha\sigma^2$ for his study of the nonlinear evolution of the Crow instability. Then the speed of a ring vortex, for example, is computed to be

$$U_R = \frac{\Gamma}{4\pi R}\left[\log\left(\frac{8R}{\sigma\sqrt{\alpha}}\right) - 1\right]. \tag{18}$$

The choice, given by Moore for zero axial flow,

$$\log\sqrt{\alpha} = -\frac{1}{2} - \int_0^\sigma \chi^2(r)\frac{dr}{r} - \int_\sigma^\infty (\chi^2(r)-1)\frac{dr}{r}, \tag{19}$$

forces (18) to agree with the general result (14) for arbitrary vorticity distributions to $O[(\Gamma\sigma/R^2)\log(\sigma/R)]$. By this time, thanks to the work of

Widnall et al. (1971) and Moore & Saffman (1972), it was known that this matching procedure produces asymptotically correct equations of motion for thin vortex filaments following an arbitrary space curve, as long as $\kappa\sigma \ll 1$ and the structure of the vortex core remains relatively undisturbed. Similarly, the cutoff scheme may be used for arbitrary vorticity distributions with the choice

$$\log (2\delta) = \frac{1}{2} - \int_0^\sigma \chi^2(r) \frac{dr}{r} - \int_\sigma^\infty (\chi^2(r)-1) \frac{dr}{r}.$$

Clearly a wide class of smoothing schemes, each having an adjustable free parameter, may be used to simulate the dynamics of vortex tubes under the restrictions mentioned above. For example, one could use the following generalized Biot-Savart law, which includes (17) and, to within the thin-filament approximation, the cutoff scheme:

$$\mathbf{u}(\mathbf{x}) = -\frac{\Gamma}{4\pi} \int_C \frac{[\mathbf{x} - \mathbf{r}(s')] \times \dfrac{\partial \mathbf{r}}{\partial s'} g(|\mathbf{x}-\mathbf{r}|/\sigma)\, ds'}{|\mathbf{x}-\mathbf{r}|^3}, \tag{20}$$

where g is arbitrary except for the required limits $g(y) \to 0$ as $y \to 0$, $g(y) \to 1$ as $y \to \infty$, and the integral constraint

$$\int_0^\infty \log (y)g'(y)\, dy = -\log 2 + \frac{1}{2} - \int_0^\sigma \chi^2(r) \frac{dr}{r} - \int_\sigma^\infty (\chi^2(r)-1) \frac{dr}{r}. \tag{21}$$

An additional term is required to accommodate possible axial flow within the filament (Widnall et al. 1971, Moore & Saffman 1972).

Applications of the Thin-Filament Approximation

With the Rosenhead-Moore approximation, one obtains evolution equations for the one-dimensional continua of space curves $\mathbf{r}_i(\xi)$ $(i = 1, 2, \ldots, N)$ for N filaments given by

$$\frac{\partial \mathbf{r}_i}{\partial t} = -\sum_j \frac{\Gamma_j}{4\pi} \int \frac{[\mathbf{r}_i(\xi, t) - \mathbf{r}_j(\xi', t)] \times \dfrac{\partial \mathbf{r}_j}{\partial \xi'} d\xi'}{(|\mathbf{r}_i - \mathbf{r}_j|^2 + \alpha\sigma_j^2)^{3/2}}, \tag{22}$$

with α given by (19). The core parameters σ_i may depend on time. In addition, it may be argued that σ_i should vary along the filament because of differing rates of vortex stretching along the filament. However, such variations in σ_i quickly produce helical vortex lines and hence axial flows that tend to eliminate these variations (Moore & Saffman 1972). These considerations suggest the model equation

$$\frac{d}{dt}(\sigma_i^2 \mathscr{L}_i) = 0, \tag{23}$$

which conserves the total volume of vorticity, with $\mathscr{L}_i(t)$ the instantaneous length of filament i. One may also wish to symmetrize the right-hand side of (22) with respect to σ_i and σ_j. These points are discussed further in Section 3.

A simple way to discretize these curves is to choose ξ to be a material coordinate and to employ a sequence of material markers. In this case, the arc-length parameterization could be computed when needed from the locations of the material coordinates. The space curves can be reconstructed to any order using splines or any other favorite interpolation scheme. For example, linear splines (piecewise linear segments) lead to analytic expressions for the ordinary differential equations in time for the material coordinates, which may be integrated forward with an explicit scheme such as Runge-Kutta or Adams-Bashforth. Cubic splines can be useful for improving the accuracy of the spatial integration and for remeshing the space curves, as dictated by the development of regions of high local curvature or highly stretched portions of the curve.

As mentioned earlier, Moore (1972) used the approximation (22) to study the nonlinear development of the most amplified unstable long wave in the trailing vortex problem. In this particular case only small deviations from linear behavior were observed—a slight increase in amplification rate and a small increase in angle between the plane of the disturbance and the horizontal, as the two disturbed vortices approach each other.

The interaction of the wing-tip vortex and flap vortices of large aircraft was studied by Leonard (1980b). Here the vortex method simulates a three-dimensional, unsteady, space-developing flow. Upstream boundary data for this flow are the circulations, core radii, and locations in the cross-flow plane of the three vortices formed in the near-wake of each wing. It had been observed experimentally that with a certain flap configuration the tip vortex and the flap outboard vortex on a model Boeing 747 would strongly interact and merge only ~ 30 spans downstream, resulting in a diffuse wake. Unfortunately, laboratory experiments as well as numerical experiments showed that this merging process is not robust. Small disturbances produce large deviations in the cross-plane trajectories of the vortices, causing a significant delay in the merger. In particular, it was observed experimentally that lowering the landing gear prevented the merger, and computationally it was found that small changes in the upstream boundary data provided by wind-tunnel measurements would upset the merging process.

Ring vortices can interact in a myriad of ways. Figure 1 illustrates the interaction of four ring vortices, initially coplanar, as computed by Y.

Nakamura (private communication). As in the case of the two-ring interaction, the rings are attracted to each other as mutual inductance and curvature effects deform and reorient the rings. The reader is encouraged to invoke the Biot-Savart law qualitatively (with appropriate arm waving and finger twisting) to obtain the results as shown. The results agree with the experiments of Oshima & Asaka (1977), which also show that the inner portions of the four vortices merge to form a single vortex while the outer portions form another vortex of opposite-sign vorticity. This process, which is not well understood, involves the relinking of vortex lines via viscous effects and probably large deformations of the vortex cores. Knowing that this change in topology does occur in nature, one can attempt to model this process in the context of thin-filament dynamics by allowing relinking to occur (Leonard 1975), but at the risk of neglecting important dynamical effects. See also the discussion below concerning the relinking of quantum vortex lines in liquid ^{4}He.

Finally, it has been observed that round jets, with only gentle perturbations applied, show interesting, erratic behavior of the jet fluid (Lee & Reynolds 1983). By pulsing the jet flow axially at a frequency $f \approx 0.5 U_{av}/D$, one obtains a steady stream of nearly circular vortex rings in the near-field.

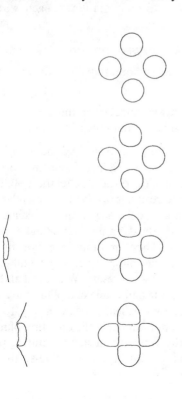

Figure 1   Vortex simulation of four vortices, initially coplanar, with time increasing downward. (Left) Side view of vortices moving left to right; (right) front view (Y. Nakamura, private communication).

Farther downstream the rings collide, merge, and become three dimensional, forming a classical (unperturbed) turbulent jet. But if, in addition, the jet nozzle is subjected to a small, steady orbital excitation around the azimuth, the included angle of the jet increases dramatically from 25° up to 80°. If the rotation period is a small integer ($= n_R$) multiple of the period of the vortex ring formation, the vorticity splits into n_R distinct trains of vortex rings equally spaced around the azimuth. The dynamics of this process for $n_R = 2$, the bifurcating jet, are being studied computationally by Parekh et al. (1983) using the thin-filament approximation. The results indicate that the axial spacing between rings, which is proportional to $1/f$, is a crucial parameter. For the spacing that yields maximum spreading, the successive rings strongly interact at a certain downstream location, launching one another along widely separated paths. If the spacing is too large, the interaction is too weak to produce wide angles. If the spacing is too small, successive rings collide and merge. Experimental and computational results for a case of large spread angle are illustrated in Figure 2. At the present time the computational results require a significantly stronger azimuthal perturbation than that used in the experiment. The discrepancy could be due to errors in the thin-filament approximation or to neglect of the additional potential-flow term required to maintain flow tangency at the nozzle wall.

Limitations of the Thin-Filament Approximation

As discussed above, the equations of motion for a thin filament are derived using the following assumptions:

1. The structure of the core remains nearly constant in time.
2. Disturbance wavelengths are much larger than the core radius.

The failure of the approximation is nicely illustrated by the analysis of linearized perturbations to a rectilinear vortex having a constant-vorticity core (Kelvin 1880). As shown in Figure 3, the dispersion relation obtained from (22) (dashed line) yields accurate results only for long wavelengths ($k\sigma \to 0$) for the unperturbed radial mode. A filament with a constant core structure is simply not able to represent the dynamics of the higher radial modes or their dispersion relations. These motions are not just of theoretical interest. A slowly rotating higher radial mode may be subject to unstable growth if an external straining flow is present (Widnall et al. 1974, Moore & Saffman 1975). For example, the amplitude of a plane wave with zero rotation rate (superposition of two helical modes, $m = \pm 1$, with $\Omega(k\sigma) = 0$) and with vector displacement oriented in the direction of maximum strain will increase exponentially. Similarly, a slowly rotating plane wave will suffer an increase in amplitude during the positive-strain part

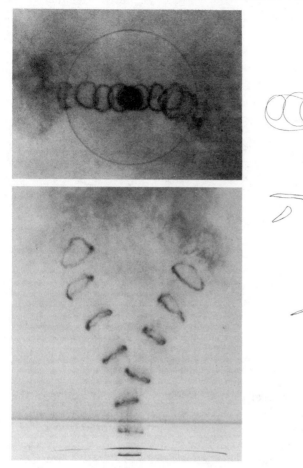

JET EXIT   JET EXIT

Figure 2 Bifurcating round jet, excited axially and azimuthally. (*Left*) Experiment of Lee & Reynolds (1983); (*right*) vortex simulation of Parekh et al. (1983); (*top*) front views; (*bottom*) side views.

of its rotation cycle that may not "heal" during the negative-strain part of the cycle, leading to exponential growth. This mechanism was shown to be responsible for the instability of a rectilinear vortex in a weak external strain field (Moore & Saffman 1975, Tsai & Widnall 1976) and in a finite-amplitude strain field (Robinson & Saffman 1984), and for the instability of an inviscid vortex ring, where the perturbation grows in the presence of the ring's own strain field (Widnall & Tsai 1977).

RADIAL MODE

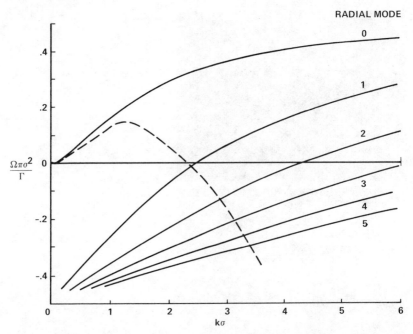

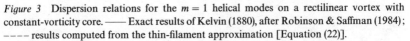

Figure 3   Dispersion relations for the $m = 1$ helical modes on a rectilinear vortex with constant-vorticity core. ——— Exact results of Kelvin (1880), after Robinson & Saffman (1984); ———— results computed from the thin-filament approximation [Equation (22)].

Another phenomenon that requires a higher-order description of the vortex core is vortex breakdown (Leibovich 1978)—a rapid change in the structure of the vortex core as one moves axially. Vortex breakdown typically involves a significant axial flow (and hence helical vortex lines). Also, it seems likely that rapid changes in curvature along a vortex tube without axial flow could also lead to breakdown.

Thus, when short-wavelength disturbances arise within the thin-filament approximation, the dynamics of the space curves are probably not well approximated, and it is likely that the assumed structure of the core is no longer correct. However, a simple way to modify the thin-filament approximation and possibly to extend its range of validity is to make a more judicious choice of the smoothing function defined in (20). For example, given the dispersion curve of Figure 3, one could determine the smoothing function g that reproduces the correct behavior for all $k\sigma$. Of course, at this point all that has been improved with certainty is the linearized motion of a rectilinear vortex with an unperturbed core structure. Further analysis is required to determine if this alternative has more general utility.

Local Induction Approximation

Early attempts to deal with the nonlinear one-dimensional continuum of a thin vortex filament encountered numerical difficulties, inspiring a further simplification to the thin-filament approximation (Hama 1962, 1963, Arms & Hama 1965). Not only did this simplification ease the numerical difficulties at the time, but it eventually led to an elegant asymptotic mathematical theory of vortex dynamics. Here one assumes that the core radius is so small that the only significant contribution to the motion of the filament is that due to local curvature. Using, for example, (12) with a cutoff we find that, to leading order,

$$\frac{\partial \mathbf{r}}{\partial t} = \log\,(L/\sigma)\,\frac{\Gamma}{4\pi}\left(\frac{\partial \mathbf{r}}{\partial s} \times \frac{\partial^2 \mathbf{r}}{\partial s^2}\right), \tag{24}$$

where L is a large-scale cutoff, as yet unspecified. Hama (1962) used this local induction approximation (LIA) to model the three-dimensional dynamics of perturbed two-dimensional Tollmien-Schlichting waves. A few years later, Betchov (1965) reformulated the dynamics given by (24) in terms of the dynamics of space curves described by a Frenet-Serret frame (O'Neill 1966)—an orthonormal frame composed of the local tangent vector $\mathbf{t}(=\partial \mathbf{r}/\partial s)$, the principal normal $\mathbf{n}$, and the binormal $\mathbf{b}(=\mathbf{t} \times \mathbf{n})$, as illustrated below.

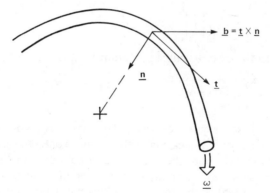

The coordinate frame varies along the curve according to the Frenet-Serret formulas

$$\mathbf{t}' = \kappa\mathbf{n}, \qquad \mathbf{n}' = -\kappa\mathbf{t}+\tau\mathbf{b}, \qquad \mathbf{b}' = -\tau\mathbf{n}, \tag{25}$$

where $\kappa(s)$ and $\tau(s)$ are the curvature and torsion, respectively, and where primes denote differentiation with respect to arc length s. The scalar functions κ and τ provide an intrinsic representation of the space curve, i.e. one without regard to its location or orientation in space.

Assuming that variations of L in (24) along the curve are negligible, we may rescale time and write this equation in the form

$$\dot{\mathbf{r}} = \kappa\mathbf{b}, \tag{26}$$

where the overdot is the derivative with respect to the new time variable. We can then obtain evolution equations for the intrinsic quantities κ and τ by repeated space differentiation of (26) and time differentiation of (25). The result, in slightly different form from that of Betchov (1965), is

$$\kappa = 2\beta \text{ sech } [\beta(s-ct)],$$

$$\dot{\tau} = \left(\frac{\kappa^2}{2} + \frac{\kappa''}{\kappa} - \tau^2\right)'. \tag{27}$$

As noted by Hasimoto (1972), these equations have a two-parameter family of solitary-wave solutions given by

$$\kappa = 2\beta \text{ sech } [\beta(s-ct)],$$

$$\tau = \alpha\beta = \text{constant}. \tag{28}$$

The wave speed c is 2τ. The parameter α fixes the shape of the curve, while β is a scaling parameter. Furthermore, Hasimoto showed that in the general case the complex function

$$\psi(s,t) = \kappa(s,t) \exp\left[i \int_0^s \tau(s',t)\,ds'\right] \tag{29}$$

evolves according to the cubic Schrödinger equation

$$\frac{1}{i}\frac{\partial\psi}{\partial t} = \frac{\partial^2\psi}{\partial s^2} + \frac{1}{2}|\psi|^2\psi, \tag{30}$$

and thus the solitary waves are, in fact, solitons (Whitham 1974). This means, among other things, that if two or more of these waves collide, they emerge unscathed from the collision and continue on their merry way. These processes are illustrated nicely by the numerical experiments of Aref & Flinchem (1985). An example of a collision is depicted in Figure 4. These authors propose that the small-scale three-dimensional waves observed experimentally in mixing layers may result from the spanwise spreading of localized disturbances on concentrated vortex tubes subjected to a mean shear, i.e. LIA [Equation (25)] plus a steady shear-flow component.

The local induction approximation (LIA) has also been useful in treating the dynamics of dense tangles of quantized vortex filaments in superfluid ^{4}He (Schwarz 1982). Using the measured speed-energy relation of quantized vortex rings and assuming the classical hydrodynamics of thin vortex

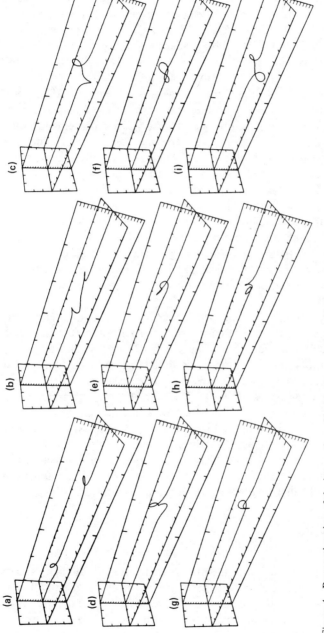

Figure 4 Perspective view of the interaction of two solitons on a rectilinear vortex as a high-torsion (fast) soliton overtakes a low-torsion (slow) soliton. Time increases from (*a*) to (*i*). Interaction computed under the local induction approximation by Aref & Flinchem (1985).

rings, i.e. $E_R = 2\pi R^2 \Gamma U_R - 3R\Gamma^2/4$ and Equation (14), one finds that the effective core radius is about 1 Å, or about one fourth the mean atomic spacing. Consequently, Schwarz has determined that LIA accounts for about 90% of the total induced velocity contribution. An additional velocity component, depending on temperature, arises from a local frictional force due to the relative velocity between the normal and superfluid components. This component tends to smooth the twists in the vortex filaments. On the other hand, as two filaments collide, severe local distortions take place if one takes into account the full Biot-Savart law. (LIA of course ignores the presence of other vortex filaments.) Rather than follow the detailed contortions of such collisions, especially since classical theory is suspect at this point, Schwarz assumes that the vortex tubes reconnect, changing the topology in such a way as to preserve the direction of vorticity. These collision events then become point sources of curvature and torsion singularities along each of the newly formed space curves, which in turn become smoothed by the frictional force. Schwarz finds that a statistical equilibrium is established in which there is a balance between these two competing phenomena. The statistical steady state that is achieved depends on temperature. The resulting mutual-friction-force density as a function of temperature is in very good agreement with experiment.

Recently, Hopfinger et al. (1982) investigated twisting distortions that propagate along concentrated tubes of vorticity in a rotating tank. The waveforms for these distortions appeared to resemble Hasimoto solitons, although the core sizes were somewhat large for LIA to be valid and collisions between solitary waves usually disrupted the vortex cores, producing vortex breakdown. Indeed, further analysis and experiments under more controlled conditions by Maxworthy et al. (1985) have shown that if the approximation (24) is used to explain the experimental results, then the large-scale cutoff length L must be a strong function of wave number or curvature. That this strong dependence is correct, at least for low-amplitude long waves for a rectilinear vortex, is demonstrated by computing the dispersion relation for this case using (24). We find that LIA gives

$$\Omega = \pm \frac{\Gamma k^2}{4\pi} \log (L/\sigma). \tag{31}$$

Comparing this result with the dispersion relation of full linear theory and $k\sigma \to 0$ [i.e. Equation (16) with an $O(k^2\sigma^2)$ correction for axial flow within the core], we see that agreement is achieved if $L \sim 1/k$ and if the constant of proportionality depends on the structure of the vortex core. Indeed, Maxworthy et al. (1985) found such a relation between L and k for traveling helical waves, standing sinusoidal waves, and solitary kink waves, in agreement with the dispersion relation (16) of linear theory. Therefore, we conclude that for these experiments the assumption of low amplitude is a better approximation than LIA with L/σ = constant, and that, in general, the use of LIA as a predictive tool must be made with caution. In a related effort, Leibovich & Ma (1983) have studied weakly nonlinear, nonaxisymmetric wave propagation on vortices and have shown how the results may be used to calibrate the Hasimoto soliton, i.e. to determine the correct cutoff length L for a given soliton. However, the results appear to be dependent upon the particular primary wave chosen for the analysis. Further work is under way to investigate nonlinear wave propagation assuming other primary waves and to determine the relevance of a particular choice to experiment (S. Leibovich, private communication).

In the above discussions concerning concentrated tubes of vorticity, it has been useful to use a Frenet-Serret frame on the space curve of the tube. With some restrictions, a similar construction of coordinates can be made for each vortex line in a continuum field of vorticity. Thus κ, τ, $\mathbf{t}$, $\mathbf{n}$, and $\mathbf{b}$ are now functions of three independent variables. By defining six additional scalar quantities, in addition to κ and τ one can write the gradients of $\mathbf{t}$, $\mathbf{n}$, and $\mathbf{b}$ in terms of $\mathbf{t}$, $\mathbf{n}$, and $\mathbf{b}$ and the eight scalar fields (Marris & Passman 1969).

The intrinsic dynamical equations for the eight scalar fields for general inviscid fluid motion are of course not local in space, but they might be useful in certain situations. P. Schatzle (private communication) is currently investigating the relinking of vortex lines in a viscous fluid using this coordinate system. Special classes of magnetohydrodynamic flows have been investigated with this coordinate system (e.g. Singh & Babu 1983).

3. SMOOTH DISTRIBUTIONS OF VORTICITY

We now turn our attention to the problem of representing smooth fields of distributed vorticity and determining accurately the evolution of such fields. Motivated by the fact that a smoothing or filtering of the point-vortex field, yielding vortex "blobs," appeared to be a promising approach in two dimensions (Chorin & Bernard 1973), Leonard (1980a,b) proposed an analogous filtering of three-dimensional space curves to yield the representation

$$\omega(\mathbf{x}, t) = \sum_i \Gamma_i \int \gamma_i[\mathbf{x} - \mathbf{r}_i(\xi, t)] \frac{\partial \mathbf{r}_i}{\partial \xi} \, d\xi, \tag{32}$$

where γ_i is a smoothing function having the normalization

$$\int \gamma_i(\mathbf{x}) \, d\mathbf{x} = 1. \tag{33}$$

It is assumed that γ_i has the form

$$\gamma_i(\mathbf{x}-\mathbf{r}_i) = \frac{1}{\sigma_i^3} \, p\!\left(\frac{|\mathbf{x}-\mathbf{r}_i|}{\sigma_i}\right), \tag{34}$$

where σ_i is the core radius of filament i. Inserting the representation (32) into (9) reduces the Biot-Savart integral to a sum of line integrals over each of the space curves representing the filaments:

$$\mathbf{u}(\mathbf{x}, t) = -\frac{1}{4\pi} \sum_j \Gamma_j \int_{C_j} \frac{[\mathbf{x}-\mathbf{r}_j(\xi, t)] \times \dfrac{\partial \mathbf{r}_j}{\partial \xi} \, q(|\mathbf{x}-\mathbf{r}_j|/\sigma_j) \, d\xi}{|\mathbf{x}-\mathbf{r}_j|^3}, \tag{35}$$

where q is defined by

$$q(y) = 4\pi \int_0^y p(r) r^2 \, dr. \tag{36}$$

From the normalization of γ_i, it is noted that $q(y) \to 1$ as $y \to \infty$, so that at a distance large compared with the core radius, the induced velocity may be calculated as if all the circulation were concentrated on the curves $\mathbf{r}_j(\xi, t)$. If $p(r) < \infty$, then q is $O(y^3)$ for small y; therefore, $\mathbf{u}(\mathbf{x}, t)$ remains bounded.

The function $p(r)$ defines the underlying basis for the expansion of the vorticity field and plays a role analogous to that of a B spline function used as a basis for a finite-element representation of a field quantity. Particularly simple choices are the Gaussian core $p(r) = (1/\pi^{3/2}) \exp(-r^2)$ used by Leonard (1980b) and the function $p(r) = (3\alpha/4\pi)(r^2+\alpha)^{-5/2}$ used by Nakamura et al. (1983), which is related to the Rosenhead-Moore smoothing function $q(y) = y^3/(y^2+\alpha)^{3/2}$ through (36). Questions of convergence and accuracy are discussed below.

It appears that for a dense swarm of vortex filaments (significant overlapping of cores), there are a number of choices for specifying the dynamics of the space curves. One possibility is to move the curves with the local velocity,

$$\frac{\partial \mathbf{r}_i}{\partial t} = \mathbf{u}(\mathbf{r}_i(t), t). \tag{37}$$

This choice is used in the convergence proofs discussed below, and any other choice should converge to this one as the density of filaments increases without limit.

If differing core radii are present, it is recommended that the right-hand sides of (35) and (22) for $\mathbf{u}(\mathbf{r}_i)$ be symmetrized with respect to σ_i and σ_j, for example, by replacing σ_j with $(\sigma_i^2+\sigma_j^2)^{1/2}$. Then one obtains exact conservation of linear and angular momentum (Leonard 1980b). Another choice, which touches base with the thin-filament approximation discussed in Section 2, is to replace q in (35) with another function, say g, when computing the right-hand side of (35). To reproduce long-wavelength thin-filament dynamics, g must satisfy the integral constraint (21). The circulation fraction $\chi(r)$ is given in terms of the function p by

$$\chi(r) = \left[1 - 4\pi \int_{r/\sigma}^{\infty} p(s)s^2 \left(1 - \frac{r^2}{s^2} \right)^{1/2} ds \right].$$ (38)

Finally, one could average the velocity field using the filter γ_i to obtain the filament velocity as follows:

$$\frac{\partial \mathbf{r}_i}{\partial t} = \int \gamma_i(\mathbf{r}_i - \mathbf{x}')\mathbf{u}(\mathbf{x}', t) \, d\mathbf{x}'.$$ (39)

This choice automatically yields conservation of linear and angular momentum and conservation of energy if the σ_i are independent of time. Further discussion can be found in Leonard (1980b).

Alternate Representations and Convergence

It would be comforting to know that vortex-blob methods converge to an exact solution of the Euler equations as the density of the blobs is increased. Such proofs might also point the way to more efficient and more accurate schemes. Indeed, Hald (1979) was first to present such a proof for two-dimensional flows. Later Beale & Majda (1982a,b) established stronger convergence results in two dimensions and proved convergence for a new three-dimensional algorithm that they proposed. Subsequently Cottet (1982) and Anderson & Greengard (1984) have simplified these proofs, and the latter propose simplified algorithms for three-dimensional flows. A typical smooth-core distribution function, such as a Gaussian, leads to a scheme that is approximately second-order accurate, but Beale & Majda (1982a,b) showed that schemes of arbitrarily high order in two and three dimensions are possible using core distribution functions satisfying certain constraints, as discussed by Hald (1979) and Leonard (1980b).

In these three-dimensional schemes, the vorticity field is represented by a collection of vortex "arrows" or vector-valued δ functions centered at points $\mathbf{x}_k(t)$ initially within cells of volume h^3 and smoothed by a spherically symmetric function γ defined by (33) and (34):

$$\omega(\mathbf{x}) = \sum_k \gamma(\mathbf{x} - \mathbf{x}_k(t))\omega_k(t)h^3.$$ (40)

The core parameter of $\gamma \, (= \sigma)$ must be large enough so that the elements overlap, i.e. $\sigma \approx h$. Inserting (40) into the Biot-Savart integral (9), one obtains a smooth velocity field analogous to the expression (35) for vortex filaments:

$$\mathbf{u}(\mathbf{x}, t) = -\frac{1}{4\pi} \sum_k \frac{(\mathbf{x} - \mathbf{x}_k(t)) \times \omega_k(t) h^3 q(|\mathbf{x} - \mathbf{x}_k(t)|/\sigma)}{|\mathbf{x} - \mathbf{x}_k(t)|^3}. \tag{41}$$

The vortex locations $\mathbf{x}_k(t)$ move with the local velocity. The vorticity vectors must also be updated in response to the local strain field. (In the filament approach, this is achieved automatically as one follows a sequence of points on the space curves defining the filaments.) In one approach (Anderson & Greengard 1984), the expression (41) for $\mathbf{u}(\mathbf{x}, t)$ is differentiated analytically to obtain the required velocity gradients. The evolution of the $\omega_k(t)$ vectors is then given by

$$\frac{d\omega_k(t)}{dt} = (\omega_k(t) \cdot \nabla_\mathbf{x}) \mathbf{u}(\mathbf{x})|_{\mathbf{x} = \mathbf{x}_k}. \tag{42}$$

Alternatively, one can update the vorticity according to the discrete equivalent of Cauchy's vorticity formula (Truesdell 1954):

$$\begin{aligned} \omega(\mathbf{x}, t) &= [\nabla_\alpha \mathbf{x}(\alpha, t)] \cdot \omega(\alpha, 0), \\ \mathbf{x}(\alpha, 0) &= \alpha, \end{aligned} \tag{43}$$

where the α are Lagrangian coordinates. Related methods have been used by Rehbach (1978) and Chorin (1980) that are a compromise between the filament representation and the "arrow" representation.

The method using (42) is attractive because the connectivity of the filaments of vorticity is not explicitly maintained. Thus, for example, the initialization process is often simpler because the initial vorticity vectors $\omega_k(0)$ are just values of the vorticity field on a regular array of lattice points. Also, the amalgamation of several neighboring vortex "arrows" into a fewer number to simulate viscous diffusion or to save computation time is simply achieved. Reconnection is possible in the vortex-filament representation (Leonard 1975), but it is less flexible and requires more coding effort. In addition, this independence simplifies generation of new vorticity at solid boundaries, where circulation is created to maintain the no-slip condition for viscous flows. For example, at one point in the computational cycle of such a scheme, one would determine the wall slip velocity on a grid of surface points and create new vortex "arrows" to eliminate the slip velocity. Chorin (1980) has implemented this method to study three-dimensional boundary-layer turbulence, using a random-walk algorithm to simulate the effects of viscous diffusion. In the vortex-filament method the connectivity of newly generated vorticity at the surface would have to be determined. In this regard, the fact that this vorticity is aligned with the level lines of surface pressure (Leonard 1980b) should be useful.

However, in acquiring the above-mentioned advantages by using the "arrow" representation, one loses the possibilities of higher-order interpolation along a given space curve, i.e. in the filament representation a single point can represent a particular "wiggle" rather than just a straight arrow. And, while the solenoidal condition for the velocity field is satisfied, the vorticity field is only approximately divergence-free. Furthermore, it appears that linear and angular momentum are not exactly conserved. Nevertheless, its ease of application and its generality make the "arrow" representation very attractive for further development, testing, and application.

Suitable test cases for three-dimensional vortex methods are not in abundance. Axisymmetric test problems include the family of vortex rings computed by Norbury (1973) and the unsteady motion of a vortex ring with a thin elliptical core analyzed by Moore (1980). In three dimensions one could compare with the linear stability analyses of strained rectilinear vortices with elliptical cross section (Robinson & Saffman 1984), of vortex rings (Widnall & Tsai 1977), and of Stuart's family of periodic shear layers (Pierrehumbert & Widnall 1982). The last test case has the virtue of having a smooth base flow. Other established numerical methods may be used to provide "exact" solutions to nonlinear, unsteady flows. For example, Couët (1979) tested his three-dimensional vortex-in-cell algorithm on a Taylor-Green problem and compared his results with those computed by a spectral code for early times.

Applications to Unsteady Vorticity Fields

The rapid nonlinear evolution of three-dimensional vortical flows in which viscous effects are relatively unimportant is a recurrent theme in fluid mechanics. In this section, we present the results from vortex simulations of several such flows in which viscous effects establish the initial conditions or the unperturbed state but appear to play a relatively minor role in the subsequent evolution of the perturbed flow. Flow structures of importance are represented by a number of computational vortex filaments.

Our first example is the evolution of a turbulent spot in a laminar boundary layer (Leonard 1980a,b, 1981). A related theoretical effort has been reported recently by Russell & Landahl (1984). Each infinite tube of vorticity in the boundary layer is decomposed into its straight, unperturbed configuration and a loop representing its contribution to the perturbed vorticity field, as shown in Figure 5. Image vorticity is used to maintain flow tangency at the wall. The no-slip condition is not enforced. The velocity field is decomposed, accordingly, as

$$\mathbf{u}(\mathbf{x}) = U(y)\hat{e}_x + \mathbf{u}'(\mathbf{x}), \tag{44}$$

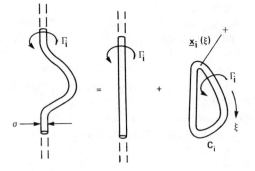

Figure 5   Decomposition of the vorticity field.

where $U(y)$ is the laminar profile and $\mathbf{u}'$ is given by a sum of Biot-Savart contributions over all the filament curves and their images. Thus, exact boundary conditions at infinity in all directions are satisfied, and computational points are required only where the vorticity is perturbed away from its laminar state.

In Figure 6 the space curves defining the vortex filaments are shown at three different times after an initial perturbation consisting of small displacements of the filaments within a three-dimensional volume of order $O(\delta^{*3})$. The results shown are typical. Cell-like structures are observed of $O(\delta^*)$ in extent in the spanwise and vertical directions but elongated in the streamwise direction. Within these cells, perturbation velocities on the order of several tenths of U_∞ are observed in all three directions, implying a vigorous mixing process at these scales. Instantaneous streamwise velocity profiles within the cells do *not* resemble ensemble-average results from experiments (Cantwell et al. 1978); however, rather good agreement is achieved if the simulation results are span-averaged over a cell width of $4\delta^*$, as shown in Figure 7. The good comparison leads to the speculation that the internal structure of a single realization of a turbulent spot ($\geqq 100\delta^*$ in streamwise extent) is dominated by the presence of many cell-like substructures such as those revealed in the numerical simulation. The initial spacing between the filaments is $0.41\delta^*$, so that a cell cross section of $4\delta^* \times 4\delta^*$ in a spanwise-vertical plane will be pierced by roughly 100 filaments—certainly adequate to represent considerable detail within each cell.

Thanks to theoretical efforts within the last decade, the inviscid instability of a vortex ring is now reasonably well understood. As discussed in Section 2, the analysis must take into account nonuniform radial motion within the vortex core to predict this short-wavelength phenomenon. The number of azimuthal waves in the most amplified eigenfunction increases as the ratio of core radius to ring radius decreases. Krutzsch (1939) studied these instabilities experimentally nearly 50 years ago. More recently, B. Sturtevant (private communication; see Van Dyke 1982) has obtained as many as 40 waves in shock-tube-generated vortex rings. Computer experiments can now be used to probe the nonlinear regime. In Figure 8 the evolution of a vortex ring composed of five computational filaments is shown (Ashurst 1981). Each of the initial 40 node points in each filament was independently subjected to a random perturbation of $0.01R$. As seen in the figure, a mode with five waves appears most amplified in this case. As vortex stretching takes place, more points are added to each filament. In the final view, 267 points are used to represent the flow.

Ashurst (1983) has also studied the evolution of the time-developing round jet with similar initial conditions (i.e. an infinite cylindrical shell of circumferential vorticity, periodic in the axial direction, with small random displacements applied initially to each node). The results are displayed in Figures 9 and 10. The number of points used range from 4000 in the initial state to approximately 8000 at the end of the simulation. The filaments merge to form snaking tubes of vorticity having a preferred length scale. Three-dimensionality is noticeable at the outset, in contrast to implications of linear stability theory; this theory usually suggests a progression through an amplified two-dimensional state onto three-dimensional instabilities of that finite-amplitude two-dimensional state. Is this simply a matter of initial amplitude? What is the origin of the preferred length scale? These questions and others suggest the need for further studies to build on the interesting preliminary results shown here.

Vortex breakdown is another important flow where multiple computational filaments are required to represent essential flow features. In this case one must deal with the rapidly changing core structures encountered in these flows. Figure 11 shows a simple demonstration of the effect of initial variation in axial flow along the vortex induced by an axial section of helical vortex filaments. As shown at the top (a) of the figure, a gradual rise in (negative) axial flow along the axis of a column vortex produces axisymmetric traveling waves emanating from the location of the initial variation, while a larger initial variation (helical filaments with a smaller pitch) leads to a wave breaking of the vorticity distribution shown at the bottom (b).

In a serious attempt to study experimentally observed vortex breakdown, Nakamura et al. (1983) used helical vortex filaments, with circulation and pitch depending on radius, to represent a given swirling flow upstream of breakdown. At this time, only a few preliminary numerical experiments have been made. In one experiment a strong initial perturbation appears to be evolving into an axisymmetric breakdown, followed downstream by a three-dimensional spiral disturbance. In this simulation the velocity

$tU_\infty/\delta^* = 8$ 28   Top Views

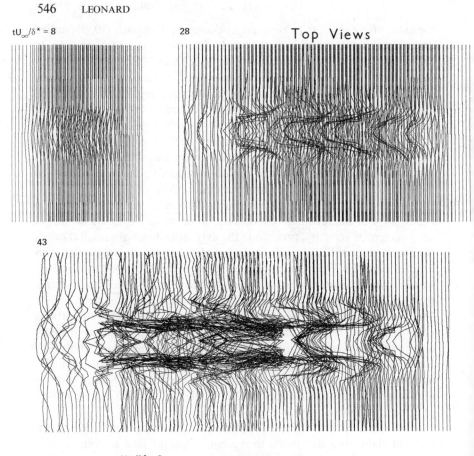

43

$tU_\infty/\delta^* = 8$ Rear Views

28

43

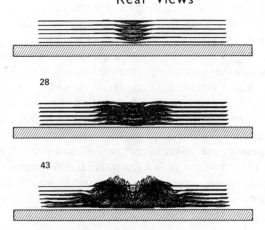

contribution from vorticity upstream and downstream of the computational domain was computed from systems of semi-infinite vortex filaments with predetermined geometry.

Finally, we make some additional comments on the possibilities of simulating turbulence with vortex methods. It is well established that the velocity fluctuations in high-Reynolds-number turbulent flows are highly intermittent in space and time, which implies that intense vorticity is confined to a small fraction of the fluid volume. Thus, Lagrangian vortex elements should provide an economical representation of these complex flows. In addition, the Fourier transform of the vorticity field, and therefore the energy spectrum, may be easily obtained from the filament representation (32). Even the simplest of curved filaments, the vortex ring, has a rich energy spectrum. This spectrum is plotted in Figure 12 for a core with zero cross-section. The $1/k$ behavior for large k is characteristic for isolated smooth tubes of vorticity and $k\sigma \ll 1$. The k^2 behavior for small k is a consequence of the nonzero impulse of the vortex ring.

Degani & Leonard (1976, and unpublished) studied the evolution of a patch of turbulence in an infinite domain consisting of five vortex filaments that initially were elliptical space curves randomly positioned and randomly oriented. After the flow developed for a time, energy spectra were computed and compared with experimental spectra of low-Reynolds-number homogeneous turbulence at the same ratio of integral scale to

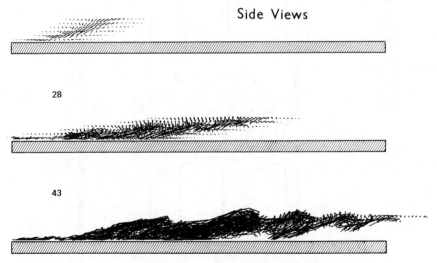

$tU_{\infty}/\delta^* = 8$

Side Views

28

43

Figure 6 Vortex filaments showing the evolution of a localized disturbance in a laminar boundary layer.

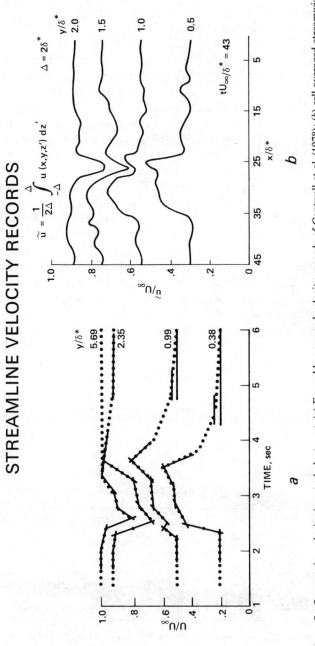

$$\tilde{u} = \frac{1}{2\Delta} \int_{-\Delta}^{\Delta} u(x,y,z')\, dz'$$

$\Delta = 2\delta^*$

Figure 7 Streamwise velocity in a turbulent spot. (*a*) Ensemble-averaged velocity records of Cantwell et al. (1978); (*b*) cell-averaged streamwise velocity from the vortex simulation of Leonard (1981).

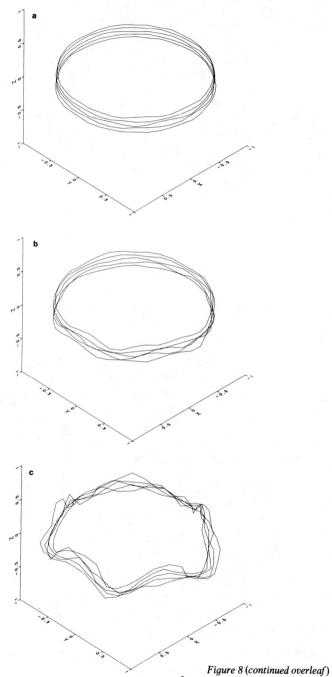

Figure 8 (continued overleaf)

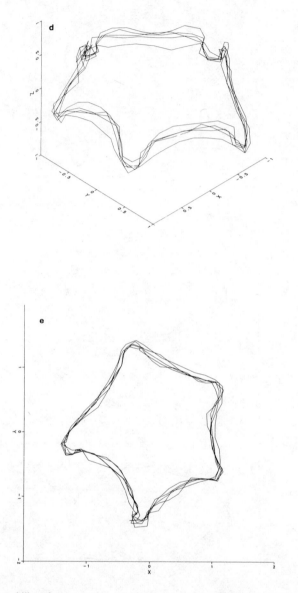

Figure 8   Instability of a vortex ring. (*a*)–(*d*) Perspective views with time increasing; (*e*) top view at final time (Ashurst 1981).

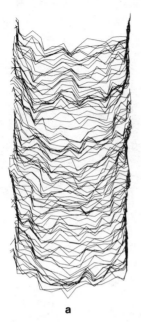

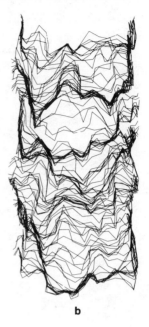

a

b

Figure 9 Perspective views of a time-developing round jet. Only the front half of the filaments is shown. The time increases from (*a*) to (*c*) (Ashurst 1983).

c

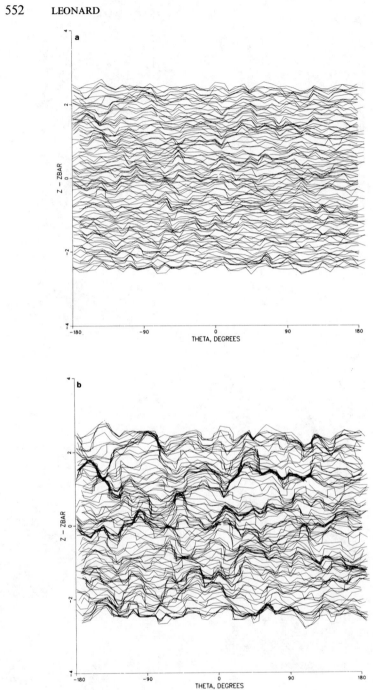

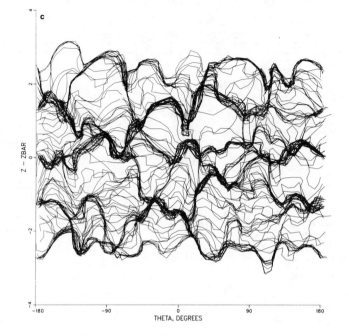

Figure 10   Views of the filaments of Figure 9 as seen from inside the jet at corresponding times (Ashurst 1983).

Taylor microscale (Figure 13). The agreement is good for the two decades in wave number shown, even though the simulation had only 500 node points at the time of the comparison. Chorin (1981, 1982) has studied inertial-range turbulence using Lagrangian vortex methods. In Chorin (1981), a rescaling technique was used to allow the tracking of increasingly finer scales of motion during the evolution. The results suggest a number of important conclusions concerning the structure and dimensionality of inertial-range vorticity. However, these results and those of Degani & Leonard need further verification, as they were obtained essentially under the thin-filament approximation. Consequently one has to assume, for example, that the cores of the isolated tubes of vorticity do not break up into other structures. Refined calculations are under way (A. Chorin, private communication; W. T. Ashurst & A. Leonard, in preparation).

4.   SUMMARY AND CONCLUSIONS

We have shown how a wide variety of three-dimensional vortex flows may be simulated numerically with Lagrangian vortex elements. If the flow

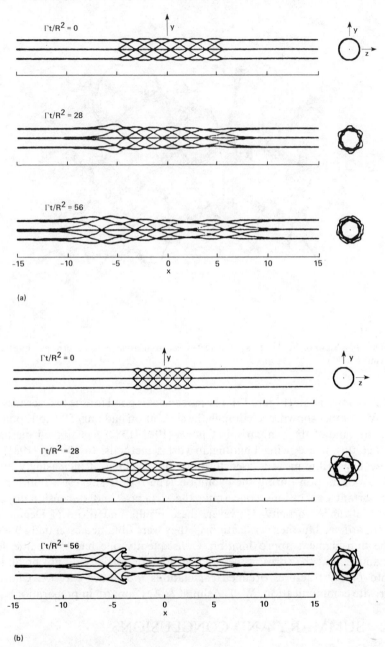

Figure 11   Evolution of a localized region of axial flow in a vortex tube. (*a*) Gradual initial variation in axial flow; (*b*) larger initial variation.

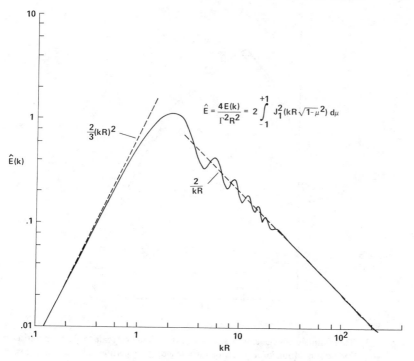

$$\hat{E} = \frac{4E(k)}{\Gamma^2 R^2} = 2 \int_{-1}^{+1} J_1^2(kR\sqrt{1-\mu^2})\, d\mu$$

$$\frac{2}{3}(kR)^2$$

$$\frac{2}{kR}$$

$\hat{E}(k)$

kR

Figure 12   Energy spectrum of a vortex ring.

consists of a few isolated vortex tubes (or is modeled as such), one can assign a computational filament to each tube and generally obtain useful results quickly and easily with a relatively small number of vortex elements. Such efforts could be quite helpful in obtaining at least a qualitative understanding of the flow under study in terms of vortex dynamics. The filament representation described in Section 2 appears to be the preferred choice for these applications because of the relative importance of self-induction due to local curvature.

If the flow is more complex or if more detailed quantitative information is desired, then a larger number of computational elements will be required. In this situation, one is quickly up against machine speed limitations. In their simplest form, vortex methods require $O(N^2)$ operations per time step for N vortex elements. Consequently the more powerful machines that are coming along do not have the impact one might hope for in using these methods. For example, a hundredfold increase in computer speed allows only a tenfold increase in N. On the other hand, more elaborate schemes can be devised that significantly lower the operation count. An example is the vortex-in-cell method (Couët et al. 1981).

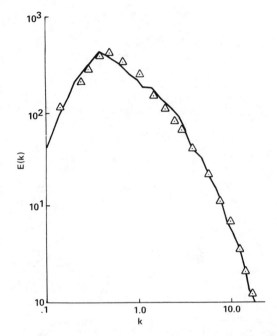

Figure 13   Three-dimensional energy spectra. The ratio of integral scale to Taylor microscale is 5. —— Vortex simulation of a patch of turbulence; △ experimental measurements in homogeneous turbulence (Comte-Bellot & Corrsin 1971).

The above pessimistic remarks concerning the computational require-ments of vortex methods of course must be tempered by the fact that vortex elements are required only where the vorticity is nonzero, often only a small fraction of the total fluid vólume. The economy of the flow representation is a significant advantage. One might well ask if the notion of economy could be pushed even further. It would seem that we would have to determine vortex structures that are, in some sense, preferred by the Navier-Stokes equations. Lundgren's (1982) spiral vortex core is in this spirit and would be a candidate for inertial-range turbulence. The vortex ring is a preferred state in isolation, but its utility in a more general setting is not known. It will be interesting to watch for future developments in this direction. The payoffs could be substantial.

As discussed in Section 3, methods employing vortex "arrows" (discon-nected vector-valued vortex blobs) enjoy the advantage of having high flexibility, in that the generation of new vorticity at a solid surface, the remeshing of the vorticity field, and the generation of prescribed initial vorticity fields can be done with relative ease. It appears that a sufficiently high density of vortex "arrows" will yield simulations of high quality. Further development and testing is required, however, to establish the practical limitations of this approach.

Vortex stretching, a hallmark of three-dimensional transitional or turbulent flows, implies an increasing total line length of vortex elements with time. In some situations the filament length/volume or "arrow" density may increase beyond a tolerable upper limit and remeshing must be done. It is clear that the process of line-length generation is connected with the transfer of energy to small scales, and that remeshing is a dissipation mechanism that removes small-scale energy from the representation. However, the effectiveness and accuracy of remeshing and other techniques for dissipating small-scale energy, such as smoothing the space curves (in the filament representation) and increasing the core parameter, have not been investigated adequately.

As work progresses to develop and improve the basic vortex methods, extensions of the method should be pursued. There have been a number of recent extensions of two-dimensional vortex methods that could be moved into three dimensions in the near future, but not without considerable effort and ingenuity. We have in mind vortex simulations of bluff-body flows, flows with nonuniform density, and chemically reacting flows. In addition, the use of Lagrangian vortex elements to represent the rotational part of the velocity field in the computation of compressible flows is being studied by several investigators. In the above-mentioned extensions the use of a fixed grid system, in addition to vortex elements, may be desirable or even necessary.

ACKNOWLEDGMENTS

The author would like to thank P. E. Dimotakis and R. S. Rogallo for many helpful comments on a draft of this paper. Linda Malaby's assistance in preparing the manuscript is gratefully acknowledged.

Literature Cited

Anderson, C., Greengard, C. 1984. On vortex methods. *SIAM J. Numer. Anal.* In press

Aref, H. 1983. Integrable, chaotic, and turbulent vortex motion in two-dimensional flows. *Ann. Rev. Fluid Mech.* 15:345–89

Aref, H., Flinchem, E. P. 1985. Dynamics of a vortex filament in a shear flow. *J. Fluid Mech.* In press

Arms, R. J., Hama, F. R. 1965. Localized-induction concept on a curved vortex and motion of an elliptic vortex ring. *Phys. Fluids* 8:553–59

Ashurst, W. T. 1981. Vortex ring instability. *Bull. Am. Phys. Soc.* 26:1267

Ashurst, W. T. 1983. *Large eddy simulation via vortex dynamics.* Presented at the AIAA Comput. Fluid Dyn. Conf., 6th, Danvers, Mass. *AIAA Pap. No. 83-1879-CP*

Batchelor, G. K. 1967. *An Introduction to Fluid Dynamics.* Cambridge: Cambridge Univ. Press. 615 pp.

Beale, J. T., Majda, A. 1982a. Vortex methods. I: Convergence in three dimensions. *Math. Comput.* 39:1–27

Beale, J. T., Majda, A. 1982b. Vortex methods. II: Higher order accuracy in two and three dimensions. *Math. Comput.* 39:29–52

Belotserkovskii, S. M. 1977. Study of the unsteady aerodynamics of lifting surfaces using the computer. *Ann. Rev. Fluid Mech.* 9:469–94

Betchov, R. 1965. On the curvature and torsion of an isolated vortex filament. *J. Fluid Mech.* 22:471–79

Cantwell, B. J., Coles, D., Dimotakis, P. E. 1978. Structure and entrainment in the plane of symmetry of a turbulent spot. *J. Fluid Mech.* 87:641–72

Chorin, A. J. 1980. Vortex models and boundary layer instability. *SIAM J. Sci. Stat. Comput.* 1:1–21

Chorin, A. J. 1981. Estimates of intermittency, spectra, and blow-up in developed turbulence. *Commun. Pure Appl. Math.* 34:853–66

Chorin, A. J. 1982. The evolution of a turbulent vortex. *Commun. Math. Phys.* 83:517–35

Chorin, A. J., Bernard, P. S. 1973. Discretization of a vortex sheet, with an example of roll-up. *J. Comput. Phys.* 13:423–29

Comte-Bellot, G., Corrsin, S. 1971. Simple Eulerian time correlation of full- and narrow-band velocity signals in grid-generated "isotropic" turbulence. *J. Fluid Mech.* 48:273–337

Cottet, G. H. 1982. PhD thesis. Univ. Pierre et Marie Curie, Paris

Couët, B. 1979. *Evolution of turbulence by three-dimensional numerical particle-vortex tracing.* PhD thesis. Stanford Univ., Calif. (Inst. Plasma Res. Rep. No. 793)

Couët, B., Buneman, O., Leonard, A. 1981. Simulation of three-dimensional incompressible flows with a vortex-in-cell method. *J. Comput. Phys.* 39:305–28

Crow, S. C. 1970. Stability theory for a pair of trailing vortices. *AIAA J.* 8:2172–79

Degani, D., Leonard, A. 1976. Statistical studies of turbulence using vortex filaments. *Bull. Am. Phys. Soc.* 21:1223

Fraenkel, L. E. 1970. On steady vortex rings of small cross-section in an ideal fluid. *Proc. R. Soc. London Ser. A* 316:29–62

Hald, O. H. 1979. Convergence of vortex methods for Euler's equations. II. *SIAM J. Numer. Anal.* 16:726–55

Hama, F. R. 1962. Progressive deformation of a curved vortex filament by its own induction. *Phys. Fluids* 5:1156–62

Hama, F. R. 1963. Progressive deformation of a perturbed line vortex filament. *Phys. Fluids* 6:526–34

Hasimoto, H. 1972. A soliton on a vortex filament. *J. Fluid Mech.* 51:477–85

Hoeijmakers, H. W. M. 1983. Computational vortex flow aerodynamics. *AGARD Conf. Proc.* 342:18-1–18-35

Hopfinger, E. J., Browand, F. K., Gagne, Y. 1982. Turbulence and waves in a rotating tank. *J. Fluid Mech.* 125:505–34

Kelvin, Lord. 1880. Vibrations of a columnar vortex. *Philos. Mag.* 10:155–68

Krutzsch, C. H. 1939. Über eine experimentell beobachtete Erscheinung an Wirbelringen bei ihrer translatorischen Bewegung in wirklichen Flüssigkeiten. *Ann. Physik* 35:497–523

Lamb, H. 1932. *Hydrodynamics.* Cambridge: Cambridge Univ. Press. 738 pp. 6th ed.

Lee, M. J., Reynolds, W. C. 1983. Structure of the bifurcating jet. *Bull. Am. Phys. Soc.* 28:1362

Leibovich, S. 1978. The structure of vortex breakdown. *Ann. Rev. Fluid Mech.* 10:221–46

Leibovich, S., Ma, H. Y. 1983. Soliton propagation on vortex cores and the Hasimoto soliton. *Phys. Fluids* 26:3173–79

Leonard, A. 1975. Numerical simulation of interacting, three-dimensional vortex filaments. In *Proc. Int. Conf. Numer. Meth. Fluid Dyn., 4th, Colo.,* pp. 245–50. Heidelberg: Springer-Verlag

Leonard, A. 1980a. Vortex simulation of three-dimensional spotlike disturbances in a laminar boundary layer. In *Turbulent Shear Flows 2,* pp. 67–77. Berlin/Heidelberg: Springer-Verlag

Leonard, A. 1980b. Vortex methods for flow simulation. *J. Comput. Phys.* 37:289–335

Leonard, A. 1981. Turbulent structures in wall-bonded shear flows observed via three-dimensional numerical simulations. In *The Role of Coherent Structures in Modelling Turbulence and Mixing,* pp. 119–45. Heidelberg: Springer-Verlag

Lugt, H. J. 1983. *Vortex Flow in Nature and Technology.* New York: Wiley. 297 pp.

Lundgren, T. S. 1982. Strained spiral vortex model for turbulent fine structure. *Phys. Fluids* 25:2193–2203

Marris, A. W., Passman, S. L. 1969. Vector fields and flows on developable surfaces. *Arch. Ration. Mech. Anal.* 32:29–86

Maxworthy, T., Hopfinger, E. J., Redekopp, L. G. 1985. Wave motions on vortex cores. Submitted for publication

Moore, D. W. 1972. Finite amplitude waves on aircraft trailing vortices. *Aeronaut. Q.* 23:307–14

Moore, D. W. 1980. The velocity of a vortex ring with a thin core of elliptical cross section. *Proc. R. Soc. London Ser. A* 370:407–15

Moore, D. W., Saffman, P. G. 1972. The motion of a vortex filament with axial flow. *Philos. Trans. R. Soc. London Ser. A* 272:403–29

Moore, D. W., Saffman, P. G. 1975. The instability of a straight vortex filament in a strain field. *Proc. R. Soc. London Ser. A* 346:413–25

Nakamura, Y., Leonard, A., Spalart, P. R. 1983. Numerical simulation of vortex breakdown by the vortex-filament method. *AGARD Conf. Proc.* 342:27-1–27-13

Norbury, J. 1973. A family of steady vortex rings. *J. Fluid Mech.* 57:417–31

O'Neill, B. 1966. *Elementary Differential Geometry*. New York/London: Academic. 411 pp. 3rd ed.

Oshima, Y., Asaka, S. 1977. Interaction of multi-vortex rings. *J. Phys. Soc. Jpn.* 42:1391–95

Parekh, D. E., Leonard, A., Reynolds, W. C. 1983. A vortex filament simulation of a bifurcating jet. *Bull. Am. Phys. Soc.* 28:1353

Pierrehumbert, R. T., Widnall, S. E. 1982. The two- and three-dimensional instabilities of a spatially periodic shear layer. *J. Fluid Mech.* 114:59–82

Rehbach, C. 1978. *Numerical calculation of unsteady three-dimensional flows with vortex sheets.* Presented at the AIAA Aerospace Sci. Meet., 16th, Huntsville, Ala. *AIAA Pap. No. 78-111*

Robinson, A. C., Saffman, P. G. 1984. Three-dimensional stability of an elliptical vortex in a straining field. *J. Fluid Mech.* 142:451–66

Rogallo, R. S., Moin, P. 1984. Numerical simulation of turbulent flows. *Ann. Rev. Fluid Mech.* 16:99–137

Rosenhead, L. 1930. The spread of vorticity in the wake behind a cylinder. *Proc. R. Soc. London Ser. A* 127:590–612

Russell, J. M., Landahl, M. T. 1984. The evolution of a flat eddy near a wall in an inviscid shear flow. *Phys. Fluids* 27:557–70

Saffman, P. G. 1970. The velocity of viscous vortex rings. *Stud. Appl. Math.* 49:371–80

Saffman, P. G., Baker, G. R. 1979. Vortex interactions. *Ann. Rev. Fluid Mech.* 11:95–122

Schwarz, K. W. 1982. Generation of superfluid turbulence deduced from simple dynamical rules. *Phys. Rev. Lett.* 49:283–85

Serrin, J. 1959. Mathematical principles of classical fluid mechanics. In *Encyclopedia of Physics, Vol. VIII/1*, pp. 125–263. Berlin/Göttingen/Heidelberg: Springer-Verlag

Singh, S. N., Babu, R. 1983. On the geometry of vortex lines in magnetofluid flows. *Nuovo Cimento* 76B:47–52

Tsai, C.-Y., Widnall, S. E. 1976. The stability of short waves on a straight vortex filament in a weak externally imposed strain field. *J. Fluid Mech.* 73:721–33

Truesdell, C. 1954. *The Kinematics of Vorticity.* Bloomington: Ind. Univ. Press. 232 pp.

Van Dyke, M. 1982. *An Album of Fluid Motion.* Stanford, Calif: Parabolic. 176 pp.

Whitham, G. B. 1974. *Linear and Nonlinear Waves.* New York: Wiley. 636 pp.

Widnall, S. E., Bliss, D. B., Tsai, C.-Y. 1974. The instability of short waves on a vortex ring. *J. Fluid Mech.* 60:35–47

Widnall, S. E., Bliss, D., Zalay, A. 1971. Theoretical and experimental study of the stability of a vortex pair. In *Aircraft Wake Turbulence and Its Detection*, pp. 305–38. New York: Plenum

Widnall, S. E., Tsai, C.-Y. 1977. The instability of the thin vortex ring of constant vorticity. *Philos. Trans. R. Soc. London Ser. A* 287:273–305

Ann. Rev. Fluid Mech. 1985. 17 : 561–608

MANTLE CONVECTION AND VISCOELASTICITY

W. R. Peltier

Department of Physics, University of Toronto, Toronto,
Ontario M5S 1A7, Canada

1. INTRODUCTION

The ongoing revolution in the Earth sciences, which began more than twenty years ago, was originally based upon the increasingly widespread acceptance of the idea that continental masses have moved horizontally with respect to one another throughout geological time. This hypothesis of continental "drift," or at least the form of the hypothesis that came to be called seafloor spreading, now serves as the basic paradigm for the organization of most geological and geophysical research.

At the center of this guiding principle is the recognition that the solid outer shell of the planet—its iron-magnesium silicate "mantle," which occupies roughly half Earth's radial extent—must be able to deform as a viscous fluid when it is subjected to an applied shear stress over geological intervals of time. To the extent that this rheological ansatz is correct, it is clear that a thermally induced convective circulation must appear in the mantle in response to a radial temperature gradient that is sufficiently in excess of adiabatic. The observed spreading of the seafloor away from hot mid-oceanic ridges is presumably a surface manifestation of such deep-seated mantle convection. Likewise, the deep ocean trenches are understood to be regions where cold surface material returns to the mantle to complete the circulation. These and other aspects of the pattern of surface motions associated with mantle convection have been described kinematically within the framework of a set of ideas that has come to be called "plate tectonics." The development of this set of ideas has consummated the revolution at a descriptive level and has delivered as its main product a clear view of the velocity field of material at the Earth's surface at the present epoch of geological time. In so doing, it has also contributed in an

561

0066–4189/85/0115–0561$02.00

important way to our understanding of all the major geological processes contributing to the formation and modification of the surface features of the planet. These processes include volcanism, mountain building, and earthquake activity, each of which is a surface manifestation of the thermally induced circulation that fills the planetary mantle.

In spite of the fact that the revolution has been consummated at the descriptive and conceptual level, there is as yet no generally accepted and detailed dynamical explanation of the kinematic picture that has emerged. The degree of contact that presently exists between theory and observation is therefore modest and limited to certain qualitative features, such as the observed variations of ocean-floor heat flow and bathymetry as a function of ocean-floor age. Major questions remain outstanding. For example, there is no general consensus as to the source of energy that maintains the circulation against frictional dissipation; the candidates are radioactive heating and/or secular cooling of the planet in bulk. Nor is there any degree of unanimity concerning the radial structure of the circulation; here the competing scenarios include a "whole-mantle" model, in which radial advection dominates the radial heat transport everywhere in the mantle, and a "layered" model, in which a thermal boundary layer exists at a particular mid-mantle depth through which heat is transported by diffusion between two separately convecting regions. A third important issue that persists in the literature on this problem concerns the extent to which the velocity field at the surface reflects that in the mantle beneath—that is, the extent to which a Stokes-flow extrapolation to depth may provide a reasonable first-order approximation to the interior velocity field.

Because these unresolved issues are crucial ones, it is clearly impossible in this brief review to do more than air them further; in doing so, I focus upon those recent advances that have contributed most to the development of the dynamical theory with which we eventually hope to more fully reconcile the kinematic observations. The paper is organized into three major sections. The present-day surface-velocity field is discussed in Section 2, following a brief historical review of the development of plate tectonics. We demonstrate that the surface velocity field possesses equally intense toroidal and poloidal constituents. A representation of the surface flow in terms of radial vorticity and horizontal divergence provides a compact scalar summary of the surface kinematics, and furthermore it can be usefully exploited to study the connection between the surface flow and the large-scale gravitational field of the planet, as measured from the orbits of artificial Earth satellites. The apparent connection between the surface velocity and gravity fields and the internal lateral heterogeneity inferred from free-oscillation splitting data are also discussed, as is the variation of surface heat flow. Section 3 reviews the past decade of theoretical and observational work, which has led to reasonably well-constrained models of the variation of mantle viscosity with depth. Hydrodynamic models of the mantle circulation are discussed in Section 4, beginning with a brief analysis of the linear stability of a spherical-shell model containing the solid-solid phase transformations, which have been inferred to exist on the basis of high-pressure laboratory experiments on the dominant mantle minerals. Very high Rayleigh number circulations computed numerically are employed to illustrate the fact that the main features of the surface observations can be generally understood in terms of the same boundary-layer scaling relations that govern constant-viscosity flows in the laboratory. A brief discussion is also provided of recent efforts to incorporate the combined effects of chemical heterogeneity and phase transformations in numerical models of the circulation, and of some further hydrodynamic complexities. The article concludes with an outlook for future developments of the dynamical theory.

2.   MANTLE CONVECTION: THE OBSERVATIONAL EVIDENCE

The earliest motivations for consideration of the notion of continental drift were entirely due to the fact that the coastlines of Africa and South America are complementary, a relationship clear even to the earliest geographers. Wegener (1929) and Holmes (e.g. 1931) were among the first to suggest that these continents were once connected and later split by the action of a thermal convective circulation in the Earth's mantle, but the idea was considered disreputable at the time. Its reemergence in the modern literature was signaled by the appearance of a paper by Hess (1962), who referred to the mechanism that split the original supercontinent of Pangea as "seafloor spreading" and believed that the circulation filled the entire volume of the mantle. Bullard et al. (1965) employed statistical methods to reconstruct the original supercontinent as it must have looked prior to breakup 200×10^6 years ago and to provide an analysis of the goodness of fit of the reconstructed continental fragments. The first quantitative measurements suggesting that the present continents were connected in the past were obtained using paleomagnetic methods (Runcorn 1956, McElhinny 1971), which showed that the paleopole paths of Europe and North America were similar until the Triassic, when they diverged, an event interpreted as coinciding with continental breakup.

2.1   The Emergence of Plate Tectonics: A Brief Historical Review

Paleomagnetic analyses played a dominant role in establishing the validity of Hess' hypothesis of seafloor spreading. In 1963 Vine & Mathews (1963) and Morley (unpublished), who were studying the patterns of magnetic anomalies over ocean ridges, realized independently that the alternate linear anomalies of normal and reversely magnetized material that parallel all the major oceanic ridges could be understood in terms of Hess' hypothesis. Hot material rising beneath the ridge crest toward the Earth's surface acquires a magnetization as it cools below the Curie point, and this has normal or reversed polarity depending upon the sense of the Earth's dipole field (which undergoes random reversals through geological time; see, for example, Busse 1978) at the time the material cools. The seafloor is therefore a magnetic tape recorder of its own history of spreading, the rates of which may be inferred by using the paleomagnetically established reversal time scale established by radioactive dating of successive sequences of volcanic rocks on land (e.g. Cox et al. 1963, Cox 1969). Using these methods, one infers present-day spreading rates for the major oceanic ridges of 1–10 cm yr^{-1} (e.g. Vine 1966).

Although paleomagnetic analysis was responsible for demonstrating that hot, mantle-derived material was ascending to the surface along the mid-oceanic ridges, it was seismological research that established equally clearly that cold material, prone to elastic fracture (earthquakes), was elsewhere descending into the mantle to complete the circulation. In fact, prior to the advance of the seafloor spreading hypothesis by Hess, Benioff (1954) had suggested that the deep ocean trenches were regions where old and cold ocean floor was downthrust into the mantle along what are today called "Benioff zones," which are the only locations in the planet where earthquakes occur to depths in excess of a few tens of kilometers. In these regions, such as the Chile, Tonga, and Japan trenches, earthquakes may occur to depths of 670 km, and their foci lie in well-defined (downgoing) "slabs" in which new events continually recur. Benioff's suggestion was first established as fact following the 1963 Alaska earthquake (Plafker 1965, Stauder & Bollinger 1965). The analysis of elastic-wave first-motion observations has played an important recent role in the development of the theory of mantle convection (e.g. Isacks & Molnar 1971), since the inferred earthquake source mechanisms appear to imply that the downgoing slabs encounter some resistance to their vertical motion at depths near 670 km, where the deepest earthquakes occur.

On the basis of global analyses of the distribution of near-surface seismicity (e.g. Isacks et al. 1968), it was realized that the global system of mid-ocean ridges and deep ocean trenches [and their crucial interconnections via transform faults (Wilson 1965)] was continuous over the surface and divided it into 12 local regions called "plates," the interiors of which were almost completely devoid of seismicity. The idea of "plate tectonics" is basically to describe all of the major geological processes such as volcanism, mountain building, and earthquake activity as consequences of the interactions between the 12 major plates at the boundaries between them. Seminal papers marking the consolidation of this set of ideas include those by Wilson (1965), McKenzie & Parker (1967), Morgan (1968), and McKenzie & Morgan (1969).

The ultimate product of the plate-tectonic revolution has been the inference of the horizontal velocity field at the Earth's surface for the present geological epoch. Using paleomagnetic data to control the directions of present relative motion among the plates, and the magnetic reversal time scale to constrain the relative drift speeds, the relative velocity field is inferred on the basis of the assumption of a constant angular velocity for each plate around a unique pole of rotation. That the individual plates do move with constant angular velocity is demanded by their rigidity. The most recent and complete analysis of the current pattern of plate velocities is that of Minster & Jordan (1978), whose results are illustrated in Figure 1, where the major plates are depicted along with their present directions of motion. In order to make the relative velocity field absolute, one is obliged to introduce an additional assumption—for example, that a frame of reference defined by a series of surface "hotspots" (e.g. Hawaii) is fixed in the mantle, or that the surface velocity field has no net rotation. Convergent boundaries on Figure 1 are trenches, whereas divergent boundaries are ridges. In order to understand the implications of this picture hydrodynamically, it proves useful to develop a description of the complex geographic pattern of surface velocities in terms of scalar fields.

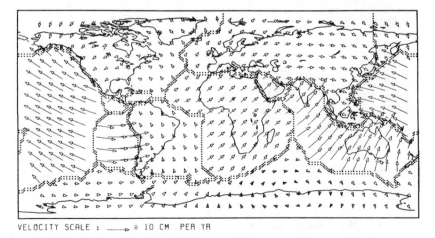

VELOCITY SCALE : ———▸ ≡ 10 CM PER YR

Figure 1 Plate boundaries and velocities from the model of Minster & Jordan (1978) in the no-net-rotation frame.

2.2   A "Two-Scalar" Representation of the Surface Kinematics: Correlations With the Gravitational Field

Two scalar fields that can be employed to characterize the vector field of surface velocity $\mathbf{u}$ are the horizontal divergence $\mathbf{V}_H \cdot \mathbf{u}$ and the vertical component of the curl, $\hat{\mathbf{z}} \cdot (\mathbf{V}_H \times \mathbf{u})$. Clearly, both these scalars are required to describe plate-tectonic motions, since mid-ocean ridges and deep ocean trenches are predominantly boundaries on which $\mathbf{V}_H \cdot \mathbf{u} \neq 0$, whereas transform faults like the San Andreas are predominantly boundaries on which $\hat{\mathbf{z}} \cdot (\mathbf{V}_H \times \mathbf{u}) \neq 0$. This representation is also useful dynamically, since convection with constant viscosity and essentially infinite Prandtl number develops in such a way that $\hat{\mathbf{z}} \cdot (\mathbf{V}_H \times \mathbf{u}) \equiv 0$ if it is zero initially (e.g. Busse 1975). We are therefore able to infer, on the basis of the importance of the $\hat{\mathbf{z}} \cdot (\mathbf{V}_H \times \mathbf{u})$ component of the surface flow, the extent to which effects due to the temperature dependence of viscosity or to the presence on the surface of chemically distinct continental material that cannot be reingested into the mantle (which couple the two scalars) are important in the Earth. This representation is also a rather elegant one, since $\mathbf{V}_H \cdot \mathbf{u}$ and $\hat{\mathbf{z}} \cdot (\mathbf{V}_H \times \mathbf{u})$ are determined entirely by the spheroidal (poloidal) and toroidal scalars, in terms of which any solenoidal field may be completely represented. From an absolute plate-motion model such as that of Minster & Jordan (1978), one may calculate the coefficients in the spherical harmonic expansion of the spheroidal and toroidal scalars from

$$
S_n^{m(_s^c)} = \frac{1}{4\pi n(n+1)} \iint \left[v_\theta \frac{\partial Y_n^{m(_s^c)}}{\partial \theta} + \frac{v_\lambda}{\sin \theta} \frac{\partial Y_n^{m(_s^c)}}{\partial \lambda} \right] d\lambda \, d\cos \theta, \tag{1}
$$

$$
T_n^{m(_s^c)} = \frac{1}{4\pi n(n+1)} \iint \left[\frac{v_\theta}{\sin \theta} \frac{\partial Y_n^{m(_s^c)}}{\partial \lambda} - v_\lambda \frac{\partial Y_n^{m(_s^c)}}{\partial \theta} \right] d\lambda \, d\cos \theta,
$$

where $v_\theta \, (\theta, \lambda)$ and $v_\lambda \, (\theta, \lambda)$ are the north-south and east-west components, respectively, of the velocity field delivered by the plate-motion model. The coefficients in the spherical-harmonic expansion of the divergence $(\mathbf{V}_H \cdot \mathbf{u})_n^m$ and of the curl $(\hat{\mathbf{z}} \cdot \mathbf{V}_H \times \mathbf{u})_n^m$ are then

$$
(\mathbf{V}_H \cdot \mathbf{u})_n^m = \frac{-n(n+1)}{a\sqrt{2-\delta_{mo}}} \cdot S_n^m,
$$

$$
(\hat{\mathbf{z}} \cdot (\mathbf{V}_H \times \mathbf{u}))_n^m = \frac{n(n+1)}{a\sqrt{2-\delta_{mo}}} \, T_n^m. \tag{2}
$$

Plates a and b of Figure 2 show the fields $\mathbf{V}_H \cdot \mathbf{u}$ and $\hat{\mathbf{z}} \cdot \mathbf{V}_H \times \mathbf{u}$ for the present epoch constructed from a spherical harmonic expansion truncated at degree and order 32 (sufficient to give $5° \times 5°$ resolution). Inspection of

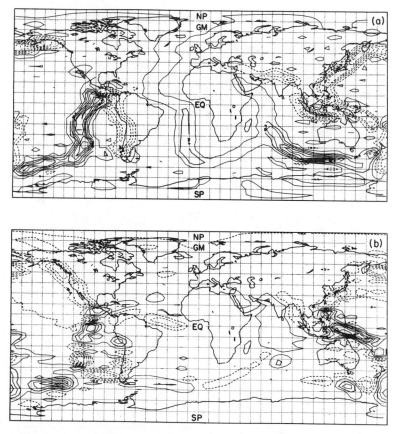

Figure 2   (*a*) Surface divergence to degree and order 32. The contour interval is 2×10^{-8} rad yr^{-1}. (*b*) Vertical component of the curl to degree and order 32. The contour interval is 2 $\times 10^{-8}$ rad yr^{-1}. The absolute plate-motion model is that of Minster & Jordan (1978) in the "hotspot" frame. From A. Forte & W. R. Peltier (in preparation).

Figure 2*a* shows that the main regions of present-day divergence are coincident with the East Pacific Rise off the west coast of South America and the East Indian Ridge in the eastern Indian Ocean. The most intense regions of convergence are in the northwestern Pacific and beneath South America. In contrast (Figure 2*b*), the most intense regions of negative vorticity are located along the west coast of North America (the San Andreas fault) and the southern tip of South America, whereas the regions of most intense positive vorticity are located north of Australia and in the South Central Pacific.

In order to represent the energy in the spheroidal and toroidal components of the flow, we may compute degree variances from

$$\sigma_n(\text{spheroidal}) = \left[\sum_{m=0}^{n} S_n^m S_n^{m*} \right]^{1/2},$$

$$\sigma_n(\text{toroidal}) = \left[\sum_{m=0}^{n} T_n^m T_n^{m*} \right]^{1/2},$$

(3)

which are shown in Figure 3a. This illustrates the interesting fact that convection in the Earth is characterized by a near-equipartition of kinetic

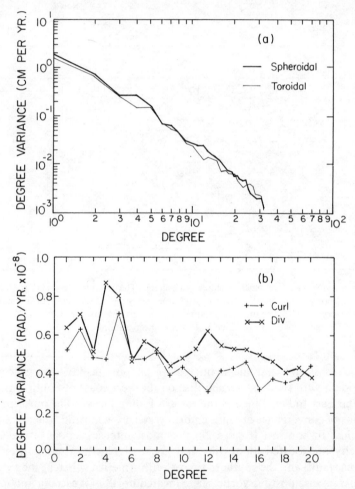

Figure 3   (a) Degree variances for the spheroidal and toroidal components of the velocity spectra. The absolute plate-motion model is that of Minster & Jordan (1978) in the "hotspot" frame. (b) Degree variances for divergence and curl. From A Forte & W. R. Peltier (in preparation).

energy between the nondivergent and irrotational components of the flow, which extends across the entire spectrum. This point was first made by Hager & O'Connell (e.g. 1978), although it was not further exploited except in the context of a downward extrapolation through the mantle of the surface velocities using a Stokes-flow approximation to the dynamics. Although the temperature and pressure dependence of viscosity may contribute to the maintenance of this equipartition, the crucial role is probably played by the presence of the continents, whose chemical buoyancy ensures that they cannot be remixed into the underlying mantle. Degree variances for divergence and curl are shown in Figure 3b. The well-defined peak at degrees 4 and 5 is associated with the global system of ridges and trenches and is thus to be seen as determined by the dominant scale of the near-surface circulation.

When the scalars $\mathbf{V}_H \cdot \mathbf{u}$ and $\hat{\mathbf{z}} \cdot \mathbf{V}_H \times \mathbf{u}$ are computed from the spheroidal and toroidal coefficients, they may be employed in correlation analyses to measure the degree of interrelation between these representations of the velocity field and other Earth properties that we might reasonably expect to be strongly connected to the nature of the internal convective circulation. One of the most promising of these other fields is the gravitational field, a useful measure of which is the field of geoid heights, which gives the geometric shape of the surface of constant potential defined by the surface of the oceans. If we call this field $Ge(\theta, \lambda)$, we may compute the degree correlation between it and either $\mathbf{V}_H \cdot \mathbf{u}$ or $\hat{\mathbf{z}} \cdot \mathbf{V}_H \times \mathbf{u}$ from

$$
\rho_n = \frac{\sum\limits_{m=-n}^{n} (Ge)_n^{m*} V_n^m}{\left[\sum\limits_{m=-n}^{n} V_n^{m*} V_n^m\right]^{1/2} \left[\sum\limits_{m=-n}^{n} (Ge)_n^{m*}(Ge)_n^m\right]^{1/2}}, \tag{4}
$$

where V_n^m is either $(\mathbf{V}_H \cdot \mathbf{n})_n^m$ or $(\hat{\mathbf{z}} \cdot \mathbf{V}_H \times \mathbf{u})_n^m$. Using the GEM10B model (Lerch et al. 1979) to deliver the coefficients Ge_n^m, we have computed (A. Forte & W. R. Peltier, in preparation) the degree correlations between geoid height and each of the scalars $\mathbf{V}_H \cdot \mathbf{u}$ and $\hat{\mathbf{z}} \cdot \mathbf{V}_H \times \mathbf{u}$; these are shown in Figure 4. For degree $n \gtrsim 9$ there is no apparent connection between Ge and either of the kinematic scalars; however, for degree $\gtrsim 9$ a striking pattern emerges. Between Ge and $\mathbf{V}_H \cdot \mathbf{u}$ an increasingly strong negative correlation develops with decreasing degree, which disappears at degree 3 and then reappears at degree 2. Between Ge and $\hat{\mathbf{z}} \cdot \mathbf{V}_H \times \mathbf{u}$, on the other hand, an increasingly positive correlation develops with decreasing degree, which peaks sharply at degree 3 and then drops at degree 2. It may be significant that the strongest correlation between Ge and $\hat{\mathbf{z}} \cdot (\mathbf{V}_H \times \mathbf{u})$ occurs at the same degree (3) at which the otherwise negative correlation between Ge and $\mathbf{V}_H \times \mathbf{u}$ disappears. Clearly, the coupling between the spheroidal and

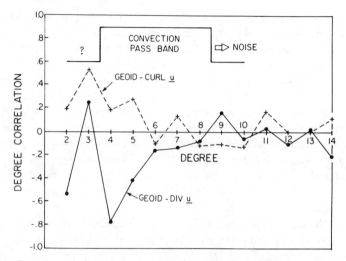

Figure 4   Degree correlations between geoid height Ge and surface divergence and curl over the spectral range 2 ≤ *n* ≤ 14. From A. Forte & W. R. Peltier (in preparation).

toroidal components of the flow is most intense at degree 3, which happens (see Section 4.1) to be the most unstable scale for a spherical shell of mantle dimensions, according to the predictions of linear theory.

In order to understand the results of such correlation analyses more fully, we must compare the geographic form of the fields over the range of wave numbers for which consistent correlations exist. Figures 5*a* and 5*b* show maps of $V_H \cdot u$ and Ge in the spectral range 3 ≤ *l* ≤ 8; inspection of these figures shows that through this band of wave numbers in which Ge and $V_H \cdot u$ are negatively correlated, positive geoid height anomalies are generally coincident with regions of convergence (trenches), whereas negative geoid anomalies are associated with regions of divergence (ridges) except in the North Atlantic (where spreading rates are extremely low). Since ridges are regions of positive topography and trenches regions of negative topography, it is clear that the Ge signal in both regions must be dominated by internal density variations and not by the topographic relief. The satellite-derived field Ge may therefore be proxy for the internal buoyancy and the data taken to suggest that there is deep thermal support in the Earth for *both* trenches *and* ridges. Plates *c* and *d* of Figure 5 show the degree-2 component of the pattern of surface divergence and the same component of the spectrum of lateral density heterogeneity, which has recently been inferred (Masters et al. 1982, Woodhouse & Dziewonski 1984) from the observed frequency splitting of the Earth's normal modes of elastic gravitational free oscillation. The two patterns are almost identical. If

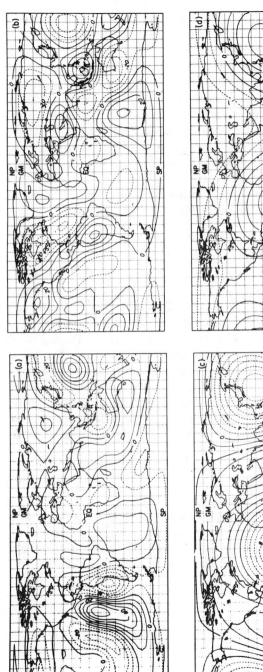

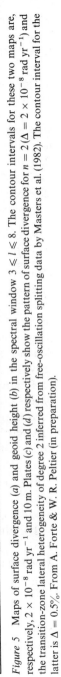

Figure 5 Maps of surface divergence (*a*) and geoid height (*b*) in the spectral window $3 \leqslant l \leqslant 8$. The contour intervals for these two maps are, respectively, 2×10^{-8} rad yr^{-1} and 10 m. Plates (*c*) and (*d*) respectively show the pattern of surface divergence for $n = 2$ ($\Delta = 2 \times 10^{-8}$ rad yr^{-1}) and the transition-zone lateral heterogeneity of degree 2 inferred from free-oscillation splitting data by Masters et al. (1982). The contour interval for the latter is $\Delta = 0.5\%$. From A. Forte & W. R. Peltier (in preparation).

Masters et al. (1982) are correct in their claim that the source of this lateral heterogeneity resides at the depth of the transition zone (420–670 km), then it is clear that the deep thermal support beneath the major regions of subduction (Chile, Kurile, and Java trenches) extends to very great depth indeed and remains extraordinarily well correlated with the surface kinematics. There is some suggestion, however, (e.g. Kawakatsu 1983) that the lateral heterogeneity observed from the free-oscillation frequency splittings may be at least partly located within the lithosphere. In this case its strong correlation with the surface velocity field follows from the fact that at least the oceanic component of the lithosphere (upper ~ 100 km of the Earth in which temperatures are sufficiently low that viscosity is effectively infinite) is the cold upper thermal boundary layer of the mantle convective circulation itself (Peltier 1980).

2.3 Ocean-Floor Heat Flow and Topography: The Lithosphere as Thermal Boundary Layer

Figure 6 shows the topography of and heat flow through the seafloor as a function of seafloor age from Parsons & Sclater (1977), based upon data from the Deep Sea Drilling Project (DSDP). The oceanic depth data (Figure 6a) are for the North Pacific, whereas the heat-flow information (Figure 6b) is a combination of measurements from the Pacific, South Atlantic, and Indian Oceans. The solid lines on each plate are a best-fitting curve through the data, while the dashed lines (which overlap the solid line on Figure 6b) are curves that follow a $t^{1/2}$ law. The fact that both of these observations closely follow the $t^{1/2}$ behavior may be understood on the basis of the assumption that vertical diffusion through the oceanic plate is precisely balanced by horizontal advection. Subject to this boundary-layer approximation, we have

$$u \frac{\partial T}{\partial x} = \kappa \frac{\partial^2 T}{\partial z^2}, \tag{5}$$

where u is the horizontal component of velocity, T the temperature field, and κ the thermal diffusivity. Under the similarity transformation $\eta' = z/x^{1/2}$, Equation (5) is reduced to an ordinary differential equation in η', which may be nondimensionalized by writing $\eta' = 2\eta(\kappa/u)^{1/2}$ to give

$$\frac{d^2 T}{d\eta^2} + 2\eta \frac{dT}{d\eta} = 0. \tag{6}$$

Subject to the boundary conditions $T = T_s$ at $x = 0$ and $T \to T_m$ as $z \to \infty$, the dimensional solution is

$$T = T_s + (T_m - T_s) \, \text{erf} \left[\frac{z}{2} \left(\frac{u}{\kappa x} \right)^{1/2} \right]. \tag{7}$$

The surface heat flow associated with this temperature solution is

$$q_s = K \left. \frac{\partial T}{\partial z} \right|_{z=0} = K(T - T_s) \left(\frac{u}{\pi \kappa x} \right)^{1/2}. \tag{8}$$

Since the age t of the lithosphere is just x/u, this boundary-layer cooling model very nicely explains the heat-flow observations shown in Figure 6b. From (7) we may also calculate the topography of the seafloor on the basis of the assumptions that it is isostatically compensated and that density variations through the boundary layer are associated only with temperature variations. Assuming compensation, Turcotte & Oxburgh (1972) showed that the depth of the seafloor beneath the surface of the ocean should be

$$D = D_{ref} - \frac{\alpha}{1 - \rho_w/\rho} \int_0^\infty (T - T_{ref}) \, dz, \tag{9}$$

where D_{ref} is the reference depth at the ridge crest where $T_{ref} = T_s$, α is the coefficient of thermal expansion, and ρ_w and ρ are respectively the densities of water and of the Earth. Substitution from (7) into (9) gives

$$D = D_{ref} - \frac{2\alpha(T_m - T_s)}{1 - \rho_w/\rho} \left(\frac{\kappa}{\pi} \right)^{1/2} \left[\left(\frac{x_{ref}}{u} \right)^{1/2} - \left(\frac{x}{u} \right)^{1/2} \right], \tag{10}$$

which again predicts a $t^{1/2}$ dependence and agrees with the observations shown in Figure 6a. The slight flattening of the topography away from the $t^{1/2}$ prediction for ages greater than ~ 60 Myr has been invoked as evidence for the existence of a second scale of convection supported by boundary-layer instability under old seafloor (Richter & Parsons 1975); however, this idea has proven to be rather contentious (e.g. Yuen et al. 1981), since there appear to be several different physical processes that could produce such flattening. This point is discussed further in Section 4.3. The fact that horizontal advection balances vertical diffusion in the oceanic lithosphere establishes that this part of the Earth's surface is intimately involved in the mantle convective circulation. In order to construct a hydrodynamic theory for this circulation, however, we need to establish values for the properties of the mantle fluid. Specifically, we require a measurement of the effective viscosity of its constituent material and some clear indication as to whether the internal density stratification is adiabatic or nonadiabatic. To infer these properties we analyze the nature of the slow gravitationally forced free-surface flows that are driven by surface loading of the planet by the large ice masses that existed on its surface 18,000 years ago.

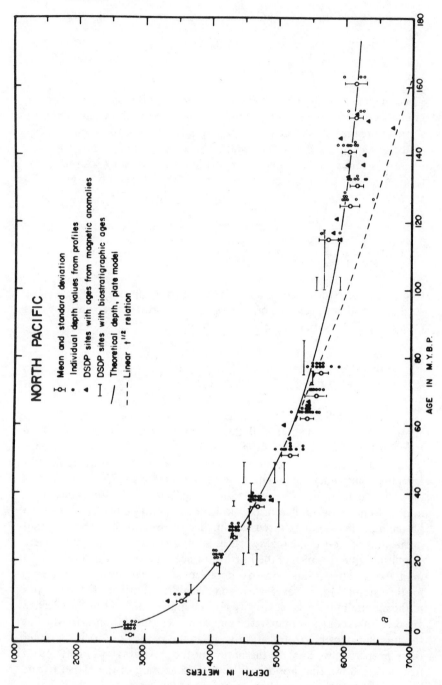

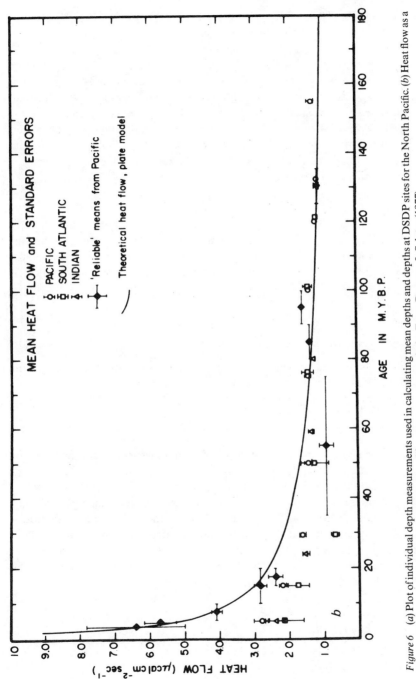

Figure 6 (*a*) Plot of individual depth measurements used in calculating mean depths and depths at DSDP sites for the North Pacific. (*b*) Heat flow as a function of the ocean-floor age. This follows a $t^{1/2}$ reference curve precisely. From Parsons & Sclater (1977).

3. THE MANTLE AS A FLUID: INFERENCES FROM GLACIAL ISOSTASY

Over the past decade, beginning with the article by Peltier (1974), a complete linear viscoelastic field theory has been developed to interpret the observed responses of the planet to gravitational interaction with the huge ice sheets that existed on the Northern Hemisphere continents at ice-age maximum (18,000 years ago). A fairly complete review of the results obtained with this theory up to 1981 can be found in Peltier (1982). In this section I briefly review those most recent results that have direct bearing on the mantle-convection issue.

3.1 The Linear Viscoelastic Field Theory

The small-amplitude creeping flow that is gravitationally induced in the Earth in response to surface loading is one that satisfies the following Laplace-transform domain forms of the equations of momentum balance and continuity, and for the perturbation of the gravitational potential:

$$\nabla \cdot \tau - \nabla(\rho_0 g_0 \mathbf{u} \cdot \hat{\mathbf{e}}_r) - \rho_0 \nabla \phi_1 - g_0 \rho_1 \hat{\mathbf{e}}_r = 0, \tag{11a}$$

$$\rho_1 = -\rho_0 \nabla \cdot \mathbf{u} - \mathbf{u} \cdot (\partial_r \rho_0) \hat{\mathbf{e}}_r, \tag{11b}$$

$$\nabla^2 \phi_1 = 4\pi G \rho_1, \tag{11c}$$

which have been linearized in small deviations from a background hydrostatic equilibrium configuration (ρ_0, p_0, ϕ_0) that satisfies

$$\nabla p_0 = -\rho_0 g_0 \hat{\mathbf{e}}_r, \tag{12a}$$

$$\nabla^2 \phi_0 = 4\pi G \rho_0. \tag{12b}$$

In (11), τ is the stress tensor, $\mathbf{u}$ the displacement vector, ϕ_1 the perturbation of the gravitational potential, ρ_1 the density perturbation, and G the gravitational constant. In general, ϕ_1 is the sum of two parts, ϕ_2 and ϕ_3, which are respectively the potential of any externally applied gravitational force field (the load) and the potential due to the internal redistribution of planetary mass forced by the load-induced deformation. In the Laplace-transform domain of the variable s, the stress-strain relation which relates τ to the strain tensor $\mathbf{e}$ is that for a three-dimensional Maxwell solid with no bulk dissipation (Peltier 1974); it has the form

$$\tau_{ij} = \lambda(s) e_{kk} \delta_{ij} + 2\mu(s) e_{ij}, \tag{13}$$

where the moduli are

$$\lambda(s) = \frac{\lambda s + \mu K/\nu}{s + \mu/\nu}, \tag{14a}$$

$$\mu(s) = \frac{\mu s}{s + \mu/v}. \tag{14b}$$

Clearly, in the limit $s \to \infty$ $(t \to 0)$ the moduli reduce to the usual elastic Lamé parameters, and the constitutive relation (13) reduces to Hooke's law. In the limit $s \to 0$ $(t \to \infty)$, on the other hand, we have $\lambda(s) \to K$ and $\mu(s) \to sv$, so that the constitutive relation effectively reduces to that for a Newtonian viscous fluid with molecular viscosity v. In this model the bulk modulus K is given by $K = \lambda + 2\mu/3$. Our problem is to solve (11) for the ice-sheet-loading-induced deformations of the Earth and, by fitting predictions of the model to the observations, to infer $v(r)$ in the interior of the planet.

It is crucial to our implementation of the so-called principle of correspondence, which is used to obtain the time-domain solution of (11) from that in the Laplace-transform domain, that the quasi-static momentum-balance equation (11a) contain the second term on the left-hand side. This describes the extra body force to which the system is subject in consequence of the fact that it is in a state of hydrostatic prestress prior to the time the surface load is applied. It is through the presence of this term that effectively nonadiabatic density discontinuities in the interior intro-duce extra internal buoyancy when they are deflected from their equilib-rium positions. As well as employing the theory to infer the viscosity of the mantle, we can also use it to infer the presence of any such internal buoyancy-producing discontinuities. We do not discuss here the technical details of the methods by which the above field equations are solved; rather, we focus on the most relevant recent results of application of this theory. The interested reader will find a recent summary of mathematical methods in Peltier (1982).

3.2   Mantle Viscosity and Postglacial Sea Levels

The primary data base for the study of mantle viscosity consists of relative sea-level histories over the past 10–15 kyr since the melting of the major ice sheets began about 18,000 years ago. The observed variation of sea level at any location is obtained by measuring the height above or depth below present-day sea level of a sequence of relict beach horizons of age determined by application of ^{14}C dating techniques. Typical examples of such data are shown in Figure 7, which is from Wu & Peltier (1983). At sites that were well within the ice margins like the Ottawa Islands in present-day Hudson Bay, the oldest beach is found at the greatest height (> 100 m) above the present level of the sea, and the uplift of the region (glacial rebound) has proceeded at an exponentially decreasing rate through time. At sites near the ice-sheet margin like the city of Boston, on the other hand, the relative sea-level histories are strikingly nonmonotonic, with initial

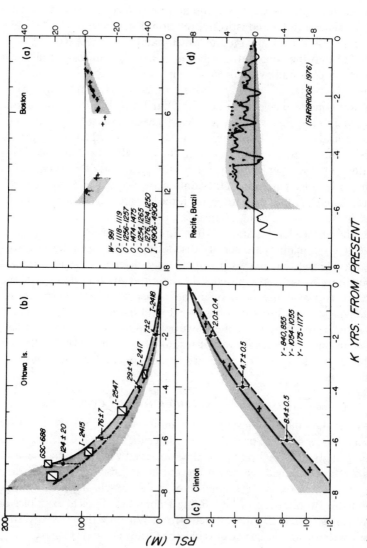

Figure 7   Examples of relative sea-level data from four different geographic locations. The Ottawa Islands are in Hudson Bay, which was under the Laurentian ice sheet; Boston and Clinton were near the ice margin; while Recife, Brazil was in the far field of the ice loads.

emergence followed by submergence that is ongoing today. Further south along the US east coast at Clinton, Connecticut, the sea-level history is one of monotonic submergence, while at a very distant site like Recife, Brazil, the sea-level record is marked by the appearance of a raised beach at about 6000 years ago, after which the coast has been continuously uplifted, but at a rate that has been a decreasing function of time.

For purposes of determining mantle viscosity, the most useful sea-level records are those from sites that were once ice covered. Since the disintegration of the ice sheet over Canada (centered on Hudson Bay) was essentially complete by about 6000 years ago, it is clear from the fact that the land has continued to rise out of the sea since that time that the Earth's mantle is not Hookean elastic. If it were, then deformation would have ceased with the cessation of melting. We seek to determine $v(r)$ by fitting the predictions of our viscoelastic field theory to the sea-level data. Using a well-constrained a priori model for the glaciation history, such as that presented in Peltier & Andrews (1976) and recently refined in Wu & Peltier (1983), we predict local sea-level histories by solving an integral equation that directly predicts the time-dependent separation of the geoid and the surface of the solid Earth. This integral equation follows simply by application of the constraint that the meltwater produced by glacial disintegration be distributed over the surface of the global ocean in such a way that the ocean surface remains equipotential. The solution of this integral equation was discussed in detail in Peltier et al. (1978), while a preliminary analysis of its predictions appeared in Clark et al. (1978) and Farrell & Clark (1976). An extensive sequence of more recent solutions is provided in Wu & Peltier (1983). Figure 8 shows the results of one such sequence of analyses for six sites in the high Canadian Arctic at which good ^{14}C-controlled relative-sea-level (RSL) curves exist. Comparisons between theory and observations are shown for three different models of the internal viscoelastic layering that differ from one another only in $v(r)$, with the elastic structure fixed to that of model 1066B of Gilbert & Dziewonski (1975). The models all have a surface lithospheric thickness of $L = 120.7$ km (in which $v = \infty$) and an upper-mantle viscosity of 10^{21} Pa s to a depth of 670 km. The different predictions shown are for lower-mantle viscosity $v_{LM} = 10^{21}, 10^{22}$, and 5×10^{22} Pa s. The data are best fit with the uniform-viscosity model, since although the model with $v_{LM} = 10^{22}$ Pa s fits the amplitudes at 8000 years ago reasonably well, it predicts present-day emergence rates that are high by almost a factor of two. On the other hand, the model with $v_{LM} = 5 \times 10^{22}$ Pa s fits the present-day emergence rates well but predicts far too little emergence over the past 8000 years. These data constrain the upper-mantle viscosity close to the value $v_{LM} = 10^{21}$ Pa s and require $10^{21} \leqslant v_{LM} \leqslant 10^{22}$ Pa s. In the next section, we address the question of

ICE 2

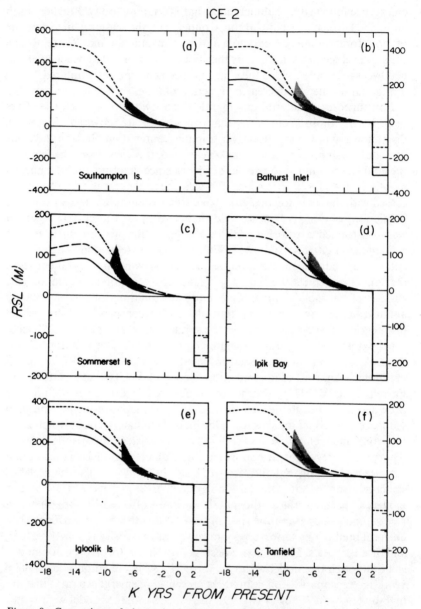

Figure 8 Comparison of observed relative sea level (hatched regions) and theoretically predicted RSL at six sites in the Canadian Arctic. The models have 1066B elastic structure, an upper-mantle viscosity of 10^{21} Pa s. The short-dashed, long-dashed, and solid curves are predictions for models with lower-mantle viscosities of 10^{21}, 10^{22}, and 5×10^{22} Pa s, respectively.

whether a theory of plate tectonics based on the thermal-convection hypothesis is compatible with this (rather large!) value of the viscosity. Before doing so, however, we briefly discuss several of the other signatures of glacial isostasy that may be invoked to constrain more tightly the above viscosity estimate.

3.3   Mantle Viscosity, $\dot{J}_2$, and the Nontidal Acceleration of Planetary Rotation

Given the mass of the ice sheets, whose melting caused the sea-level variations discussed in the last section [$\sim 2 \times 10^{19}$ kg from the Laurentian (Canadian) ice sheet, $\sim 0.6 \times 10^{19}$ kg from the Fennoscandian (northwestern European) ice sheet, and $\sim 0.7 \times 10^{19}$ kg from the West Antarctic ice sheet), it should not be too surprising to learn that the disintegration of these large ice masses induced substantial changes in the Earth's rotation, causing both the rate of rotation to change and the pole itself to wander with respect to the surface geography. Even though the ice sheets have long since disappeared, their influences on these two components of the planetary rotation are still evident today because of the slow viscous flow in the mantle induced by the associated gravitational imbalance. In the past several years, it has been increasingly recognized that the nontidal acceleration of planetary rotation that had been inferred from analyses of ancient eclipse data (e.g. see Morrison 1973, Muller & Stephenson 1975, and Morrison & Stephenson 1982) was due, as originally suggested by Dicke (1966), to the influence of deglaciation. Recent analyses of this effect include those by Sabadini & Peltier (1981), Peltier (1982), and Peltier & Wu (1983). Very recently this datum has been improved as a consequence of a new direct method of measurement using laser ranging data to the *LAGEOS* satellite, which has been shown to require (Yoder et al. 1983, Peltier 1983, 1984a, Rubincam 1984) a present-day secular variation of J_2 (the degree-2 zonal coefficient in the spherical harmonic expansion of the Earth's gravitational potential field) in the amount of $(-3.5 \pm 0.3) \times 10^{-11}$ yr^{-1}.

The theoretical prediction of $\dot{J}_2$ or, equivalently, of the nontidal acceleration of planetary rotation $\dot{\omega}_3 / \Omega$, since the two are linearly related, takes the form (e.g. Peltier 1982)

$$\frac{\dot{\omega}_3}{\Omega} = -\frac{2 m_e a^2}{3C} \dot{J}_2 = -\frac{I_{33}^R}{C} \left[D_1 \dot{f}(t) + \sum_{j=1}^{M} r_j^2 \frac{d}{dt} (f * e^{-s_j^2 t}) \right], \tag{15}$$

where C is the Earth's principal component of inertia, I_{33}^R is the axial inertia perturbation that would be produced by the surface ice and water load if the Earth were rigid, $D_1 = 1 + k_2^E$, and $f(t)$ is a function that describes the history of glaciation and deglaciation (equal to 1 at glacial maximum and 0 during an interglacial), while r_j^2 and s_j^2 are the amplitudes and inverse relaxation times, respectively, of the M normal modes that are required to synthesize the time dependence of the Love number k of degree 2 (Peltier 1976). The load history $f(t)$ is taken to be a periodic sawtooth function of period 10^5 yr, with individual pulses characterized by a 90-kyr linear rise and a linear 10-kyr collapse. This function is well constrained by $^{18}O/^{16}O$ data from deep-sea cores (Hays et al. 1976), which are proxy for past fluctuations in the volume of continental ice (Shackleton 1967). Figure 9, from Peltier (1983), shows the results of a sequence of predictions of the $LAGEOS$ observation for a suite of models, all of which (again) have 1066B elastic structure and which differ from one another only in their viscosities beneath 670-km depth (labeled ν_{LM}). Calculations are shown for models including 1 (Laurentia), 2 (Laurentia + Fennoscandia), and all 3 (+ Antarctica) ice sheets. With the upper-mantle viscosity fixed at 10^{21} Pa s, we require $2.7 \times 10^{21} \leqslant \nu_{LM} \leqslant 4.4 \times 10^{21}$ Pa s in order to fit the observed $\dot{J}_2$. This is consistent with the result from the sea-level data discussed in the last section. Although the $\dot{J}_2$ observation can also be fit with an extremely high value for the lower-mantle viscosity (near 10^{23} Pa s), this possibility is ruled out by the sea-level data. The above bounds on the viscosity of the

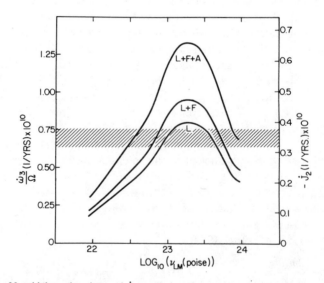

Figure 9   Nontidal acceleration and $\dot{J}_2$ predictions for models with 1, 2, or 3 ice sheets as a function of the lower-mantle viscosity ν_{LM}. Model L includes the Laurentian (Canadian) ice sheet only, L + F the Fennoscandian ice sheet also, and L + F + A the large-scale deglaciation of West Antarctica as well. The $\dot{J}_2$ inferred from the orbit of the $LAGEOS$ satellite is shown as the hatched region. The elastic Earth model is 1066B, and the upper-mantle viscosity is 10^{21} Pa s.

deep mantle are the best currently available and are extremely important, as we shall see, insofar as the construction of mantle-convection models is concerned.

3.4 Mantle Viscosity and the Anomalous Gravitational Field

In all three regions of the Earth that were once covered by large continental ice masses, the surface of the Earth remains depressed somewhat below its height of gravitational equilibrium. In consequence, if one subtracts from the local vertical component of the gravitational acceleration the value for an equilibrium reference spheroid at the same latitude, one finds a negative anomaly at each of these locations. Over Hudson Bay the peak anomaly Δgm is $-40 \leqslant \Delta gm \leqslant -30$ mgal (Walcott 1970), for Fennoscandia the range is $-20 \leqslant \Delta gm \leqslant -15$ mgal (Balling 1980), while for West Antarctica (Bentley et al. 1982) it is near that in Fennoscandia. Our theoretical prediction of the free air anomaly makes use of Green's function G for this signal as (e.g. Longman 1963)

$$G(\theta, t) = \frac{g}{m_e} \sum_{l=0}^{\infty} [l + 2h_l - (l+1)k_l]P_l (\cos \theta), \tag{16}$$

in which h_l and k_l are the surface local Love numbers of degree l. To predict Δg we simply convolve this Green function with the space- and time-dependent surface load. This calculation, because the observation provides an absolute measure of the current degree of gravitational disequilibrium, is found to be particularly sensitive to the degree of gravitational disequilibrium that existed prior to deglaciation at 18 kyr BP (before present), and it is also extremely sensitive to the existence of any internal buoyancy-generating interfaces in the mantle (as we shall see).

With the history of loading again constrained by $^{18}O/^{16}O$ data from deep-sea cores, and 20 prior 10^5-yr glacial cycles employed in the calculation, Figure 10 (Peltier & Wu 1982, Wu & Peltier 1982) compares the observed free-air gravity anomaly with the predicted value (for disk-load models of the ice sheets) as a function of the mantle viscosity beneath 670-km depth. The upper-mantle viscosity is fixed at 10^{21} Pa s, and the 1066B structure is employed to constrain the elastic component of the model. Results are shown for both Fennoscandia and Laurentia. Three different calculations are shown on each plate: The dash-dotted line is the prediction based upon the assumption that isostatic equilibrium prevails initially; the dashed line includes the influence of initial disequilibrium but assumes that the ice-sheet radius remains fixed while the volume fluctuates; and the most accurate prediction, the solid line, incorporates the effect of simultaneous volume and radius variations under the assumption that the

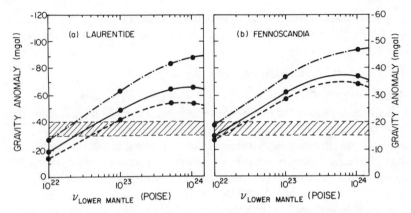

Figure 10 Observed peak free-air gravity anomalies over Fennoscandia and Laurentia (hatched regions) and predictions as a function of lower-mantle viscosity v_{LM}.

ice sheet maintains a plastic profile at all times. The lower-mantle viscosity preferred by the data is again in accord with that deduced on the basis of the previously discussed relative sea-level fluctuations and $\dot{J}_2$ observation. The ability of this model to fit the gravity data, however, is due to the fact that the free-air-anomaly prediction contains an important contribution from very long time-scale modes of viscous gravitation relaxation; these modes are supported by internal density discontinuities in the mantle, which are able to produce a buoyant restoring force when they are deflected from their equilibrium positions by ice-sheet loading (Peltier 1981b, 1982, Peltier & Wu 1982, Wu & Peltier 1983). In the 1066B structure the crucial mode is that called M1 by Peltier (1976), which is supported by the $\sim 6.2\%$ density discontinuity at 670-km depth in this model. This may be of critical importance to the construction of models of the mantle convective circulation, as is argued in the next section.

3.5 Mantle Discontinuities and Internal Buoyancy

The strong sensitivity of the Δg prediction to the presence of long-relaxation-time modes supported by small internal-mantle density discontinuities is due to the fact (mentioned previously) that the gravity anomaly provides an absolute measure of the current degree of disequilibrium and thus of the amount of viscous relaxation that has yet to occur before equilibrium will be achieved. The relative sea-level and nontidal acceleration data, on the other hand, are relatively insensitive to this physics. In a relaxation spectrum containing a range of relaxation times it is clear that the greater the elapsed time over which free relaxation has been occurring, the greater the fraction of the remaining relaxation carried by the modes with longest relaxation time. Figure 11 shows a relaxation diagram for the elastic Earth model 1066B with a 120.7-km-thick lithosphere and a constant mantle viscosity of 10^{21} Pa s. At degree $n = 6$, the M1 mode has a relaxation time of about 300,000 yr. The other modes in this diagram are supported at the Earth's surface (M0), the core-mantle boundary (C0), the mantle-lithosphere interface (L0), and the small discontinuity at 420-km depth (M2) associated with the olivine-spinel phase transformation. All of the density variation in 1066B is treated as nonadiabatic, and all discontinuities of density then support unique modes of relaxation. Table 1 shows the relative strength of each mode in the spectrum for a range of spherical harmonic degrees. Note that M2 is relatively unimportant compared with M1, which carries about 14% of the relaxation at degree 2 and 10% at degree 6. Since this is the dominant scale of the Laurentian ice sheet (i.e. $n = 6$), it should be clear that this internal mode will make a strong contribution to the observed Δg.

In order to demonstrate the importance of this effect explicitly, we show in Figure 12 a sequence of peak free-air gravity anomalies from calculations for simplified Earth models that are incompressible in their interiors [see

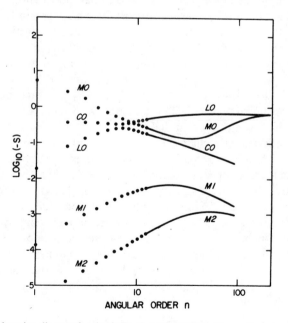

Figure 11   Relaxation diagram for elastic Earth model 1066B with a uniform mantle viscosity of 10^{21} Pa s and an elastic lithosphere of thickness 120.7 km. The five main modal branches are discussed in the text. The inverse relaxation times $-S$ have been nondimensionalized with a characteristic time of 10^3 yr.

Table 1   Strengths (in %) of viscous modes for h and k Love numbers in Earth model 1066B

l	M0		M1		M2		L0		C0		C1	
	h	k	h	k	h	k	h	k	h	k	h	k
2	39.1	57.5	15.6	4.4	2.8	0.3	2.5	2.3	37.0	35.0	3.0	0.5
3	49.3	62.7	14.0	4.4	2.4	0.4	1.5	1.5	30.9	30.6	1.9	0.4
4	48.4	59.9	13.1	4.6	2.2	0.4	1.1	1.2	25.1	25.5	1.1	0.3
5	60.8	71.1	12.5	4.9	2.1	0.5	1.1	1.2	18.9	19.4	0.6	0.2
6	69.2	78.4	11.9	5.1	2.0	0.5	1.8	1.9	12.2	12.5	0.3	0.1
7	74.7	83.0	11.3	5.2	1.9	0.5	7.2	7.4	2.8	3.0	0.2	0.1
8	71.7	78.8	10.6	5.2	1.9	0.6	2.0	2.1	12.1	12.6	0.1	0.0
9	85.9	91.8	9.9	5.1	1.8	0.6	1.4	1.5	1.0	1.0	0.0	0.0
10	86.6	92.6	9.3	5.0	1.7	0.6	0.7	0.8	0.4	0.5	0.0	0.0
15	91.6	95.1	6.2	3.9	1.4	0.7	0.0	0.0	0.0	0.0	0.0	0.0
20	94.6	96.6	3.8	2.6	1.1	0.6	0.0	0.0	0.0	0.0	0.0	0.0
25	96.7	97.8	2.2	1.5	0.8	0.5	—	—	—	—	—	—
50	99.7	99.8	—	—	—	—	0.1	0.1	—	—	—	—
90	99.7	99.8	—	—	—	—	0.1	0.1	—	—	—	—

Figure 5 of Wu & Peltier (1984) for additional information]. The elastic structures are regional averages of 1066B for the upper and lower mantles and differ from one another in their lower-mantle viscosities and in the density jump $\Delta\rho$ at 670-km depth [$\Delta\rho/\rho$ (670) is labeled in percent adjacent to the appropriate curve on the figure]. Forcing is from a disk-load approximation to the Laurentian ice sheet (mass 2×10^{19} kg and radius $15°$), and the load history contains 20 prior 10^5-yr glacial cycles as for the calculations shown on Figure 11 for 1066B. Dashed curves are predictions based on the assumption of initial (i.e. $-18,000$ yr) isostatic equilibrium, while the solid curves include the influence of initial disequilibrium on the basis of the assumption that the ice-sheet radius remains fixed while the volume fluctuates. Inspection of these results shows that it is apparently not possible to fit the observed gravity anomaly over Hudson Bay with a model that contains no source of internal mantle buoyancy. Indeed, in the context of the incompressible homogeneous layered models employed here, we need $\Delta\rho/\rho$ (670) $\approx$ 12% in order to fit the observed Δgm over Hudson Bay with a viscosity of the deep mantle near that demanded by the previously discussed observations of relative sea level and nontidal acceleration of rotation. With $\Delta\rho/\rho$ (670) equal to the seismically constrained value near 6.2%, however, the same effect is achieved by adding the buoyancy due to the 420-km discontinuity.

The above analysis is important for the following reason. Although high-pressure experiments in the diamond anvil apparatus (Yagi et al. 1979, Jeanloz & Thompson 1983) imply that the 670-km seismic discontinuity

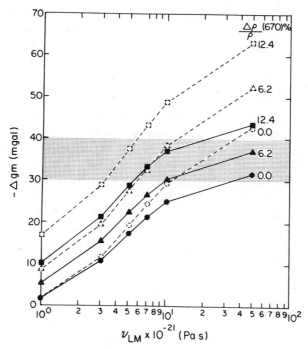

Figure 12 Observed free-air gravity anomaly over Hudson Bay and predictions for a sequence of homogeneous two-layered mantle models as a function of lower-mantle viscosity and the density increased at 670-km depth. The solid curves include the effect of initial isostatic disequilibrium, while the dashed curves are predictions based upon the assumption that isostatic equilibrium prevails initially.

could be entirely explained as an equilibrium phase transformation, seismic reflections from this boundary apparently require it to be significantly sharper than is easily explicable in these terms (Anderson 1981, Lees et al. 1983). It would clearly be useful if it were possible to distinguish, on the basis of the rebound data alone, whether the above demonstrated requirement for the existence of a buoyancy-generating interface in the mantle could be construed as also requiring that the interface be of chemical origin. Clearly, if material could change phase immediately after the boundary were displaced, then a phase boundary would not be able to generate any buoyant restoring force subsequent to displacement. Although the time scale for the candidate phase transformation in laboratory diamond anvil experiments *is* rather short, the time scale in the Earth turns out to be extremely long. The reason for this is that in order for material to change phase, the latent heat of transformation must be removed, and this occurs only on the thermal-diffusion time scale. The solution to an appropriate Stefan problem (e.g. O'Connell 1976, Mareschal & Gangi 1977) shows that the time scale on which the phase boundary actually migrates is

$$\tau = \left[\frac{T_m \sqrt{k \Delta \rho / \rho_1 \rho_2}}{2\gamma^{-1} K(\gamma^{-1} g \rho - \beta)} \right]^2,$$

in which T_m is the temperature at the interface, γ the Clapeyron slope, β the local vertical temperature gradient, and $\Delta \rho = \rho_2 - \rho_1$ the density difference between the phases. For the spinel $\rightarrow$ perovskite + magnesiowustite transition at 670-km depth (which has $\gamma \sim -3$ MPa K^{-1}) this time scale is approximately 500,000 yr. Since this is long compared with the loading time scale of $\sim 10^5$ yr and with the time scale of $\sim 10^4$ yr that governs the dominant modes of viscous gravitational relaxation, it seems clear that even if the internal mantle discontinuities are phase boundaries, they would act as though they were chemical on the time scale of rebound processes. In either event they would produce the buoyant restoring force required by the Δg observations. We may not therefore expect to understand whether the required layering is adiabatic or nonadiabatic by studying glacial isostatic adjustment processes.

3.6   Mantle Viscoelasticity and Deglaciation-Induced True Polar Wander

The last of the signatures of glacial isostasy that might be invoked to constrain the internal mantle viscoelastic structure is the presently occurring wander of the rotation pole at the rate of $0.95°/10^6$ yr along the $76 \pm 5°$ West meridian toward Greenland (e.g. Dickman 1977). That this is also a response to the Pleistocene glacial cycle has been argued in several recent articles (Nakiboglu & Lambeck 1980, 1981, Sabadini & Peltier 1981, Peltier 1982, Peltier & Wu 1983, Wu & Peltier 1984). The first mathematically correct solution for the forced polar wander of a layered viscoelastic Earth was obtained in Peltier (1982) and is discussed at length in Wu & Peltier (1984). The form of the theoretical prediction of this quantity in this new theory is

$$\frac{\dot{\omega}_i}{\Omega} = \frac{\Omega}{A\sigma_0} I_{j3}^R \left[D_1 \dot{f}(t) + D_2 f(t) + \sum_{i=1}^{M-1} E_i \frac{d}{dt} (f * e^{-\lambda_i t}) \right], \tag{17}$$

where the numbers $D_1, D_2, E_i,$ and λ_i are defined in Peltier (1982). The main difference between the mathematical structure of the polar-wander solution (17) and that for the nontidal acceleration (15) is in the fact that the history-dependent term in the present case involves a sum over $M-1$ modes rather than the M modes that govern the evolution of $\dot{J}_2$. This is a consequence of

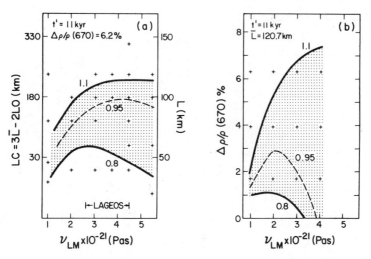

Figure 13   Trade-off curves in (L, v_{LM}) and $(\Delta\rho/\rho, v_{LM})$ space for fits to the polar wander observed in the pole path computed by the International Latitude Service for the period since 1900 A.D. Any point within the boundaries fits the observed speed of $0.95°/10^6$ yr within a standard error.

the fact, first pointed out by Peltier (1982), that when the surface load is zero ($f \equiv 0$), $\dot{\omega}_{1,2}/\Omega$ differs from zero only because the centrifugal and isostatic adjustment contributions to the forcing just fail to cancel one another. In fact, for a homogeneous Earth model that has only one mode of relaxation, they do *exactly* cancel, and such a model therefore cannot explain the present-day observed wander of the rotation pole. The nonzero polar-wander speed is therefore carried by a term involving fewer (by one) exponential time constants than are required to describe the isostatic adjustment process itself. Figure 13 (from Peltier 1984b) shows a series of fits to the $\dot{\omega}_{1,2}/\Omega$ datum in terms of trade-off curves in (L, v_{LM}) and $(\Delta\rho/\rho$ (670), $v_{LM})$ space and illustrates the marked sensitivity of this datum to all three of these parameters. (Upper-mantle viscosity v_{UM} is again held fixed at 10^{21} Pa s, and L is lithospheric thickness.) Elastic structures are the same flat models employed in Section 3.5. Clearly the previously derived viscous and elastic structure is also compatible with this datum, but the datum is not a terribly useful one for constraining the parameters of this structure.

4.   SIMPLE HYDRODYNAMIC MODELS OF MANTLE CONVECTION

Given that the mantle viscosity is close to the value $v = 10^{21}$ Pa s and that its other physical properties are known to be close to the values listed in

Table 2   Physical properties of the mantle

$$2 \times 10^{-5} \leqslant \alpha \leqslant 3.5 \times 10^{-5} \text{ K}^{-1}$$
$$10^{-6} \leqslant \kappa \leqslant 2 \times 10^{-6} \text{ m}^2 \text{ s}^{-1}$$
$$10^{21} \leqslant \nu \leqslant 5 \times 10^{21} \text{ Pa s}$$
$$C_p \approx 1.25 \times 10^3 \text{ J kg}^{-1} \text{ K}^{-1}$$
$$\rho \approx 4500 \text{ kg m}^{-3}$$

Olivine-spinel phase change	Spinel-perovskite phase change
$0.046 \leqslant \Delta\rho/\rho \leqslant 0.05$	$\Delta\rho/\rho \approx 0.11$
$1800 \leqslant T \leqslant 1900$ K	$2300 \leqslant T \leqslant 2400$ K
$-0.3 \times 10^{-3} \leqslant \beta \leqslant -1.3 \times 10^{-3}$ K m^{-1}	$-0.3 \times 10^{-3} \leqslant \beta \leqslant -1.3 \times 10^{-3}$ K m^{-1}
$3 \times 10^6 \leqslant \gamma \leqslant 6.2 \times 10^6$ Pa K^{-1}	$-3 \times 10^6 \leqslant \gamma \leqslant -1 \times 10^6$ Pa K^{-1}
$0.03 \leqslant S \leqslant 0.10$	$-0.08 \leqslant S \leqslant -0.03$
$7.3 \times 10^5 \leqslant R_Q \leqslant 4.0 \times 10^6$	$-1.6 \times 10^5 \leqslant R_Q \leqslant 8.6 \times 10^5$
$(d \simeq 2.9 \times 10^6$ m$)$	$(d \simeq 2.9 \times 10^6$ m$)$

Table 2, we are in a reasonable position to attempt to test the ability of the convection hypothesis to reconcile the data discussed in Section 2. We perform this test in a sequence of steps.

4.1   Linear Stability of the Mantle Shell: The Influence of Endothermic and Exothermic Phase Transformations

The theory for the hydrodynamic stability of spherical shells is provided in Chandrasekhar (1961). If we use the nondimensionalization of Peltier (1973), the stability equation for the neutral boundaries may be cast in the form of a set of six simultaneous ordinary differential equations as

$$\frac{dY}{dr} = \mathbf{A}\mathbf{Y}, \tag{18}$$

where the solution 6-vector is given by $\mathbf{Y} = (w, w^1, w^{11}, \pi/\delta, \Theta, \Theta^1)$. Here, w is the radial-velocity perturbation, π/δ the scaled pressure fluctuation, and Θ the temperature perturbation. The matrix of coupling coefficients is a function of the Rayleigh number Ra ($= g\alpha\Delta T d^3/\kappa\nu$ when the shell is heated from below) and of the spherical harmonic degree l. The explicit form of the matrix is given by I. Kay & W. R. Peltier (in preparation), who focus their analysis on the influence of the exothermic and endothermic olivine-spinel and spinel-postspinel (spinel $\rightarrow$ perovskite + magnesiowustite) phase transitions at 420-km and 670-km depth. As first demonstrated by Busse & Schubert (1971) and further elaborated on by Schubert & Turcotte (1971), Peltier (1973), and Schubert et al. (1975), the effect of such boundaries on small-amplitude motions may be incorporated by imposing the following boundary conditions of discontinuity at the undisplaced position of the phase interface

$$\Theta_2 - \Theta_1 = \frac{1}{r} R_Q w(r),$$ (19a)

$$\frac{\pi_2}{\delta} - \frac{\pi_1}{\delta} = S\Theta,$$ (19b)

where region 1 (2) is the region of the light (heavy) phase of density ρ_1 (ρ_2). The nondimensional parameters R_Q and S are defined as follows (γ the Clapeyron slope and α the coefficient of thermal expansion):

$$R_Q = \frac{g\alpha d^3 Q}{\kappa v c_p}, \qquad S = \frac{\Delta\rho/\rho}{d\alpha\left(\dfrac{g\rho}{\gamma} + \dfrac{dT}{dr}\right)}.$$ (20)

Values for these parameters are estimated in Table 2 and may be compared with previous estimates by Schubert et al. (1975). Our estimates of R_Q are clearly several orders of magnitude higher than those of Schubert et al., and our values of S are similarly one or two orders of magnitude lower. There are two reasons for this. Firstly, the radial length scale d is chosen equal to the mantle thickness (the only natural length scale in the problem), and the value of v is taken compatible with that inferred from the isostatic adjustment data in Section 3. This new scaling has a profound effect because it ensures that the importance of the phase boundaries (in linear theory) will be governed by the latent-heat release (as measured by R_Q) and only marginally affected by the pressure difference arising from the density difference between the phases (as measured by S). For a spinel-postspinel phase change that has negative Clapeyron slope the latent heat release effect is strongly destabilizing, while for the olivine-spinel transition it is equally strongly stabilizing. Results from such analyses are illustrated in Figure 14, which shows the critical system Rayleigh number (minimum on the neutral curve) as a function of R_Q for two different values of S. The calculations are performed with a single olivine–spinel-type transition in the center of the shell (i.e. not its physical position). In the scaling of Schubert et al., we would be in the lower-left corner of this figure at small R_Q and S, where the critical wave number (l) is 3. In the new scaling we are in the upper-right corner, where R_Q is large and the critical wave number is 6, since the influence of the phase transition has caused the most unstable mode to degenerate to a two-layered radial structure. The destabilizing effect of the spinel-postspinel transition is so intense, however, that it may dominate the system stability when both transitions are in place to produce a single-layered structure once more (I. Kay & W. R. Peltier, in preparation). Analyses of the effects of

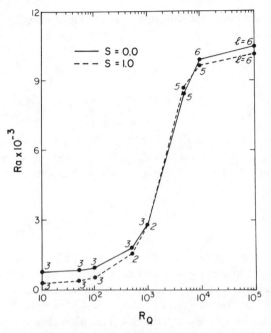

Figure 14 Critical Rayleigh number as a function of R_Q for two values of S. The phase change is of olivine–spinel-type and is located in the center of the spherical shell. The numbers adjacent to the curves denote the critical wave numbers at the onset of instability. From I. Kay and W. R. Peltier (in preparation).

phase boundaries upon finite-amplitude convection include those by Richter (1973), Christensen (1982), and Olson & Yuen (1982). All of these calculations have been done in plane layers of vertical length scale small compared with the full depth of the mantle, and so they are not properly scaled to deliver a correct assessment of the ability of a phase transition to modify the radial structure of the circulation in a shell of mantle depth. Further finite-amplitude calculations in more appropriate geometries are required to better assess the influence that such boundaries might have on the long-time-scale thermal circulation. We have already shown that they play a crucial role in the isostatic adjustment process.

4.2 *The Boundary-Layer Structure of High-Rayleigh-Number Flows*

Using the values for the physical parameters of the mantle listed in Table 2, we may estimate the Rayleigh number of the shell to be Ra $\simeq 10^7$ if the length scale d is taken equal to the mantle thickness (Sharpe & Peltier 1978), or Ra $\simeq 10^5$ for a length scale $d \simeq 670$ km such as would be appropriate to an upper-mantle-confined flow (e.g. McKenzie et al. 1974, Oxburgh & Turcotte 1978). Here we have used estimates of ΔT in excess of adiabatic from the melting point of iron at core-mantle boundary pressures and from the properties of the spinel-postspinel phase change for the whole-mantle and layered-mantle flows, respectively. Since these estimates of Ra are at least two orders of magnitude in excess of the critical value for the onset of convection, it is clear on a priori grounds that the mantle must be filled with a thermally forced circulation and that this circulation should be in the boundary-layer regime. Since the Prandtl number of the mantle, $\text{Pr} = \nu/\kappa$, is effectively infinite (from Table 2), we obtain reasonable qualitative estimates of the expected structure of the convective circulation by solving the following field equations in the Boussinesq limit for a two-dimensional plane layered Cartesian geometry:

$$\nabla \cdot \mathbf{u} = 0, \tag{21a}$$

$$-\nabla p = -\rho \hat{\mathbf{k}} + \alpha T_1 \nabla^2 \mathbf{u}, \tag{21b}$$

$$\frac{DT}{Dt} = \kappa^1 \nabla^2 T + \varepsilon^1. \tag{21c}$$

This form [e.g. see Jarvis & McKenzie (1980), Jarvis & Peltier (1982)] is useful for analysis of flows that are partly heated from below and partly from within, but it is important to note that effects expected because of the temperature dependence of viscosity and of melting have been entirely neglected. In this nondimensionalization, velocity is measured in units of $U = g\alpha T_1 d^2/\nu$, time in units of d/U, and pressure in units of $\rho_d g d$ (where ρ_d is the density adjacent to the top surface). The temperature is measured in units of $T_1 = \Delta T = (T_0 - T_d)$ if heating is from below; if a fixed heat flow through the base is imposed, then T_0 is undetermined and $T_1 = 1°C$ is chosen arbitrarily. The nondimensional parameters are $\kappa^1 = K\nu/g\alpha T_1 d^3 \rho_d C_p$ and $\varepsilon^1 = H\nu/g\alpha T_1^2 d\rho_d C_p$, in which H is the rate of internal heating per unit volume. Two more useful nondimensional parameters are

$$\mu = \frac{Hd}{F + Hd} \tag{22a}$$

and

$$R_B = \frac{g\alpha\Delta T d^3}{K\nu} \quad \text{or} \quad R_R = \frac{g\alpha d^4(F + Hd)}{\rho_d C_p K^2 \nu}, \tag{22b}$$

where (22a) is the ratio of the internal heating to the total heat flow through the upper surface, and (22b) is an appropriate Rayleigh number, R_B for heating from below (the Benard problem) and R_R (see Roberts 1967) for the case in which a particular heat flux through the lower boundary is maintained. Solutions to (21) may be constructed using a stream-function–vorticity formulation (McKenzie et al. 1974, Jarvis & Peltier 1982), and two examples of steady solutions at $R_B = 10 \, R_c$ and $R_B = 10^4 \, R_c$ are shown in Figure 15 for heating entirely from below. Solutions are represented by isotherms T', streamlines ψ', and horizontally averaged temperature and horizontal velocity $\langle T' \rangle$ and $\langle u' \rangle$, respectively. Clearly, the boundary-layer structure in the temperature field becomes very sharply defined at high R_B, whereas no boundary layer develops in the velocity field, in consequence of the infinite Prandtl number. On the basis of a large number of integrations of this sort, we may compute the variation of average flow properties such as $\bar{\delta}$ (the thermal boundary-layer thickness) and $\bar{u}$ and $\bar{w}$ (the characteristic horizontal and vertical speeds at the surface and in the vertical thermal plumes, respectively); we can also calculate the variation of the heat transfer represented by the Nusselt number $\text{Nu} = (d/\Delta T)/(d\langle T \rangle / dz)_{z=d}$, which is just the ratio of the actual heat transfer to that which would be effected by

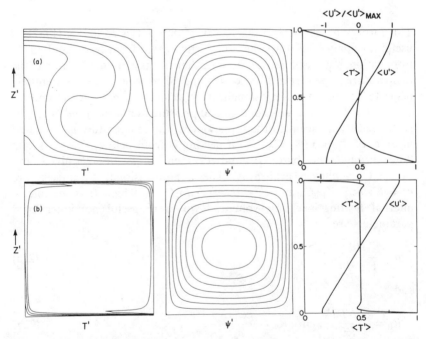

Figure 15 Steady-state heated-from-below convective circulation at (a) $R_B = 10 \, R_c$ and (b) $R_B = 10^4 \, R_c$. Note the sharp thermal boundary layers that are characteristic of the high-Rayleigh-number regime. From Jarvis & Peltier (1982).

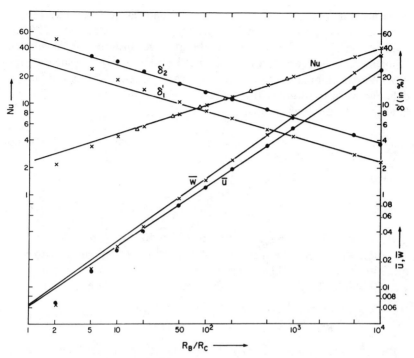

Figure 16   Power-law fits to $\bar{u}$, $\bar{w}$, $\bar{\delta}$, and Nu as a function of Rayleigh number for convective circulations that are heated from below. From Jarvis & Peltier (1982).

conduction alone. These data are shown as a function of R_B in Figure 16 and are observed to fit power-law relations of the form

$$\bar{\delta} = \beta_1(a)d(R_B/R_c)^{S_1}, \tag{23a}$$

$$\bar{u} = \beta_2(a)(K/d)(R_B/R_c)^{S_2}, \tag{23b}$$

$$\bar{w} = \beta_3(a)(K/d)(R_B/R_c)^{S_3}, \tag{23c}$$

$$\bar{q} = \beta_4(a)(K\Delta T/d)(R_B R_c)^{S_4} \tag{23d}$$

$$= (K\Delta T/d)\text{Nu}.$$

Jarvis & Peltier (1982) obtain $\mathbf{S} = (-0.278, 0.645, 0.684, 0.313)$ and $\boldsymbol{\beta} = (0.300, 8.468, 8.671, 2.30)$ for the aspect ratio $a = 1$, which is near the preferred ratio for natural flows. The numerical results are close to those obtained from boundary-layer theory (e.g. Turcotte & Oxburgh 1967, Roberts 1979, Olson & Corcos 1980).

These predictions of the boundary-layer structure of high-Rayleigh-number flows may be directly employed to test the ability of the convection hypothesis to reconcile the observations discussed in Section 2. This is because the numbers $\bar{\delta}$, $\bar{u}$, and $\bar{q}$ on the left-hand sides of relations (23) are geophysical observables, where $\bar{\delta}$ is the average thickness of the surface lithosphere, $\bar{u}$ the characteristic plate speed, and $\bar{q}$ the average surface heat flow. We can eliminate the unknown ΔT from the set (23a, 23b, 23d) in two different ways to estimate the depth of convection d in terms of known quantities. Combining (23b) and (23d), we find

$$d_1 = \frac{\beta_4^{1/2}}{\beta_2} \left[\frac{C_p v}{R_c^{1/3} \bar{q} \alpha g} \right]^{1/2},$$ (24a)

whereas combining (23a) and (23b) yields

$$d_2 = \frac{(\bar{\delta}^2/\kappa)}{\beta_2 \beta_1^2 R_c^{2/3}} \cdot \bar{u},$$ (24b)

as in Peltier (1980). With $\bar{q} \simeq 80$ mW m^{-2}, $\bar{\delta} \simeq 100$ km, and $\bar{u} \simeq 4$ cm yr^{-1}, we obtain $d_1 \simeq 3 \times 10^3$ km and $d_2 \simeq 4 \times 10^3$ km (i.e. compatible with each other and with a radial length scale close to the thickness of the mantle). If we set (24a) = (24b) and solve for the viscosity v, we obtain

$$v = \bar{\delta}^4 \bar{q} \frac{(\alpha g)}{(C_p \kappa^2)} \cdot \frac{1}{\beta_1^4 \beta_4 R_c}.$$ (24c)

It should come as no surprise that the viscosity of the mantle "fluid" implied by (24c) is near 10^{21} Pa s, the same value as that inferred on the basis of the isostatic adjustment analyses discussed in Section 3. Clearly if the viscosity of the mantle were much different from that suggested by these data, it would be very difficult to explain the surface observations in terms of the convection hypothesis. As it is, the observations are fitted by this hypothesis extremely well.

4.3   The Influence of Internal Heating

The above described successful test of the compatibility of the convection hypothesis with geophysical observations was based on the assumption that the circulation was heated entirely from below, i.e. by the flux of heat out of the core in a whole-mantle scenario. The above discussion then shows that this scenario is internally self-consistent. Although geochemical data (e.g. O'Nions et al. 1979, Smith 1977) very strongly suggest that the observed surface heat flow is far too high (approximately by a factor of two) to be explained by the amount of radioactivity in the mantle, the Earth nevertheless does contain such elements (Th, U, and K are the most important long-lived radioactive elements). The deficit between the observed heat flow and the rate of radioactive heating in the interior of the mantle must of course be made up by the loss of internal energy due to the secular cooling of the Earth in bulk or by radioactive heating in the core (most likely due to K). Various thermal-history calculations based upon the idea of parameterized convection first developed in Sharpe & Peltier (1978, 1979) have been performed that attempt to infer the so-called Urey ratio (Ur = heat flow/heat production) directly (e.g. Schubert et al. 1980, Davies 1980, Stacey 1980, Cook & Turcotte 1981, McKenzie & Richter 1981, Spohn & Schubert 1982, Richter 1984). These applications of the Sharpe & Peltier method have led to estimates of the fraction of the observed surface heat flow due to secular cooling of from 10 to 60%; only the latter estimate is compatible with the a priori geochemical constraints. If the core were devoid of radioactivity, then since its effective heat capacity (including the latent heat of freezing of the inner core) is near that of the mantle, only about 25% of the surface heat flow should be crossing the core-mantle boundary if the system were in a steady state. This is an a priori estimate of the extent to which the mantle convective circulation is forced by heating from below. Since it is highly unlikely that the system *is* in a steady state and that the core is devoid of radioactivity, however, this estimate could be enormously in error. Heat loss from the core could be a highly unsteady process, which would have important implications for the maintenance of the geodynamo (e.g. see Jones 1977). Although the use of the Sharpe-Peltier parameterization scheme to estimate Ur *may* deliver a reasonable long-time average, the use of this average to estimate the extent to which the present-day circulation is forced by heating from below could be extremely misleading. As argued in Peltier (1981a), if the heat flow across the core-mantle boundary is as important as suggested in Sharpe & Peltier (1978, 1979), the problem of the energy source of the geodynamo is no longer pressing, since the outer core would be stirred by cooling from above at a rate determined by the strength of the mantle convective circulation. In particular, one may need no longer invoke the existence of a chemically induced circulation in the outer core due to freezing out of the inner core to sustain the geodynamo (e.g. Loper & Roberts 1983). Some consensus that this argument is correct appears to be developing (e.g. Mollett 1984).

One way to obtain a more direct estimate of the present-day magnitude of the forcing that is due to heating from below is to use the observed variations of heat flow and bathymetry on the ocean floor as a function of seafloor age. Figure 17 documents an attempt to do this by Jarvis & Peltier (1980, 1981, 1982). The figure shows a calculation of surface bathymetry and heat flow across the top of an aspect ratio 1 connection cell for a value of $R_R = 5 \times 10^6$ and $\mu = 0.2$ (20% heating from within). The calculations marked "plate" and "no plate" differ from one another in that in the former case a boundary condition of constant horizontal velocity at the upper surface was employed to mimic the influence of lithospheric rigidity, whereas in the latter case the usual free-slip condition was invoked. The dashed lines are $t^{1/2}$ reference curves pinned at the arrows. These calculations show that convective flows that are mostly heated from below at high Rayleigh number predict the observed $t^{1/2}$ variations discussed in Section 2.3 very well. On the other hand, flows that are heated entirely from within predict no such variation with age because the rising hot plume is absent as a result of the absence of a hot lower boundary layer. If the oceanic lithosphere *is* the cold upper thermal boundary layer of the mantle general circulation, then heat flow and topography should be diagnostic of the present mode of heating. Since small μ produces a flattening of the bathymetry (but not of the heat flow) away from the $t^{1/2}$ reference curve (flattening = 0 for $\mu = 0$), we may use the flattening shown on Figure 6 to estimate μ. Such an estimation is shown in Figure 18 (Jarvis & Peltier 1982), which implies that in the whole-mantle scenario for which $R_R \simeq 10^7$ a value of only $\mu \simeq 0.2$ would be required to deliver the observed flattening (see Figure 17). This is in good accord with a priori geochemical constraints if

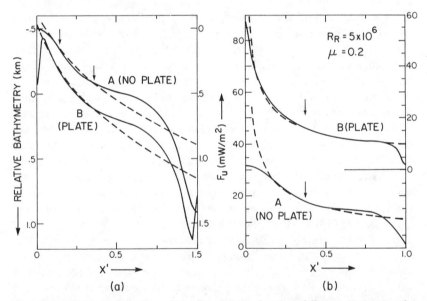

Figure 17   Relative bathymetry and heat flow (F_u) as a function of distance (age) across the top surface of a steady-state, aspect ratio 1 convection cell with $R_R = 5 \times 10^6$ and $\mu = 0.2$. The curves marked "plate" illustrate the influence of lithospheric rigidity upon the predicted variations of the two quantities. Note that this effect is very important insofar as the variation of heat flow is concerned and acts to ensure that F_u follows the $t^{1/2}$ law across the entire width of the cell. The effect of internal heating (finite μ) is such as to flatten the predicted bathymetry variations away from the $t^{1/2}$ (dashed) reference curve. The heat-flow variation is not flattened by this effect, in agreement with the observations of Figure 6.

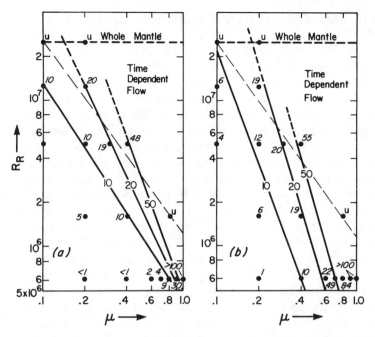

Figure 18 Lines of constant percentile of bathymetric flattening in Rayleigh number–fractional internal heating (μ) space. Plates *a* and *b* respectively employ an exact and approximate method of computing the bathymetry. From Jarvis & Peltier (1982).

the core contains sufficient radioactivity but should not be taken too literally because we have employed steady-state constant-viscosity models to infer a property of what is most probably a highly unsteady flow in which sharp variations of viscosity undoubtedly exist. At the very least, however, this analysis demonstrates that it is unnecessary to invoke the existence of a secondary flow forced by boundary-layer instability to explain the observed flattening. The reason why internal heating leads to this bathymetric flattening has to do with the asymmetry between upgoing and downgoing streams that it induces. This effect was made evident in the two-dimensional numerical calculations of Peckover & Hutchinson (1974) and McKenzie et al. (1974) and was demonstrated in the laboratory experiments of de la Cruz-Reyna (1976).

4.4   The Influence of Viscosity Variations

In Section 2.2 we showed that the surface velocity field induced by convection was characterized by a near-equipartition between the spheroidal and toroidal components of the flow. For a flow forced by thermal buoyancy alone, it is well known that in a constant-viscosity fluid the toroidal component of the flow will remain zero if it is zero initially. In order to couple the radial forcing to the toroidal component of the flow, we require either strong temperature and/or pressure variations of viscosity that then link the spheroidal and toroidal fields together or the asymmetry associated with the presence of light continental material at the Earth's surface. Viscosity variations must be strong in the mantle convective circulation, since the viscosity of the "solid" is ultimately due to the temperature and pressure activation of the migration of dislocations and defects in the solid lattice, and also since precipitous decreases of viscosity accompany the pressure-release melting that occurs near the surface at mid-oceanic ridges and in the back-arc regions landward of the major ocean trenches where volcanic activity is endemic. Because melting is such a crucial ingredient of the mantle convective circulation, the two-dimensional numerical calculations that have included only temperature and pressure variations of viscosity (e.g. Foster 1969, Torrance & Turcotte 1971, DeBremaecker 1977, McKenzie 1977, Kopitzke 1979) are not of much direct use in helping to understand the mantle-convection process. Such calculations inevitably produce a layer of virtually stagnant fluid near the cold upper surface because of the high values of viscosity that obtain at low temperature. This immobile surface is nothing like the plate tectonics described in Section 2. The reason why the surface of the Earth is mobile in spite of the $v(T)$ effect is that as hot material rising in the mantle toward a ridge crest nears the surface, it melts and its viscosity decreases precipitously, rather than increasing precipitously as would be the case if only the temperature (and normal pressure) dependence were important. This was discussed in detail by Peltier (1980). Schmeling & Jacoby (1981) included a simple representation of this effect in a two-dimensional numerical calculation, but a great deal remains to be done in this area. Christensen (1983) argues that non-Newtonian effects in the near-surface region could also contribute to the ability of the flow to activate the motion of material near the surface. Of equal importance to the variation of v with T, p, and melting is the question of the actual rheology of the mantle. Although the success of the Newtonian viscoelastic model in reconciling the isostatic adjustment data discussed in Section 3 provides circumstantial evidence that the solid-state creep of mantle material may in fact satisfy a Newtonian law, laboratory data on the creep of olivine single crystals suggest a non-Newtonian rheology with a power-law exponent near 3. Calculations of the two-dimensional thermally induced circulation of such material, however, show surprisingly small deviations from those of an appropriately scaled Newtonian analogue (e.g. Parmentier et al. 1976, Parmentier & Morgan 1982, Christensen 1983). Whether the similarity persists in three-dimensional time-dependent flow is a question that has yet to be answered.

In spite of the fact that such viscosity variations as those discussed above might be of importance to understanding the observed strength of the toroidal component of the surface velocity field (and, indeed, the ability of the surface material to move at all), it would appear that once the surface has been broken into its characteristic distribution of interacting plates, the fact that these plates have extraordinarily high viscosity (i.e. are essentially elastic) does not play any crucial active role in determining flow speed. This is suggested by virtue of the fact that the boundary-layer relations for constant-viscosity convection between stress-free boundaries provide a good reconciliation of the observed relationship between the parameters $\bar{u}$, $\bar{\delta}$, and $\bar{q}$ discussed in Section 4.2. However, much further effort needs to be expended to understand the effects of strong viscosity variations on high-Rayleigh-number flows. One of these effects is undoubtedly to induce a marked asymmetry in the stability of the cold boundary layer at the surface and the hot boundary layer at the core-mantle boundary. The upper boundary layer is stabilized against local convective instability by its high viscosity (Peltier 1981a, Yuen et al. 1981), while the hot lower boundary layer will be strongly destabilized convectively because of the low viscosity attending the high temperatures. Peltier (1980) and Yuen & Peltier (1980a,b) have argued that this lower boundary-layer instability could be the mechanism that produces the thermal plumes required to understand the origin of oceanic islands like Hawaii, and Peltier (1981a) has suggested that local melting may also play a crucial role in this mechanism.

4.5   Determinants of the Radial Mixing Length

Until quite recently, the question of the depth extent of individual mantle convection cells appeared to be one that might be unanswerable. The totality of the geophysical evidence, as recently discussed in some detail by Peltier & Jarvis (1982), was conflicting. The observation that earthquake activity in the downgoing slab ceases at the depth of the 670-km seismic discontinuity, coupled with the compressive nature of deep earthquake focal mechanisms, suggests the existence of some barrier to mass transfer at this level. On the other hand, the deep focal mechanism is itself not sufficiently well understood to rule out the possibility that such deep events, which do tend to cluster around this depth, may be triggered by processes *due* to material penetration of the interface. Certainly it is true that whatever the barrier to mass transfer might be, it is not associated with any drastic increase of the viscosity from the upper to the lower mantle as has often been suggested in the past (McKenzie 1966, 1967, McKenzie & Weiss 1975). This possibility appears to be ruled out by the analyses presented in Section 2. It also seems clear, on the basis of the new analyses of the influence of phase transformations on convection, that mass transfer across the 670-km interface could not be impeded by this boundary if it were entirely produced by the spinel $\rightarrow$ perovskite + magnesiowustite phase transition that has been predicted to exist at this depth on the basis of high-pressure laboratory experiments on the olivine binary. Is the location of the 670-km boundary at the depth predicted for this phase change to be considered entirely coincidental? The possibility apparently cannot be ruled out entirely (e.g. Jeanloz & Thompson 1983).

The only mechanism that appears plausible at this juncture that explains how mass transfer across 670-km depth might be inhibited is the action of an increase of mean atomic weight from the upper to the lower mantle across this boundary. Richter & Johnson (1974) showed that a chemical increase of ρ of only a few percent would be sufficient, in linear theory, to prevent mixing. The efficiency of this effect at preventing mixing has been investigated at finite amplitude by Olson & Yuen (1982) and Christensen & Yuen (1984), who treat the boundary as part phase change and part chemical and show that anything is possible. Olson (1984) has completed an interesting laboratory study of the ability of a two-layer system to remain stable in the presence of the entrainment processes that are bound to occur across the internal interface in such a system; he finds that it might be difficult to maintain chemically separate layers over a time scale equal to the age of the Earth. What we require are direct methods of inferring whether any chemical component of the 670-km discontinuity is in fact present or, better still, a direct demonstration that material either is or is not penetrating the boundary.

Recent analyses of seismic travel-time anomalies from deep-focus earthquakes in the northwestern Pacific by Jordan (1977) and Creager & Jordan (1984) have in fact suggested that the cold material of the downgoing slab *does* exist on a downward extension through the 670-km discontinuity to depths of at least 900–1000 km and therefore that at least some upper-mantle material does penetrate the boundary. On the other hand, Giardini & Woodhouse (1984), who have analyzed the deep seismicity of the Tonga trench, find some evidence that the slab might not penetrate at this location. The direct seismic inferences therefore remain ambiguous. Further analyses of the travel-time and free-oscillation data to reveal more clearly the internal lateral heterogeneity of the mantle, such as those by Masters et al. (1982), Dziewonski (1984), and Woodhouse & Dziewonski (1984), may eventually provide the required information.

In Section 3.5 I pointed out an alternate avenue of investigation that might in fact have proven to be equally useful in attempting to resolve this issue. To the extent that the 670-km discontinuity is a chemical boundary, it will be capable of inducing a buoyancy force when it is deflected from equilibrium by the gravitational field due to an imposed surface load. If the boundary were a phase boundary able to migrate quickly in response to a surface-load-induced pressure change, then no buoyant restoring force would be produced by its deflection. Material would simply change phase so as to keep the boundary flat. I have shown that the free-air gravity anomaly is a particularly sensitive measure of the requirement for the presence of internal buoyancy in the mantle. The M1 mode associated with the $\Delta\rho$ across the 670-km interface seems to be required to fit the observed gravity anomaly over Hudson Bay but perhaps not for the anomaly associated with the smaller-scale Fennoscandian load. It is unfortunately not possible to infer on the basis of the requirement that M1 exist that the mantle must be chemically layered. This is because the time scale on which an internal phase boundary is actually able to migrate in response to an applied surface load is effectively governed by thermal diffusion, and thus insofar as glacial rebound is concerned, the candidate phase transitions would act as though they were chemical discontinuities. Precisely why the mantle general circulation might adopt a vertically layered style, as has been argued on the basis of geochemical trace-element abundances (Wasserburg & De Paolo 1979, O'Nions et al. 1979) and from noble-gas concentrations (Allègre et al. 1983), therefore remains obscure, although chemical layering remains the most plausible candidate. A completely convincing argument for its existence, however, has yet to be produced. It seems probable to me that neither of the whole-mantle or layered-mantle scenarios upon which attention has recently been focused are sufficiently complex to serve adequately as a representation of the general circulation in this region of the Earth, although the former is more plausible than the latter on the basis of the present evidence. The strongest argument in favor of the whole-mantle scenario is that based upon the absence of layering in the observed radial variation of viscosity (e.g. Peltier & Jarvis 1982).

5. CONCLUDING REMARKS

Our understanding of the convective circulation in the Earth's mantle is in a state of rapid evolution. However, there can be no doubt at this point that the surface processes on Earth of mountain building, volcanism, earthquake activity, and continental drift are all caused by the presence of such a circulation. The application of geochemical tracer analyses to the understanding of internal mixing processes is providing important new constraints on the nature of the flow. Seismological data also appear capable of eventually delivering explicit three-dimensional tomographs of the internal lateral heterogeneity. Analyses of the slow-creeping free-surface flows induced by the gravitational imbalance produced by deglaciation appear to be capable of explicitly revealing the presence of buoyancy-producing interfaces in the interior. Such analyses, however, may not help us determine whether the 670-km interface is a phase boundary or a chemical discontinuity and thus explicitly reveal the radial mixing length of the flow. This outstanding question is extremely important in our attempts to understand the impact of mantle convection on planetary thermal history and the coupling between core and mantle, which is of much concern in connection with the mechanism of magnetic-field generation by magnetohydrodynamic processes in the core.

ACKNOWLEDGMENTS

I have enjoyed discussions with several colleagues on the subject of this paper, including Gary Jarvis, Ian Kay, Alex Forte, Mark Tushingham, Patrick Wu, and Jean-Claude Mareschal. My research at the University of Toronto is sponsored through NSERC Grant A9627.

Literature Cited

Allègre, C. J., Straudacher, T., Sarda, P., Kurz, M. 1983. Constraints on evolution of Earth's mantle from rare gas systematics. *Nature* 303 : 762–66

Anderson, D. L. 1981. Hotspots, basalts, and the evolution of the mantle. *Science* 213 : 82–89

Balling, N. 1980. The land uplift in Fennoscandia, gravity field anomalies and isostasy. In *Earth Rheology, Isostasy and Eustasy,* ed. N.-A. Mörner, pp. 297–321. New York : Wiley. 599 pp.

Benioff, H. 1954. Orogenesis and deep crustal structure : additional evidence from seismology. *Bull. Geol. Soc. Am.* 65 : 385–400

Bentley, C. R., Robertson, J. D., Greischer, L. L. 1982. Isostatic gravity anomalies on the Ross Ice Shelf, Antarctica. In *Antarctic Geoscience, Proc. Symp. Antarct. Geol. Geophys., 3rd,* ed. C. Craddock, pp. 1077–81. Madison : Univ. Wisc. Press

Bullard, E. C., Everett, J. E., Smith, A. G. 1965. The fit of the continents around the Atlantic. *Philos. Trans. R. Soc. London Ser. A* 258 : 41–51

Busse, F. H. 1975. Patterns of convection in spherical shells. *J. Fluid Mech.* 72 : 67–85

Busse, F. H. 1978. Magnetohydrodynamics of the Earth's dynamo. *Ann. Rev. Fluid Mech.* 10 : 435–62

Busse, F. H., Schubert, G. 1971. Convection in a fluid with two phases. *J. Fluid Mech.* 46 : 801–12

Chandrasekhar, S. 1961. *Hydrodynamic and Hydromagnetic Stability.* Oxford : Clarendon. 652 pp.

Christensen, U. 1982. Phase boundaries in finite amplitude mantle convection. *Geophys. J. R. Astron. Soc.* 68 : 487–97

Christensen, U. 1983. Convection in a variable viscosity fluid : Newtonian versus power law rheology. *Earth Planet. Sci.*

Lett. 64 : 153–62

Christensen, U., Yuen, D. A. 1984. The interaction of a subducting lithospheric slab with a chemical or phase boundary. *J. Geophys. Res.* In press

Clark, J. A., Farrell, W. E., Peltier, W. R. 1978. Global changes in postglacial sea level : a numerical calculation. *Quat. Res. (NY)* 9 : 265–87

Cook, F. A., Turcotte, D. L. 1981. Parameterized convection and the thermal evolution of the Earth. *Tectonophysics* 75 : 1–17

Cox, A., Doell, R. R., Dalrymple, G. B. 1963. Geomagnetic polarity epochs and Pleistocene geochronometry. *Nature* 198 : 1049–51

Cox, A. 1969. Geomagnetic reversals. *Science* 163 : 237–45

Creager, K. C., Jordan, T. H. 1984. Slab penetration into the lower mantle. *J. Geophys. Res.* 89(B5) : 3031–49

Davies, G. F. 1980. Thermal histories of convective Earth models and constraints on radiogenic heat production in the Earth. *J. Geophys. Res.* 85(B5) : 2517–30

DeBremaecker, J.-C. 1977. Convection in the Earth's mantle. *Tectonophysics* 41 : 195–208

de la Cruz-Reyna, S. 1976. The thermal boundary layer and seismic focal mechanisms in mantle convection. *Tectonophysics* 35 : 149–60

Dicke, R. H. 1966. The secular acceleration of the Earth's rotation and cosmology. In *The Earth-Moon System,* ed. B. G. Marsden, A. G. W. Cameron, pp. 98–164. New York : Plenum

Dickman, S. R. 1977. Secular trend of the Earth's rotation pole : consideration of motion of the latitude observatories. *Geophys. J. R. Astron. Soc.* 51 : 229–44

Dziewonski, A. M. 1984. Mapping the lower

mantle: determination of lateral heterogeneity in P-velocity up to degree 6. *J. Geophys. Res.* 89(B7):5929–52

Farrell, W. E., Clark, J. A. 1976. On postglacial sea level. *Geophys. J. R. Astron. Soc.* 46:647–67

Foster, T. D. 1969. Convection in a variable viscosity fluid heated from within. *J. Geophys. Res.* 74:685–93

Giardini, D., Woodhouse, J. H. 1984. Deep seismicity and modes of deformation in the Tonga subduction zone. *Nature* 307:505–9

Gilbert, F., Dziewonski, A. M. 1975. An application of normal mode theory to the retrieval of structural parameters and source mechanisms from seismic spectra. *Philos. Trans. R. Soc. London Ser. A* 278:187–269

Hager, B. H., O'Connell, R. J. 1978. Subduction zone dip angles and flow driven by plate motion. *Tectonophysics* 50:111–33

Hays, J. D., Imbrie, J., Shackleton, N. J. 1976. Variations in the Earth's orbit: pacemaker of the ice ages. *Science* 194:1121–32

Hess, H. H. 1962. History of ocean basins. In *Petrologic Studies: A Volume to Honor A. F. Buddington*, ed. A. E. J. Engel, H. L. James, B. F. Leonard, pp. 599–620. New York: Geol. Soc. Am.

Holmes, A. 1931. Radioactivity and Earth movements. *Trans. Geol. Soc. Glasgow* 18(Part 3):559–606

Isacks, B., Molnar, P. 1971. Distribution of stresses in the descending lithosphere from a global survey of focal-mechanism solutions of mantle earthquakes. *Rev. Geophys. Space Phys.* 9:103–74

Isacks, B., Oliver, J., Sykes, L. R. 1968. Seismology and the new global tectonics. *J. Geophys. Res.* 73:5855–99

Jarvis, G. T., McKenzie, D. P. 1980. Convection in a compressible fluid with infinite Prandtl number. *J. Fluid Mech.* 96:515–83

Jarvis, G. T., Peltier, W. R. 1980. Oceanic bathymetry profiles flattened by radiogenic heating in a convecting mantle. *Nature* 285:649–51

Jarvis, G. T., Peltier, W. R. 1981. Effects of lithospheric rigidity on ocean floor bathymetry and heat flow. *Geophys. Res. Lett.* 8:857–60

Jarvis, G. T., Peltier, W. R. 1982. Mantle convection as a boundary layer phenomenon. *Geophys. J. R. Astron. Soc.* 68:389–427

Jeanloz, R., Thompson, A. B. 1983. Phase transitions and mantle discontinuities. *Rev. Geophys. Space Phys.* 21:51–74

Jones, G. M. 1977. Thermal interaction of the core and the mantle and long-term behavior of the geomagnetic field. *J. Geophys. Res.* 82:1703–9

Jordan, T. H. 1977. Lithospheric slab penetration into the lower mantle beneath the Sea of Okhotsk. *J. Geophys.* 43:473–96

Kawakatsu, H. 1983. Can pure path models explain free oscillation data? *Geophys. Res. Lett.* 10:186–89

Kopitzke, U. 1979. Finite element convection models: comparison of shallow and deep mantle convection, and temperature in the mantle. *J. Geophys.* 46:97–121

Lees, A. C., Bukowiski, M. S. T., Jeanloz, R. 1983. Reflection properties of phase transition and compositional change models of the 670-km discontinuity. *J. Geophys. Res.* 88(B10):8145–59

Lerch, F. J., Klosko, S. M., Laubscher, R. E., Wagner, C. A. 1979. Gravity model improvement using Geos 3 (GEM 9 and 10). *J. Geophys. Res.* 84(B8):3897–3916

Longman, I. M. 1963. A Green's function for determining the deformation of the Earth under surface mass loads. 2. Computations and numerical results. *J. Geophys. Res.* 68:485–96

Loper, D. E., Roberts, P. H. 1983. Compositional convection and the gravitationally powered dynamo. In *Stellar and Planetary Magnetism*, ed. A. M. Soward, pp. 297–327. New York: Gordon & Breach

Mareschal, J.-C., Gangi, A. F. 1977. Equilibrium position of phase boundary under horizontally varying surface loads. *Geophys. J. R. Astron. Soc.* 49:757–72

Masters, G., Jordan, T. H., Silver, P. G., Gilbert, F. 1982. A spherical Earth structure from fundamental spheroidal mode data. *Nature* 298:609–13

McElhinny, M. W. 1971. Geomagnetic reversals during the Phanerozoic. *Science* 172:157–59

McKenzie, D. P. 1966. The viscosity of the Lower Mantle. *J. Geophys. Res.* 71:3995–4010

McKenzie, D. P. 1967. The viscosity of the Mantle. *Geophys. J. R. Astron. Soc.* 14:297–305

McKenzie, D. 1977. Surface deformation, gravity anomalies, and convection. *Geophys. J. R. Astron. Soc.* 48:211–38

McKenzie, D. P., Morgan, W. J. 1969. Evolution of triple junctions. *Nature* 224:125–33

McKenzie, D., Parker, R. L. 1967. The North Pacific: an example of tectonics on a sphere. *Nature* 216:1276–80

McKenzie, D. P., Richter, F. M. 1981. Parameterized thermal convection in a layered region and the thermal history of the Earth. *J. Geophys. Res.* 86(B12):667–80

McKenzie, D. P., Weiss, N. O. 1975. Speculations on the thermal and tectonic history of the Earth. *Geophys. J. R. Astron. Soc.* 42 : 131–74

McKenzie, D. P., Roberts, J. M., Weiss, N. O. 1974. Convection in the Earth's mantle: towards a numerical simulation. *J. Fluid Mech.* 62 : 465–538

Minster, J. B., Jordan, T. H. 1978. Present day plate motions. *J. Geophys. Res.* 83(B11) : 5331–54

Mollett, S. 1984. Thermal and magnetic constraints on the cooling of the Earth. *Geophys. J. R. Astron. Soc.* 76 : 653–66

Morgan, W. J. 1968. Rises, trenches, great faults, and crustal blocks. *J. Geophys. Res.* 73 : 1959–82

Morrison, L. V. 1973. Rotation of the Earth and constancy of G. *Nature* 241 : 519–20

Morrison, L. V., Stephenson, F. R. 1982. In *Sun and Planetary System*, ed. W. Fricke, G. Teleki, pp. 173–78. Dordrecht : Reidel. 538 pp.

Muller, P. M., Stephenson, F. R. 1975. The acceleration of the Earth and Moon from early astronomical observations. In *Growth Rhythms and History of the Earth's Rotation*, ed. G. D. Rosenberg, S. K. Runcorn, pp. 459–534. New York : Wiley. 559 pp.

Nakiboglu, S. M., Lambeck, K. 1980. Deglaciation effects upon the rotation of the Earth. *Geophys. J. R. Astron. Soc.* 62 : 49–58

Nakiboglu, S. M., Lambeck, K. 1981. Corrections to "Deglaciation effects upon the rotation of the Earth." *Geophys. J. R. Astron. Soc.* 64 : 559

O'Connell, R. J. 1976. The effects of mantle phase changes on postglacial rebound. *J. Geophys. Res.* 81 : 971–74

Olson, P. 1984. An experimental approach to convection in a two layered mantle. *J. Geophys. Res.* In press

Olson, P., Corcos, G. M. 1980. A boundary layer model for mantle convection with surface plates. *Geophys. J. R. Astron. Soc.* 62 : 195–219

Olson, P., Yuen, D. A. 1982. Thermochemical plumes and mantle phase transitions. *J. Geophys. Res.* 87(B5) : 3993–4002

O'Nions, R. K., Evensen, N. M., Hamilton, P. J. 1979. Geochemical modelling of mantle differentiation and crustal growth. *J. Geophys. Res.* 84(B11) : 6091–6101

Oxburgh, E. R., Turcotte, D. L. 1978. Mechanisms of continental drift. *Rep. Prog. Phys.* 41 : 1249–1312

Parmentier, E. M., Morgan, J. III. 1982. Thermal convection in non-Newtonian fluids: volumetric heating and boundary layer scaling. *J. Geophys. Res.* 87(B9) : 7757–62

Parmentier, E. M., Turcotte, D. L., Torrance, K. E. 1976. Studies of finite amplitude non-Newtonian thermal convection with application to convection in the earth's mantle. *J. Geophys. Res.* 81 : 1839–46

Parsons, B., Sclater, J. G. 1977. An analysis of the variation of ocean floor bathymetry and heat flow with age. *J. Geophys. Res.* 82 : 803–27

Peckover, R. S., Hutchinson, I. H. 1974. Convective rolls driven by internal heat sources. *Phys. Fluids* 17 : 1369–71

Peltier, W. R. 1973. Penetrative convection in the planetary mantle. *Geophys. Fluid Dyn.* 5 : 47–88

Peltier, W. R. 1974. The impulse response of a Maxwell Earth. *Rev. Geophys. Space Phys.* 12 : 649–69

Peltier, W. R. 1976. Glacial isostatic adjustment. II. The inverse problem. *Geophys. J. R. Astron. Soc.* 46 : 669–706

Peltier, W. R. 1980. Mantle convection and viscosity. In *Physics of the Earth's Interior*, ed. A. M. Dziewonski, E. Boschi, pp. 362–431. Amsterdam : North Holland

Peltier, W. R. 1981a. Surface plates and thermal plumes: separate scales of the mantle convective circulation. In *The Evolution of the Earth*, ed. R. J. O'Connell, W. Fyfe, pp. 229–48. Washington DC : Am. Geophys. Union

Peltier, W. R. 1981b. Ice age geodynamics. *Ann. Rev. Earth Planet. Sci.* 9 : 199–225

Peltier, W. R. 1982. Dynamics of the ice age Earth. *Adv. Geophys.* 24 : 1–146

Peltier, W. R. 1983. Constraint on deep mantle viscosity from LAGEOS acceleration data. *Nature* 304 : 434–36

Peltier, W. R. 1984a. The thickness of the continental lithosphere. *J. Geophys. Res.* In press

Peltier, W. R. 1984b. The LAGEOS constraint on deep mantle viscosity. *J. Geophys. Res.* In press

Peltier, W. R., Andrews, J. T. 1976. Glacial isostatic adjustment. I. The forward problem. *Geophys. J. R. Astron. Soc.* 46 : 605–46

Peltier, W. R., Jarvis, G. T. 1982. Whole mantle convection and the thermal evolution of the earth. *Phys. Earth Planet. Inter.* 29 : 281–304

Peltier, W. R., Wu, P. 1982. Mantle phase transitions and the free air gravity anomalies over Fennoscandia and Laurentia. *Geophys. Res. Lett.* 9 : 731–34

Peltier, W. R., Wu, P. 1983. Continental lithospheric thickness and deglaciation induced true polar wander. *Geophys. Res. Lett.* 10 : 181–84

Peltier, W. R., Farrell, W. E., Clark, J. A. 1978. Glacial isostasy and relative sea level: a global finite element model. *Tectonophysics* 50:81–110

Plafker, G. 1965. Tectonic deformation associated with the 1964 Alaska earthquake. *Science* 148:1675–87

Richter, F. M. 1973. Dynamical models for sea-floor spreading. *Rev. Geophys. Space Phys.* 11:223–87

Richter, F. M. 1984. Regionalized models for the thermal evolution of the Earth. *Earth Planet. Sci. Lett.* In press

Richter, F. M., Johnson, C. 1974. Stability of a chemically layered mantle. *J. Geophys. Res.* 79:1635–39

Richter, F. M., Parsons, B. 1975. On the interaction of two scales of convection in the mantle. *J. Geophys. Res.* 80:2529–41

Roberts, G. O. 1979. Fast viscous Bénard convection. *Geophys. Astrophys. Fluid Dyn.* 12:235–72

Roberts, P. H. 1967. Convection in horizontal layers with internal heat generation. Theory. *J. Fluid Mech.* 30:33–49

Rubincam, D. P. 1984. Postglacial rebound observed by Lageos and the effective viscosity of the lower mantle. *J. Geophys. Res.* 89(B2):1077–87

Runcorn, S. K. 1956. Palaeomagnetic comparisons between Europe and North America. *Proc. Geol. Assoc. Can.* 8:77–85

Sabadini, R., Peltier, W. R. 1981. Pleistocene deglaciation and the Earth's rotation: implications for mantle viscosity. *Geophys. J. R. Astron. Soc.* 66:552–78

Schmeling, H., Jacoby, W. R. 1981. On modelling the lithosphere in mantle convection. *J. Geophys.* 50:89–100

Schubert, G., Turcotte, D. L. 1971. Phase changes and mantle convection. *J. Geophys. Res.* 76:1424–32

Schubert, G., Yuen, D. A., Turcotte, D. L. 1975. Role of phase transitions in a dynamic mantle. *Geophys. J. R. Astron. Soc.* 42:705–35

Schubert, G., Stevenson, D., Cassen, P. 1980. Whole planet cooling and the radiogenic heat source contents of the Earth and Moon. *J. Geophys. Res.* 85(B5):2531–38

Shackleton, N. J. 1967. Oxygen isotope analyses and Pleistocene temperature readdressed. *Nature* 215:15–17

Sharpe, H. N., Peltier, W. R. 1978. Parameterized mantle convection and the Earth's thermal history. *Geophys. Res. Lett.* 5:737–40

Sharpe, H. N., Peltier, W. R. 1979. A thermal history model for the Earth with parameterized convection. *Geophys. J. R. Astron. Soc.* 59:171–203

Smith, J. V. 1977. Possible controls on the bulk composition of the Earth: implications for the origin of the Earth and Moon. *Proc. Lunar Sci. Conf.* 8:333–69

Spohn, T., Schubert, G. 1982. Modes of mantle convection and the removal of heat from the Earth's interior. *J. Geophys. Res.* 87(B6):4682–96

Stacey, F. D. 1980. The cooling Earth: a reappraisal. *Phys. Earth Planet. Inter.* 22:89–96

Stauder, W., Bollinger, G. A. 1965. The S-wave project for focal mechanism studies—earthquakes of 1963. *Air Force Off. Sci. Res., Grant AF-AFOSR 62-458, Rep.*

Torrance, K. E., Turcotte, D. L. 1971. Thermal convection with large viscosity variations. *J. Fluid Mech.* 47:113–25

Turcotte, D. L., Oxburgh, E. R. 1967. Finite amplitude convection cells and continental drift. *J. Fluid Mech.* 28:29–42

Turcotte, D. L., Oxburgh, E. R. 1972. Mantle convection and the new global tectonics. *Ann. Rev. Fluid Mech.* 4:33–68

Vine, F. J. 1966. Spreading of the ocean floor: new evidence. *Science* 154:1405–15

Vine, F. J., Matthews, D. H. 1963. Magnetic anomalies over oceanic ridges. *Nature* 199:947–49

Walcott, R. I. 1970. Isostatic response to loading of the crust in Canada. *Can. J. Earth Sci.* 7:716–27

Wasserburg, G. J., De Paolo, D. J. 1979. Models of Earth structure inferred from neodymium and strontium isotopic abundances. *Proc. Natl. Acad. Sci. USA* 76:3594–98

Wegener, A. 1929. *Die Enstehung der Kontinente und Ozeane.* Braunschweig: Vieweg. 4th Ed. Transl., 1929, as *The Origins of Continents and Oceans.* London: Methuen. Reprinted, 1966, by Dover

Wilson, J. T. 1965. A new class of faults and their bearing on continental drift. *Nature* 207:343–47

Woodhouse, J. H., Dziewonski, A. M. 1984. Mapping the upper mantle: three dimensional modelling of Earth structure by inversion of seismic waveforms. *J. Geophys. Res.* 89(B7):5953–86

Wu, P., Peltier, W. R. 1982. Viscous gravitational relaxation. *Geophys. J. R. Astron. Soc.* 70:435–86

Wu, P., Peltier, W. R. 1983. Glacial isostatic adjustment and the free air gravity anomaly as a constraint on deep mantle viscosity. *Geophys. J. R. Astron. Soc.* 74:377–450

Wu, P., Peltier, W. R. 1984. Pleistocene deglaciation and the Earth's rotation: a new analysis. *Geophys. J. R. Astron. Soc.* 76:753–91

Yagi, T., Bell, P. M., Mao, H. K. 1979. Phase relations in the system MgO-FeO-SiO_2 between 150 and 700 kbar at $1000°C$. *Carnegie Inst. Washington Yearb.* 78:614–18

Yoder, C. F., Williams, J. G., Dickey, J. O., Schultz, B. E., Eanes, R. J., Tapley, B. D. 1983. Secular variation of earth's gravitational harmonic J_2 coefficient from Lageos and nontidal acceleration of earth rotation. *Nature* 303 : 757–62

Yuen, D. A., Peltier, W. R. 1980a. Mantle plumes and the thermal stability of the D^{11} layer. *Geophys. Res. Lett.* 7 : 625–28

Yuen, D. A., Peltier, W. R. 1980b. Temperature dependent viscosity and local instabilities in mantle convection. In *Physics of the Earth's Interior*, ed. A. Dziewonski, E. Boschi, pp. 432–63. Amsterdam : North Holland

Yuen, D. A., Peltier, W. R., Schubert, G. 1981. On the existence of a second scale of convection in the upper mantle. *Geophys. J. R. Astron. Soc.* 65 : 171–90

SUBJECT INDEX

A

Accelerated crucible rotation
technique
crystal growth and, 206
Ackeret formulas, 1
Acoustics
ocean bottom, 232-33
oceanic, 219-24
Acoustic tomography
oceanic, 233-34
Acoustic waves
Kutta condition and, 412-14
Adaptive grid techniques, 490
Adiabatic flat plates
boundary layers on, 322
Advection, 128
Aerodynamics
see Sports ball aerodynamics
Aeroengine exhaust
Kutta condition and, 414
Aircraft
wing-tip vortex/flap vortices
interaction of, 530
Air entrainment
coating flows and, 84-86
Airfoil flow
wind-tunnel walls and, 494-95
Airfoils
Kutta condition and, 432-35
shearing transformations and,
492-93
Air pollutants
dispersion over complex terrain, 239-65
Algebraic grid generation, 490-
503
Arrhenius kinetics
combustion waves and, 270
Astrophysics
multicomponent convection
and, 32-34
Atlantic Ocean
equatorial undercurrent and,
359
Rossby-gravity waves in, 361
subsurface countercurrents in,
360
Atmospheric processes
modeling of
criteria for, 241-44
Axial compressors, 1
multistage, 3

B

Bagley end correction, 50
Bagley plots, 50

Ball aerodynamics
see Sports ball aerodynamics
Baseball aerodynamics, 167-77
nonspinning baseball tests in,
173-77
principles of, 167-71
spinning baseball tests in,
171-73
Bidisperse suspensions
lateral segregation and convection in, 100
Biot-Savart integral, 541
Biot-Savart law, 526, 527, 528,
538
Bismuth-silicon-oxide
Czochralski flow of, 198
Black box
perturbed boundary layers
and, 322-23
Blade coating, 74-75
Bluff bodies
Kutta condition and, 411
separated flows around
Kutta condition and, 427-
31
wake of
thermal effects in, 3
Bluff obstacles
turbulent flow around, 458
Bodies of revolution
at subsonic Mach numbers, 3-
4
Boltzmann's superposition principle, 46
Boolean operations
grid generation and, 500-2
Boundary effects
equatorial ocean circulation
and, 375-77
Boundary layers
calculation methods for, 324-
28
regions of, 325-26
shock waves and, 3
surface roughness in, 328-32
see also Perturbed boundary
layers
Boundary layer suction, 4
Boussinesq approximation
Czochralski bulk flow and,
199
Box model, 132
Boycott effect, 106
Brillouin point, 427, 430
Brownian diffusion, 104
Bulk dilatation
perturbed boundary layers
and, 341, 347

shock wave/boundary layer interactions and, 348

C

Capillary number
dip coating and, 71-74
film formation and, 67-69
Cauchy's vorticity formula, 542
Caustics, 221
Centrifugal flow
crystal-growth melts and, 192-
93
Centrifuge experiments
flux dependence on gravity in,
31
Chapman-Jouguet detonation,
272
Clarification zone, 93
Coating flows, 65-87
air entrainment and, 84-86
capillary number and, 67-69
computer simulation of, 82-84
contact angles and, 84-86
instability of, 79-82
metered by coating meniscus,
71-79
premetered, 69-71
rectilinear, 66-67
Coating meniscus
coating flow metering by, 71-
79
curvature restrictions on, 70-
71
surface tension and, 68
Combustion waves
doubly infinite plane model
for, 273-74
mathematical model for, 274-
76
paradigm models for, 268-73
see also Subsonic combustion
waves
Complex flows
narrow plumes in, 454-67
Compositional convection, 36-37
Earth's core and, 39-40
Compressible flows
bulk dilatation and, 341, 347
lateral divergence and, 341,
346-47
longitudinal curvature and,
341-46
Compression corners
shock wave/boundary layer interactions and, 350
Compression zone, 94

609

Computer simulation
of coating flows, 82-84
Concave curvature
perturbed boundary layers
and, 342-46
Connective tissue
transport phenomena in, 31
Constitutional supercooling, 35
Continuity equation
for turbulent boundary layers,
324
Continuum hypothesis, 121
Convection
see specific type
Convective fractionation, 37
Convergence zones, 221
Convex curvature
perturbed boundary layers
and, 342-46
Coordinate transformations
grid generation and, 488-90
Coriolis accelerations, 241
Cricket ball aerodynamics, 152-
67
dynamic tests in, 158-64
meteorological conditions and,
166-67
optimum swing conditions in,
164-66
principles of, 152-54
static tests in, 154-58
Crystal growth
accelerated crucible rotation
and, 206
double-diffusive convection
and, 34-35
normal freezing and, 212
in space, 210-11
Crystal growth melts, 191-212
Czochralski bulk flow and,
193-94
numerical simulation of,
198-206
Czochralski process and, 191-
93
variants of, 206-11
flow visualization in, 194-98
zone melting and, 211-12
Crystallization
sidewall, 37
Crystallization convection, 36-37
Curveball
aerodynamics of, 171-73
Czochralski bulk flow, 193-94
boundary conditions for, 200-
1
dependent variables of, 199
magnetic fields and, 207-8
numerical simulation of, 198-
206
similarity parameters for, 196-
97
vorticity-stream function formulation and, 201-6

Czochralski process, 191-93
variants of, 206-11

D

Deep Sea Drilling Project, 572
Deflagration waves
see Combustion waves
Dendritic interfaces
growth rates of, 35-36
Density stratifications
layer formation in
eddy viscosity and, 33
Diffusion
see specific type
Diffusion coefficients
multicomponent convection
and, 30
Dip coating, 71-74
Dirac delta functions, 451
Dispersion
in estuaries, 141-45
longitudinal, 129
in estuaries, 142
Taylor's model of, 130-32
mechanisms of, 128-30
of nonconserved substances,
133-35
in rivers, 135-41
Dispersion models, 119-45
for conserved substances, 132-
33
ensemble means and time
averages in, 122-27
fundamental equation of, 121-
22
instrumentation and, 127-28
Dividing-streamline height
air pollutant dispersion and,
243-44
Doppler effect, 2, 4-9
Doppler-Mach cone
discovery of, 2
Double-diffusive convection, 11,
27
crystal growth and, 34-35
magma chambers and, 36-39
magnetic fields and, 34
see also Multicomponent convection
Dynamic wetting
coating flows and, 84-85

E

Earth's core
compositional convection and,
39-40
solidification at, 35
Eddies
planetary waves and, 228
sound transmission and, 229-
30

Eddy diffusivity, 130
oceanic layer formation and,
30, 33
Eddy simulation
turbulent shear layers and,
322
Eddy viscosity
layer formation in density
stratifications and, 33
oceanic layer formation and,
30, 33
perturbed flows and, 336-37
Effective longitudinal dispersion
coefficient, 131
Eikonal equation
oceanic acoustics and, 220
Ekman divergence, 380, 382,
387
El Niño, 359, 399-404
sea-surface-temperature anomalies and, 361
Elastic liquids
hole-pressure errors and, 49
Elliptic methods
grid generation and, 503-10
Energy conversion
Kutta condition and, 439-43
Ensemble means
dispersion models and, 122-27
Equatorial currents, 359-60
Equatorial ocean circulation,
359-406
boundary effects and, 375-77
currents and, 377-78
El Niño and, 399-404
equations of motion for, 362-
71
equatorially trapped waves
and, 372-74
islands and, 377
linear continuously stratified
models for, 367-71
solutions to, 382-89, 391-
95
long-wavelength approximation and, 374-75
nonlinear continuously stratified models for, 371
solutions to, 389-91
reduced-gravity models for,
364-66
surface-layer models for, 363-
67
solutions to, 379-82
thermodynamic
mixed-layer models for,
366-67
Equatorial oceans
sea-surface-temperature anomalies in, 359, 361
Equatorial resonances, 392
Equatorial undercurrent, 359
Estuaries
dispersion in, 141-45

dispersion of nonconserved
   substances in, 133-35
dispersion of pollutants in,
   132-33
gravitational circulation in,
   144-45
longitudinal dispersion in, 142
Extrudate swell, 50-51
Extrusion rheometry, 48-51

F

Fermat's principle
   oceanic acoustics and, 220
Feynman path integrals
   underwater acoustics and, 227
Flaring, 201
Floating-zone melting
   crystal growth and, 211-12
Fluid element displacement
   equations for, 453-54
   turbulent diffusion and, 448-
      54
Fluid layers
   superimposed
      multicomponent convection
         in, 14-17
Fluids
   multicomponent convection
      in, 13-14
Forcing terms, 504
Fourier transforms
   underwater acoustics and, 225
Frenet-Serret formulas, 535
Froude number, 241-44, 256,
   259-60

G

Gas-liquid drops
   see Multiphase drops
Gaussianity
   dispersion and, 135-39
Geology
   multicomponent convection
      and, 36-39
Geophysics
   multicomponent convection
      and, 39-40
Glacial isostasy
   mantle convection and, 576-
      603
Golf ball aerodynamics, 177-88
   dimple geometry and, 184-88
   principles of, 177-79
Golf balls
   forces on
      measurements of, 180-83
   trajectories of
      computation of, 183-84
Gothert rule, 3
Gravitational fields
   mantle viscosity and, 583-84

multicomponent convection
   and, 13-14
Gravity
   coating flows and, 67
   multicomponent convection
      and, 31-32
Greenland Sea bottom water
   double-diffusive convection
      in, 27
Grid generation, 487-520
   adaptive grid movement and,
      510-19
   algebraic, 490-503
   Boolean operations and, 500-2
   concavity controls and, 503-
      10
   coordinate transformation and,
      488-90
   elliptic methods and, 503-10
   Hermite transformation and,
      491-93
   hyperbolic methods and, 503
   multisurface transformation
      and, 493-500
   partial differential equations
      methods and, 503-10
   shearing transformation and,
      491-93
Grids
   supersonic flow and, 3

H

Hadamard-Rybczynski formula,
   310
Head-wave phenomenon, 2
Heat-salt fingers, 14-17
Heavy gases
   atmospheric dispersion of
      ensemble means and, 123
Helmholtz equation
   underwater acoustics and, 224
Hermite-Birkhoff interpolation,
   493
Hermite interpolation
   defective, 493
Hermite transformations
   grid generation and, 491-93
High-speed flames
   subsonic, 282-84
Hilbert transform, 418, 420
Hills
   three-dimensional
      air pollutant dispersion and,
         247-60
      flow structure around, 247-
         51
   two-dimensional
      towing tank and, 263-64
Hindered settling, 94-98
Hodge theory, 518
Hole-pressure errors
   elastic liquids and, 49

Horizontal mixing
   equatorially trapped waves
      and, 374
Hugoniot diagram
   discontinuous deflagration
      waves and, 269
Humidity
   swing bowling and, 166-67
Hyperbolic methods
   grid generation and, 503

I

Inclined channels
   enhanced sedimentation in,
      105-14
Incompressible flows
   Crighton-Leppington problem
      and, 436
   lateral divergence and, 341,
      346-47
   longitudinal curvature and,
      341-46
   vorticity and, 523-57
Indian Ocean
   Rossby-gravity waves in, 361
Internal waves
   oceanic sound transmission
      and, 230-31
Islands
   equatorial waves and, 377

K

Kelvin waves
   dispersion relations for, 394
   equatorial currents and, 378
   equatorially trapped, 360,
      373-74, 384
   propagation of, 386
   reflection of, 376
Knuckleball
   aerodynamics of, 173-77
Kutta condition, 411-44
   characteristics of, 412-16
   energy-conversion mechanisms
      and, 439-43
   leading-edge flows and, 427
   nozzle flows and, 431-32
   oscillating airfoils and, 432-35
   separated flows and, 427-31
   shear-layer receptivity and,
      435-39
   steady trailing-edge flows and,
      416-19
   unsteady trailing-edge flows
      and, 419-26
Kutta-Joukowsky hypothesis,
   411

L

Lagrangian transport method,
   133, 135

Lakes
multicomponent convection in, 24-25
Laminar flows
sedimentation and, 107-14
shear stress at capillary wall for, 49
Langmuir cells
windrow formation and, 30
Laplace-Beltrami operator, 518
Lateral divergence
perturbed boundary layers and, 341, 346-47
Lava
convective fractionation and, 37
komatiite
top cooling and, 37
Leading-edge flows
Kutta condition and, 427-31
Linear density gradients
Froude number and, 244
Linear viscoelastic field theory
mantle convection and, 576-77
Linear viscoelasticity
polymer melts and, 46-48
polymer solutions and solids and, 45-46
Liquid-liquid drops
see Multiphase drops
Liquid natural gas tanks
multicomponent convection in, 25
Lithosphere
as thermal boundary layer, 572-73
Local induction approximation, 535-39
Log law
turbulent boundary layers and, 326
Longitudinal curvature
perturbed boundary layers and, 341-46
shock wave/boundary layer interactions and, 348-49
Longitudinal dispersion, 129
in estuaries, 142
Taylor's model of, 130-32
Low-speed flames
propagation of, 276-79

M

Mach cone, 2
Mach number
origin of, 1-2
subsonic
bodies of revolution and, 3-4
subsonic combustion waves and, 267-68

Magma chambers
basaltic
convective heat transfer in, 38
double-diffusive convection in, 36-39
Magnetic fields
Czochralski growth in, 207-8
thermal convection and, 33-34
Magnus effect, 151-52, 160, 167
Mantle
physical properties of, 590
Mantle convection, 561-604
glacial isostasy and, 576-603
high-Rayleigh-number flows and, 592-96
hydrodynamic models of, 589-603
internal heating and, 596-99
ocean floor and, 572-73
plate tectonics and, 39
radial mixing length and, 601-3
surface kinematics and, 566-72
viscosity variations and, 599-601
Mantle discontinuities
internal buoyancy and, 584-88
Mantle shell
linear stability of, 590-92
Rayleigh number of, 592-93
Mantle viscoelasticity
deglaciation-induced polar wander and, 588-89
Mantle viscosity
gravitational field and, 583-84
planetary rotation and, 581-83
postglacial sea levels and, 577-81
Marangoni effects, 35
Marangoni flows, 193
Marangoni number, 197
Materials science
multicomponent convection and, 34-36
Melt fracture, 51
Mesoscale, 219
oceanic sound transmission and, 228-31
Metallurgy
multicomponent convection and, 34-36
Meteorological flows
step changes in, 333
Molecular diffusion, 128
coupled, 21-22
gravity and, 31-32
turbulent dispersion and, 450
Momentum equation
boundary-layer approximation to, 324

Monodisperse suspensions
sedimentation of, 93-98
Multicomponent convection, 11-40
astrophysics and, 32-34
chemical studies and, 30-32
direct applications of, 24-25
geology and, 36-39
geophysics and, 39-40
laboratory experiments in, 14-18
metallurgy and, 34-36
oceanography and, 26-30
stability analyses in, 18-24
Multiphase drops, 289-318
stability of, 312-16
static, 292-302
translation of, 302-11
vapor explosions and, 316-18
Multisurface transformation
grid generation and, 493-500

N

Navier-Stokes equations, 418, 425-26, 438
Czochralski bulk flow and, 199
turbulent shear layers and, 322
Near-field flow, 194
Newtonian cooling
equatorially trapped waves and, 374
Newtonian liquids
shear rate at capillary wall for, 49
two-dimensional flow of, 66
Nonorthogonality
fluid-dynamic equations and, 491
Normal freezing
crystal growth and, 212
Nozzle flows
Kutta condition and, 431-32

O

Ocean floor
heat flow and topography of, 572-73
Oceanography
multicomponent convection and, 26-30
Ocean sound transmission, 217-35
acoustic environment and, 219-24
analytic models of, 224-28
inverse techniques and, 231-34
mesoscale and internal-wave effects and, 228-31

Ocean tomography, 233-34
Ocean waves
  equatorially trapped, 371-74
  unstable, 360-61
Orr-Sommerfeld equation, 112

P

Pacific Ocean
  equatorial undercurrent and,
    359
  Rossby-gravity waves in, 361
  subsurface countercurrents in,
    360
Parabolic equation
  underwater acoustics and,
    224-27
Partial differential equations
  grid generation and, 503-10
Peclet number, 104, 449
Perturbed boundary layers, 321-
    53
  black box and, 322-23
  levels of complexity of, 322-
    24
  multiple, 347-52
  outer layer disturbances and,
    340-47
  pressure gradient changes and,
    337-40
  rapid-distortion theory and,
    327
  surface roughness and, 328-33
  wall blowing and, 336-37
  wall heat flux and, 333-36
  wall region disturbances and,
    328-37
  see also Turbulent shear layers

Planetary rotation
  nontidal acceleration of
    mantle viscosity and, 581-
    83
Planetary waves
  mesoscale effects and, 228
Plate tectonics, 561, 563-65
  mantle convection and, 39
Plumes
  homogeneous turbulence and,
    467-82
  steady complex flows and,
    454-67
PNK theory, 106
Polar wander
  deglaciation-induced
    mantle viscoelasticity and,
    588-89
Pollutant dispersion
  over complex terrain, 239-65
  in estuaries, 132-33
Polydisperse suspensions
  sedimentation of, 98-104

Polyethylene
  flow curves from, 50
  linear
    melt fracture and, 51
    strain hardening in elongation
      and, 58
    uniaxial stretching and, 55
Polyethylene melts
  steady shear rate flow of
    superimposed oscillations
      on, 53
Polyisobutylene
  strain hardening in elongation
    and, 58
Polyisobutylene solutions
  shear rate flow of
    superimposed oscillations
      on, 53
Polymer melts
  boundary conditions for, 49
  extrusion rheometry and, 48-
    51
  linear viscoelasticity and, 46-
    48
  multiaxial stretching and, 58-
    61
  rheogoniometer for, 52-53
  rheometry of, 45-61
  shear-flow behavior of, 53-54
  uniaxial stretching and, 54-58
  viscometric flows in, 51-54
Polymers
  amorphous
    molten state of, 46
Polymer solutions
  cross diffusion in, 30-31
Polystyrene
  uniaxial stretching and, 55
Polystyrene solutions
  shear rate flow of
    superimposed oscillations
      on, 53
Polyvinylchloride
  boundary conditions for, 49
Prandtl number
  turbulent, 333, 335
  transport of heat and, 325
Pressure gradients
  perturbed boundary layers
    and, 337-40
  shock wave/boundary layer in-
    teractions and, 348-49
Pycnocline
  vertical movement of, 364-65

R

Rabinowitsch-Weissenberg correction, 49
Rapid-distortion theory
  perturbed boundary layers
    and, 327

Rayleigh drag
  equatorially trapped waves
    and, 374
Reynolds number, 3
  origin of, 1
Reynolds-Schmidt product, 241
Rheogoniometers
  for polymer melts, 52-53
Rheometers
  multiaxial stretching, 58-61
  Munstedt's, 55
  uniaxial stretching, 54-58
Rheometry
  extrusion, 48-51
  polymer melt, 45-61
Ribbing lines, 79-82
Richardson number, 241
Rivers
  dispersion in, 135-41
Roll coating, 75-79
Rosenhead-Moore approximation, 529
Rossby-gravity waves, 361, 372-
  74, 376
  dispersion relations for, 394
  equatorial currents and, 378
  equatorial resonances and, 392
  equatorially trapped, 384
  propagation of, 386
  ray theory and, 392-95
Rossby number, 241
Rosseland approximation
  Czochralski bulk flow and,
    199

S

Salinity
  gravitational circulation in
    estuaries and, 144
  longitudinal flux of, 142-44
Salinity gradients
  layer formation from, 17-18
Salt fingers, 14-17
  existence in the ocean, 26-27
Schlieren method, 2
Seafloor
  heat flow and topography of,
    572-73
Seafloor spreading
  Hess' hypothesis of, 563-64
Sea levels
  postglacial
    mantle viscosity and, 577-
      81
Sea-surface-temperature anoma-
  lies, 359, 361, 366-67,
    370, 389
  El Niño and, 400-1
Sedimentation, 91-116
  in inclined channels, 105-14
  Kynch's theory of, 94

of monodisperse suspensions, 93-98
of polydisperse suspensions, 98-104
Semiconvection, 33
Shadow zones, 221
Shear dispersion, 129
Shear flow, 51-54
   instability of
      Kutta condition and, 412
      polymer melts and, 53-54
      straining and, 478
      thin clouds in, 466-67
      turbulent, 474-78
Shearing transformations
   grid generation and, 491-93
Shear layers
   Kutta condition and, 435-39
Shear stress
   surface roughness and, 331-32
Shock wave/boundary layer interactions, 347-52
   pressure and temperature variations in, 321
Shock waves
   boundary layers and, 3
Sidewall crystallization, 37
Silicon
   Czochralski flow of, 198
Slide coater, 71
Slot coater, 69-70
Snell's law
   oceanic acoustics and, 220
Solar ponds
   multicomponent convection in, 24-25
Solitons, 536
Sommerfeld problem, 414
Sonar
   development of, 218
Sound transmission
   see Ocean sound transmission
Southern Oscillation, 401-2
Space
   crystal growth in, 210-11
Spatial-instability theory
   Kutta condition and, 415
Sports ball aerodynamics, 151-88
   baseball, 167-77
   cricket ball, 152-67
   golf ball, 177-88
Stall flutter, 4
Stars
   diffusion rates in, 12
   heat/salt diffusive convection and, 32-33
   rotating
      multicomponent convection and, 33
      thermal convection in magnetic fields and, 33-34

Stefan-Boltzmann radiation
   Czochralski bulk flow and, 200-1
Stokes flow, 310
Stokes law, 91-92
Straining flows
   normal diffusion in, 455-56
   turbulent diffusion in, 478
Strouhal number, 432, 442
Subsonic combustion waves, 267-86
Sulfide ore deposits
   oceanic
      flow phenomena and, 36
Sunspots
   convection theory and, 34
Superposition principle
   Boltzmann's, 46
   time-temperature, 48
Supersonic airfoils
   Ackeret formulas and, 1
Supersonic flow
   adverse pressure gradients and, 348-49
   grids in, 3
Supersonic projectiles
   head wave of, 2
Surface curvature
   perturbed boundary layers and, 341-46
Surface roughness
   step changes in
      boundary layers and, 328-32
Surfaces
   turbulent flow onto, 479-81
   turbulent flow parallel to, 481-82
Surface tension
   coating films and, 68
   multicomponent convection and, 35
Suspensions
   see specific type
Svanberg vorticity, 202
Sverdrup balance, 385-86, 388
Swift-Steiber approximation, 77

T

Taylor-Green problem, 543
Tensor product interpolant, 501
Thermal convection
   magnetic fields and, 33-34
Thermocapillary flow
   crystal-growth melts and, 193
Thermohaline convection, 11
Thickening zone, 94
Thin clouds
   in shear flows, 466-67
Thin-filament approximation, 526-29

applications of, 529-32
limitations of, 532-34
Time averages
   dispersion models and, 122-27
Tollmien-Schlichting waves, 415
   local induction approximation and, 535
Towing tank
   air pollutant dispersion and, 244-47
   two-dimensional hills and, 263-64
Trade winds
   reversal of, 360
Trailing-edge flows
   Kutta condition and, 416-26
Transitional-speed flames
   propagation of, 280-82
Trouton ratio, 58
Turbines, 1
Turbulence
   homogeneous
      broad plumes in, 467-82
   inhomogeneous
      diffusion calculations in, 471
Turbulent diffusion, 128, 447-83
   analysis in terms of fluid-element displacement, 448-54
   homogeneous, 467-82
   near the source, 467-73
   steady complex flows and, 454-67
   onto surfaces, 479-82
   over time, 473-78
Turbulent diffusivity, 461
Turbulent energy transport equation, 325
Turbulent flows
   around bluff obstacles, 458
Turbulent shear layers
   Navier-Stokes equations and, 322
   see also Perturbed boundary layers

U

Underwater acoustics
   parabolic equation and, 224-27

V

Vapor explosions
   multiphase drops and, 316-18
Vertical mixing
   equatorially trapped waves and, 374

Viscometric flows, 51-54
Volcanoes
  andesitic
    double-diffusive convection
      and, 39
  lava from
    convective fractionation
      and, 37
von Kármán-Cochran centrifugal
  flow, 192-93
Vortex breakdown, 545-47
Vortex flows
  three-dimensional, 523-57
Vortex ring
  energy spectrum of, 555
  instability of, 550
Vortex tubes
  axial flow in, 554
  isolated
    motion of, 526-39

Vorticity
  smooth distributions of, 539-
    53

W

Wall heat flux
  boundary layers and, 333-36
Water-methanol mixtures
  diffusion in, 21
Waves
  see Ocean waves
Weather
  swing bowling and, 166-67
Wiener-Hopf problem, 414, 420
Wind profiles
  Froude number and, 244
Windrows
  Langmuir cells and, 30
Winds

equatorial currents and, 379-
    99
  see also specific type
Wind tunnels
  supersonic, 3
Wings
  ground effects on, 4
Wing-tip vortices
  flap vortex interaction with,
    530

Y

Yanai wave, 374
Yoshida jet, 382-83

Z

Zone melting
  crystal growth and, 211-12

CUMULATIVE INDEXES

CONTRIBUTING AUTHORS, VOLUMES 1–17

A

Acosta, A. J., 5:161–84
Acrivos, A., 17:91–118
Adamson, T. C. Jr., 12:103–38
Alksne, A. Y., 2:313–54
Allen, C. M., 17:119–49
Allen, J. S., 12:389–433
Antonia, R. A., 13:131–56
Aref, H., 15:345–89
Arndt, R. E. A., 13:273–328
Ashley, H., 4:431–72
Ashton, G. D., 10:369–92

B

Baker, G. R., 11:95–122
Barenblatt, G. I., 4:285–312
Batchelor, G. K., 6:227–55
Bearman, P. W., 16:195–222
Becker, E., 4:155–94
Belotserkovskii, S. M., 9:469–94
Benton, E. R., 6:257–80
Berger, E., 4:313–40
Berger, S. A., 15:461–512
Berman, N. S., 10:47–64
Binnie, A. M., 10:1–10
Bird, G. A., 10:11–31
Bird, R. B., 8:13–34
Bogy, D. B., 11:207–28
Bradshaw, P., 9:33–54
Brennen, C., 9:339–98
Brenner, H., 2:137–76
Brooks, N. H., 7:187–211
Browand, F. K., 7:273–305
Brown, S. N., 1:45–72
Buchhave, P., 11:443–503
Burgers, J. M., 7:1–11
Busemann, A., 3:1–12
Busse, F. H., 10:435–62

C

Callander, R. A., 10:129–58
Canny, M. J., 9:275–96
Cantwell, B. J., 13:457–515
Caughey, D. A., 14:261–83
Cermak, J. E., 8:75–106

Chatwin, P. C., 17:119–49
Cheung, F. B., 15:293–319
Chiang, A. S., 13:351–78
Christensen, J., 12:139–58
Christiansen, W. H., 7:115–39
Clark, A. Jr., 6:257–80
Comte-Bellot, G., 8:209–31
Corcos, G. M., 10:267–88
Cox, R. G., 3:291–316
Crighton, D. G., 11:11–33;
  17:411–45
Crochet, M. J., 15:241–60
Csanady, G. T., 7:357–86

D

Davidson, J. F., 9:55–86
Davis, R. H., 17:91–118
Davis, S. H., 8:57–74
Denn, M. M., 12:365–87
De Vries, O., 15:77–96
Dickinson, R. E., 10:159–95
Donnelly, R. J., 6:179–225
Dowson, D., 11:35–66
Drew, D. A., 15:261–91
Dussan V., E. B., 11:371–400
Dwyer, H. A., 13:217–29

E

Eichelbrenner, E. A., 5:339–60
Eiseman, P. R., 17:487–522
Eliassen, A., 14:1–11
Emmons, H. W., 2:15–36;
  12:223–36
Engelund, F., 14:13–37
Epstein, M., 15:293–319
Evans, D. V., 13:157–87

F

Falcón, M., 16:179–93
Fay, J. A., 5:151–60
Fenton, J. D., 14:39–60
Ferri, A., 5:301–38
Ffowcs Williams, J. E., 1:197–222; 9:447–68
Field, J. E., 15:97–122

Fischer, H. B., 5:59–78; 8:107–33
Fletcher, N. H., 11:123–46
Flick, R. E., 8:275–310
Flügge, W., 5:1–8
Flügge-Lotz, I., 5:1–8
Friedman, H. W., 3:63–88
Fung, Y. C., 3:189–210

G

Garrett, C., 11:339–69
Gebhart, B., 5:213–46
Gence, J. N., 15:201–22
George, W. K. Jr., 11:443–503
Goldsmith, H. L., 7:213–47
Goldstein, M. E., 16:263–85
Goldstein, S., 1:1–28
Griffith, W. C., 10:93–105
Grimshaw, R., 16:11–44
Guedes de Carvalho, J. R. F., 9:55–86

H

Hall, M. G., 4:195–218
Hamblin, P. F., 14:153–87
Hanratty, T. J., 13:231–52
Harleman, D. R. F., 4:7–32
Harrison, D., 9:55–86
Hart, J. E., 11:147–72
Hasimoto, H., 12:335–63
Hawthorne, W. R., 1:341–66
Hayes, W. D., 3:269–90
Hendershott, M., 2:205–24
Herczyński, R., 12:237–69
Hertzberg, A., 7:115–39
Hill, J. C., 8:135–61
Ho, C. -M., 16:365–424
Holt, M., 8:187–214
Hopfinger, E. J., 15:47–76
Horikawa, K., 13:9–32
Horlock, J. H., 5:247–80
Hoskins, B. J., 14:131–51
Hoult, D. P., 4:341–68
Howard, L. N., 4:473–94
Huerre, P., 16:365–424
Hunt, J. C. R., 3:37–62;
  17:447–85

Hunter, C., 4:219–42
Hutter, K., 14:87–130
Hutter, U., 9:399–419

I

Imberger, J., 10:267–88;
    14:153–87
Inman, D. L., 8:275–310
Israeli, M., 6:281–318

J

Jaffrin, M. Y., 3:13–36
Jahn, T. L., 4:93–116
Jenkins, J. T., 10:197–219
Jerskey, T., 5:9–30
Johnson, R. E., 17:289–320
Jones, R. T., 1:223–44; 9:1–11

K

Kassoy, D. R., 17:267–87
Kazhikhov, A. V., 13:79–95
Keller, H. B., 10:417–33
Kennedy, J. F., 1:147–68
Kerrebrock, J. L., 5:281–300
Kogan, M. N., 5:383–404
Koh, R. C. Y., 7:187–211
Korbacher, G. K., 6:319–58
Korobeinikov, V. P., 3:317–46
Kovasznay, L. S. G., 2:95–112
Krylov, V. S., 1:293–316

L

Ladyzhenskaya, O. A., 7:249–
    72
Laguna, G. A., 16:139–77
Lai, W. M., 11:247–88
Lake, B. M., 12:303–34
Lakshminarayana, B., 5:247–80
Landweber, L., 11:173–205
Lanford, O. E. III, 14:347–64
Langlois, W. E., 17:191–215
Laufer, J., 7:307–26
Launder, B. E., 15:429–59
Lauterborn, W., 16:223–44
Laws, E. M., 10:247–66
Leal, L. G., 12:435–76
Lebovitz, N. R., 11:229–46
Lee, J. H. S., 16:311–36
Leibovich, S., 10:221–46;
    15:391–427
Leith, C. E., 10:107–28
Leonard, A., 17:523–59
Lesser, M. B., 15:97–122
Levich, V. G., 1:293–316
Libby, P. A., 8:351–76
Lick, W., 2:113–36
Liepmann, H. W., 16:139–77
Lightfoot, E. N., 13:351–78
Lighthill, M. J., 1:413–46

Lin, C. C., 13:33–55
Lin, J. -T., 11:317–88
Linson, L. M., 3:63–88
Lissaman, P. B. S., 15:223–39
List, E. J., 14:189–212
Livesey, J. L., 10:247–66
Loitsianskii, L. G., 2:1–14
Lomax, H., 7:63–88; 11:289–
    316
Long, R. R., 4:69–92
Lugt, H. J., 15:123–47
Lumley, J. L., 1:367–84;
    11:443–503

M

Macagno, E. O., 12:139–58
MacCormack, R. W., 11:289–
    316
Macken, N. A., 5:185–212
Marble, F. E., 2:397–446
Martin, S., 13:379–97
Mason, S. G., 3:291–316
Maxworthy, T., 7:273–305;
    13:329–50
McCreary, J. P. Jr., 17:359–409
McCroskey, W. J., 14:285–311
McCune, J. E., 5:281–300
McIntire, L. V., 12:159–79
Mehta, R. D., 17:151–89
Mei, C. C., 10:393–416
Meissner, J., 17:45–64
Melcher, J. R., 1:111–46
Messiter, A. F., 12:103–38
Michael, D. H., 13:189–215
Mikhailov, V. V., 3:371–96
Miles, J. W., 6:17–35; 12:11–43
Milgram, J. H., 4:397–430
Millsaps, K., 16:1–10
Moin, P., 16:99–137
Mollo-Christensen, E., 5:101–18
Monin, A. S., 2:225–50
Morel-Seytoux, H. J., 8:233–74
Mow, V. C., 11:247–88
Munk, M. M., 13:1–7
Munk, W., 2:205–24; 11:339–69
Mysak, L. A., 12:45–76

N

Naudascher, E., 11:67–94
Neiland, V. Ya., 3:371–96
Newman, J. N., 2:67–94
Nickel, K., 5:405–28
Nieuwland, G. Y., 5:119–50
Noble, P. T., 13:351–78
Nordstrom, C. E., 8:275–310
Novak, R. A., 1:341–66

O

Oppenheim, A. K., 5:31–58
Orszag, S. A., 6:281–318

Ostrach, S., 14:313–45
Oxburgh, E. R., 4:33–68

P

Palm, E., 7:39–61
Panofsky, H. A., 6:147–77
Pao, Y. -H., 11:317–38
Parlange, J. -Y., 5:79–100;
    12:77–102
Patel, V. C., 11:173–205
Patrick, R. M., 3:63–88
Patterson, G. S. Jr., 10:289–300
Payatakes, A. C., 14:365–93
Peake, D. J., 14:61–85
Pearson, J. R. A., 8:163–81
Pedley, T. J., 9:229–74
Peltier, W. R., 17:561–608
Penner, S. S., 5:9–30
Peregrine, D. H., 15:149–78
Peskin, C. S., 14:235–59
Peterlin, A., 8:35–55
Petschek, H. E., 3:63–88
Philip, J. R., 2:177–204
Phillips, N. A., 2:251–92
Phillips, O. M., 1:245–64; 6:93–
    110
Pieńkowska, I., 12:237–69
Pipkin, A. C., 9:13–32
Plesset, M. S., 9:145–85
Prosperetti, A., 9:145–85

R

Raichlen, F., 7:327–56
Raizer, Yu. P., 1:385–412
Rallison, J. M., 16:45–66
Rand, R. H., 15:29–45
Raupach, M. R., 13:97–129
Reethof, G., 10:333–67
Reichenbach, H., 15:1–28
Reshotko, E., 8:311–49
Reynolds, W. C., 8:183–208
Rhines, P. B., 11:401–41
Rich, J. W., 2:355–96
Rivlin, R. S., 3:117–46
Roberts, P. H., 4:117–54;
    6:179–225
Roberts, W. W. Jr., 13:33–55
Robinson, A. R., 2:293–312
Rockwell, D., 11:67–94
Rodden, W. P., 4:431–72
Rodi, W., 15:429–59
Rogallo, R. S., 16:99–137
Rohsenow, W. M., 3:211–36
Rott, N., 17:1–9
Rouse, H., 8:1–12
Rusanov, V. V., 8:377–404
Ruschak, K. J., 17:65–89
Russel, W. B., 13:425–55
Russell, D. A., 7:115–39
Ryzhov, O. S., 10:65–92

618   CONTRIBUTING AUTHORS

S

Sadhal, S. S., 17:289–320
Saffman, P. G., 11:95–122
Saibel, E. A., 5:185–212
Sano, O., 12:335–63
Saville, D. A., 9:321–37
Sawyers, K. N., 3:117–46
Schowalter, W. R., 16:245–61
Schwartz, L. W., 14:39–60
Sears, M. R., 11:1–10
Sears, W. R., 11:1–10
Seebass, A. R., 12:181–222;
  16:337–63
Shapiro, A. H., 3:13–36
Shen, S. -F., 9:421–45
Shercliff, J. A., 3:37–62
Sherman, F. S., 1:317–40;
  10:267–88
Simpson, J. E., 14:213–34
Sivashinsky, G. I., 15:179–99
Skalak, R., 7:213–47
Smits, A. J., 17:321–58
Snyder, W. H., 17:239–66
Sobieczky, H., 16:337–63
Solonnikov, V. A., 13:79–95
Soloukhin, R. I., 5:31–58
Soward, A. M., 4:117–54
Spee, B. M., 5:119–50
Spielman, L. A., 9:297–319
Spindel, R. C., 17:217–37
Spreiter, J. R., 2:313–54
Steger, J. L., 7:63–88
Stewartson, K., 1:45–72
Stolzenbach, K. D., 4:7–32
Streeter, V. L., 6:57–73
Stuart, J. T., 3:347–70
Sychev, V. V., 3:371–96

T

Takahashi, T., 13:57–77
Talbot, L., 15:461–512
Tani, I., 1:169–96; 9:87–111
Tanner, R. I., 9:13–32
Taub, A. H., 10:301–32
Taylor, C. M., 11:35–66
Taylor, G. I., 1:111–46; 6:1–16
Thom, A. S., 13:97–129
Thomas, J. H., 15:321–43
Tien, C. L., 7:167–85
Tijdeman, H., 12:181–222
Tobak, M., 14:61–85
Traugott, S. C., 3:89–116
Treanor, C. E., 2:355–96
Truesdell, C., 6:111–46
Tuck, E. O., 10:33–46
Turcotte, D. L., 4:33–68
Turner, J. S., 1:29–44; 6:37–56;
  17:11–44

U

Uhlenbeck, G. E., 12:1–9

V

Van Atta, C. W., 6:75–91
Van Dyke, M., 1:265–92;
  16:287–309
van Wijngaarden, L., 4:369–96
Veronis, G., 2:37–66
Villat, H., 4:1–5
Vincenti, W. G., 3:89–116
Vogel, A., 16:223–44
Votta, J. J., 4:93–116

W

Walters, K., 15:241–60
Wegener, P. P., 5:79–100
Wehausen, J. V., 3:237–68
Werlé, H., 5:361–82
Widnall, S. E., 7:141–65
Wieghardt, K., 7:89–114
Wille, R., 4:313–40
Williams, F. A., 3:171–88;
  8:351–76
Williams, J. C. III, 9:113–44
Willmarth, W. W., 3:147–70;
  7:13–38
Winant, C. D., 12:271–301
Winet, H., 9:339–98
Wood, D. H., 17:321–58
Wooding, R. A., 8:233–74
Wu, T. Y., 4:243–84
Wylie, E. B., 6:57–73
Wyngaard, J. C., 13:399–423

Y

Yaglom, A. M., 11:505–40
Yao, L. -S., 15:461–512
Yen, S. M., 16:67–97
Yeung, R. W., 14:395–442
Yih, C. -S., 1:73–110
Yuen, H. C., 12:303–34

Z

Zel'dovich, Ya. B., 1:385–412;
  4:285–312; 9:215–28
Zeman, O., 13:253–72
Zweifach, B. W., 3:189–210

CHAPTER TITLES, VOLUMES 1–16

HISTORY

Fluid Mechanics in the First Half of This Century	S. Goldstein	1:1–28
The Development of Boundary-Layer Theory in the USSR	L. G. Loitsianskii	2:1–14
Compressible Flow in the Thirties	A. Busemann	3:1–12
As Luck Would Have It—A Few Mathematical Reflections	H. Villat	4:1–6
Ludwig Prandtl in the Nineteen-Thirties: Reminiscences	I. Flügge-Lotz, W. Flügge	5:1–8
The Interaction Between Experiment and Theory in Fluid Mechanics	G. I. Taylor	6:1–16
Some Memories of Early Works in Fluid Mechanics at the Technical University of Delft	J. M. Burgers	7:1–11
Hydraulics' Latest Golden Age	H. Rouse	8:1–12
Recollections From an Earlier Period in American Aeronautics	R. T. Jones	9:1–11
History of Boundary-Layer Theory	I. Tani	9:87–111
Some Notes on the Study of Fluid Mechanics in Cambridge, England	A. M. Binnie	10:1–10
The Kármán Years at GALCIT	W. R. Sears, M. R. Sears	11:1–10
Some Notes on the Relation Between Fluid Mechanics and Statistical Physics	G. E. Uhlenbeck	12:1–9
My Early Aerodynamic Research—Thoughts and Memories	M. M. Munk	13:1–7
Vilhelm Bjerknes and his Students	A. Eliassen	14:1–12
Contributions of Ernst Mach to Fluid Mechanics	H. Reichenbach	15:1–28
Karl Pohlhausen, As I Remember Him	K. Millsaps	16:1–10
Jakob Ackeret and the History of the Mach Number	N. Rott	17:1–9

FOUNDATIONS

Nonlinear Continuum Mechanics of Viscoelastic Fluids	R. S. Rivlin, K. N. Sawyers	3:117–46
Bounds on Flow Quantities	L. N. Howard	4:473–94
Self-Similar Solutions as Intermediate Asymptotics	G. I. Barenblatt, Ya. B. Zel'dovich	4:285–312
Prandtl's Boundary-Layer Theory From the Viewpoint of a Mathematician	K. Nickel	5:405–28
The Meaning of Viscometry in Fluid Dynamics	C. Truesdell	6:111–46
Mathematical Analysis of Navier-Stokes Equations for Incompressible Liquids	O. A. Ladyzhenskaya	7:249–72
Steady Non-Viscometric Flows of Viscoelastic Liquids	A. C. Pipkin, R. I. Tanner	9:13–32
Existence Theorems for the Equations of Motion of a Compressible Viscous Fluid	V. A. Solonnikov, A. V. Kazhikhov	13:79–95
Topology of Three-Dimensional Separated Flows	M. Tobak, D. J. Peake	14:61–85
Wave Action and Wave-Mean Flow Interaction, With Application to Stratified Shear Flows	R. Grimshaw	16:11–44
Nonlinear Interactions in the Fluid Mechanics of Helium II	H. W. Liepmann, G. A. Laguna	16:139–77

619

NON-NEWTONIAN FLUIDS, RHEOLOGY

Drag Reduction by Additives	J. L. Lumley	1:367–84
Rheology of Two-Phase Systems	H. Brenner	2:137–76
Nonlinear Continuum Mechanics of Viscoelastic Fluids	R. S. Rivlin, K. N. Sawyers	3:117–46
The Meaning of Viscometry in Fluid Dynamics	C. Truesdell	6:111–46
Useful Non-Newtonian Models	R. B. Bird	8:13–34
Instability in Non-Newtonian Flow	J. R. A. Pearson	8:163–81
Steady Non-Viscometric Flows of Viscoelastic Liquids	A. C. Pipkin, R. I. Tanner	9:13–32
Drag Reduction by Polymers	N. S. Berman	10:47–64
Numerical Methods in Non-Newtonian Fluid Mechanics	M. J. Crochet, K. Walters	15:241–60
Nonlinear Interactions in the Fluid Mechanics of Helium II	H. W. Liepmann, G. A. Laguna	16:139–77
Stability and Coagulation of Colloids in Shear Fields	W. R. Schowalter	16:245–61
Rheometry of Polymer Melts	J. Meissner	17:45–64

INCOMPRESSIBLE INVISCID FLUIDS

Vortex Breakdown	M. G. Hall	4:195–218
Periodic Flow Phenomena	E. Berger, R. Wille	4:313–40
Bounds on Flow Quantities	L. N. Howard	4:473–94
Secondary Flows: Theory, Experiment, and Application in Turbomachinery Aerodynamics	J. H. Horlock, B. Lakshminarayana	5:247–80
The Structure and Dynamics of Vortex Filaments	S. E. Widnall	7:141–65
The Structure of Vortex Breakdown	S. Leibovich	10:221–46
Vortex Interactions	P. G. Saffman, G. R. Baker	11:95–122
Low-Gravity Fluid Flows	S. Ostrach	14:313–45
Autorotation	H. J. Lugt	15:123–47
Integrable, Chaotic, and Turbulent Vortex Motions in Two-Dimensional Flows	H. Aref	15:345–89
Wave Action and Wave-Mean Flow Interaction, With Application to Stratified Shear Flows	R. Grimshaw	16:11–44
Vortex Shedding from Oscillating Bluff Bodies	P. W. Bearman	16:195–222
Computing Three-Dimensional Incompressible Flows With Vortex Elements	A. Leonard	17:523–59

COMPRESSIBLE FLUIDS

Shock Waves and Radiation	Ya. B. Zel'dovich, Yu. P. Raizer	1:385–412
Vibrational Relaxation in Gas-Dynamic Flows	J. W. Rich, C. E. Treanor	2:355–96
Dynamics of Dusty Gases	F. E. Marble	2:397–446
Compressible Flow in the Thirties	A. Busemann	3:1–12
The Coupling of Radiative Transfer and Gas Motion	W. G. Vincenti, S. C. Traugott	3:89–116
Sonic Boom	W. D. Hayes	3:269–90
Gas Dynamics of Explosions	V. P. Korobeinikov	3:317–46
The Theory of Viscous Hypersonic Flow	V. V. Mikhailov, V. Ya. Neiland, V. V. Sychev	3:371–96
Mixing-Controlled Supersonic Combustion	A. Ferri	5:301–38
Transonic Airfoils: Recent Developments in Theory, Experiment, and Design	G. Y. Nieuwland, B. M. Spee	5:119–50
Experiments in Gasdynamics of Explosions	A. K. Oppenheim, R. I. Soloukhin	5:31–58
A Blunt Body in a Supersonic Stream	V. V. Rusanov	8:377–404
Compressible Turbulent Shear Layers	P. Bradshaw	9:33–54
Viscous Transonic Flows	O. S. Ryzhov	10:65–92

Transonic Flow Past Oscillating Airfoils | H. Tijdeman, R. Seebass | 12:181–222
Existence Theorems for the Equations of
Motion of a Compressible Viscous Fluid | V. A. Solonnikov, A. V. Kazhikhov | 13:79–95
The Computation of Transonic Potential Flows | D. A. Caughey | 14:261–83
The Impact of Compressible Liquids | M. B. Lesser, J. E. Field | 15:97–122
Supercritical Airfoil and Wing Design | H. Sobieczky, A. R. Seebass | 16:337–63

RAREFIED GAS FLOW
Transition from Continuum to Molecular Flow | F. S. Sherman | 1:317–40
Molecular Gas Dynamics | M. N. Kogan | 5:383–404
Numerical Solution of the Nonlinear
Boltzmann Equation for Nonequilibrium
Gas Flow Problems | S. M. Yen | 16:67–97

MAGNETOHYDRODYNAMICS, PLASMA FLOW, ELECTROHYDRODYNAMICS
Electrohydrodynamics: A Review of the Role
of Interfacial Shear Stresses | J. R. Melcher, G. I. Taylor | 1:111–46
Solar-Wind Flow Past Objects in the Solar
System | J. R. Spreiter, A. Y. Alksne | 2:313–54
Magnetohydrodynamics at High Hartmann
Number | J. C. R. Hunt, J. A. Shercliff | 3:37–62
Collisionless Shocks in Plasmas | H. S. Friedman, L. M. Linson, R.
 | M. Patrick, H. E. Petschek | 3:63–88
Magnetohydrodynamics of the Earth's Core | P. H. Roberts, A. M. Soward | 4:117–54
Electrokinetic Effects with Small Particles | D. A. Saville | 9:321–37
Magnetohydrodynamics of the Earth's
Dynamo | F. H. Busse | 10:435–62
Magneto-Atmospheric Waves | J. H. Thomas | 15:321–43

VISCOUS FLUIDS
Laminar Separation | S. N. Brown, K. Stewartson | 1:45–72
The Theory of Viscous Hypersonic Flow | V. V. Mikhailov, V. Ya. Neiland,
 | V. V. Sychev | 3:371–96
Prandtl's Boundary-Layer Theory from the
Viewpoint of a Mathematician | K. Nickel | 5:405–28
The Fluid Mechanics of Lubrication | E. A. Saibel, N. A. Macken | 5:185–212
The Meaning of Viscometry in Fluid
Dynamics | C. Truesdell | 6:111–46
Spin-Up | E. R. Benton, A. Clark Jr. | 6:257–80
Mathematical Analysis of Navier-Stokes
Equations for Incompressible Liquids | O. A. Ladyzhenskaya | 7:249–72
Steady Non-Viscometric Flows of Viscoelastic
Liquids | A. C. Pipkin, R. I. Tanner | 9:13–32
Electrokinetic Effects with Small Particles | D. A. Saville | 9:321–37
Viscous Transonic Flows | O. S. Ryzhov | 10:65–92
Stokeslets and Eddies in Creeping Flow | H. Hasimoto, O. Sano | 12:335–63
Particle Motions in a Viscous Fluid | L. G. Leal | 12:435–76
Existence Theorems for the Equations of
Motion of a Compressible Fluid | V. A. Solonnikov, A. V. Kazhikhov | 13:79–95
Flow in Curved Pipes | S. A. Berger, L. Talbot, L. -S. Yao | 15:461–512
Vortex Shedding from Oscillating Bluff
Bodies | P. W. Bearman | 16:195–222
Coating Flows | K. J. Ruschak | 17:65–89
Sedimentation of Noncolloidal Particles at
Low Reynolds Numbers | R. H. Davis, A. Acrivos | 17:91–118
Mantle Convection and Viscoelasticity | W. R. Peltier | 17:561–608

BOUNDARY-LAYER THEORY
Laminar Separation | S. N. Brown, K. Stewartson | 1:45–72
Boundary-Layer Transition | I. Tani | 1:169–96
Higher-Order Boundary-Layer Theory | M. Van Dyke | 1:265–92

The Development of Boundary-Layer Theory
  in the USSR   L. G. Loitsianskii   2:1–14
Turbulent Boundary Layer   L. S. G. Kovasznay   2:95–112
Atmospheric Boundary Layer   A. S. Monin   2:225–50
Boundary Layers in Ocean Circulation Models   A. R. Robinson   2:293–312
The Theory of Viscous Hypersonic Flow   V. V. Mikhailov, V. Ya. Neiland,
                                          V. V. Sychev   3:371–96
Three-Dimensional Boundary Layers   E. A. Eichelbrenner   5:339–60
Prandtl's Boundary-Layer Theory from the
  Viewpoint of a Mathematician   K. Nickel   5:405–28
Pressure Fluctuations Beneath Turbulent
  Boundary Layers   W. W. Willmarth   7:13–37
Boundary-Layer Stability and Transition   E. Reshotko   8:311–49
Compressible Turbulent Shear Layers   P. Bradshaw   9:33–54
History of Boundary-Layer Theory   I. Tani   9:87–111
Incompressible Boundary-Layer Separation   J. C. Williams III   9:113–44
Study of the Unsteady Aerodynamics of
  Lifting Surfaces Using the Computer   S. M. Belotserkovskii   9:469–94
Numerical Methods in Boundary-Layer Theory   H. B. Keller   10:417–33
Ship Boundary Layers   L. Landweber, V. C. Patel   11:173–205
Analysis of Two-Dimensional Interactions
  Between Shock Waves and Boundary
  Layers   T. C. Adamson Jr., A. F. Messiter   12:103–38
Some Aspects of Three-Dimensional Laminar
  Boundary Layers   H. A. Dwyer   13:217–29
Progress in the Modeling of Planetary
  Boundary Layers   O. Zeman   13:253–72
The Turbulent Wall Jet—Measurements and
  Modeling   B. E. Launder, W. Rodi   15:429–59
Aerodynamics of Sports Balls   R. D. Mehta   17:151–89
The Response of Turbulent Boundary Layers
  to Sudden Perturbations   A. J. Smits, D. H. Wood   17:321–58
The Kutta Condition in Unsteady Flow   D. G. Crighton   17:411–45

STABILITY OF FLOW
Nonlinear Stability Theory   J. T. Stuart   3:347–70
Vortex Breakdown   M. G. Hall   4:195–218
Bounds on Flow Quantities   L. N. Howard   4:473–94
The Stability of Time-Periodic Flows   S. H. Davis   8:57–74
Instability in Non-Newtonian Flow   J. R. A. Pearson   8:163–81
Boundary-Layer Stability and Transition   E. Reshotko   8:311–49
The Structure of Vortex Breakdown   S. Leibovich   10:221–46
Self-Sustained Oscillations of Impinging Free
  Shear Layers   D. Rockwell, E. Naudascher   11:67–94
Finite Amplitude Baroclinic Instability   J. E. Hart   11:147–72
Meniscus Stability   D. H. Michael   13:189–215
Stability of Surfaces That Are Dissolving or
  Being Formed by Convective Diffusion   T. J. Hanratty   13:231–52
Instabilities, Pattern Formation, and
  Turbulence in Flames   G. I. Sivashinsky   15:179–99
The Form and Dynamics of Langmuir
  Circulations   S. Leibovich   15:391–427
Stability and Coagulation of Colloids in Shear
  Fields   W. R. Schowalter   16:245–61
Perturbed Free Shear Layers   C. -M. Ho, P. Huerre   16:365–424

TURBULENCE
Shear-Flow Turbulence   O. M. Phillips   1:245–64
The Turbulent Boundary Layer   L. S. G. Kovasznay   2:95–112
Bounds on Flow Quantities   L. N. Howard   4:473–94
Intermittency in Large-Scale Turbulent Flows   E. Mollo-Christensen   5:101–18
Sampling Techniques in Turbulence
  Measurements   C. W. Van Atta   6:75–91

Pressure Fluctuations Beneath Turbulent
  Boundary Layers   W. W. Willmarth   7:13–37
New Trends in Experimental Turbulence
  Research   J. Laufer   7:307–26
Homogeneous Turbulent Mixing with
  Chemical Reaction   J. C. Hill   8:135–61
Computation of Turbulent Flows   W. C. Reynolds   8:183–208
Hot-Wire Anemometry   G. Comte-Bellot   8:209–31
Turbulent Flows Involving Chemical Reactions   P. A. Libby, F. A. Williams   8:351–76
Compressible Turbulent Shear Layers   P. Bradshaw   9:33–54
Aeroacoustics   J. E. Ffowcs Williams   9:447–68
Turbulence and Mixing in Stably Stratified
  Waters   F. S. Shermann, J. Imberger, G. M.
                                                       Corcos   10:267–88
Geostrophic Turbulence   P. B. Rhines   11:401–41
The Measurement of Turbulence With the
  Laser-Doppler Anemometer   P. Buchhave, W. K. George Jr., J.
                                                       L. Lumley   11:443–503
Similarity Laws for Constant-Pressure and
  Pressure-Gradient Turbulent Wall Flows   A. M. Yaglom   11:505–40
Turbulence In and Above Plant Canopies   M. R. Raupach, A. S. Thom   13:97–129
Conditional Sampling in Turbulence
  Measurement   R. A. Antonia   13:131–56
Cup, Propeller, Vane, and Sonic
  Anemometers in Turbulence Research   J. C. Wyngaard   13:399–423
Organized Motion in Turbulent Flow   B. J. Cantwell   13:457–515
Turbulent Jets and Plumes   E. J. List   14:189–212
The Strange Attractor Theory of Turbulence   O. E. Lanford III   14:347–64
Homogeneous Turbulence   J. N. Gence   15:201–22
Integrable, Chaotic, and Turbulent Vortex
  Motions in Two-Dimensional Flows   H. Aref   15:345–89
The Turbulent Wall Jet—Measurements and
  Modeling   B. E. Launder, W. Rodi   15:429–59
Numerical Simulation of Turbulent Flows   R. Rogallo, P. Moin   16:99–137
Aeroacoustics of Turbulent Shear Flows   M. E. Goldstein   16:263–85
Perturbed Free Shear Layers   C. -M. Ho, P. Huerre   16:365–424
The Response of Turbulent Boundary Layers
  to Sudden Peturbations   A. J. Smits, D. H. Wood   17:321–58
Turbulent Diffusion From Sources in Complex
  Flows   J. C. R. Hunt   17:447–85

CONVECTION
Buoyant Plumes and Thermals   J. S. Turner   1:29–44
Instability, Transition, and Turbulence in
  Buoyancy-Induced Flows   B. Gebhart   5:213–46
Nonlinear Thermal Convection   E. Palm   7:39–61
Multicomponent Convection   J. S. Turner   17:11–44
Buoyancy-Driven Flows in Crystal-Growth
  Melts   W. E. Langlois   17:191–215

HEAT TRANSFER
Boiling   W. M. Rohsenow   3:211–36
Fluid Mechanics of Heat Disposal from Power
  Generation   D. R. F. Harleman, K. D.
                                                       Stolzenbach   4:7–32
Fluid Mechanics of Heat Pipes   C. L. Tien   7:167–85
Complex Freezing-Melting Interfaces in Fluid
  Flow   M. Epstein, F. B. Cheung   15:293–319

COMBUSTION, FLOWS WITH CHEMICAL REACTION
Theory of Combustion of Laminar Flows   F. A. Williams   3:171–88
Chemically Reacting Flows   E. Becker   4:155–94
Mixing-Controlled Supersonic Combustion   A. Ferri   5:301–38

Flow Lasers	W. H. Christiansen, D. A. Russell, A. Hertzberg	7:115–39
Homogeneous Turbulent Mixing with Chemical Reaction	J. C. Hill	8:135–61
Turbulent Flows Involving Chemical Reactions	P. A. Libby, F. A. Williams	8:351–76
Scientific Progress on Fire	H. W. Emmons	12:223–36
Instabilities, Pattern Formation, and Turbulence in Flames	G. I. Sivashinsky	15:179–99
Dynamic Parameters of Gaseous Detonations	J. H. S. Lee	16:311–36
Mathematical Modeling for Planar, Steady, Subsonic Combustion Waves	D. R. Kassoy	17:267–87

SHOCK WAVES, EXPLOSIONS
Shock Waves and Radiation	Ya. B. Zel'dovich, Yu. P. Raizer	1:385–412
Sonic Boom	W. D. Hayes	3:269–90
Gas Dynamics of Explosions	V. P. Korobeinikov	3:317–46
Collisionless Shocks in Plasmas	H. W. Friedman, L. M. Linson, R. M. Patrick, H. E. Petschek	3:63–88
Experiments in Gasdynamics of Explosions	A. K. Oppenheim, R. I. Soloukhin	5:31–58
A Blunt Body in a Supersonic Stream	V. V. Rusanov	8:377–404
Underwater Explosions	M. Holt	9:187–214
Dust Explosions	W. C. Griffith	10:93–105
Analysis of Two-Dimensional Interactions Between Shock Waves and Boundary Layers	T. C. Adamson Jr., A. F. Messiter	12:103–38
Dynamic Parameters of Gaseous Detonations	J. H. S. Lee	16:311–36

AERO- AND HYDRODYNAMIC SOUND, ACOUSTICS
Hydrodynamic Noise	J. E. Ffowcs Williams	1:197–222
Sonic Boom	W. D. Hayes	3:269–90
Noise from Aircraft Turbomachinery	J. E. McCune, J. L. Kerrebrock	5:281–300
Aeroacoustics	J. E. Ffowcs Williams	9:447–68
Turbulence-Generated Noise in Pipe Flow	G. Reethof	10:333–67
Model Equations of Nonlinear Acoustics	D. G. Crighton	11:11–33
Air Flow and Sound Generation in Musical Wind Instruments	N. H. Fletcher	11:123–46
Aeroacoustics of Turbulent Shear Flows	M. E. Goldstein	16:263–85
Sound Transmission in the Ocean	R. C. Spindel	17:217–37
The Kutta Condition in Unsteady Flow	D. G. Crighton	17:411–45

FLOWS IN HETEROGENEOUS AND STRATIFIED FLUIDS, ROTATING FLOWS
Stratified Flows	C. -S. Yih	1:73–110
Analogy Between Rotating and Stratified Fluids	G. Veronis	2:37–66
Finite Amplitude Disturbances in the Flow of Inviscid Rotating and Stratified Fluids Over Obstacles	R. R. Long	4:69–92
Experiments in Rotating and Stratified Flows: Oceanographic Application	T. Maxworthy, F. K. Browand	7:273–305
Turbulence and Mixing in Stably Stratified Waters	F. S. Sherman, J. Imberger, G. M. Corcos	10:267–88
Wakes in Stratified Fluids	J. -T. Lin, Y. -H. Pao	11:317–38
Geostrophic Turbulence	P. B. Rhines	11:401–41
Fluid Modeling of Pollutant Transport and Diffusion in Stably Stratified Flows Over Complex Terrain	W. H. Snyder	17:239–66

FREE-SURFACE FLOWS (WATER WAVES, CAVITY FLOWS)
Tides	M. Hendershott, W. Munk	2:205–24
Applications of Slender-Body Theory in Ship Hydrodynamics	J. N. Newman	2:67–94
The Motion of Floating Bodies	J. V. Wehausen	3:237–68
Cavity and Wake Flows	T. Y. Wu	4:243–84

Hydrofoils and Hydrofoil Craft	A. J. Acosta	5:161–84
Harbor Seiching	J. W. Miles	6:17–35
The Effect of Waves on Rubble-Mound Structures	F. Raichlen	7:327–56
Numerical Methods in Water-Wave Diffraction and Radiation	C. C. Mei	10:393–416
Solitary Waves	J. W. Miles	12:11–43
Topographically Trapped Waves	L. A. Mysak	12:45–76
Instability of Waves on Deep Water	H. C. Yuen, B. M. Lake	12:303–34
Power From Water Waves	D. V. Evans	13:157–87
Strongly Nonlinear Waves	L. W. Schwartz, J. D. Fenton	14:39–60
Numerical Methods in Free-Surface Flows	R. W. Yeung	14:395–442
The Impact of Compressible Liquids	M. B. Lesser, J. E. Field	15:97–122
Breaking Waves on Beaches	D. H. Peregrine	15:149–78
The Form and Dynamics of Langmuir Circulations	S. Leibovich	15:391–427
Wave Action and Wave-Mean Flow Interaction, With Application to Stratified Shear Flows	R. Grimshaw	16:11–44
Secondary Flow in Curved Open Channels	M. Falcón	16:179–93

BUBBLES, FILMS, SURFACE TENSION, BUBBLY FLOWS, CAVITATION

Surface-Tension-Driven Phenomena	V. G. Levich, V. S. Krylov	1:293–316
Boiling	W. M. Rohsenow	3:211–36
One-Dimensional Flow of Liquids Containing Small Gas Bubbles	L. van Wijngaarden	4:369–96
Oil Spreading on the Sea	D. P. Hoult	4:341–68
Spherical Cap Bubbles	P. P. Wegener, J. -Y. Parlange	5:79–100
Bubble Dynamics and Cavitation	M. S. Plesset, A. Prosperetti	9:145–85
Cavitation in Bearings	D. Dowson, C. M. Taylor	11:35–66
Drop Formation in a Circular Liquid Jet	D. B. Bogy	11:207–28
On the Spreading of Liquids on Solid Surfaces: Static and Dynamic Contact Lines	E. B. Dussan V.	11:371–400
Meniscus Stability	D. H. Michael	13:189–215
Cavitation in Fluid Machinery and Hydraulic Structures	R. E. A. Arndt	13:273–328
Low-Gravity Fluid Flows	S. Ostrach	14:313–45
Mathematical Modeling of Two-Phase Flow	D. A. Drew	15:261–91
The Deformation of Small Viscous Drops and Bubbles in Shear Flows	J. M. Rallison	16:45–66
Rheometry of Polymer Melts	J. Meissner	17:45–64
Coating Flows	K. J. Ruschak	17:65–89
Fluid Mechanics of Compound Multiphase Drops and Bubbles	R. E. Johnson, S. S. Sadhal	17:289–320

DIFFUSION, FILTRATION, SUSPENSIONS

Flow in Porous Media	J. R. Philip	2:177–204
Rheology of Two-Phase Systems	H. Brenner	2:137–76
Dynamics of Dusty Gases	F. E. Marble	2:397–446
Suspended Particles in Fluid Flow Through Tubes	R. G. Cox, S. G. Mason	3:291–316
Longitudinal Dispersion and Turbulent Mixing in Open-Channel Flow	H. B. Fischer	5:59–78
Double-Diffusive Phenomena	J. S. Turner	6:37–56
Transport Properties of Two-Phase Materials With Random Structure	G. K. Batchelor	6:227–55
Multiphase Fluid Flow Through Porous Media	R. A. Wooding, H. J. Morel-Seytoux	8:233–74
On the Liquidlike Behavior of Fluidized Beds	J. F. Davidson, D. Harrison, J. R. F. Guedes de Carvalho	9:55–86
Particle Capture From Low-Speed Laminar Flows	L. A. Spielman	9:297–319
Drag Reduction by Polymers	N. S. Berman	10:47–64
Water Transport in Soils	J. -Y. Parlange	12:77–102

Toward a Statistical Theory of Suspension
Coastal Sediment Processes | R. Herczyński, I. Pieńkowska
K. Horikawa | 12:237–69
13:9–32
Stability of Surfaces That Are Dissolving or
Being Formed by Convective Diffusion | T. J. Hanratty | 13:231–52
Brownian Motion of Small Particles
Suspended in Liquids | W. B. Russel | 13:425–55
Dynamics of Oil Ganglia During Immiscible
Displacement in Water-Wet Porous Media | A. C. Payatakes | 14:365–93
Multicomponent Convection | J. S. Turner | 17:11–44
Sedimentation of Noncolloidal Particles at
Low Reynolds Numbers | R. H. Davis, A. Acrivos | 17:91–118
Mathematical Models of Dispersion in Rivers
and Estuaries | P. C. Chatwin, C. M. Allen | 17:119–49

MATHEMATICAL METHODS
Nonlinear Wave Propagation in Fluids | W. Lick | 2:113–36
Self-Similar Solutions as Intermediate
Asymptotics | G. I. Barenblatt, Ya. B. Zel'dovich | 4:285–312
Periodic Flow Phenomena | E. Berger, R. Wille | 4:313–40
Prandtl's Boundary-Layer Theory From the
Viewpoint of a Mathematician | K. Nickel | 5:405–28
Nonlinear Dispersive Waves | O. M. Phillips | 6:93–110
Mathematical Analysis of Navier-Stokes
Equations for Incompressible Liquids | O. A. Ladyzhenskaya | 7:249–72
Existence Theorems for the Equations of
Motion of a Compressible Viscous Fluid | V. A. Solonnikov, A. V. Kazhikhov | 13:79–95
The Strange Attractor Theory of Turbulence | O. E. Lanford III | 14:347–64
Mathematical Modeling of Two-Phase Flow | D. A. Drew | 15:261–91
Computer-Extended Series | M. Van Dyke | 16:287–309

NUMERICAL METHODS
Critique of Numerical Modeling of
Fluid-Mechanics Phenomena | H. W. Emmons | 2:15–36
Sampling Techniques in Turbulence
Measurements | C. W. Van Atta | 6:75–91
Numerical Simulation of Viscous
Incompressible Flows | S. A. Orszag, M. Israeli | 6:281–318
Relaxation Methods in Fluid Mechanics | H. Lomax, J. L. Steger | 7:63–88
A Blunt Body in a Supersonic Stream | V. V. Rusanov | 8:377–404
Finite-Element Methods in Fluid Mechanics | S. -F. Shen | 9:421–45
Study of the Unsteady Aerodynamics of
Lifting Surfaces Using the Computer | S. M. Belotserkovskii | 9:469–94
Monte Carlo Simulation of Gas Flows | G. A. Bird | 10:11–31
Prospects for Computational Fluid Mechanics | G. S. Patterson Jr. | 10:289–300
Numerical Methods in Boundary-Layer
Theory | H. B. Keller | 10:417–33
Numerical Solution of Compressible Viscous
Flows | R. W. MacCormack, H. Lomax | 11:289–316
The Computation of Transonic Potential
Flows | D. A. Caughey | 14:261–83
Numerical Methods in Free-Surface Flows | R. W. Yeung | 14:395–442
Numerical Methods in Non-Newtonian Fluid
Mechanics | M. J. Crochet, K. Walters | 15:241–60
Numerical Solution of the Nonlinear
Boltzmann Equation for Nonequilibrium
Gas Flow Problems | S. M. Yen | 16:67–97
Numerical Simulation of Turbulent Flows | R. S. Rogallo, P. Moin | 16:99–137
Computer-Extended Series | M. Van Dyke | 16:287–309
Grid Generation for Fluid Mechanics
Computations | P. R. Eiseman | 17:487–522
Computing Three-Dimensional Incompressible
Flows With Vortex Elements | A. Leonard | 17:523–59

EXPERIMENTAL METHODS

Unsteady Force and Pressure Measurements W. W. Willmarth 3:147–70
Use of Lasers for Local Measurement of
  Velocity Components, Species Densities,
  and Temperatures S. S. Penner, T. Jerskey 5:9–30
Hydrodynamic Flow Visualization H. Werlé 5:361–82
Sampling Techniques in Turbulence
  Measurements C. W. Van Atta 6:75–91
Optical Effects in Flow A. Peterlin 8:35–55
Hot-Wire Anemometry G. Comte-Bellot 8:209–31
The Measurement of Turbulence With the
  Laser-Doppler Anemometer P. Buchhave, W. K. George Jr., J.
    L. Lumley 11:505–40
Conditional Sampling in Turbulence
  Measurement R. A. Antonia 13:131–56
Field-Flow Fractionation (Polarization
  Chromatography) E. N. Lightfoot, A. S. Chiang, P.
    T. Noble 13:351–78
Cup, Propeller, Vane, and Sonic
  Anemometers in Turbulence Research J. C. Wyngaard 13:399–423
Modern Optical Techniques in Fluid
  Mechanics W. Lauterborn, A. Vogel 16:223–44

BIOLOGICAL FLUID DYNAMICS

Blood Flow R. T. Jones 1:223–44
Hydromechanics of Aquatic Animal
  Propulsion M. J. Lighthill 1:413–46
Peristaltic Pumping M. Y. Jaffrin, A. H. Shapiro 3:13–36
Microcirculation: Mechanics of Blood Flow in
  Capillaries Y. C. Fung, B. W. Zweifach 3:189–210
Locomotion of Protozoa T. L. Jahn, J. J. Votta 4:93–116
Hemodynamics H. L. Goldsmith, R. Skalak 7:213–47
Pulmonary Fluid Dynamics T. J. Pedley 9:229–74
Flow and Transport in Plants M. J. Canny 9:275–96
Fluid Mechanics of Propulsion by Cilia and
  Flagella C. Brennen, H. Winet 9:339–97
Mechanics of Animal Joints V. C. Mow, W. M. Lai 11:247–88
Fluid Mechanics of the Duodenum E. O. Macagno, J. Christensen 12:139–58
Dynamic Materials Testing: Biological and
  Clinical Applications in Network-Forming
  Systems L. V. McIntire 12:159–79
The Fluid Dynamics of Insect Flight T. Maxworthy 13:329–50
The Fluid Dynamics of Heart Valves:
  Experimental, Theoretical, and
  Computational Methods C. S. Peskin 14:235–59
Fluid Mechanics of Green Plants R. H. Rand 15:29–45

FLUID DYNAMICS OF MACHINERY

Aerodynamics of Turbo-Machinery W. R. Hawthorne, R. A. Novak 1:341–66
Secondary Flows J. H. Horlock, B. Lakshminarayana 5:247–80
The Fluid Mechanics of Lubrication E. A. Saibel, N. A. Macken 5:185–212
Noise From Aircraft Turbomachinery J. E. McCune, J. L. Kerrebrock 5:281–300
Optimum Wind-Energy Conversion Systems U. Hütter 9:399–419
Cavitation in Fluid Machinery and Hydraulic
  Structures R. E. A. Arndt 13:273–328
On the Theory of the Horizontal-Axis Wind
  Turbine O. De Vries 15:77–96
Flow in Curved Pipes S. A. Berger, L. Talbot, L. -S. Yao 15:461–512

FLUID DYNAMICS OF AIRBORNE VEHICLES

Wing-Body Aerodynamic Interaction H. Ashley, W. P. Rodden 4:431–72
Transonic Airfoils: Recent Developments in
  Theory, Experiment, and Design G. Y. Nieuwland, B. M. Spee 5:119–50

Aerodynamics of Powered High-Lift Systems   G. K. Korbacher   6:319–58
The Structure and Dynamics of Vortex
  Filaments   S. E. Widnall   7:141–65
Recollections From an Earlier Period in
  American Aeronautics   R. T. Jones   9:1–11
Study of the Unsteady Aerodynamics of
  Lifting Surfaces Using the Computer   S. M. Belotserkovskii   9:469–94
Transonic Flow Past Oscillating Airfoils   H. Tijdeman, R. Seebass   12:181–222
Unsteady Airfoils   W. J. McCroskey   14:285–311
Autorotation   H. J. Lugt   15:123–47
Low-Reynolds-Number Airfoils   P. B. S. Lissaman   15:223–39
Supercritical Airfoil and Wing Design   H. Sobieczky, A. R. Seebass   16:337–63

FLUID DYNAMICS OF WATERBORNE VEHICLES
Hydromechanics of Aquatic Animal
  Propulsion   M. J. Lighthill   1:413–46
Applications of Slender-Body Theory in Ship
  Hydrodynamics   J. N. Newman   2:67–94
The Motion of Floating Bodies   J. V. Wehausen   3:237–68
Sailing Vessels and Sails   J. H. Milgram   4:397–430
Hydrofoils and Hydrofoil Craft   A. J. Acosta   5:161–84
Hydrodynamic Problems of Ships in Restricted
  Waters   E. O. Tuck   10:33–44
Ship Boundary Layers   L. Landweber, V. C. Patel   11:173–205
Power From Water Waves   D. V. Evans   13:157–87

FLUID DYNAMICS OF HYDRAULIC STRUCTURES AND OF THE ENVIRONMENT
Formation of Sediment Ripples, Dunes and
  Antidunes   J. F. Kennedy   1:147–68
Oil Spreading on the Sea   D. P. Hoult   4:341–68
Fluid Mechanics of Heat Disposal From Power
  Generation   D. R. F. Harleman, K. D.
    Stolzenbach   4:7–32
Buoyant Plumes and Wakes   J. A. Fay   5:151–60
Longitudinal Dispersion and Turbulent Mixing
  in Open-Channel Flow   H. B. Fischer   5:59–78
Harbor Seiching   J. W. Miles   6:17–35
Waterhammer and Surge Control   V. L. Streeter, E. B. Wylie   6:57–73
Fluid Mechanics of Waste-Water Disposal in
  the Ocean   R. C. Y. Koh, N. H. Brooks   7:187–211
The Effect of Waves on Rubble-Mound
  Structures   F. Raichlen   7:327–56
Hydraulics' Latest Golden Age   H. Rouse   8:1–12
Aerodynamics of Buildings   J. E. Cermak   8:75–106
On the Liquidlike Behavior of Fluidized Beds   J. F. Davidson, D. Harrison, J. R.
    F. Guedes de Carvalho   9:55–86
River Meandering   R. A. Callander   10:129–58
River Ice   G. D. Ashton   10:369–92
Water Transport in Soils   J. -Y. Parlange   12:77–102
Coastal Sediment Processes   K. Horikawa   13:9–32
Debris Flow   T. Takahashi   13:57–77
Turbulence In and Above Plant Canopies   M. R. Raupach, A. S. Thom   13:97–129
Cavitation in Fluid Machinery and Hydraulic
  Structures   R. E. A. Arndt   13:273–328
Sediment Ripples and Dunes   F. Engelund, J. Fredsøe   14:13–37
Dynamics of Lakes, Reservoirs, and Cooling
  Ponds   J. Imberger, P. F. Hamblin   14:153–87
Dynamics of Oil Ganglia During Immiscible
  Displacement in Water-Wet Porous Media   A. C. Payatakes   14:365–93
Secondary Flow in Curved Open Channels   M. Falcón   16:179–93
Mathematical Models of Dispersion in Rivers
  and Estuaries   P. C. Chatwin, C. M. Allen   17:119–49

GEOPHYSICAL FLUID DYNAMICS
Stratified Flows   C. -S. Yih   1:73–110
Formation of Sediment Ripples, Dunes, and
  Antidunes   J. F. Kennedy   1:147–68
Analogy Between Rotating and Stratified
  Fluids   G. Veronis   2:37–66
Tides   M. Hendershott, W. Munk   2:205–24
Atmospheric Boundary Layer   A. S. Monin   2:225–50
Models for Weather Prediction   N. A. Phillips   2:251–92
Boundary Layers in Ocean Circulation Models   A. R. Robinson   2:293–312
Finite Amplitude Disturbances in the Flow of
  Inviscid Rotating and Stratified Fluids Over
  Obstacles   R. R. Long   4:69–92
Mantle Convection and the New Global
  Tectonics   D. L. Turcotte, E. R. Oxburgh   4:33–68
Magnetohydrodynamics of the Earth's Core   P. H. Roberts, A. M. Soward   4:117–54
Buoyant Plumes and Wakes   J. A. Fay   5:151–60
The Atmospheric Boundary Layer Below 150
  Meters   H. A. Panofsky   6:147–77
Hydrodynamics of Large Lakes   G. T. Csanady   7:357–86
Mixing and Dispersion in Estuaries   H. B. Fischer   8:107–33
Currents in Submarine Canyons: An
  Air-Sea-Land Interaction   D. L. Inman, C. E. Nordstrom, R.
                                                       E. Flick   8:275–310
Objective Methods for Weather Prediction   C. E. Leith   10:107–28
Rossby Waves—Long-Period Oscillations of
  Oceans and Atmospheres   R. E. Dickinson   10:159–95
Magnetohydrodynamics of the Earth's
  Dynamo   F. H. Busse   10:435–62
Internal Waves in the Ocean   C. Garrett, W. Munk   11:339–69
Geostrophic Turbulence   P. B. Rhines   11:401–41
Coastal Circulation and Wind-Induced
  Currents   C. D. Winant   12:271–301
Models of Wind-Driven Currents on the
  Continental Shelf   J. S. Allen   12:389–433
Progress in the Modeling of Planetary
  Boundary Layers   O. Zeman   13:253–72
Frazil Ice in Rivers and Oceans   S. Martin   13:379–97
Sediment Ripples and Dunes   F. Engelund, J. Fredsøe   14:13–37
Dynamics of Glaciers and Large Ice Masses   K. Hutter   14:87–130
The Mathematical Theory of Frontogenesis   B. J. Hoskins   14:131–51
Gravity Currents in the Laboratory,
  Atmosphere, and Ocean   J. E. Simpson   14:213–34
Snow Avalanche Motion and Related
  Phenomena   E. J. Hopfinger   15:47–76
The Form and Dynamics of Langmuir
  Circulations   S. Leibovich   15:391–427
Fluid Modeling of Pollutant Transport and
  Diffusion in Stably Stratified Flows Over
  Complex Terrain   W. H. Snyder   17:239–66
Modeling Equatorial Ocean Circulation   J. P. McCreary, Jr.   17:359–409
Turbulent Diffusion From Sources in Complex
  Flows   J. C. R. Hunt   17:447–85
Mantle Convection and Viscoelasticity   W. R. Peltier   17:561–608

ASTRONOMICAL FLUID DYNAMICS
Solar-Wind Flow Past Objects in the Solar
  System   J. R. Spreiter, A. Y. Alksne   2:313–54
Self-Gravitating Gaseous Disks   C. Hunter   4:219–42
Spin-Up   E. R. Benton, A. Clark Jr.   6:257–80
Hydrodynamics of the Universe   Ya. B. Zel'dovich   9:215–28
Relativistic Fluid Mechanics   A. H. Taub   10:301–32

Rotating, Self-Gravitating Masses   N. R. Lebovitz   11:229–46
Some Fluid-Dynamical Problems in Galaxies   C. C. Lin, W. W. Roberts Jr.   13:33–55
Progress in the Modeling of Planetary
    Boundary Layers   O. Zeman   13:253–72
Magneto-Atmospheric Waves   J. H. Thomas   15:321–43

OTHER APPLICATIONS
Flow Lasers   W. H. Christiansen, D. A. Russell,
                                                                        A. Hertzberg   7:115–39
Fluid Mechanics of Heat Pipes   C. L. Tien   7:167–85
Flow Through Screens   E. M. Laws, J. L. Livesey   10:247–66
Continuous Drawing of Liquids to Form
    Fibers   M. M. Denn   12:365–87
Coating Flows   K. J. Ruschak   17:65–89
Aerodynamics of Sports Balls   R. D. Mehta   17:151–89
Buoyancy-Driven Flows in Crystal-Growth
    Melts   W. E. Langlois   17:191–215

MISCELLANEOUS
Superfluid Mechanics   P. H. Roberts, R. J. Donnelly   6:179–225
Experiments in Granular Flow   K. Wieghardt   7:89–114
On the Liquidlike Behavior of Fluidized Beds J. F. Davidson, D. Harrison, J. R.
                                                                        F. Guedes de Carvalho   9:55–86
River Meandering   R. A. Callander   10:129–58
Flows of Nematic Liquid Crystals   J. T. Jenkins   10:197–219
Relativistic Fluid Mechanics   A. H. Taub   10:301–32
River Ice   G. D. Ashton   10:369–92
Debris Flow   T. Takahashi   13:57–77
Frazil Ice in Rivers and Oceans   S. Martin   13:379–97
Brownian Motion of Small Particles
    Suspended in Liquids   W. B. Russel   13:425–55
Sediment Ripples and Dunes   F. Engelund, J. Fredsøe   14:13–37
Nonlinear Interactions in the Fluid Mechanics
    of Helium II   H. W. Liepmann, G. A. Laguna   16:139–77

ORDER FORM

Annual Reviews Inc.

A NONPROFIT SCIENTIFIC PUBLISHER

4139 EL CAMINO WAY • PALO ALTO, CA 94306-9981 • (415) 493-4400

s for Annual Reviews Inc. publications may be placed through your bookstore; subscription agent; par-
ting professional societies; or directly from Annual Reviews Inc. by mail or telephone (paid by credit
or purchase order). Prices subject to change without notice.

iduals: Prepayment required in U.S. funds or charged to American Express, MasterCard, or Visa.

ational Buyers: Please include purchase order.

nts: Special rates are available to qualified students. Refer to Annual Reviews *Prospectus* or contact
al Reviews Inc. office for information.

ssional Society Members: Members whose professional societies have a contractural arrangement with
al Reviews may order books through their society at a special discount. Check with your society for
nation.

ar orders: When ordering current or back volumes, please list the volumes you wish by volume number.

ing orders: (New volume in the series will be sent to you automatically each year upon publication. Can-
on may be made at any time.) Please indicate volume number to begin standing order.

blication orders: Volumes not yet published will be shipped in month and year indicated.

rnia orders: Add applicable sales tax.

ge paid (4th class bookrate / surface mail) by Annual Reviews Inc.

NNUAL REVIEWS SERIES		Prices Postpaid per volume USA/elsewhere	Regular Order Please send:	Standing Order Begin with:
			Vol. number	Vol. number
l Review of ANTHROPOLOGY				
ols. 1-10	(1972-1981)	$20.00/$21.00		
ol. 11	(1982)	$22.00/$25.00		
ols. 12-13	(1983-1984)	$27.00/$30.00		
ol. 14	(avail. Oct. 1985)	$27.00/$30.00	Vol(s). _____	Vol. _____
l Review of ASTRONOMY AND ASTROPHYSICS				
ols. 1-19	(1963-1981)	$20.00/$21.00		
ol. 20	(1982)	$22.00/$25.00		
ols. 21-22	(1983-1984)	$44.00/$47.00		
ol. 23	(avail. Sept. 1985)	$44.00/$47.00	Vol(s). _____	Vol. _____
l Review of BIOCHEMISTRY				
ols. 29-34, 36-50	(1960-1965; 1967-1981)	$21.00/$22.00		
ol. 51	(1982)	$23.00/$26.00		
ols. 52-53	(1983-1984)	$29.00/$32.00		
ol. 54	(avail. July 1985)	$29.00/$32.00	Vol(s). _____	Vol. _____
l Review of BIOPHYSICS				
ols. 1-10	(1972-1981)	$20.00/$21.00		
ol. 11	(1982)	$22.00/$25.00		
ols. 12-13	(1983-1984)	$47.00/$50.00		
ol. 14	(avail. June 1985)	$47.00/$50.00	Vol(s). _____	Vol. _____
l Review of CELL BIOLOGY				
ol. 1	(avail. Nov. 1985)	est. $27.00/$30.00	Vol. _____	Vol. _____
l Review of EARTH AND PLANETARY SCIENCES				
ols. 1-9	(1973-1981)	$20.00/$21.00		
ol. 10	(1982)	$22.00/$25.00		
ols. 11-12	(1983-1984)	$44.00/$47.00		
ol. 13	(avail. May 1985)	$44.00/$47.00	Vol(s). _____	Vol. _____
Review of ECOLOGY AND SYSTEMATICS				
ols. 1-12	(1970-1981)	$20.00/$21.00		
ol. 13	(1982)	$22.00/$25.00		
ols. 14-15	(1983-1984)	$27.00/$30.00		
ol. 16	(avail. Nov. 1985)	$27.00/$30.00	Vol(s). _____	Vol. _____

1

RDERING INFORMATION ON PAGE 4

		Prices Postpaid per volume USA/elsewhere	Regular Order Please send:	Standing Begin w
			Vol. number	Vol. nu

Annual Review of **ENERGY**

Vols. 1-6	(1976-1981)	$20.00/$21.00		
Vol. 7	(1982)	$22.00/$25.00		
Vols. 8-9	(1983-1984)	$56.00/$59.00		
Vol. 10	(avail. Oct. 1985)	$56.00/$59.00	Vol(s)._____	Vol.____

Annual Review of **ENTOMOLOGY**

Vols. 8-16, 18-26	(1963-1971; 1973-1981)	$20.00/$21.00		
Vol. 27	(1982)	$22.00/$25.00		
Vols. 28-29	(1983-1984)	$27.00/$30.00		
Vol. 30	(avail. Jan. 1985)	$27.00/$30.00	Vol(s)._____	Vol.____

Annual Review of **FLUID MECHANICS**

Vols. 1-5, 7-13	(1969-1973; 1975-1981)	$20.00/$21.00		
Vol. 14	(1982)	$22.00/$25.00		
Vols. 15-16	(1983-1984)	$28.00/$31.00		
Vol. 17	(avail. Jan. 1985)	$28.00/$31.00	Vol(s)._____	Vol.____

Annual Review of **GENETICS**

Vols. 1-15	(1967-1981)	$20.00/$21.00		
Vol. 16	(1982)	$22.00/$25.00		
Vols. 17-18	(1983-1984)	$27.00/$30.00		
Vol. 19	(avail. Dec. 1985)	$27.00/$30.00	Vol(s)._____	Vol.____

Annual Review of **IMMUNOLOGY**

Vols. 1-2	(1983-1984)	$27.00/$30.00		
Vol. 3	(avail. April 1985)	$27.00/$30.00	Vol(s)._____	Vol.____

Annual Review of **MATERIALS SCIENCE**

Vols. 1-11	(1971-1981)	$20.00/$21.00		
Vol. 12	(1982)	$22.00/$25.00		
Vols. 13-14	(1983-1984)	$64.00/$67.00		
Vol. 15	(avail. Aug. 1985)	$64.00/$67.00	Vol(s)._____	Vol.____

Annual Review of **MEDICINE: Selected Topics in the Clinical Sciences**

Vols. 1-3, 5-15	(1950-1952; 1954-1964)	$20.00/$21.00		
Vols. 17-32	(1966-1981)	$20.00/$21.00		
Vol. 33	(1982)	$22.00/$25.00		
Vols. 34-35	(1983-1984)	$27.00/$30.00		
Vol. 36	(avail. April 1985)	$27.00/$30.00	Vol(s)._____	Vol.____

Annual Review of **MICROBIOLOGY**

Vols. 17-35	(1963-1981)	$20.00/$21.00		
Vol. 36	(1982)	$22.00/$25.00		
Vols. 37-38	(1983-1984)	$27.00/$30.00		
Vol. 39	(avail. Oct. 1985)	$27.00/$30.00	Vol(s)._____	Vol.____

Annual Review of **NEUROSCIENCE**

Vols. 1-4	(1978-1981)	$20.00/$21.00		
Vol. 5	(1982)	$22.00/$25.00		
Vols. 6-7	(1983-1984)	$27.00/$30.00		
Vol. 8	(avail. March 1985)	$27.00/$30.00	Vol(s)._____	Vol.____

Annual Review of **NUCLEAR AND PARTICLE SCIENCE**

Vols. 12-31	(1962-1981)	$22.50/$23.50		
Vol. 32	(1982)	$25.00/$28.00		
Vols. 33-34	(1983-1984)	$30.00/$33.00		
Vol. 35	(avail. Dec. 1985)	$30.00/$33.00	Vol(s)._____	Vol.____

2

SEE ORDERING INFORMATION ON PAGE 4